ADVANCED NANOMATERIALS FOR POINT OF CARE DIAGNOSIS AND THERAPY

ADVANCED NANOMATERIALS FOR POINT OF CARE DIAGNOSIS AND THERAPY

Edited by

SUSHMA DAVE
Department of Applied Sciences JIET Jodhpur, Rajasthan, India

JAYASHANKAR DAS
Valnizen Healthcare Vile Parle West Mumbai, Maharashtra, India

SOUGATA GHOSH
Department of Microbiology, School of Science, RK University,
Rajkot, Gujarat, India

ELSEVIER

Elsevier
Radarweg 29, PO Box 211, 1000 AE Amsterdam, Netherlands
The Boulevard, Langford Lane, Kidlington, Oxford OX5 1GB, United Kingdom
50 Hampshire Street, 5th Floor, Cambridge, MA 02139, United States

ISBN: 978-0-323-85725-3

For Information on all Elsevier publications visit our website at https://www.elsevier.com/books-and-journals

Publisher: Susan Dennis
Editorial Project Manager: Emerald Li
Production Project Manager: Sruthi Satheesh
Cover Designer: Christian Bilbow

Typeset by Aptara, New Delhi, India

CONTENTS

Contributors

Laxmi R. Adil
Department of Chemistry, Indian Institute of Technology Guwahati, Guwahati, Assam, India

Rashmi Ahire
Department of Chemistry, Savitribai Phule Pune University (Formerly Pune University), Pune, Maharashtra, India

Anshebo G. Alemu
Faculty of Natural & computational science, Department of Physics, Samara University, Samara, Ethiopia

Anshebo T. Alemu
Faculty of Engineering, Adama Science and Technology University, Ethiopia

Pallab K. Bairagi
Ordinance Factory Nalanda, Department of Defence Production, Ministry of Defence, Rajgir, Bihar, India

Shahani Begum
Plant Biotechnology Laboratory, Post Graduate Department of Botany, Utkal University, Bhubaneswar, Odisha, India

Kamalakanta Behera
Department of Chemistry, Biochemistry and Forensic sciences, Amity School of Applied Sciences, Amity University Haryana, Haryana, India

Dhiraj Bhatia
Biological Engineering Discipline, Indian Institute of Technology Gandhinagar, Palaj, Gujarat, India

S.P. Bhatnagar
Department of Physics, Maharaja Krishnakumarsihji Bhavnagar University, Bhavnagar, Gujarat, India

Satabdi Bhattacharjee
Department of Microbiology, Assam University, Silchar, Assam, India

Debasis Bisoi
Pradyumna Bal Memorial Hospital, Kalinga Institute of Medical Sciences, Bhubaneswar, Odisha, India

Prakash Bobde
Department of Research and Development, University of Petroleum & Energy Studies, Dehradun, Uttarakhand, India

Ch. G. Chandaluri
Faculty of Chemistry, Humanities and Sciences Division Indian Institute of Petroleum and Energy, Visakhapatnam, India

Moirangthem A. Chanu
Department of Chemistry, Indian Institute of Technology Guwahati, Guwahati, Assam, India

Naveen K. Dandu
Material Science Division, Argonne National Lab, Argonne, IL

Anirban Das
Department of Chemistry, Biochemistry and Forensic sciences, Amity School of Applied Sciences, Amity University Haryana, Haryana, India

Jayashankar Das
Valnizen Healthcare Vile Parle West Mumbai, Maharashtra, India

Hemlata Das
Department of Medical Oncology and Department of Pathology, IMS and Sum Hospital, Bhubaneswar, India

Dwiti K. Das
National Centre for Nanoscience and Nanotechnology, University of Mumbai, Mumbai, Maharashtra, India

Sushma Dave
Department of Applied Sciences JIET Jodhpur, Rajasthan, India

Vishakha Dave
Department of Physics, Maharaja Krishnakumarsihji Bhavnagar University, Bhavnagar, Gujarat, India

Ayushman Gadnayak
IMS and SUM Hospital, Siksha 'O' Anusandhan (Deemed to be University), Bhubaneswar, Odisha, India

Sayantani Garai
Department of Biotechnology, University of Engineering & Management, Kolkata

Goutam Ghosh
UGC-DAE Consortium for Scientific Research, Mumbai Centre, CFB Building, BARC Campus, Mumbai, India

Sreejita Ghosh
Microbiology Research Laboratory, Department of Biotechnology, Maulana Abul Kalam Azad University of Technology, Simhat, Haringhata, Nadia, West Bengal

Sougata Ghosh
Department of Microbiology, School of Science, RK University, Rajkot, Gujarat, India

Swati Goswami
Department of Microbiology, School of Science, RK University, Kasturbadham, Rajkot, Gujarat, India

Parameswar K. Iyer
Department of Chemistry, Indian Institute of Technology Guwahati, Guwahati, Assam, India; Centre for Nanotechnology, Indian Institute of Technology Guwahati, Guwahati, Assam, India

Yogesh Jadhav
School of energy studies, Savitribai Phule Pune University (Formerly Pune University), Pune, Maharashtra, India

Sandesh Jadkar
School of energy studies, Savitribai Phule Pune University (Formerly Pune University), Pune, Maharashtra, India

Fahmida Khan
Department of Chemistry, National Institute of Technology Raipur, India

Prateek Khare
Department of Chemical Engineering, Madan Mohan Malaviya University of Technology, Gorakhpur, Uttar Pradesh, India

Mst N. Khatun
Department of Chemistry, Indian Institute of Technology Guwahati, Guwahati, Assam, India

Sindhu Kilaru
Department of Medical Oncology and Department of Pathology, IMS and Sum Hospital, Bhubaneswar, India

Byoung-Suhk Kim
Department of Organic Materials & Fiber Engineering, Jeonbuk National University, Jeollabuk, South Korea

Kisan Kodam
Department of Chemistry, Savitribai Phule Pune University (Formerly Pune University), Pune, Maharashtra, India

Spoorthy Kolluri
Department of Medical Oncology and Department of Pathology, IMS and Sum Hospital, Bhubaneswar, India

Vijay Kumar
Department of Microbiology, School of Science, RK University, Kasturbadham, Rajkot, Gujarat, India

Dibyajit Lahiri
Department of Biotechnology, University of Engineering & Management, Kolkata

Lidong Li
Centre for Nanotechnology, Indian Institute of Technology Guwahati, Guwahati, Assam, India; School of Materials Science and Engineering, University of Science and Technology Beijing, Beijing, China

N. Mahender Reddy
Department of Chemistry, Chaitanya Bharathi Institute of Technology (A), Gandipet, Hyderabad, Telangana, India

Indrani Medhi
Centre for Nanotechnology, Indian Institute of Technology Guwahati, Guwahati, Assam, India

Rasbindu Mehta
Department of Physics, Maharaja Krishnakumarsihji Bhavnagar University, Bhavnagar, Gujarat, India

A.M. Vinu Mohan
Electrodics & Electrocatalysis Division, CSIR-Central Electrochemical Research Institute (CECRI), Karaikudi, Tamil Nadu, India

Padmaja Mohanty
Valnizen Healthcare Vile Parle West Mumbai, Maharashtra, India

Jatindra N. Mohanty
IMS and SUM Hospital, Siksha 'O' Anusandhan (Deemed to be University), Bhubaneswar, Odisha, India

Neeta Mohanty
Department of Oral Pathology & Microbiology, Institute of Dental Sciences, Siksha 'O' Ansuandhan deemed to be University, Bhubaneswar, Odisha, India

Maheswata Moharana
Department of Chemistry, National Institute of Technology Raipur, India

Subrata Mondal
Department of Chemistry, Indian Institute of Technology Guwahati, Guwahati, Assam, India

Vinod Morya
Biological Engineering Discipline, Indian Institute of Technology Gandhinagar, Palaj, Gujarat, India

Dipro Mukherjee
Department of Biotechnology, University of Engineering & Management, Kolkata

Moupriya Nag
Department of Biotechnology, University of Engineering & Management, Kolkata

Vinod Nandre
Department of Chemistry, Savitribai Phule Pune University (Formerly Pune University), Pune, Maharashtra, India

Rahul Narasimhan
Centre for Nanotechnology, Indian Institute of Technology Guwahati, Guwahati, Assam, India

Ananya Nayak
Centre for Genomics and Molecular Therapeutics, Siksha 'O' Anusandhan (Deemed to be University), Bhubaneswar, Odisha, India

Silpa P A
Department of ECE, Sahrdaya College of Engineering & Technology, Thrissur, Kerala, India

Sukdeb Pal
Wastewater Technology Division, CSIR-National Environmental Engineering Research Institute (CSIR-NEERI), Nagpur, India; Academy of Scientific and Innovative Research (AcSIR), Ghaziabad, India

Deepak Panchal
Wastewater Technology Division, CSIR-National Environmental Engineering Research Institute (CSIR-NEERI), Nagpur, India; Academy of Scientific and Innovative Research (AcSIR), Ghaziabad, India

Soumya S. Panda
Department of Medical Oncology and Department of Pathology, IMS and Sum Hospital, Bhubaneswar, India

Saroj Prasad Panda
Department of Medical Oncology and Department of Pathology, IMS and Sum Hospital, Bhubaneswar, India

Swagatika Panda
Department of Oral Pathology & Microbiology, Institute of Dental Sciences, Siksha 'O' Ansuandhan deemed to be University, Bhubaneswar, Odisha, India

Medha Pandya
The KPES Science Collage, Maharaja Krishnakumarsihji Bhavnagar University, Bhavnagar, Gujarat, India

Sabyasachi Parida
Pradyumna Bal Memorial Hospital, Kalinga Institute of Medical Sciences, Bhubaneswar, Odisha, India

Retwik Parui
Department of Chemistry, Indian Institute of Technology Guwahati, Guwahati, Assam, India

Ravi Patel (Kumar)
Department of Research and Development, University of Petroleum & Energy Studies, Dehradun, Uttarakhand, India

Snigdha Pattanaik
Department of Orthodontics & Dentofacial Orthopaedics, Institute of Dental Sciences, Siksha O Anusandhan (Deemed to be University), Bhubaneshwar, Odisha, India

Subrat Kumar Pattanayak
Department of Chemistry, National Institute of Technology Raipur, India

Om Prakash
Wastewater Technology Division, CSIR-National Environmental Engineering Research Institute (CSIR-NEERI), Nagpur, India

Anulipsa Priyadarshini
School of Applied Sciences, Kalinga Institute of Industrial Technology, Deemed to be University, Bhubaneswar, Odisha, India

Sivaprakasam Radhakrishnan
Department of Organic Materials & Fiber Engineering, Jeonbuk National University, Jeollabuk, South Korea

Varun Rai
School of Materials Science and Engineering, Nanyang Technological University, Singapore

Pravat Rajbanshi
Department of Chemical Engineering, Indian Institute of Technology Kanpur, Kanpur, Uttar Pradesh, India

Anjali Rajwar
Biological Engineering Discipline, Indian Institute of Technology Gandhinagar, Palaj, Gujarat, India

Kola Ramesh
Department of Chemistry, Chaitanya Bharathi Institute of Technology (A), Gandipet, Hyderabad, Telangana, India

Gubbala V. Ramesh
Department of Chemistry, Chaitanya Bharathi Institute of Technology (A), Gandipet, Hyderabad, Telangana, India

Satya Ranjan Misra
Department of Oral Medicine & Radiology, Institute of Dental Sciences, Siksha 'O' Ansuandhan deemed to be University, Bhubaneswar, Odisha, India

Rakesh Rawal
Department of Life Sciences, University School of Sciences, Gujarat University, Ahmedabad, Gujarat, India

Rina Rani Ray
Microbiology Research Laboratory, Department of Biotechnology, Maulana Abul Kalam Azad University of Technology, Simhat, Haringhata, Nadia, West Bengal

Swayamprabha Sahoo
IMS and SUM Hospital, Siksha 'O' Anusandhan (Deemed to be University), Bhubaneswar, Odisha, India

Maheswata Sahoo
IMS and SUM Hospital, Siksha 'O' Anusandhan (Deemed to be University), Bhubaneswar, Odisha, India

Tejaswini Sahoo
School of Applied Sciences, Kalinga Institute of Industrial Technology, Deemed to be University, Bhubaneswar, Odisha, India

Satya Narayan Sahu
School of Applied Sciences, Kalinga Institute of Industrial Technology (KIIT), Deemed to be University, Bhubaneswar, India

Jnana R. Sahu
School of Applied Sciences, Kalinga Institute of Industrial Technology, Deemed to be University, Bhubaneswar, India

Rojalin Sahu
School of Applied Sciences, Kalinga Institute of Industrial Technology, Deemed to be University, Bhubaneswar, Odisha, India

Priyanka Samal
Department of Clinical Hematology- Hemato-oncology & Stem Cell Transplant, IMS & SUM Hospital, Bhubaneswar, Odisha, India

Debarchita Sarangi
Department of Prosthodontics, Crown & Bridge, Institute of Dental Sciences, Siksha O Anusandhan (Deemed to be University), Bhubaneshwar, Odisha, India

D. Saritha
Department of Chemistry, Chaitanya Bharathi Institute of Technology (A), Gandipet, Hyderabad, Telangana, India

Deepak Senapati
School of Applied Sciences, Kalinga Institute of Industrial Technology, Deemed to be University, Bhubaneswar, Odisha, India

Abhishek Sharma
Wastewater Technology Division, CSIR-National Environmental Engineering Research Institute (CSIR-NEERI), Nagpur, India; Academy of Scientific and Innovative Research (AcSIR), Ghaziabad, India

Indu Sharma
Department of Microbiology, Assam University, Silchar, Assam, India

Ram S. Singh
Department of Physics, O P Jindal University, Raigarh, Chhattisgarh, India

Vishal Singh (K.)
Department of HSE and Civil Engineering, University of Petroleum & Energy Studies, Dehradun, Uttarakhand, India

Purusottam Tripathy
Wastewater Technology Division, CSIR-National Environmental Engineering Research Institute (CSIR-NEERI), Nagpur, India

Suresh Waghmode
Department of Chemistry, Savitribai Phule Pune University (Formerly Pune University), Pune, Maharashtra, India

Shu Wang
Institute of Chemistry, Chinese Academy of Sciences, Beijing, P.R. China

Ramesh B. Yathirajula
Centre for Nanotechnology, Indian Institute of Technology Guwahati, Guwahati, Assam, India

CHAPTER 1

Nanomaterials-based biosensors

Anirban Das[a], Varun Rai[b], Kamalakanta Behera[a], Ram S. Singh[c]
[a]Department of Chemistry, Biochemistry and Forensic sciences, Amity School of Applied Sciences, Amity University Haryana, Haryana, India
[b]School of Materials Science and Engineering, Nanyang Technological University, Singapore
[c]Department of Physics, O P Jindal University, Raigarh, Chhattisgarh, India

1.1 Introduction

An entity that can be utilized to detect and/or quantify a biochemical molecule (analyte) is known as a biosensor [1]. The variation in physical or chemical properties of either the analyte or the biosensor when they come in proximity or form a bond with each other is exploited for the biosensing process. The applications of biosensing are manifold (1) biomarkers, biomolecules that are physiological indicators of a disease condition need to be analyzed *in vivo* or *in vitro* for therapeutics and diagnostics (2) in environmental conservation and monitoring, markers or disruptive biomolecules like hormones and pesticides as well as in (3) agriculture and food processing, contaminant(s) have to be identified and quantified. For example, biosensors may be used to study the cellular environment to investigate signaling mechanisms, metabolic activity, ion concentration or pH, and changes in the microenvironment. The "sensing" process involves conversion of bioanalyte interactions to a detectible and reproducible response that may be detected by electrochemical [2], photochemical [3], photo (electro) chemical [4], colorimetric [5], fluorometric [6], mechanical motion [7], magnetic [8], or chemical [9] means. The detection techniques may require the attachment of another molecule often called a label or tag. The major thrust areas that are for improvement in design of biosensors as well as biosensing techniques are lowering the detection limit and multiplexing, the ability to detect multiple analytes in a mixture [10].

Nanoparticles are molecular or atomic aggregations that are 100 nm or less along any direction. The physical and chemical properties vary from their bulk (larger) counterparts of same composition due to the phenomenon of quantum confinement. Additionally, the manifold increase in surface area resulting from the small size of the particles make these materials catalytically much more active than their bulk counterparts. Another property of nanoparticles that make them very useful in several fields, including sensing is the versatility endowed on these particles due to our ability of tailoring their properties by variation in their shape, size and anisotropy. This variation may lead to variation in surface area or surface termination leading to distinct physical or chemical response to external stimuli which is utilized in sensing applications. The

Advanced Nanomaterials for Point of Care Diagnosis and Therapy
DOI: https://doi.org/10.1016/B978-0-323-85725-3.00020-9

ability to attach functional groups on the surface of these nanoparticles also contributes to the versatility of these materials. Nanotechnology is significantly influencing the field of biosensors primarily due to the ability of nanostructures to multiplex in real time. The development of nanotechnology led to the miniaturization of devices which in turn simultaneously led to growth of the field of nanobiosensors. For example, techniques like photolithography, a surface patterning method directed by light, laid the foundation of microarrays and the surface-enhanced laser desorption/ionization time-of flight technique in proteomics. In this chapter, the principles used in nanobiosensing as well as the mechanism of action of nanosized sensors utilizing nanomaterials or nanostructures would be discussed and classified. Representative examples of nanosized biosensors would also be discussed. The chapter is classified according to the detection method. At the end of the chapter a specialized topic nanozymes is discussed [11].

1.2 Classification based on detection techniques

1.2.1 Colorimetric biosensors

Colorimetric biosensors are able detect and/or quantify biological parameters or activity by visually observing a color change either using the naked eye or using simple optical detectors. Thus this is a viable technique to develop sensors that may be used in point-of care biomedical devices [5]. Au NPs are widely utilized in nanobiosensors using the colorimetric principle as their color varies according to their size, shape, and degree of aggregation (Fig. 1.1) [12]. Additionally these NPs can be easily functionalized. There are reports of unmodified AuNPs being used to colorimetrically detect RNA of various viruses, for example, HCV (hepatitis C virus), SVCV (spring viraemia of carp virus), CGMMV (cucumber green mottle mosaic virus), BVDV, (bovine viral diarrhea virus), MCMV (maize chlorotic mottle virus), and CyHV-3 (cyprinid herpes-virus-3). The color of the Au NP arises due to the phenomenon of surface plasmon resonance (SPR). As the size of the NPs increases or smaller NPs coagulate there is red shift of the absorption and scattering peaks and the color of the colloidal solution changes from red to purple to blue [13]. In the presence of the RNA of the virus, the Au NP aggregate and this leads to the visually observable color change. In the absence

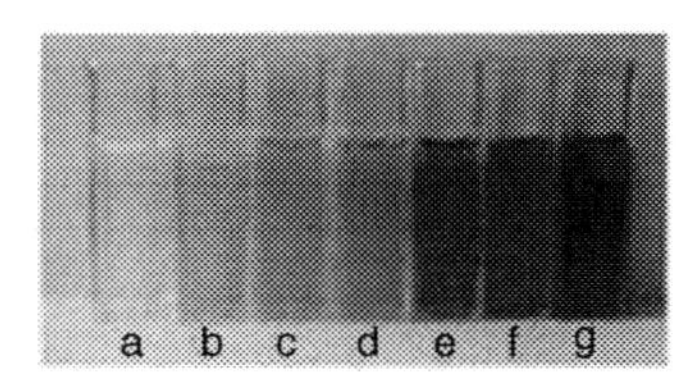

Fig. 1.1 Color changes of the AuNPs formed in the presence of different concentrations of dopamine (A) 0, (B) 2.5×10^{-6}, (C) 5×10^{-6}, (D) 8×10^{-6}, (E) 1×10^{-5}, (F) 1.5×10^{-5}, and (G) 2×10^{-5} M. *(Reproduced from Ref. [12] with permission of American Chemical Society).*

of virus RNA, the single strand DNA primers that are added to the bioassay have their nitrogenous bases uncoiled thus facilitating attractive electrostatic interaction between the Au NP and the bases.

Functionalized Au NPs have been reported to show enhanced specificity toward the analyte of interest. The nature of the functional group influences the sensing properties. For example, antibody functionalized Au NPs, in presence of antigen for the influenza A virus coagulate to change color toward longer wavelength. Engineered peptides that have higher binding affinity toward specific viruses than antibodies result in enhanced sensitivity of Au NPs functionalized by these peptides than the ones functionalized by the antibodies. The colorimetric response of Au NPs has also been employed to implement lab-on-chip. An immunochromatographic strip combined with a reverse transcription PCR was used to detect the HINI virus. Ag NPs also undergo color change on aggregation (yellow to orange) and there is report of the use of Ag NPs along with Au NPs for detection of the Kaposi's sarcoma-associated herpes virus.

1.2.2 Electrochemical biosensors

Electrochemical biosensors relies on electrochemical signal readout derived from redox processes following successful detection of analytes (proteins, DNAs, RNAs, metal ions) by probe molecules attached on electrode substrate. Nanomaterials-based electrodes are very promising for electrochemical sensing owing to the unique physical, chemical properties and electron transport properties [14–17]. Moreover, nanomaterials offer large surface area to –volume ratios, that's very critical for immobilization of large number of probe molecules to improve sensitivity, specificity, and lead to miniaturization of sensing devices. Electrochemical biosensors based on nanomaterials involve less sample volume, show fast quantitative analysis, and can be customized in sizes for power requirements and onsite application. Molecular biorecognition is coupled with redox process in electrochemical biosensor, and successful recognition event changes the physical proximity of redox centers tagged with probe molecule, with the electrode surface. The changed proximity of redox centers with electrode affects the electron transfer resistance and subsequently current or impedance is recorded by electrochemical working station - potentiostat. Evolution of electrochemical biosensors also depends on advancement in robust electronic circuitry, size, and portability of potentiostat. Therefore, advancement in materials for electrochemical biosensors development and electrochemical working station is very important for wide spread biomedical application in epidemics.

Nanoporous alumina membrane based electrochemical biosensor utilizes submicrometer thick alumina membrane over platinum conducting wire electrode, as template for attaching probe molecules (dengue cDNA or antiWNV-DIIIimmunoglobulinM). The electrochemical signal is recorded by introducing a redox molecules (ferrocyanide or ferrocenemethanol), which shows faradic current with electrode during potential scan. Binding of analytes (target DNA or West Nile virus protein domain III) with probes attached on alumina membrane substrate, blocks the nanopores and access of redox molecules toward platinum electrode tip that subsequently reduces the electrochemical signal readout. The electrochemical nanobiosensor based on nanoporous alumina shows detection limits of 4 pg mL^{-1} with logarithmic linearity up to 53 pg mL^{-1} for West Nile virus protein domain III and detection limit of 9.55×10^{-12} M with 6 order of linearity for the ss-31 merc DNA of Dengue Virus RNA as shown in Fig. 1.2 [18,19].

Various signal amplification strategies have been employed to improve detection limit of electrochemical biosensor [20]. Primary capture probe ss DNA molecules are immobilized on the electrode and successful binding of analyte DNAs leads to formation of ds DNA, where positively charged redox active intercalators (hexa-ammineruthenium [II]) are intercalated across backbone and electrochemical signal is recorded. Gold nanoparticles attached with multiple secondary reporter probe ss DNA are introduced to amplify electrochemical signal involving positively charged redox intercalator [21]. Single binding of analyte ss DNA with secondary reporter probe attached on gold nanoparticles, and subsequently with capture probe DNA attached on electrode surface, brings significantly large number of redox active intercalator in proximity of the electrode, resulting in enhanced electrochemical signal.

1.2.3 Photoelectrochemical biosensors

In the PEC technique, light energy falls on a semiconductor catalyst material that had been deposited on an optically transparent conducting electrode (e.g., indium tin oxide – ITO coated glass slide) called the working electrode (WE). This causes the excitation of electrons from the valence band to the conduction band of the semiconductor and thus generating electron-hole pairs (excitons). These excitons drive redox reactions thus facilitating conversion of light energy to chemical energy (Fig. 1.3). The circuit is completed using a counter electrode (CE) which usually consists of a Pt wire or a mesh. A differential potential is applied to prevent the recombination of these excitons and thus a higher yield of chemical energy is obtained from the incident light. The electrochemical redox reactions result in measurable change in voltage or generation of current which is recorded and quantified against a standard reference electrode (RE) and used as a parameter to quantify efficiency of the process [4,22].

The fine-tuning the properties of the photoactive materials are key to increasing the efficiency of the PEC process and this is often achieved by engineering the band gap of the semiconductor. Apart from the composition, in nano-sized materials this engineering may be achieved by variation in shape, size and anisotropy. The various types of photoactive nanomaterials that have been reported include (1) individual semiconductors, for example, (a) TiO_2 in various shapes, sizes and morphologies, (b) ZnO nanowire arrays, (c) quantum dots of various compositions (e.g., CdS and PbS), and (d) g-C_3N_4 (2) semiconductor-semiconductor heterojunctions (a) p-n heterojunction, for example,

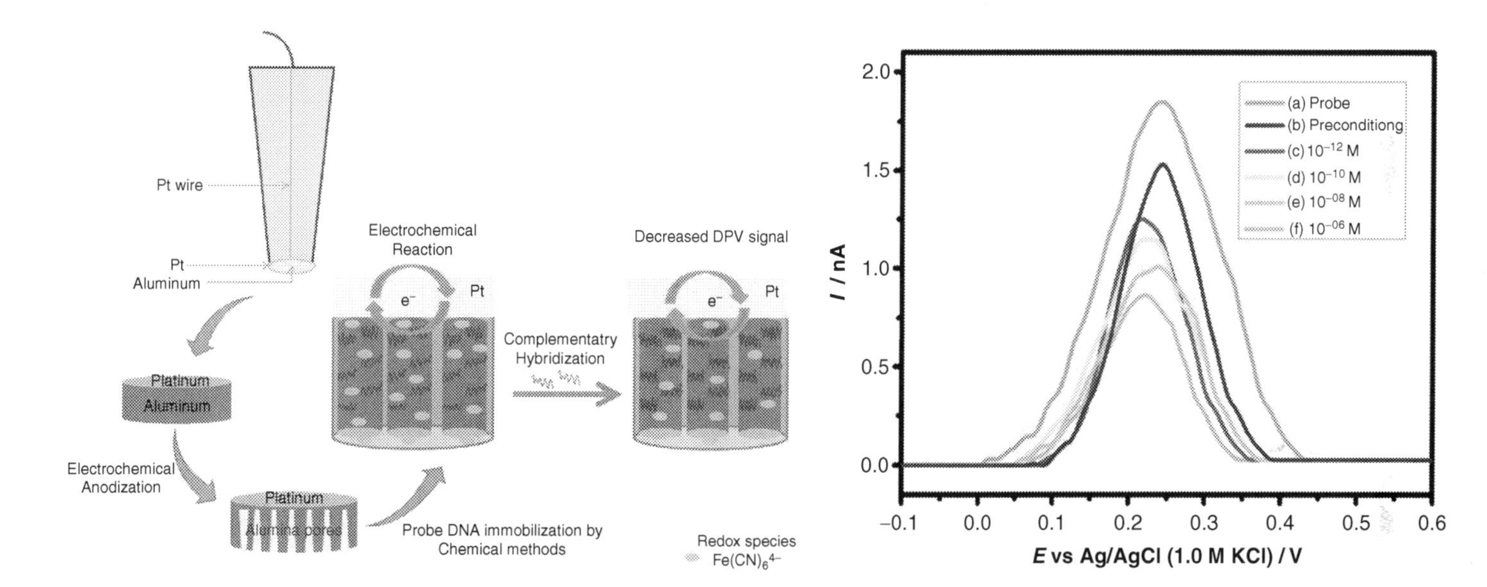

Fig. 1.2 (A) Scheme of construction and operation for nanoporous alumina membrane based DNA biosensor. (B) Differential pulse voltammetry current signal response of an electrochemical probe cDNA biosensor, and toward increasing concentrations of analyte complementary ssDNA to probe cDNA [19].

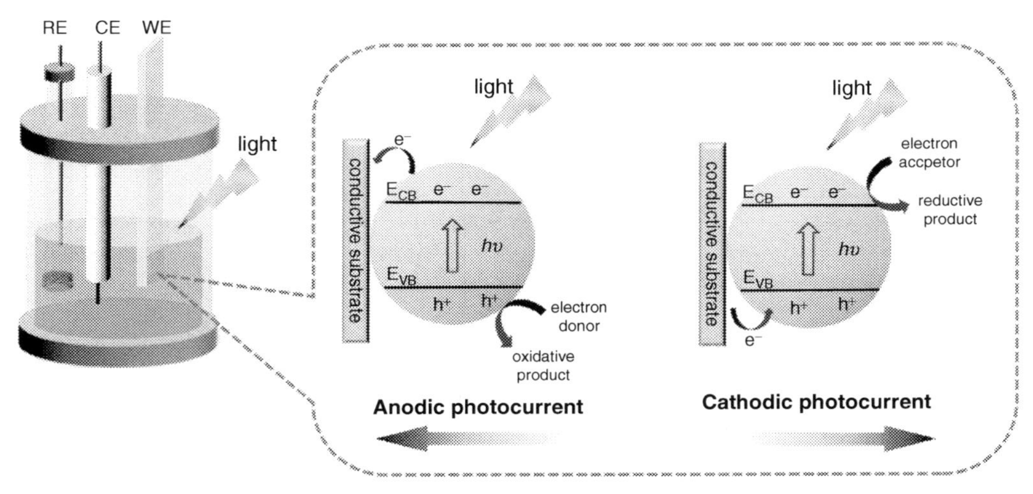

Fig. 1.3 ***Diagram of PEC sensing with the traditional three electrode system and the photocurrent generation mechanism.*** *(Reproduced from Ref. [22] with permission of Royal Society of Chemistry).*

CuO/ZnO, MnO_2/g-C_3N_4, (b) n-n heterojunctions, for example, TiO_2/g-C_3N_4, and (c) z-scheme system, for example, $BiVO_4$/CdS systems, (3) semiconductor metal heterojunctions, for example, Au NPs on 3D TiO_2 nanorods, and (4) upconversion-based semiconductor composites, for example, core-shell NaYF4:Yb, Tm@TiO2.

Due to its stability, environmental friendliness and cost-effectiveness, TiO_2, an n-type semiconductor, is the most widely used single semiconductor material for PEC applications. However, it has a very wide band gap and the various techniques used to overcome this shortcoming include tailoring its morphology or preparing composites. Using Cu_2O–TiO_2 as a nanoctalyst, a PEC based biosensor was reported to simultaneously sense glucose and generate biohydrogen [23]. The catalyst was coated on an optically transparent conducting material to comprise the WE (Fig. 1.4). Irradiation with light causes the electron in valence band of Cu_2O to move to the conduction band leaving behind holes in the valence band that oxidize the glucose to gluconic acid. The electrons in the conduction band travel through the external circuit to the CE (Pt) where the water is reduced to generate H_2 gas. The photocurrent generated (normalized per surface area- photocurrent density) increased with the increase in glucose concentration as shown in Fig. 1.4B and thus this process is used to "sense" glucose concentration. The curve has a linear response of photocurrent between 375 mM and 1.5 mM glucose that reaches saturation at about 2.5 mM glucose. There is negligible interference from ascorbic acid and dopamine. Pure Cu_2O though active as a photocatalyst is rapidly photocorroded. A thin TiO_2 layer prevents the photocorrosion of TiO_2 and increases the light absorption range.

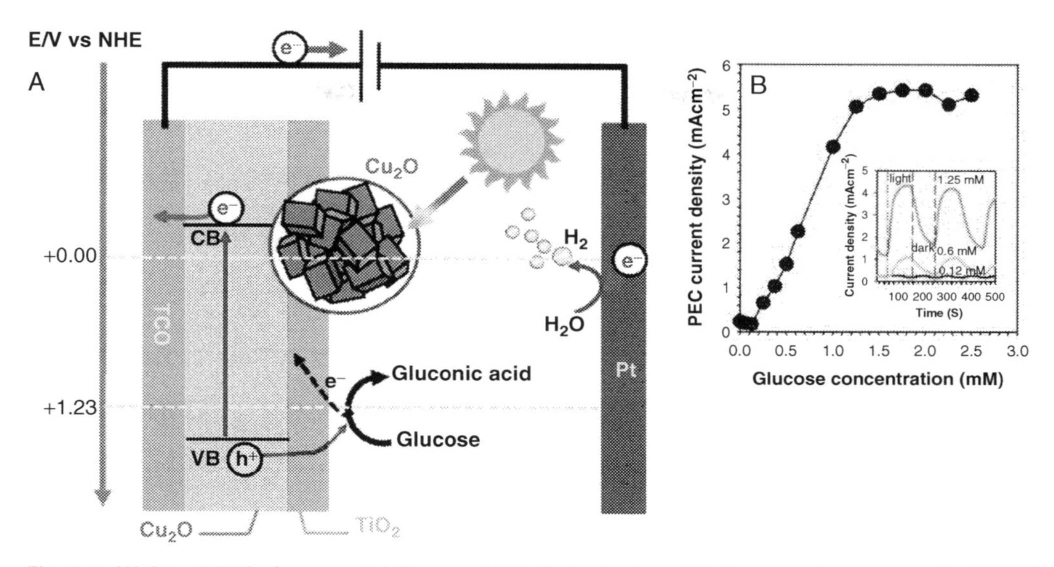

Fig. 1.4 (A) Direct PEC glucose oxidation, and (B) photoelectrocatalytic current response at a Cu2O/TiO2 electrode in 0.1 M NaOH with different concentrations of glucose (inset: the photocurrent response of glucose measured at an applied potential of 0.7 V). *(Reproduced from Ref. [23] with permission of Elsevier).*

Semiconductor quantum dots have high extinction coefficient and large dipole moments resulting in high absorption of light and high rate of charge separation of the excitons. Thus QDs are extensively used as photoactive materials for PEC applications, either by themselves or as sensitizers for wide band gap semiconductors. Examples of semiconductors used in PEC based biosensors include PbS and CdS. Individual semiconductors suffer from weak visible light absorption efficiency and fast exciton recombination. Engineering of heterojunctions that are formed by coupling of two semiconductors with different band structures, are one of the strategies employed to overcome these disadvantages. Typically Z-scheme, p-n and semiconductor-metal heterojuctions have been reported for PEC biosensor applications. Coupling of a p-type and an n-type semiconductor results in a p-n junction. A space charge region created by diffusion of holes and electrons aids in charge separation by guiding these in opposite directions. The holes move to the valence band of the p-type semiconductor and the electrons move to the conduction band of the n-type semiconductor resulting in efficient charge separation and electron transfer. Some examples of p-n semiconductor based PEC biosensors include a Cu_2O/ZnO based sensor used detect biomolecules and a $MnO_2/g\text{-}C_3N_4$ based sensor used to detect glucose and lactose. The Z-scheme, designed to imitate the process of photosynthesis comprises of two separate systems, photosystem I (PS I) and photosystem II (PS II) connected *via* a medium (Fig. 1.5). On absorption of light energy, the photoinduced electrons of PSII system migrate to the

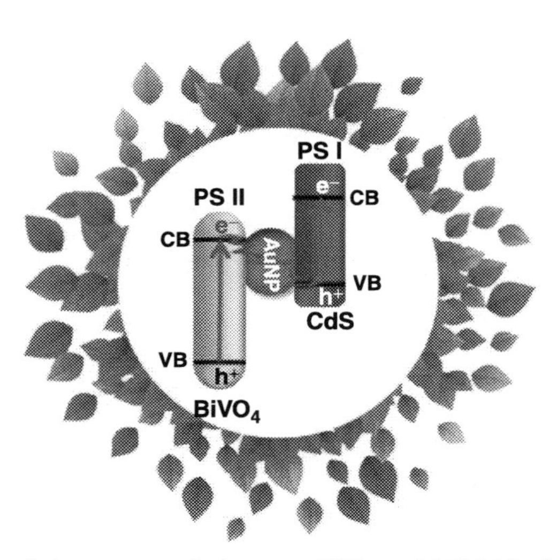

Fig. 1.5 ***The photogenerated electron transfer between $BiVO_4$ and CdS QD in the Z-scheme heterojunctions.*** *(Reproduced from Ref. [24] with permission of American Chemical Society).*

valence band of PS I and accumulate there, while the holes accumulate in the valence band of PS II, resulting in very efficient charge separation for the overall system and optimum capability to carry out redox reactions [24].

A prostate-specific antigen (PSA) biosensor based on the PEC "Z-scheme" was reported. In this system PS I consisted of by CdS quantum dots while PS II consisted of $BiVO_4$. Au NPs deposited on (010) facets of $BiVO_4$ are used as a bridge between the two photosystems to facilitate transfer of electrons to the valence band of PS I from the conduction band of PS II (Fig. 1.6). A PSA aptamer conjugated magnetic bead was designed. A DNA walker strand that was partially complementary to it was attached. Hairpin DNA1 was conjugated onto the AuNPs while hairpin DNA2 labeled with CdS quantum dot (QD-H2) was synthesized. In presence of the target analyte, PSA, the DNA walker strand is released and induces the opening of DNA1. Further, in the presence of QD-H2, the DNA walker facilitates the hybridization of DNA1 with DNA2 on the Au NP in a stepwise fashion ultimately leading to the formation of a Z-scheme photosystem by assembly of CdS QDs on the AuNPs@BiVO4. This Z-scheme PEC sensing system, under optimum conditions displays good photocurrent responses toward the target PSA in the working range of 0.01–50 ng mL^{-1} at a low detection limit of 1.5 pg mL^{-1} .

Various strategies have been reported to increase the photocurrent response hence the sensitivity of PEC based biosensors. One of the techniques is introduction of plasmonic nanoparticles (e.g., Au NPs). As these when incorporated on a metal oxide-based

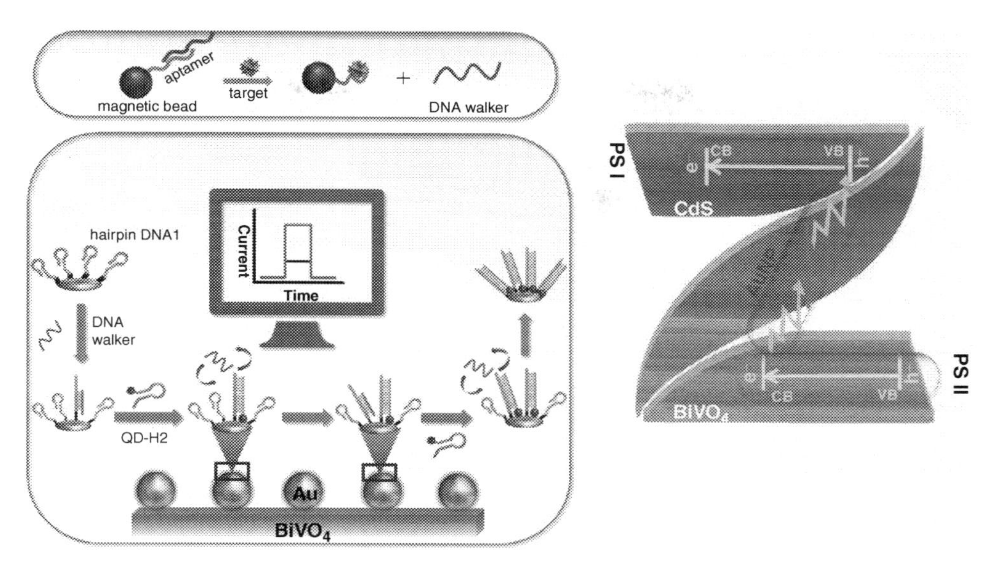

Fig. 1.6 ***Schematic illustration of the photoelectrochemical sensing platform based on the Z-scheme heterojunction.*** *(Reproduced from Ref. [24] with permission of American Chemical Society).*

photocatalyst may result in a Schottky barrier, which hinders exciton recombination. The visible light activity and the conductivity may also increase. Additionally, interactions between the exciton and the plasmon may be utilized to increase biosensing efficiency. In some cases, the electrocatalytic activity of these metal NPs aids the PEC biosensing process. Incorporation of carbon-based nanomaterials, for example, carbon nanotubes that have high surface area and stability along with high thermal and electrical conductivities has also been reported for increasing the sensitivity of PEC based biosensors. Other techniques are modification of the nanocatalyst surface by organic molecules that facilitate electron transport and design of dual sensitizers that involve coupling of wide and small band gap semiconductors for wider utilization of light energy and preventing recombination of charge. Another technique is the process of upconversion using primarily lanthanide based compounds and semiconductors wherein low energy infra-red radiation is converted into higher energy photons (ultraviolet or visible light), thus increasing the energy absorption range of the sensors.

1.2.4 Acoustic biosensors

Acoustic biosensors are based on inverse piezoelectric effect, that is, electric power applied to a piezoelectric material (e.g., quartz) produces acoustic waves. The biological analytes to be sensed are adsorbed on the piezoelectric crystal to alter the physical quantities such as resonant frequency, dissipation, and acoustic velocity, which are monitored with respect to time and analyzed further to extract useful biological information. It is categorized mainly into two groups: bulk acoustic wave and surface acoustic wave biosensors. In bulk acoustic wave biosensors, electric power is applied across the top and bottom of the crystal, which causes atoms of the crystal to oscillate side-by-side in the plane of crystal. In case of surface acoustic wave biosensors, electric power is applied across the surface of the crystal, which causes atoms of the crystal to oscillate up-and-down in the out of plane of the crystal. Generally, AT quartz is used in bulk acoustic wave biosensors while ST quartz is used in surface acoustic biosensors [25]. Nanomaterials can be utilized into surface acoustic biosensors to improve their performances and also in miniaturization of the devices. For example, coating of Au (gold) nanoparticles composited with a SiO_2 layer on ST quartz reduces the detection limit (from 9.4 ng mL^{-1} to 37 pg mL^{-1}) of carcinoembriyonic antigen (a biomarker for cancer) in a surface acoustic biosensor. The improved sensitivity of the device is attributed to the mass amplification of the biological analytes on the surface of Au nanoparticles coated ST quartz [26]. Graphene is another fascinating nanomaterial that can be effectively utilized in surface acoustic biosensors [27]. Graphene used in a surface acoustic biosensor to detect endotoxin is shown in Fig. 1.7. The sensing layer in the device is fabricated as follows. Single layer graphene grown by chemical vapor deposition is coated on the quartz substrate followed by improvement in surface hydrophilicity using chitosan and immobilization of the aptamer on the surface by

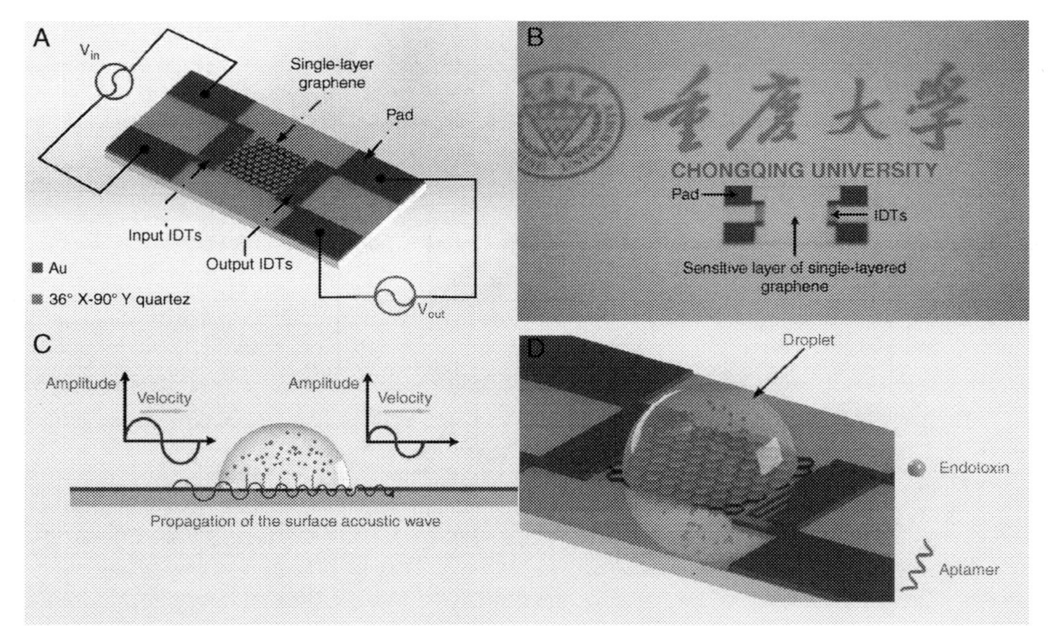

Fig. 1.7 ***Graphene used in a surface acoustic biosensor.*** (A) The surface acoustic biosensor with quartz crystal coated with graphene. (B) Practically fabricated surface acoustic biosensor. (C and D) Detection of endotoxin by the surface acoustic biosensor [28].

cross linking with glutaraldehyde. Detection of endotoxin is observed by monitoring the phase shift of the device due to the change in mass on the sensitive area. This graphene based surface acoustic biosensor has high sensitivity for detection of endotoxin with a detection limit of 3.53 ng mL^{-1} [28].

1.3 Nanozymes

For biomedical sensing applications, functional biomaterials with enzyme like characteristics, called nanozymes have been reported. The first report of use of nanozymes in sensing was in 2008 where magnetic Fe_3O_4 nanoparticles were used to mimic the enzyme peroxidase and were used to detect H_2O_2 and glucose. Since that report various colorimetric, fluorescent, chemiluminescent, and electrochemical nanozyme biosensors have been developed for in-vitro sensors. The detection of H_2O_2 was carried out by peroxidase mimics, various nanomaterials have been used for example, Fe_3O_4 NPs and its composites with protein, silica, graphene, and graphene oxide as well as hybrids with AgNPs, FeTe, iron phosphate, CuO, Au NPs and NC, Au @ Pt core shell nanorods, NiTe nanowires, carbon-based materials including C nanodots, carbon nitride dots, tungsten carbide nanorods, CoFe, and CoFe oxide nanoparticles, PtPd nanodendrites on graphene nanosheets, MgSe, Prussian blue, MWCNT-Prussian blue NPs polyoxometalate and polyoxometallate - Fe_3O_4 nanoparticles, $BiFeO_3$ NPs, FeS Co_3O_4 CdS, and layered double hydroxides [11]. However despite nanozymes being promising materials for the development of biosensors they currently have the following disadvantages (1) efficiency of nanozymes are lower than natural enzymes, (2) the 3D complexity of the natural enzymes cannot be mimicked by nanozymes, (3) the current range of artificial enzymes is narrow, limited to just mimics of peroxidase and oxidase, and (4) biosafety of the nanozymes has not been adequately considered.

1.4 Conclusion

A biosensor is an entity that can be utilized to detect and/or quantify a biochemical molecule called an analyte. The biosensing process undergoes the alteration in physical or chemical properties of either the analyte or the biosensor when they come in proximity or form a bond with each other. Nanomaterials are the ideal candidates to be utilized in biosensors as their sizes are compatible to the sizes of the biological molecules. In addition, the large surface area of nanomaterials is advantageous in attaching biological molecules for sensor development. The binding/molecular recognition event is translated into a sensing signal. Various important biosensors categorized based on the detection techniques are systematically described with their advantages and limitations. In particular, functional biomaterials with enzyme like characteristics, called nanozymes have been thoroughly discussed.

References

[1] AP.F. Turner, Biosensors: sense and sensibility, Chem. Soc. Rev. 42 (8) (2013) 3184–3196, https://doi.org/10.1039/C3CS35528D.

[2] J.L. Hammond, N. Formisano, P. Estrela, S. Carrara, J. Tkac, Electrochemical biosensors and nanobiosensors, Essays Biochem. 60 (1) (2016) 69–80, https://doi.org/10.1042/EBC20150008.

[3] K. Mahato, P. Kumar Maurya, P. Chandra, Fundamentals and commercial aspects of nanobiosensors in point-of-care clinical diagnostics, 3 Biotech 8 (3) (2018) 149, https://doi.org/10.1007/s13205-018-1148-8.

[4] A. Devadoss, P. Sudhagar, C. Terashima, K. Nakata, A. Fujishima, Photoelectrochemical biosensors: new insights into promising photoelectrodes and signal amplification strategies, J. Photochem. Photobiol. C 24 (2015) 43–63.

[5] V.X.T. Zhao, TIt Wong, X.T. Zheng, Y.N. Tan, X. Zhou, Colorimetric biosensors for point-of-care virus detections, Mater. Sci. Energy Technol. 3 (2020) 237–249.

[6] M. Schäferling, Fluorescence-based biosensors. In: Encyclopedia of Analytical Chemistry, John Wiley & Sons, Ltd, 2016, pp. 1–52.

[7] A. Munawar, Y. Ong, R. Schirhagl, M.A. Tahir, WS. Khan, SZ. Bajwa, Nanosensors for diagnosis with optical, electric and mechanical transducers, RSC Adv. 9 (12) (2019) 6793–6803, https://doi.org/10.1039/C8RA10144B.

[8] D.J. Denmark, X. Bustos-Perez, A. Swain, M.-.H. Phan, S. Mohapatra, SS. Mohapatra, Readiness of magnetic nanobiosensors for point-of-care commercialization, J. Electron. Mater. 48 (8) (2019) 4749–4761, https://doi.org/10.1007/s11664-019-07275-7.

[9] S. Kurbanoglu, N.K. Bakirhan, A. Shah, S.A. Ozkan, Chapter 5 - Chemical nanosensors in pharmaceutical analysis, in: S.A. Ozkan, A. Shah (Eds.), New Developments in Nanosensors for Pharmaceutical Analysis, Academic Press, 2019, pp. 141–170.

[10] M. Holzinger, A.L.e. Goff, S. Cosnier, Nanomaterials for biosensing applications: a review, Front. Chem. 2 (2014) 63, pp. 1–10. https://doi.org/10.3389/fchem.2014.00063.

[11] S. Lin, J. Wu, J. Yao, W. Cao, F. Muhammad, H. Wei, Chapter 7 - Nanozymes for Biomedical Sensing Applications: From In Vitro to Living Systems. in: Sarmento, B.; das Neves, J. (Eds.). In Biomedical Applications of Functionalized Nanomaterials, Elsevier: 2018, pp. 171–209. https://doi.org/10.1016/B978-0-323-50878-0.00007-0.

[12] R. Baron, M. Zayats, I. Willner, Dopamine-, L-DOPA-, adrenaline-, and noradrenaline-induced growth of Au nanoparticles: assays for the detection of neurotransmitters and of tyrosinase activity, Anal. Chem. 77 (6) (2005) 1566–1571, https://doi.org/10.1021/ac048691v.

[13] H. Aldewachi, T. Chalati, M.N. Woodroofe, N. Bricklebank, B. Sharrack, P. Gardiner, Gold nanoparticle-based colorimetric biosensors, Nanoscale 10 (1) (2018) 18–33, https://doi.org/10.1039/C7NR06367A.

[14] V. Rai, J. Deng, C.-.S. Toh, Electrochemical nanoporous alumina membrane-based label-free DNA biosensor for the detection of Legionella sp, Talanta 98 (2012) 112–117.

[15] V. Rai, Y.T. Nyine, H.C. Hapuarachchi, H.M. Yap, L.C. Ng, C.-.S. Toh, Electrochemically amplified molecular beacon biosensor for ultrasensitive DNA sequence-specific detection of Legionella sp, Biosens. Bioelectron. 32 (1) (2012) 133–140.

[16] R. Sheervalilou, O. Shahraki, L. Hasanifard, M. Shirvaliloo, S. Mehranfar, H. Lotfi, Y. Pilehvar-Soltanahmadi, Z. Bahmanpour, S.S. Zadeh, Z. Nazarlou, Electrochemical nano-biosensors as novel approach for the detection of lung cancer-related microRNAs, Curr. Mol. Med. 20 (1) (2020) 13–35.

[17] P.D. Sinawang, V. Rai, R.E. Ionescu, R.S. Marks, Electrochemical lateral flow immunosensor for detection and quantification of dengue NS1 protein, Biosens. Bioelectron. 77 (2016) 400–408.

[18] B.T.T. Nguyen, G. Koh, HSi Lim, A.J.S. Chua, M.M.L. Ng, C.-.S. Toh, Membrane-based electrochemical nanobiosensor for the detection of virus, Anal. Chem. 81 (17) (2009) 7226–7234.

[19] V. Rai, H.C. Hapuarachchi, L.C. Ng, S.H. Soh, Y.S. Leo, C.-.S. Toh, Ultrasensitive c DNA detection of dengue virus RNA using electrochemical nanoporous membrane-based biosensor, PLoS One 7 (8) (2012) e42346.

[20] V. Rai, C.-.S. Toh, Electrochemical amplification strategies in DNA nanosensors, Nanosci. Nanotechnol. Lett. 5 (6) (2013) 613–623.

[21] J. Zhang, S. Song, L. Wang, D. Pan, C. Fan, A gold nanoparticle-based chronocoulometric DNA sensor for amplified detection of DNA, Nat. Protoc. 2 (11) (2007) 2888–2895.
[22] Z. Qiu, D. Tang, Nanostructure-based photoelectrochemical sensing platforms for biomedical applications, J. Mater. Chem. B 8 (13) (2020) 2541–2561, https://doi.org/10.1039/C9TB02844G.
[23] A. Devadoss, P. Sudhagar, C. Ravidhas, R. Hishinuma, C. Terashima, K. Nakata, T. Kondo, I. Shitanda, M. Yuasa, A. Fujishima, Simultaneous glucose sensing and biohydrogen evolution from direct photoelectrocatalytic glucose oxidation on robust Cu 2 O–TiO 2 electrodes, PCCP 16 (39) (2014) 21237–21242.
[24] S. Lv, K. Zhang, Y. Zeng, D. Tang, Double photosystems-based 'Z-Scheme'photoelectrochemical sensing mode for ultrasensitive detection of disease biomarker accompanying three-dimensional DNA walker, Anal. Chem. 90 (11) (2018) 7086–7093.
[25] R. Fogel, J. Limson, A.A. Seshia, Acoustic biosensors, Essays Biochem. 60 (1) (2016) 101–110.
[26] S. Li, Y. Wan, Y. Su, C. Fan, V.R. Bhethanabotla, Biosens. Bioelectron. 95 (2017) 48–54.
[27] R.S. Singh, D. Li, Q. Xiong, I. Santoso, X. Yu, W. Chen, A. Rusydi, A.T.S. Wee, Anomalous photoresponse in the deep-ultraviolet due to resonant excitonic effects in oxygen plasma treated few-layer graphene, Carbon 106 (2016) 330–335.
[28] J. Ji, Y. Pang, D. Li, Z. Huang, Z. Zhang, N. Xue, Yi Xu, X. Mu, An aptamer-based shear horizontal surface acoustic wave biosensor with a CVD-grown single-layered graphene film for high-sensitivity detection of a label-free endotoxin, Microsyst. Nanoeng. 6 (1) (2020) 1–11.

CHAPTER 2

Emerging technology for point-of-care diagnostics: Recent developments

Subrata Mondal[a], Rahul Narasimhan[b], Ramesh B. Yathirajula[b], Indrani Medhi[b], Lidong Li[b,c], Shu Wang[d], Parameswar K. Iyer[a,b]
[a]Department of Chemistry, Indian Institute of Technology Guwahati, Guwahati, Assam, India
[b]Centre for Nanotechnology, Indian Institute of Technology Guwahati, Guwahati, Assam, India
[c]School of Materials Science and Engineering, University of Science and Technology Beijing, Beijing, China
[d]Institute of Chemistry, Chinese Academy of Sciences, Beijing, P.R. China

2.1 Introduction

In earlier days, it was required to send patient's sample away to the laboratories and wait for weeks to know the results. Consequently, mortality rate was comparatively higher because of prolonged diagnosis period. However, testing at 'Point of Care' can speed up clinical decision making and fasten the process of diagnosis, prognosis and operational treatment choice. POC testing makes use of smarter technology that simplifies testing process and offer lab quality diagnostic results within a very short period of time to improve productivity and reduce staff burden. In parallel to the improved clinical care, there may be economic benefits arising from adoption of POC based diagnosis. Expanses involving specimen packaging, transportation and reduced staff burden can significantly shrink total cost of care. In a survey, it has been found that improved diagnosis time and reduced staff burden due to adoption of POC technology decreases total laboratory expense by 8-20 % and waiting time by 46 minutes per patient which in turn helps reducing morbidity rate [1]. Consequently, entire medical regimes are being motivated toward point of care based diagnostics technologies. The invention of first wearable device, pacemaker in the year of 1958, is believed the pioneer in the world of wearable devices [2]. Since then, it has gone through a series of advancements. Present PoC based sensors functions like a "microcomputer" fitted with their "own sensors" and empowered by a microbattery with a communication tool. This chapter briefly discusses about a range of PoC devices and their futuristic applications in healthcare regime.

2.2 Power systems

The power systems play an essential role in point-of-care testing (POCT) systems. Power is highly essential for sensing, data processing and wireless communications. Sustainable energy can be the best choice for POC sensors which can be optimized by designing

Advanced Nanomaterials for Point of Care Diagnosis and Therapy
DOI: https://doi.org/10.1016/B978-0-323-85725-3.00021-0

self-powered sensors. They are developed to produce power from the environment like human body or the solar cells [3,4]. The main feature of the development of intravenous applications consists of integrating the instrumentation and the powering part in the same device. Hence two main problems can be overcome. The first one consists of combining the necessary instrumentation and communication systems to control the biosensors and to send the information through human skin, while the second one lies in the way to transfer adequate energy to empower the device. POC sensors communicate to other smart tools via Wi-Fi, Bluetooth and near-field communication (NFC) technology. Altogether this innovation has led to the development of POC sensors for remote monitoring of the patients in ambulatory environment which were previously impossible [5].

2.2.1 Lithium batteries

Lithium possesses the greatest electrochemical potential and provides the largest energy density with respect to their weight. The energy density of lithium-ion is almost twice than that of the standard nickel-cadmium. The load characteristics are appreciably good and perform similarly to nickel-cadmium in terms of discharge. The main advantages of lithium ion batteries are high energy density, relatively low self-discharge which is less than half of that of nickel-based batteries [6].

A Li-ion battery is fabricated by connecting primary Li-ion cells in series (to increase voltage), in parallel (to increase current) or combined configurations. Multiple battery cells are often integrated into a module. Usually, a basic Li-ion cell consists of an anode (negative electrode), which is contacted by an electrolyte containing lithium ions and a cathode (positive electrode). The both the electrodes remain isolated from each other by a separator, usually micro porous polymeric membrane, that allows the exchange of lithium ions between the two electrodes. In addition to liquid electrolyte, gel, polymer and ceramic electrolyte are also being investigated for applications in Li-ion batteries [7].

2.2.2 Solar cells

The solar cell is an electrical system which transforms the energy of sunlight into electricity through the photovoltaic effect. Three types of solar panels are known such as, mono crystalline, polycrystalline, and thin-film based solar cells. Each of these cells unique from the fabrication perspective.

POC based biosensors developed using nanomaterial offers promising approaches to develop high performance sensors with enhanced resolution and detection limits. As renewable energy is an promising alternative to other traditional sources such as fossil fuels, POC sensors integrated with photovoltaic systems which can detect several bioanalytes and empower itself by solar energy harvesting in lower cost. Such kind of devices can be eco-friendly as well as cost effective in nature [4,8].

Meanwhile, several biomedical equipments with complex structures require high power consumption. Even for smaller POC sensors, batteries are essential for the power supply. The disposal of batteries may cause environmental pollutions. Hence, development in renewable energy and energy harvesting are attracting immense scientific interests as an promising alternative to traditional energy sources like fossil fuels and nuclear power [9]. Photovoltaic (PV) devices which can convert sunlight into electricity may be viable solution for energy these days. Recently, the silicon wafer-based technology has gained significant industrial attention, but many other new technologies based on thin-film solar cell, dye-sensitized solar cell (DSSC) and organic materials solar cell offer multiple advantages including reduced cost and appreciable energy conversion efficiency [4,10,11]. The DSSC possess several advantages, which includes traditional roll printing techniques, flexible, transparent, and low-cost. Although the conversion efficiency of such devices is lower than the silicon solar cell, its potential efficiency was estimated to be good enough to consider them to compete with fossil fuel. Therefore, it is extremely important to further develop POC biosensor systems integrated with advanced nanomaterials to improve performance and reduce costs. Hence, the photovoltaic based power systems can be integrated with POC biosensors to reduce the cost following a green strategy [8].

2.2.3 Triboelectric nanogenerators (TENGs)

TENG is being extensively used in self-powered sensing devices. In some cases a combination of multiple energy generators are also being used. For example, a tribogenerator are integrated with a supercapacitor for energy storage in a transparent stretchable material to power a strain sensor [12].

Power is essential for the sensors for detecting bioanalytes, data processing, and wireless communication. Sustainable energy is essential for POC sensors that may be optimized by designing self-powered sensors. They are made to produce power from the environment like solar cells or from the human body (Fig. 2.1A) [13]. Thermogenerators are known to harvest energy from body heat (Fig. 2.1C) [15]. Electrical energy may be harvested from biological fluids like sweat which are known as biofuels [19,20]. Mechanical energy of a human body can be transformed to electrical energy through electromagnetic nanogenerators (ENGs), piezoelectric nanogenerators (PENGs), and triboelectric nanogenerators (TENGs) [14]. TENG is being widely used in self-powered POC sensors [16]. In some cases, a combination of these energy generators are also used [12,21]. For example, a supercapacitor for energy storage and a tribogenerator were integrated in a transparent stretchable substrate to power a strain sensor [16]. Dye sensitized solar cell, supercapacitor, and TENG have been integrated for self-powered textile where SC acts as the power storage bottom layer (Fig. 2.1B–E). The electrical energy stored in SC is collected from solar energy and converted by DSSC. The mechanical energy of the human body converted to electrical energy by the TENG [12]. Piezoelectric based nanogenerator systems is also a rising technology that can be used in POC sensors (Fig. 2.1F) [18].

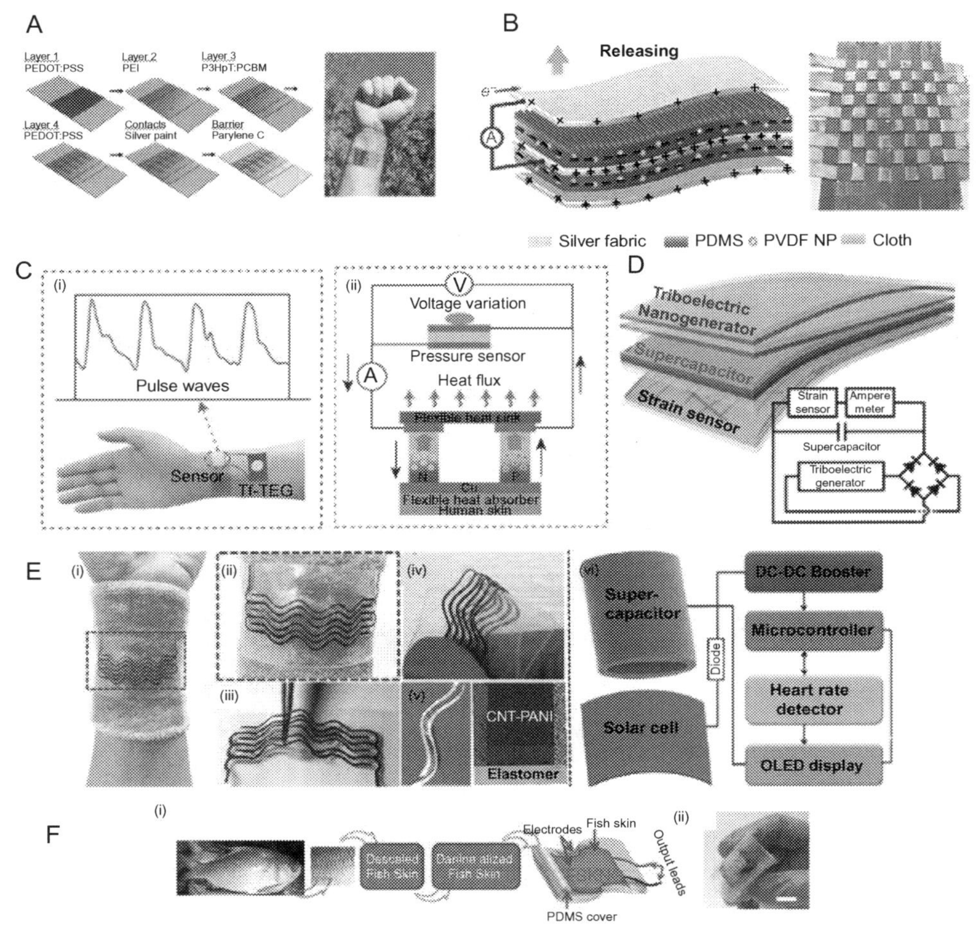

Fig. 2.1 (A) Schematic summary of the fabrication processes of wearable organic solar cells (left) and a digital image of the wearable solar cell equipped on skin (right). *Reprinted with permission from ref [13]. Copyright 2015 Elsevier B.V.* (B) Schematic representation of the working mechanism of the flexible woven (FW) TENG (FW-TENG, left), a digital image of the 2D woven structure interlacing the silver fabric and double-sided contact-separation structure (right). *Reprinted with permission from ref [14]. Copyright 2020 Elsevier Ltd.* (C) Schematic descriptions and morphology of the self-powered wearable pressure sensing system integrating thin film thermoelectric generator (tf-TEG) with a flexible pressure sensor. (i) Schematic illustration of tf-TEG and pressure sensor worn on the wrist. (ii) Equivalent circuit diagram of the complete system for sensing. *Reprinted with permission from ref [15]. Copyright 2020 Elsevier.* (D) Schematic illustration of the TENG, SC, and strain sensor. Transparent and stretchable strain sensor equipped on the neck. Circuit diagram of strain sensor with SC for charge storage which is charged by TENG. *Reprinted with permission from ref [16]. Copyright 2015 American Chemical Society.* (E) Digital images of (i) stretchable sweat band incorporated with interdigited solid state (IS) SC, (ii) close-up of ISSC, (iii) ISSC showing its freestanding nature, (iv) ISSC showing its flexibility for bending deformation (180°), and (v) close-up section of ISSC electrode. (vi) Schematic representation of the integration of the flexible solar cell and ISSC to a DC-DC booster to power a pulse rate sensor. *Reprinted with permission from ref [17]. Copyright 2019 Elsevier.* (F) (i) Fabrication of nanogenerator (NG) based on raw fish skin (FSK) and (ii) digital image of the FSKNG showing the flexibility. Scale bar ~20 mm. *Reprinted with permission from ref [18]. Copyright 2017 American Chemical Society.*

2.3 Technologies involved

2.3.1 Transistor

A transistor is a three terminal system where the channel current is regulated by the gate voltage. A FET amplifier is nothing but an amplifier which makes use of one or more field-effect transistors (FETs). The MOSFET amplifier is the most common type of FET amplifier which uses metal–oxide–semiconductor FETs (MOSFETs). The main features of a FET used for amplification is that it shows very high level of input impedance and low degree of output impedance. It can give one of the best impedance matching. A transistor functions as an amplifier by improving the strength of a weak signal. The DC bias voltage fixed in the emitter base junction, maintains it in forward biased condition. This forward bias is upheld irrespective of the signal polarity [22].

FET-based biosensing, also known as BioFET (or FEB), ChemFET, or ISFET (ion-sensitive FET) based on the analyte and mechanism under investigation, is an detection approach that makes use of a FET system that has been developed for biological sensing applications [24–26]. The FET biosensing is immensely important for a number of reasons such as it can be compatible with the "complementary metal oxide semiconductor" (CMOS) process, indicates FEBs can be embedded as a "system on a chip" (SoC) for electronic digital processing, and in a miniaturized fashion. The idea of such embedding for biosensors has been termed as "internet of biology" [24]. To use in the channel of the semiconductor, a wide range of materials have been studied, including organic semiconductors (OSCs), nanowires (NWs), one-dimensional (1D) materials such as CNTs and two-dimensional (2D) materials like graphene and other transition metal dichalcogenide (TMD) monolayers like MoS_2 [27–38]. Given such a wide range of options for channel substrate, and in addition several techniques for functionalization, there can be many different types of analytes that can be detected, including: proteins, DNA, chemicals, pH, antibodies, and other complex molecules [25]. The channel can also be modified using nanoparticles [39–41] or nanopores [39–44]. FEB systems provide ultra-high sensitivity, selectivity, stability, scalability, and can be fabricated on flexible substrates [45–48].

Particularly, interest in FET based biosensors was mainly inspired by the advantage of portability, rapid electrical detection without the need for labeling the biomolecules, low power consumption, and inexpensive mass production. In a typical FET device, both the electrodes (source and drain) are used to connect a semiconductor material (channel). Current passing through the channel is electrostatically regulated by a third electrode called the gate. In case of an FET biosensor (Fig. 2.2A–D), the gate present in a logic transistor is removed and the dielectric coating is functionalized with targeting receptors for selective detection of the desired biomolecules. On capturing a charged bioanalytes, produces an electrostatic effect, which is then transduced into a output signal in the form of change in electrical parameters of the FET such as channel conductance or drain-to-source current. In parallel to the dielectric layer, different polymers/lipids may also be used to in the

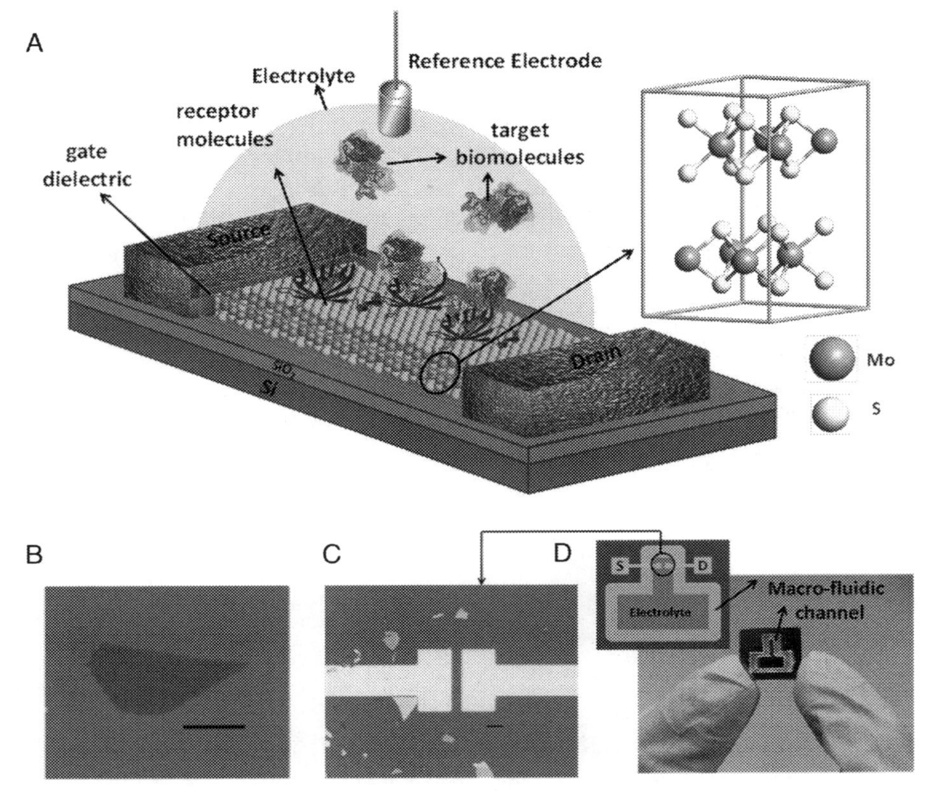

Fig. 2.2 ***MoS_2-based FET biosensor device.*** (A) Schematic diagram of MoS_2-based FET biosensor. For biosensing, the dielectric layer covering the MoS2 channel is functionalized with receptors for specifically capturing the target biomolecules. The charged biomolecules after being captured induce a gating effect, modulating the device current. An electrolyte gate in the form of an Ag/AgCl reference electrode is used for applying bias to the electrolyte. The source and drain contacts are also covered with a dielectric layer to protect them from the electrolyte (not shown in this figure). (B) Optical image of a MoS_2 flake on 270 nm SiO_2 grown on degenerately doped Si substrate. Scale bar, 10 μm. (C) Optical image of the MoS_2 FET biosensor device showing the extended electrodes made of Ti/Au. Scale bar, 10 μm. (D) Image and schematic diagram (inset figure) of the chip with the biosensor device and macrofluidic channel for containing the electrolyte. Inlet and outlet pipe for transferring the fluid and the reference electrode are not shown in the figure. *Reprinted with permission from ref [23]. Copyright 2014 American Chemical Society.*

channel, and in many cases direct functionalization of the channel using specific linkers/groups/receptors has shown significant improvements [23].

2.3.2 Electrochemical

The most simple and common electrochemical detection system based on the measurement of the current in presence of a constant potential, often known as amperometry.

Measurement using a constant potential has the benefit of avoiding unwanted effects of changes of the electrochemical double-layer charging at the interface between the solution and working electrode producing nonfaradaic currents. Electrochemical detection systems have shown application in flow analysis as well. The main features of flow analysis are enhanced selectivity and rapid detection carried out during the flow.

Electron transfer is believed to be the fundamental process in the chemistry of biosensors and more generally, in living systems [49]. Electron transfer (ET) is believed to be one of the simplest chemical processes as no new bonds are formed or broken. The fact that ET is present all over in biology and chemistry can be evident from several observations such as ET is the central to all the electrochemical processes, redox chemistry, photosynthesis, and respiration pathways [50].

2.3.3 Microfluidics

The main aim of a POC diagnostic research is to invent a proto-type, chip-based and miniaturized tool box that may be used to examine multiple analytes in complex biological samples. Moreover, the integration of microfluidics with improved biosensing system may result in improved POC diagnostics [51].

Point-of-care (POC) diagnosis systems have shown many advantages such as rapid response precise results and ease of data analysis. These microfluidics based POC systems have immense potential in providing improved healthcare including early detection of disease, easy monitoring and increased personalization [52].

Glass is a transparent and amorphous solid that exhibits properties such as chemical inertness, surface stability, robustness, and solvent compatibility. Furthermore, it is known for its biocompatibility, hydrophilicity, and yields a homogeneous coating, which makes it an ideal material for using in biomedical devices [53]. Nowadays, silicon is being as a preferred substrate for the developing MF channels for its high tolerance to varying conditions and necessity of low bonding temperature. In parallel to this, polymeric materials are also used in MF devices and immensely influenced the commercial manufacturers because of their cost effectiveness and easy fabrication steps, with compared to silicon and glass substrates [54]. Poly (methyl methacrylate) (PMMA) and poly (dimethylsiloxane) (PDMS) are being mostly used in MF devices since they result intransparency and excellent chemical/electrical resistivity indicating significant application for large-scale fabrication of MF devices [55].

2.4 Monitoring health parameters

The real time monitoring of physiological conditions of human beings monitoring has evolved to improve the longevity and quality of human life and considerably reduced the burden on the medical professionals. Sensor technology has advanced along with the information technology which gives human beings accurate and faster information on various physiological condition, which is crucial in avoiding larger health complications in future. The monitoring devices can be embedded into various places or objects in home or as a wearable device by integrating the sensor into watches, apparels, or jewelery.

2.4.1 Pedometer

Pedometer is an inexpensive physical activity assessment tool that can monitor the physical movement like number of steps taken during walking, jogging or running. And over the past few years the pedometers have gained the attention among the fitness enthusiast due to its increased accuracy in the ambulatory measurement and an effective tool to fight against the rise in obesity problem. The pedometer contains a motion sensor which can provide various performance output like distance travelled, number of steps, average speed, average pace and calories burnt. The performance metric can also be presented using an audio output along with the visual display. The earlier form of pedometers are tightly strapped to the hip, leg or shoe which was highly inconvenient before the advent of pedometers that are integrated into wrist watches. Classical pedometers are called mechanical pedometers which require a pendulum to detect the physical mobility. As the person walks, the swing created by movement creates inertia in the pendulum which can be sensed by a mechanical stop with a mechanical counter which is later converted as step count. Electronic pedometer consists of an open circuit, which can become a closed circuit with the movement of the pendulum. The closed circuit causes the current to flow thereby switching the electrical transducer for a momentary step.

In recent years, the moving pendulum of an electronic pedometer can be replaced with triaxial MEMS accelerometers due to its advantage like smaller size, lower power consumption and cost, higher accuracy. In addition, accelerometers can also record the intensity of the walking steps along with the frequency and pattern of the physical activity [56]. Accelerometers are a piezoelectric pedometer device which can detect the displacement in all the vertical, anteroposterior, and lateral planes. The accelerator contains a piezoelectric element along with the seismic mass. During the physical activity, the seismic mass creates a temporary pressure on the pressure sensitive piezoelectric material, which is transduced into an electrical signal. The accuracy of the pedometers can be determined by the rate of data acquisition or sampling frequency at which the sensor detects the step count. A sampling frequency ranging from 1 to 64 Hz is followed by the commercially available pedometer products.

2.4.2 Blood pressure

Detection in the fluctuation of blood pressure from the normal range (120/80mmHg) helps in the early diagnosis of various conditions in human body like diabetics, Heart problems, parathyroid disease, hypoglycemia, anaphylaxis, anemia, vascular disease, kidney problems, and hormonal imbalance. Blood pressure detector measures the pressure created due to the circulation blood upon the wall of the blood vessels. The waveform produced while measuring the blood pressure varies between high systolic blood pressure (SBP), which occurs during the end of the cardiac cycle and low diastolic blood pressure (LBP) during the beginning of the cardiac cycle. Blood pressure can be measured by both invasive and non-invasive monitoring. Invasive monitoring is the most accurate and reliable blood pressure measurement technique which can be performed only by the medical or healthcare professionals. Invasive measurement involves penetrating a cannula needle in the artery of the patient and is widely used in surgical application to continuously monitor the blood pressure of the patient. Noninvasive measurement can be used to have a tentative measurement with a little compromise in the accuracy, and these sophisticated methods do not involve penetrating into the blood vessels. Mercury sphygmomanometer is a classical non-invasive instrument used successfully for the last 100 years by placing an air-cuff over the upper arm. When the air-cuff is inflated, it temporarily obstructs the blood flow in the artery and the blood pressure can be detected by observing the Korotkoff sound through stethoscope or a Doppler when the air-cuff is slowly deflated. In contrast to observing the Korotkoff sound, oscillometry blood pressure monitor use a pressure sensor device to detect the pressure oscillation. The blood pressure is then computed from the oscillation using various empirical algorithms. Recently, an oscillometry finger pressing blood pressure monitor was also developed that can be deployed in any smart phones Fig. 2.3A–B [57]. The oscillometry setup uses the existing strain gauge available in the smart phone as a pressure sensitive element to monitor the BP with an accuracy of 4.0 ± 11.4 mmHg for (SBP) and -9.4 ± 9.7 mmHg for SBP and LBP respectively. Also, ultrasound is used to measure the clinical arterial blood pressure using the same principle used in ultrasound elastography (measures the tissue stiffness). The procedure measures the elasticity of the arterial wall and the tissue surrounding the artery from the obtained ultrasound images and further the data is mapped to the corresponding systolic and diastolic pressure based on the look up table. The studies from this procedure has shown an absolute relative relative error of 6.4% for SBP and 6.0% for LBP. [58] In addition, signals obtained from electrocardiogram (ECG) or photoplethysmogram (PPG) can also used to measure the blood pressure by extracting the pulse transition time (PTT). PTT refers to the propagation of the pulse pressure wave between two locations in the cardiovascular system which can be used to estimate the systolic blood pressure [59,60].

2.4.3 Galvanic skin response

Galvanic skin response (GSR) also called as skin conductance or electrodermal activity used to measure the physiological data by observing the variation in the electrical characterization of skin by placing two electrodes at the tip of the fingers. The primary reason for the change in the electrical conductivity of the skin is due to the variation in the human

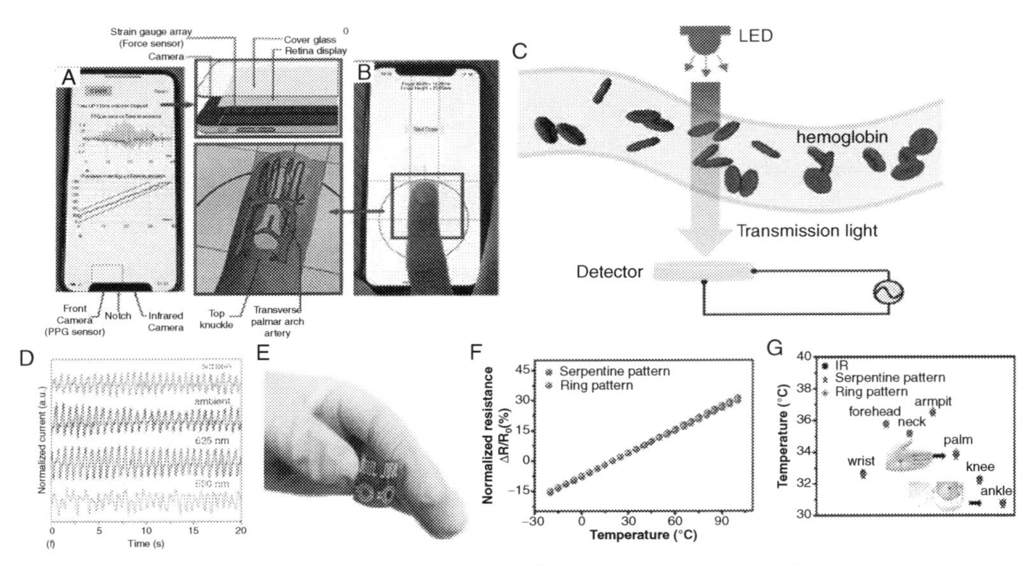

Fig. 2.3 (A) Mobile application that implement scillometric finger pressing blood pressure sensor which uses the strain gauge under the mobile display as the pressure sensing element and (B) photograph of the user making a measurement by placing the finger over the specific area on the screen of the mobile which also measures the height fingertip width and height from the artery in the middle of the finger to the top of the fingertip. *Reprinted with permission from ref [57]. Copyright 2018 Nature.* (C) Schematic representation of the heart rate detection system. (D) Photocurrent of the heart beat sensor under various illumination of light. (E) Flexible temperature sensor. *Reprinted with permission from ref [68]. Copyright 2020 American Chemical Society.* (F) Responsivity of the temperature sensor in the temperature range −20 to 100°C (g). Temperature sensed at various location of the body using ring and serpentine shaped sensor. *Reprinted with permission from ref [71]. Copyright 2019 Elsevier.*

body sweating. The sweat glands are controlled by the arousal of the sympathetic branch of the autonomic nervous system. Therefore, GSR can monitor the emotional behavior of the human under various circumstances like stress, pain and emotional trauma. The GSR monitor applies a constant voltage which is very small that the individual cannot be felt. The output current of GSR is monitored with a high sensitive parameter analyzer. The characterization of skin conductance is of two types, tonic and phasic. Tonic skin conductance is the base or reference data, where the individual is not provoked by any discrete environment. The phasic skin conductance is measured when the individual is provoked by external discrete stimuli which results in change in the conductance with respect to the tonic skin conductance. Tonic and phasic skin conductance level changes from person to person and also dependent on the individual's current psychological state.

2.4.4 Pulse oximeter

Pulse oximeter is a noninvasive diagnostic device that measures the oxygen saturation level in the blood without inserting a needle or extracting a sample of patient's blood.

The device monitors, how well the oxygen is pumped by the heart throughout the body. The device is a flexible component which can be used to give instantaneous results during intensive care, emergency aid, and recovery process. The instrument is effectively being used along with infrared temperature sensor as a preliminary testing method during the COVID-19 pandemic [61]. During the measurement, the device is clipped over the finger, toe or earlobe where a beam of light pass through the skin and the hemoglobin level is calculated by measuring the absorbance of light by the blood and thereby the arterial oxygen saturation of the blood is derived from the hemoglobin level. The pulse oximeter contains a photoplethysmogram (PPG) sensor which is categorized into transmission and reflectance type. In a transmission pulse oximetry, the LED, and the photodetector are placed facing each other, such that the light from the LED gets detected by photodetector after passing through the tissue. Whereas in a reflectance pulse oximetry, the LED and photodetector are placed on the same side. Most PPG sensors use green or infrared LED since it can penetrate through the tissue easily. Lee et al developed a wristwatch-type PPG sensor which specifically measures the PPG waves from the radial artery and the ulnar artery of the wrist [62]. Instead Kim et al. has developed wearable pulse oximetry system which can be embodied on any location of the body including the fingernail by measuring the PPG wave from the blood capillaries [63]. The device uses a near-field communication technology where the measured data can be transmitted wireless and also act as a power supply for the system.

2.4.5 Heart rate monitor

Electrocardiogram (ECG) is a technique that measures the electrical activity of the heart by placing an electrode on the chest, arms and legs of the person. The ECG test records the time varying signal produced due to the successive contract and relax of the atrial and ventricular fibers by applying a potential difference between the electrodes. The electrical signals can be approximated in a waveform containing peak and trough labeled as P, Q, R, S and T as shown in the figure. Various signal processing techniques can be used to extract accurate results like QT interval, RR Interval from the ECG waveform. Similar to ECG, PPG sensors can also monitor the irregular heartbeats (cardiac arrhythmias) [64–67]. Chen et al reported using an Antimony selenide in a photodetector to sense the heartbeat of the individual Fig. 2.3C–D. The sensor was able to sense the heartbeat with the presence of the room light and can eliminate the need of an external LED. The devices shows a difference of 0.002 bpm difference with the commercially available heart beat sensor [68].

2.4.6 Temperature sensor

Temperature sensor has become one of the vital device used throughout the healthcare sector to monitor the temperature, which is the primary response of the human body during any infections or other conditions like allergies, cardiovascular condition, pulmonological diagnostics and allergies. Also, it has played a pivotal role in monitoring the public during the COVID-19 pandemic. Clinical temperature measurement is of two types, one is the surface temperature and the other is the core temperature. Surface temperature can be measured using a semiconducting transducer like thermistor, resistive temperature detectors (RTD), pyroelectric detectors. Thermistors are resistors which changes its resistance with the change in the temperature. Thermistors are highly sensitivity to the temperature with a minimum 4% change in the resistance per degree. Increase in the resistance of the thermistor with the increase in the temperature is called Positive temperature coefficient (PTC) thermistor and decrease in the resistance of the thermistor with the increase in the temperature is called negative temperature coefficient (NTC) thermistor. NTC thermistors are widely used among the medical industry to monitor the temperature of the patient due to its better precision measurement and long term stability. Patients temperature can be accurately measured using thermistor by placing the probe on the patients rectal or the esophageal. Apart from body temperature, NTC thermistors are also used during dialysis and incubation period to maintain the appropriate temperature. Commercially available thermistors contain ceramic oxide material with a various disadvantage like poor integrability. Meanwhile electrical conducting polymers have emerged as an active material of the thermistor with a good sensitivity, sinter ability and also stretchablity. Recently an ultra-sensitive stretchable temperature sensor has been developed using a self-healing organohydrogels with a sensitivity of 19.6% [69]. Also thermal sensors can be fabricated using semi-crystalline polymer matrix with a conductive fillers graphite, CNT, graphene oxide. The use of a silver fractal as a conductive filler in a matrix of polyacrylate which exhibits a superior PTC effect for a temperature interval of 34°C to 37°C has been reported [70]. In contrast to thermistors, RTD sensors use metals like nickel, copper or platinum as the active material in the form of a fine wire which changes its resistance with respect to the temperature. A flexible RTD containing a patterned silver film with a shape of a ring and serpentine Fig. 2.3E–G was reported to have a sensitivity of 0.36 % and a good linearity of 0.16% [71]. Another RTD sensor developed can distinguish the temperature on different parts of the body which makes it a viable candidate for wearable and portable thermal sensitive device.

To have a contactless remote sensing, infrared temperature sensor is widely used to avoid further spread of the infection without compromising the accuracy and speed. Infrared temperature sensor detects the surface temperature of the individual by sensing the infrared energy emitted by the individual. The detector contains a lens which detects the IR energy (heat) of the person with respect to the absolute temperature of the surrounding environment. The gradients in the IR energy are converted into an electrical signal for the digital display of the temperature. Pyroelectric thermal detectors are a type of infrared sensor which can detects the incident radiation of electromagnetic frequency. Pyroelectric thermal detectors consist of a thin foil of crystalline material with pyroelectricity property where the charges are induced on the surface of the electrode due to the change in the polarization of the material under the vicinity of thermal flux.

2.5 Biomarkers for disease diagnosis

Biomarkers are biomolecules expressed in cells and can be a useful tool in disease detection. Not only biomarkers facilitate better understanding of diseases, but also aid in predicting or assessing the effects of a therapeutic intervention. Currently, many diagnostic tests and imaging technologies are in use that utilizes the detection or quantitative measurement of biomarkers. Some of the biomarkers and their detection systems are listed in the following sections.

2.5.1 Ion monitoring

2.5.1.1 Sodium

Sodium (Na) maintains the homeostatic state of the body through the regulation of salt. The sodium level of bodily fluids ranges from 135–145 mM. Blood and lymph fluid contains approximately 85% of total sodium content. Higher levels of sodium in the body is filtered by the kidneys and passed as urine. Sweating also releases smaller amounts of sodium. A few other biological mechanisms are also in place for sodium regulation and sensing. For example, adrenal glands secrete the hormone aldosterone that regulates sodium levels. In addition, the gene Nax is responsible for sodium sensing in the brain [72].

Imbalance of sodium can have deleterious effects such as improper functioning of nerves and muscles. In addition, low sodium levels may cause heart failure, malnutrition, or diarrhea [73]. Therefore, it is important to keep sodium levels under control. Nowadays, quantitative measurement of sodium in blood is possible by a simple sodium blood test.

2.5.1.2 Potassium

Similar to sodium, potassium (K^+) is another electrolyte with key roles in the body. Some important benefits of balanced potassium levels include the regulation of osmotic pressure and intracellular pH, maintenance of transmembrane potential and hormone secretion. Healthy potassium levels in the urine are in the range of 25–125 mM [74]. Like sodium, excretion of potassium is through urine filtered by the kidney [75]. Potassium sensing can occur via the transmembrane enzyme KdpD which either acts as a kinase when external potassium levels are low or as a phosphatase when levels are high [76].

Potassium imbalance can result in many complex diseases. For example, high potassium levels can cause renal hypertension, Cushing's syndrome, aldosteronism, etc. On the other hand, low potassium levels can cause chronic renal failure, Addison's disease, pyelonephritis among many others [77]. Several methods are available to selectively detect potassium ions that employ fluorescent, electrical or electrochemical techniques.

2.5.1.3 Chloride

Balance in chloride concentrations is also important for homeostasis. Increase or decrease in concentrations of such ions is indicative of a multitude of diseases such as cystic fibrosis (CF), and amyotrophic lateral sclerosis (ALS) among many others [78,79].

Water disinfection systems utilize introduction of chlorine. However, due to the toxic nature of high doses of chlorine, it is essential to monitor chlorine levels. Household water should contain no more than 0.2 ppm of free chlorine as per the recommendations of World Health Organization [80]. Abnormal high chlorine concentrations has a greater potential to adversely affect human health. For example, many cellular properties such as pH, volume and membrane potential may be affected by dysregulated chloride levels.

2.5.1.4 Zinc

Zinc ranks second in the catalogue of most abundant trace element with critical roles in cell growth. Impaired growth, decreased immunity, and abnormal brain functions are some of the consequences of zinc deficiency. Zinc is administered to treat tissue inflictions, diarrhea and ulcerative colon disease [81–83].

Zinc sensing is achieved by the extracellular receptor ZnR, which in turn activates the inositol 1,4,5–trisphosphate (IP3) pathway. In colonocytes, ZnR upregulates the Na+/H+ exchanger isoform 1 (NHE1) and consequently Na+/H+ exchange increases. This interaction has important roles in cellular ion homeostasis and cell proliferation. Dysregulation zinc concentrations and its related pathway is also identified to be one of the factors responsible for prostate cancer. Prostate cancer samples have a tenfold decrease in zinc ion concentrations in comparison to normal healthy samples. Thus, zinc has important roles in control of cell proliferation and overall survival [84]. Zinc levels can be measured from body samples such as urine and increase or decrease can be indicative of neoplastic growth or tumor.

2.5.1.5 Copper

In neurodegenerative diseases such as Alzheimer's, copper ion concentrations are not properly regulated and imbalance in Cu^+ ions is indicative of such diseases. As Cu^+ is an essential cation for many enzymes, fluctuations in its levels may affect oxidative metabolism. During catalysis, copper changes its oxidative state and can actively switch between Cu^+ and Cu^{2+} states. This, in turn, may result in conformational changes of the enzyme structure. However, such cycles between the two redox states also result in the accumulation of reactive oxygen species or ROS. ROS radicals are one of the major sources of cellular damage. Dysregulation of copper levels is detected using bodily fluids such as blood and urine as test samples [85–87].

2.5.1.6 Cadmium

Another important element involved in enzymatic activities is cadmium. However, unlike copper, cadmium is detrimental for enzymatic activities especially in DNA repair mechanisms. Cadmium is harmful for health and causes oxidative stress. Evolutionarily, organisms have adapted to intake less cadmium and has protective mechanisms in place against cadmium toxicity. Examples of protective proteins against cadmium are metallothionein, glutathione (GSH) and phytochelatin (PC) pathway proteins, antioxidant enzymes, and heat shock proteins [88].

Diseases associated with cadmium toxicity are respiratory diseases, cardiovascular diseases, neurodegenerative diseases and also certain types of cancer [89]. Exposure to cadmium can be detected from urine and blood samples.

2.5.1.7 Lead

Lead is one of the most toxic heavy metal pollutant with a severe detrimental impact to animal health and even the environment. Once introduced in the food chain, lead persists and accumulates and can have harmful toxicological effects. Lead toxicity can be evaluated using a blood test [90]. Currently, various methods for the removal of lead from water samples are under development and are of particular significance.

2.5.1.8 Mercury

Mercury (II) is another example of a heavy metal pollutant which is highly poisonous. Exposure to mercury is particularly dangerous as tissues readily absorb mercury even at lower concentrations [91]. Similar to cadmium, mercury ions or Hg^{2+} binds enzymes that are involved in critical processes of energy production, cell growth and nucleic acid synthesis [92]. Consequently, Hg^{2+} binding negatively impacts enzyme function. Hg^{2+} levels in drinking water are strictly recommended to be limited to 1 μg/L in China and 2 μg/L in USA. Mercury poisoning is detected from urine samples and levels exceeding 0.05 mg/L is usually indicative of lead poisoning [93].

2.5.2 pH

Changes in pH are a good indicator of impact in cellular functions and can be a useful tool in early diagnosis of diseases. Maintenance of intracellular pH is a requirement for optimal performance of many enzymes involved in a range of different cellular functions such as signal transduction, cell cycle, inflammation, ion transport, etc., and fluctuations can promote cancerous growth. One important reason for optimal pH for enzyme function is because of structural conformation of the protein and increase or decrease in pH can cause misfolding. Diseases related to abnormal pH fluctuations include neurodegenerative diseases such as Alzheimer's, gastrointestinal disorders and cancer [94,95].

Owing to the significance of monitoring pH values, many methods have been developed to quantify pH, which includes the use of organic dyes and fluorescent proteins. However, existing methods are not accurate as many factors may bias the measurement and thus, decrease the reliability of the detection system [96]. Therefore, there is a growing need to develop more accurate and real-time measurement systems for intracellular pH.

2.5.3 Glucose

Blood glucose levels are deregulated especially in metabolic diseases such as diabetes in which blood glucose levels can increase beyond 230 mg/dL [97,98]. Defects in regulation of glucose levels in diabetes is due to defective usage of the hormone insulin. The main function of insulin is in cellular glucose intake to be subsequently used as a source of energy in various metabolic pathways. Inability to use insulin by the body or decreased production due to pancreas malfunction can result in the manifestation of diabetes. Progressively, diabetic patients acquire several medical complications such as kidney disease, stroke, cardiovascular diseases. Diabetes is a complex disease with currently no cure. However, the deleterious consequences of this disease can be kept under control by frequently monitoring the glucose levels [99].

In addition to blood, glucose levels can also be detected and measured using plasma or serum samples. Glucose monitoring techniques can be based on enzymatic or hexokinase methods [100]. Enzymatic methods for glucose monitoring is the used mode for glucose measurement for home use. In this method, the signal is due to oxidation of hydrogen peroxide and is proportional to the amount of glucose present. As shown in Fig. 2.4, the enzyme glucose oxidase (GOx) converts glucose to gluconic acid in presence of water (H_2O) and oxygen (O_2). This reaction also results in the production of hydrogen peroxide (H_2O_2). Further oxidation of H_2O_2 at the electrode produces current as signal and acts as an indirect measurement of glucose levels [100].

The glucose detection method based on hexokinase involves a multistep process or chemical reactions. In presence of adenosine triphosphate (ATP) and magnesium ions, the enzyme hexokinase converts glucose to glucose-6-phosphate (G6P). Subsequent, catalysis of G6P and nicotinamide adenine dinucleotide (NAD) by the enzyme glucose-6-phosphate dehydrogenaseproduces NADH. Absorbance at 340 nm wavelength of NADH is then measured and is an indirect measurement of the amount of glucose present [100].

2.5.4 Lactate

Lactate levels can be indicative of a range of conditions such as high stress, shock or conditions leading to heart failure. High lactate produces an acidic environment that

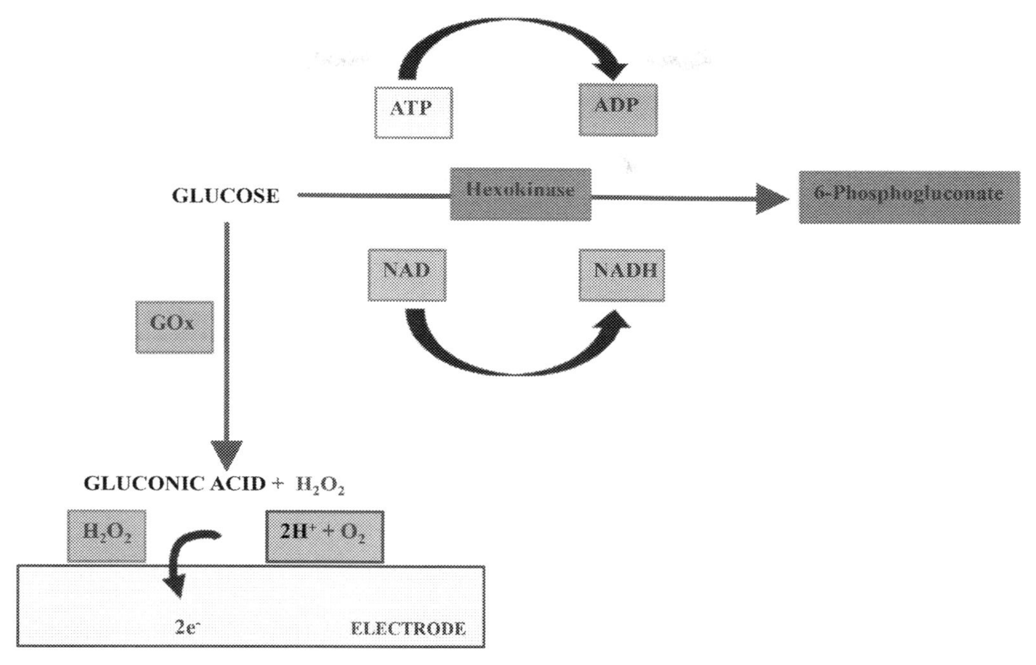

Fig. 2.4 ***Schematic diagram of glucose detection techniques.***

supports cancer growth. Therefore, early detection of elevated lactate levels may help to prevent the progression of conditions harmful to health.

Lactate levels can be measured using many bodily fluids such as blood, saliva, etc. Enzymes such as lactase oxidase (LOX) are utilized to measure blood lactate levels. Physical activity can also modulate blood lactate levels and can cause levels to increase to 25 mmol/L during strenuous exercise. However, normal levels is between 0.5–1.5 mmol/L. Human saliva has ~0.2 mM of lactate [101–103].

2.5.5 Bacteria

Microscopic organisms such as certain bacteria can have serious health consequences. Examples include *Salmonella* that causes food poisoning and pneumonia causing bacteria *Streptococcus pneumoniae*. Several microbiological techniques exists to detect bacteria with the most traditional one being bacteria grown on an agar plate. However, such techniques overlook the heterogeneity that exists at a bulk population level [104]. pH plays an important role in bacterial growth. Bacteria can efficiently maintain a pH gradient between the intracellular and external environments. For example, neutralophilic bacteria can survive and grow in the pH range or 5.5-9 while maintaining their internal pH in the range of 7.5-7.7. The bacteria *Streptococcus mutans* are present in dental plaques with a pH of ~ 4.8. Even in conditions of pH outside their growth range, bacteria can adapt to survive by using different strategies. Certain bacteria can cease their growth in order to survive and resume their growth when the pH is in their optimal growth range. Enteric bacteria such as *E. coli* and *Salmonella* survive the acidic stomach environment and ceases growth. Similarly, extremely alkaliphilic bacteria can survive high pH ranges of 10–13 [105].

As bacteria has evolved several strategies to survive at extreme conditions and can also develop antibiotic resistance, it is important to have effective strategies to detect several bacterial species for effective disease diagnosis and treatment decisions.

2.5.6 Virus

Pathogens such as viruses are responsible for a plethora of diseases worldwide. Being contagious in nature, viruses pose a serious threat to human health and are one of the major causes of mortality. Therefore, having effective strategies in place for control of viral infections is extremely crucial.

There are many methods for viral detection such as PCR-based amplification. Newer approaches based on nanoparticles (NP) have also been developed for viral detection. Additionally, development of biosensors that utilizes antibodies, nucleic acids, or enzymes as sensing element can also diagnose viral infections. Furthermore, aptosensors or biosensors that utilize aptamers as sensing elements are under development. Aptamers are three-dimensional structures formed of single stranded DNA or RNA with high affinity toward their target. Examples of targets of aptamers include proteins or their building blocks aminoacids, chemical compounds or other molecules. Aptamers are developed using the technique systematic evolution of ligands by exponential enrichment (SELEX). SELEX involves an iterative step-wise selection process from a large collection of random sequences of RNA or DNA oligonucleotides which are exposed to the target ligand and only the bound oligonucleotide is PCR amplified for further selection. As an alternative to antibodies, aptamers are highly preferred in biosensors owing to its high sensitivity and selectivity. For example, several biosensors are available that uses aptamers for the detection of avian influenza virus (AIV) subtype H5N1. Specificity against H5N1 is due to capture of viral particles by a DNA aptamer fixed on the electrode surface and integrated into different transducers [106,107].

2.5.7 Cancer

Cancer is a complex disease which results in deregulation of multiple molecular pathways and progressively leads to accumulation of deleterious mutations. Cancer promoting oncogenes help in the progression of the disease while certain other genes such as tumor suppressors inhibit tumor growth. Lack of or impairment of DNA repair mechanisms and enhanced cell proliferation results in increase in mutations and subsequent inactivation of tumor suppressor genes [108]. Advanced stages of cancer are difficult to treat and leads to low cell survival. Therefore, understanding and assessing cancer progression by utilizing known cancer biomarkers can tremendously help in overall survival. A range of biomolecules such as proteins, lipids, nucleic acids and sugars have been identified as cancer biomarkers. Extensive understanding of these biomarkers can help to understand the disease progression and also aid in deciding the course of treatment [108,109].

Cancer biomarkers vary depending on the type of cancer or the tissue of origin. Table 2.1 provides examples of cancer biomarkers depending on the cancer type and their method of detection. The catalogue of cancer biomarkers is increasing due to rapid advancement of technologies such as DNA/RNA high throughput sequencing. Advancement in non-invasive imaging technology helps to monitor cancer progression or response to treatment in patients effectively. Further identification and discovery of cancer biomarkers can serve useful for a more in depth understanding of such a complex disease. Along with this, improving the design and enhancing the detection sensitivity of biosensors for early detection and diagnosis can significantly help the effectiveness or success of cancer treatment [110].

2.5.8 Sleep

Sleep is one of the most important biological state that allows body and mind to take rest and recover [112]. Adequate amount of good quality sleep is essential to improve memory, mood and overall growth of the body. Sleep deprivation may often lead to many psychological and neurological disorders. A study in United States suggests that inappropriate sleep in young adults results in diabetes, obesity, hypertension, and many cognitive disorders [113]. Hence, several sleep parameters including amount, quality and phases of sleep is of great clinical concern for health professionals. Presently, polysomnography is believed to be the gold standard in determining sleep quality and stages. The technology makes use of EEG, ECG, EMG, and EOG all together to monitor sleep stages, overall sleep time, wake/arousal cycle [114]. Although, the extracted from PSG is highly reliable, it involves numerous wired probes attached to patient body, critical instrumentation and trained technician that limits its application in 'Point of Care'. To address the issue several portable, wireless and user friendly smart sleep sensor has been developed [115]. The wrist worn actigraphs can be an excellent choice to diagnose sleep related disorders like depression, dementia circadian rhythm sleep disorders, etc. [116]. The probe makes use of a simple concept that a person is more prone to make periodic movement in awake stage rather than in sleep stage [117]. Wrist actigraph (WA) through a accelerometer monitor body movements followed by an algorithm that determines sleep stages, quality and circadian rhythm [118]. However, the technique is not reliable to compute rapid eye movement (REM) and non REM sleep. However, several other devices that record additional physiological parameters like electro-dermal activities, heart rate and respiration rate also has been developed in order to improve accuracy in monitoring sleep architecture. Such devices comprised of accelerometer,

Table 2.1 Cancer biomarkers.

Type of cancer	Biomarkers	Detection technique
Acute myeloid leukemia (AML)	EGR1 (early growth response 1)	FISH
	IDH1/IDH2 (isocitrate dehydrogenase)	PCR-based mutation detection
	FLT3 (fms-like tyrosine kinase 3)	PCR
Acute promyelocyticleukemia	PML-RARa	FISH, PCR, cytogenetics
B-cell chronic lymphocytic leukemia (CLL)	TP53 (tumor protein p53)	FISH
	ATM	FISH
Breast cancer	HER2 (human epidermal growth factor receptor 2)	FISH, IHC
	ER/PR (estrogen receptor/progesterone receptor)	IHC, LBA
	PIK3CA (phosphatidylinositol 3-kinase catalytic subunit alpha)	Real time PCR
	TOP2A (DNA topoisomerase II alpha)	FISH
Chronic myeloid leukemia (CML)	BCR-ABL	FISH, PCR, cytogenetics
Colorectal cancer	KRAS (Kirsten rat sarcoma viral oncogene homolog)	Real time PCR
	BRAF (B-Raf)	Real time PCR
Follicular lymphoma (FL)	EZH2 (enhancer of zeste homologue 2)	Allele specific PCR
Gastrointestinal cancer (GIST)	KIT	IHC
Melanoma	BRAF (B-Raf)	Real time PCR
Lung cancer (NSCLC)	EGFR (epidermal growth factor receptor)	Sequencing, IHC
	AKL (anaplastic lymphoma kinase)	FISH, IHC
Ovarian cancer	BRCA1/BRCA2 (breast cancer gene)	Multiplex PCR, Sequencing
Prostate cancer	PSA (prostate specific antigen)	PCR
	PCA3 (prostate cancer antigen 3)	PCR
Urothelial cancer	FGFR3 (fibroblast growth factor receptor 3)	Real time PCR for mutational detection

Table 2.1 data obtained from reference [111].

gyroscope and ECG electrode altogether [114]. The Oura ring is one of the most popular and accurate sleep-activity trackers available in market that track sleep activities with 48% specificity and 96% sensitivity with compare to PSG [119]. Watchpat is another sleep monitoring system that measures peripheral arterial tone (PAT) [116,120]. It allows users to non-invasively measure arterial pulsatile volume changes at the fingertip which is a crucial parameter to track respiratory events. The sleep profiler is another wearable smart device to be worn on forehead and records ECG, EMG, EEG, head displacement through a series of sensors attached to it [116,121]. However, sleep trackers based on contactless technology are gaining high popularity among the users because of their ease of use. ResMed developed S+ sleep sensor that is not required to be worn. It makes use of low frequency radio waves to detect body movements of the user followed by a proprietary algorithm to produce data related to sleep stages, quality and architecture [116]. But probably, the most popular contactless sleep tracker is Beddit-3 developed by Apple Inc. The device is to be placed under the bed of the user. The device through its piezoelectric sensor records small forces exerted by body and decodes it to ECG, pulse rate and other respiratory activities [122]. However, all these devices are the best choices for personal sleep trackers but need robust validation before clinical application.

2.5.9 Seizure and epilepsy

Nowadays, epilepsy has emerged as one of the most abundant neurological disorders that results in sudden loss of awareness, pain and abnormal brain activities. It often appears with a bundle of involuntary movements involving a partial part of the body along with loss of consciousness. It has been established that the excessive electrical discharge in a specific neuronal cells results in seizure activities in humans. Table 2.2 sums up currently available devices for seizure detection. Epileptic patients are often advised to maintain a seizure log book as an integral part of the treatment [123]. Nowadays, video EEG is being used as the gold standard for detection of epileptic seizures [123]. However, it suffers from complex instrumentation and difficult to use in ambulatory environment. B-Alert has been developed as a wireless wearable system based on EEG for detection of seizure [124]. The Emotiv EPOC is also a smart wearables used for similar purpose [124]. In parallel to the EEG, some devices are also being developed based on electrodermal activities (EDA) that measures regulation of the skin conductance with respect to activities of the autonomic nervous system [125]. The Affectiva Q-sensor is a smart wearable devices which quantify EDA and skin temperature. In spite of good detection rate, results are affected by body movements and prone to produce artefacts. In addition to this, an accelerometer can also be used for detection of the epileptic seizures. An armband with commercial name Epicare free quantifies vibrations and convulsion fitted with an alarm is an excellent choice to detect epileptic seizure during sleep [125].

Table 2.2 Currently available devices for seizure detection. Adapted with permission from ACS Appl. Bio Mater. 4 (2021) 47–70. [126]

Company	Brand Name	Device type
ActiGraph	wGT3X, wActiSleep, GT3X, ActiSleep	Triaxis accelerometer watch
Advanced Brain Monitoring	X series – EEG wireless monitoring, B-Alert wireless EEG systems	EEG wireless headset to interpret health signals
Affectiva	Affectiva Q-Sensor	Wearable and wireless sensor to quantify EDA, activity and temperature and
Alert-It	Ep-It Companion Monitor	Bed motion sensor, accelerometer placed below mattress
Apple Inc.	EpiWatch	Apple Watch App with Health App for measuring heart rate, accelerometer and gyroscope. data acquired through iPhone and Apple Watch
Ashametrics Company	Wrist LifeBand, Ankle LifeBand, Chest LifeBand	Soft, wearable bands to monitor EDA, and temperature along with an accelerometer
D.C.T.Associates Pty Ltd.	Vigil-Aide	Detects initiation of seizure by movement sensor and buzzes an alarm
Emfit	Emfit Seizure Monitor	Bed motion detector along with accelerometer below mattress
Empatica	E3 Wristband	Wristband with accelerometer, EDA monitor, temperature sensor, gyroscope and seizures sensors
Hexoskin	Smart Shirt	Monitors pulse rate, respiratory rate, and body activities: sleep, and breathing volume
Holst Centre/IMEC Hobo Heeze BV	Epilepsy Seizure Monitoring System	Sensors for epileptic seizures. Associated with an algorithm. Also measures EEG, ECG and accelerometer sensors
Misfit	Beddit Sleep Monitor	Mattress strap to fit entire width, monitors heart rate, respiration, snoring and movement.
Neuro:On	Neuro:On Smart Sleep Mask	Monitord brain waves, eye movement, muscle tension, body temp, pulse, body temp and body movement.
Sensorium	Sensealert-102/EP200	Bed motion probes (accelerometer below mattress)
Smart Monitor Corp.	SmartWatch	Motion sensors along with GPS inferable in wristband
Vahlkamp	Epi-Watcher	Bed motion probes (accelerometer below the mattress) with timer to outweigh false positives signals. comprises a wireless alarm and a spoken message to pre-defined numbers

2.6 Conclusion and future perspective

Point of care devices are futuristic and have potential to transform medical diagnostics by altering the method of data acquisition, processing, and analyzing health information. The real applications of such devices are extensive and expected to grow in faster rate as they have become an integral part of everyday wearables. In last decade, the PoC devices have undergone multi-level advancements in the area of device engineering to replace conventional wearable diagnostic devices with the patient friendly non-invasive tools. However, there is enough space for further improvements in accuracy, selectivity and calibration. Also, stability of such devices is a big challenge as PoC application of such devices requires long term stability, biocompatibility and rapid response. Moreover, these devices need stability and accuracy in highly competitive biological environments. In addition to this, the devices must be free from calibration and interferences from other biological signals. Continuous supply of power in such devices remained another potential challenge. Nowadays, coin cell batteries are being used in most of the point of care devices. Such batteries are heavy in terms of weight and suffer from rigidity. The detail analysis of composition of various biofluids and their levels are essential for disease diagnosis. The use of PoC devices has made it possible to monitor levels of biomarkers in real time. Hence, the PoC devices may have the ability to revolutionize the healthcare scenario, however, further advancements and intensive optimization is needed before clinical applications.

References

[1] B. Nørgaard, C.B. Mogensen, Blood sample tube transporting system versus point of care technology in an emergency department; effect on time from collection to reporting? A randomised trial, Scand. J. Trauma Resusc. Emerg. Med. 20 (2012) 71.

[2] K. Jeffrey, V. Parsonnet, Cardiac pacing, 1960–1985, Circulation 97 (1998) 1978–1991.

[3] Z. Yang, J. Deng, X. Sun, H. Li, H. Peng, Stretchable, wearable dye-sensitized solar cells, Adv. Mater. 26 (2014) 2643–2647.

[4] S. Cai, X. Hu, Z. Zhang, J. Su, X. Li, A. Islam, L. Han, H. Tian, Rigid triarylamine-based efficient DSSC sensitizers with high molar extinction coefficients, J. Mater. Chem. A 1 (2013) 4763–4772.

[5] K. Guk, G. Han, J. Lim, K. Jeong, T. Kang, E.K. Lim, J. Jung, Evolution of wearable devices with real-time disease monitoring for personalized healthcare, Nanomaterials *29* (2019) 813.

[6] R. Gracia, D. Mecerreyes, Polymers with redox properties: materials for batteries, biosensors and more, Polym. Chem. 4 (2013) 2206–2214.

[7] W. Xing, A.M. Wilson, G. Zank, J.R. Dahn, Pyrolysed pitch-polysilane blends for use as anode materials in lithium ion batteries, Solid State Ion 93 (1997) 239–244.

[8] B. O'Regan, M. Grätzel, A low-cost, high-efficiency solar cell based on dye-sensitized colloidal TiO2 films, Nature 353 (1991) 737–740.

[9] A.N. Menegaki, Growth and renewable energy in Europe: Benchmarking with data envelopment analysis, Renew. Energy 60 (2013) 363–369.

[10] B. Shin, O. Gunawan, Y. Zhu, N.A. Bojarczuk, S.J. Chey, S. Guha, Thin film solar cell with 8.4% power conversion efficiency using an earth-abundant Cu_2ZnSnS_4 absorber, Prog. Photovoltaics Res. Appl. 21 (2013) 72–76.

[11] J. Krantz, T. Stubhan, M. Richter, S. Spallek, I. Litzov, G.J. Matt, E. Spiecker, C.J. Brabec, Spray-coated silver nanowires as top electrode layer in semitransparent P3HT:PCBM-based organic solar cell devices, Adv. Funct. Mater. 23 (2013) 1711–1717.

[12] Z. Wen, M.-H. Yeh, H. Guo, J. Wang, Y. Zi, W. Xu, J. Deng, L. Zhu, X. Wang, C. Hu, L. Zhu, X. Sun, Z.L. Wang, Self-powered textile for wearable electronics by hybridizing fiber-shaped nanogenerators, solar cells, and supercapacitors, Sci. Adv. 2 (2016) e1600097.
[13] T.F. O'Connor, A.V. Zaretski, S. Savagatrup, A.D. Printz, C.D. Wilkes, M.I. Diaz, E.J. Sawyer, D.J. Lipomi, Wearable organic solar cells with high cyclic bending stability: materials selection criteria, Sol. Energy Mater. Sol. Cells 144 (2016) 438–444.
[14] L. Liu, X. Yang, L. Zhao, W. Xu, J. Wang, Q. Yang, Q. Tang, Nanowrinkle-patterned flexible woven triboelectric nanogenerator toward self-powered wearable electronics, Nano Energy 73 (2020) 104797.
[15] Y. Wang, W. Zhu, Y. Deng, B. Fu, P. Zhu, Y. Yu, J. Li, J. Guo, Self-powered wearable pressure sensing system for continuous healthcare monitoring enabled by flexible thin-film thermoelectric generator, Nano Energy 73 (2020) 104773.
[16] B.-U. Hwang, J.-H. Lee, T.Q. Trung, E. Roh, D.-I. Kim, S.-W. Kim, N.-E. Lee, Transparent stretchable self-powered patchable sensor platform with ultrasensitive recognition of human activities, ACS Nano 9 (2015) 8801–8810.
[17] V. Rajendran, A.M.V. Mohan, M. Jayaraman, T. Nakagawa, All-printed, interdigitated, freestanding serpentine interconnects based flexible solid state supercapacitor for self powered wearable electronics, Nano Energy 65 (2019) 104055.
[18] S.K. Ghosh, D. Mandal, Sustainable energy generation from piezoelectric biomaterial for noninvasive physiological signal monitoring, ACS Sustain. Chem. Eng. 5 (2017) 8836–8843.
[19] Y. Su, J. Wang, B. Wang, T. Yang, B. Yang, G. Xie, Y. Zhou, S. Zhang, H. Tai, Z. Cai, G. Chen, Y. Jiang, L.-Q. Chen, J. Chen, Alveolus-inspired active membrane sensors for self-powered wearable chemical sensing and breath analysis, ACS Nano 14 (2020) 6067–6075.
[20] A.J. Bandodkar, J.-M. You, N.-H. Kim, Y. Gu, R. Kumar, A.M.V. Mohan, J. Kurniawan, S. Imani, T. Nakagawa, B. Parish, M. Parthasarathy, P.P. Mercier, S. Xu, J. Wang, Soft, stretchable, high power density electronic skin-based biofuel cells for scavenging energy from human sweat, Energy Environ. Sci. 10 (2017) 1581–1589.
[21] Y. Zi, L. Lin, J. Wang, S. Wang, J. Chen, X. Fan, P.-K. Yang, F. Yi, Z.L. Wang, Triboelectric–pyroelectric–piezoelectric hybrid cell for high-efficiency energy-harvesting and self-powered sensing, Adv. Mater. 27 (2015) 2340–2347.
[22] D. Kahng, Silicon-Silicon dioxide surface device, in: S.M. Sze (Eds.), Semiconductor devices: Pioneering papers, World Scientific (1991) 583–596.
[23] D. Sarkar, W. Liu, X. Xie, A.C. Anselmo, S. Mitragotri, K. Banerjee, MoS2 field-effect transistor for next-generation label-free biosensors, ACS Nano 8 (2014) 3992–4003.
[24] B.R. Goldsmith, L. Locascio, Y. Gao, M. Lerner, A. Walker, J. Lerner, J. Kyaw, A. Shue, S. Afsahi, D. Pan, J. Nokes, F. Barron, Digital biosensing by foundry-fabricated graphene sensors, Sci. Rep. 9 (2019) 434.
[25] Y.-C. Syu, W.-E. Hsu, C.-T. Lin, Review—Field-Effect Transistor Biosensing: Devices and Clinical Applications, ECS J. Sol. State Sc. 7 (2018) Q3196–Q3207.
[26] M.S. Makowski, A. Ivanisevic, Molecular analysis of blood with micro-/nanoscale field-effect-transistor biosensors, Small 7 (2011) 1863–1875.
[27] Q. Li, N. Lu, L. Wang, C. Fan, Advances in nanowire transistor-based biosensors, Small Methods 2 (2018) 1700263.
[28] S.G. Surya, H.N. Raval, R. Ahmad, P. Sonar, K.N. Salama, V.R. Rao, Organic field effect transistors (OFETs) in environmental sensing and health monitoring: a review, Trends Analyt. Chem. 111 (2019) 27–36.
[29] L. Torsi, M. Magliulo, K. Manoli, G. Palazzo, Organic field-effect transistor sensors: a tutorial review, Chem. Soc. Rev. 42 (2013) 8612–8628.
[30] S.M. Goetz, C.M. Erlen, H. Grothe, B. Wolf, P. Lugli, G. Scarpa, Organic field-effect transistors for biosensing applications, Org. Electron. 10 (2009) 573–580.
[31] F. Maddalena, M.J. Kuiper, B. Poolman, F. Brouwer, J.C. Hummelen, D.M. de Leeuw, B. De Boer, P.W.M. Blom, Organic field-effect transistor-based biosensors functionalized with protein receptors, J. Appl. Phys. 108 (2010) 124501.
[32] J. Lei, H. Ju, Nanotubes in biosensing, Nanomed. Nanobiotechnol. 2 (2010) 496–509.
[33] G. Gruner, Carbon nanotube transistors for biosensing applications, Anal. Bioanal. Chem. 384 (2006) 322–335.

[34] A. Nag, A. Mitra, S.C. Mukhopadhyay, Graphene and its sensor-based applications: a review, Sens. Actuator A Phys. 270 (2018) 177–194.
[35] M. Pumera, Graphene in biosensing, Mater. Today 14 (2011) 308–315.
[36] C. Zhu, D. Du, Y. Lin, Graphene-like 2D nanomaterial-based biointerfaces for biosensing applications, Biosens. Bioelectron. 89 (2017) 43–55.
[37] P.K. Kannan, D.J. Late, H. Morgan, C.S. Rout, Recent developments in 2D layered inorganic nanomaterials for sensing, Nanoscale 7 (2015) 13293–13312.
[38] K. Shavanova, Y. Bakakina, I. Burkova, I. Shtepliuk, R. Viter, A. Ubelis, V. Beni, N. Starodub, R. Yakimova, V. Khranovskyy, Application of 2D non-graphene materials and 2D oxide nanostructures for biosensing technology, Sensors 16 (2016) 223.
[39] S. Mao, G. Lu, K. Yu, J. Chen, Specific biosensing using carbon nanotubes functionalized with gold nanoparticle–antibody conjugates, Carbon 48 (2010) 479–486.
[40] S. Myung, A. Solanki, C. Kim, J. Park, K.S. Kim, K.-B. Lee, Graphene-encapsulated nanoparticle-based biosensor for the selective detection of cancer biomarkers, Adv. Mater. 23 (2011) 2221–2225.
[41] X. Chen, Z. Guo, G.-M. Yang, J. Li, M.-Q. Li, J.-H. Liu, X.-J. Huang, Electrical nanogap devices for biosensing, Mater. Today 13 (2010) 28–41.
[42] J. Basu, C. RoyChaudhuri, Graphene nanoporous FET biosensor: influence of pore dimension on sensing performance in complex analyte, IEEE Sens. J. 18 (2018) 5627–5634.
[43] Y. Wang, Q. Yang, Z. Wang, The evolution of nanopore sequencing, Front. Genet. 5 (2015) 449.
[44] R. Ren, Y. Zhang, B.P. Nadappuram, B. Akpinar, D. Klenerman, A.P. Ivanov, J.B. Edel, Y. Korchev, Nanopore extended field-effect transistor for selective single-molecule biosensing, Nat. Commun. 8 (2017) 586.
[45] M. Xu, D. Obodo, V.K. Yadavalli, The design, fabrication, and applications of flexible biosensing devices, Biosens. Bioelectron 124-125 (2019) 96–114.
[46] Y.H. Kwak, D.S. Choi, Y.N. Kim, H. Kim, D.H. Yoon, S.-S. Ahn, J.-W. Yang, W.S. Yang, S. Seo, Flexible glucose sensor using CVD-grown graphene-based field effect transistor, Biosens. Bioelectron. 37 (2012) 82–87.
[47] O.S. Kwon, S.J. Park, J.-Y. Hong, A.R. Han, J.S. Lee, J.S. Lee, J.H. Oh, J. Jang, Flexible FET-Type vegf aptasensor based on nitrogen-doped graphene converted from conducting polymer, ACS Nano 6 (2012) 1486–1493.
[48] D. Lee, T. Cui, Low-cost, transparent, and flexible single-walled carbon nanotube nanocomposite based ion-sensitive field-effect transistors for pH/glucose sensing, Biosens. Bioelectron. 25 (2010) 2259–2264.
[49] S. F. Nelson, Electron transfer reactions in organic chemistry, in: V. Balzani (Ed.), Electron Transfer in Chemistry, Wiley, 2001, pp. 342–392.
[50] H. Taube, H. Myers, R.L. Rich, Observations on the mechanism of electron transfer in solution, J. Am. Chem. Soc. 75 (1953) 4118–4119.
[51] S. Solanki, C.M. Pandey, Biological applications of microfluidics system, in: C. Dixit, A. Kaushik (Eds.), Microfluidics for biologists, Springer, Cham, 2016, pp. 191–221.
[52] K. Ren, Y. Chen, H. Wu, New materials for microfluidics in biology, Curr. Opin. Biotechnol. 25 (2014) 78–85.
[53] L. Kulinsky, Z. Noroozi, M. Madou, Present technology and future trends in point-of-care microfluidic diagnostics, Methods Mol. Biol. 949 (2013) 3–23.
[54] M.J. Schöning, H. Lüth, Novel concepts for silicon-based biosensors, Phys. Status Solidi 185 (2001) 65–77.
[55] H. Becker, L.E. Locascio, Polymer microfluidic devices, Talanta 56 (2002) 267–287.
[56] J.E. Berlin, K.L. Storti, J.S. Brach, Using activity monitors to measure physical activity in free-living conditions, Phys. Ther. 86 (2006) 1137–1145.
[57] A. Chandrasekhar, C.-S. Kim, M. Naji, K. Natarajan, J.-O. Hahn, R. Mukkamala, Smartphone-based blood pressure monitoring via the oscillometric finger-pressing method, Sci. Transl. Med. 10 (2018) eaap8674.
[58] A.M. Zakrzewski, A.Y. Huang, R. Zubajlo, B.W. Anthony, Real time blood pressure estimation from force-measured ultrasound, IEEE Trans. Biomed. Eng. 65 (2018) 2405–2416.
[59] X. Ding, B.P. Yan, Y.-T. Zhang, J. Liu, N. Zhao, H.K. Tsang, Pulse transit time based continuous cuffless blood pressure estimation: a new extension and a comprehensive evaluation, Sci. Rep. 7 (2017) 11554.

[60] X. Ding, N. Zhao, G. Yang, R.I. Pettigrew, B. Lo, F. Miao, Y. Li, J. Liu, Y. Zhang, Continuous blood pressure measurement from invasive to unobtrusive: celebration of 200th birth anniversary of carl ludwig, IEEE J. Biomed. Health Inform. 20 (2016) 1455–1465.

[61] S. Shah, K. Majmudar, A. Stein, N. Gupta, S. Suppes, M. Karamanis, J. Capannari, S. Sethi, C. Patte, Novel use of home pulse oximetry monitoring in covid-19 patients discharged from the emergency department identifies need for hospitalization, Acad. Emerg. Med. 27 (2020) 681–692.

[62] Y. Lee, H. Shin, J. Jo, Y. Lee, Development of a wristwatch-type PPG array sensor module, IEEE Int. Conf. Consum. Electron, 2011 168–171.

[63] J. Kim, P. Gutruf, A.M. Chiarelli, S.Y. Heo, K. Cho, Z. Xie, A. Banks, S. Han, K.-I. Jang, J.W. Lee, K.-T. Lee, X. Feng, Y. Huang, M. Fabiani, G. Gratton, U. Paik, J.A. Rogers, Miniaturized battery-free wireless systems for wearable pulse oximetry, Adv. Funct. Mater. 27 (2017) 1604373.

[64] S. Baek, Y. Ha, and H.-w. Park. 2020. Accuracy of wearable devices for measuring heart rate during conventional and nordic walking. PM&R Online ahead of print.

[65] A. Hochstadt, E. Chorin, S. Viskin, A.L. Schwartz, N. Lubman, R. Rosso, Continuous heart rate monitoring for automatic detection of atrial fibrillation with novel bio-sensing technology, J. Electrocardiol. 52 (2019) 23–27.

[66] K. Ishii, N. Hiraoka, Nail tip sensor: toward reliable daylong monitoring of heart rate, Trans. Electr. Electron. Eng. 15 (2020) 902–908.

[67] H. Sharma, Heart rate extraction from PPG signals using variational mode decomposition, Biocybern. Biomed. Eng. 39 (2019) 75–86.

[68] C. Chen, K. Li, F. Li, B. Wu, P. Jiang, H. Wu, S. Lu, G. Tu, Z. Liu, J. Tang, One dimensional Sb_2Se_3 enabling a highly flexible photodiode for light-source-free heart rate detection, ACS Photonics 7 (2020) 352–360.

[69] J. Wu, Z. Wu, Y. Wei, H. Ding, W. Huang, X. Gui, W. Shi, Y. Shen, K. Tao, X. Xie, Ultrasensitive and stretchable temperature sensors based on thermally stable and self-healing organohydrogels, ACS Appl. Mater. Interfaces 12 (2020) 19069–19079.

[70] J. Kim, D. Lee, K. Park, H. Goh, Y. Lee, Silver fractal dendrites for highly sensitive and transparent polymer thermistors, Nanoscale 11 (2019) 15464–15471.

[71] L. Kang, Y. Shi, J. Zhang, C. Huang, N. Zhang, Y. He, W. Li, C. Wang, X. Wu, X. Zhou, A flexible resistive temperature detector (RTD) based on in-situ growth of patterned Ag film on polyimide without lithography, Microelectron. Eng. 216 (2019) 111052.

[72] M. Noda, T.Y. Hiyama, Sodium sensing in the brain, Pflugers Arch 467 (2015) 465–474.

[73] M. Ankarcrona, F. Mangialasche, B. Winblad, Rethinking alzheimer's disease therapy: are mitochondria the key?, J. Alzheimer's Dis. 20 (2010) S579–S590.

[74] G. Song, R. Sun, J. Du, M. Chen, Y. Tian, A highly selective, colorimetric, and environment-sensitive optical potassium ion sensor, Chem. Commun. 53 (2017) 5602–5605.

[75] C.A. Cuevas, X.-T. Su, M.-X. Wang, A.S. Terker, D.-H. Lin, J.A. McCormick, C.-L. Yang, D.H. Ellison, W.-H. Wang, Potassium sensing by renal distal tubules requires Kir4.1, J. Am. Soc. Nephrol. 28 (2017) 1814–1825.

[76] M.K. Ali, X. Li, Q. Tang, X. Liu, F. Chen, J. Xiao, M. Ali, S.-H. Chou, J. He, Regulation of inducible potassium transporter KdpFABC by the KdpD/KdpE two-component system in Mycobacterium smegmatis, Front. Microbiol. 8 (2017) 570.

[77] X. Liu, C. Ye, X. Li, N. Cui, T. Wu, S. Du, Q. Wei, L. Fu, J. Yin, C.-T. Lin, Highly sensitive and selective potassium ion detection based on graphene hall effect biosensors, Materials 11 (2018) 399.

[78] L. Trnkova, V. Adam, J. Hubalek, P. Babula, R. Kizek, Amperometric sensor for detection of chloride ions, Sensors 8 (2008) 5619–5636.

[79] J.P. Kim, Z. Xie, M. Creer, Z. Liu, J. Yang, Citrate based fluorescent materials for low cost chloride sensing in the diagnosis of cystic fibrosis, Chem. Sci. 8 (2017) 550–558.

[80] Y.-K. Yen, K.-Y. Lee, C.-Y. Lin, S.-T. Zhang, C.-W. Wang, T.-Y. Liu, Portable nanohybrid paper based chemiresistive sensor for free chlorine detection, ACS Omega 5 (2020) 25209–25215.

[81] Y. Xu, Y. Zhou, W. Ma, S. Wang, A fluorescent sensor for zinc detection and removal based on core-shell functionalized Fe_3O_4@SiO_2 nanoparticles, J. Nanomater. 2013 (2013) 178138.

[82] S. Chen, T. Sun, Z. Xie, D. Dong, N. Zhang, A fluorescent sensor for intracellular Zn^{2+} based on cylindrical molecular brushes of poly(2-oxazoline) through ion-induced emission, Polym. Chem. 11 (2020) 6650–6657.

[83] Y.-S. Yang, C.-M. Ma, Y.-P. Zhang, Q.-H. Xue, J.-X. Ru, X.-Y. Liu, H.-C. Guo, A highly selective "turn-on" fluorescent sensor for zinc ion based on a cinnamyl pyrazoline derivative and its imaging in live cells, Anal. Methods 10 (2018) 1833–1841.
[84] M. Hershfinkel, W.F. Silverman, I. Sekler, The zinc sensing receptor, a link between zinc and cell signaling, Mol. Med. 13 (2007) 331–336.
[85] J. Jo, H.Y. Lee, W. Liu, A. Olasz, C.-H. Chen, D. Lee, Reactivity based detection of copper(II) ion in water: oxidative cyclization of azoaromatics as fluorescence turn on signaling mechanism, J. Am. Chem. Soc. 134 (2012) 16000–16007.
[86] B. Ram, S. Jamwal, S. Ranote, G.S. Chauhan, R. Dharela, Highly selective and rapid naked-eye colorimetric sensing and fluorescent studies of Cu^{2+} ions derived from spherical nanocellulose, ACS Appl. Polym. Mater. 2 (2020) 5290–5299.
[87] P. Verwilst, K. Sunwoo, J.S. Kim, The role of copper ions in pathophysiology and fluorescent sensors for the detection thereof, Chem. Commun. 51 (2015) 5556–5571.
[88] S.A. Winter, R. Dölling, B. Knopf, M.N. Mendelski, C. Schäfers, R.J. Paul, Detoxification and sensing mechanisms are of similar importance for Cd resistance in Caenorhabditis elegans, Heliyon 2 (2016) e00183.
[89] R.F. Aglan, M.M. Hamed, H.M. Saleh, Selective and sensitive determination of Cd(II) ions in various samples using a novel modified carbon paste electrode, J. Anal. Sci. Technol. 10 (2019) 7.
[90] H. Ebrahimzadeh, M. Behbahani, A novel lead imprinted polymer as the selective solid phase for extraction and trace detection of lead ions by flame atomic absorption spectrophotometry: synthesis, characterization and analytical application, Arab. J. Chem. 10 (2017) S2499–S2508.
[91] S. Bhatt, G. Vyas, P. Paul, A new molecular probe for colorimetric and fluorometric detection and removal of Hg2+ and its application as agarose film-based sensor for on-site monitoring, J. Fluoresc. 30 (2020) 1531–1542.
[92] D. Dai, J. Yang, Y. Wang, Y.-W. Yang, Recent progress in functional materials for selective detection and removal of mercury(II) ions, Adv. Funct. Mater. 31 (2021) 2006168.
[93] W. Zhong, L. Wang, D. Qin, J. Zhou, H. Duan, Two novel fluorescent probes as systematic sensors for multiple metal ions: focus on detection of hg^{2+}, ACS Omega 5 (2020) 24285–24295.
[94] Y. Chen, C. Wang, Y. Xu, G. Ran, Q. Song, Red emissive carbon dots obtained from direct calcination of 1,2,4-triaminobenzene for dual-mode pH sensing in living cells, New J. Chem. 44 (2020) 7210–7217.
[95] S.-L. Yang, W.-S. Liu, G. Li, R. Bu, P. Li, E.-Q. Gao, A pH-sensing fluorescent metal–organic framework: pH-triggered fluorescence transition and detection of mycotoxin, Inorg. Chem. 59 (2020) 15421–15429.
[96] J. Shangguan, D. He, X. He, K. Wang, F. Xu, J. Liu, J. Tang, X. Yang, J. Huang, Label-free carbon-dots-based ratiometric fluorescence ph nanoprobes for intracellular pH sensing, Anal. Chem. 88 (2016) 7837–7843.
[97] M. Shokrekhodaei, S. Quinones, Review of non-invasive glucose sensing techniques: optical, electrical and breath acetone, Sensors 20 (2020) 1251.
[98] A.L. Rinaldi, E. Rodríguez-Castellón, S. Sobral, R. Carballo, Application of a nickel hydroxide gold nanoparticles screen-printed electrode for impedimetric sensing of glucose in artificial saliva, J. Electroanal. Chem. 832 (2019) 209–216.
[99] H. Teymourian, A. Barfidokht, J. Wang, Electrochemical glucose sensors in diabetes management: an updated review (2010–2020), Chem. Soc. Rev. 49 (2020) 7671–7709.
[100] W. Villena Gonzales, A.T. Mobashsher, A. Abbosh, The progress of glucose monitoring—a review of invasive to minimally and non-invasive techniques, devices and sensors, Sensors 19 (2019) 800.
[101] I. Olaetxea, A. Valero, E. Lopez, H. Lafuente, A. Izeta, I. Jaunarena, A. Seifert, Machine learning-assisted raman spectroscopy for pH and lactate sensing in body fluids, Anal. Chem. 92 (2020) 13888–13895.
[102] Z. Zhang, X. Xu, K. Chen, Lactate clearance as a useful biomarker for the prediction of all cause mortality in critically ill patients: a systematic review study protocol, BMJ Open 4 (2014) e004752.
[103] A. Kuşbaz, İ. Göcek, G. Baysal, F.N. Kök, L. Trabzon, H. Kizil, B. Karagüzel Kayaoğlu, Lactate detection by colorimetric measurement in real human sweat by microfluidic-based biosensor on flexible substrate, J. Text. Inst. 110 (2019) 1725–1732.
[104] G. Pitruzzello, D. Conteduca, T.F. Krauss, Nanophotonics for bacterial detection and antimicrobial susceptibility testing, Nanophotonics 9 (2020) 4447–4472.

[105] T.A. Krulwich, G. Sachs, E. Padan, Molecular aspects of bacterial pH sensing and homeostasis, Nat. Rev. Microbiol. 9 (2011) 330–343.
[106] M.S. Draz, H. Shafiee, Applications of gold nanoparticles in virus detection, Theranostics 8 (2018) 1985–2017.
[107] R. Jalandra, A.K. Yadav, D. Verma, N. Dalal, M. Sharma, R. Singh, A. Kumar, P.R. Solanki, Strategies and perspectives to develop SARS-CoV-2 detection methods and diagnostics, Biomed. Pharmacother. 129 (2020) 110446.
[108] B.J. Czowski, R. Romero-Moreno, K.J. Trull, K.A. White, Cancer and pH dynamics: transcriptional regulation, proteostasis, and the need for new molecular tools, Cancers 12 (2020) 2760.
[109] A.N. Bhatt, R. Mathur, A. Farooque, A. Verma, B.S. Dwarakanath, Cancer biomarkers - current perspectives, Indian J. Med. Res. 132 (2010) 129–149.
[110] J. Vallamkondu, E.B. Corgiat, G. Buchaiah, R. Kandimalla, P.H. Reddy, Liquid crystals: a novel approach for cancer detection and treatment, Cancers 10 (2018) 462.
[111] N. Goossens, S. Nakagawa, X. Sun, Y.J.T.C.R. Hoshida, Cancer biomarker discovery and validation, Transl. Cancer Res. 4 (2015) 256–269.
[112] D.W. Carley, S.S. Farabi, Physiology of Sleep, Diabetes Spectr 29 (2016) 5.
[113] O.M. Buxton, E. Marcelli, Short and long sleep are positively associated with obesity, diabetes, hypertension, and cardiovascular disease among adults in the United States, Soc. Sci. Med. 71 (2010) 1027–1036.
[114] A.J. Boe, L.L. McGee Koch, M.K. O'Brien, N. Shawen, J.A. Rogers, R.L. Lieber, K.J. Reid, P.C. Zee, A. Jayaraman, Automating sleep stage classification using wireless, wearable sensors, npj Digit. Med. 2 (2019) 131.
[115] G. Medic, M. Wille, M.E. Hemels, Short- and long-term health consequences of sleep disruption, Nat. Sci. Sleep 9 (2017) 151–161.
[116] B. Byrom, M. McCarthy, P. Schueler, W. Muehlhausen, Brain monitoring devices in neuroscience clinical research: the potential of remote monitoring using sensors, wearables, and mobile devices, Clin. Pharmacol. Ther. 104 (2018) 59–71.
[117] C.H. Schenck, M.W. Mahowald, R.L. Sack, Assessment and management of insomnia, JAMA 289 (2003) 2475–2479.
[118] K.L. Stone, S. Ancoli-Israel, Actigraphy, in: M. Kryger, T. Roth, W.C. Dement (Eds.), Principles and Practice of Sleep MedicineSixth Edition, Elsevier, 2017, pp. 1671–1678.
[119] J. Dunn, R. Runge, M. Snyder, Wearables and the medical revolution, Per. Med. 15 (2018) 429–448.
[120] J.A. Pinto, L.B.M. d. Godoy, R.C. Ribeiro, E.I. Mizoguchi, L.A.M. Hirsch, L.M. Gomes, Accuracy of peripheral arterial tonometry in the diagnosis of obstructive sleep apnea, Brazilian J. Otorhinolaryngol. 81 (2015) 473–478.
[121] C. Stepnowsky, D. Levendowski, D. Popovic, I. Ayappa, D.M. Rapoport, Scoring accuracy of automated sleep staging from a bipolar electroocular recording compared to manual scoring by multiple raters, Sleep Med. 14 (2013) 1199–1207.
[122] J. Paalasmaa, H. Toivonen, M. Partinen, Adaptive heartbeat modeling for beat-to-beat heart rate measurement in ballistocardiograms, IEEE J. Biomed. Health Inform. 19 (2015) 1945–1952.
[123] K. Vandecasteele, T. De Cooman, Y. Gu, E. Cleeren, K. Claes, W.V. Paesschen, S.V. Huffel, B. Hunyadi, Automated epileptic seizure detection based on wearable ECG and PPG in a hospital environment, Sensors 17 (2017) 2338.
[124] T.S. Grummett, R.E. Leibbrandt, T.W. Lewis, D. DeLosAngeles, D.M.W. Powers, J.O. Willoughby, K.J. Pope, S.P. Fitzgibbon, Measurement of neural signals from inexpensive, wireless and dry EEG systems, Physiol. Meas. 36 (2015) 1469–1484.
[125] A. Ulate-Campos, F. Coughlin, M. Gaínza-Lein, I.S. Fernández, P.L. Pearl, T. Loddenkemper, Automated seizure detection systems and their effectiveness for each type of seizure, Seizure Euro. J. Epilep. 40 (2016) 88–101.
[126] S. Mondal, N. Zehra, A. Choudhury, P. K. Iyer, Wearable sensing devices for point of care diagnostics, ACS Appl. Bio Materials 4 (2021) 47–70.

CHAPTER 3

Quantum dots enabled point-of-care diagnostics: A new dimension to the nanodiagnosis

Swayamprabha Sahoo[a], Ananya Nayak[b], Ayushman Gadnayak[a], Maheswata Sahoo[a], Sushma Dave[c], Padmaja Mohanty[d], Jatindra N. Mohanty[a], Jayashankar Das[d]

[a]IMS and SUM Hospital, Siksha 'O' Anusandhan (Deemed to be University), Bhubaneswar, Odisha, India
[b]Centre for Genomics and Molecular Therapeutics, Siksha 'O' Anusandhan (Deemed to be University), Bhubaneswar, Odisha, India
[c]Department of Applied Sciences JIET Jodhpur, Rajasthan, India
[d]Valnizen Healthcare Vile Parle West Mumbai, Maharashtra, India

3.1 Introduction

Point-of-care (POC) diagnosis is an important device for preventing and eliminating parasitic diseases that include leishmaniasis, malaria and lymphatic filariasis. Most of the POC assay use immunochromatography and subsidiary strategies to evaluate flow with gold nanoparticle as per reporters [1]. Analyzes totally based on gold nanoparticles commonly provide semiquantitative or qualitative consequences and relatively low sensitivity. Upward conversion phosphorus nanoparticles, and super-paramagnetic particles [2]. A quantum dot (Qdot) based assay in a site flow analysis layout with a mobile reader to trace antibody reaction the use of neurocysticercosis (NCC) as a diseased model was developed, and discovered that it executed further to standard reaction detection platforms antibodies in patients with NCC [3]. The mobile reader set up gives the gain of portability and flexibility to be used in region where the laboratory is not immediately accessible. The new POC assay with mobile reader is a possible option for detecting antibody responses [4]. Fig. 3.1 shows implications and barriers while using POCT device.

3.2 Characteristics of quantum dots-based point-of-care testing device

The Qdot-based POCT assay could also be a side stream immune-chromatographic sandwich analysis utilizing Qdots as a following material, which has the quality structure of an immune-chromatographic test strip. The antibody pair utilized within the tape was advanced by Nanjing Vazyme Medical Co., Ltd. This antibody epitope is practically to the BRAHMS antibody against procalcitonin. For this test 80 μL serum or plasma, blood required as test material and a reaction time of quarter hour. [5]

Advanced Nanomaterials for Point of Care Diagnosis and Therapy
DOI: https://doi.org/10.1016/B978-0-323-85725-3.00005-2

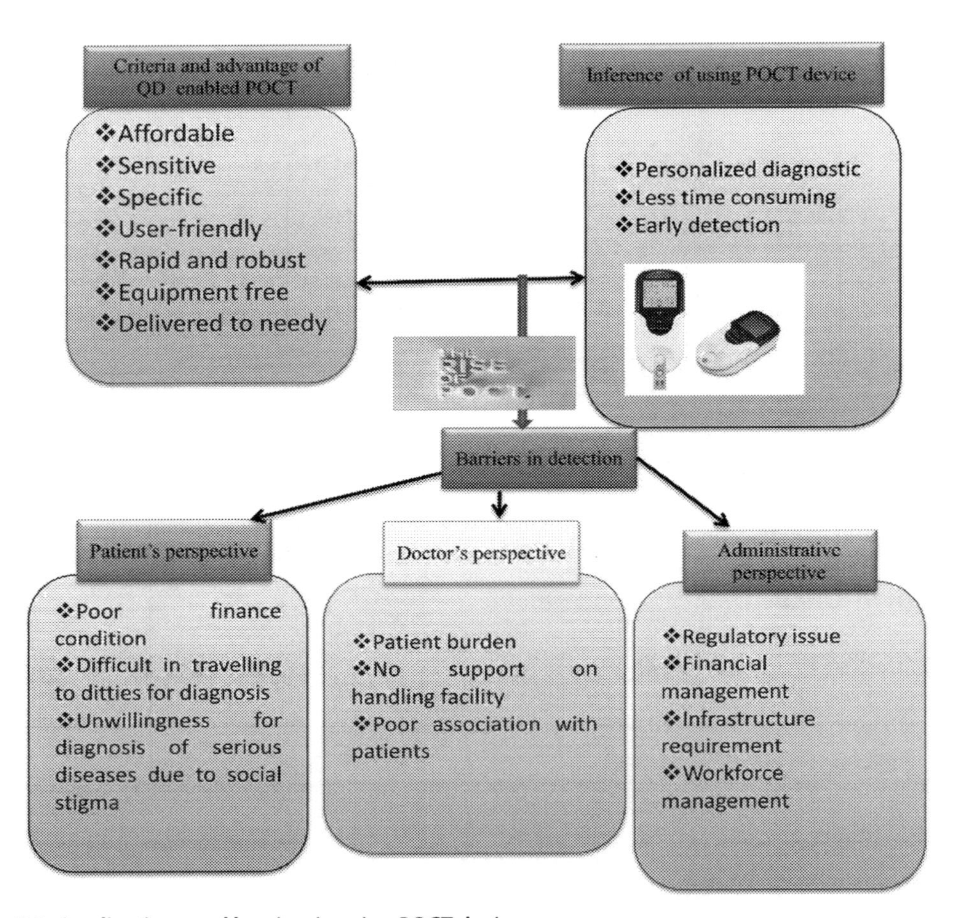

Fig. 3.1 ***Implications and barriers in using POCT device.***

A wide range of samples are dispensed precisely into the sampling opening without dilution. The sample is then reached to the top point of the strip on account of the capillary effect, since the liquid passes through the conjugated carrier, the Qdot of the antiprocalcitonin sensitive antibody are derived from the fluid and tie to the procalcitonin protein inside the sample Fig. 3.2. The procalcitonin antibody Qdot complex at that point moves with the liquid to the test line and control line, which is then captured by another antiprocalcitonin antibody on the test line [6]. At last, the extrasensitized Qdot that aren't certain to the procalcitonin protein are captured by the immobilized recombinant procalcitonin protein on the control line. The fluorescence signal from the test line and control line was recorded by an adjusted fluorescence immunoassay, and in this way the procalcitonin fixation inside the sample was determined by the proportion of sign power on the test line and control line. Each pile of tape is recalibrated and hence the alignment bend is put away on the memory card. Calibration could likewise be a 7-point alignment inside the reach 0.03 to 45.00 ng mL^{-1}. Prior to estimation, the

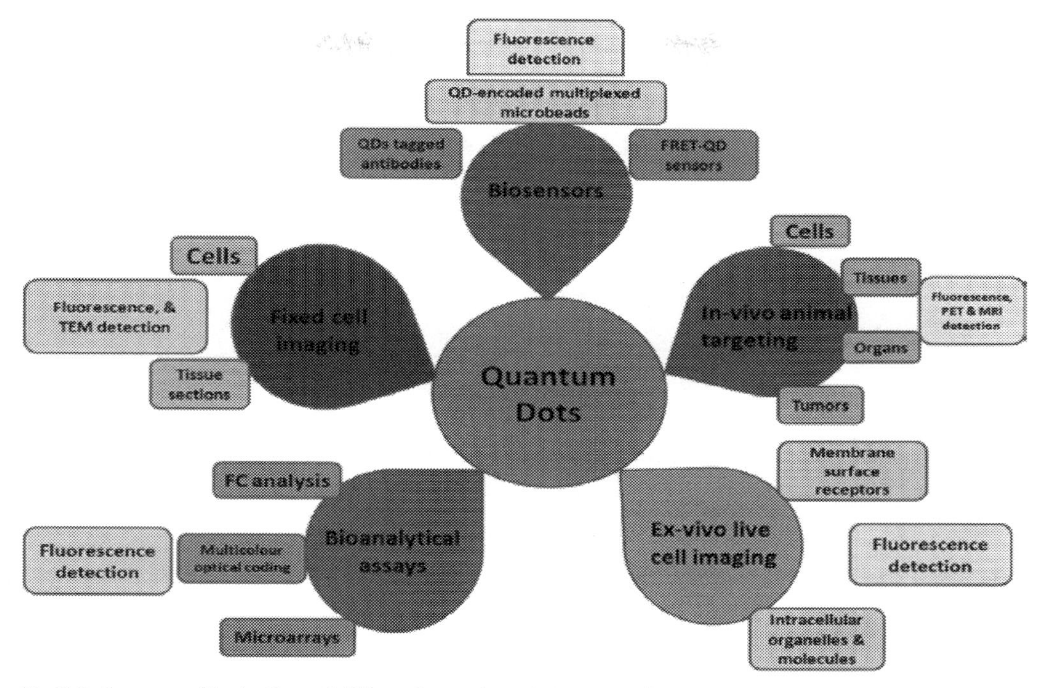

Fig 3.2 ***A sum up illustration of different quantum dot approach.***

calibration curve was moved to a fluorescent immunoassay [7]. This examination has been affirmed by the Chinese Food and Drug Administration. We bought all the test strips and fluorescent immunoassays used in this investigation from Nanjing Vazyme Medical Co., Ltd.

3.3 Nanobiosensors for point-of-care diagnostics

To ensure better health care, it is imperative to achieve world-class health management standards through timely solutions supported by rapid diagnosis, intelligent data analysis & information analysis [8] POC tests make certain rapid detection of analytes in the vicinity of the patient and allow better diagnosis, monitoring and treatment of disease. It also allows for quick medical resolution as diseases are often diagnosed early, which is leading to better health conditions for patients and permits them to begin treatment earlier [9–11]. Various potential treatment devices have been developed in the past to pave the way for testing next generation treatments. Biosensors are a very important part of the service kit because they are directly responsible for the presentation of the items bioanalytically. Therefore, they were screened for potential nursing applications, which are essential for personalized patient health management, as they frequently measure the contented of biological markers or chemical reactions, generating signals

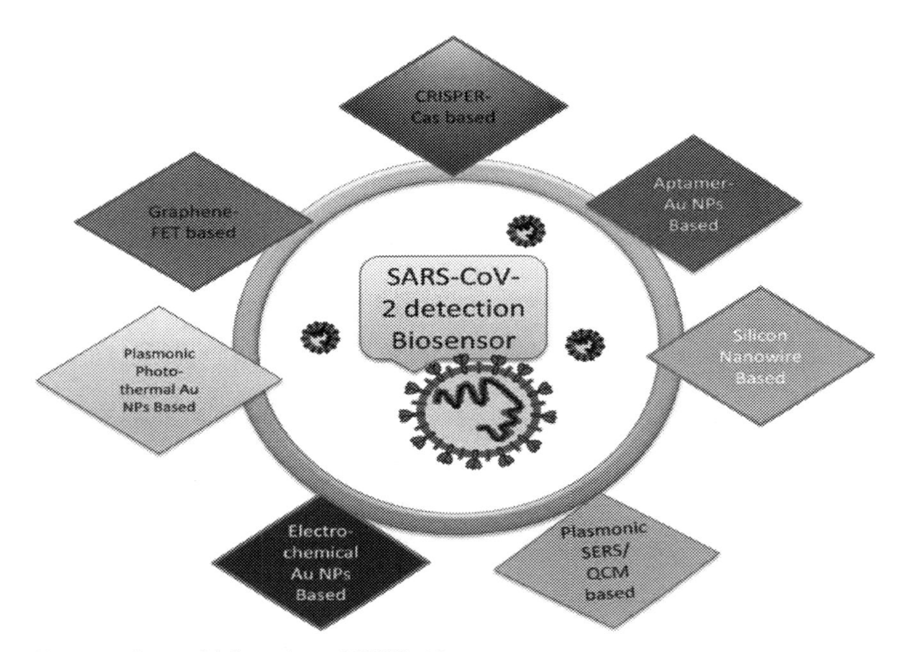

Fig 3.3 ***Biosensor for rapid detection of COVID-19.***

that are primarily related with analytes concentration and should therefore recognize markers [12–14]. They cause diseases like body fluids. High selectivity and susceptibility allow early diagnosis and treatment of the target disease. Therefore, facilitating timely therapeutic solutions and integrating them along with nanotechnology be capable of improving the evaluation process of disease onset and development as well as support in planning treatments for a wide range of diseases [15]. In this review, it was revealed that how nanotechnology is used in development of nanosensors and what could be the possible trends are currently being observed in these nanosensors to diagnose various diseases in specific objects [16–17] as shown in Fig. 3.3.

3.3.1 Nanosensors for point-of-care diagnostics of cancer

Malignancy is one of the main sources of death, in developing nation, however represents to one of every seven deaths around the world. There are in excess of 200 species cancer. Yet the most widely recognized sort is breast malignancy, esophageal malignant growth, ovarian cancer, colon malignant growth, prostate malignancy, cellular breakdown in the lungs, bladder cancer, kidney cancer, lymphoma, thyroid cancer, skin cancer, liver cancer, and pancreatic cancer. Breast cancer and ovarian malignancy are two perilous diseases that are often detailed in women. Around 180,000 new diagnosed cases are analyzed every year while 238,000 women are determined to have ovarian malignancy around the world, of which 151,000 deaths occurred. Chances of cancer death can be managed by only early detection or diagnosed way that could reduce both death rate and early recovery of patient falling in cancer. The ideal diagnostic tool is portable and ensures reliability [18]. It is seeking research to develop new tools that enable continuous, inexpensive, and real-time in vivo cancer monitoring, which ensuring early diagnosis, drug efficiency as well as effective drug delivery [19]. The utilizing of nanotechnology against drug diagnosis, cancer therapy, & drug delivery has made it possible to use nanomaterials to extract and detect specific tumor biomarkers, circulating tumor cells, or extracellular vesicles secreted by tumors [20–22]. Recent research results show the emergence of nano- & micro-based technologies that are integrated into various sensory and molecular communication platforms. Several authors reported the sensors for cancer biomarkers. This section describes recent advances in nanosensors for the treatment of cancer biomarkers [19,23].

3.3.2 Nanosensors for point-of-care diagnostics of diabetes

The development of affordable and easy and simple to use diagnostic tools remains a significant objective for global health issue [24]. Thus, new industrial products for nonstop glucose monitoring are being developed that have separate sampling points throughout the day to provide acceptable blood sugar control data for diabetic patients [30]. A different potential approach is the discovery of biomarkers from available body fluids using biosensors for POC as this can improve patient care by monitoring health status in real time and remotely [25]. The nanotechnology research, which includes nanomaterials and nanosensors, is also focused on the constant monitoring & these efforts also impact through increasing the sensor region, enhancing the catalytic property of electrodes and providing the nanoscale sensors. Nanotechnology research now days are used as these points of care for diagnostic equipment. For example, diabetes, where catalytic nanotubes such as graphene, carbon nanotubes, Qdots, and electrospine nanofibers are included in the biosensor to expand sensitivity detection limits and response time [26,27]. A number of new biosensors with nanomaterials are currently being developed. Second one is laboratory on-chip and embedded nanosensors for detecting glucose *in vivo* and *in vitro* [29]. The real-time monitoring device or tool is a powerful monitoring and diagnostic tool for measuring glucose in diabetes testing and field diagnosis [28].

3.3.3 Current trends of nanosensors for point-of-care diagnostics in infectious diseases

Infectious diseases, for example, dengue fever, cholera, hepatitis, severe respiratory syndrome malaria, and avian flu are usually caused by pathogenic microbes like bacteria, viruses, fungi, and parasites, which create a serious impact on human's health because of the characteristics such as reproductive speed and uncertainty. The salient vogue of malaria, TB, HIV, and outbreak of rising infectious diseases like Ebola, Influenza A and MERS poses major challenges for patient care in resource limited settings (RLS). Despite of advance technologies for diagnostic can't be achieved in RLS mainly for economic constrains. Inexpensive and easy point-of-care testing (POCT) that is rely less on operator training and environmental context, have been broadly studied to provide rapid diagnosis and treatment monitoring in settings outside the laboratory [31–34] what makes it different from other diseases. World Health Organization (WHO) announced that an appropriate tool for diagnosing infectious diseases must occurred specificity, sensitivity, accuracy, performance, ease of use, and low cost [35]. Conventional diagnostic techniques for this disease include microscopy and culture strategies, immunology as well as polymerase chain reaction (PCR). Although these techniques have made significant contributions to the recognition and diagnosis of infectious diseases, and subsequently the prevention and treatment of various infectious diseases, they should have proven impediments, for example, delays and inaccuracies since they are costly and, particularly in developing nations, also require ability [36,37]. Because of the extraordinary properties of nanomaterials in mechanical, optical, magnetic, electrical terms, and catalytic advances in nanotechnology have obtained numerous utilizations, particularly in biomedical applications, for example, drug delivery, tissue engineering, bioimaging, and also in nanodiagnostics [38–41]. Due to its specific early detection properties as well as high sensitivity, nanodiagnostics, the potential for on-site treatment tests, have gotten the best consideration in the diagnostics of infectious diseases. Mainly it's due to their potential to ensure portability, stability, and accessibility. During this investigation, we will specialize in various nanodiagnostic devices designed to diagnose diverse infectious diseases [42].

The rapid increase in research in chemical and biochemical sensors, on-chip laboratories, cellular technologies and wearable electronic devices offers unprecedented opportunities to test manual and mobile POC system for self-testing [43]. Successful implementation of POCT technology results in minimal user intervention during the process to reduce user error, easy-to-use, user-friendly and clear recognition platform; high diagnostic sensitivity and specificity; immediate clinical evaluation; and low production and consumption costs.

3.3.4 Recent trends in biosensor for SARS-CoV-2 detection

Switching to a diagnostic test for nCOV2 from a laboratory setting for a point of care could prospectively alter the speed & quality of the tests that can be performed. This describes 11 diagnostic tests which may be appropriate for testing of Corona (nCOV2) at POCT. Out of the six molecular tests & five antibody linked tests, some devices have got elevated diagnostic precision throughout controlled testing. However, there is currently a lack of data on the effectiveness of clinical settings & also a clear comprehending of the optimal population & the role of these tests in treatment pathways. Despite the major contributions made by biomedical engineering, nanotechnology and materials science to the development of POC diagnostics, there are still many technical challenges that must be overcome. New technologies based on advanced nanomaterials and micro fluids, improved analysis sensitivity, device miniaturization, cost reduction, and powerful multiplex detection could impact the future development of POC diagnostics.

The COVID 19 pandemic is now becoming increasingly severe because of its continued universal spread and a lack of suitable therapeutic & diagnostic systems. Global health authorities are working hard to combat the Corona (COVID 19) epidemic, studying all aspects of therapeutic development and specializing in studying the smart diagnostic tools required for the selective and rapid detection of the COVID 19 proteins [44–46]. The push for rapid mass population analysis for COVID 19 is filed by innovative techniques in biosensor development [47]. For the SARS-CoV-2 test of viral genomic RNA, membrane proteins and glycoprotein peaks requiring an immediate immune response if safe from the host to the ACE-2 receptor [48]. Humoral responses are mediated by IgG and IgM antibodies, which do not usually detect COVID-19 disease and are also utilized for a potential therapy called plasma therapy. Rapid COVID-19 POC tests can supply a result as "while you wait," ideally within 2 h of giving a sample. Probably this could help everyone to isolate in early stage that can diminish the spread of infection. For rapid testing people are interested for point of care diagnosis, antigen and molecular tests [49]. The SARS-Cov-2 virus biosensor includes CRISPR-Cas-based RNA detection; aptamer Au NP supports detection of viral proteins; detection of viral RNA is supported by AuNP electrochemistry; transistor biosensor with graphene field effect for virus detection, virus protein protein detection, silicon nanowires supported; nonlabeling plasmonic techniques such as surface plasmon resonance, surface enhanced Raman scattering, & microbalance quartz crystal biosensors for COVID-19 [50–51].

3.3.5 Future perspectives of quantum dot com point-of-care diagnosis

India's health system is a case of unprecedented budget constraints, different population groups and different health needs. This example requires unique solutions that are affordable, accessible, and effective, rather than minor technological advances that are at least unrelated or meet country-specific requirements [50]. The ROS diagnostic device meets the requirements of the individual situation, doctor, patient, and rural population. In addition, ROS testing is becoming more complex, with the potential to create chip technology laboratories that multitask and increase the effectiveness of timely patient care [51]. Despite the various barriers to using the POC diagnostic tool, scenarios are changing and encouragement is needed to expand its use to maximize its utility in meeting the health needs of rural populations.

References

[1] A. Kaushik, A.M. Mubarak, Point of care sensing devices: better care for everyone, Sensors 18 (12) (2018) 4303.
[2] A. Kaushik, A. Yndart, S. Kumar, R.D. Jayant, A. Vashist, A.N. Brown, C.Z. Li, M. Nair, A sensitive electrochemical immunosensor for label-free detection of Zika-virus protein, Sci. Rep. 8 (1) (2018) 1–5.
[3] Y.-C. Wang, Y.-T. Lee, T. Yang, J.-R. Sun, C.-F. Shen, C.-M. Cheng, Current diagnostic tools for coronaviruses–From laboratory diagnosis to POC diagnosis for COVID-19, Bioeng. Transl. Med. 5 (3) (2020) e10177.

[4] A. Kaushik, A. Vasudev, S.K. Arya, S.K. Pasha, S. Bhansali, Recent advances in cortisol sensing technologies for point-of-care application, Biosens. Bioelectron. 53 (2014) 499–512.
[5] R. Karmakar. Quantum dots and it method of preparations-revisited, Prajnan O Sadhona 2 (2015) 116–142.
[6] P. Malik, S. Gulia, R. Kakkar, Quantum dots for diagnosis of cancers, Adv. Mat. Lett. 4 (2013) 811–822.
[7] D.E. Zhang, X.M. Ni, H.G. Zheng, Y. Li, X.J. Zhang, Z.P. Yang, Synthesis of needle-like nickel nanoparticles in water-in-oil microemulsion, Mater. Lett. 59 (16) (2005) 2011–2014.
[8] E. Gatebes. Nanotechnology: The magic bullet towards attainment of Kenya's Vision 2030 on industrialization. (2016).
[9] G. Bagherzade, M.M. Tavakoli, MH. Namaei, Green synthesis of silver nanoparticles using aqueous extract of saffron (Crocus sativus L.) wastages and its antibacterial activity against six bacteria, Asian Pac. J. Trop. Biomed 7 (3) (2017) 227–233.
[10] N.M. Noah, P.M. Ndangili, Current trends of nanobiosensors for point-of-care diagnostics, J. Anal. Methods Chem. 2019 (2019).
[11] A. Verma, MS. Mehata, Controllable synthesis of silver nanoparticles using neem leaves and their antimicrobial activity, J. Radiat. Res. Appl. Sci. 9 (1) (2016) 109–115.
[12] A.I. Usman, A.A. Aziz, OA. Noqta, Application of green synthesis of gold nanoparticles: a review, Jurnal Teknologi 81 (1) (2019).
[13] X. Luo, A. Morrin, A.J. Killard, MR. Smyth, Application of nanoparticles in electrochemical sensors and biosensors, Electroanalysis 18 (4) (2006) 319–326.
[14] K.A. Gray, L. Zhao, M. Emptage, Bioethanol, Curr. Opin. Chem. Biol 10 (2006) 141–146.
[15] S. Eustis, MA. El-Sayed, Why gold nanoparticles are more precious than pretty gold: noble metal surface plasmon resonance and its enhancement of the radiative and nonradiative properties of nanocrystals of different shapes, Chem. Soc. Rev. 35 (3) (2006) 209–217.
[16] X. Zhang, Gold nanoparticles: recent advances in the biomedical applications, Cell Biochem. Biophys 72 (3) (2015) 771–775.
[17] G. Doria, J. Conde, B. Veigas, L. Giestas, C. Almeida, M. Assunção, J. Rosa, P.V. Baptista, Noble metal nanoparticles for biosensing applications, Sensors 12 (2) (2012) 1657–1687.
[18] R. Mosayebi, A. Ahmadzadeh, W. Wicke, V. Jamali, R. Schober, M. Nasiri-Kenari, Early cancer detection in blood vessels using mobile nanosensors, IEEE Trans. Nanobiosci 18 (2) (2018) 103–116.
[19] P.S. Gaikwad, R. Banerjee, Advances in point-of-care diagnostic devices in cancers, Analyst 143 (6) (2018) 1326–1348.
[20] C.G. Siontorou, G.P. Nikoleli, D.P. Nikolelis, S. Karapetis, N. Tzamtzis, S. Bratakou, Point-of-care and implantable biosensors in cancer research and diagnosis. In: Next Generation Point-of-Care Biomedical Sensors Technologies for Cancer Diagnosis, Springer, Singapore, 2017, pp. 115–132.
[21] Y. Okaie, T. Nakano, T. Hara, S. Nishio, Target Detection and Tracking by Bionanosensor Networks, Springer, 2020 Springer Nature Switzerland AG, 2016.
[22] E. Salvati, F. Stellacci, S. Krol, Nanosensors for early cancer detection and for therapeutic drug monitoring, Nanomedicine 10 (23) (2015) 3495–3512.
[23] P. Mohanty, C. Yu, W. Xihua, K. Mi Hong, C.L. Rosenberg, D.T. Weaver, E. Shyamsunder. Field effect transistor nanosensor for breast cancer diagnostics. arXiv preprint arXiv:1401.1168 (2014).
[24] K.J. Cash, HA. Clark, Nanosensors and nanomaterials for monitoring glucose in diabetes, Trends Mol. Med. 16 (12) (2010) 584–593.
[25] V. Kumar, S. Hebbar, A. Bhat, S. Panwar, M. Vaishnav, K. Muniraj, V. Nath, R.B. Vijay, S. Manjunath, B. Thyagaraj, C. Siddalingappa, Application of a nanotechnology-based, point-of-care diagnostic device in diabetic kidney disease, Kidney Int. Rep. 3 (5) (2018) 1110–1118.
[26] B.R. Azamian, J.J. Davis, K.S. Coleman, C.B. Bagshaw, ML. Green, Bioelectrochemical single-walled carbon nanotubes, J. Am. Chem. Soc. 124 (43) (2002) 12664–12665.
[27] B. Song, D. Li, W. Qi, M. Elstner, C. Fan, H. Fang, Graphene on Au (111): a highly conductive material with excellent adsorption properties for high-resolution bio/nanodetection and identification, Chem. Phys. Chem. 11 (3) (2010) 585–589.
[28] M. Taguchi, A. Ptitsyn, E.S. McLamore, J.C. Claussen, Nanomaterial-mediated biosensors for monitoring glucose, J. Diabetes Sci. Technol. 8 (2) (2014) 403–411.
[29] X. Kang, J. Wang, H. Wu, I.A. Aksay, J. Liu, Y. Lin, Glucose oxidase–graphene–chitosan modified electrode for direct electrochemistry and glucose sensing, Biosens. Bioelectron. 25 (4) (2009) 901–905.

[30] S. Alwarappan, C. Liu, A. Kumar, CZ. Li, Enzyme-doped graphene nanosheets for enhanced glucose biosensing, J. Phys. Chem. C. 114 (30) (2010) 12920–12924.
[31] Y. Wang, L. Yu, X. Kong, L. Sun, Application of nanodiagnostics in point-of-care tests for infectious diseases, Int. J. Nanomed. 12 (2017) 4789.
[32] B. Ray, E. Ghedin, R. Chunara, Network inference from multimodal data: a review of approaches from infectious disease transmission, J. Biomed. Inform 64 (2016) 44–54.
[33] A.S. Fauci, DM. Morens, The perpetual challenge of infectious diseases, N. Engl. J. Med. 366 (5) (2012) 454–461.
[34] W.G. Lee, Y.G. Kim, C. BG, U. Demirci, A. Khademhosseini, Nano/microfluidics for diagnosis of infectious diseases in developing countries, Adv. Drug. Deliv. Rev. 62 (4-5) (2010) 449–457.
[35] P. Yager, T. Edwards, E. Fu, K. Helton, K. Nelson, M.R. Tam, B.H. Weigl, Microfluidic diagnostic technologies for global public health, Nature 442 (7101) (2006) 412–418.
[36] P. Yager, G.J. Domingo, J. Gerdes, Point-of-care diagnostics for global health, Annu. Rev. Biomed. Eng. 10 (2008).
[37] K.L. Kotloff, J.P. Nataro, W.C. Blackwelder, D. Nasrin, T.H. Farag, S. Panchalingam, Y. Wu, S.O. Sow, D. Sur, R.F. Breiman, AS. Faruque, Burden and aetiology of diarrhoeal disease in infants and young children in developing countries (the Global Enteric Multicenter Study, GEMS): a prospective, case-control study, Lancet North Am. Ed. 382 (9888) (2013) 209–222.
[38] T. Laksanasopin, T.W. Guo, S. Nayak, A.A. Sridhara, S. Xie, O.O. Olowookere, P. Cadinu, F. Meng, N.H. Chee, J. Kim, C.D. Chin, A smartphone dongle for diagnosis of infectious diseases at the point of care, Sci. Transl. Med. 7 (273) (2015).
[39] Y. Wang, S. Yi, L. Sun, Y. Huang, M. Zhang, Charge-selective fractions of naturally occurring nanoparticles as bioactive nanocarriers for cancer therapy, Acta Biomater. 10 (10) (2014) 4269–4284.
[40] L. Sun, Y. Huang, Z. Bian, J. Petrosino, Z. Fan, Y. Wang, K.H. Park, T. Yue, M. Schmidt, S. Galster, J. Ma, Sundew-inspired adhesive hydrogels combined with adipose-derived stem cells for wound healing, ACS Appl. Mater. Interfaces 8 (3) (2016) 2423–2434.
[41] F. Chai, L. Sun, Y. Ding, X. Liu, Y. Zhang, T.J. Webster, C. Zheng, A solid self-nanoemulsifying system of the BCS class IIb drug dabigatran etexilate to improve oral bioavailability, Nanomedicine 11 (14) (2016) 1801–1816.
[42] P. Zhang, H. Lu, J. Chen, H. Han, W. Ma, Simple and sensitive detection of HBsAg by using a quantum dots nanobeads based dot-blot immunoassay, Theranostics 4 (3) (2014) 307.
[43] P. Tallury, A. Malhotra, L.M. Byrne, S. Santra, Nanobioimaging and sensing of infectious diseases, Adv. Drug. Deliv. Rev. 62 (4-5) (2010) 424–437.
[44] G.K. Darbha, U.S. Rai, A.K. Singh, PC. Ray, Gold-nanorod-based sensing of sequence specific HIV-1 virus DNA by using hyper-Rayleigh scattering spectroscopy, Chemistry. 14 (13) (2008) 3896–3903.
[45] V.K. Iglič, R. Dahmane, T.G. Bulc, P. Trebše, S. Battelino, M.B. Kralj, M. Benčina, K. Bo-hinc, D. Božič, M. Debeljak, D. Dolinar, From extracellular vesicles to global environment: a cosmopolite Sars-Cov-2 virus, IJCMCR. 4 (1) (2020) 004, http://doi.org/10.46998/IJCMCR 2020;79.
[46] C. Huang, Y. Wang, X. Li, L. Ren, J. Zhao, Y. Hu, L. Zhang, G. Fan, J. Xu, X. Gu, Z. Cheng, Clinical features of patients infected with 2019 novel coronavirus in Wuhan, China, Lancet North Am. Ed. 395 (10223) (2020) 497–506.
[47] B.B. Munnink, R.S. Sikkema, D.F. Nieuwenhuijse, R.J. Molenaar, E. Munger, R. Molenkamp, A. van der Spek, P. Tolsma, A. Rietveld, M. Brouwer, N. Bouwmeester-Vincken, Jumping back and forth: anthropozoonotic and zoonotic transmission of SARS-CoV-2 on mink farms, bioRxiv. (2020).
[48] T. Yang, Y.-C. Wang, C.-F. Shen, C.-M. Cheng, Point-of-care RNA-based diagnostic device for COVID-19, Diagnostics 10 (3) (2020) 165.
[49] B. Udugama, P. Kadhiresan, H.N. Kozlowski, A. Malekjahani, M. Osborne, V.Y. Li, H. Chen, S. Mubareka, J.B. Gubbay, WC. Chan, Diagnosing COVID-19: the disease and tools for detection, ACS nano 14 (4) (2020) 3822–3835.
[50] J. Xiang, M. Yan, H. Li, T. Liu, C. Lin, S. Huang, C. Shen, Evaluation of enzyme-linked immunoassay and colloidal gold-immunochromatographic assay kit for detection of novel Coronavirus (SARS-Cov-2) causing an outbreak of pneumonia (COVID-19), MedRxiv. (2020).
[51] A.N. Konwar, V. Borse, Current status of point-of-care diagnostic devices in the Indian healthcare system with an update on COVID-19 pandemic, Sensors International 1 (2020) 100015.

CHAPTER 4

Nanomaterials-based disposable electrochemical devices for point-of-care diagnosis

A.M. Vinu Mohan
Electrodics & Electrocatalysis Division, CSIR-Central Electrochemical Research Institute (CECRI), Karaikudi, Tamil Nadu, India

4.1 Introduction

The development of Lab-on-a-chip devices for clinical diagnostics has seen rapid growth over the past decade. Human blood plasma provides crucial information pertaining to disease diagnostics [1]. The conventional diagnostic tests are usually performed at centralized laboratories having expensive bench-top analyzers and operated by well trained personnel. Because of this, patients usually need to wait for several days to receive their diagnostic results. Hence, there has been a growing need to provide diagnostic results at the point of care, for prompt and accurate treatment of diseases in hospitals and for homecare diagnostics [2]. Point-of-care testing (POCT) is playing a vital role in improving the clinical outcome in health care management [3]. POCT diagnostic systems are portable instruments that can rapidly provide real-time diagnostic information by untrained personnel at a patient site in the home, the field, critical care unit, an ambulance, or a hospital. The POCT systems provide rapid and accurate test results allowing patients to ensure follow-up treatment immediately [4].

Biosensors are vital elements of a POCT device which detects the presence or the level of biomarkers from body fluids. Biosensors consist of biological recognition components like enzymes, antibodies, nucleic acids, animal or vegetable tissues, which are closely bonded with an appropriate transducer [5]. These immobilized recognition constituents selectively interact with the target molecules and produce physicochemical changes on the transducer's surface [6]. The changes are recognized by the transducer, and converted into quantifiable signals and subsequently determine the amount of analyte present in the sample. Depending on the type of transducers, biosensors can be classified into optical, electrochemical, mass-based, or piezoelectric sensors [7]. Depending on the mode of interactions between the analytes and the biological entities, biosensors can be divided into two types such as catalytic and affinity biosensor. In the former one, the interactions result in the formation of a new reaction product, but the latter one causes binding of analyte onto the transducer surface [8].

Advanced Nanomaterials for Point of Care Diagnosis and Therapy
DOI: https://doi.org/10.1016/B978-0-323-85725-3.00025-8

Electrochemical sensors are devices that deliver information about the analyte concentration by generating analytical signal from the recognition constituent coupled with an electrochemical transducer. The electron transfer kinetics at the electrode-electrolyte interface is crucial for sensitive electrochemical detection of analytes. The sensitivity of electrochemical transduction can be enhanced by appropriate surface modification at the electrode surface. The use of nanostructured materials enhances the electrochemically active surface area of the electrodes and provides higher response current even for low concentration of analytes. Selectivity is the most fundamental requirement for any type of sensor modalities for quantitatively detecting analyte of interest in the presence of structurally similar interferents, particularly in complex real samples. Functionalizing electrodes with conducting polymers and other electrically conductive nanomaterials increase the electron transfer kinetics and thus reduce the overpotential required for electrochemical oxidation/reduction of analytes. Several electroanalytical techniques are widely used for developing electrochemical sensors such as cyclic voltammetry, amperometry, differential pulse voltammetry, square wave voltammetry, impedance spectroscopy, and many more.

Majority of the commercially available electrochemical sensors are disposable in nature as they offers the advantages of eliminating surface fouling that can results in loss of reproducibility and sensitivity. The most critical factor defining disposability has been the economic efficiency which can be achieved via high-throughput scalable fabrication at very low costs, and by using minimum quantities of active materials [9]. Cost efficiency has been a key factor for the most successful commercial advances in disposable sensors; particularly glucose strips based diabetes monitoring, and pregnancy test kits for reproductive health assessment. When designing an electrochemical sensor the choice of the material for the working electrode is one of the important factors. The disposable electrochemical sensors use mostly carbon based electrodes as working and counter electrodes and Ag/AgCl traces as reference electrodes. Recently biodegradable and compostable electrodes using activated charcoal [10] and melanin [11] have been developed for diverse electrochemical applications so as to reduce the environmental impact and manufacturing cost of disposable sensors.

The development of nanostructured materials has paved the way for developing high-performance electrochemical sensors for medical diagnostics, environmental monitoring and food safety [12]. The design of nanomaterials based electrochemical sensors has gained enormous attention due to their high sensitivity and selectivity, real-time monitoring and ease of use [13]. Carbon based nanomaterials such as carbon nanotubes (CNTs), fullerenes and graphene have been widely used in electroanalytical and electrocatalytic applications due to their ability to promote electron transfer in electrochemical reactions [14,15]. Owing to the unique size and shape dependent physical, chemical and electrochemical properties, metal nanoparticles play various key roles in designing the electrochemical sensor and biosensor platforms. Metal nanoparticles offer effortless synthesis, ease of surface functionalization, stress-free sensor fabrication, facile catalysis of electro-chemical reactions and the enhancement of electron transfer rate [16,17]. With recent advances on the design of the metal nanoparticles, noble metallic nanomaterials such as gold (Au), silver (Ag), platinum (Pt), palladium (Pd) and their bimetallic alloys and core-shell nanoparticles are predominantly attracted for sensor and biosensor applications [18,19]. The chemical composition, surface condition, crystal structure quality, crystallographic axis orientation, etc. are critical parameters of nanomaterials, which cumulatively influence electron transport mechanisms. This chapter addresses how nanomaterials are utilized in the design of high-performance electrochemical sensors and biosensors for POCT applications. The recent advances in disposable electrode fabrication methods and their various applications in enzymatic, non-enzymatic and aptamer based sensors are discussed. Moreover, the burgeoning wearable sensors for non-invasive continuous monitoring of biomarkers and the associated feedback controlled drug delivery systems are detailed.

4.2 Paper-based disposable sensors for point-of-care testing

A POCT device that explores paper as a testing platform is termed as paper based analytical device (PAD), where cellulose fiber contribute a large component of the paper substrate. Cellulose based paper possesses distinctive characteristics for sensing application such as flexibility, low-cost, light-weight, biocompatible, eco-friendly, and hydrophilic in nature. Besides, the paper surface can be tirelessly functionalized by chemical or physical approaches and can be folded or stacked. The hydrophilic and porous nature of paper substrate enables adsorbing and percolating fluids by capillary action, and the autonomous flow of solution is independent on any external pumping sources. Recently PADs have attracted increased attention for detecting biomarkers in clinical, pharmaceutical and environmental surroundings. The healthcare monitoring necessitates rapid, reliable and simpler analytical method with augmented signal response and low limit of detection (LOD). The common analytical techniques used in PADs are colorimetric, fluorescent, chemiluminescent, electrochemical, photo-electrochemical, and surface plasmon resonance. The traditional colorimetric detection mode offers real-time naked-eye detection of analytes in a qualitative or semi quantitative manner. But the advent of microelectrodes and nanomaterials based electrocatalysts realizes highly sensitive quantitative assessment of markers by various electrochemical methods.

The advanced electrochemical techniques possess the potential for miniaturization, low-power consumption, low fabrication cost, portability, high sensitivity and selectivity, which enables effortless design and fabrication of the paper based analytical tools. The versatile electrochemical methods like amperometry, potentiometry, coulometry, and advanced pulse voltametric techniques can be used to harvest sensitive electrochemical signals from PADs. The studies showed that the electrochemical analysis on paper substrate is well comparable to results obtained by the conventional bench-top techniques such as UV-Vis spectroscopy, liquid/gas chromatography and inductively coupled plasma-mass spectrometry, with respect to sensitivity and selectivity. In addition, PADs allow multiplexed monitoring of multiple analytes, simultaneously from a single platform. The sample pre-treatment processes like collection, separation, extraction and concentration can be carried out within a single multiplexed platform, which further reduces cost, analysis time, and improves the efficiency of clinical diagnosis.

The choice of paper material from the available variety of paper materials is purely based on the fabrication steps involved in developing the prototype, and also depends on the required application. Chromatographic filter papers have been widely utilized for developing sensors and microfluidic devices, due to its inherent wicking ability. The Whatman® cellulose paper is extensively explored owing to the significant parameters such as porosity, particle retention and flow rate. Even though filter paper is a good choice of material, the drawbacks associated with its physical characteristics necessitate other types of paper materials for PAD application. For instance, nitrocellulose membranes are hydrophobic and possess a high degree of non-specific interaction toward biomolecules like enzymes, proteins and DNA. Glossy paper also employed as an appropriate sensor platform as it is a flexible substrate made of cellulose fiber blended with inorganic fillers. The non-degradability of glossy paper, and its moderately smooth surface over the filter paper offers facile entrapment of nanomaterials. Other paper materials and substrates used for electrochemical detection are cellulose acetate filter paper, art paper, office paper, nitrocellulose membrane and PVDF filter membrane [20].

Development of electrochemical PADs (ePADs) includes constructing a barrier wall, electrode fabrication and sensor functionalization. The wall formation specifically defines the hydrophilic and hydrophobic surfaces of the biosensor which is required to conduct the electrochemical reactions in a defined area and also to restrict the reverse flow of liquids. The barrier also required to avoid possibility of electrical contacts to participate in the electrochemical reaction. The sample volume can be controlled by judiciously defining the sensing area, and thus the size of the active transducer part also can be established accordingly. The wall preparation can be performed by either manual cutting or wet etching methods. In general, barriers are realized by physical clogging of the pores present in the cellulose substrate with polymers. The resolution and accuracy of the boundary can be improved by advanced equipment based approaches such as wax printing, ink-jet printing, plasma irradiation, corona treatment, screen-printing, flexographic printing, laser printing, ink stamping, vapor phase deposition, and photolithography. The widely explored method is wax coating owing to their low cost, preserved flexibility of the patterned surface, and nontoxic nature of the reagents. After the wax deposition process, devices need to be heated to facilitate penetration of the material inside pores of the paper fibers.

Microelectrodes possess vital importance for ePAD performance, and a variety of electrode materials including carbon, metals and nanoparticles are explored for their fabrication. Several electrode fabrication techniques have been utilized in the past decade, including screen printing, pencil/pen drawing, inkjet-printing, soft lithography and many more. In electrochemical biosensors the molecular recognition process happens on the working electrode surface, therefore in order to enhance the sensitivity, nanostructured materials and electrocatalysts are generally functionalized. Some of the examples are metal nanoparticles, based on gold, platinum, palladium, and their composites, graphene, carbon nanotubes, fullerene, iron oxide, and so on. In order to achieve selective molecular sensing, the transducer's surface is generally modified with antibodies, immobilized enzymes [21], DNA [22], aptamers [23], molecularly imprinted polymers [24–26], and so on. The paper fibers are well suited for chemical or physical immobilization of recognition components. The standard electrode modification methods for paper sensors include physical adsorption, layer-by-layer deposition, covalent bonding, cross-linking, self-assembly electrodeposition, chemically or electrochemically induced polymerization, and many more.

Paper-based electrochemical sensors were first introduced by Dungchai and coworkers in 2009 [27]. The microfluidic channels were realized on filter paper using photolithography, and screen-printing was utilized to fabricate electrodes (Fig. 4.1A). Enzyme based biosensors were constructed for determining glucose, lactate, and uric acid in biological samples by functionalizing the working electrode with corresponding oxidase enzymes. This analytical method demonstrates the feasible integration of electrochemical sensor and paper-based microfluidic technology for portable, miniaturized point of care analysis. During early stages of invention, 2D sensor design was introduced for simultaneous monitoring of multiple analytes from a single sample reservoir. In 2D sensors, three electrodes are fabricated on a same surface of the paper, but in a 3D pattern, the paper is folded like an origami configuration, where the working electrode is printed on one segment of the paper, and the reference and counter electrodes are patterned on another. The major advantage of 3D design is that the fluid can move tirelessly in both vertical and horizontal directions to occupy the maximum reaction zone. It is significant to note that the counter electrode needs to fabricate bigger than the working and reference electrodes which facilitates unlimited current transfer among the electrodes. Besides, the working electrode should configure maximum closely to the reference electrode to minimize the resistance between them.

The common capillary-based lateral flow systems offer good selectivity, rapid parallel assessment of multiple analytes, minimal sample pre-treatment procedures, and low cost. But, the lateral flow has bottlenecks, such as the need for large sample volume, contamination during the transport and sample evaporation. The vertical and other three dimensional flow of solution facilitates co-optimization of the fluidic pathway to eliminate interference or competition among designed assays, and requires small sample volume. Stackable layers can be fabricated by these three dimensional flow methods, and thus the flow can be controlled without extending the fluid pathways. Seong et al.

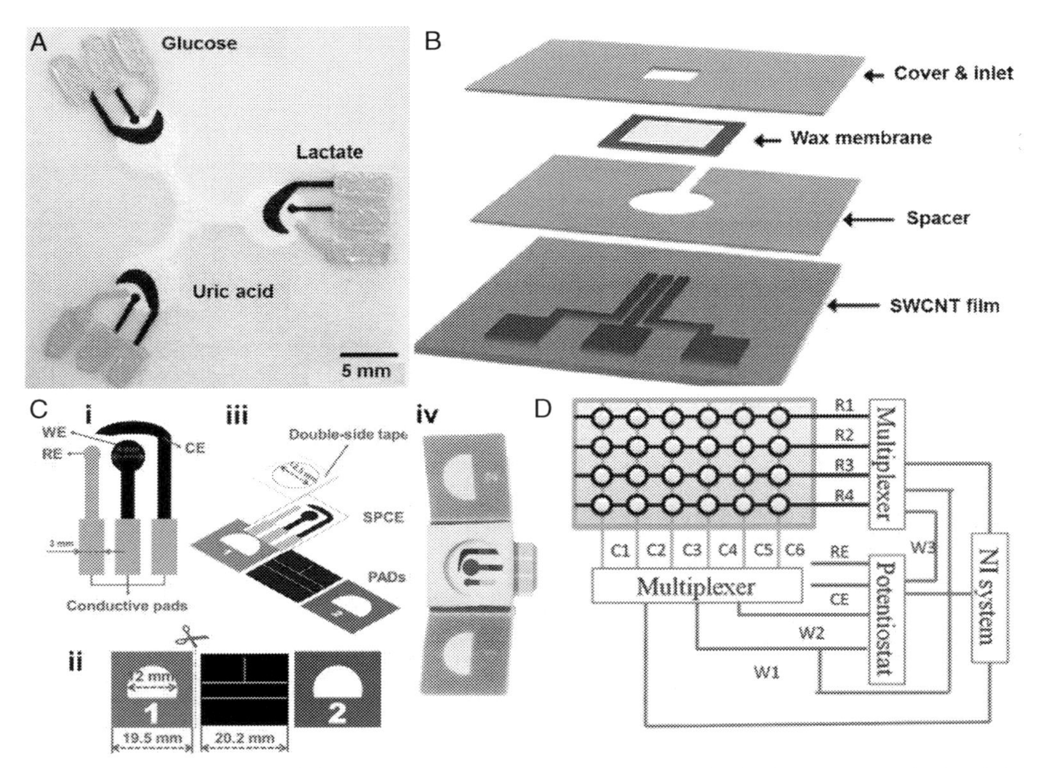

Fig. 4.1 (A) Optical image of the three electrode paper-based microfluidic device having hydrophilic part at the center of the device that wicks sample into the three separate test zones (reproduced with permission from ref. [27]). The silver electrodes and contact pads are realized by Ag/AgCl paste and PB modified carbon was used as working electrodes. (B) Schematic diagram of functional paper fluidic device for acetaminophen determination (reproduced with permission from ref. [28]). (C) Schematically representing the design of (i) printed electrode, (ii) paper based device, and (iii) the assembly of electrode and PADs. (iv) The digital image of the printed PAD device (reproduced with permission from ref. [29]). (D) Schematic diagram of the addressable electrode array detection platform for cancer screening (reproduced with permission from ref. [30]).

developed a paper fluidic device for analyzing acetaminophen in the presence of ascorbic acid [28]. The device possesses a vertical flow configuration, where the sample solution is flowed vertically through the paper substrate. Acetaminophen is an antipyretic analgesic commonly used for treating headaches, backache, arthritis and fever. Acetaminophen poisoning resulted from consuming excessive repeated doses can be rapidly monitored by a simple POCT device. The sensitivity of the acetaminophen detection was increased by using micro-patterned SWCNT film as the electrode, and AuNPs were deposited on it (Fig. 4.1B). The device showed a good sensitivity of 13.3 mA/M and a detection limit of 15.0 μM along with a relative standard deviation of 3.3%.

Human C-reactive protein (CRP) is a nonspecific pentameric protein produced by stimulation of endogenous proinflammatory cytokines in the liver. The infection, cell damage, or other inflammatory conditions cause abnormal level of CRP in the circulation system and CRP is a critical risk factor for cardiovascular or other heart diseases. Hence, monitoring the accurate level of CRP in patients is significant for early therapeutic treatment. In general, immunonephelometric or immunoturbidimetric assays are using for determining the CRP level, but the method is costly, time-consuming, and expert personnel are required. Chailapakul et al. developed a paper based disposable electrochemical sensor for sensitive and selective monitoring of CRP [29]. The label free sensor utilizes thiol-terminated poly (2-methacryloyloxyethyl phosphorylcholine) (PMPC-SH) and electrodeposited AuNPs, and the electrochemical responses were measured by differential pulse voltammetry (DPV) (Fig. 4.1C). The PAD was successfully used for detecting the precise CRP level in simulated body fluids.

Huang *et al* reported a disposable ePAD based on an addressable electrode array for early cancer screening and evaluation [30]. It is imperative to develop a reliable point-of-care sensor for sensitive measurement of tumor markers. Such ePADs facilitate timely diagnosis and treatment, alleviate patient stress and reduce costs. Monitoring a single marker is not advisable due to its limited specificity, and analyzing group of tumor markers in complex serum samples improves the diagnostic accuracy. Also, in comparison with parallel single-marker analysis, multiplex monitoring provides shortened assay time and reduced sample consumption. The sensing sites are fabricated on a square paper and Ag/AgCl reference and carbon counter electrodes are shared with another piece of stacked paper. The wax patterns constituted reservoir for the electrochemical cell, and the disposable array possesses 24 sensing sites that could simultaneously sense four tumor markers from 6 samples (Fig. 4.1D). The performance of the multienzymatic system was augmented by the use of gold nanoparticles (AuNPs) and MWCNTs, which results in a wide linear detection range and appreciable LOD for biomarkers.

The recent research related with medical diagnosis is particularly focused on paper based microfluidic devices owing to their simplicity, low cost and high commercial perspectives. A major breakthrough in this area is the commercialization of paper-based pregnancy test kit, which is one of the first lateral flow assays. PADs facilitate POC detection, predominantly in low income countries, where the people are severely affected by infectious diseases such as diarrhea, respiratory infections, HIV/AIDS, malaria, tuberculosis, and many more. Moreover, early prognosis is primarily important for diseases like cancer, diabetes and liver malfunctions. The quantification of biomarkers like glucose, lactate, nitrite, nucleic acid, uric acid, alcohol, and many more, can be conducted from the physiological fluids like blood and urine by paper based devices.

4.3 Screen-printed disposable sensors for point-of-care analysis

The POCT devices mandate portable, lightweight and low-cost sensing devices, which can replace the bulky instrument as the analytical tool. In general, electrochemical sensors are used by disc/rod type electrodes immersed in electrolytes containing large beaker. The process is tedious and the use of expensive electrodes necessitates surface regeneration of electrodes (polishing) for its repeated use. The advent of novel fabrication technologies eases the sensor development by miniaturizing planar electrodes in an inexpensive substrate. Advanced printing technologies enable the construction of three electrode system having working, reference and counter electrodes by exploiting the corresponding conductive inks. Screen printing technology is ideal for large scale production of disposable electrodes in a planar substrate [31]. The electrodes can be printed on rigid ceramic/glass substrate or flexible plastic or paper substrates. The emerging wearable technologies require sensors to be constructed on wearable fabrics or other textile based substrates. As the screen-printing offers coating of viscous slurries over any fabric or flexible substrates, the soft, flexible and stretchable electrochemical sensors can be efficiently constituted for diverse wearable applications.

The glucose biosensor strips represent one of the major commercialized screen printed biosensor, used for diabetes monitoring. The global pandemic threat, especially the severe COVID-19 caused by severe acute respiratory syndrome coronavirus 2 (SARS-CoV-2) is highly transmissible, pathogenic and life threatening. Society is in constant demand for developing rapid testing kits for monitoring the disease spread, and thus reducing the infection rate. The screen printing technique offers decentralized analyses by cost-effective construction of rapid antigen/antibody based sensing kits. The preparation of sensing strip involves multilayer printing of electrodes and insulator layer, and functionalization of the working electrode surfaces. Fig. 4.2 illustrates the process

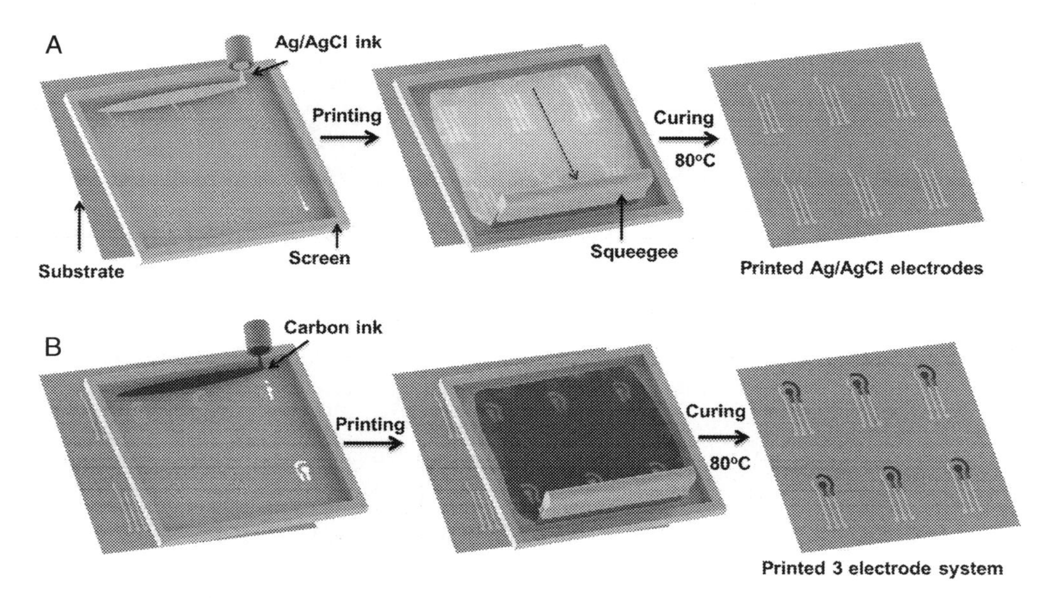

Fig. 4.2 Schematically representing the fabrication of screen-printed electrochemical sensors, including printing of (A) Ag/AgCl ink followed by (B) the carbon ink for realizing typical three electrode system.

of printing (a) Ag/AgCl ink followed by (b) carbon ink for realizing typical planar 3 electrodes for electrochemical sensing. The printing process necessitates development of screen/stencil with desired geometry and dimensions. The stencils are usually fabricated by laser cutting, chemical etching or electronic cutting methods. While printing, the stencils are positioned on the required substrates, and the printable inks are rolled over the pattern using a soft squeegee having sharp edged rubber blades. The printing process transfers the stencil configuration to the underlying substrate and the printed traces needs to be cured in a convection oven at desired temperature. The electrodes are then functionalized with suitable modifiers in order to enhance the sensitivity and selectivity of the sensors. For some applications, nanostructured catalysts or electrocatalysts are blending with the carbon based conductive inks, but the ink formulation must be homogenous to maintain the batch reproducibility of the electrodes. Screen-printed electrodes are widely explored for enzyme based amperometric sensors, aptamer based immunosensors and ion selective membrane based potentiometric sensors.

4.3.1 Enzyme-based disposable amperometric sensors

In enzyme-based amperometric sensors, the generation of current signal is monitored when a constant potential is applied between the working and reference electrodes. The enzyme sensors can be work with minimal sample volumes, and miniaturized easily to operate with complex matrices [32]. The enzymatic sensors have been witnessed significant research advancements, and based on the sensing mechanisms they can be classified into three major types. Clark and Lyons introduced the first generation of enzyme-based sensors by monitoring the hydrogen peroxide generation or oxygen consumption during the enzymatic reaction [33]. The variation of oxygen level in the background causes lack of precision for the sensors following oxygen utilization. Updike and Hicks addressed this issue by using a dual cathode system having one of them is functionalized with glucose oxidase (GOx) enzyme and the signal difference between the electrodes are measured [34]. The first-generation sensors are suffering from poor selectivity because the presence of various electroactive species like ascorbic acid, uric acid, dopamine and other drugs in the real samples could interfere with the measurements. The second generation of enzyme sensors overcomes this problem by eliminating the overpotential required for the electrochemical reaction. The use of artificial redox mediators enhances the electron transfer rates and the product of enzymatic reaction can be detected at low applied potential. For instance, redox active conducting polymers, metal heaxcyanoferrates, ferrocene derivative and many more are widely utilized as redox mediators. The third generation sensors facilitate direct electron transfer and avoid the use of mediators. The enzymes are directly immobilizing on the electrode surface and thus decreases the distance between the recognition moiety and the transducer. Recently, nano-structured conductive materials are modified onto the electrodes for increasing the electrochemically active surface area and to augment the electron transfer kinetics [35].

The enzyme-based sensors are mainly used in three ways such as *off-line* detectors, *in vivo* detectors and *on-line* detectors. In *off-line* detectors, samples are collected and introduced in the sensing chamber of the biosensor which measures the level of the analyte of interest. For instance, all the commercially available glucometers are functioning in the same way, where the screen-printed strips measure the glucose concentration from the blood droplet of diabetic patients, and the portable meter displays the glucose level. I*n vivo* biosensors are implanted in the body for continuously monitoring the biomarker concentration. The biocompatibility of these implantable sensors is crucial for clinical trials, including the biosensor constituents and substrates. The *on-line* detectors are integrated into a flow line coming from a sample extractor in the body or biological matter. This flow through detector usually consists of an implanted micro dialysis probe, which is a minimally-invasive sampling tool used for continuous measurement of free, unbound biomarker levels in the extracellular fluid.

Researchers paid immense efforts on developing nanostructured materials in order to boost up the sensitivity of enzymatic sensors. Nanostructured Pt is an outstanding electrode material owing to their distinctive properties such as high surface area, catalytic activity, conductivity and favorable electrochemical stability. Pd is another metallic catalyst, widely explored for electrochemical sensing, fuel cell and organic coupling reactions [36]. Kalcher et al. reported a disposable glucose sensor based on a composite material of MWCNTs and palladium [37]. The as developed screen-printed enzymatic sensor showed acceptable linearity in sub-millimolar concentration range with LOD 0.14 mM. The multi-metallic alloys are more promising catalysts compared to monometallic catalysts due to their synergistic effects. Pt–Pd composites showed excellent electro-catalytic response toward oxygen [38], H_2O_2 [39], nitrite [40], and many more [41]. Recently, three-dimensional (3D) nanostructures of noble metal catalysts have witnessed great attention for various electrocatalytic applications. The 3D skeleton provides abundant active sites, and the conductive framework offers good electron transport properties. Lan et al. developed snow flake-like Pt–Pd bimetallic nanoclusters and functionalized on screen-printed gold nano film electrode [42]. The sensor possesses good electrocatalytic behavior toward H_2O_2, and when GOx was immobilized, the biosensor showed favorable properties toward glucose.

The detection of biomarkers like glucose, lactate, tumor markers, and many more directly from whole blood samples are becoming increasingly important in modern clinical diagnostics. The whole blood is mainly complex in nature which consists of a mixture of substances like protein, glucose, hormones and inorganic salts. Wang et al. developed a low-cost disposable glucose sensor by coupling screen-printed electrode with a paper disc [43]. The electrode was modified with graphene/polyaniline/AuNPs/GOx, and the assay is based on the direct electrochemistry of GOx, which consists of two flavin adenine dinucleotide (FAD) cofactors with the ability to restore enzymatic activity. Graphene, an aromatic carbon monolayer nanomaterial, is promising for sensitive electrochemical detection owing to its advantages such as excellent electron transfer kinetics, electrical conductivity, light weight, high mechanical strength and large surface area. The assay measures the decrease in current response from FAD when inflamed by the enzyme-substrate reaction using DPV technique. This analytical device is suitable for POC analysis of glucose in resource limited settings.

The quantification of glucose and cholesterol levels in human blood are crucial for diagnosing the risks of several serious diseases such as diabetes mellitus, arteriosclerosis, hypertension, cerebral thrombosis and coronary artery disease. Diabetes mellitus is accompanied with long-term complications including kidney failure, blindness, heart disease, and cardiovascular disease. Recently, the composites of graphene with other nanostructures such as metal nanoparticles, metal oxides and/or conducting polymers are attracted for enhancing the sensing characteristics. Ounnunkad et al. developed a platinum/reduced graphene oxide (rGO)/poly (3-aminobenzoic acid) film on a screen-printed electrode, and utilized for sensitive amperometric detection of glucose and cholesterol [44]. At a working potential of +0.50 V, the sensor showed linear current responses to glucose and cholesterol in the concentration ranges of 0.25–6.00 mM and 0.25–4.00 mM, respectively. Both sensors exhibited a good anti-interference capability and recoveries in human serum trials.

Drop casting method is the simplest for all the electrode modification techniques, where physical adsorption plays a vital role to immobilize biological entities as transducers. The reversibility of the sensor is crucial for continuous monitoring of biomarkers using implanted or wearable devices. The robust immobilization of enzymes is crucial for achieving better electron transfer rate and thus augments the sensitivity of detection. The rGO nanosheets are well known for its ability to provide facile adsorption of enzymes via electrostatic interaction [45]. The amine-terminated MWCNTs are yet another platform for stable enzyme immobilization, where the hydrogen bonding interaction of amino group and carboxyl group of enzyme establish adequate anchoring. The use of conducting polymers like polyaniline, PEDOT:PSS, polypyrrole, and many more boost up the electron transfer kinetics and shorten the ion diffusion path. Another hand, the AuNPs provide excellent electrocatalytic activity, and when stabilized with citrate, AuNPs could present efficient negative charges for anionic doping. Maity et al utilized layer by layer deposition of these functional materials for sensitive amperometric detection of glucose [46]. The biosensor exhibited good reproducibility, high stability, wide linear range (1–10 mM), lowest detection limit (64 μM) and appreciable sensitivity (246 μ Acm^{-2} mM^{-1}). Different types of biosensors have been recently reported based on immobilization of enzymes on screen-printed electrodes. Table 4.1 shows the electrode modifiers, detection limits and linear range for various enzyme based sensors [47–59].

Table 4.1 Enzyme-based disposable sensors fabricated on screen-printed electrodes.

Sensor	Analyte	Enzyme	Linear range	LOD/sensitivity	References
AuNPs–MWCNT–Nafion	Alcohol	Alcohol dehydrogenase	0.2 to 25 mM	50 μM	[47]
Graphite nanosheets zinc oxide nanoparticles	Glucose	GOx	0.3 to 4.5 mM	30.07 μA mM^{-1} cm^{-2}	[48]
MWCNT	Lactose	Cellobiose dehydrogenase	0.5 to 200 μM	250 nM	[49]
SWCNT/PVI-Os[a]	Glucose	GOx	0.5 to 8.0 mM	32 A mM^{-1} cm^{-2}	[50]
AuNPs/graphene	Glucose	GOx	0 to 40 mg dL^{-1}	0.3 mg dL^{-1}	[51]
Magnetic Fe_3O_4 nanoparticles	Glucose	GOx	0 to 33.3 mM	1.74 μA mM^{-1}	[52]
Iridium nanoparticles	Glucose	GOx	0 to 15 mM	-	[53]
pTBA[b]-AuZn alloy oxide	Glucose	GOx	30 to 500 mg/dL	17.23±0.32 mg/dL	[54]
Pt nanocluster	Cholesterol	Cholesteroloxidase	2 to 486 μM	2 μM	[56]
Rhodium–graphite/MWCNT	Cholesterol	Cytochrome P450scc	0 to 80 μM	1.12 μA/ (mM mm^2)	[55]
AuNPs	Cholesterol	Cholesterol esterase, Cholesterol oxidase	5 to 5000 μg/mL	3.0 μg/mL	[57]
rGO-CS-Fc/Pt NPs[c]	Cholesterol	Cholesterol esterase, Cholesterol oxidase	0.5 to 4.0 mg/mL	0.871 nA/mM/cm^2	[58]
Ferrocene-capped AuNPs	Cholesterol	Cholesterol esterase, Cholesterol oxidase	50 μM to 15 mM	12 μM	[59]

[a]Osimium (bpy)2-complexed poly (1-vinylimidazole).
[b]Poly (terthiophene benzoic acid.
[c]reduced graphene oxide-chitosan-ferrocene carboxylic acid/platinum nanoparticles.

4.3.2 Nonenzymatic disposable amperometric sensors

The sensitivity of enzyme based sensors largely related to the activity of the immobilized enzyme, thus reproducibility remains a critical challenge in precise sensing and quality control. As the robust enzyme entrapment is an inevitable step during the biosensor construction, complicated advanced engineering of biological entities on the electrode surfaces is required. In addition, the stability of the enzyme at ambient conditions is challenging. The factors such as pH, temperature and moisture possess a critical role in the long term stability of enzymes. Furthermore, the performance of enzyme sensors is affected by the deficiency of oxygen, which causes non-linear and poor sensor responses in the absence of mediator. Even in the presence of mediator, it is difficult to completely eliminate the competition of dissolved oxygen at the electron-binding site. These limitations of enzyme based sensors can be addressed by non-enzymatic sensors, advantageous in terms of structural simplicity and easy functionalization for mass production. In contrast, non-enzymatic sensors are free from oxygen dependence and they generate electrical currents by directly catalyzing the oxidation/reduction of analyte on the electrode surface. The non-enzymatic sensors are highly stable, and their performances are not much altered with respect to pH and temperature. The rapid developments in nanotechnology fueled the use of various nanostructures for non-enzymatic sensing, and the sensors can be easily constructed by mass production techniques such as vapor deposition, injection, thermosetting, photocuring, and polymer coating.

Metal nanoparticles based on Au, Pt, Ag, Ni, Cu and Pd have been widely used for non-enzymatic estimation of glucose [60–64] due to their better electro catalytic properties. Electrochemically deposited nickel nanoparticles on screen printed electrodes are good candidates for realizing disposable non-enzymatic sensors [62]. Nickel is a ferromagnetic, rigid and ductile metal, extensively used as a cathode material in fuel-cell. Nickel functions as an efficient alternative for expensive noble metals. Nickel nanoparticles (NiNPs) form high-valent oxyhydroxide species on the electrode surface in alkaline medium which catalyzes the electrochemical oxidation of analytes. In addition, NiNPs functionalized electrodes offer excellent stability and anti-fouling performance. Su et al utilized NiNPs for glucose detection from human serum and urine samples [65]. NiNPs were electrochemically deposited on a screen-printed electrode, and the fabricated sensor showed a wide range of detection from 5 μM to 1.5 mM and high sensitivity of 1.9134 μA μM CM^{-1}.

Core-shell nanostructures are composite nanomaterials having an inner layer of core material and an outer layer of another shell material. The distinctive characteristics of the core and the shell can be tuned by increasing the number of shells. Core-shell nanostructures provide broad multifunctional properties and the choice of shell material plays a vital role in enhancing the optical/electrical/magnetic properties. These nanostructures often exhibit synergistic effect, by combining the benefits of individual core and shell layers. The protective outer coating of shell around the core material eliminates migration and aggregation, and thus preserves the chemical activity and the stability of nanoparticles. In biosensors, the core-shell nanostructures amplify the signal due to their large surface area and catalytic properties [66]. Sadasivuni et al. reported a non-enzymatic glucose sensor by modifying the screen-printed electrode with CeO_2@CuO core shell nanostructures [67]. The amperometric studies demonstrated that the modified electrode can detect glucose at +0.4 V, with a sensitivity of 3319.83 μA mM^{-1} cm^{-2} and detection limit of 0.019 μM. More importantly, the sensors possess anti-interference and anti-poisoning activity.

Metal oxide nanoparticles having different morphologies have been prepared in the last decade via various versatile methods. These particles offer high surface area, stability, electrical and photochemical properties. The electron transfer kinetics between the electrode and analyte can be enhanced with metal oxide nanoparticles and thus they can consider as electrocatalysts or electronic wires. The facile bonding of metal oxide nanostructures with the electrode surface can be realized by several approaches like electrodeposition, physical adsorption, electropolymerization and covalent bonding. The major disadvantage of these nanostructures is its wide band gap, which makes them as semiconductor and poor ion transport materials. These difficulties can be resolved by hybridizing with other metal nanoparticles, carbonaceous materials, or polymers [68]. For instance, zinc oxide (ZnO) possesses high chemical stability, electrochemical activity and exceptional isoelectric point which are adequate for electrochemical applications. Doping ZnO nanoparticles with Fe enhances the conductivity as well as the surface area. Ahmad et al utilized such Fe@ZnO modified disposable screen-printed electrode for glucose sensing application. The sensor showed remarkable electro-oxidation of glucose with a LOD of 0.30 μM [69].

4.3.3 Aptamer-based electrochemical immunosensors

Immunosensors use antibodies or antibody fragments as molecular recognition component which form stable complexes with specific analytes, called as antigens. Depending on the signal transduction methods, immunosensors can be classified into different groups such as optical, electrochemical, piezoelectric and thermometric. For each kind of transducers, particular labelling is required either on the antigen or on the antibody, and some of the labelling methods can be utilized to different detection techniques. Majority of the existing immunosensors are either related to a competitive or sandwich assay, which usually applies to the detection of low and high molecular weight molecules, respectively. There are mainly two methods for the development of direct competitive immunoassays, in the first approach, immobilized antibodies reacts with free antigens which are in competition with labelled antigens. In another method, the immobilized antigens compete with free antigens for labelled free antibodies. In a sandwich assay, the immobilized antibodies are endowed to interact with free antigens, and subsequently labelled antibodies are directed to a second binding spot of the antigen [70]. In immunosensors, the use of antibodies potentially improves the selectivity of the sensors, and the possibility of enzymatic amplification provides high sensitivity. In addition, the 96-well plate typically employed in enzyme linked immunosorbent assay (ELISA) tests is appropriate for highly sensitive multianalyte detection [71].

The electrochemical methods allow easy miniaturization of immunoassays by exploiting the advances in microelectronics and microelectrode fabrication. Hence, electrochemical immunoassays are the most promising substitute for optical methods, and they are useful for multiplexed monitoring with minimal sample requirements (microliters to nanoliters of samples). For instance, Kakabakos et al. developed a disposable screen-printed immunosensor for sensitive detection of C−reactive protein (CRP) in human serum [72]. CRP is an acute-phase protein manufactured in the liver, which is a valuable biomarker for inflammation plays, and useful for the possibility of myocardial infection, peripheral arterial disease, stroke, and sudden cardiac death. In this work, a sandwich-type immunoassay was constructed using CRP capturing antibodies attached on bismuth citrate-modified screen-printed electrode. The estimation of the analyte involves acidic dissolution of the quantum dots and anodic stripping voltammteric detection of the released Pb (II), and the developed immunosensor exhibits extraordinary sensitivity, selectivity and analytical simplicity.

Nanostructured materials have been widely utilized for amplifying the electrode responses of immunosensors. CNTs have attracted much attention due to their impeccable properties such as high strength and flexibility, low density and high thermal and electrical conductivity. Ho et al. utilized MWCNTs for enhancing the signal responses for carcinoembryonic antigen [73]. Monoclonal anti-CEA antibodies are covalently immobilized on CNT modified disposable screen-printed electrode and a sandwich immunoassay was developed with tagged ferrocene carboxylic acid encapsulated liposomes. The redox responses from the released ferrocene carboxylic acid are measured using square wave voltammetry. Pingarrón et al. developed AuNPs- poly (amidoamine) nanostructured screen-printed carbon electrodes for tau protein detection from human plasma [74]. A sandwiched immunoassay was prepared, and the amperometric signals were measured at −200 mV against the Ag pseudo-reference electrode, upon adding hydroquinone as electron transfer mediator and H_2O_2 as the enzyme substrate. The immunocomplex formation was estimated and the sensor showed good selectivity and low LOD (1.7 pg mL^{-1}). The promising performances of these analytical devices offer simple operation, low-cost and the possibility to miniature into pocket-size electrochemical transducers. This enables the development of automated POCT platforms for on-site determination of variety of healthcare biomarkers.

4.4 Ink-jet printed disposable sensors

Ink-jet printing is one of the non-contact deposition method, recently exploited for large scale fabrication of flexible electronic devices [75], disposable biosensors [76], light emitting diodes [77] and so on. In inkjet-printing, the functional inks are directly writing on flexible substrates, and the ink can be tuned by incorporating metal nanoparticles, redox polymers or carbon based conducting materials for multifunctional applications. There are mainly two important modes of ink-jet printing, one is drop-on-demand and the other is continuous printing processes. The former approach depend on the thermally induced droplets, and the latter exploits a continuous electro-conductive stream of fluid which is delivered via a nozzle based on piezoelectric vibration [78]. The factors like viscosity, surface tension, electrical conductivity and durability of the ink patterns should be optimized to realize ink-jet printed electrodes in high performance and resolution. Furthermore, the surface pre-treatment of substrates are often necessary for better ink adherence and cartridge compatibility, and ink sintering temperature are crucial for establishing robust printed traces. The high temperature requirement of metal nanoparticles based inks are challenging for ink-jet printing as many of the common plastic based substrates are unstable at elevated temperature [79].

Carbon based materials such as CNTs, graphene, graphene oxide, and reduced graphene oxide possess broad applications in flexible printed electronics owing to their versatile properties such as high mechanical strength, electrical conductivity, and optical transparency. Metal nanoparticles are attracting tremendous attraction for flexible electronics as their thermodynamic size effect decreases the melting points compared to the bulk materials [80]. For instance, Peltonen et al. developed a paper based ink-jet printed electroanalytical system for glucose monitoring [81]. A planar three-electrode system is constituted on paper substrate by ink-jet printing of AuNPs based working and counter electrodes, and silver nanoparticle (AgNPs) based reference electrode. The direct printing of highly conductive electrodes on flexible substrates is compatible for roll-to-roll fabrication. The conductivity of the printed traces remains constant upon bending or twisting deformations, and the partial penetration of the nanoparticles ink endows good adhesion and scratch resistance. The GOx was functionalized by entrapping with electropolymerized PEDOT films and used as an amperometric highly sensitive glucose sensor.

Silver nanoparticles also have drawn huge interest for biosensor preparation owing to its lower oxidation state, high surface to volume ratio and low-cost compared to noble metals like Au and Pt [82]. Benvidi et al. developed flexible biosensor systems using inkjet printing of silver nanoparticles [83]. AgNPs was chemically deposited by expulsion of a mixture of silver nitrate and ascorbic acid onto different flexible substrates like paper and textile fabrics. Bhansali et al. developed a disposable aptamer sensor, enriched by functionalized magnetic nanoparticles, for detecting salivary cortisol level. The sensors are realized by ink-jet printing of CNT-copper porphyrin ink on a photo-paper substrate. The sensor could detect the salivary cortisol variations which is useful for POCT monitoring of patients having obstructive sleep apnea [84].

4.5 Laser-scribed graphene based disposable sensors

Graphene is a single-atom thick, two-dimensional sheet of sp^2 bonded carbon. Graphene possesses unique electronic, mechanical, optical, thermal, and electrochemical properties. Graphene provides two-dimensional environment for electron transport, and facilitates rapid heterogeneous electron transfer at their edges. Graphene possess larger surface area of 2630 $m^2\ g^{-1}$ compared to that of graphite (~10 $m^2\ g^{-1}$) and carbon nanotubes (1315 $m^2\ g^{-1}$). Graphene shows excellent electrical conductivity and are mechanically flexible in nature. Graphene based electrodes have uniformly distributed electrochemically active sites, and it is typically prepared from three dimensional graphite [85]. Several methods have been adopted for the synthesis of graphene such as mechanical exfoliation [86], chemical vapor deposition [87] and reduction of exfoliated graphene oxide [88].

The laser induced preparation of reduced graphite oxide was initially reported for supercapacitor application [89]. Later, this process is advanced to the production of laser scribed graphene by thermally reducing graphite oxide using precisely controlled high energy CO_2 laser [90]. Such expanded graphene films with limited restacked configuration possess good electrical conductivity, large surface area (1520 $m^2\ g^{-1}$) and porosity. Tour et al. fabricated flexible laser induced graphene electrodes onto a thin film of polyimide (PI) [91]. This direct pattering of electrochemically active graphene layers on the flexible substrates offers roll-to-roll manufacturing of disposable electrochemical sensors [92]. Zhao et al. anchored copper nanoparticles on laser induced graphene and used for non-enzymatic detection of glucose [93]. The combination of graphene and copper drastically improves the performance of electrochemical sensors because of the synergistic effect between both the components. This results in augmented electrocatalytic active surface area and electron transfer kinetics for glucose oxidation. The developed glucose sensor illustrates appreciable sensitivity of 495 mA $mM^{-1}\ cm^{-2}$ and the LOD of 0.39 mM. Park et al. utilized Pt-Au nanoparticles for amplifying the signal response of laser scribed graphene toward dopamine [94]. The results showed that the incorporation of the bimetallic catalyst greatly enhances the electrocatalytic activity toward the oxidation of dopamine. The as fabricated disposable sensor exhibited an exceptional sensitivity of 865.80 $\mu A\ mM^{-1}\ cm^{-2}$ and a lowest detection limit of 75 nM.

The direct writing of laser scribed graphene on flexible PI film provides binder free, porous, three dimensional conductive carbon configuration, rich in defects and edge plane domains. These self-supporting flexible electrodes possess large specific surface area and electrolyte accessibility which is favorable for electrochemical reactions. Alshareef et al. developed such a flexible platform and used for electrochemical oxidation of ascorbic acid, dopamine, and uric acid [95]. The sensitivity of detection was amplified by decorating the electrode surface with PtNPs via simple electrodeposition process (Fig. 4.3). The electrode modification greatly improves the electron transfer kinetics and sensitivity of the electrochemical sensor. Hence, the laser induced graphene fabrication from PI substrate is an excellent method for realizing high through-put biosensors, and the sensitivity of the transducer can be further boosted up by incorporating nanostructured materials.

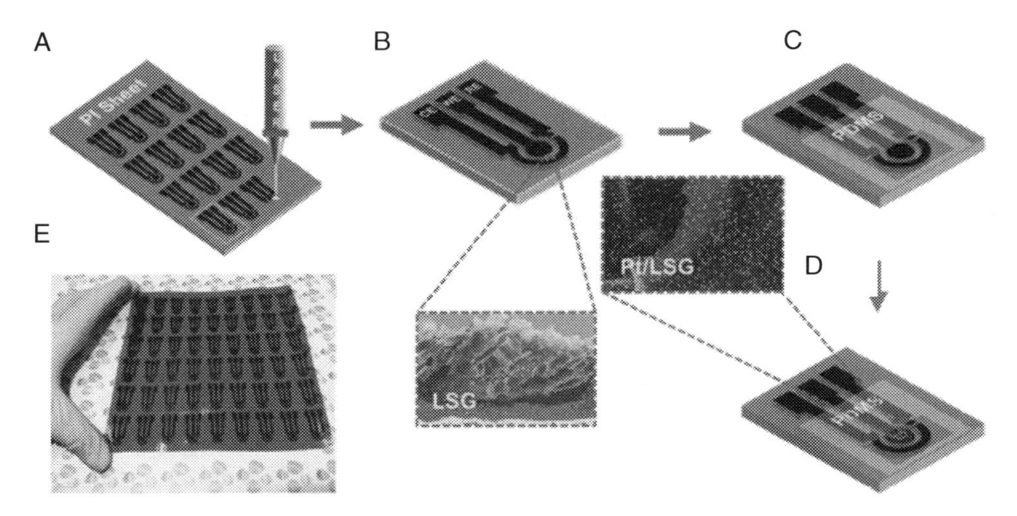

Fig. 4.3 Schematic illustration of (A) the fabrication of arrays of lased scribed graphene electrodes on PI sheet, (B) 3D view of the electrode pattern, (C) selective passivation of the electrode area by PDMS, and (D) electrodeposition of PtNPs on the working electrode. The projected images show the vertical cross-sectional SEM image of an electrode showing porous and protruded morphology of graphene, and the anchored PtNPs over graphene sheets. (E) Optical image of patterned electrode arrays on a PI sheet (reproduced with permission from ref. [95]).

4.6 Recent advances and future perspectives

The rapid growth in technological innovations results emerging of a new trend of POCT platforms, which enables self-evaluation of biomarkers and transferring of the data to patients or healthcare providers via wireless connectivity [96]. Such POCT system can be placed directly on the epidermis or other wearable substrates with minimal user intervention, and the device can continuously monitor the relevant biomarkers without interrupting routine activities. The wearable POCT device monitors the underlying physiology of the human body by measuring health or fitness level indicators from body fluids like sweat, interstitial fluid, saliva or tears fluid. Monitoring multiple analytes simultaneously from the body fluids necessitate fruitful amalgamation of several components such as sample handling, multiplexed recognition system, signal transducers, signal analysis and readout, and power sources [12]. The non-invasive prognosis implemented by wearable sensors offer new opportunities for remote and continuous healthcare assessment in non-clinical settings. The use of user-friendly wearable healthcare devices encourages people to invest greater care and interest in their own healthcare in a smarter and cheaper way.

4.6.1 Flexible and stretchable sensors for wearable applications

The three dimensional and curvilinear nature of human body necessitate the sensors to be soft and flexible to have conformal integration onto the skin surface. The use of rigid sensors and its forceful integration disturb wearer's routine activities and adversely affect the real-time monitoring. Moreover, the textile fabric substrate also possesses mechanical flexibility and stretchability. Hence, the mechanical resiliency of the wearable sensor, including the active sensing part, electrical contact and the associated electronics are potentially important [97]. The use of soft, secondary skin like sensors with mechanical properties similar to the skin provides seamless integration with the non-linear surfaces and ensures robust electrochemical performances when subjected to multiaxial strains. The irregular body motion and muscle movements impart high level of strain on the devices, and the robust bonding of the transducer endows noise-free signal collection. As the skin surface possesses ~30% stretchability, the large strain may cause micro crack formation at the electrodes, which adversely affects the electrical conductivity [98].

Introducing stretchable properties to the system facilitate complete flexibility and ensure robust performances under harsh environmental conditions. The stretchability can be accomplished by two ways such as deterministic and random composite routes. The deterministic approach exploits the design induced stretchability, via wavy shape or island-bridge configuration. The three dimensional wavy structure is realized by prestretching the substrate before establishing the conductive traces [99]. Upon releasing the strain, the conductive layer forms out-of-plane wrinkled structure that can eliminate the strain related problems. In island-bridge pattern, the active rigid islands are interconnected by stretchable curve (serpentine, coil, etc.) shaped bridges [100]. Upon stretching, the unwinding of curvy region accommodates most of the strain and protects the active transducers from damage [101]. The random composite route provides intrinsic stretchability by exploiting the hyper elastic polymer based stretch enduring conductive inks. The facile incorporation of nanostructured materials such as carbon nanoparticles, nanoparticles, metal nanowires, nanosheets and flakes with the elastic binders like cross-linked block copolymers of polystyrene [102], PDMS [103], polyurethane [104], and latex [105] realizes intrinsically conductive electrodes. The assimilation of both the deterministic and composite routes is another versatile approach for establishing highly stretchable and conductive wearable electrochemical systems [106-108].

In general, the level of biomarkers in human body fluids is very low and the use of nanostructured materials for signal amplification is essential for their sensitive monitoring. Cheng et al. utilized vertically aligned gold nanowires (v-AuNWs) for multiplexed, potentiometric analysis of Na^+, K^+ and pH in sweat [109]. The stretchable "tattoo-like" epidermal sensors can be patterned in-plane design, and when functionalized with specific ion-selective membranes, the v-AuNWs array shows good reproducibility, sensitivity, and stability (Fig. 4.4). The performance of the sensor could be maintained even under 30% strains, which is ideal for epidermal monitoring of electrolytes from human

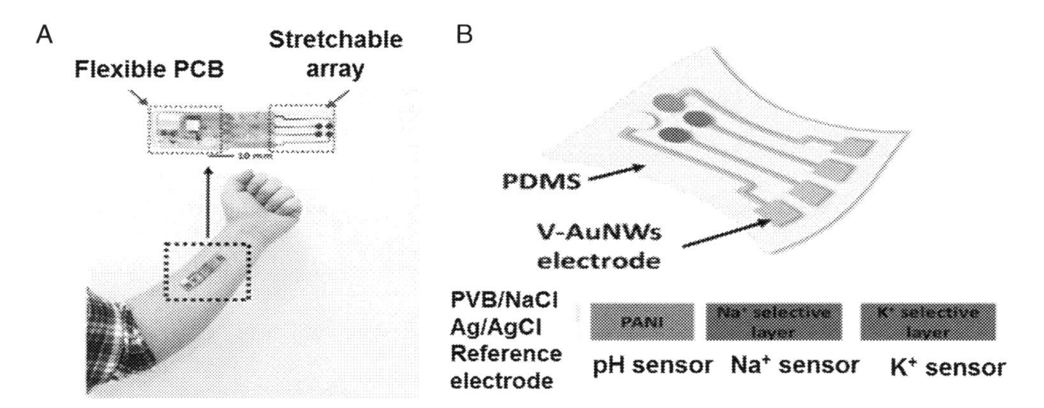

Fig. 4.4 (A) An integrated wearable multiplexed sweat sensing device on a subject's wrist including a flexible PCB and stretchable electrodes. (B) Schematic of a stretchable ion sensing electrodes based on v-AuNWs containing for pH, Na^+ and K^+ monitoring (reproduced with permission from ref. [109]).

sweat. Wang et al. developed a wearable lactate biosensor based on AgNWs molecularly imprinted polymers (MIPs) deposited on screen-printed electrode [110]. The MIPs were prepared by electropolymerizing 3-aminophenylboronic acid on AgNWs modified electrode, and the biosensor exhibited high sensitivity and selectivity toward lactate from 10^{-6} M to 0.1 M, with the LOD of 0.22 μM. The flexible sensors showed robust electrochemical performance even after bended and twisted for 200 times. Such wearable sensor could non-invasively detect sweat lactate level during exercise activities.

Self-powered sensors are useful for wearable sensing as the continuous signal collection and wireless transmission requires high amount of energy over a prolonged period of time [101]. Self-powered systems are functioning by generating electricity from various renewable sources such as thermoelectric, photovoltaic, piezoelectric, and triboelectric generators. The sensitivity of these self-powered sensing systems can be enhanced by introducing nanoparticles. Lin et al. developed a self-powered wearable electrochemical system for lactate detection by directly growing monometallic and bimetallic nanoparticles on conductive carbon fibers [111]. The discharged capacity of a capacitor was used for electrochemically reducing metal ions like Ag, Au, Pt and Pd to form monometallic nanoparticles. Similarly bimetallic PdAu nanoparticles were prepared and used for bandage based electrochemical sensing of lactate from sweat samples. A triboelectric nanogenerator was integrated, which can power the lactate sensor from one minute of its operation.

Fiber based sensors are getting enormous attention as they can be weaved into a fabric substrate and function as textile based wearable smart sensors. Fiber based devices offer high level of flexibility and they can be stretched while maintaining its performances. Cheng et al. developed gold nanowires coated fiber based smart textile for sweat pH monitoring. Gold nanowires were vertically grown on elastomeric fiber and functionalized with PANI for pH analysis. The pH sensor demonstrated a typical sensitivity of 60.6 mV/pH and the sensing activity remains unaffected even under 100% strain. These fiber based sensors can be tirelessly weaved into textiles, which offer more comfort to the wearer for personalized health and fitness monitoring.

4.6.2 Wearable devices for diagnosis and therapy

The recent advances in point-of-care analysis involve the real-time biomarker analysis coupled with feed-back controlled drug-delivery system. For instance, diabetes is one of the most chronic diseases, and the patients are required to check their blood glucose levels periodically. The continuous management of diabetes necessitates having precise shots of insulin proportional to the glucose concentration. The repetitive blood analysis and invasive drug delivery is cumbersome and accompanied with intense stress. Therefore, a unique method for painless and stress-free glucose assessment and accurate management of homeostasis via controlled drug delivery is greatly enviable.

Recently, the wearable transdermal drug delivery is focused on microneedle based minimally invasive systems. Microneedles possess sharp tips having high aspect ratio and made of rigid materials to realize successful skin penetration. Biocompatible polymers with high mechanical strength are generally preferred for these medical devices [112–114]. Kim et al. developed a wearable hybrid device for sweat-based glucose evaluation and multistage transdermal drug delivery [115]. A patch-based wearable system was developed for noninvasive glucose analysis and microneedle-based therapy. The system allows mass fabrication of porous metal electrodes and reliable sweat analysis with minimal amount of samples. The presence of multiple sweat uptake layers control the efficiency of sweat sampling and the use of porous gold nanostructure maximizes the electrochemically active surface area. The transdermal multistage drug delivery platform consists of biocompatible hyaluronic acid based hydrogel microneedles loaded with drug incorporated phase change nanoparticles. Such systems facilitate continues real-time monitoring of health and fitness status and feedback-controlled therapy without pain and stress.

4.7 Summary

The chapter elaborated the efficacy of nanostructured materials for improving the sensitivity of disposable electrochemical sensors by amplifying the electrode responses toward analyte concentration. The size and shape dependant physical, catalytic and electrochemical properties of nanomaterials play a vital role in the biosensor development. The nanomaterials significantly enhance the electrochemically active surface area and the electron transfer kinetics between the electrode-electrolyte interfaces. The growing demand for disposable biosensors for POC analyses was realized by advanced mass fabrication approaches like screen-printing, ink-jet printing and laser induced writing.

Advanced POCT systems offer fast, accessible and reliable information from human body fluids, and the smart devices wirelessly transfers data to the smartphone or any cloud devices. The recent advances in the wearable POCT devices include continuous real-time monitoring of health and fitness status of individuals. Furthermore, these wearable systems can be used for smart therapy with respect to the feedback from transdermal sensors. Apart from clinical diagnostics and wearable sensing, the disposable electrochemical sensors are widely exploited for several applications, ranging from environmental, pharmaceutical, forensic, agricultural, and food sciences.

Acknowledgment

Vinu Mohan A.M. acknowledges the Department of Science and Technology (DST) for providing the INSPIRE Faculty award and the research grant (DST/INSPIRE/04/2016/001601), and Council of Scientific & Industrial Research (CSIR) for financial support.

References

[1] M. Medina-Sánchez, S. Miserere, A. Merkoçi, Nanomaterials and lab-on-a-chip technologies, Lab Chip 12 (11) (2012) 1932–1943.
[2] V. Gubala, L.F. Harris, A.J. Ricco, M.X. Tan, D.E. Williams, Point of care diagnostics: status and future, Anal. Chem. 84 (2) (2012) 487–515.
[3] S. Nayak, N.R. Blumenfeld, T. Laksanasopin, S.K. Sia, Point-of-care diagnostics: recent developments in a connected age, Anal. Chem. 89 (1) (2017) 102–123.
[4] J. Hu, S. Wang, L. Wang, F. Li, B. Pingguan-Murphy, T.J. Lu, et al., Advances in paper-based point-of-care diagnostics, Biosens. Bioelectron. 54 (2014) 585–597.
[5] A.J. Bandodkar, S. Imani, R. Nunez-Flores, R. Kumar, C. Wang, A.V. Mohan, et al., Re-usable electrochemical glucose sensors integrated into a smartphone platform, Biosens. Bioelectron. 101 (2018) 181–187.
[6] A.P. Turner, Biosensors: sense and sensibility, Chem. Soc. Rev. 42 (8) (2013) 3184–3196.
[7] T. Ng, W. Liao, Sensitivity analysis and energy harvesting for a self-powered piezoelectric sensor, J. Intell. Mater. Syst. Struct. 16 (10) (2005) 785–797.
[8] H.A. Alhadrami, Biosensors: classifications, medical applications, and future prospective, Biotechnol. Appl. Biochem. 65 (3) (2018) 497–508.
[9] A.J. Killard, Disposable sensors, Curr. Opin. Electrochem 3 (1) (2017) 57–62.
[10] J. Kim, I. Jeerapan, B. Ciui, M.C. Hartel, A. Martin, J. Wang, Edible electrochemistry: food materials based electrochemical sensors, Adv. Healthc. Mater 6 (22) (2017) 1700770.
[11] Y.J. Kim, W. Wu, S.-E. Chun, J.F. Whitacre, C.J. Bettinger, Biologically derived melanin electrodes in aqueous sodium-ion energy storage devices, Proc. Natl. Acad. Sci. 110 (52) (2013) 20912–20917.
[12] A. VinuMohan, Screen-printed electrochemical sensors for environmental contaminants. In: Nanosensor Technologies for Environmental Monitoring, Springer, Switzerland, 2020, pp. 85–108.
[13] A.V. Mohan, K. Aswini, V. Biju, Electrochemical codeposition of gold particle–poly (2- (2-pyridyl) benzimidazole) hybrid film on glassy carbon electrode for the electrocatalytic oxidation of nitric oxide, Sens. Actuators B 196 (2014) 406–412.
[14] A.V. Mohan, B. Brunetti, A. Bulbarello, J. Wang, Electrochemical signatures of multivitamin mixtures, Analyst 140 (22) (2015) 7522–7526.
[15] V.M.A. Mohanan, A.K. Kunnummal, V.M.N. Biju, Electrochemical sensing of hydroxylamine using a wax impregnated graphite electrode modified with a nanocomposite consisting of ferric oxide and copper hexacyanoferrate, Microchim. Acta 183 (6) (2016) 2013–2021.

[16] H. Li, D. Xu, Silver nanoparticles as labels for applications in bioassays, Trends Anal. Chem. 61 (2014) 67–73.
[17] A.V. Mohan, G. Rambabu, K. Aswini, V. Biju, Electrocatalytic behaviour of hybrid cobalt–manganese hexacyanoferrate film on glassy carbon electrode, Thin Solid Films 565 (2014) 207–214.
[18] A. Chen, S. Chatterjee, Nanomaterials based electrochemical sensors for biomedical applications, Chem. Soc. Rev. 42 (12) (2013) 5425–5438.
[19] C. Fenzl, T. Hirsch, A.J. Baeumner, Nanomaterials as versatile tools for signal amplification in (bio) analytical applications, Trends Anal. Chem. 79 (2016) 306–316.
[20] V.B.C. Lee, N.F. Mohd-Naim, E. Tamiya, M.U. Ahmed, Trends in paper-based electrochemical biosensors: from design to application, Anal. Sci. 34 (1) (2018) 7–18.
[21] A.V. Mohan, K. Aswini, A.M. Starvin, V. Biju, Amperometric detection of glucose using Prussian blue-graphene oxide modified platinum electrode, Anal. Methods 5 (7) (2013) 1764–1770.
[22] J. Zhai, H. Cui, R. Yang, DNA based biosensors, Biotechnol. Adv. 15 (1) (1997) 43–58.
[23] R. Kirby, E.J. Cho, B. Gehrke, T. Bayer, Y.S. Park, D.P. Neikirk, et al., Aptamer-based sensor arrays for the detection and quantitation of proteins, Anal. Chem. 76 (14) (2004) 4066–4075.
[24] K. Aswini, A.V. Mohan, V. Biju, Molecularly imprinted poly (4-amino-5-hydroxy-2, 7-naphthalenedisulfonic acid) modified glassy carbon electrode as an electrochemical theophylline sensor, Mater. Sci. Eng. C 65 (2016) 116–125.
[25] K. Aswini, A.V. Mohan, V. Biju, Molecularly imprinted polymer based electrochemical detection of L-cysteine at carbon paste electrode, Mater. Sci. Eng. C 37 (2014) 321–326.
[26] V.M.A. Mohanan, A.K. Kunnummal, V.M.N. Biju, Selective electrochemical detection of dopamine based on molecularly imprinted poly (5-amino 8-hydroxy quinoline) immobilized reduced graphene oxide, J. Mater. Sci. 53 (15) (2018) 10627–10639.
[27] W. Dungchai, O. Chailapakul, CS. Henry, Electrochemical detection for paper-based microfluidics, Anal. Chem. 81 (14) (2009) 5821–5826.
[28] S.H. Lee, J.H. Lee, V.-.K. Tran, E. Ko, C.H. Park, W.S. Chung, et al., Determination of acetaminophen using functional paper-based electrochemical devices, Sens. Actuators B 232 (2016) 514–522.
[29] C. Pinyorospathum, S. Chaiyo, P. Sae-Ung, V.P. Hoven, P. Damsongsang, W. Siangproh, et al., Disposable paper-based electrochemical sensor using thiol-terminated poly (2-methacryloyloxyethyl phosphorylcholine) for the label-free detection of C-reactive protein, Microchim. Acta 186 (7) (2019) 1–10.
[30] S. Ge, L. Ge, M. Yan, X. Song, J. Yu, J. Huang, A disposable paper-based electrochemical sensor with an addressable electrode array for cancer screening, Chem. Commun. 48 (75) (2012) 9397–9399.
[31] A.J. Bandodkar, C.S. López, A.M.V. Mohan, L. Yin, R. Kumar, J. Wang, All-printed magnetically self-healing electrochemical devices, Sci. Adv. 2 (11) (2016) e1601465.
[32] G. Rocchitta, A. Spanu, S. Babudieri, G. Latte, G. Madeddu, G. Galleri, et al., Enzyme biosensors for biomedical applications: Strategies for safeguarding analytical performances in biological fluids, Sensors 16 (6) (2016) 780.
[33] L.C. Clark Jr, C. Lyons, Electrode systems for continuous monitoring in cardiovascular surgery, Ann. N.Y. Acad. Sci. 102 (1) (1962) 29–45.
[34] S.J. Updike, GP. Hicks, The enzyme electrode, Nature 214 (5092) (1967) 986–988.
[35] J. Castillo, S. Gáspár, S. Leth, M. Niculescu, A. Mortari, I. Bontidean, et al., Biosensors for life quality: design, development and applications, Sens. Actuators B 102 (2) (2004) 179–194.
[36] M. Iqbal, Y.V. Kaneti, J. Kim, B. Yuliarto, Y.M. Kang, Y. Bando, et al., Chemical design of palladium-based nanoarchitectures for catalytic applications, Small 15 (6) (2019) 1804378.
[37] V. Guzsvány, J. Anojčić, E. Radulović, O. Vajdle, I. Stanković, D. Madarász, et al., Screen-printed enzymatic glucose biosensor based on a composite made from multiwalled carbon nanotubes and palladium containing particles, Microchim. Acta 184 (7) (2017) 1987–1996.
[38] J. Li, H.-.M. Yin, X.-.B. Li, E. Okunishi, Y.-.L. Shen, J. He, et al., Surface evolution of a Pt–Pd–Au electrocatalyst for stable oxygen reduction, Nat. Energy 2 (8) (2017) 1–9.
[39] X. Sun, S. Guo, Y. Liu, S. Sun, Dumbbell-like PtPd–Fe3O4 nanoparticles for enhanced electrochemical detection of H2O2, Nano Lett. 12 (9) (2012) 4859–4863.
[40] R. Xi, S.-.H. Zhang, L. Zhang, C. Wang, L.-.J. Wang, J.-.H. Yan, et al., Electrodeposition of Pd-Pt nanocomposites on porous GaN for electrochemical nitrite sensing, Sensors 19 (3) (2019) 606.

[41] X. Niu, M. Lan, C. Chen, H. Zhao, Nonenzymatic electrochemical glucose sensor based on novel Pt–Pd nanoflakes, Talanta 99 (2012) 1062–1067.
[42] X. Niu, C. Chen, H. Zhao, Y. Chai, M. Lan, Novel snowflake-like Pt–Pd bimetallic clusters on screen-printed gold nanofilm electrode for H2O2 and glucose sensing, Biosens. Bioelectron. 36 (1) (2012) 262–266.
[43] F.-.Y. Kong, S.-.X. Gu, W.-.W. Li, T.-.T. Chen, Q. Xu, W. Wang, A paper disk equipped with graphene/polyaniline/Au nanoparticles/glucose oxidase biocomposite modified screen-printed electrode: Toward whole blood glucose determination, Biosens. Bioelectron. 56 (2014) 77–82.
[44] S. Phetsang, J. Jakmunee, P. Mungkornasawakul, R. Laocharoensuk, K. Ounnunkad, Sensitive amperometric biosensors for detection of glucose and cholesterol using a platinum/reduced graphene oxide/poly (3-aminobenzoic acid) film-modified screen-printed carbon electrode, Bioelectrochemistry 127 (2019) 125–135.
[45] G. Bharath, R. Madhu, S.-.M. Chen, V. Veeramani, A. Balamurugan, D. Mangalaraj, et al., Enzymatic electrochemical glucose biosensors by mesoporous 1D hydroxyapatite-on-2D reduced graphene oxide, J. Mater. Chem. B 3 (7) (2015) 1360–1370.
[46] D. Maity, C. Minitha, R.K. RT, Glucose oxidase immobilized amine terminated multiwall carbon nanotubes/reduced graphene oxide/polyaniline/gold nanoparticles modified screen-printed carbon electrode for highly sensitive amperometric glucose detection, Mater. Sci. Eng. C 105 (2019) 110075.
[47] S. Zhen, Y. Wang, C. Liu, G. Xie, C. Zou, J. Zheng, et al., A novel microassay for measuring blood alcohol concentration using a disposable biosensor strip, Forensic Sci. Int. 207 (1-3) (2011) 177–182.
[48] C. Karuppiah, S. Palanisamy, S.-.M. Chen, V. Veeramani, P. Periakaruppan, Direct electrochemistry of glucose oxidase and sensing glucose using a screen-printed carbon electrode modified with graphite nanosheets and zinc oxide nanoparticles, Microchim. Acta 181 (15-16) (2014) 1843–1850.
[49] G. Safina, R. Ludwig, L. Gorton, A simple and sensitive method for lactose detection based on direct electron transfer between immobilised cellobiose dehydrogenase and screen-printed carbon electrodes, Electrochim. Acta 55 (26) (2010) 7690–7695.
[50] Q. Gao, Y. Guo, W. Zhang, H. Qi, C. Zhang, An amperometric glucose biosensor based on layer-by-layer GOx-SWCNT conjugate/redox polymer multilayer on a screen-printed carbon electrode, Sens. Actuators B 153 (1) (2011) 219–225.
[51] Z. Pu, R. Wang, J. Wu, H. Yu, K. Xu, D. Li, A flexible electrochemical glucose sensor with composite nanostructured surface of the working electrode, Sens. Actuators B 230 (2016) 801–809.
[52] B.-.W. Lu, W-C. Chen, A disposable glucose biosensor based on drop-coating of screen-printed carbon electrodes with magnetic nanoparticles, J. Magn. Magn. Mater. 304 (1) (2006) e400–e402.
[53] J. Shen, L. Dudik, C-C. Liu, An iridium nanoparticles dispersed carbon based thick film electrochemical biosensor and its application for a single use, disposable glucose biosensor, Sens. Actuators B 125 (1) (2007) 106–113.
[54] D.-.M. Kim, S.J. Cho, C.-.H. Cho, K.B. Kim, M.-.Y. Kim, Y-B. Shim, Disposable all-solid-state pH and glucose sensors based on conductive polymer covered hierarchical AuZn oxide, Biosens. Bioelectron. 79 (2016) 165–172.
[55] S. Carrara, V.V. Shumyantseva, A.I. Archakov, B. Samorì, Screen-printed electrodes based on carbon nanotubes and cytochrome P450scc for highly sensitive cholesterol biosensors, Biosens. Bioelectron. 24 (1) (2008) 148–150.
[56] K.S. Eom, Y.J. Lee, H.W. Seo, J.Y. Kang, J.S. Shim, SH. Lee, Sensitive and non-invasive cholesterol determination in saliva via optimization of enzyme loading and platinum nano-cluster composition, Analyst 145 (3) (2020) 908–916.
[57] Y. Huang, L. Cui, Y. Xue, S. Zhang, N. Zhu, J. Liang, et al., Ultrasensitive cholesterol biosensor based on enzymatic silver deposition on gold nanoparticles modified screen-printed carbon electrode, Mater. Sci. Eng. C 77 (2017) 1–8.
[58] G. Li, J. Zeng, L. Zhao, Z. Wang, C. Dong, J. Liang, et al., Amperometric cholesterol biosensor based on reduction graphene oxide-chitosan-ferrocene/platinum nanoparticles modified screen-printed electrode, J. Nanopart. Res. 21 (7) (2019) 1–16.
[59] B. Feng, Y-N. Liu, A disposable cholesterol enzyme biosensor based on ferrocene-capped gold nanoparticle modified screen-printed carbon electrode, Int. J. Electrochem. Sci. 10 (2015) 4770–4778.
[60] H. Zhu, X. Lu, M. Li, Y. Shao, Z. Zhu, Nonenzymatic glucose voltammetric sensor based on gold nanoparticles/carbon nanotubes/ionic liquid nanocomposite, Talanta 79 (5) (2009) 1446–1453.

[61] D. Rathod, C. Dickinson, D. Egan, E. Dempsey, Platinum nanoparticle decoration of carbon materials with applications in non-enzymatic glucose sensing, Sens. Actuators B 143 (2) (2010) 547–554.
[62] Y. Liu, H. Teng, H. Hou, T. You, Nonenzymatic glucose sensor based on renewable electrospun Ni nanoparticle-loaded carbon nanofiber paste electrode, Biosens. Bioelectron. 24 (11) (2009) 3329–3334.
[63] L.-.M. Lu, X.-.B. Zhang, G.-.L. Shen, R-Q. Yu, Seed-mediated synthesis of copper nanoparticles on carbon nanotubes and their application in nonenzymatic glucose biosensors, Anal. Chim. Acta 715 (2012) 99–104.
[64] H.-.F. Cui, J.-.S. Ye, W.-.D. Zhang, C.-.M. Li, J.H. Luong, F.-.S. Sheu, Selective and sensitive electrochemical detection of glucose in neutral solution using platinum–lead alloy nanoparticle/ carbon nanotube nanocomposites, Anal. Chim. Acta 594 (2) (2007) 175–183.
[65] S. Kubendhiran, S. Sakthinathan, S.-.M. Chen, C.M. Lee, B.-.S. Lou, P. Sireesha, et al., Electrochemically activated screen printed carbon electrode decorated with nickel nano particles for the detection of glucose in human serum and human urine sample, Int. J. Electrochem. Sci 11 (2016) 7934–7946.
[66] P.K. Kalambate, Z. Huang, Y. Li, Y. Shen, M. Xie, Y. Huang, et al., Core@ shell nanomaterials based sensing devices: a review, Trends Anal. Chem. 115 (2019) 147–161.
[67] T. Dayakar, K.V. Rao, K. Bikshalu, V. Malapati, KK. Sadasivuni, Non-enzymatic sensing of glucose using screen-printed electrode modified with novel synthesized CeO2@ CuO core shell nanostructure, Biosens. Bioelectron. 111 (2018) 166–173.
[68] J.M. George, A. Antony, B. Mathew, Metal oxide nanoparticles in electrochemical sensing and biosensing: a review, Microchim. Acta 185 (7) (2018) 1–26.
[69] W. Raza, K. Ahmad, A highly selective Fe@ ZnO modified disposable screen printed electrode based non-enzymatic glucose sensor (SPE/Fe@ ZnO), Mater. Lett. 212 (2018) 231–234.
[70] F. Ricci, G. Volpe, L. Micheli, G. Palleschi, A review on novel developments and applications of immunosensors in food analysis, Anal. Chim. Acta 605 (2) (2007) 111–129.
[71] F. Ricci, G. Adornetto, G. Palleschi, A review of experimental aspects of electrochemical immunosensors, Electrochim. Acta 84 (2012) 74–83.
[72] C. Kokkinos, M. Prodromidis, A. Economou, P. Petrou, S. Kakabakos, Disposable integrated bismuth citrate-modified screen-printed immunosensor for ultrasensitive quantum dot-based electrochemical assay of C-reactive protein in human serum, Anal. Chim. Acta 886 (2015) 29–36.
[73] S. Viswanathan, C. Rani, A.V. Anand, Ho J-aA, Disposable electrochemical immunosensor for carcinoembryonic antigen using ferrocene liposomes and MWCNT screen-printed electrode, Biosens. Bioelectron. 24 (7) (2009) 1984–1989.
[74] C.A. Razzino, V. Serafín, M. Gamella, M. Pedrero, A. Montero-Calle, R. Barderas, et al., An electrochemical immunosensor using gold nanoparticles-PAMAM-nanostructured screen-printed carbon electrodes for tau protein determination in plasma and brain tissues from Alzheimer patients, Biosens. Bioelectron. 163 (2020) 112238.
[75] S.P. Chen, H.L. Chiu, P.H. Wang, Y.C. Liao, Inkjet printed conductive tracks for printed electronics, Ecs. J. Solid State Sc 4 (4) (2015) P3026–P3033.
[76] E. Bihar, S. Wustoni, A.M. Pappa, K.N. Salama, D. Baran, S. Inal, A fully inkjet-printed disposable glucose sensor on paper, npj Flex. Electron 2 (1) (2018) 1–8.
[77] L. Zhou, L. Yang, M. Yu, Y. Jiang, C.F. Liu, W.Y. Lai, et al., Inkjet-printed small-molecule organic light-emitting diodes: halogen-free inks, printing optimization, and large-area patterning, ACS Appl. Mater. Interfaces 9 (46) (2017) 40533–40540.
[78] A. Al-Halhouli, H. Qitouqa, A. Alashqar, J. Abu-Khalaf, Inkjet printing for the fabrication of flexible/ stretchable wearable electronic devices and sensors, Sens. Rev. 38 (4) (2018) 438–452.
[79] A. Moya, G. Gabriel, R. Villa, FJ. del Campo, Inkjet-printed electrochemical sensors, Curr. Opin. Electrochem 3 (1) (2017) 29–39.
[80] L. Nayak, S. Mohanty, S.K. Nayak, Ramadoss A. A review on inkjet printing of nanoparticle inks for flexible electronics, J. Mater. Chem. C 7 (29) (2019) 8771–8795.
[81] A. Määttänen, U. Vanamo, P. Ihalainen, P. Pulkkinen, H. Tenhu, J. Bobacka, et al., A low-cost paper-based inkjet-printed platform for electrochemical analyses, Sens. Actuators B 177 (2013) 153–162.
[82] M.A. Abrar, Y. Dong, P.K. Lee, W.S. Kim, Bendable electro-chemical lactate sensor printed with silver nano-particles, Sci. Rep. 6 (1) (2016) 1–9.
[83] Z. Abadi, V. Mottaghitalab, M. Bidoki, A. Benvidi, Flexible biosensor using inkjet printing of silver nanoparticles. Sensor Review. 2014.

[84] R.E. Fernandez, Y. Umasankar, P. Manickam, J.C. Nickel, L.R. Iwasaki, B.K. Kawamoto, et al., Disposable aptamer-sensor aided by magnetic nanoparticle enrichment for detection of salivary cortisol variations in obstructive sleep apnea patients, Sci. Rep. 7 (1) (2017) 1–9.
[85] M. Pumera, Electrochemistry of graphene: new horizons for sensing and energy storage, Chem. Rec. 9 (4) (2009) 211–223.
[86] K.S. Novoselov, A.K. Geim, S.V. Morozov, D. Jiang, Y. Zhang, S.V. Dubonos, et al., Electric field effect in atomically thin carbon films, Science 306 (5696) (2004) 666–669.
[87] S. Bae, H. Kim, Y. Lee, X. Xu, J.S. Park, Y. Zheng, et al., Roll-to-roll production of 30-inch graphene films for transparent electrodes, Nat. Nanotechnol. 5 (8) (2010) 574–578.
[88] W. Choi, I. Lahiri, R. Seelaboyina, YS. Kang, Synthesis of graphene and its applications: a review, Crit. Rev. Solid State 35 (1) (2010) 52–71.
[89] W. Gao, N. Singh, L. Song, Z. Liu, A.L. Reddy, L. Ci, et al., Direct laser writing of micro-supercapacitors on hydrated graphite oxide films, Nat. Nanotechnol. 6 (8) (2011) 496–500.
[90] V. Strong, S. Dubin, M.F. El-Kady, A. Lech, Y. Wang, B.H. Weiller, et al., Patterning and electronic tuning of laser scribed graphene for flexible all-carbon devices, ACS Nano 6 (2) (2012) 1395–1403.
[91] Z. Peng, J. Lin, R. Ye, E. Samuel, J. Tour, Flexible and stackable laser-induced graphene supercapacitors, ACS Appl. Mater. Interfaces 7 (5) (2015) 3414–3419.
[92] S.R. Nambiar, A.V. Mohan, Metal nanoparticles-based disposable sensors, Disposable Electrochemical Sensors for Healthcare Monitoring, Royal Society of Chemistry, United Kingdom, 2021, pp. 170–203.
[93] Y. Zhang, N. Li, Y. Xiang, D. Wang, P. Zhang, Y. Wang, et al., A flexible non-enzymatic glucose sensor based on copper nanoparticles anchored on laser-induced graphene, Carbon 156 (2020) 506–513.
[94] X. Hui, X. Xuan, J. Kim, J.Y. Park, A highly flexible and selective dopamine sensor based on Pt-Au nanoparticle-modified laser-induced graphene, Electrochim. Acta 328 (2019) 135066.
[95] P. Nayak, N. Kurra, C. Xia, HN. Alshareef, Highly efficient laser scribed graphene electrodes for on-chip electrochemical sensing applications, Adv. Electron. Mater 2 (10) (2016) 1600185.
[96] S. Imani, A.J. Bandodkar, A.V. Mohan, R. Kumar, S. Yu, J. Wang, et al., A wearable chemical–electrophysiological hybrid biosensing system for real-time health and fitness monitoring, Nat. Commun. 7 (1) (2016) 1–7.
[97] P. Bocchetta, D. Frattini, S. Ghosh, A.M.V. Mohan, Y. Kumar, Y. Kwon, Soft materials for wearable/flexible electrochemical energy conversion, storage, and biosensor devices, Materials 13 (12) (2020) 2733.
[98] A.V. Mohan, V. Rajendran, R.K. Mishra, M. Jayaraman, Recent advances and perspectives in sweat based wearable electrochemical sensors, Trends Anal. Chem. 131 (2020) 116024.
[99] K.K. Kim, S. Hong, H.M. Cho, J. Lee, Y.D. Suh, J. Ham, et al., Highly sensitive and stretchable multidimensional strain sensor with prestrained anisotropic metal nanowire percolation networks, Nano Lett. 15 (8) (2015) 5240–5247.
[100] A.V. Mohan, N. Kim, Y. Gu, A.J. Bandodkar, J.M. You, R. Kumar, et al., Merging of thin-and thick-film fabrication technologies: toward soft stretchable "island–bridge" devices, Adv. Mater. Technol. 2 (4) (2017) 1600284.
[101] A.J. Bandodkar, J.-.M. You, N.-.H. Kim, Y. Gu, R. Kumar, A.V. Mohan, et al., Soft, stretchable, high power density electronic skin-based biofuel cells for scavenging energy from human sweat, Energy Environ. Sci. 10 (7) (2017) 1581–1589.
[102] I. You, M. Kong, U. Jeong, Block copolymer elastomers for stretchable electronics, Acc. Chem. Res. 52 (1) (2018) 63–72.
[103] A. Larmagnac, S. Eggenberger, H. Janossy, J. Vörös, Stretchable electronics based on Ag-PDMS composites, Sci. Rep. 4 (1) (2014) 1–7.
[104] D. Wang, H. Li, M. Li, H. Jiang, M. Xia, Z. Zhou, Stretchable conductive polyurethane elastomer in situ polymerized with multi-walled carbon nanotubes, J. Mater. Chem. C 1 (15) (2013) 2744–2749.
[105] Y. Huang, C. Hao, J. Liu, X. Guo, Y. Zhang, P. Liu, et al., Highly stretchable, rapid-response strain sensor based on SWCNTs/CB nanocomposites coated on rubber/latex polymer for human motion tracking, Sens. Rev. 39 (2) (2019) 233–245.
[106] A.J. Bandodkar, R. Nuñez-Flores, W. Jia, J. Wang, All-printed stretchable electrochemical devices, Adv. Mater. 27 (19) (2015) 3060–3065.

[107] A.J. Bandodkar, I. Jeerapan, J.-.M. You, R. Nuñez-Flores, J. Wang, Highly stretchable fully-printed CNT-based electrochemical sensors and biofuel cells: combining intrinsic and design-induced stretchability, Nano Lett 16 (1) (2016) 721–727.
[108] V. Rajendran, A.V. Mohan, M. Jayaraman, T. Nakagawa, All-printed, interdigitated, freestanding serpentine interconnects based flexible solid state supercapacitor for self powered wearable electronics, Nano Energy 65 (2019) 104055.
[109] Q. Zhai, L.W. Yap, R. Wang, S. Gong, Z. Guo, Y. Liu, et al., Vertically aligned gold nanowires as stretchable and wearable epidermal ion-selective electrode for noninvasive multiplexed sweat analysis, Anal. Chem. 92 (6) (2020) 4647–4655.
[110] Q. Zhang, D. Jiang, C. Xu, Y. Ge, X. Liu, Q. Wei, et al., Wearable electrochemical biosensor based on molecularly imprinted Ag nanowires for noninvasive monitoring lactate in human sweat, Sens. Actuators B 320 (2020) 128325.
[111] C.-.H. Chen, P.-.W. Lee, Y.-.H. Tsao, Z.-H. Lin, Utilization of self-powered electrochemical systems: metallic nanoparticle synthesis and lactate detection, Nano Energy 42 (2017) 241–248.
[112] H. Wang, G. Pastorin, C. Lee, Toward self-powered wearable adhesive skin patch with bendable microneedle array for transdermal drug delivery, Adv. Sci. 3 (9) (2016) 1500441.
[113] R.K. Mishra, A.V. Mohan, F. Soto, R. Chrostowski, J. Wang, A microneedle biosensor for minimally-invasive transdermal detection of nerve agents, Analyst 142 (6) (2017) 918–924.
[114] A.V. Mohan, J.R. Windmiller, R.K. Mishra, J. Wang, Continuous minimally-invasive alcohol monitoring using microneedle sensor arrays, Biosens. Bioelectron. 91 (2017) 574–579.
[115] H. Lee, C. Song, Y.S. Hong, M.S. Kim, H.R. Cho, T. Kang, et al., Wearable/disposable sweat-based glucose monitoring device with multistage transdermal drug delivery module, Sci. Adv. 3 (3) (2017) e1601314.

CHAPTER 5

Fabrication of nanomaterials for biomedical imaging

Abhishek Sharma[a,b], Deepak Panchal[a,b], Om Prakash[a], Purusottam Tripathy[a], Prakash Bobde[c], Sukdeb Pal[a,b]
[a]Wastewater Technology Division, CSIR-National Environmental Engineering Research Institute (CSIR-NEERI), Nagpur, India
[b]Academy of Scientific and Innovative Research (AcSIR), Ghaziabad, India
[c]Department of Research and Development, University of Petroleum & Energy Studies, Dehradun, Uttarakhand, India

5.1 Introduction

Nanotechnology has evolved as an interdisciplinary research initiative to understand and modify materials by integrating engineering, chemistry, medicine, etc. [1]. It is not only limited to synthesis and characterization of nanomaterials, but have also found upscaled applications in various sectors such as environmental remediation, optics, electrical and electronics, energy, communication, biomedicine, etc. [2]. Nanomaterials are identified containing at least one dimension in the nanometre scale and exhibit properties distinct from their bulk counterparts. The distinctive characteristics are concealed in their incredibly nano-scaled structures [3,4]. In particular, nanoscale dimensions render novel and tunable electronic and optoelectronic characteristics with high luminescence efficiency, low exciton binding energy and thermal conductivity that make nanomaterials extremely favorable for FET (field-effect transistor), solid-state lighting/display, memory device, photodetector, and lasing applications in the biomedical world [5]. When nanomaterials are applied in medical sciences, it gives rise to a branch of medicine called "nanomedicine" resulting in substantial preclinical, transitional and clinical applications. Nanomedicine requires the use of nano-scaled materials, such as nanorobots and biocompatible nanoparticles that helps in easy diagnosis, drug delivery, actuation or sensing purposes in a living organism [6]. These characteristics would accelerate the advancement and the production of nanomaterials. Additionally, the ongoing research will also evolve unparalleled possibilities in individualized diagnostics and treatment methods.

Among the enormous application of nanoparticles in medicine, biomedical imaging is the most attractive and booming research field. Bioimaging is also referred to as molecular imaging and is defined as "visualization, characterization, and measurement of biological processes at the molecular and cellular levels in humans and other living systems" [7]. It is an evolving interdisciplinary research field combining various areas (genetics, molecular biology, cytology, physics, nuclear medicine, medicine, chemistry, pharmacology, etc.) which help in the detection of physiological processes and therefore provides useful clinical data

Advanced Nanomaterials for Point of Care Diagnosis and Therapy
DOI: https://doi.org/10.1016/B978-0-323-85725-3.00023-4

for treatment strategies of various diseases and post-treatment condition [2,8]. In comparison to the conventional imaging probes (organic molecules or metal–organic compounds), the inorganic nanomaterials have facilitated outstanding performance due to their unique physicochemical characteristics based on their small size [9]. Attributes such as magnetism, optical activity and X-ray attenuation have been used to develop nanoparticle-based probes for bioimaging. For example, quantum dots with improved optical stability, tunable emission wavelength and chemical constancy have been considered as a robust and better alternative than fluorescent dyes for fluorescent tagging in optical imaging [10]. Similarly, superparamagnetic iron oxide nanoparticles showed enhanced detection sensitivity as strong T_2 MRI (magnetic resonance imaging) contrast agents that have been preferred over conventional Gd^{3+}-based MRI contrast agents [11]. These nanomaterials can be conjugated to specific targeting ligands that increase the selected binding capability to diseased sites. This enables the easy penetration with enhanced permeability of nanocarriers through micro vessels and can be taken up easily by cells offering highly-selective payload accumulation at target sites. Even after the excellent characteristics of nanomaterials, the nanomaterial-based probes have many drawbacks that avert their substantial use in clinical settings and therefore, a few are approved. However, their applicability can be improved with several surface modification techniques to enhance biocompatibility and functionalities such as stimuli-responsiveness, targeted imaging [12].

Among the enormous nanomaterials, inorganic nanoparticles (quantum dots, gold nanoparticles, silica nanoparticles, magnetic nanoparticles, carbon nanotubes, fullerenes, and graphene) based molecular imaging technologies (as shown in Fig. 5.1) have vast

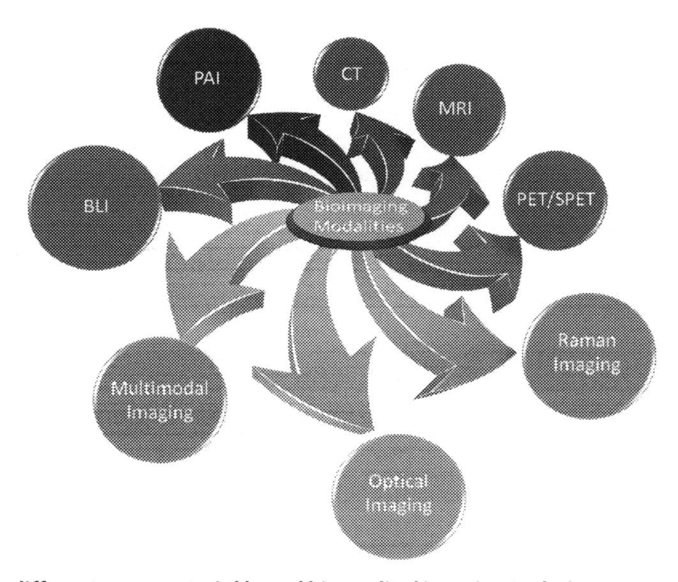

Fig. 5.1 ***Illustrates different nanomaterial based biomedical imaging techniques.***

potential than conventional imaging technologies. Therefore, in this book chapter, we discuss nanomaterials' application and their recent progress in bioimaging. Moreover, their design considerations and various techniques are also addressed from diagnostic and imaging perspectives.

5.2 Bioimaging modalities

A non-invasive tomography process renders a multiplanar and 3-D detailed in-vivo images termed Magnetic resonance imaging (MRI). MRI works on the principle of contrast mechanism that developed due to the signal alterations between abutting regions such as tissue, bone, vessels and by nearby magnetic materials. During the process of MRI, the contrast enhancement is achieved through the interaction between contrast agents and neighbouring water protons. The fundamental concept of MRI depends on the nuclear magnetic resonance (NMR) and the relaxation of proton spins in a magnetic field [13]. Upon exposure to a strong magnetic field, ^{1}H protons (water protons) adsorb resonant radiofrequency (RF) pulses leading to an immediate excitation and subsequent returning to the ground state by emission of the RF energy. In this process, the applied magnetic field coordinate with either like or unlike the spin of nuclei. There can be two different ways for relaxation (returning) of excited nuclei as, longitudinal (T_1) and transverse (T_2) [9,14]. Using these pathways, superparamagnetic iron oxide NPs were comprehensively employed as contrast agents (CAs) [15]. Therefore, the magnetic materials act as strong MRI contrast agents that helps producing MR signals. The receiver's coils are employed to enhance the sensing of the radiated signal across the body. Further, the amplitude of signals is sensed and plotted on greyscale for imaging. [16]. The intrinsic contrast can be achieved through the process. While some external CAs are used in the process for good spatiotemporal, specific, sensitive and responsive MRI images [2]. In recent advances, many nanoparticles such as magnetic, high Z element and quantum dots are being used as a contrast agent in MRI to enhance the image's potential. Computed tomography (CT) is derived from terms, computed (with a computer), tomo (to cut) and graph(y) (image). This is the most advanced tool for diagnosing imaging focussed on X-ray encounters with organs or contrast agents. CT works on the principle of attenuation coefficient of X-ray beam that passes through density of *in-vivo* organs. CT enables a two-dimensional segment perpendicular to the axis of the acquisition mechanism to rebuild the density of the body [2,17]. Chemical-based contrast agents are considered safest, but severe adverse effects have also been observed due to their high osmolality and viscosity [17,18]. Consequently, nanoparticulate CT agents have been expected to have enormous potentials to generate an adequate image of body. Optical imaging (OI) is an evolving and advanced modality of imaging that is expected to have a significant effect on lethal disease prevention and care. It is a non-invasive and non-toxic technique that renders almost real-time visualization without removing or altering the observed tissues. It can detect both functional and structural deviation with high resolution images owing to pico or femtomolar concentration [19]. In optical imaging, fluorescence and bioluminescence imaging are common procedures to analyze the non-ionizing radiation and provide the high-resolution 2D or multidimensional image information [2,20]. Fluorescence imaging system contains a source of light (appropriate wavelength) which excites the target fluorescent molecules and tissues. Subsequently, they emitted back a longer-wavelength and lower energy photons that are used for imaging [19]. Other modalities in bioimaging are Bioluminescence Imaging (BLI) using the enzymatic reaction between an enzyme and its substrate that has been used for animal molecular imaging. Photoacoustic imaging (PAI) (also called optoacoustic and thermoacoustic imaging) is an evolving imaging modality based on the photoacoustic effect [21]. It can conduct real-time as well as cross-sectional imaging of tissue in vivo without peeling of tissues [22]. Similarly, Positron emissions tomography (PET) is another diagnostic imaging technology intended to use positron-emitting radio isotopic compounds as molecular image tests and *in-vivo* assessment of biochemical processes. None of the single modality is ideal that fills all the medical application's necessities. Because they often exhibit low sensitivity for detection and produce unitary information. To overcome these drawbacks, a combination of two or more modalities can considerably enhance the sensitivity of detection to get more detailed information. As a consequence, there is an apparent proneness to hybrid imaging approaches, namely optical/MR, PET/CT, MR/PET and optical/CT imaging etc. The synergistic advantages of dual and multi-modals could provide more helpful diagnosis and treatment policies [23]. Due to the easy synthesis, numerous modification procedures and unique characteristics of nanomaterials are viable in bioimaging and supporting traditional bioimaging processes and techniques. The application of nanomaterials in bioimaging is one of the most promising approaches in dual and multimodal imaging systems because of its high surface area and multi-functional groups. Morphological modifications and surface functionalization with ligands turn NPs into multifunctional entities. These multifunctional NPs are viable to use in the field of bioimaging in both dual and multimodal imaging system [2]. For analysis and therapy of diseases, an enormous number of NPs based dual and multimodal imaging probes have been advanced and applied for *in vivo* functional imaging.

5.3 Fabrication of nanomaterials for bioimaging

Nanomaterial's (or nanomedicine) applications in the biomedical world are rooted in advanced functional design and have been realised in pre-clinical experimental diagnosis. In due course, nanomaterials will facilitate customized clinical care based on molecular profiles of each patient. Nanomaterial production in medical science is

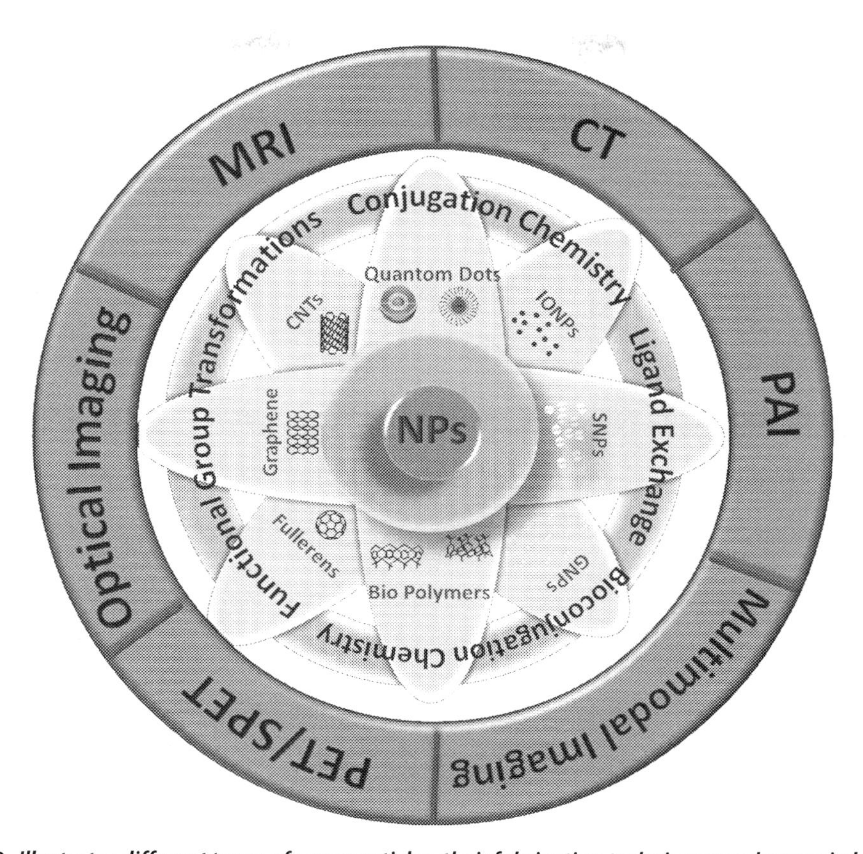

Fig. 5.2 ***Illustrates different types of nanoparticles, their fabrication techniques and usage in biomedical imaging.***

quick and multidirectional, but currently their uses are under developing stages. The advantages of using nanoparticles can be broken down into various directions such as diagnostic drug and gene delivery, molecular imaging, and targeted gene therapy [24,25]. We plan to study the evolution of various nanomaterials in biomedical imaging, as shown in Fig. 5.2.

5.3.1 Iron oxide nanoparticles (IONPs)

Iron based nanomaterials, specifically, iron oxide nanoparticles (IONPs) are considered as typical contrast agents that can be used in MRI (as T-2- and T-2-weighted) [26]. Among several synthetic methods, coprecipitation of ferrous/ferric ions in ammonical aqueous media is the easiest way that usually hindered by poor crystallinity and polydispersed. Therefore, other methods such as decomposition at high heat (thermal), can produces the iron nanomaterials with hight monodispersity and uniform crystallinity. The hydrophobic IONPs surface are subsequently coated with polymers of silica, phospholipids, or amphiphilic to increase the *in-vivo* solubility and biocompatibility. The IONPs can be categorized into three subtypes in terms of particle size distributions: particles with diameter > 30 nm known as superparamagnetic iron oxide nanoparticles (SPIO), diameter > 20 nm known as ultrasmall superparamagnetic iron oxide nanoparticles (USPIO) and diameter < 10 nm known as very small superparamagnetic iron oxide (VSOP). Dual system of using different modalities can lead to improved imaging tool in in-vivo, ex-vivo and in-vitro for living subjects. This multimodal system can be easily explored using the combination of MRI along with optical dual modal imaging [26]. A Cy5.5-arginyl peptide conjugated nanoparticle was developed from crosslinked iron oxide (CLIO) by the group of researchers for in vivo near-infrared fluorescence (NIRF) and MR imaging [27]. An innovative imaging probe containing of CLIO-Cy5.5 in conjugation with peptide EPPT was also developed, which explicitly recognised under glycosylated mucin-1 antigen (uMUC-1) on different tumor cells [28]. This probe showed high specificity toward the number of *in-vitro* uMUC-1-positive human adenocarcinomas. Furthermore, in uMUC-1-positive tumors an accumulation of the developed probe was displayed using NIRF and MR imaging. On the contrary, no such signals were observed in control tumors. Therefore, the probe could be extended not only to the early staging and detection of tumor recurrence, but also shown helpful in monitoring of therapeutic efficacy. Chlorotoxin, a larger peptide (36-amino acid) isolated from the Israeli scorpion venom exhibit high affinity for molecules like MMP-2 endopeptidase, have the capability for conjugating with USPIO for glioma tumor imaging. A multifunctional nanoprobe containing motifs of Cy5.5-Cltx on USPIO's surface using PEGylation linkage was synthesized that provide easy MR imaging and fluorescence microscopy [29]. This nanoprobe was extremely stable and has prolonged retention time period inside targeted cells (at least 24 h) as compared to traditional optical fluorophores, which is beneficial for intraoperative imaging applications. Cheon's group synthesized $DySiO_2$-$(Fe_3O_4)_n$ nanoparticle that showed excellent performance ofT2-weighted contrast mediated enhanced MR images along with strong fluorescent features detecting polysialic acids that expressed on neuroblastoma and other cell lines [30]. A parallel concept have also been used for MR imaging of targeted specific adenovirus gene delivery [31]. The same cross-linker sulfo-SMCC may be introduced into the Lys residue-containing adenovirus and MnMEIO (manganese-doped magnetism engineered iron oxide) that facilitated the formation of darker MR distincted contrast as compared to control nontreated and free MnMEIO-treated cell lines upon their application on targeted region.

5.3.2 Gold nanoparticles (GNPs)

Gold is one of the most potent nanoparticles/material in the field of bioimaging, sensing and therapy because of vivid colorant, morphology and tuneable plasmon nature [32–34]. A large number of GNPs exhibiting different morphologies have been used such as nano rods, shells, cages, webs and surface-enhanced Raman scattering (SERS) NPs for the investigation of *in-vitro* and *in-vivo* cancer imaging and diagnosis [2]. There are some enhanced modalities of GNPs in the field of bioimaging such as surface plasmon resonance, optoelectronic cohesion, photothermal and photoacoustic reflexes. Morphological variations in GNPs have changed characteristics and nature of imaging, nano-spheres and rods of GNPs are appropriate for dark optical and photothermal imaging. Whereas, photoacoustic imaging is provided by nano-spheres, shells, cages which surprisingly attained the deep tissues, lymph nodes and circulatory systems [35]. These variety of gold nanostructure were synthesized via seed-mediated growth with satisfactory reproducibility [36]. With the development of quantum flock synthesis and bio-coupling, gold nanoparticles have entered the golden age of bioimaging based on fluorescence [37]. Synthesis of water-dispersible gold nanoparticles with reactive surface properties allowed the various molecules/ligands to be attached, including drugs, contrast agents and other moieties [38]. Synthesis of dumbbell shaped nanoparticles composed gold and magnetic nanoparticles was used for dual MR and optical imaging prepared via magnetic Fe_3O_4 growth on Au in oleic acid and oleyl-amine environment [39]. Hyeon group proposed the magnetic gold (Mag-GNS) nanoshells that could be used as a medium for MRI imaging [40]. PLGA-Au-Mn nanocomposite embodied with rhodamine was also developed for imagery of T-2 MRI [41]. Gold and manganese shells have made photothermal properties and MR contrast improvement possible.

5.3.3 Silica nanoparticles (SNPs)

Silica nanomaterials are one of the crucial categories of inorganic nanoparticles for bioimaging due to their size-controllable morphology, ease of functionalization, biocompatibility and hydrophilic surface. Silica is also accepted and has been commonly used in cosmetics by the USFDA [42]. Silica NPs have enormous pertinence in bioimaging due to their adjustable surface and pores as well as they cope with fluorescent/MRI/PET/SPECT as contrast agents [43]. Also, Cornell dots (silica nanoparticles) encapsulated with fluorescent dyes have also been evaluated for stage I human clinical trials as cancer-targeted imaging probes [44]. [45] prepared the silica nanoparticles (50-2000 nm) under the controlled hydrolysis of silyl ethers into silanols followed by silanols condensation in the mixture of ammonia, water and alcohol. The silica NPs sizes were depending on the concentration of silyl ether and alcohol. Similarly, mesoporous silica NP (SNP) was also synthesized via simple sol-gel method by the *in-situ* polymerization of silyl ethers. Various of non-ionic surfactants, polymers, triethanolamine and propanetriol were added in an optimized concentration of silyl ether to avert the accumulation of SNPs. The hydrophilic character of SNPs make them promising material in the field of biomedical such as drugs delivery, bioimaging and diagnosis [46]. A number of alterations are frequently required for biocompatibility and for coupling with contrasting agents involving incorporation of functional groups onto the nanoparticles using co-condensation and post-synthesis surface modification process [47]. Amino-functionalized SNPs are easily conjugated with biomolecules, nucleic acid and peptides with various functional fragment such as alkyl halide, NHS esters and carboxylic acids [12]. Fluorescent samples such as rhodamine isothiocyanate, fluorescein isothiocyanate, methyl violegene, alexa-fluorine dyes, ruthenium, terbium and europium coordination complexes can be introduced with modified silica nanoparticles that effectively delivered to different cell lines. As a result, one or two photon fluorescence microscopes can visualise those cells which have been labelled. Lin et al. [48] prepared the silica-coated core-shell NPs (SPIO) via thermal decomposition process with precursors oleic acid and oleyl amine. Further, the hydrophobic SPIO NPs were modified with dye-doped silica shells via reversed microemulsion process. The modified SPIO was able to monitor *in-vivo* with MRI contrast agents. Similarly, (C. [49]) synthesised silica- and alkoxysilane-coated nanoparticles and applied in MRI for cell labelling and *in-vitro* cell tracking.

5.3.4 Carbon nanotubes (CNTs)

CNTs are most prominently used nanomaterials in the field of health care such as diagnostic, imaging and drug delivery [50]. They formed by sp^2 hybridization of carbon (graphite) in the form of hollow cylindrical tubes with high aspect ratio. Depending upon the number of wall layers of carbon, CNTs are divided into three groups such as:

- Single-walled nanotubes (SWNTs),
- Double-walled nanotubes (DWNTs) and
- Multiwalled nanotubes (MWNTs).

Synthesized CNTs are hydrophobic and are found mostly in aggregate or packaged form with uniform suspension in aqueous solutions. This feature facilitates the use of CNTs in bioimaging, drug delivery, photothermal therapy, and other applications related to health care. In the aqueous phase, the dispersion of CNTs can be carried out by controlled acoustic cavitation of CNTs aqueous solution with amphiphilic molecules or surfactants. Covalent and non-covalent functionalization processes are the most efficient method for stable dispersion of CNTs in both the organic and aqueous phases. Other fundamental properties like broad absorption band, outstanding photoacoustic response, NIR photoluminescence and unique Raman/surface-enhanced Raman scattering are also responsible for bioimaging [51]. CNTs very well show these properties as the most prevalent Raman fingerprint band for bioimaging is the Graphite mode (G-band). NIR excitation of CNTs helps in minimizing the autofluorescence of biospecimen and photobleaching of CNTs [51]. Further, CNT's optoacoustic imaging property offers another propitious bioimaging technique for tumor by providing excellent photo-to-acoustic conversion efficiency and photothermal-acoustic response.

CNTs can conjugate with radionuclide, fluorescent dyes, and other organic/inorganic nanoparticles and serve as contrasting agents for different bioimaging techniques like MRI, CT, PET, SPECT (single-photon emission computed tomography), etc. [12].

Conjugation of CNTs could be achieved by different targeting molecules like RGD peptide, EGF, transferrin, aptamers, herceptin, folic acid, and many antibodies. The biodistribution and targeted delivery of such bio conjugated CNTs by simple incubation in living cells or by intravenous or intratumoral injection in tumors are significant. Coating CNTs with PEG polymer accomplish very good biodistribution that are effective in vivo tumor targeting. CNT's wide absorption band allows using a far red/NIR light source for excitation, without inducing autofluorescence of cell or tissue. The effective penetration of both the excited and emitted NIR photoluminescence (1 to 1.8 μm) through organelles, despite the low photoluminescence quantum efficiency of CNT, enables one to achieve best signal-to-noise ratios. However, difference in photoluminescence quantum efficiency of CNTs have been observed due to surface coating and metallic concentration of CNT. Coating of CNTs with PEGylated phospholipids or sodium cholate gives the quantum efficiency of photoluminescence higher than that stabilized in normal surfactants. Further for improving the photoluminescence strength of CNTs, resonance coupling with GNPs, accompanied by plasmon-assisted energy transfer is used.

CNTs based probes supports Raman imaging as one of the most promising and effective modality of bioimaging [52]. Another promising and efficient modality for bioimaging is the photoacoustic response of CNT. The basic theory of photoacoustic imaging is the ultrasonic transducer that are used to detect the optical excitation energy transformed by a biospecimen into high magnitude sound waves. However, without the use of any contrasting agent, the separation of the conditions such as tumor from normal tissues by optoacoustic imaging is difficult. CNTs deliver outstanding photoacoustic conversion and photothermal-acoustic response efficiency, making these materials one of the most promising contrast agents in tumor optoacoustic imaging. CNTs can combine with tumor targeting molecules such as anticancer antibodies and RGD peptides, thus developing into promising biomedical optoacoustic imaging materials [53].

5.3.5 Graphene

Graphene is a carbon allotrope with hybridization state of sp^2 and 2D honeycomb lattice. It also serves as raw material for the production of other carbon forms like CNTs and fullerene. Naturally, graphene is a ring-shaped molecule with several unsaturated carbon-carbon single bonds in a plane, that provide surface functionalization due to reactive sites and free π electrons. It has unique physical properties like reasonable biocompatibility, strong mechanical strength, ease of processing, and modification due to its flexible surface functionalization and ultrahigh surface area characteristics attracting researchers' interest [54]. They also explored the different forms of graphene like graphene quantum dots, graphene oxide and reduced graphene oxide for bioimaging. For the first time graphene was recovered by exfoliation of graphite using adhesive tape [55].

Graphene have some essential properties, such as visible/NIR photoluminescence, distinctive Raman bands, and optoacoustic and photothermal responses for bioimaging [56,57]. In particular, graphene's NIR photoluminescence is an important characteristic to reduce autofluorescence during *in vitro* and *in vivo* bioimaging. In addition, the conjugation of fluorescent moieties such as gold quantum clusters, radionuclides such as ^{66}Ga, ^{99m}Tc, ^{125}I and ^{64}Cu, organic dyes, semiconductor quantum dots, magnetic materials such as Gd (III) complexes and SPIONs to graphene or its derivatives will prepare multimodal imaging probes. Further, conjugation of peptides, antibodies or ligands toward cancer markers resulted in imaging of targeted cancer cells and tumor milieu. Use of photothermal therapy by graphene for Alzheimer's disease, cancers and its effects have recently been examined [57,58]. Interestingly, graphene oxide-Fe_3O_4-PEG composite under NIR laser illumination (*in vivo* photothermal treatment) showed a substantial decrease in the mice's tumor size [59]. In addition, the special 2D graphene structure enables photothermal therapy to be coupled with chemotherapy and photodynamic therapy by covalent or non-covalent conjugation. In such graphene-based combination therapy, excellent therapeutic impacts are observed. In short, versatile surface chemistry, wide surface area and special photoluminescence, Raman and graphene's photothermal properties demonstrate its great potential for bioimaging. However, it could be further enhanced by producing multimodal and versatile graphene arsenals for image-guided NIR phototherapy [12].

5.3.6 Fullerenes

Fullerenes(C_{60}) are the 0-D form of graphitic carbon which found in various forms such as hollow sphere, ellipsoid or tubes. C_{60} (a carbon buckyball) was discovered by Harold Kroto and group in 1985, while synthesised in huge amount by Huffman in 1990. Fullerene is made up of symmetry of twenty hexagonal and twelve pentagonal rings. There are 6:6 rings double bonds (among two hexagons) and 6:5 rings double bonds (among hexagons and pentagons). Thus, the [6,6] bonds are used by nucleophilic, radical additions as well as cycloadditions to functionalize fullerene(C_{60}) [60]. A number of atoms or ions can be accommodated by the fullerenes in their hollow center. The compounds in the hollow center that contain atoms or ions are known as endohedrals. Endohedrals metallofullerenes are functionalized fullerenes that can encapsulate the transition metal atom inside the carbon cage that forms a electronic interaction between C-cage and metals. Endohedrals metallofullerenes are sparingly soluble in polar solvents. It can be used in the application of biomedical research field [61,62]. To enhance the MRI quality, Gadolinium (Gd) encapsulated fullerenes have been used as contrast agents. Gd-DTPA is widely used for the improving of relaxation rates of water protons in MRI as a contrasting agent [63]. But there is a risk of release of residual Gd^{3+} ions from the chelate which may cause toxicity. Howeve, there are very less chances of discharge of Gd^{3+} ions from fullerene cage. Therefore, Gd @ X2n is appropriate in MRI as contrasting agent. [64] reported that the NMR contour of both water-soluble functionalized fullerenes such as Gd@C60(OH)*x* and Gd@C_{60}[C(COOH)$_2$]. They concluded that these functionalized fullerenes were perfect candidates for MIR contrasts with pH stimulation due to their strong pH dependence on the proton relaxivity.

It has been found that functionalized fullerenes with peptides and amino acids significantly activate the enzymes involved in the oxidative deamination of biogenic amines significantly [65]. A fullerene functionalised material is able to focused on specific tissue [66]. A tissue-vectorized bisphosphonate fullerene was synthesized and assessed in vitro to bone tissue. Amide bisphosphonate interacted multiple hydroxyl groups, which has a affinity to the calcium phosphate mineral hydroxyapatite of bone. Cagle et al. [67] found that the holmium-fullerenol resided (> 1h) in blood, while metal chelates (Na_2[166Ho(DTPA)(H_2O)]) can reside more than 4 days in liver and bones, showing slow but steady clearance. This unique characteristic of endohedral makes it prominent non-toxic radiotracer. Similarly, β-emitter (133Xe) was deceived into fullerenes via ion implantation process [68].

5.3.7 Quantum dots (QDs)

Semiconductor quantum dots (SCQDs) are one of the prominent QDs. With optimized time, temperature and substrates during synthesis can control the morphology and structure of QDs. Generally, QDs are formed by groups II and group VI or group III and group V elements, with size one to ten nanometer. This contributes to the effect of quantum confinement with narrow emission bands, large absorption bands and also provide luminescence features. QDs have many unique functional properties such as extended photostability, high quantum efficacy and ability to synchronized multicolour imaging via single wavelength excitation [69,70]. Therefore, use of QDs in imaging applications have grown rapidly compared to traditional fluorescence dyes due to the different attractive features of these nanomaterials [71]. In organic phase, the excellent quality QDs are prepared and also capped with extremely hydrophobic aliphatic ligands. For biological applications of QDs, reactions including ligand exchange and surface alterations are essential requisites. During, modification of the surface of QDs, properties of QDs should be unchanged and equitably distributed and the conjugating species should also be biocompatible for biological applications. The capped QDs by organic solvent such as mercaptopropionic acid and thioglycolic acid can assist in direct cells interaction and translocation to the cytosol. QDs in association with various peptides such as nucleus localization signal or mitochondrial localization signal halps in targeted intracellular labelling and mitochondrial. However, for imaging different targeting sites such as lymph nodes, tumors, blood vessels, etc., non-targeted QDs modified with surface molecules such as polyarginine, polymers, carboxylic acids, and gelatin are also explored. In addition, quantum dot – D-lactose conjugates have also been used in in vivo applications for selective labelling and imaging of leukocy [72,73].

The water soluble and paramagnetic micellular coated, PEGylated and cyclic RGD modified QDs was synthesized which was demonstrated as molecular imaging probe for fluorescence microscopy and MRI [74]. PEGylated and functionalized Gd^{3+} ion phospholipid micelles were performed to make the hydrophobic CdSe/ZnS QDs water soluble, bio-consistent and MR detectable. The coupling of RGD peptide and maleimide on the QDs surface specially targets the $\alpha v\beta 3$-integrins, which displayed on the angiogenic-endothelial cells and tumors. In the in vitro test assessment for HUVECs (angiogenic human umbilical vein endothelial cells) adequate fluorescent imaging ability and substantial T_1-weighted contrast enhancement were observed [2].

A PEGylated and polyethyleneimine coupled QDs was prepared, which could penetrate cell membranes and interrupt endosomal organelles in cells [75]. By the use of near ID QDs (emission λ 800nm), Cao et al. [76] detected the fluorescent images carcinoma cells after 6 h by cell endocytosis. A coupled aptamer- (Apt-) doxorubicin (Dox) QDs was synthesised with the ability of sensing the prostate cancer cells that express the PSMA (prostate-specific membrane antigen) protein [77].

5.4 Nanoparticle design considerations: Core and surface fabrication

Certain parameters should be taken into consideration while devising of the therapeutic probes or multimodal imaging agents with nanoparticles: (i) toxicity of nanomaterials toward living beings, (ii) formation of possible metabolites after cell uptake or vascular circulation, (iii) biodegradability and biocompatibility to prevent detrimental accumulations in organelles, (iv) convenience for chemical modification of nanomaterial, and (v) complete investigation of synthesized nanomaterials *in-vitro* and *in-vivo* before their actual applications on living organisms. Here, some of the common technologies for modification in structural design and functionalization strategies of nanomaterials are summarized:

5.4.1 Fabrication of nanoparticles core

The type of nanomaterial to be core, fabrication method of imaging probes is generally determined by the imaging purposes and specific imaging modality.

5.4.2 Shell structure synthesis

Designing a shell structure is complex than synthesizing a core or nanomaterial. The shell provides more protection to the core from outer microenvironment and improves its physical property and stability. Biocompatibility of nanomaterial is largely depending upon shell material and therefore, with improved characteristics the unexpected immunological response can be prevented. The shell structure plays a key role in the drug delivery and therapy by bursting the shell and releasing the therapeutic drugs in a controlled fashion on the targeted site. The shell could also be considered as drug payload, contrast or therapeutic agents' incorporating to strengthen their functionality.

5.4.3 Surface modifications

The outermost surface of shell (nanomaterial) may be extremely perceptive to the micro bio environment like plasma, blood or receptor at binding sites, their surface modification using stabilizers (surfactants) is vital to maintain durability. For achieving the targeted molecular imaging, specific surface ligands of nanomaterials are expected to be in conjugation with biomolecules (proteins, antibodies, or peptides). Therefore, modification protocols for designing of inner core or on the surface can facilitate various functionalities like, imaging probes and therapeutic payloads.

5.5 Fabrication techniques of nanoparticles

Various new production approaches have been developed that have many advantages for the production of desired structure of nanomaterials. As shown in Table 5.1 top-down synthetic process is able to provide desired shapes and sizes of NPs, while it's tough to attain desirable structures in bottom-up approaches. This section focuses in the development of different synthetic processes and their advantages/disadvantages for shape-specific nanomaterial drug nanocarriers.

One recent fabrication technique that generates homogeneous nanoparticles with greater yield and enhanced size is microfluidic devices. They have been engineered

Table 5.1 Advanced nanofabrication technique for synthesis of substantial nanoparticles.

Shape	Size	Charge	Techniques	Reference
Rods Bipyramids Decahedral	6.1 ± 0.2 nm	Neutral	Thermally-induced seed twinning	[88]
Cubic cylinder	Cubic: 2, 3, 5 μm cylindrical nanoparticles (diameter: height 200 nm: 200 nm, 100 nm: 300 nm, 150 nm: 450 nm)	Positive	Top-down lithographic fabrication method (PRINT)	[89]
Spherical	Usually 422 nm but after prednisolone loading: 448~660 nm)		Electrohydrodynamic atomization	[90]
Rod			Imprint lithography	[91]
Discoidal Soft vs rigid properties	1 μm (1000 × 400 nm)	Negative	Top-down fabrication technique (electron beam lithography)	[92]
Spherical	10 nm	Neutral	Bottom-up chemical methods	[93]
Spherical	17, 28, and 48 nm	Hydrophobic	Langmuir-Blodgett method	[94]

to synthesize particles of various diameters from nm to μm by means of materials like metals and polymers. Xu et al. [78] focused on the flow system for production of monodisperse polymeric particles. These monodisperse particles were manufactured with small size variations up to 30 μm with an average radius of 14 μm. A comparison was also done for the monodisperse microparticles against the microparticles of the same size produced by conventional method fabricated on the basis of release kinetics of bupivacaine [78]. Higher yields were shown by monodisperse particles produced from the microfluidic system and could encapsulate larger molecules. Qin et al. [79] manufactured micro-fabricated devices by using design software such as photolithography and etching. They used soft lithography to render unique structures on photomasks' surface by micro- and nano-scale patterns. Soft lithography could be used to emulsify polymeric tools and particles because it can regulate the form and structure more efficiently in terms of dimension. It renders very high performance as well as economical. But synthesis issues remain persist with techniques such as replica, micro and nano transfer moldings. Nanoparticles generated by an injectable nanoparticle (ING) can be combined with molecular imaging probes for tumor imaging and treatment. These nanoparticles were successfully accumulated in mice tumors [80].

Currently, an advanced top-down method-based photolithography process introduced which render designed photomask's polymer sheets. It also renders highly defined and pollution free synthesis. Near-infrared wavelength light is used to mitigate the photodamage related issues. Due to above advantages, photolithography has contributed in the study of deeper tissues and large-scale patterning. However, this latest technique exhibits a number of advantages over other available techniques, it requires high maintenance cost. Electrospray is a feasible method of microencapsulation built to resolve the limitations of solvent extraction methods. The sub-micro sized monodisperse particles were synthesized by the application of high voltage to the needle and the field [81,82].

For multimodal imaging and image-guided therapy, coaxial electrospray was used [83]. Experimental and theoretical studies on coaxial electrospray of poly(lactide-co-glycolide) microparticles sought to address the limitations of loss of bioactivity and low encapsulation during nanofabrication. In the presence of solvent like oleic acid, naturally occurring ore of high purity was restructured to synthesize the magnetite nanoparticles [84]. Various methods have been reported for the synthesis of magnetic nanoparticles, including sol-gel techniques, precipitation, and thermal decomposition. The nanoparticles made using the method of wet grinding are smaller in size, more smooth, less agglomerated and have a fast melting time [84]. Merkel et al. [85] said the top-down manufacturing methods would easily control the particles' shape and size. Thorough analysis of the flexibility of

Particle Replication in Non-wetting Templates (PRINT ®) is also possible with this method allowing formation of particle with good shape and size, it can easily be increased at a low price.

It is costly to use physiochemical processes for synthesis nanoparticles and requires the usage of harmful chemicals that constitute risks to the environment and living health. Roy et al. [86] used the term 'green chemistry' and as he used starch from boiled raw rice for the preparation of various silver nanoparticles. The particles formed involved spherical, rod-like, hexagonal and floral processes which needed no hazardous agents and materials and were simple, cost-effective and environmentally friendly. Green synthesis is also used with plant extracts in single-celled organisms like bacteria and fungi [86].

However, there are a range of drawbacks, including high operating costs, poor efficiency, corrosive etchants exposed surfactants, high energy radiation, relatively low temperature and the wave length used in the manufacture process. For several researchers, these limitations have proven to be averse for using such conventional techniques. Therefore, the discovery and development of new nanofabrication technique was inspired to overcome from these disadvantages. Recent nano-manufacturing techniques have created possibilities for the use of non-planar surfaces and large surface area. These methods are also applicable to biological, sensitive organic and organometallic materials. In addition, they allow low-cost fabrication and offer high-throughput compared to traditional techniques [87].

5.6 Conclusions

Nanotechnology and has witnessed a huge development and is still undergoing evolutionary developments and growth in the biomedical world especially the bio imaging. Nanomaterials have improved and inflate the proficiencies of conventional methods for bio-imaging due to their huge surface are, controlled surface modifications (core and shell) and ease of synthesis. Different nanomaterials such as, noble (gold and silver), carbon based (CNT and fullerene), quantum dots (QDs) and most importantly the magnetic particles have allowed the researchers for an enhanced imaging and diagnostic performance. Despite the huge development, a few of the nanomaterials have got clinical successes and many are in their beginning stage. Therefore, a more advance research is needed for further extension of nanoscience in biomedical world. For in-vivo applications of nanomaterials, their biological compatibility, bioaccumulation and systemic toxicity in the body have to be considered. Moreover, the scarce knowledge of induced toxicity of and their health risk for living tissue of nanomaterials needs to be updated. Thus, more research work is still needed for the acceptance and promotion of nanomaterial advances in bioimaging applications.

Acknowledgment

AS and DP acknowledge the University Grant Commission, New Delhi, India for providing Junior & Senior Research Fellowship, respectively. Director, CSIR-NEERI is thankfully acknowledged for giving the opportunity to pursue the work in CSIR-NEERI, Nagpur, India. The article is checked for plagiarism using the iThenticate software and recorded in the Knowledge Resource Center, CSIR-NEERI, Nagpur for anti-plagiarism (KRC No.: CSIR-NEERI/KRC/2020/DEC/WWTD/1).

References

[1] C.M. Niemeyer, Nanoparticles, proteins, and nucleic acids: biotechnology meets materials science, Angew. Chem. Int. Ed. 40 (22) (2001) 4128–4158.
[2] Z. Liu, F. Kiessling, J. Gätjens, Advanced nanomaterials in multimodal imaging: design, functionalization, and biomedical applications, J. Nanomater. 2010 (2010), doi:https://doi.org/10.1155/2010/894303.
[3] S.K. Nune, P. Gunda, P.K. Thallapally, Y.-.Y. Lin, M. Laird Forrest, C.J. Berkland, Nanoparticles for biomedical imaging, Expert Opin. Drug Deliv. 6 (11) (2009) 1175–1194.
[4] Y. Xu, A. Karmakar, W.E. Heberlein, T. Mustafa, A.R. Biris, A.S. Biris, Multifunctional magnetic nanoparticles for synergistic enhancement of cancer treatment by combinatorial radio frequency thermolysis and drug delivery, Adv. Healthc. Mater. 1 (4) (2012) 493–501.
[5] T. Zhu, S.G. Cloutier, I. Ivanov, K.L. Knappenberger, I. Robel, F. Zhang, Nanocrystals for electronic and optoelectronic applications, 2012.
[6] J.K. Patra, G. Das, L.F. Fraceto, E.V.R. Campos, M.P. Rodriguez-Torres, L.S. Acosta-Torres, L.A. Diaz-Torres, et al., Nano based drug delivery systems: recent developments and future prospects, J. Nanobiotechnol. 16 (1) (2018) 1–33.
[7] D.A. Mankoff, A definition of molecular imaging, J. Nucl. Med. 48 (6) (2007) 18N–21N.
[8] W. Jin, D.-.H. Park, Functional layered double hydroxide nanohybrids for biomedical imaging, Nanomaterials 9 (10) (2019) 1404.
[9] D. Kim, J. Kim, YIl Park, N. Lee, T. Hyeon, Recent development of inorganic nanoparticles for biomedical imaging, ACS Cent. Sci. 4 (3) (2018) 324–336.
[10] X. Gao, Y. Cui, R.M. Levenson, L.W.K. Chung, S. Nie, In vivo cancer targeting and imaging with semiconductor quantum dots, Nat. Biotechnol. 22 (8) (2004) 969–976.
[11] N. Lee, H.R. Cho, M.H. Oh, S.H. Lee, K. Kim, B.H. Kim, K. Shin, et al., Multifunctional Fe3O4/TaO x Core/Shell nanoparticles for simultaneous magnetic resonance imaging and X-ray computed tomography, J. Am. Chem. Soc. 134 (25) (2012) 10309–10312.
[12] V. Biju, Chemical modifications and bioconjugate reactions of nanomaterials for sensing, imaging, drug delivery and therapy, Chem. Soc. Rev. 43 (3) (2014) 744–764.
[13] M.A. Brown, R.C. Semelka, MRI: Basic Principles and Applications, John Wiley & Sons, New Jersy, 2011.
[14] H.B. Na, C.S. In, H. Taeghwan, Inorganic nanoparticles for MRI contrast agents, Adv. Mater. 21 (21) (2009) 2133–2148.
[15] M.M.J. Modo, J.W.M. Bulte, Molecular and Cellular MR Imaging, CRC Press, Boca Raton, 2007.
[16] F. Rabai, R. Ramani, Magnetic resonance imaging: anesthetic implications. In: Essentials of Neuroanesthesia, Academic Press, London, 2017, pp. 519–532.
[17] N. Lee, S.H. Choi, T. Hyeon, Nano-sized CT contrast agents, Adv. Mater. 25 (19) (2013) 2641–2660.
[18] S. Namasivayam, M.K. Kalra, W.E. Torres, W.C. Small, Adverse reactions to intravenous iodinated contrast media: an update, Curr. Probl. Diagn. Radiol. 35 (4) (2006) 164–169.
[19] G.D. Luker, K.E. Luker, Optical imaging: current applications and future directions, J. Nucl. Med. 49 (1) (2008) 1–4.
[20] G. Pirovano, S. Roberts, S. Kossatz, T. Reiner, Optical imaging modalities: principles and applications in preclinical research and clinical settings, J. Nucl. Med. 61 (10) (2020) 1419–1427.

[21] M. Xu, L.V. Wang, Photoacoustic imaging in biomedicine, Rev. Sci. Instrum. 77 (4) (2006) 041101.
[22] C. Lee, M. Jeon, C. Kim, Photoacoustic imaging in nanomedicine. In: Applications of Nanoscience in Photomedicine, Chandos Publishing, Cambridge, 2015, pp. 31–47.
[23] J. Cheon, J.-.H. Lee, Synergistically integrated nanoparticles as multimodal probes for nanobiotechnology, Acc. Chem. Res. 41 (12) (2008) 1630–1640.
[24] R. Sinha, G.J. Kim, S. Nie, D.M. Shin, Nanotechnology in cancer therapeutics: bioconjugated nanoparticles for drug delivery, Mol. Cancer Ther. 5 (8) (2006) 1909–1917.
[25] Y. Liu, H. Miyoshi, M. Nakamura, Nanomedicine for drug delivery and imaging: a promising avenue for cancer therapy and diagnosis using targeted functional nanoparticles, Int. J. Cancer 120 (12) (2007) 2527–2537.
[26] S. Laurent, D. Forge, M. Port, A. Roch, C. Robic, L. Vander Elst, R.N. Muller, Magnetic iron oxide nanoparticles: synthesis, stabilization, vectorization, physicochemical characterizations, and biological applications, Chem. Rev. 108 (6) (2008) 2064–2110.
[27] L. Josephson, M.F. Kircher, U. Mahmood, Yi Tang, R. Weissleder, Near-infrared fluorescent nanoparticles as combined MR/optical imaging probes, Bioconjug. Chem. 13 (3) (2002) 554–560.
[28] A. Moore, Z. Medarova, A. Potthast, G. Dai, In vivo targeting of underglycosylated MUC-1 tumor antigen using a multimodal imaging probe, Cancer Res. 64 (5) (2004) 1821–1827.
[29] O. Veiseh, C. Sun, J. Gunn, N. Kohler, P. Gabikian, D. Lee, N. Bhattarai, et al., Optical and MRI multifunctional nanoprobe for targeting gliomas, Nano Lett. 5 (6) (2005) 1003–1008.
[30] J.-.H. Lee, Y.-. Jun, S-In Yeon, J.-.S. Shin, J. Cheon, Dual-mode nanoparticle probes for high-performance magnetic resonance and fluorescence imaging of neuroblastoma, Angew. Chem. Int. Ed. 45 (48) (2006) 8160–8162.
[31] Y.-M. Huh, E-S. Lee, J.-H. Lee, Y.-W. Jun, P.-H. Kim, C.-O. Yun, J.-H. Kim, J.-S. Suh, J. Cheon, Hybrid nanoparticles for magnetic resonance imaging of target-specific viral gene delivery, Adv. Mater. 19 (20) (2007) 3109–3112.
[32] K. Saha, S.S. Agasti, C. Kim, X. Li, V.M. Rotello, Gold nanoparticles in chemical and biological sensing, Chem. Rev. 112 (5) (2012) 2739–2779.
[33] P. Shi, K. Qu, J. Wang, M. Li, J. Ren, X. Qu, pH-responsive NIR enhanced drug release from gold nanocages possesses high potency against cancer cells, Chem. Commun. 48 (61) (2012) 7640–7642.
[34] S. Liang, Q. Zhou, M. Wang, Y. Zhu, Q. Wu, X. Yang, Water-soluble l-cysteine-coated FePt nanoparticles as dual MRI/CT imaging contrast agent for glioma, Int. J. Nanomed. 10 (2015) 2325.
[35] M.W. Knight, N.J. Halas, Nanoshells to nanoeggs to nanocups: optical properties of reduced symmetry core–shell nanoparticles beyond the quasistatic limit, New J. Phys. 10 (10) (2008) 105006.
[36] E.C. Dreaden, A.M. Alkilany, X. Huang, C.J. Murphy, M.A. El-Sayed, The golden age: gold nanoparticles for biomedicine, Chem. Soc. Rev. 41 (7) (2012) 2740–2779.
[37] E.S. Shibu, S. Sugino, K. Ono, H. Saito, A. Nishioka, S. Yamamura, M. Sawada, Y. Nosaka, V. Biju, Singlet-oxygen-sensitizing near-infrared-fluorescent multimodal nanoparticles, Angew. Chem. Int. Ed. 52 (40) (2013) 10559–10563.
[38] A.C. Templeton, W. Peter Wuelfing, R.W. Murray, Monolayer-protected cluster molecules, Acc. Chem. Res. 33 (1) (2000) 27–36.
[39] C. Xu, J. Xie, D. Ho, C. Wang, N. Kohler, E.G. Walsh, J.R. Morgan, Y.E. Chin, S. Sun, Au–Fe3O4 dumbbell nanoparticles as dual-functional probes, Angew. Chem. Int. Ed. 47 (1) (2008) 173–176.
[40] J. Kim, S. Park, JiE Lee, S.M. Jin, J.H. Lee, InSu Lee, I. Yang, et al., Designed fabrication of multifunctional magnetic gold nanoshells and their application to magnetic resonance imaging and photothermal therapy, Angew. Chem. 118 (46) (2006) 7918–7922.
[41] H. Park, J. Yang, S. Seo, K. Kim, J. Suh, D. Kim, S. Haam, K.-.H. Yoo, Multifunctional nanoparticles for photothermally controlled drug delivery and magnetic resonance imaging enhancement, Small 4 (2) (2008) 192–196.
[42] C. Contado, L. Ravani, M. Passarella, Size characterization by sedimentation field flow fractionation of silica particles used as food additives, Anal. Chim. Acta 788 (2013) 183–192.
[43] T.-.J. Yoon, K.N. Yu, E. Kim, J.S. Kim, B.G. Kim, S.-.H. Yun, B.-.H. Sohn, M.-.H. Cho, J.-.K. Lee, S.B. Park, Specific targeting, cell sorting, and bioimaging with smart magnetic silica core–shell nanomaterials, Small 2 (2) (2006) 209–215.
[44] H. Ow, D.R. Larson, M. Srivastava, B.A. Baird, W.W. Webb, U. Wiesner, Bright and stable core– shell fluorescent silica nanoparticles, Nano Lett. 5 (1) (2005) 113–117.

[45] B.G.Trewyn, II. Slowing, S. Giri, H.-.T. Chen,VS-Y. Lin, Synthesis and functionalization of a mesoporous silica nanoparticle based on the sol–gel process and applications in controlled release, Acc. Chem. Res. 40 (9) (2007) 846–853.
[46] R. Rani, K. Sethi, G. Singh, Nanomaterials and their applications in bioimaging. In: Plant Nanobionics, Springer, Cham, 2019, pp. 429–450.
[47] Si-H. Wu, C.-.Y. Mou, H.-.P. Lin, Synthesis of mesoporous silica nanoparticles, Chem. Soc. Rev. 42 (9) (2013) 3862–3875.
[48] Yu-S. Lin, Si-H Wu, Y. Hung, Yi-H. Chou, C. Chang, M.-.L. Lin, C.-.P. Tsai, C.-.Y. Mou, Multifunctional composite nanoparticles: magnetic, luminescent, and mesoporous, Chem. Mater. 18 (22) (2006) 5170–5172.
[49] C. Zhang, B. Wängler, B. Morgenstern, H. Zentgraf, M. Eisenhut, H. Untenecker, R. Krüger, et al., Silica-and alkoxysilane-coated ultrasmall superparamagnetic iron oxide particles: a promising tool to label cells for magnetic resonance imaging, Langmuir 23 (3) (2007) 1427–1434.
[50] S. Kumar, R. Rani, N. Dilbaghi, K. Tankeshwar, Ki-H. Kim, Carbon nanotubes: a novel material for multifaceted applications in human healthcare, Chem. Soc. Rev. 46 (1) (2017) 158–196.
[51] K. Kostarelos, A. Bianco, M. Prato, Promises, facts and challenges for carbon nanotubes in imaging and therapeutics, Nat. Nanotechnol. 4 (10) (2009) 627–633.
[52] A. Bianco, K. Kostarelos, M. Prato, Making carbon nanotubes biocompatible and biodegradable, Chem. Commun. 47 (37) (2011) 10182–10188.
[53] A. De La Zerda, C. Zavaleta, S. Keren, S. Vaithilingam, S. Bodapati, Z. Liu, J. Levi, et al., Carbon nanotubes as photoacoustic molecular imaging agents in living mice, Nat. Nanotechnol. 3 (9) (2008) 557–562.
[54] J. Lin, Y. Huang, P. Huang, Graphene-based nanomaterials in bioimaging. In: B. Sarmento, J. das Neves, (Eds.), Biomedical Applications of Functionalized Nanomaterials. Elsevier, Amsterdam, 2018 pp. 247–287.
[55] A. Geim, K. Novoselov, The rise of graphene, Nat. Mater 6 (2007) 183–191. https://doi.org/10.1038/nmat1849.
[56] W. Choi, I. Lahiri, R. Seelaboyina, Y.S. Kang, Synthesis of graphene and its applications: a review, Crit. Rev. Solid State Mater. Sci. 35 (1) (2010) 52–71.
[57] K. Yang, L. Feng, X. Shi, Z. Liu, Nano-graphene in biomedicine: theranostic applications, Chem. Soc. Rev. 42 (2) (2013) 530–547.
[58] X. Sun, Z. Liu, K. Welsher, J.T. Robinson, A. Goodwin, S. Zaric, H. Dai, Nano-graphene oxide for cellular imaging and drug delivery, Nano Res. 1 (3) (2008) 203–212.
[59] K. Yang, L. Hu, X. Ma, S. Ye, L. Cheng, X. Shi, C. Li, Y. Li, Z. Liu, Multimodal imaging guided photothermal therapy using functionalized graphene nanosheets anchored with magnetic nanoparticles, Adv. Mater. 24 (14) (2012) 1868–1872.
[60] I. Rašović, K. Porfyrakis, Functionalisation of fullerenes for biomedical applications, 2019, pp. 109-122.
[61] R. Partha, J.L. Conyers, Biomedical applications of functionalized fullerene-based nanomaterials, Int. J. Nanomed. 4 (2009) 261.
[62] A. Kumar, Fullerenes for biomedical applications, J. Environ. Appl. Biores. 03 (04) (2015) 175–191.
[63] P. Caravan, J.J. Ellison, T.J. McMurry, R.B. Lauffer, Gadolinium (III) chelates as MRI contrast agents: structure, dynamics, and applications, Chem. Rev. 99 (9) (1999) 2293–2352.
[64] É. Tóth, R.D. Bolskar, A. Borel, G. González, L. Helm, A.E. Merbach, B. Sitharaman, L.J. Wilson, Water-soluble gadofullerenes: toward high-relaxivity, pH-responsive MRI contrast agents, J. Am. Chem. Soc. 127 (2) (2005) 799–805.
[65] A. Bianco, T. Da Ros, M. Prato, C. Toniolo, Fullerene-based amino acids and peptides., J. Pept. Sci. 7 (4) (2001) 208–219.
[66] K.A. Gonzalez, L.J. Wilson, W. Wu, GH. Nancollas, Synthesis and in vitro characterization of a tissue-selective fullerene: vectoring C60 (OH) 16AMBP to mineralized bone, Bioorg. Med. Chem. 10 (6) (2002) 1991–1997.
[67] D.W. Cagle, S.J. Kennel, S. Mirzadeh, J. Michael Alford, L.J. Wilson, In vivo studies of fullerene-based materials using endohedral metallofullerene radiotracers, Proc. Natl. Acad. Sci. 96 (9) (1999) 5182–5187.

[68] S. Watanabe, N. Ishioka, T. Sekine, A. Osa, M. Koizumi, H. Shimomura, K. Yoshikawa, H. Muramatsu, Production of endohedral 133Xe-fullerene by ion implantation, J. Radioanal. Nucl. Chem. 255 (3) (2003) 495–498.
[69] R. Freeman, I. Willner, Optical molecular sensing with semiconductor quantum dots (QDs), Chem. Soc. Rev. 41 (10) (2012) 4067–4085.
[70] J. Li, F. Cheng, H. Huang, L. Li, J.-J. Zhu, Nanomaterial-based activatable imaging probes: from design to biological applications, Chem. Soc. Rev. 44 (21) (2015) 7855–7880.
[71] N. Hildebrandt, Biofunctional quantum dots: controlled conjugation for multiplexed biosensors, ACS Nano 5 (7) (2011) 5286–5290.
[72] D. Ficai, A.M. Grumezescu (Eds.), Nanostructures for Novel Therapy: Synthesis, Characterization and Applications, Elsevier, Amsterdam, 2017.
[73] I.V. Martynenko, A.P. Litvin, F. Purcell-Milton, A.V. Baranov, A.V. Fedorov, Y.K. Gun'Ko, Application of semiconductor quantum dots in bioimaging and biosensing, J. Mater. Chem. B 5 (33) (2017) 6701–6727.
[74] W.J.M. Mulder, R. Koole, R.J. Brandwijk, G. Storm, P.T.K. Chin, G.J. Strijkers, C.M. Donegá, K. Nicolay, A.W. Griffioen, Quantum dots with a paramagnetic coating as a bimodal molecular imaging probe, Nano Lett. 6 (1) (2006) 1–6.
[75] H. Duan, S. Nie, Cell-penetrating quantum dots based on multivalent and endosome-disrupting surface coatings, J. Am. Chem. Soc. 129 (11) (2007) 3333–3338.
[76] Yu' Cao, K. Yang, Z. Li, C. Zhao, C. Shi, J. Yang, Near-infrared quantum-dot-based non-invasive in vivo imaging of squamous cell carcinoma U14, Nanotechnology 21 (47) (2010) 475104.
[77] V. Bagalkot, L. Zhang, E. Levy-Nissenbaum, S. Jon, P.W. Kantoff, R. Langer, O.C. Farokhzad, Quantum dot− aptamer conjugates for synchronous cancer imaging, therapy, and sensing of drug delivery based on bi-fluorescence resonance energy transfer, Nano Lett. 7 (10) (2007) 3065–3070.
[78] Q. Xu, M. Hashimoto, T.T. Dang, T. Hoare, D.S. Kohane, G.M. Whitesides, R. Langer, D.G. Anderson, Preparation of monodisperse biodegradable polymer microparticles using a microfluidic flow-focusing device for controlled drug delivery, Small 5 (13) (2009) 1575–1581.
[79] D. Qin, Y. Xia, G.M. Whitesides, Soft lithography for micro-and nanoscale patterning, Nat. Protoc. 5 (3) (2010) 491–502.
[80] R. Xu, G. Zhang, J. Mai, X. Deng, V. Segura-Ibarra, S. Wu, J. Shen, et al., An injectable nanoparticle generator enhances delivery of cancer therapeutics, Nat. Biotechnol. 34 (4) (2016) 414–418.
[81] J. Xie, W.J. Ng, L.Y. Lee, C.-.H. Wang, Encapsulation of protein drugs in biodegradable microparticles by co-axial electrospray, J. Colloid Interface Sci. 317 (2) (2008) 469–476.
[82] J. Yao, L.K. Lim, J. Xie, J. Hua, C.-.H. Wang, Characterization of electrospraying process for polymeric particle fabrication, J. Aerosol Sci. 39 (11) (2008) 987–1002.
[83] S. Yuan, F. Lei, Z. Liu, Q. Tong, T. Si, R.X. Xu, Coaxial electrospray of curcumin-loaded microparticles for sustained drug release, PLoS One 10 (7) (2015) e0132609.
[84] G. Priyadarshana, N. Kottegoda, A. Senaratne, A.De Alwis, V. Karunaratne, Synthesis of magnetite nanoparticles by top-down approach from a high purity ore, J. Nanomater 2015 (2015), doi:http://dx.doi.org/10.1155/2015/317312 317312.
[85] T.J. Merkel, K.P. Herlihy, J. Nunes, R.M. Orgel, J.P. Rolland, J.M. DeSimone, Scalable, shape-specific, top-down fabrication methods for the synthesis of engineered colloidal particles, Langmuir 26 (16) (2010) 13086–13096.
[86] E. Roy, S. Patra, S. Saha, D. Kumar, R. Madhuri, P.K. Sharma, Shape effect on the fabrication of imprinted nanoparticles: Comparison between spherical-, rod-, hexagonal-, and flower-shaped nanoparticles, Chem. Eng. J. 321 (2017) 195–206.
[87] B.D. Gates, Q. Xu, M. Stewart, D. Ryan, C.G Willson, G.M. Whitesides, New approaches to nanofabrication: molding, printing, and other techniques, Chem. Rev. 105 (4) (2005) 1171–1196.
[88] A. Sánchez-Iglesias, N. Winckelmans, T. Altantzis, S. Bals, M. Grzelczak, L.M. Liz-Marzán, High-yield seeded growth of monodisperse pentatwinned gold nanoparticles through thermally induced seed twinning, J. Am. Chem. Soc. 139 (1) (2017) 107–110.
[89] S.E.A. Gratton, P.A. Ropp, P.D. Pohlhaus, J. Christopher Luft, V.J. Madden, M.E. Napier, J.M. DeSimone, The effect of particle design on cellular internalization pathways, PNAS 105 (33) (2008) 11613–11618.
[90] K. Huanbutta, T. Sangnim, S. Limmatvapirat, J. Nunthanid, P. Sriamornsak, Design and characterization of prednisolone-loaded nanoparticles fabricated by electrohydrodynamic atomization technique, Chem. Eng. Res. Des. 109 (2016) 816–823.

[91] C.J. Bowerman, J.D. Byrne, K.S. Chu, A.N. Schorzman, A.W. Keeler, C.A. Sherwood, J.L. Perry, et al., Docetaxel-loaded PLGA nanoparticles improve efficacy in taxane-resistant triple-negative breast cancer, Nano Lett. 17 (1) (2017) 242–248.
[92] J. Key, A.L. Palange, F. Gentile, S. Aryal, C. Stigliano, D.Di Mascolo, E.De Rosa, et al., Soft discoidal polymeric nanoconstructs resist macrophage uptake and enhance vascular targeting in tumors, ACS Nano 9 (12) (2015) 11628–11641.
[93] Li Zhang, C. Guan, Y. Wang, J. Liao, Highly effective and uniform SERS substrates fabricated by etching multi-layered gold nanoparticle arrays, Nanoscale 8 (11) (2016) 5928–5937.
[94] T. Ishida, Y. Tachikiri, T. Sako, Y. Takahashi, S. Yamada, Structural characterization and plasmonic properties of two-dimensional arrays of hydrophobic large gold nanoparticles fabricated by Langmuir-Blodgett technique, Appl. Surf. Sci. 404 (2017) 350–356.

CHAPTER 6

Surface modification with nanomaterials for electrochemical biosensing application

Sivaprakasam Radhakrishnan[a], Byoung-Suhk Kim[a], Sushma Dave[b]
[a]Department of Organic Materials & Fiber Engineering, Jeonbuk National University, Jeollabuk, South Korea
[b]Department of Applied Sciences JIET Jodhpur, Rajasthan, India

6.1 Introduction

Biosensors are contributing vital role to solve many problems in the fast growing world [1]. Biosensors are analytical equipment that comprises a bioreceptor (such as enzyme, antibody, protein, DNA etc.), transducer (electrical, optical, colorimetric, etc.) and signal amplifier/outputs (Fig. 6.1). In commonly, transducer was integrated with a bio-receptor to produce a signal (proportional to a concentration of analyte) and then transferred to a detector. The development of biosensor devices for identification and quantification of various substances are becoming more crucial and highly desirable in different sectors including medical monitors, diagnostics, environmental monitors, forensic application, warfare agents detection, food safety and so on [2].

The conventional analytical techniques such as chromatography, spectrophotometers, and capillary electrophoresis have delivered high sensitivity. But the major drawbacks aroused from expensive instrumentations, tedious sample pre-treatment, highly skilled person to operate the instrument, occupation of the large space, etc. [3]. Alternatively, the biosensors devices are having attractive properties including rapid detection, high sensitivity and selectivity, high precise, analytical simplicity both in and outside the laboratories and so on [4]. These special attractive features of biosensors devices are extensively applied in several sectors [5–10]. It is evident, the market size of the biosensors are significantly increased from 7.3 billon \$ in the year 2003 to 10.2 billion \$, 19.2 billion \$ in the year 2007, and 2019, respectively. And also it is expected to increase into 31.5 billion \$ in the year 2024. The value of the biosensors market significantly increased which demonstrated the huge demand of biosensors devices in different areas [11].

The biosensors could be classified majorly into two categories (i) based on types of transducers (electrochemical, optical, electrical, piezoelectric, thermal, and colorimetric) and (ii) nature of biological recognition elements (enzyme → enzymatic biosensor; antibodies → immunosensors; DNA → DNA sensor; microbial cell → microbial sensors and etc.) [12]. Among the various types of biosensors, the electrochemical biosensors has special attention because of their unique advantages such as low cost, easy

Advanced Nanomaterials for Point of Care Diagnosis and Therapy
DOI: https://doi.org/10.1016/B978-0-323-85725-3.00002-7

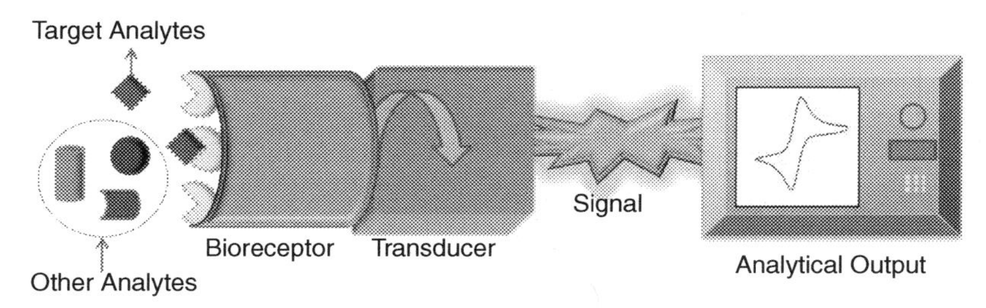

Fig. 6.1 ***Schematic representation of sensor construction.***

miniaturization, point-of-care (POC) testing in low resource settings, continuous monitoring, ultra-low concentration detection [13]. The electrochemical biosensors further classified into three categories according to the working principle governing their method of measurement including potentiometric, amperometric, and impedimetric.

The electrochemical biosensors was first successfully commercialized and clearly proposed their working mechanism into the scientific communities. It is well-known that Prof. Leland C. Clark was first proposed the enzyme electrode with glucose oxidase in 1962 [14]. The Clark developed methodologies were further applied and produced the first of many biosensors based on enzymatic approach by Yellow Springs Instruments Company, Ohio, USA and established their biosensors company around the world. Followed by several researchers and companies have been developed various biosensors system for different applications. Current trend on the development of biosensor device was based on engineered nanomaterials (as surface modifier, mediator, immobilizing matrix, and so on), flexible and wearable biosensors, and self-powered biosensors [15]. However, in this chapter, it has emphasized only an overview of the electrochemical biosensors design, advantages, and role of nanomaterials in the biosensor fabrication for different analyte detection.

6.2 Essential requirement and role of nanomaterials in ideal biosensors development

In order to design ideal biosensor for commercial application there are several analytical parameters should be evaluate. For example, the proposed biosensors should be high specific, stable under normal storage condition, stability for large number of measurements, the biosensor chemical reaction should not affect with physical parameters (electrolyte stirring, solution pH, and temperature as manageable), the sensor probe should be biocompatible for invasive measurements and the electrochemical response should be rapid, accurate and reproducible [16,17]. To satisfy all the above parameters, researchers are deeply investigating numerous approaches such as different surface functionalization with nanomaterials, new types of transducer design, and new types of biomolecules as bio-recognition layer, novel electrode fabrication and so on [18–22]. The surface modification with nanomaterials is one of the key approaches to enhance the analytical parameters for biosensor development [23–26]. The new type of materials with attractive physical and chemical properties may precisely response to single target molecule detection with good sensitivity and selectivity [27–29]. The identification of ultra low concentration of target analyte in a solution was a tough job in the biosensors development. Biosensors research communities are deeply working toward achieve this ultra-low concentration and single molecule detection of various analytes. It is due to the fact that the very minimum amount target analyte molecules were available for the detection in the low volume of solution. Further, the low concentration of target analytes will consume considerable time to reach the bio-recognition/transducer surface. In order to detect the low concentration of analyte in solution and decrease the analyte diffusion time could be attain by engineering the transducer surface with good specificity to the target analyte molecules. In any biosensors design, the tuning of bio-recognition layer is vital steps to increase the specificity of the sensor device for specific target analyte detection in complex matrix [30,31]. The high specificity provides an efficient transmission of biological interaction into a detectable signal, which is important parameter to achieve good sensitivity, selectivity and to bring the short response time. In addition to that, the fabrication of biosensors device should be compatible with biological matrix for invasive usage and can also useful to detect multiple analytes. To achieve good analytical performance (selectivity, sensitivity, long-term stability, reproducibility, and so on) of the fabricated sensors device, various type of nanomaterials (noble metal nanoparticles, transition metal based oxide/sulfide/phosphate and etc., conducting polymers, carbon nanomaterials, biomaterials and their composite of these) with different size and structure (0, 1, 2 and 3-D dimensional) have been extensively used in the past several years [32–36].

Because, the development of new types of nanomaterials are significantly interest and give paramount important among the scientific community in the past few years owing to their attractive properties through alteration of size into the nanoscale [37–39]. These special properties of nanomaterials are making them much useful in sensor design for various purposes such as a catalyst, surface modifier, immobilization matrix for biomolecules attachment, redox mediator, and signal amplifier and so on. For example, several biomolecules (enzyme, antibody, DNA, etc.) were anchored with nanomaterials as bio-recognition layer in the sensor design [40–43]. These bio-recognition layer could be provided a selective interaction with target analyte in the presence of complex matrix and hence provide an excellent analytical performance of the sensor device. Further, instead of bio-recognition layer, specific porous materials were also used to detect various analyte molecules. It is well-known that the molecularly imprinted polymers (MIPs) are fascinating methodology for the selective and sensitive identification of target analyte molecules [44,45]. The MIPs was fabricated by simply polymerization of specific monomer in presence of the analyte molecule as template. Followed by, the co-immobilized template molecule (analyte) will be evacuated from the polymer through suitable treatment to get the desirable cavities. This cavity could be providing a sufficient space for the sitting of target molecule in polymer and yield a great selectivity. Further, the MIPs could be attach or modify over the several substrates including glassy carbon, platinum, gold, graphite, silicon and nanoparticles [44]. It is well-know that the powder forms of the nanomaterials are also extensively used for biosensor application to enhance sensitivity of the sensor device. In most commonly, the powder samples were obtained from different synthesis approach such as solvothermal, sol-gel method, precipitation, microwave, sonochemical, thermolysis and etc. [46–49]. The powder nanomaterials was modified over electrode substrate involve mainly two methods (i) immersion of the desire electrode in nanomaterials dispersion in solvents and (ii) drop-cast of the nanomaterials dispersion over the substrate. In contrast, electrodeposition of nanomaterials is also more popular for sensor applications. Because, the electrodeposition has distinct advantages including low cost with green approach, simple room temperature preparation in large area of substrate, tuning the film thickness from nanometer to several micron thicknesses, control the morphologies *via* varying the chemical composition of bath solution and adding the suitable additives, high purity, less time and deposition over any substrate with short time. Several nanomaterials (conducing polymers, noble metals, transition metal and alloy, carbon materials and so on) could be prepared by electrodeposition method.

Now, it is clear that the fabrication of nanomaterials with good physical and chemical properties through simple synthesis process is highly desirable for various sensor applications. Because, the nanomaterials modified surface was provide a good analytical sensing performance, miniaturized sensor device and also used to avoid from the surface fouling of the modified electrode surface. It is evident from the significant number of review articles published using nanomaterials in past couple of years [50–56]. In the following section mainly focused on different biosensors design using different nanomaterials.

6.3 Fabrication of electrochemical biosensors using nanomaterials

6.3.1 Nucleic acid biosensors

The detection of specific DNA sequences is much important in gene profiling, medical diagnosis, biological warfare detection, food industries and forensic applications [57]. The DNA sensing methodologies relies on heterogeneous hybridization between a tethered nucleic acid with known genomic sequences and the free target nucleic acid with a complimentary genomic sequence in the solution [58]. The attachment of DNA on transducer surface is critical steps in fabrication of DNA sensors. Because, the DNA attachment should be vertical from the electrode surface, whereas it is horizontal bases of the DNA interact with electrode surface they restrict the DNA double helix formation. Hence, researchers are using different immobilization approach over the nanomaterials and transducer. For example, in this work, polyaniline coated polypyrrole nanotubes (PPy-PANi) was used as surface modifier in the DNA sensor fabrication (Fig. 6.2) [59]. Initially, PPy-PANi nanoutbes are covered with gold electrode by drop-caste method and then glutaraldehyde (GA) used as cross-linker to create linkage between amine group of the polyaniline and aldehyde group of the GA (Fig. 6.2; step-C). Followed by 5′-amine modified DNA immobilized over the GA treated polymer substrate through bonding between NH2 group of DNA and -CHO group of GA (Fig. 6.2; step-D). The probe ssDNA has been successfully immobilized over the GA treated PPy-PANi composite nanomaterials and then probe ssDNA modified electrode has been used to distinguish the hybridization efficiency with its corresponding complimentary, non-complimentary, and mismatched target DNA sequences. Here, methylene blue (MB) was used as redox indicator and differential pulse voltammery

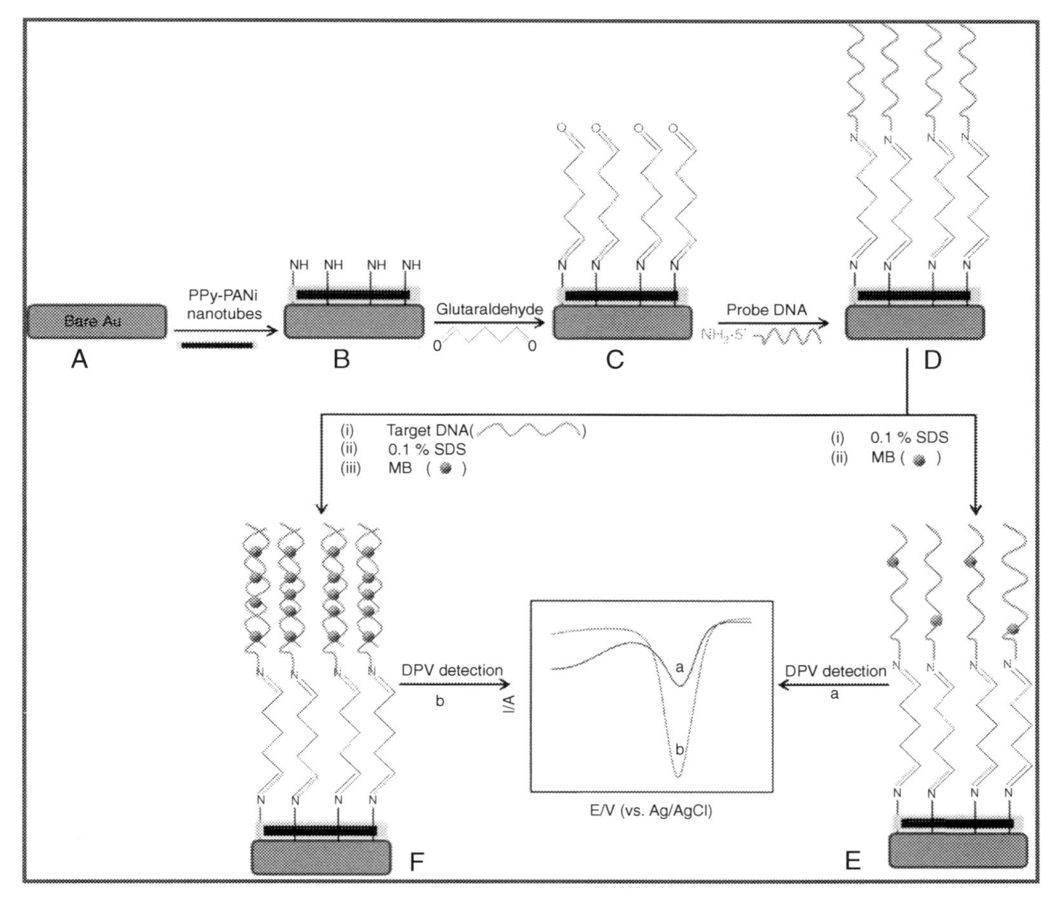

Fig. 6.2 ***Schematic illustration of construction and DPV detection of DNA using polyaniline (PANi) coated polyyrrole nanotubes (PPy)****. Reprinted with permission from Ref. [59]. Copyright (2013), RSC.*

used for the DNA hybridization detection. The PPy-PANi-GA substrate exhibited good linear range (1×10^{-13} and 1×10^{-09} M) for target DNA hybridization detection with detection limit of 5.0×10^{-14} M.

Similarly, electrodeposited poly (3,4-ethylene-dioxythiophene) (PEDOT) thin film was used for DNA construction as shown in Fig. 6.3. The EDOT monomers initially electropolymerized into PEDOT by cyclic voltammogram method over the glassy carbon (GC) electrode surface and then immersed into the gold nanoparticles (Au NPs) solution to form PEDOT/AuNPs through affinity interaction between the thiol group of PEDOT and gold nanoparticles. Followed by, thiolated probe ssDNA used for the immobilization over the PEDOT-AuNP film by similar affinity interaction of AuNPs and –SH-ssDNA. Then ssDNA modified surface used for the hybridization discrimination study with its corresponding complimentary, non-complimentary, and mismatched target DNAs. The hybridization event has been monitored by chronocoulometry in presence of ruthenium hexamine as indicator. The PEDOT-AuNP-S-ssDNA modified could detect the target DNA concentration ranging from 1.0×10^{-15} to 1.0×10^{-13} M with detection limit of 2.6×10^{-16} M [60].

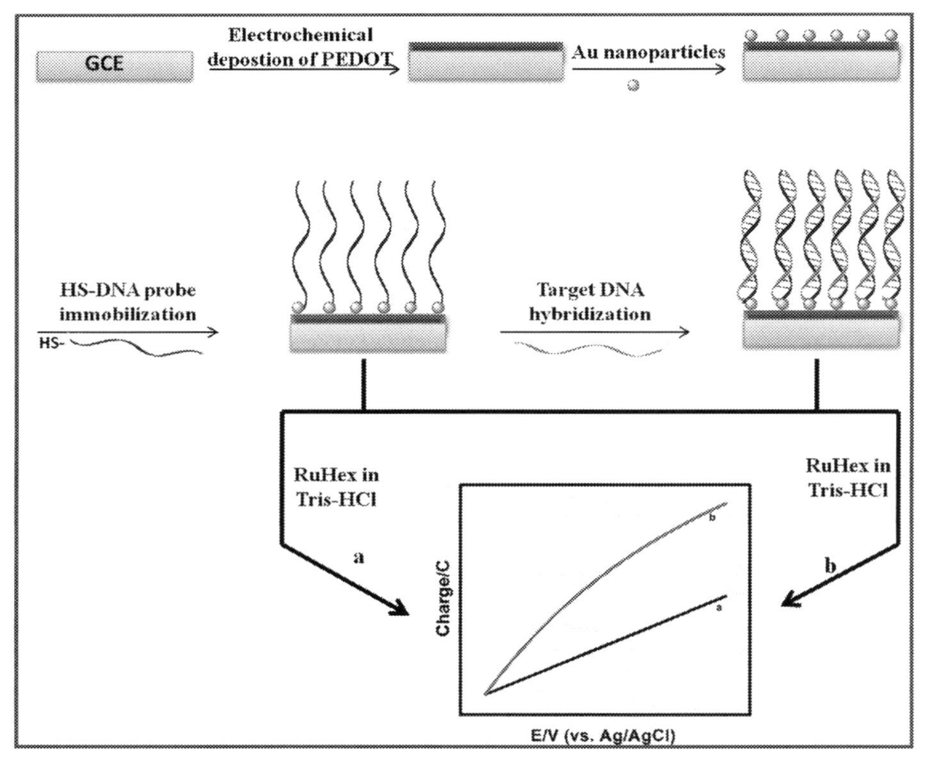

Fig. 6.3 ***Schematic diagrams for DNA sensor fabrication and chronocoulometry detection using PEDOT-gold nanoparticles [60].*** *Reprinted with permission from Ref. [60]. Copyright (2013), RSC.*

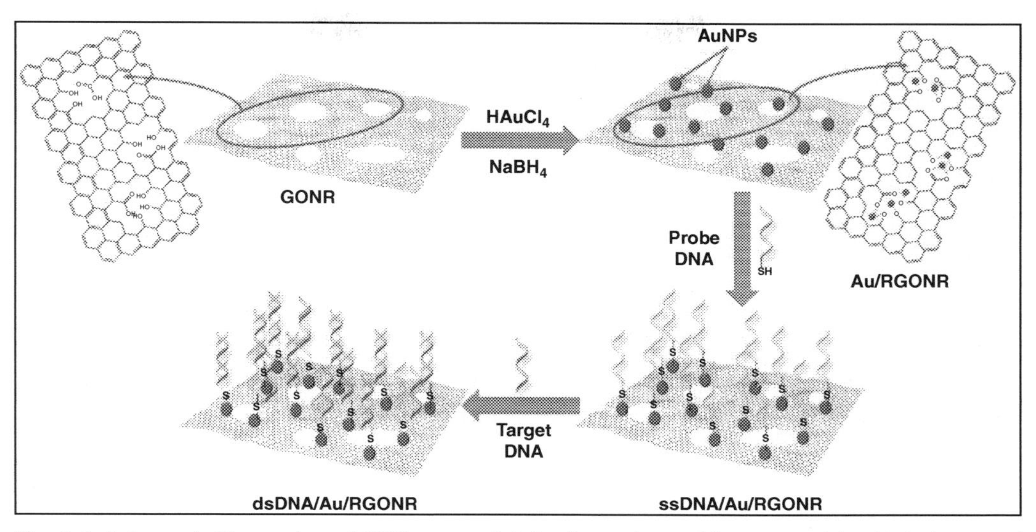

Fig. 6.4 ***Schematic illustration of DNA sensor fabrication using gold nanoparticles (Au) decorated graphene.*** *Reprinted with permission from Ref. [61]. Copyright (2013), RSC.*

Similar way, gold nanoparticles were anchored over the graphene sheets and then thiol modified DNA immobilized over the AuNP-graphene sheets (Fig. 6.4) [61]. Followed by, the ssDNA modified surface used for hybridization detection with corresponding complimentary target DNA by amperometric method. The AuNP-graphene modified electrode showed good linear concentration (0.1 fM to 10^{-6} M) of complimentary target DNA [61].

6.3.2 Environmental sensors

The environmental pollution is one of major issues in world-wide due to the main reason of rapid urbanization and industrialization. The environmental pollutions are severely affecting the quality of air, soil and water. Further, this polluted environment will be created a several acute and chronic diseases for human being and strongly affects the economics of the nation. The reduction and detection of these pollutions are highly tough task for all the developed and developing nations. Hence, the developments of efficient system to detect and prevent these environmental pollutions are high desirable to save the peoples from several health issues [62,63]. Currently, several analytical techniques have been applied for the detection of environmental pollutions from air, soil and water. The electrochemical detection is one of the important analytical methods because of their unique merits. For example, bisphenols (BPs) is highly toxic compound and excessive intake will create adverse health effect to human. In most commonly, this BP enters into the environment from different ways including polycarbonate plastics, plastic industries and their consumer products. The BP mainly contaminated in soil and water through industries waste waters, disposed plastic products (toys, bucket, bags and so on) [64]. The BP creates a several diseases for human when consumption of BP such as infertility, prostate cancer, birth illness. Hence, development simple system to detection BP is much important role in analytical chemistry. Ali et al. developed bispheneol A (BA-A) sensors using molecularly imprinted polymer (MIP) methodologies using polyacrylate, β-cyclodextrin, and graphene oxide [64]. The developed MIP based sensors exhibited good linear range (0.02-1.0 μM) and detection limit (8 nM). Further, the optimized process was extended to the detection of BP-A in water samples. Very recently, Baghayeri et al. fabricated BP-A sensor using MWCNT/$CuFe_2O_4$ modified GC [65]. The nanocomposite electrode demonstrated an excellent sensing performance toward BP-detection and delivered a good linear range from 0.01 to 120 μM with detection limit of 3.2 nM. The proposed electrode exhibited good selectivity, reproducibility, and applicability in different real samples. Similarly, Cong et al. developed a new composite materials containing $BiVO_4$-MoS_2-Co_3O_4 and used for the detection of bisphenol A [66]. The new composite materials modified surface demonstrated good analytical performances for the electrochemical detection of BP-A. Similarly, Wang et al. was fabricated BP-A sensor using AuNPs/MoS_2/IL-graphene composite [67]. The sensor exhibited excellent electro-catalytic properties for the oxidation of BP-A due to their synergistic effect during combination of AuNPs, MoS_2, IL-graphene. Further, the optimized protocol was used to BPA detection in lake water sample. Xu et al. designed unique 3D structures with support of 2D Ni-MOF and CNTs [68]. This special 3D structure materials used for the electrochemical detection of BP-A and observed an excellent performance for BPA sensing. The proposed electrode exhibited an ultra-low detection limit of 0.35 nM and good sensitivity of 284.64 μA μM^{-1} cm^{-2} for BPA. Amiri et al. prepared magnetic nanocomposite (ZnO/$ZnCo_2O_4$) through Glycyrrhiza glabra extract [69]. The green mediated prepared magnetic materials modified electrode delivered a good sensing performance for BP-detection and also successfully applied for the detection BP-A in water samples. It is well-known that hydroquinone (HQ) is one of the phenolic compounds and used for various chemical and biological industries. Due to their extensive usages of HQ was contaminated in environment largely. The HQ was more toxic to human health and animals. Further, it is not easy to degrade through ecological condition. The detection of HQ is important for various applications. Radhakrishnan et al. fabricated electrochemical sensors for HQ detection using polyaniline-iron oxide- reduced graphene oxide (PANi-Fe_2O_3-rGO) ternary composite [70]. The combinations of three nanomaterials were provided a attractive catalytic properties toward electro oxidation of HQ. The ternary composite modified electrode exhibited good analytical performance including wide linear range, detection limit, sensitivity and reproducibility. Further the optimized sensors used for the determination of HQ in tab water samples.

Nitrite is one of important toxic substances in drinking water. The excess consumption of nitrite was created several illnesses such as methemoglobinemia, stomach cancer

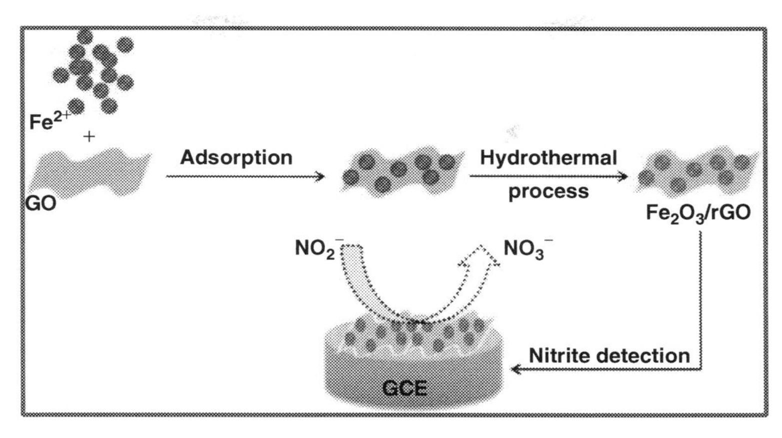

Fig. 6.5 ***Schematic illustrations of Fe2O3-rGO formation and their nitrite detection.*** *Reprinted with permission from Ref. [62]. Copyright (2013), Elsevier.*

and etc. This nitrite contamination was mainly occurred in the ground water through anthropogenic process and waste water from chemical industries.

Radhakrishnan et al. developed an electrochemical sensor for effective detection of nitrite by differential pulse voltammetry method based on Fe_2O_3 nanoparticles decorated graphene sheets modified electrode [71].

The Fe_2O_3-rGO hybrid composite modified electrode displayed a good electrochemical activity of the electro-oxidation of nitrite as shown in Fig. 6.5. The synergetic activity for Fe_2O_3-rGO hybrid composite is due to the restriction of graphene sheet aggregation in the presence of Fe_2O_3 and hence the surface area and conductivity of the materials was improved greatly. The Fe_2O_3-rGO hybrid composite modified electrode displayed a good linear range 5.0×10^{-8} M to 7.8×10^{-4} M with a detection limit of 1.5×10^{-8} M. The well-optimized electrode was further verified for real detection of nitrite in water samples. Hydrogen peroxide (H_2O_2) detection is very important in several industrial and biomedical applications. Recently, Riaz et al. developed a electrochemical H_2O_2 sensor using cobalt/carbon nanocomposite [72]. The modified electrode displayed a good sensing performance with low limit of detection of 10 μM and good sensitivity of 300 μA/mM cm^2. The optimized method was enabled on commercial screen printed electrode with integrated with a portable electrochemical workstation. Radhakrishnan et al. fabricated H_2O_2 sensors through enzymatic route based on horseradish peroxidase (HRP) enzyme modified with cerium oxide-reduced graphene oxide (CeO_2-rGO) (Fig. 6.6) [73]. The HRP/CeO_2-rGO modified electrode exhibited a good selective and sensitive response toward H_2O_2 detection. The enzyme loading over the CeO_2-rGO modified electrode was found to be 4.270×10^{-10} mol cm^{-2}. The fabricated enzymatic electrode delivered a good linear range from 0.1 to 500 μM with detection limit of 21 nM. Chen et al. designed a terbuthylazine electrochemical sensor based on MIP method [74]. Here, 3-thiophenemalonic acid used as monomer to

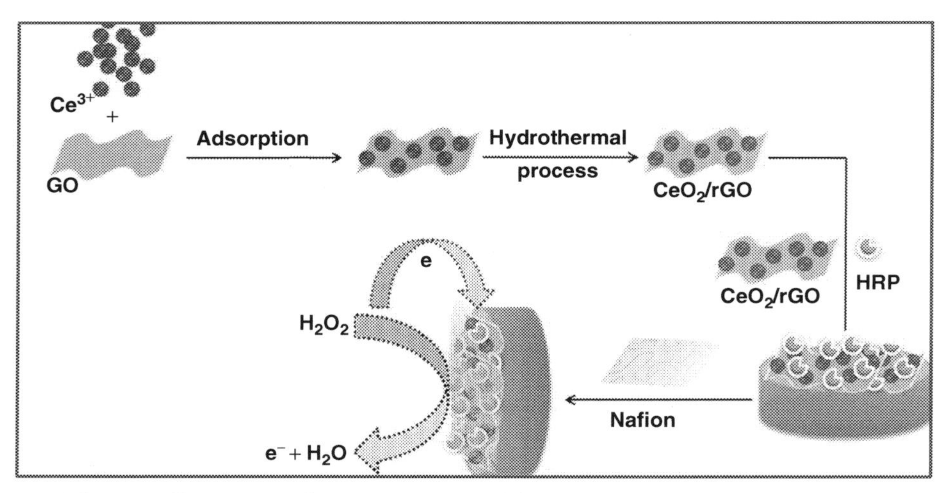

Fig. 6.6 ***Schematic illustrations of HRP/CeO_2/rGO modification and H_2O_2 detection.*** *Reprinted with permission from Ref. [73]. Copyright (2013), Elsevier.*

prepare MIP. The fabricated electrode was detecting the terbuthylazine greatly with good linear range from 2.5×10^{-7} mol/L to 1.2×10^{-4} mol/L, sensitivity, stability and reproducibility.

6.3.3 Food adulteration sensors

The prevention of food adulteration is one of critical problem in universe. The food adulteration problem is seriously affecting the human health with numerous acute and chronic diseases. The adulterations are persists at every stages of food with mixture of several harmful chemicals and artificial colors. So, prevention of food adulteration and development of efficient cost-effective analytical system will be urgently required to save people from various illnesses. Currently, electrochemical techniques were broadly used for the detection of various food adulteration and additives based on different nanomaterials modified electrodes. For example, Acrylamide (AA) is a commercially produced chemical compounds and used for several applications including polymer industries, water treatment, gel electrophoresis, ore, oil and fabrics industries [75]. However, AA is one of the important toxic chemicals due to its high neurotoxicity, potential carcinogenicity, genotoxicity, reproductive, and developmental toxicity [76]. Further, the AA was released higher concentration in fired and grilled starchy foods and creates several health issues during consumption of fired food items [77]. In past few years, several researchers demonstrated for the electrochemical detection of AA using various approach. Suchman et al. demonstrated AA electrochemical sensor using adsorption stripping voltammetry for the detection of AA through complex formation between AA and Ni2+. The developed protocol has been successfully verified for the detection of AA in various real samples including bake rolls, potato chips, and crackers [78].

Tong et al. fabricated AA electrochemical sensor using DNA-graphene oxide. The fabricated sensors showed a good linear range, stability and reproducibility [79]. Silva et al. developed first electrochemical biosensor for AA screening based on a direct biochemical interaction between the analyte and intact bacterial cell and demonstrated an excellent analytical outputs [80]. Radecka et al. developed a simple electrochemical AA sensors using single-walled carbon nanotubes/hemoglobin. The fabricated electrode showed very low detection limit as 1 nM and developed sensor verified with direct determination of AA in potato crisps [81]. Jalavand et al. designed a new type of sensing platform for the effective detection of AA using HG-DDAB/Pt-Au-Pd nanoparticles/Ch-IL/MWCNT-IL. The fabricated new sensing platform exhibited a two linear concentration from 0.03–39.0 nM and 39–150 nM with detection limit of 0.01 nM and also verified with real detection of AA in potato chips [82]. Caffeine (1,3,7-trimethylxanthine) and vanillin (4-hydroxy-3-methoxybenzaldehyde) are major additives in various foods, drinks and other products. However, excess consumption these food additives for long-time can create to severe health problem including gastric acid secretion, blood pressure, heart diseases, and cancer. For example, Sun et al. developed vanillin sensor using Fe@Fe3C-C composite as electrode materials for the sensor design [83]. The designed electrode displayed a good analytical performance including good linear range from 10 nM to 50 μM and detection limit was 2.63 nM. Further, the developed sensor used for the accurate detection of vanillin with good recoveries in various real samples including Chocolate, cooky, yogurt, jelly, and milk powder. Zeng et al. reported vanillin detection based on AuPd nanoparticles-graphene composite [84]. The composite modified electrode exhibited good response toward vanillin detection in the concentration ranges from 0.1 to 7 and 10-40 μM. The fabricated electrode was further extended to the detection of vanillin in vanilla bean, vanilla tea and biscuit samples. In contract, Wang et al. developed a electrochemical imprinted sensor using graphene oxide/MWCNT/IL/AuNP/MIP as sensing platform for the effective detection of vanillin [85]. The imprinted electrochemical sensors showed good linear range (1.0×10^{-8} to 2.5×10^{-6} mol L^{-1}) for the vanillin detection with detection limit of 6.23×10^{-10} mil L^{-1}. Deng et al. designed a new material for vanillin detection using carbon black paste electrode with graphene-PVP composite [86]. The vanillin detection was optimized with support of several experimental including pH value, accumulation potential, and time. The modified electrode demonstrated a good linear range for vanillin detection from 0.02–2.0 μM, 2.0–40 μM. and 40–100 μM. Sivakumar et al. developed CoS/Nafion modified electrode for the electrochemical detection of vanillin and obtained good analytical performance with wide linear range from 0.5 – 56 μM and detection limit 70 nM [87]. Further, the developed protocol further used for the detection of vanillin in food samples such as chocolate and milk powder. Very recently, we have fabricated electrochemical caffeine sensors using Cu-MOF and graphene hybrid composite modified glassy carbon electrodes [88]. The fabricated hybrid composite electrode demonstrated a good linear range from 5 μM to 450 μM and detection limit 710 nM. The modified electrode provided a good stability, selectivity, sensitivity and could be used for the accurate detection of caffeine in tea and coffee samples. Another important compound in food industry is sulfite, which is used as preservative. It is also extensively found in non-alcoholic beverages, alcoholic beverages, dried fruits, jams and pickles and etc. Jost et al. developed a sulfite sensor using AuNPs/silsequioxane composite as sensing platform [89]. The designed electrode delivered a good linear range from 2.54 to 48.6 mg L^{-1} with limit of detection for sulfite was 0.88 mg L^{-1}. Recently, Saleh et al. developed a eugenol through electrochemical method based on graphene oxide/SnO_2 [90]. Eugnol is major additive in food industries and consumption of high doses could create lot of health effects including kidney damage, digestive problems, blood pressure and liver failure. Based on the graphene oxide/SnO_2 composite modified electrode could be possible to detection eugnol effectively. The proposed modified electrode was detecting eugnol as 20 nM. It is well-known that the trimethylamine (TMA) is a typical fish-odor substance and the concentration is directly linked to freshness of the fish. Choi et al. developed a electrochemical sensor for the identification of trimethylamine compound based on enzyme immobilized over the electrodeposited gold nanoparticles electrode. The trimethylamine dehydrogenase-Au NPs modified electrode exhibited good sensing performance toward electrochemical detection of TMA with good linear range (0 to 2.5 mM) and detection limit (1 μM) [91].

6.3.4 Small biomolecules sensors

It is well-known that our human body has several biomolecules and these molecules are playing an impressive role for the proper function of the human body. In particularly, small biomolecules such as glucose, ascorbic acid, dopamine, vitamins, uric acid and so forth are playing vital role for normal activities of human body and resisting illness of human from various diseases [92]. The electrochemical method is more flexible, reliable, accurate, and rapid for detection of these small biomolecules over the other analytical techniques. The detection of glucose in human blood serum is important in diabetic issues. There are several commercial glucose sensor kits available for accurate determination of glucose in blood samples. However, the working nature of these available meters based on enzymatic method, which has several drawbacks including enzyme stability, cost, and complicated preparation. In order to solve these issues, researchers are recently focusing on the development of non-enzymatic glucose sensors based on various nanomaterials [93,94]. For example, nickel sulfide (NiS)-reduced graphene oxide hybrid composite has been used for the non-enzymatic glucose detection [95]. The fabricated composite electrode showed good response toward electro-oxidation of glucose and obtained wide linear range from 5.0×10^{-5} to 1.7×10^{-3} M and detection limit of 10 μM. Similarly, CuS nanostructures with different morphologies (microflower, cauliflower, and nanoparticles interconnected network-like structures) were prepared by changing the polarity of solvent during the hydrothermal synthesis [96]. The synthesized different morphologies of CuS was further utilized for electrochemical non-enzymatic glucose sensors applications. Among three different morphologies of CuS, nanoparticles inter-connected network-like structures showed good sensing performance for glucose oxidation. It showed good linear range, detection limit, sensitivity, reproducibility, and repeatability. Recently, polypyrrole (PPy) coated CuS@SiO_2 porous sphere was prepared and applied for the glucose detection [97]. The PPy-CuS@SiO_2 modified electrode demonstrated excellent catalytic behavior due to their combination of electroactive conducting polymers and CuS. It showed excellent sensing performance toward electro-oxidation of glucose with good linear range, detection limit, and sensitivity. Recently, Lotfi et al. fabricated g-C_3N_4/NiO/CuO ternary composite for the non-enzymatic electrochemical glucose sensor application [98]. The ternary composite modified glassy carbon electrode displayed a wide glucose concentration from 0.4 μM to 8.5 mM with a detection limit of 0.1 μM. Further, the proposed electrode has good stability, sensitivity, reproducibility. Ascorbic acid or vitamin C is one of the most important water soluble vitamins. It is present in biological systems and various foods. It was used as antioxidant in food, beverage, pharmaceutical formulations, and cosmetic application. Further, the AA was played a key role in biological system and extensively used for common cold, mental illness, infertility, cancer and AIDS. Hence, there are lots of opportunities on the development of AA sensor in various areas [51]. Zhang et al. developed AA electrochemical sensor using branch-trunk Ag nanostructure modified electrode [99]. The Ag nanostructures were achieved by galvanic replacement method. The Ag modified electrode demonstrated a good linear range 0.17 μM to 1.80 mM and a detection limit of 60 nM. A new type of electrochemical AA sensors was designed by Li et al. based on MOF-carbon naotubes. The MOF based electrode displayed a good catalytic performance for the electro-oxidation of AA [100]. The fabricated electrode showed a good working range, limit of detection, sensitivity, selectivity and reproducibility. Recently, Co_3O_4.Fe_2O_3 nanosphere materials have been used for the fabrication of electrochemical AA sensor [101]. The metal oxide modified electrode exhibited a good analytical performance for AA detection. The present modified electrode delivered a linear range from 0.1 nM to 0.01 mM and limit of detection 96.02 ± 4.80 pM. A Mesoporous $CuCo_2O_4$ rod was used as surface modifier for the fabrication of electrochemical sensor for AA detection [102]. The $CuCo_2O_4$ modified electrode performed good sensing behavior toward the electro-catalytic oxidation of AA. Further, the optimized $CuCo_2O_4$ modified electrode verified for the detection of AA in vitamin C tablets. Wang et al. recently proposed MIP based electrochemical sensor for AA detection using MXene [103]. The MIP based electrode delivered a good response for the detection of AA with good linear range from 0.5 μM to 10 μM and detection limit of 0.27 μM. The MIP based electrode further extended to the detection of AA in vitamin C tablets. Butein ($C_{15}H_{12}O_5$) is important flavonoid and broadly used for antioxidant, anti-inflammatory, anti-cancer properties. Recently, Kanagavalli et al. fabricated electrochemical sensors for butein detection using carbon nanoparticles as sensing platform [104]. The designed carbon electrode was showed good response for electro-oxidation of butein with wide linear range of 10 to 100 μM and detection limit of 7.6 μM. Uric acid is one of the important biomolecules in our body. The concentration of UA raise in our body will lead several illnesses like renal failure, gout, cardiovascular, pregnancy disease. Karthika et al. designed uric acid sensor using SrWO4 as surface modifier. The SrWO4 modified electrode displayed a good response for uric acid detection with detection limit of 33 ppm and linear range 1 nM to 500 nM. The optimized sensor was further utilized for successfully detection of UA in urine and serum samples with good recovery [105]. Guan et al. developed uric acid sensor based on IL COF-1 and MWCNT/AuNPs used as sensing platform. The composite modified electrode showed good response for simultaneous detection of dopamine and uric acid with good linear range, sensitivity and detection limit [106]. Zhang et al. fabricated electrochemical sensor for simultaneous detection of ascorbic acid (AA), dopamine (DA) and uric acid (UA) using nitrogen-doped reduced graphene oxide (N-rGO). The N-rGO modified electrode demonstrated a good catalytic behavior for the simultaneous electro-oxidation of UA, AA, and DA and exhibited a good linear range, stability, reproducibility [107]. Similarly, Reddy et al. fabricated electrode for detection of uric acid in the presence of ascorbic acid and dopamine. The poly(DPA)/SiO_2@Fe_3O_4 modified carbon paste electrode used for the detection of these analytes. The combination of polydopamine and metal oxide was provided a synergetic effect toward electro-oxidation of uric acid in presence of ascorbic acid, and dopamine [108]. The hybrid composite modified electrode showed good linear range from 1.2 to 1.8 μM with a good limit of detection. Further, the designed electrode verified with real sample to accurate determination of uric acid.

6.4 Conclusions, future prospects, and challenges

A comprehensive overview of various types of nanomaterials and different method of sensor fabrication for electrochemical detection of nucleic acid, environmental hazardous chemicals, food adulterations and small biomolecules has been examined. These detailed studies were clearly demonstrated that the sensing performance was mainly linked with the surface modification over the transducer using various nanomaterials, bioreceptors, and so on. These nanomaterials and bi-receptors modified surface have provided a good sensing platform for the selective detection of various analytes with good response time, sensitivity, selectivity, reproducibility, stability and real-sample applicability. Currently, different types of electrochemical sensor for detection of different target molecules have been achieved using nanomaterials as surface modifiers. Although the improved performance after the blending of nanomaterials as hybrid composite materials including binary, ternary and so forth, there is still good opportunity to further improve the electrochemical sensor performance for detection of various analyte molecules. The attractive properties of these nanomaterials and their vast application in different sensors will make them in future to bring real analytical device with cost-effective, portable, and accurate detection for various analyte molecules detection.

Acknowledgments

Dr. S. Radhakrishnan acknowledges the DST, New Delhi, India for the DST-Inspire Faculty Award (DST/INSPIRE/04/2015/002259).

References

[1] M.K. Sezginturk, Commercial biosensors and their application Chapter one Introduction to commercial biosensor, Clinical Food Beyond (2020) 1–28.
[2] E.B. Bahadir, M.K. Sezginturk, Application of commercial biosensors in clinical, food, environmental, and biothreat/biowarfare analysis, Anal. Chem. 478 (2015) 107–120.
[3] G-F. Leonor, M. Novell, P. Blondeau, F.J. Andrade, A disposable, simple, fast, and low-cost paper-based biosensor and its application to the determination of glucose in commercial orange juices, Food Chem 265 (2018) 64–69.
[4] F. Mustafa, A. Othman, S. Andreescu, Cerium oxide-based hypoxanthine biosensor for fish spoilage monitoring, Sens. Actuat. B: Chem. 332 (2021) 129435.
[5] B. Ince, M.K. Sezginturk, A high sensitive and cost-effective disposable biosensor for adiponectin determination in real human serum samples, Sens. Actuat. B: Chem. 328 (2021) 129051.
[6] N.J. Forrow, S.W. Bayliff, A commercial whole blood glucose biosensor with a low sensitivity to hematocrit based on an impregnated porous carbon electrode, Biosens. and Bioelectron. 21 (2005) 581–587.
[7] A. Kirchhain, A. Bonini, F. Vivaldi, N. Poma, F.D. Francesco, Latest developments in non-faradic impedimetric biosensors: towards clinical applications, Trend Anal. Chem. 133 (2020) 116073.
[8] A. Hashem, M.A.M. Hossain, Ab.R. Marlinda, M.Al. Mamun, K. Simarani, M.R. Johan, Nanomaterials based electrochemical nucleic acid biosensors for environmental monitoring: A review, Appl. Surf. Sci. Adv. 4 (2021) 100064.
[9] C. Griesche, A.J. Baeumner, Biosensors to support sustainable agriculture and food safety, Trend. Anal. Chem. 128 (2020) 115906.
[10] L.D. Mello, L.T. Kubota, Review of the use of biosensors as analytical tools in the food and drink industries, Food Chem 77 (2002) 237–256.
[11] J.H.T. Luong, K.B. Male, J.D. Glennon, Biosensor technology: technology push versus market pull, Biotech. Adv. 26 (2008) 492–500.
[12] X. Lu, M. Cui, Q. Yi, A. Kamrani, Detection of mutant genes with different types of biosensor method, Trend. Anal. Chem. 126 (2020) 115860.
[13] R. Gupta, N. Raza, S.K. Bhardwaj, K. Vikrant, K-H. Kim, N. Bhardwaj, Advances in nanomaterials-based electrochemical biosensors for the detection of microbial toxins, pathogenic bacteria in food matrices, J. Hazard. Mater. 401 (2021) 123379.
[14] V. Scognamiglio, F. Arduini, The technology tree in the design of glucose biosensors, Trend. Anal. Chem. 120 (2019) 115642.
[15] R.C. Reid, I. Mahbub, Wearable self-powered biosensors, Curr. Opin. Electrochem. 19 (2020) 55–62.
[16] S. Andreescu, J-L. Marty, Twenty years research in cholinesterase biosensors: From basic research to practical applications, Biomol. Eng. 23 (2006) 1–15.
[17] D.V. Voort, C.A. Mcneil, R. Renneberg, J. Korf, W.T. Hermens, J.F.C. Glatz, Biosensors: basic features and application for fatty acid-binding protein, an early plasma marker of myocardial injury, Sens. and Actuat. B: Chem. 105 (2005) 50–59.

[18] A. Mokhtarzadeh, E.-K. Reza, P. Paria, M. Hejazi, G. Nasrin, H. Mohammad, B. Behzad, M. de le. Guardia, Nanomaterials-based biosensors for detection of pathogenic virus. TrAC trend, Anal. Chem. 97 (2017) 445–457.

[19] L. Qian, S. Durairaj, S. Prins, A. Chen, Nanomaterial-based electrochemical sensors and biosensors for the detection of pharmaceutical compounds, Biosens. Bioelectron. 175 (2021) 112836.

[20] M. Bahri, A. Baraket, N. Zine, M.B. Ali, J. Bausells, A. Errachid, Capacitance electrochemical biosensor based on silicon nitride transducer for TNF-α cytokine detection in artificial human saliva: heart failure (HF), Talanta 209 (2020) 120501.

[21] A. Bensana, F. Achi, Analytical performance of functional nanostructured biointerfaces for sensing phhenolic ompounds, Colloids Surf. B: Biointerf. 196 (2020) 111344.

[22] K. Takasu, K. Kushiro, K. Hayashi, Y. Iwasaki, S. Inoue, E. Tamechika, M. Takai, Polymer brush biointerfaces for highly sensitive biosensors that preserve the structure and function of immobilized proteins, Sensors Actuators B: Chem 216 (2015) 428–433.

[23] F.G. Nejad, S. Tajik, H. Beitollahi, I. Sheikhshoaie, Magnetic nanomaterials based electrochemical (bio) sensors for food analysis, Talanta (2021), Article In Press, doi:10.1016/j.talanta.2020.122075.

[24] K. Murtada, V. Moreno, Nanomaterials-based electrochemical sensors for the detection of aroma compounds-towards analytical approach, J. Electroanal. Chem. 861 (2020) 113988.

[25] T. Xiao, J. Huang, D. Wang, T. Meng, X. Yang, Au and Au-based nanomaterials: synthesis and recent progress in electrochemical sensor application, Talanta 206 (2020) 120210.

[26] K. Shrivas, A. Ghosale, P.K. Bajpai, T. Kant, K. Dewangan, R. Shankar, Advances in flexible electronics and electrochemical sensors using conducting nanomaterials: a review, Microchem. J. 156 (2020) 104944.

[27] B. Liu, J. Liu, Sensors and biosensors based on metal oxide nanomaterials, Trend. in Anal. Chem. 121 (2019) 115690.

[28] M. Hromadova, F. Vavrek, Electrochemical electron transfer and its relation to charge transport in single molecule junction, Curr. Opin. Electrochem. 19 (2020) 63–70.

[29] R.J. Nichols, S.J. Higgins, Charge transfer in single molecules at electrochemical interface, Encyclopedia Interf. Chem. (2018) 15–23.

[30] J. Lee, S. Kim, H.Y. Chung, A. Kang, S. Kim, H. Hwang, S.I. Yang, W.S. Yun, Electrochemical microgap immunosensors for selective detection of pathogenic Aspergillus niger, J. Hazard. Mater. 411 (2021) 125069.

[31] S.A. Hira, D. Annas, S. Nagappan, Y.A. Kumar, S. Song, H-J. Kim, S. Park, K.H. Park, Electrochemical sensor based on nitrogen-enriched metal-organic framework for selective and sensitive detection of hydrazine and hydrogen peroxide, J. Environ. Chem. Eng. (2021), In Press, doi:10.1016/j.jece.2021.105182.

[32] Q. Wang, X. Wang, M. Xu, X. Lou, F. Xia, One-dimensional and two-dimensional nanomaterials for the detection of multiple biomolecules, Chin. Chem. Lett. 30 (2019) 1557–1564.

[33] Y. Wang, L. Wang, X. Zhang, X. Liang, Y. Feng, W. Feng, Two-dimensional nanomaterials with engineered bandgap: synthesis, properties, applications. Nanotoday, 37 (2021) 101059.

[34] S. Su, X. Gu, Y. Xu, J. Shen, D. Zhu, J. Chao, C. Fan, L. Wang, Two-dimensional nanomaterials for biosensing applications, Trends Anal. Chem. 119 (2019) 115610.

[35] B. Zhang, J.-Y. Sun, M.-Y. Ruan, P.-X. Gao, Tailoring two-dimensional nanomaterials by structural engineering for chemical and biological sensing, Sens. Actuat. Rep. 2 (2020) 100024.

[36] D. Jiang, Z. Chu, J. Peng, J. Luo, Y. Mao, P. Yang, W. Jin, One-step synthesis of three-dimensional $Co(OH)_2$/rGO nano-flowers as enzyme-mimic sensors for glucose detection, Electrochim. Acta 270 (2018) 147–155.

[37] X. Yang, R. Wu, N. Xu, X. Li, N. Dong, G. Ling, Y. Liu, P. Zhang, Application and prospect of antimonene: a new two-dimensional nanomaterials in cancer theranostics, J. Inorg. Biochem. 212 (2020) 111232.

[38] L. Reverte, P.-S. Beatriz, M. Campas, New advances in electrochemical biosensors for the detection of toxins: nanomaterials, magnetic beads and microfluidics systems. A review, Anal. Chim. Acta 908 (2016) 8–21.

[39] K.M. Koo, N. Soda, M.J.A. Shiddiky, Magnetic nanomaterials-based electrochemical biosensors for the detection of diverse circulating cancer biomarkers. Curr. Opin. Electrochem. 25 (2021) 100645.

[40] A. Kannan, S. Radhakrishnan, Fabrication of an electrochemical sensor based on gold nanoparticles functionalized polypyrrole nanotubes for the highly sensitive detection of L-dopa, Mater. Today Commun 25 (2020) 101330.
[41] S. Radhakrishnan, S. Prakash, C.R.K. Rao, M.Vijayan, Organically soluble bifunctional polyaniline-magnetite composites for sensing and supercapacitor application, Electrochem. Solid-State Lett. 12 (A84) (2009).
[42] X. Lin, X. Lian, B. Luo, X.-C. Huang, A highly sensitive and stable electrochemical HBV DNA biosensor based on ErGO-supported Cu-MOF, Inorg. Chem. Commun. 119 (2020) 108095.
[43] U. Eletxigerra, J. Martinez-Perdiguero, S. Merino, Disposable microfluidic immune-biochip for rapid electrochemical detection of tumor necrosis factor alpha biomarker, Sens. Actuat. B 221 (2015) 1406–1411.
[44] J.W. Lowdon, H. Dillen, P. Singla, M. Peeters, T.J. Cleij, B.V. Grinsven, K. Eersels, MIPs for commercial application in low-cost and assay-an overview of the current status quo, Sens. Actuat. B 325 (2020) 128973.
[45] A. Raziqu, A. Kidakova, R. Boroznjak, R.A. Jekaterina, V. Syritski, Development of a portable MIP-based electrochemical sensor for detection of SARS-CoV-2 antigen, Biosens. Bioelectron. 178 (2021) 113029.
[46] G. Manjari, S. Saran, S. Radhakrishnan, P. Rameshkumar, A. Pandikumar, S.P. Devipriya, Facile green synthesis of Ag-Cu decorated ZnO nanocomposite for effective removal of toxic organic compounds and an efficient detection of nitrite ions, J. Environ. Manag. 262 (2020) 110282.
[47] P. Kanchana, S. Radhakrishnan, M. Navaneethan, M. Arivanandhan, Y. Hayakawa, C. Sekar, Electrochemical sensor based on Fe doped hydroxyapatite-carbon nanotubes composite for L-dopa detection in the presence of uric acid, J. Nanosci. Nanotech. 16 (2016) 6185–6192.
[48] S. Radhakrishnan, H.-Y. Kim, B.-S. Kim, Expeditious and eco-friendly fabrication of highly uniform microflower superstructures and their application in highly durable methanol oxidaiton and high performance supercapacitors, J. Mater. Chem. 4 (2016) 12253–12262.
[49] N. Lavanya, S. Radhakrishnan, C. Sekar, Fabrication of hydrogen peroxide biosensor based on Ni doped SnO2 nanoparticles, Biosens. Bioelectron. 36 (2012) 41–47.
[50] W. Wang, X. Wang, N. Chen, Y. Luo, Y. Lin, W. Xu, D. Du, Recent advances in nanomaterials-based electrochemical (bio)sensors for pesticides detection, Trends in Anal. Chem. 132 (2020) 116041.
[51] K. Dhara, R.M. Debiprosad, Review on nanomaterials-enabled electrochemical sensors for ascorbic acid detection, Anal. Biochem. 586 (2019) 113415.
[52] F. Arduini, S. Cinti, V. Mazzaracchio, V. Scognamiglio, A. Amine, D. Moscone, Carbon black as an outstanding and affordable nanomaterials for electrochemical (bio)sensor design, Biosens. Bioelectron. 156 (2020) 112033.
[53] K.V. Ratnam, H. Manjunatha, S. Janardan, K. Chandra Babu Naidu, S. Ramesh, Nonenzymatic electrochemical sensor based on metal oxide, MO (M=Cu, Ni, Zn and Fe) nanomaterials for neurotransmitters: an abridged review, Sens. Inter. 1 (2020) 100047.
[54] H. Filik, A.A. Avan, Review on applications of carbon nanomaterials for simultaneous electrochemical sensing of environmental contaminant dihydroxybenzene isomers, Arabian J. Chem. 13 (2020) 6092–6105.
[55] H. Du, Y. Xie, J. Wang, Nanomaterial-sensors for herbicides detection using electrochemical techniques and prospect applications, Trends Anal. Chem. 135 (2021) 116718.
[56] E. Sheikhzadeh, V. Beni, M. Zourob, Nanomaterial application in bio/sensors for the detection of infectious diseases, Talanta (2020), In Press, doi:10.1016/j.talanta.2020.122026.
[57] J.P. Tosar, J.Laiz G.Branas, Electrohcemical DNA hybridization sensors applied to real and complex biological samples, Biosens. Bioelectron. 26 (2010) 1205–1217.
[58] F. Lucarelli, G. Marazza, A.P.F. Turner, M. Mascini, Carbon and gold electrodes as electrochemical transducers for DNA hybridization sensors, Biosens. Bioelectron. 19 (2004) 515–530.
[59] S. Radhakrishnan, C. Sumathi, V. Dharuman, J. Wilson, Polypyrrole nanotubes-polyaniline composite for DNA detection using methylene blue as intercalator, Anal. Methods 5 (2013) 1010–1015.
[60] S. Radhakrishnan, C. Sumathi, V. Dharuman, J. Wilson, Gold nanoparticles functionalized poly(3,4-ethylenedioxythiophene) thin film for highly sensitive label free DNA detection, Anal. Methods 5 (2013) 684–689.

[61] N.K. Mogha, V. Sahu, R.K. Sharma, D.T. Masram, Reduced graphene oxide nanoribbon immobilized gold nanoparticles based electrochemical DNA biosensor for the detection of Mycobacterium tuberculosis, J. Mater. Chem. B 6 (2018) 5181–5187.
[62] B. Zou, J.G. Wilson, F.B. Zhan, Y. Zeng, Air pollution exposure assessment methods utilized in epidemiological studies, J. Environ. Monit. 11 (2009) 475–490.
[63] H. Nakamura, Recent organic pollution and its biosensing methods, Anal. Methods 2 (2010) 430–444.
[64] H. Ali, S. Mukhopadhyay, N.R. Jana, Selective electrochemical detection of bisphenol A using a molecularly imprinted polymer nanocomposite, New J. Chem 43 (2019) 1536–1543.
[65] M. Baghayeri, A. Amiri, M. Fayazi, M. Nodehi, A. Esmaeelnia, Electrochemical detection of bisphenol a on a MWCNT/$CuFe_2O_4$ nanocomposite modified glassy carbon electrode, Mater. Chem. Phys. 261 (2021) 124247.
[66] Y. Cong, W. Zhang, W. Ding, T. Zhang, Y. Zhang, N. Chi, Q. Wang, Fabrication of electrochemically-modified $BiVO_4$-MoS_2-Co_3O_4 composite film for bisphenol A degradation, J. Environ. Sci. 102 (2021) 341–351.
[67] Y. Wang, Y. Liang, S. Zhang, T. Wang, X. Zhuang, C. Tian, F. Luan, S-Q. Ni, X. Fu, Enhanced electrochemical sensor based on gold nanoparticles and MoS2 nanoflowers decorated ionic liquid-functionalized graphene for sensitive detection of bisphenol A in environmental water, Microchem. J. 161 (2021) 105769.
[68] C. Xu, L. Liu, C. Wu, K. Wu, Unique 3D hetrostructures assembled by quasi-2D Ni-MOF and CNTs for ultrasensitive electrochemical sensing of bisphenol A, Sens. Actuators B 310 (2020) 127885.
[69] M. Amiri, M.-M. Hadi, Green synthesis of ZnO/ZnCo2O4 and its application for electrochemical determination of bisphenol A, Microchem. J. 160 (2021) 105663.
[70] S. Radhakrishnan, K. Karthikeyan, C. Sekar, J. Wilson, S.J. Kim, A promising electrochemical sensing platform based on ternary composite of polyaniline-Fe2O3-reduced graphene oxide for sensitive hydroquinone determination, Chem. Eng. J. 259 (2015) 594–602.
[71] S. Radhakrishnan, K. Krishnamoorthy, C. Sekar, J. Wilson, S.J. Kim, A highly sensitive electrochemical sensor for nitrite detection based on Fe2O3 nanoparticles decorated reduced graphene oxide nanosheets, Appl. Catal. B: Environ. 148-149 (2014) 22–28.
[72] M.A. Riaz, Z. Yuan, A. Mahmood, F. Liu, X. Sui, J. Chen, Q. Huang, X. Liao, L. Wei, Y. Chen, Hierarchically porous carbon nanofibers embedded with cobalt nanoparticles for efficient H_2O_2 detection on multiple sensor platform, Sens. Actuators B 319 (2020) 128243.
[73] S. Radhakrishnan, S.J. Kim, An enzymatic biosensors for hydrogen peroxide based on one-pot preparation of CeO2-reduced graphene oxide nanocomposite, RSC Adv. 5 (2015) 12937–12943.
[74] J. Chen, G.-R. Xu, L-y. Bai, J.-Z. Tao, Y.-Q. Liu, Y.-P. Zhang, A high sensitive sensor for terbuthylazine determination based on molecularly imprinted electropolymer of 3-thiophenemalonic acid, Int. J. Electrochem. Sci. 7 (2012) 9812–9824.
[75] K. Kotsiou, M. Tasioula-Margari, E. Capuano, V. Fogliano, Effect of standard phenolic compounds and olive oil phenolic extracts on acrylamide formation in an emulsion system, Food Chem. 124 (2011) 242–247.
[76] C. Pelucchi, C.L. Vecchia, C. Bosetti, P. Boyle, P. Boffetta, Exposure to acrylamide and human cancer—a review and meta-analysis of epidemiologic studies, Ann Oncol 22 (2011) 1487–1499.
[77] D.S. Mottram, B.L. Wedzicha, A.T. Dodson, Acrylamide is formed in the Maillard reaction, Nature 419 (2002) 448–449.
[78] H. Vesela, E. Sucman, Determination of acrylamide in food using adsorption stripping voltammetry, Czech J. Food Sc. 31 (2013) 401-106.
[79] D. Li, Y. Xu, L. Zhang, H. Tong, A label-free electrochemical biosensors for acrylamide based on DNA immobilized on graphene oxide-modified glassy carbon electrode, Int. J. Electrochem. Sci. 9 (2014) 7217–7227.
[80] N.A.F. Silva, M.J. Matos, A. Karmali, M.M. Rocha, An electrochemical biosensor for acrylamide determination: merits and limitations, Portugalia Electrochimica Acta 29 (2011) 361–373.
[81] A. Krajewska, J. Radecki, H. Radecka, A voltammetric biosensor based on glassy carbon electrodes modified with single-walled carbon nanotubes/hemoglobin for detection of acrylamide in water extracts from potato crisps, Sensors 8 (2008) 5832–5844.

[82] K. Varmira, O. Abdi, N-B. Gholivand, H.C. Gholivand, H.C. Giocoechea, A.R. Jalalvand, Intellectual modifying a bare glassy carbon electrode to fabricate a novel and ultrasensitive electrochemical biosensors: application to determination of acrylamide in food samples, Talanta 179 (2018) 509–517.

[83] J. Sun, T. Gan, K. Wang, Z. Shi, J. Li, L. Wang, A novel sensing platform based on a core-shell Fe@Fe3C-C nanocomposite for ultrasensitive determination of vanillin, Anal. Methods 6 (2014) 5639–5646.

[84] L. Shang, F. Zhao, B. Zeng, Sensitive voltammetric determination of vanillin with an AuPd nanoparticles-graphene composite modified electrode, Food Chem. 151 (2014) 53–57.

[85] X. Wang, C. Luo, L. Li, H. Duan, An ultrasensitive molecularly imprinted electrochemical sensor based on graphene oxide/carboxylated multiwalled carbon nanotube/ionic liquid/gold nanoparticles composite for vanillin analysis, RSC Adv. 5 (2015) 92932–92939.

[86] P. Deng, Z. Xu, R. Zeng, C. Ding, Electrochemical behavior and voltammetric determination of vanillin based on an acetylene black paste electrode modified with graphene-polyvinylpyrrolidone composite film. Food Chem. 180 (2015) 156–163.

[87] M. Sivakumar, M. Sakthivel, S.-M. Chen, Simple synthesis of cobalt sulfide nanorods for efficient electrocatalytic oxidation of vanillin in food samples, J. Colloid Interface Sci. 490 (2017) 719–726.

[88] A. Venkadesh, J. Mathiyarasu, S. Radhakrishnan, Voltammetric sensing of caffeine in food sample using Cu-MOF and graphene, Electroanalysis 32 (2020) 1–8, doi:10.1002/elan.202060488.

[89] J.P. Winiarski, R.B. Marflia, H.A. Magosso, C.L. Jost, Electrochemical reduction of sulfite based on gold nanoparticles/silsesquioxane-modified electrode, Electrochim. Acta. 251 (2017) 522–531.

[90] G. Fadillah, W.P. Wicaksono, I. Fatimah, T.A. Saleh, A sensitive electrochemical sensor based on functionalized graphene oxide/SnO2 for the determination of eugenol, Microchem. J. 159 (2020) 105353.

[91] Y-B. Choi, H.G. Kim, G.H. Han, H-H. Kim, S.W. Kim, Voltammetric detection of trimethylamine using immobilized trimethylamine dehydrogenase on an electrodeposited gold nanoparticles electrode, Biotechnol. Bioprocess Eng. 16 (2011) 631–637.

[92] Y. Manmana, T. Kubo, K. Otsuka, Recent development of point-of-care (POC) testing platform for biomolecules, Trends in Anal. Chem. 135 (2021) 116160.

[93] Q. Dong, H. Ryu, Y. Lei, Metal oxide based non-enzymatic electrochemical sensors for glucose detection, Electrochim. Acta 370 (2021) 137744.

[94] E. Sehit, Z. Altintas, Significance of nanomaterials in electrochemical glucose sensors: an updATED REVIew, Biosens. Bioelectron. (2016-2020), 159 (2020) 112165.

[95] S. Radhakrishnan, S-J. Kim, Facile fabrication of NiS and a reduced graphene oxide hybrid film for non-enzymatic detection of glucose. RSC Adv. 5 (2015) 44346–44352.

[96] A. Venkadesh, S. Radhakrishnan, J. Mathiyarasu, Eco-friendly synthesis and morphology-dependent superior electrocatalytic properties of CuS nanostructure, Electrochim. Acta 246 (2017) 544–552.

[97] S. Radhakrishnan, V. Ganesan, J. Kim, Voltammetric nonenzymatic sensing of glucose by using a porous nanohybrid composed of CuS@SiO_2 spheres and polypyrrole, Microchim. Acta 187 (260) (2020).

[98] Z. Lotfi, M.B. Gholivand, M. Shamispur, Non-enzymatic glucose sensor based on a g-C3N4/NiO/CuO nanocomposite, Anal. Biochem. 616 (2021) 114062.

[99] Y. Zhang, P. Liu, S. Xie, M. Chen, M. Zhang, Z. Cai, R. Liang, Y. Zhang, F. Cheng, A novel electrochemical ascorbic acid sensor based on branch-trunk Ag hierarchical nanostructures, J. Electroanal. Chem. 818 (2018) 250–256.

[100] Y. Li, W. Ye, Y. Cui, B. Li, Y. Yang, G. Qian, A metal-organic frameworks carbon nanotubes based electrochemical sensor for highly sensitive and selective determination of ascorbic acid, J. Mol. Struct. 1209 (2020) 127986.

[101] M.M. Alam, H.B. Balkhoyor, A.M. Asiri, M.R. Karim, M.T.S. Chani, M.M. Rahman, Fabrication of ascorbic sensor acid with Co_3O_4.Fe_2O_3 nanospehre materials by electrochemical technique, Surf. Interfaces 20 (2020) 100607.

[102] X. Xiao, Z. Zhang, F. Nan, Y. Zhao, P. Wang, F. He, Y. Wang, Mesoporous CuCo2O4 rods modified glassy carbon electrode as a novel non-enzymatic amperometric electrochemical sensors with high sensitive ascorbic acid recognition, J. Alloys Compd. 852 (2021) 157045.

[103] Q. Wang, X. Xiao, X. Hu, L. Huang, T. Li, M. Yang, Molecularly imprinted electrochemical sensor for ascorbic acid determination based on MXene modified electrode, Mater. Lett. 285 (2021) 129158.

[104] P. Kanagavalli, S. Radhakrishnan, G. Pandey, V. Ravichandiran, G.P. Pazhani, M. Veerapandian, G. Gurumurthy Hegde, Electrochemical tracing of butein using carbon nanoparticles interfaced electrode processed from biowaste, Electroanalysis 32 (2020) 1220–1225.
[105] A. Karthika, A. Suganthi, M. Rajarajan, An ultrahigh selective uric acid sensor based on $SrWO_4$ nanocomposite using pomelo leaf extract solubilized Nafion modified glassy carbon electrode, J. Sci-Adv. Mater. Dev. (2021), In Press, doi:10.1016/j.jsamd.2021.01.002.
[106] Q. Guan, H. Guo, R. Xue, M. Wang, X. Zhao, T. Fan, W. Yang, M. Xu, W. Yang, Electrochemical sensor based on covalent organic frameworks-MWCNT-NH_2/AuNPs for simultaneous detection of dopamine and uric acid, J. Electroanal. Chem. 880 (2021) 114932.
[107] H. Zhang, S. Liu, Electrochemical sensor based on nitrogen-doped reduced graphene oxide for the simultaneous detection of ascorbic acid, dopamine, and uric acid, J. Alloys Compd. 842 (2020) 155873.
[108] Y.V.M. Reddy, B. Sravani, S. Agarwal, V.K. Gupta, G. Madhavi, Electrochemical sensor for detectionof uric acid in the presence of ascorbic acid and dopamine using the poly(DPA)/SiO_2@Fe_3O_4 modified carbon paste electrode, J. Electroanal. Chem. 820 (2018) 168–175.

CHAPTER 7

Nanomaterials for sensors: Synthesis and applications

Laxmi R. Adil[a], Retwik Parui[a], Mst N. Khatun[a], Moirangthem A. Chanu[a], Lidong Li[b,d], Shu Wang[c], Parameswar K. Iyer[a,d]
[a]Department of Chemistry, Indian Institute of Technology Guwahati, Guwahati, Assam, India
[b]School of Materials Science and Engineering, University of Science and Technology Beijing, Beijing, China
[c]Institute of Chemistry, Chinese Academy of Sciences, Beijing, P.R. China
[d]Centre for Nanotechnology, Indian Institute of Technology Guwahati, Guwahati, Assam, India

7.1 Introduction of nanomaterials

Nanomaterials (NMs) chemistry is comparatively a young branch of research. NMs are building blocks for numerous nanotechnological devices. Some decades ago NMs word would have puzzled scientists regardless of the fact that in the nature NMs always existed. The term "nanomaterials" indicates particles having sizes in the range 1 to 100 nm, with at least one in out of three dimensions. The chemical and physical properties of molecules or atoms change in NMs state, than corresponding materials in bulk. NMs possess higher surface area and also exhibit quantum effects due to its very small size. Therefore, the unique properties of NMs cannot be predicted by comparing it with the bulk materials.

Developments in the field of nanotechnology prompted the design and development of novel NMs with different size and shapes for various interdisciplinary applications in material science, chemistry and biology as well as allied areas. NMs can be synthesized by various types of synthesis approaches. NMs include very diverse materials viz. metals, metal-organic framework (MOF), carbon dots, polymer nanoparticles, etc. NMs also have diversity in morphology like disk, cube, spherical, rod-like shape. In most of the cases, synthesized NMs surface have to be modified to stabilize because of NMs nano size that leads to chemically high reactivity and prevent it to form aggregates or agglomeration. The NMs can also be surface functionalized according to requirements for any desired applications. Sensing and bioimaging plays a decisive role in the biomedical and diagnostic applications (Fig. 7.1). Hence, fluorescent NMs are the most valuable materials for unravelling the secrets and development of science.

7.2 Classification of nanomaterials

On the basis of size, shape, chemical and physical properties NMs can be further classified into different classes.

Advanced Nanomaterials for Point of Care Diagnosis and Therapy
DOI: https://doi.org/10.1016/B978-0-323-85725-3.00017-9

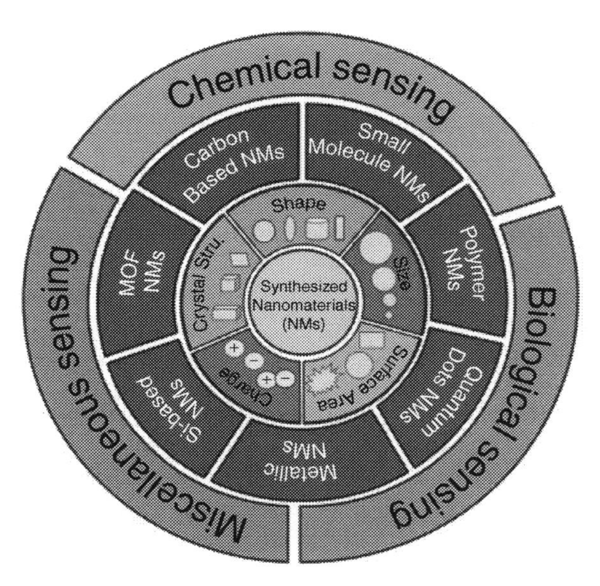

Fig. 7.1 ***Diverse nanomaterial types and their uses in different sensing applications.***

7.2.1 Organic small molecule-based nanomaterials

The designing of fluorescent probes based on the small molecule is trending exponentially and increasing as a hot research area because it is considered as a powerful tool for its application in chemo and bio-sensing. The first fluorescent organic small material-quinine, was discovered in 1845. Since then, both theoretical and experimental efforts were made to tune their color and design more fluorescent materials. Presently, many small molecules-based fluorescent materials family are developed [1]. Organic fluorescent materials (OFMs) are one of the most essential and versatile weapons in modern fluorescent techniques due to their high sensitivity and excellent bio-compatibility [2–4]. Over the last few decades, OFMs significantly impacted biomedical research in terms of their capability to achieve high spatiotemporal resolution images, which have provided exceptional clinical accuracy at the subcellular level. The incorporation of donor (D) and acceptor (A) units into the organic backbone is the most usual method to improve the optical property and results in the NIR emissive materials [5].

7.2.2 Polymer-based nanomaterials

Polymer nanoparticles have emerged as promising materials for chemical sensing, bio-imaging, drug delivery and therapy due to the molecular wire effect, high sensitivity, higher quantum yield, higher signal-to-noise ratio and protect the drug by encapsulating etc. Wide varieties of polymers have been synthesized, and they have tunable emission, ranging from blue to infrared on incorporating suitable repeating units or monomer functionalization [6].

7.2.3 Quantum dots

Development in semiconductor material and technology has enabled the need of designing a small core that contains mobile electrons. Quantum dots (QDs) are crystalline in nature and size falling in the nanoregime with an ability to transport electrons. QDs are used in bioimaging and in various commercial displays due to higher stability, cost-effectiveness, environment-friendly, narrow and symmetric emission. Over the past decade QDs along with organic luminescent materials have found significant applications in various fields and have appeared as better options compared to inorganic nanomaterials due to their high biocompatibility, less toxicity, economical and show efficient color-tunability [7].

7.2.4 Metallic nanomaterials

Metal NMs are arguably the most studied class of NMs systems. In 1857, Michael Faraday synthesized gold colloid and studied its interaction with light. NMs plasmonic effect can be tuned by chemical synthesis and NMs activity was controlled by variation in the shape and surface. Cadmium (Cd), cobalt (Co), gold (Au), silver (Ag) metal nanoparticles are the most studied NMs systems. Metallic NMs were also used for coloring the glasses a few centuries ago, such as the use of gold NPs for making beautiful stained-glass preparation as done in early days by ancient Romans [8].

7.2.5 Silicon-based nanomaterials

Silicon-based NMs (SiNMs) are commonly renowned for their size-dependent tunable optical properties along with low cytotoxicity. In addition to the low toxicity, quantum confinement effect results in superior fluorescence stability and less photo bleaching effect that have unfolded a novel avenue of application in bioimaging and other diagnostic applications. The large surface area of the Si NMs offers more analyte interactions over the surface and enhances its sensitivity to the next level. Consequently, the electrical conductivity improves upon binding with the charged particles, which has made it a strong contender for the fabrication of a nano-sensor device. More intense research has recently developed new polymer stuffed Si NMs and surface functionalized silicon-based NMs that provide better solubility in both water and organic environments. Apart from the solubility, it also advances the targeting ability, photophysical properties and stability of the probe under a wide pH range. Moreover, magnetic core encapsulation is an unusual approach for easy functionalization in Si NMs, which also plays a massive role in separating metal particles using the magnetic property [9,10].

7.2.6 Metal-organic framework nanoparticles

Metal-organic frameworks (MOFs) have specific structural and functional properties and they are considered as the porous compound in one, two or three dimensions. It's framework was created by metal ion and bridging organic linkers via coordination bonding. These can be used as gas adsorption, energy storage, sensor, drug delivery and as a catalyst. MOFs are getting explosive attention because of their exceptional properties, application and feasible synthesis methods [11].

7.2.7 Carbon-based nanoparticles

Carbon-based NMs have received enormous consideration as an efficient sensing element owing to impressive electrical conductivity, high tensile strength, chemical consistency and biostability. They have revealed innumerable advantages over the traditional sensing tools regarding rapidity, sensitivity, and detection limits. Different allotropes of carbon like graphite, diamond, etc. can be developed depending upon the several hybridization states of carbon atoms. Moreover, controlling structural conformation using various synthetic pathways helps to design a powerful sensing platform with unique electrical properties as the morphology directly influences the electron transport process. Depending upon the morphological dimension, carbon NMs have been classified into fullerenes (0D), carbon nanotubes (1D), graphene (2D). Among all, graphene and carbon nanotubes are most frequently used for sensing applications endowing excellent optical properties and facilitating easy incorporation of biomolecules with a high surface to volume ratio [12].

7.3 Synthesis of nanomaterials

Synthesis of NMs can be performed using different types of methods in the solid, liquid and gas phases.

7.3.1 Organic small molecule-based nanomaterials

Various strategies have been employed to prepare organic small molecule based NMs involving the non-covalent interaction like hydrophobic effect, hydrogen-bonding, electrostatic and van der Waals interactions etc. Among all the self-assembly strategies, nanoprecipitation, guest-host assembly, polymer encapsulation etc. are extensively used methods that show promising behavior in the diverse field of research.

a. Nanoprecipitation

Nanoprecipitation is the most simple and straight forward technique that has been widely used for assembling organic molecules, mainly based on the hydrophobic and hydrophilic interactions. Generally, in the assembling process, water-miscible organic solvents (tetrahydrofuran, ethanol, acetone etc.) are used to solubilise the hydrophobic organic dyes. The stock solution of the organic dyes is inoculated to a large amount of water or some other poor solvent, subsequently keeping under vortex. Owing to hydrophobic and hydrophilic repulsion, organic dyes accumulate into nanoparticle forms [13]. The limitation arises from the uncontrollable size distribution of the nanoparticles that were lost by the encapsulation of the organic small molecule with amphiphilic polymer matrix through the nanoprecipitation process. This also provides better photostability, anti-photobleaching property and fine colloidal quality of the probe inside the biological cell with the fast and selective response toward respective analytes. Meanwhile, the change in the preparation procedure and polymer matrix can easily tune the morphology and size of the nanoparticles. 1,2-distearoyl-sn-glycero-3-phosphoethanolamine-N-[methoxy(poly(ethylene glycol))] (DSPE-PEG) derivatives are the most common amphiphilic polymer used for the nanoprecipitation of AIEgens [14]. Guest-host chemistry has revealed significant improvement for the effective formulation of supramolecular assembly of AIEgens in chemical and biological research areas [15]. Host materials like cyclodextrins (CDs), cucurbiturils (CBs), crown ethers commonly possess a particular cavity size that can encapsulate specific guests and enhance their selectivity toward recognizing specific analytes.

b. Covalent bonding

Although the physical interaction-based nanoparticle fabrication is preferable for the effortless handling process and fast operation time, this process cannot modify the functional group. In this aspect, a covalent bonding approach is an efficient tool for the preparation of surface-modified nanoparticles. Besides, it is relatively easy to maintain the dye loading ratio at the time of fabrication. Secondly, the disadvantages of dye diffusion from the nanoparticles during application can be prevented due to covalent bonding. Few simple reactions like Schiff base type reactions, isothiocyanate- (ITC), and amine-regulated reactions, are generally followed to link organic small molecule with the polymer or bio conjugates, e.g., NMs ofTPE-CS was prepared via anchoring TPE functionalized thiocyanate (TPE-ITC) and chitosan (CS) [16].

7.3.2 Polymer-based nanomaterials

NMs of polymer are synthesized by polymerizing the monomer in the heterophase medium and dispersing the prepared polymer into suitable medium.

a. Synthesis of polymer NMs in heterophase medium

The single step of generating polymer nanoparticles by polymerising the monomers into the medium into which is a bad solvent for the polymer delivers an enormous scope in terms of nanoparticle morphology and size control. The first report for synthesis of polypyrrole polymer nanoparticles was performed in the presence of methylcellulose into a water medium in which pyrrole was readily soluble [17]. Polyaniline is used extensively because of its incredible stability and easy preparation method. Usually, polyaniline is synthesized in an acidic aqueous medium by using the chemical oxidation polymerisation method [18]. PEDOT (poly(3,4-ethylenedioxythiophene) is the most studied nanoparticles of polythiophene because it has tremendous transparency and superior thermal stability. Nanoparticle of PEDOT are prepared using aqueous oxidation polymerization of 3,4-(ethylenedioxy)- thiophene. Anionic counter ions will act as a stabiliser and PEDOT containing polyelectrolyte anionic polystyrene sulfonate (PSS) which balance the cationic PEDOT [19].

b. Synthesis of polymer NMs by precipitation technique

Polymer nanoparticle was synthesized by reprecipitation method it comprises polymer solution addition into the non-solvent that leads to sudden precipitation of the polymer. In this method, the polymer was firstly dissolved in the appropriate good solvent, then it was added to the bad solvent, which is miscible with a suitable solvent. After completion of mixing the two solvents, the polymer nanoparticles were dispersed in the bad solvent. The polymer nanoparticle size is adjusted by the polymer concentration in the solution [20].

c. Synthesis of polymer NMs by emulsion technique

NMs synthesis by emulsion technique has attracted immense attention in the past decades in the industry and academia due to higher bioavailability, intracellular uptake, less toxicity and high stability etc. Emulsion techniques such as microemulsion, miniemulsion, polymerization and process of emulsion solvent evaporation method control the size, morphology and its photophysical properties. Microemulsion polymerization process is the utmost common techniques in which monomers and oil in water emulsions are emulsified after the polymerization [21]. In miniemulsion polymerization technique, droplets of monomer were converted to polymer nanoparticles [22]. In microemulsion polymerization technique thermodynamic stable dispersions of water and oil was emulsified and it was further stabilized by cosurfactant and surfactant [23].

7.3.3 Quantum dots nanomaterials

There are numerous routes that have been developed for the synthesis of QD comprising top-down and bottom-up approach. Molecular beam epitaxy, electron beam lithography, etc., are methods used in the top-down synthesis of QD and self-assembly in the solution approach was used in the bottom-up approach method.

a. Organometallic and nonorganometallic strategy

Organometallic strategy involves two steps of QDs synthesis and it requires surfactants, precursors and solvents in which surfactants can act as a solvent. In the first step nucleation of the precursor occurs while growth of the nanocrystal happens in the second step. These two steps were separate and is controlled by various factors such as temperature, the concentration of precursors etc. When the temperature rises organometallic material will generate a highly reactive precursor to produce QDs. In the growth state surfactants are anchored on the surface of the nanocrystal and they also act as a capping ligand that control the growth and solubility of the QDs. In organometallic strategy, $Cd(CH_3)_2$ is generally used though it is moisture sensitive and high toxic. To reduce the cost, toxicity and instability of the QDs, research has been directed toward more environment friendly non-toxic methods. To overcome this hurdle, cadmium oxide, cadmium myristate was used as an alternative for the production of CdSe/ZnSe core/shell QDs [24].

b. Microwave based

In the conventional heating method whole setup was uniformly heated however, in the microwave heating method, only polar molecule will get heated because of the high absorption of microwaves energy that enhanced the QDs formation rate and yield. Nitrogen doped N-CQD was synthesized in water medium by using xylan and NH_4OH as precursor and the reaction mixture was treated in microwave at 200°C for 10 minutes [25].

c. Microemulsion based

Microwave emulsion-based approach to synthesize QDs at room temperature is the most popular strategy and is categorized in two methods as normal microemulsions (oil in water) and reverse microemulsions (water in oil). The high-quality QDs was formed due to the newly formed nanocluster, which strongly stabilized the growth of QDs. CdS QDs were synthesized in dioctyl sulfosuccinate proliniumisopropylester ([ProC3]AOT) ionic liquid by using isooctane and 2-hydroxyethylammonium formate (HO-EAF) and stirred for 12 hours [26].

7.3.4 Metal nanomaterials

Unique optical and electrical features of the metal nanoparticles (NMs) has offered an explosion of interest in their application in the biological and chemical sensor field over the last few decades [27]. Especially size and shape-dependent tunability of the optical properties have elicited more attention and expanded its application exponentially in the fundamental and technical research. NMs like gold, silver, platinum and copper are most preferential as a sensor material holding ease of surface functionalization, easy synthesis procedure and tunable photophysical properties. Moreover, these NMs offer the desired product with less toxicity and specific targets, which successfully fulfil the criteria for the biological application for further improvement in the clinical diagnosis and therapeutics. However, more intense research has paved some excellent physicochemical features like surface plasmon resonance (SPR), higher surface-enhanced Raman scattering (SERS), etc., in the development of sensing materials. This segment emphasizes the exceptional properties and the synthesis of noble gold, silver, platinum, and copper MNs developed for sensing utilization.

a. Gold NMs

Au nanoparticles (AuNPs) are among the most leading candidates encompassing excellent biocompatibility and chemical stability, utilized in diverse research fields like drug delivery, therapeutics, biosensing, chemical catalysis, etc. [28]. Additionally, the excellent redox activity of these NPs offer fabrication of nanoscale devices, improving the attributes of chemo and bio-sensors. Owing to the high chemical inertness of the AuNPs, it has been utilized as an electrochemical sensor. In terms of a colorimetric assay, the aggregation of the AuNPs into cluster generates an easily distinguishable naked eye color change [29]. Generally, the small sized AuNPs display absorption at ~520 nm having SPR, but the more resonance coupling in the cluster form shifts the absorption maxima toward higher wavelength endowing high molar extinction coefficient [30]. This intense and sharp visible color change phenomenon enhances probe sensitivity for the detection of toxic analytes and critical disease biomarkers and also helps to achieve a very low detection limit as compared to the traditional organic dyes.

However, the synthesis procedure has a considerable impact on the particle size and shape, which decides its properties leading to more versatile application opportunities. Thus, various fabrication processes have been followed to achieve unique shaped and sized AuNPs holding specific features. Compared to other traditional reduction methods, citrate-mediated reduction of gold (III) molecules has received immense attention attributing simplicity, low-cost materials and easy tunability to uniform particles in the variable size range [31]. In 1951, a simple reduction of $HAuCl_4$ was reported using citrate molecule in water medium, that produced 20 nm size uniform AuNPs [32]. After a few years, in 1973, different size particles (16 to 147 nm) were prepared using the same method by controlling the ratio of stabilizing and reducing agents [33]. Nevertheless, the difficulties in obtaining bigger size particles (>150 nm) were further uplifted by the introduction of a new one-step reaction via o-diaminobenzene based gold reduction using poly(N-vinyl-2-pyrrolidone) (PVP) [34]. Next, the Brust-Schiffrin Method had opened another way to synthesize thermally stable AuNPs within the range of 1.5-2 nm [35,36]. Initially, $AuCl_4^-$ was transferred from aqueous to the organic phase and further reduced by $NaBH_4$. The high binding affinity of sulfur atom toward gold, having soft nature, was utilized to stabilize the NPs. However, these methods are unable to obtain different morphology except for the spherical shape. On the other hand, variations in the morphology such as nanocages and nanowire can include additional chemical and physical features compared to spherical NPs. Xia et al. fulfilled this requirement by synthesizing gold nanocages using silver nano-cube as a template [37]. After this, one new single-step organic reaction was established to synthesize crystalline ultrathin gold nanowires in the presence of triisopropylsilane and octylamine under hexane solvent at room temperature [38]. The resulting nanowires self-assembled and formed a two-dimensional network that further acted as an active material for SERS application. Many more wet chemical processes were recently invented to prepare various 2D AuNPs due to unique shapes like nanobelt, [39] nanoplate, etc. [40].

b. Silver NMs

Silver nanoparticles (AgNPs) is another important topic in nanoscience. However, initially, the rich and fascinating properties of AuNPs did not allow more research on the AgNPs. Later the commercial availability and the superior properties of AgNPs like electrical conductivity, a wide variety of color tunability have enticed more and showed promising behavior in sensor application. The amalgamation of excellent localized SPR property and distinct morphology provides the best surface-enhanced Raman scattering effect related to the other metal NPs under a specific environment [41]. Moreover, the high sensitivity and selectivity of AgNPs make them an ideal contender for biological sensing, especially for the early-stage detection of biomarkers that cause life-threatening diseases. AgNPs are also recognized for their antimicrobial activity, which plays an essential role in pharmaceutical applications.

The most straightforward approach to synthesize AgNPs is the reduction of $AgNO_3$ in the presence of reducing agents like sodium citrate, $NaBH_4$ etc. [42]. The reduction of Ag(I) results in metallic Ag that leads to accumulation into oligomeric clusters in an aqueous solution. However, this method is not suitable for the synthesis of the monodispersed NPs. This challenge was overcome through the synthesis of monodispersed silver nano-cubes via the polyol process [43]. Here, the condition for reduction was controlled to obtain these monodispersed AgNPs. The concentration of $AgNO_3$ was multiplied by three times and was maintained at the ratio of poly(vinyl pyrrolidone) and $AgNO_3$ as 1.5 in polyethylene glycol. This process was further improved by employing sodium hydrosulfide for the bulk preparation of nano-cubes [44]. Later, the incorporation of silver seeds results in controlled growth of the nano-cubes and formed desired size NMs ranging from 30-200 nm [45]. As the properties of the AgNPs can be tailered by the size and shape of the particles, the synthesis of many other different shapes like nanowire [46], nanoprism [47], nanoflower [48], nanoplate [49] etc. were also developed. Meanwhile, uniformly shaped, well-distributed nano rice in bulk was also prepared [50]. These NMs exhibited plasmon resonance features having excitation at the NIR region. Overall, different synthetic approaches include more distinctive properties and broadening AgNPs applications in diverse fields day by day.

7.3.5 Silicon-based nanomaterials

Several methods, like laser mediated pyrolysis, ultrasonic dispersion, plasma-assisted synthesis, and other wet chemical techniques, have been followed to synthesize less than 10nm size SiNPs. In 1992, the ultrasonic dispersion method was introduced where the colloidally suspended zero-dimensional silicon nanoparticles were prepared via silicon wafer etching in different organic solvents [51]. However, this method resulted in highly pure SiNPs but unable to obtain a uniform structure. Later on, more efforts have been put into the functionalization to obtain attractive electronic and photophysical properties. In 2013, a new chemical-etching process was reported to synthesize highly photostable, red-emissive silicon nanocrystal (SiNC) with a 4.1 nm hydrodynamic diameter for tracking and imaging single-molecule present in the plasma membrane [52]. In this method, initially, the hydroxyl group terminated SiNCs were prepared by immersing the silicon wafer into $HF/HNO_3/H_2O$ solution. These hydroxyl groups (silanol) were further functionalized with mercaptotrimethoxysilane and liberated from the wafer through mechanical scraping using a razor blade. Next, the pure NCs were labelled with maleimide-transferrin derivative for the biological application. The next year, another novel method was proposed for preparing numerous luminescent SiNPs ranging from 3-4 nm diameter through functional group manipulation [53]. At first, hydrogen silsesquioxane (HSQ) was converted into oxide-embedded SiNCs using 1100°C under the inter atmosphere. After that, the hydride surfaced SiNCs were liberated from the SiO_2 surface via etching with 1:1:1 49% HF/EtOH/H_2O. Next, the resultant H-SiNPs were functionalized by tailoring various functionalities (amine, alkyl, and phosphene) to regulate their emission properties. However, morphological tuning of SiNPs is another crucial factor in gaining diverse properties for different applications, particularly in good-quality optical sensors. In 1964, a bottom-up strategy named vapor-liquid-solid (VLS) method was proposed for the one-dimensional growth of SiNWs in the presence of nanoparticle catalyst where the nanoparticles control the growth and diameter of the NWs. But the process was unsuccessful in getting nanoscale SiNWs [54]. In 1997 this process was further modified and a laser ablation technique was employed for the generation of good-quality, pure, crystalline and nanoscopic SiNWs [55]. Using this technique VLC method was further improvised to CVD-VLC via the introduction of chemical vapor deposition (CVD) process for a more controlled reactant source which enabled the more controlled growth of the SiNWs. Still, many innovative synthetic techniques are developing for further betterment in properties and application.

7.3.6 Metal organic framework nanomaterials

Irrespective of conventional organic and inorganic materials MOF materials synthesis methodologies development is still in its infancy. While many reports for the synthesis of MOF NPs are available, reliable preparation methods for uniform NPs yet remains a challenge.

a. Rapid nucleation

To attain rapid nucleation of NMs, rapid heating is a new approach, as conventional heating methods take several hours, however, microwave-assisted NMs were synthesized within one minute. Because of the rapid heating, energy was generated throughout the bulk of the material and boiling point of solvent was exceeded by using pressurized containers. Microwave heating is broadly used because it speedily grinds the precursors and yields MOF NMs with small size difference. Microwave-assisted MOF NMs, the cubic $Zn_4O(BDC)_3$ (IRMOF-1/MOF-5, BDC=1,4-benzenedicarboxylate) was first synthesized by solvothermal method [56]. In addition to heating by microwave method, ultrasound instrument also demonstrated very useful to fast precursor method to consume the bulk material and then the nucleation step. Ultrasound acoustic cavitation creates hot spots in size of 200 nm with transient extreme higher temperature (5000 K) and higher pressure (1000 bar). In these hot spot nucleation and growth of Fe(OH) BDC (MIL-53-Fe) NPs occurred rapidly [57].

b. Microemulsion strategy

The microemulsion strategy delivers an effectual method to control the MOF NPs shape and it also narrow the NPs size distribution. Rather than fast nucleation and transient heating procedure, in this method the nanoparticle size is controlled by the isolation of the nucleation sites. Two non-miscible solvents such as water and oil are chosen and both are mixed for create monodisperse nanoscale emulsions droplets. This droplet size of emulsion can be tuned by amphiphilic surfactant concentration. Emulsion droplets act as size limiting vessels for MOF precursors, in which the NMs nucleation was controlled by the reaction temperature. In isooctane/1-hexanol/water system by using the cetyltrimethylammonium bromide (CTAB) surfactant, the first crystalline nanorods of $Ln_2(BDC)_3(H_2O)_4$ (Ln = Tb^{3+}, Gd^{3+}, Eu^{3+}) was synthesized [58].

c. Coordination modulation

Modulation in the coordination is a most effective and adaptable strategy to regulate the metal-ligand interactions to synthesize MOF NMs and also these methods leads to the most uniform size MOF NMs. This method can be combined with the rapid nucleation and microemulsion strategies. Ligands those are monotopic are mixed in the reaction system to modulate into coordinate sites of available metal ion chemically. The shape, size, uniformity of NPs was regulated by chemical modulators. The MOF $Cu_2(ndc)_2(dabco)$ (ndc=1,4-naphthalene dicarboxylate; dabco =1,4-diazabicyclo[2.2.2]octane) NPs have two coordination modes. The Cu-dabco (amine) to bridge and Cu-ndc (carboxylate) to form 2D sheets. Acetic acid act as a modulator that selectively formed only square-rod shaped anisotropic NPs were selectively on the (100) surface [59].

7.3.7 Carbon-based nanomaterials

a. Carbon nanotube

The carbon nanotubes (CNT), a one-dimensional tube-like rolled hollow graphene sheet, have encouraged researchers with their superior properties in the developing nanotechnology area. It has shown excellent progress in the chemical and biosensor field since 1991 [60]. Commonly CNT comprises the hexagonal ring having sp^2 hybridized carbon atom as a building unit. But in some cases, sp^3 hybridization also can be found. Non-covalent interactions with a highly delocalized π cloud in the surface even feasible ease of functionalization with biological macromolecules, making them a supreme candidate for bio-sensing applications. These π bonds enable multi-walled nanotube structure formation through π-π interaction. According to the rolled layers number, CNTs are devided into two groups; Single-walled carbon nanotubes (SWCNTs) and multi-walled carbon nanotubes (MWCNTs). SWCNTs enfold into a cylindrical structure with a 0.4 nm to 2.5 nm diameter. These exhibit outstanding electrical conductivity owing to extended π conjugation. Amalgamating two features, miniature size and excellent conductivity, offer its application for electron transfer reaction as a single electrode [61].

The three most popular techniques for the economical production of CNT are arc discharge, laser ablation and chemical vapor deposition (CVD) [62,63]. In the arc discharge method, two pure graphite electrodes were employed as cathode and anode and distanced by 1-2 mm under a helium atmosphere at very low pressure (30-150 torr). Sometimes, various metals (Fe, Mo, Ni, Co) containing anode rods were utilized to achieve different metal incorporated CNT and SWCNT owing to unique features. Generally, arc production continues up to 1 min and after that the reaction is terminated by separating the two electrodes as current passing can raise the chamber's temperature too high. This arc effect consequences the sublimation of the anode graphite rod and deposition on to cathode electrode in the form of single and multiple walled CNT. The final product properties depend on several parameters like reaction atmosphere, current and the voltage used for arching, and the gas's pressure inside the reaction chamber. Although this method plays a pivotal role in the large-scale preparation of CNT, it has limited control of the product's alignment. Laser ablation is another efficient technique for the extensive synthesis of the CNT. In this method, a powerful laser tool is used as an excitation source for the vaporization of pure graphite inside a quartz chamber at 1200°C under Argon atmosphere. The resulted vapor dust is regulated in a pulse flow by the inert gas from the high-temperature compartment and accumulated on a low-temperature copper rod outside the chamber. Here, the laser power is an essential parameter that decides the diameter of the nanoparticle formed. However, it was found that the increment in the laser power produced thinner CNT [64]. Other reports on this method also demonstrate that the ultrafast laser source can generate pure and high-quality SWCNTs in large quantities [65]. The key demerit of this method is the non-uniform nanotube formation. Although this method is very potent in synthesizing less impurity-containing nanotube, it is not so commercially relaxing due to the requirement of pure graphite blocks, slow rate carbon deposition, high power laser source, etc. Finally, the chemical vapor deposition technique is the most advantageous and controllable nanotube synthesis procedure in terms of purity and quantity of the production among all these three processes. Depending on the several improvisations on the typical CVD protocol, it has been classified into water-assisted CVD, catalytic chemical vapor deposition (CCVD), plasma-enhanced CVD, radiofrequency CVD and microwave plasma-assisted CVD. Currently, CCVD is the most effective method for the preparation of CNT. The principal mechanism is based on the chemical dissociation of precursor hydrocarbon material in the presence of the catalytic amount of transition material, which is quite similar to the arc discharge method regarding the precursor excitation. Usually, the whole reaction is performed inside a horizontal quartz furnace where the catalyst is kept in the middle of the furnace under atmospheric pressure. Now hydrocarbon and inert gas comprising reaction mixtures are flown over the catalyst at a temperature varying from 500-1000°C. Further, cooling of the reaction tube leads to the precipitation of well-grown sp^2 hybridized nanotube. Some of the variable conditions like the concentration of the hydrocarbon, pressure and temperature of the tube, nature of the catalyst used, etc. can control the properties of nanotubes. Overall the CVD process is the most flexible, cost-effective, and easily handled procedure to obtain the desired dimensioned CNT.

b. Graphene NMs

Graphene is a densely packed single layer carbon atom comprising hexagonal elementary ring and forming a two-dimensional honeycomb-like planner sheet, the parental structure of the other allotropes of carbon embracing CNT, fullerene [66]. The existence of 2D crystal was proven wrong with the invention of graphene material [67]. Owing to a similar building unit, graphene exhibits similar electrical conductivity, tensile strength with CNT. Still, the structural difference produces well-discriminative data in Raman spectra. It also reveals a high surface to volume ratio, fast electron mobility, excellent band tunability, good elasticity, and porous structure forming from multiple single layer stacking, making it eligible for the sensing application with proper selectivity and sensitivity. Several methods, like chemical vapor deposition, chemical exfoliation of graphite, the mechanical breaking of graphite crystal, etc., have been established to attain good quality graphene sheets. However, chemical processes are most eligible for the economic and bulk production of graphene film, which can be utilized in industrial applications.

Although the mechanical breaking of graphite crystal is the initial procedure demonstrating the graphene film preparation, it demands more time and effort with low productivity, which causes a large barrier in industrial utilization purposes [68]. On the other hand, the chemical reduction of graphite oxide is a potential synthetic strategy for the large-scale production of graphene sheets at a lesser cost [69]. The oxidation of graphite block leads to lighter color graphite oxide, and commonly anchors with hydrophilic groups such as carboxylic and enol functionalities at the plane edges. These hydrophilic groups facilitate the colloidal dispersion of the GO in water through van der Waals interaction. However, the elimination of the hydrophilic functionalities by chemical reduction decreases their hydrophilicity, enabling the accumulation endowing π-π stacking between the restored aromatic ring and produces exfoliated well-stacked graphene sheet [70]. This π-π interaction also introduces some improved features retaining previous properties of precursor material. Nevertheless, the stability of the produced stacking sheet in the solution is also a challenging task as, in most cases, ultrasonication also failed to redisperse the precipitated portion. A synthetic pathway was recently demonstrated for the preparation of exfoliated graphite oxide by ultrasonication, which was subsequently reduced by hydrazine to suspended nano-platelets [71]. However, the initially generated nanoplatelets were not so stable in the water. Yet, the use of poly(sodium 4-styrenesulfonate) in the time of reduction gave graphene sheet with extended enhancement in its stability, which was not precipitated even after one year. Later, another strategy was reported using a single strand of DNA to stabilize a highly concentrated graphene sheet in an aqueous solution [72]. The co-addition of single-strand DNA and hydrazine at the time of reduction outcomes DNA intercalated stable, ordered nanocomposite structure.

Owing to the uncontrolled surface area and incomplete reduction in the chemical exfoliation process, has encouraged scientists to put more efforts into further developing the synthesis process to obtain high-quality controlled layered graphene sheets. Moreover, stabilizing materials used in the chemical process directly impacts the electronic properties and results in a thick film, restricting its applications in various fields. To overcome these drawbacks, a self-assembly-based bottom-up approach was established to produce high-quality film in between lamellar mesostructured silica [73]. This method exhibits several advantages like no involvement of hazardous chemicals, the requirement of a mild reaction environment, generation of one-layer pure graphene film on a large scale with lesser thickness, and long-term dispersibility.

7.4 Properties of nanomaterials

The properties of NMs have generally categorized into chemical and physical.

a. Chemical properties

The NMs have properties such as more stability and less reactivity against the external factors such as light, moisture, heat. NMs also have antifungal, antibacterial, antiviral properties and these are suitable for bioimaging and it can be employed in environmental monitoring applications.

b. Physical properties

NMs have smaller diameter and large surface area that are critical factors leading to exceptional physical properties compared to their bulk counterpart materials. Physical properties include optical, magnetic, tensile, flexible, hydrophobic, hydrophilic, cellular uptake, cytotoxicity, conductive properties, etc.

7.5 Characterization of nanoparticles

NMs have unprecedented physicochemical properties and to understand the extraordinary behavior it is essential to characterize its properties by using different techniques.

a. Particle size

NMs synthesized by using different methods in different diameters were measured by using several techniques, and the most used were DLS, AFM, TEM and SEM. The size of NMs depends or can change with the measurement methods, for example DLS determines the hydrodynamic radius of suspended particles whereas electron microscope provides the size of the isolated particles form the surroundings [74].

b. Morphology

Scanning and transmission electron microscopy (SEM and TEM) have been extensively used to acquire the shape and size of NMs. Thickness and differentiation of nanocapsule from nanosphere were widely performed by the TEM instrument. Information of surface morphology with high resolution in three dimensions was performed by atomic force microscopy (AFM). Surface topography, cavities and pores information can be determined by using above mentioned instruments [75].

c. Zeta potential or surface charge (ζ)

The surface charge of NMs expressed as the zeta potential are affected by the interface of the dispersing medium solvation effect. It is a measurement of the magnitude of the charge attraction or repulsion between the particles. Zeta potential study gives detailed information about the aggregation and dispersion. In the solution state, zeta potential can be measure by dynamic light scattering (DLS), whereas in the gaseous state, NMs charge was determined by differential mobility analyser (DMA) [76].

d. Specific surface area (SSA)

The surface area is the utmost important factor for characterization. It is defined as the surface area per mass unit of any material. This SSA characterization is the most influential and causes extraordinary effect on the chemical, physical and optical properties and performance of NMs. SSA analysis has been performed by the BET technique for measuring the surface area of NMS. In the BET technique SSA analysis was performed by adsorption of inert gas (argon, nitrogen) over the surface of NMs was studied [77].

e. Crystal structure

Powder and single crystal diffraction methods are non-destructive analytical methods that give detailed information about the molecular arrangements, bond angles, bond lengths and lattice of crystals. The powder and crystal samples can be studied by powder X-ray Diffractometers (XRD) and Single Crystal XRD technique respectively. The relation between NMs crystal structure and its toxicity was also established [78].

f. Chemical composition

In the chemical composition determination technique sample material analyzed for elemental or isotopic composition. The elemental and chemical composition of NMs identify the purity and ensure its good performance. The composition was determined using X-ray photoelectron spectroscopy (XPS), ion chromatography, atomic emission spectroscopy and energy dispersive X ray (EDX), etc. [79].

g. Optical properties

NMs exhibit diverse and tunable optical properties and their emission is sensitive to concentration, solvent, aggregation state and their size and shape. The important technique to study their emission and absorption properties are photoluminescence, phosphorescence and UV-vis. Absorption spectroscopic technique. UV vis. spectroscopy technique is also used for the transition metal and NMs quantitative determination in solution [80].

h. Magnetic properties

Charged materials with mass i.e. protons, electrons, negative ions etc. lead to the generation of magnetic effects. The NMs can be classified to paramagnetic, diamagnetic, ferromagnetic and antiferromagnetic according to their behavior in the magnetic field. The magnetic properties of NMs rely on core-shell structure, chemical composition and ionic substitutions. In order to characterize the magnetic properties vibrating sample magnetometer (VSM), and superconducting quantum interference devices (SQUID) technique can be employed [81].

7.6 Nanomaterials application in sensing

7.6.1 Chemical sensing

Sensing plays a key role in protecting the present status of our environment, which is under deep distress. The release of various hazardous chemicals and gases from industry or accidental fire explosion directly impacts human health. The large-scale use of pesticides, insecticides, and herbicides in the agricultural field is a prime environmental issue worldwide. Hence, the present sensing materials need to improve in sensitivity, selectivity, stability, and many other aspects, which is quite challenging. NMs development has unique size-dependent photophysical properties, large reactive surface area, and rapid response compared to their bulk counterparts, which opens up broad possibilities for researchers to meet the present challenges.

7.6.1.1 Nanomaterials for explosive sensing

In recent years, the research on detection and accurate quantification of explosive materials are fast-growing and concerning homeland security, environment, and health safety. The common explosive materials are compound based on nitro (TNT, DNT, RDX, picric acid), nitrates such and peroxides namely, triacetone triperoxide (TATP). Nitroexplosive compounds are common ingredients in fireworks, dye, pharmaceutical, leather, and matchbox industries. Ammonium nitrate is used as a fertilizers and as an explosive material. TATP is mostly used by the terrorist as a homemade explosive because of its easy synthesis from readily available precursor materials, i.e., acetone and hydrogen peroxide. For example, the terrorist bombings attack that took place in the London subway system in 2005 was due to TATP. Besides its explosive nature, long-time exposure to these compounds affects humans' biological systems and many other organisms, causing a mutagenic and carcinogenic effect [82,83]. Some of them are quite soluble in water and readily contaminate water and soil, thereby easily reaching its toxicity in humans and other organisms. Hence, there is a continual requirement for sensing trace amounts of such explosive compounds in the ground, air, and water to ensure the health and environmental safety and security.

Reliable sensing of trace explosive materials remains a challenging task. Peroxide and nitrate-based explosives are highly sensitive to heat, shock, and friction, so contact-free detection is highly desirable. Ensuring security from terrorism activities, the effective vapor phase detection of various hidden explosives buried in warzones or shields in various containers is highly demanding. In many cases, vapor phase detection is complicated because of their low vapor pressure at ambient temperature. Explosives wrapped in plastic reduce their vapor concentration by 1000.

Furthermore, many explosive compounds are known to be sticky in nature and will adsorb on the surface with high surface energy, resulting in the accumulation of many explosive molecules, thus reducing analyte molecules' number density per sampling volume. Consideration of all the above challenges, NMs having the potential to detect both in solution and gas-phase, ultra-sensitivity up to picomolar and even less concentration, large reactive surface area, with quick response time is more attractive. Several methods, such as trained canines, surface-enhanced Raman spectroscopy (SERS), infrared spectroscopy (I.R.), mass spectrometry (M.S.), X-ray imaging, techniques, gas chromatography (G.S.), electrochemical technique, and so on have been reported for the trace amount detection of nitro explosives. Although such techniques give reliable response, the analytical methods coupled with NMs are considered a better choice for onsite detection, low cost, reproducibility, and quick response.

Nitro explosives (NAC) are electron deficient in nature, so their detection is based on the electron-rich NMs' electron transfer mechanism toward NAC, resulting in the quenching of fluorescence. Due to the benzene ring in nitroaromatics (TNT, picric acid), π-π stacking with the fluorophore also contributed to fluorescence quenching. In 2010, for the first time, electron-rich graphene nanostructure was reported as an electrochemical platform for TNT sensing [84]. Due to its size dependent fluorescence nature, the quantum dot represents an excellent sensing platform for various explosives. Thus, CdSe QDs were used for NAC detection where its fluorescence quenches only in chloroform medium with an LOD of 10^{-6} to 10^{-7} M [85]. Again, the surface of QDs can be modified with different functionality to act as a receptor for any analyte of interest. Further significant development is known on the use of a multichannel nanoparticle array to detect and differentiate various explosives. Here, multicolor fluorescence QDs are functionalized to have a different surface receptor (OH. and OMe). These surfaces can respond to a wide range of explosives through various mechanisms such as host–guest interaction, electrostatic interaction, and π-π stacking (Fig. 7.2A) [86].

The major problem in explosive sensing is the lack of selectivity as all nitro explosives have similar electron deficiency. Picric acid having a high mutagenic effect is selectively detected from environmental water samples and in the vapor phase by using CP nanoparticles. The mechanism for this excellent selectivity is the combined effect of the "molecular-wire effect," electrostatic interaction, resonance energy transfer (RET) and photoinduced electron transfer (PET) (Fig. 7.2B). The water-dispersible polymeric NPs (HCPN-S) with hydrophobic backbone and hydrophilic sulfonate side chains could selectively sense TNT by efficient encapsulation of hydrophobic TNT inside the CP core [87]. Further, the discriminative detection of various explosives such as nitro explosives and peroxide explosives is performed using nanowire-based sensor device such as field-effect-transistor arrays (NW-FETs) [88]. Here, the pattern recognition of intrinsic

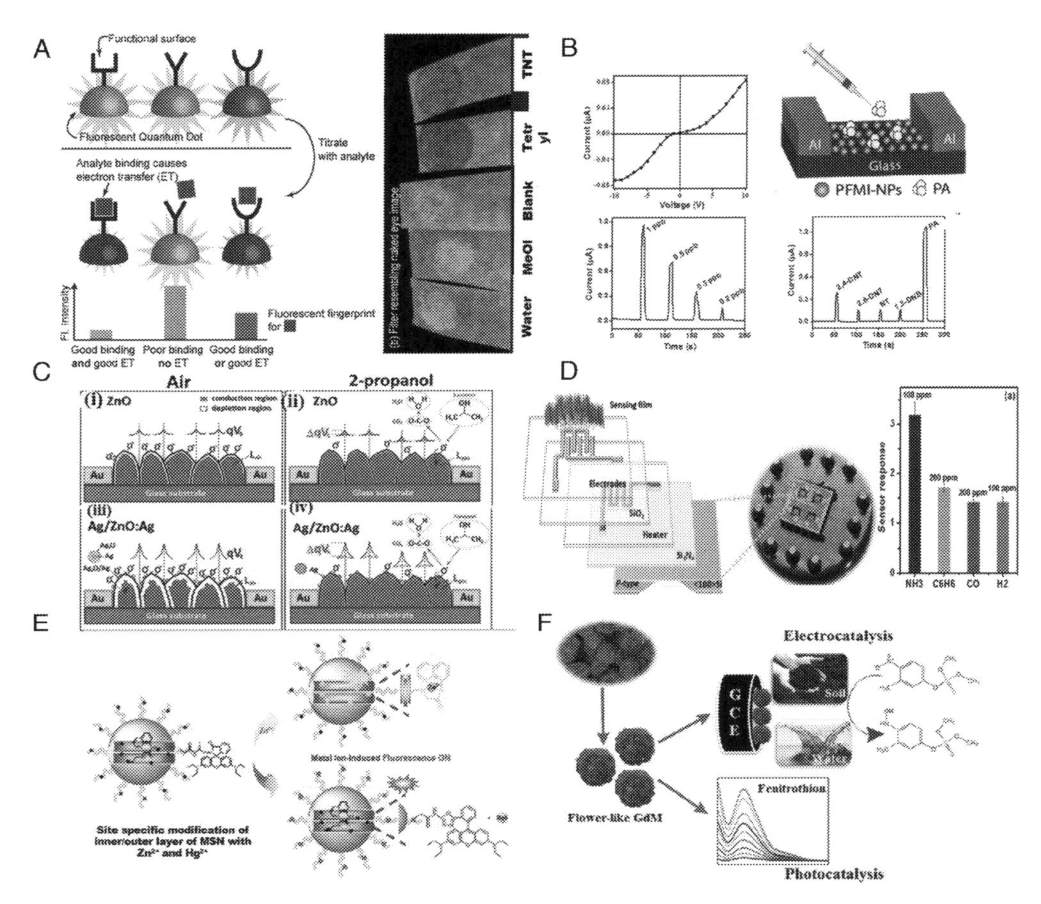

Fig. 7.2 (A) Differential analyte binding across the multicoloured QD array giving a fluorescent fingerprint for the analyte which can be detected by naked eye on a filter paper based device. Reproduced with permission [86]. Copyright 2015 American Chemical Society (B) Detection of picric acid by using a two terminal device where polymeric nanoparticles PFMT-NPs act as a active layer. Inset (i) I-V characteristics (ii) two terminal device (iii) output current response with increasing concentration of picric acid (iv) selectivity study of PFMI-NPs with various nitro explosives. Reproduced with permission [92]. Copyright 2015, American Chemical society. (C) Proposed gas sensing mechanism for ZnO, and Ag/ZnO:Ag columnar films integrated in sensor structures during exposure to: (i), (iii) ambient air and (ii), (iv) 2-propanol vapors. Reproduced with permission [104]. (D) MEMS-based PdO/WO_3 microsensors for selective sensing of H_2. Reproduced with permission [166]. Copyright 2016, American Chemical society. (E) Florescence off/on based sensing of Zn^{2+} and Hg^{+} induced by specific site modification of inner/outer surface of MSN with Zn^{2+} and Hg^{+}. Reproduced with permission [167] Copyright 2013, Royal Society of Chemistry. (F) Electrocatalytic and photocatalytic detection Fenitrothion by Gadolinium Molybdate. Reproduced with permission [123]. Copyright 2018, American Chemical Society.

thermodynamics and kinetics of the mode of interaction between the nano-sensor array and various explosives gives the explosives' fingerprint.

Moreover, Raman spectroscopy, electrical method, electrochemical method, and many more analytical techniques coupled with NMs achieved ultrasensitive detection of various explosive materials with excellent selectivity. Ammonium nitrate (AN) present in explosives is decomposed readily and detected as trace ammonia. Recently, surface-enhanced Raman scattering/spectroscopy (SERS) signals of Ag-Cu CPs were being used for AN detection because of their high sensitivity and molecular fingerprinting ability [89]. Due to a lack of nitro group and aromatic rings, sensing of peroxide explosives such as TATP is quite challenging. However, a remarkable improvement is the development of a new fluorogenic SiO_2 nanomaterial that drastically increases the fluorescence emission in the presence of triacetone triperoxide vapors (TATP) [90].

The following Table 7.1 listed different types of NMs used for the detection of various explosive materials along with mechanism and their limit of detection.

7.6.1.2 Nanomaterials for volatile organic compounds sensing

The volatile organic compounds (VOCs) are characterized by low molecular mass and very high vapor pressure at room temperature. Because of high vapor pressure, they can quickly pollute the air that we breathe and affect various biological processes of human health. For example, exposure to the high concentration (>500 ppb) of acetone, hydrogen sulfide (>100 ppb), formaldehyde (>500 ppb), hydrazine (4.0 ppm), trimethylamine and ethanol (>40 ppb) in the breath are associated with irritation in the eyes, nose, dizziness, lung cancer, and diabetes disease. Moreover, hydrazine has a mutagenic and carcinogenic ability that damages the kidney, liver, and central nervous system. Depending on functional groups present in the molecular structure, the indoor VOC can broadly be classified into seven types. These are terpenes, esters, and aldehydes, and hydrocarbons of aliphatic, aromatic, oxygenated, and halogenated. The source of such VOCs is many, but 50 to 300 different VOCs can be produced from indoor environments of daily household products such as air fresheners, cosmetics, paint, and cleansers, disinfectants, stored fuels and automotive products etc. Another important source is our body metabolism, where it releases as end products through body fluids (sweat, breathing, urine, blood, skin), and proper analysis of it is considered a non-invasive technique for early-stage diagnosis of various diseases [99].

Considering the health hazards potential of VOC even at very low concentration (sub ppb) Environmental Protection Agency (EPA) and the National Institute of Occupational Safety have established several guidelines for health safety limiting the exposure to VOCs inside and outside air [100]. But the main challenge is that the conventional technique for indoor VOCs detection is time-consuming, lacks selectivity, highly expensive, and fails to give accurate information for VOC exposure on real-time applications. So, the development of a user-friendly VOCs sensing platform with excellent sensitivity and selectivity is highly demanding. Nowadays, NMs with large reactive surface areas and sized dependent properties open up a new strategy to solve such problems.

Table 7.1 Different types of NMs used for detection of various explosive materials along with mechanism and their limit of detection.

Explosives	Types of NMs	Materials	Materials detected	Mechanism	LOD	Reference
Nitroexplosives DNT, TNT, tetryl, RDX, Picric Acid, TNT, DNT	Organic small molecule-based nanoparticles	Thiazol-substituted pyrazoline	Picric Acid	acid-base interaction induced ET	0.002 ppm	[91]
	Polymer-based nanoparticles	PFMI	Picric Acid	coulombic interaction/ PET/RET	30.9 pM	[92]
		HCPN-S	TNT	hydrophobic cavities that can encapsulate hydrophobic TNT	3.7 nM,	[87]
	QDs	Calixarene and cyclodextrin having O.H. and OMe surface	DNT, TNT, tetryl, RDX and (PETN)	E.T.	1-0.2 ug/ml	[86]
	Metallic nanoparticles	Combination of AuNPs with photoactivated TiO_2 substrate	RDX, TNT, DNT, PETN	UV light mediating migration from TiO2 to the AuNPs	3.1×10^{-8} M - 10^{-15} M	[93]
	Carbon based	N doped carbon dots	Picric acid	FRET, acid/base nteraction	0.25 mM	[94]
	Si-based nanoparticles	SOI-NW	DNP	-	1 fM	[95]
	Metal-Organic Framework Nanoparticles	MOF $[Cd_2(btc)_2(H_2O)_2]$	N.B., NT, and DNT	self-sacrificing template	18.1 ppb (DNT)	[96]
Nitrate explosives N.G., EGDN, PETN ammonium nitrate	Nanocomposite	Ag–Cu alloy nanoparticles	AN	SERS signals	5 mM	[97]
	Nanocomposite	Nanoparticles Composites on Au Surfaces	N.G., EGDN, PETN	Structure similarity of imprinted matrix	200 fM, 400 fM and 20 pM	[98]
Peroxide explosives, TATP, HMDT	silica nanoparticles	Surface functionalized silica nanoparticles	TATP	ICT	0.12 mg L^{-1}	[90]
	Nanosensors array	ZnO-SiNW	TATP, HMDT	Complexation	fM	[88]

Undoubtedly, metal-oxide semiconductors, such as SnO_2 and ZnO, are extensively used for VOCs detection due to their novel morphologies and structure. The tunable macro/meso ZnO nanostructure is used to detect ethanol and acetone because a large surface area for contact was provided for targeted VOCs by the hierarchical macro/mesoporous nanostructure [101]. But metal oxide semiconductors operate efficiently only at a temperature above 200°C.

To further improve VOCs sensing performance, great attention has been paid to the use of nanocomposites for VOCs detection, i.e., the integrated usage of two or three NMs such as CNT, graphene nanostructure, Si nanowire Ag, Pt, ZnO, TiO_2, gold nanoparticles and so on. Recently, SnO_2 nanobelts having unique morphological and functional properties are doped with Eu to have selective sensing of acetone electrochemically and conductometically [102]. Au doped MoS_2 can distinguish oxygen functionalized and non-functionalized VOCs by analyzing different VOCs' response patterns [103]. Herein, Au-MoS_2 nanocomposites' electrical response is positive for toluene, hexane and negative for oxygen-containing VOCs (ethanol and acetone). The observation was the opposite in the case of pristine MoS_2. Again Ag-doped ZnO columnar film was able to differentiate between various VOCs such as propanol, acetone, methane, and hydrogen, where preferential adsorption of these VOCs occurred at several Ag-doped ZnO surface sites (Fig. 7.2C) [104]. Another strategy for VOC detection is the use of nano heterojunction, improving the charge transfer at the interface with p-n heterojunction formation [105]. The deposition of NiO onto the ZnO nanoparticle networks forms a p-n heterojunction. This heterojunction is quite effective for room temperature selective detection of VOCs, commonly used for breath analysis i.e., acetone and ethanol. These enhanced sensory signals mechanisms are due to the formation of unique ultra-porous, p-n nanoheterojunctions, enhancing the interaction with the target analyte.

Further, CP/CP nanocomposites are known to improve the sensory signals due to the molecular wire effect. The polymeric nanoparticle's sensitivity is directly correlated with the interaction between the functional group attached to the polymeric chain and the gas molecule. The sensitivity of these polymeric materials can be further improved by blending with another NM, where it forms an unstable nano morphology, which can be easily rearranged upon exposure to VOCs. A nano P_3HT/PCBM sensing chip was recently designed that is applicable for real-time sensing wherever VOCs are used or produced. This device's detection limit shows that this device can quickly respond to several VOCs (n-octane, toluene, o-xylene, chlorobenzene, dichlorobenzene) to detect the overexposure concentration limit of these VOCs. This device can give an alarm signal within a fraction of minutes for timely remedies [106]. Another interesting thing is that a colloidal crystal-PDMS composite was developed with the outer surface with the crystallized NPs and in bulk PDMS was used as sensor for VOC based on nanoscale easy tear (NET) mechanism. Various chemical stimuli are generated for vapor phase detection within a short time when the residual PDMS layer is removed entirely from the colloidal–PDMS composite. The multiplex sensing (benzene, toluene, xylene) potential selectively for the target analyte was illustrated by patterning topography and graphics of the colloidal crystal-polymer. This proposed NET mechanism displays outstanding achievement for simple, on-site and real-time VOCs detection by using colorimetric detection [107].

7.6.1.3 Nanomaterials for toxic gases sensing

The rapid development of human civilization, such as industrialization and increasing vehicles' use, leads to the release of enormous hazardous and toxic gases such as NOx, COx, SO_2, NH_3, $COCl_2$, hydrocarbons etc., that consequently threaten human survival and many lives. In Syria, since 2013, the alleged use of chemical weapons such as chlorine, phosgene, and sarin resulted in the death of more than 3000 people with 13000 injuries [108,109]. In addition to these, the burning of hydrocarbons, fossil fuels, and plastics also contributed to releasing such toxic gases. The long-term exposure to such gases affects the respiratory system even at very low concentration, undergoing neurological disorder. However, most such gases are colorless, tasteless, and odorless, so our body sense organs cannot detect it. Hence, there is an ultimate need to develop a reliable and portable sensing platform to detect such toxic gases.

Metal oxide (ZnO, SnO_2, Ni_2O_3, CuO etc.) in its various forms such as nanoparticles, nanowires, nanoflowers, nanorods, and core-shell array nanotubes, hollow spheres, nanosheets, and so on are widely used for gas sensing purposes. The intrinsic defect in metal oxide, such as oxygen vacancies, is highly attractive because it provides the active sites for gas molecules adsorption. Metal oxide semiconductors can be n or p-type semiconductors in the case of an n-type semiconductor, and the sensor's resistance increases or decreases depending on the gas molecule in contact, which is oxidizing or reducing gas molecules. Here, gas molecules are adsorbed on the active site drain (provide) elections continuously to the adsorbing surface [110]. However, most of these NMs need to improve a lot in sensitivity, selectivity from interfering gas molecules, and working temperature. Considerable efforts by many researchers' shows that the development of hybrid NMs such as nanocomposites, metal-doped nanoparticles, and designing heterojunction and surface functionalization gives better sensing performance. The decoration of tungsten oxide nanoneedles with PdO nanoparticle enhances the gas sensing performances, suppressing humidity. The PdO doped N.N.s sensor response to hydrogen (H_2) is 755 times higher than the corresponding undoped N.N.s at 150°C temperature (Fig. 7.2D) [111,166].

A nanocomposite of Graphene Oxide and Organic Dye (R-GO and BPB) based sensor is developed for rapid and dual-mode sensing of Ammonia Gas. The devices, i.e., visual mode and electrical mode devices, display remarkable sensitivity toward ammonia gas at a concentration below the threshold of 25 ppm [112]. The good dual-mode operation in sensing helps enhance the LODs accuracy with a long operation time. Moreover, these NMs can be used as wearable chemical sensors for health and environmental safety by working at ambient conditions.

Recently, several molybdenum tungsten chalcogenide NMs have been reported for gas sensing based on chemoresistance [113]. Continuous efforts have been given to improve the gas sensing performance of MoO_3 and WO_3, and as a great achievement, Mo and W dichalcogenides (MX2 X = S, Se, and Te) were found as an alternative to them. Rh-doped $MoSe_2$ can be an excellent SO_2 gas sensor due to the nanomaterial surface's physisorption, while it acts as a promising adsorbent for gases like CO, NO, and NO_2. It is also outlined that the efforts for making metal oxides nanocomposite or nanohybrids, decoration, and surface sensitization of graphene and its derivatives, with noble metal nanoparticles, will provide promising improvement in transition metal dichalcogenides resulting in excellent performances for gas sensing.

7.6.1.4 Nanomaterials for metal ions sensing

Metal ions (As, Pb, Cd, Cr, Hg, Co, Cu, Mg, and so on) are serious environmental pollutants creating a worldwide concern because of their non-biodegradable nature. They are essential in our body only up to a specific concentration limit. If the dose of consumption or exposure exceeds the body requirements, it may cause several acute and chronic poisoning. The rapid development in industrialization and mining activities release such metal ions continuously and contaminate natural water bodies, thereby posing a health risk. Moreover, natural processes like rain, weathering, and soil erosion are also responsible for discharging metal ions into water bodies. As such specific detection and complete removal of these poisonous metal ions are in great demand.

For achieving a highly sensitive detection, continuous efforts have been made by researchers. Recently, NMs having unique properties such as large reactive surface area, robust adsorption decent catalytic efficiency and represent a better choice as a sensing platform for heavy metals. Various types of NMs based on polymer, small molecule, metal, metal oxide, carbon, silicon, as well as a hybrid in the form of nanoparticles, nanosheets, nanodots, etc. have shown promising results. Also, the hydroxyapatite-based NPs doped on silver ion-exchanged nanocrystalline ZSM-5 nanocomposite (Ag-Nano-ZSM-5) material have shown ultra-sensitivity (<1 ppb) toward toxic heavy metal ions Cd^{2+} (0.5–1600 ppb), Pb^{2+} (0.6–1600), As^{3+} (0.9–1800), and Hg^{2+} (0.8–1800 ppb). The combined effect of high adsorption affinity for hydroxyapatite NPs, mesoporous nanocrystalline zeolite, higher reactive surface area and the lowering of electron transfer resistance for the Ag-Nano-ZSM-5 nanomaterial is accountable to its efficient electrocatalytic activity. Moreover, it can detect these toxic metals in a real water sample with satisfactory results [114].

Hybrid NMs based on SiO_2 and Core-Shell of Fe_3O_4@SiO_2 also give promising results for metal ion (Hg^{2+}, Pb^{2+}, Cd^{2+}, Cu^{2+}, Zn^{2+}, Fe^{3+}, etc.) detection (Fig. 7.2E) [115,167]. These NMs improved the sensing performance of metal ions, such as selectivity, sensitivity, and binding affinity. Moreover, hybrid nano-sensors have achieved unique advantages that result in promising applications: i) being solid materials, they are applicable in solid-liquid heterogeneous phase; ii) the hybrid NMs are recyclable through some chemical treatments, after sensing studies iii) mostly the hybrid sensors are stable for a relatively long time in a wide range of pH iv) relatively low cost of raw materials.

Further, CP NMs (PEGylated SCPNs) were also developed based on a fluorescent electronic tongue for the rapid discrimination of toxic heavy metal ions in aqueous media [116]. The mechanism of sensing is based on fluorescence quenching due to electron transfer between the analyte with the distinct energy level of the conjugated nanoparticle. The pattern recognition method is successfully used for the discrimination of various metal ions. This, CP nano sensing array will have promising prospects for real-time monitoring of environmentally toxic heavy metals.

In 2020, a Metal-Organic Framework Nanoparticles (NH_2-MIL-125(Ti)) was designed to detect lead-based on reversible fluorescence switching [117]. Here, toxic lead is sensed (LOD -7.7 pM) with a concentration less than the maximum limit approved by WHO. Moreover, the NMs are recoverable and can detect lead in the river (186 pM) and tap water (152 pM).

7.6.1.5 Nanomaterials for pesticides, insecticides, and herbicides sensing

Nowadays, as control from pests, insects, or unwanted weeds, several pesticides, insecticides, and herbicides are used in agricultural fields. However, according to the United States Environment Protection Agency (US EPA), every year, around two billion pounds of pesticides are used only in North America. The problem is that only 1% of the pesticide reaches its target, which means most of the pesticides used will contaminate the environment without giving many benefits [118]. Agriculture and Food Organization defined a pesticide as "Any substance or mixture of substances intended for preventing, destroying, or controlling any pest, including vectors of human or animal disease, or unwanted species of plants or animals causing harm during or otherwise interfering with the production, processing, and storage or marketing of food agriculture commodities, wood or wood products, or animal foodstuffs, or which may be administered to animals to control insects arachnids or other pests in or on their bodies." This materials are used as a desiccants, defoliants, growth regulator, fruit thinning agents or agents for avoiding the premature fall of fruits, and matters that are added to crops before or afterward harvest to prevent deterioration during storage or transport [119]. Despite all these advantages, the long-term exposure or consumption of food containing pesticide residue has the capability to harm to harm entire ecology other than the target organism.

The most common types of pesticides available in the market are Organochlorine pesticides (OCPs), organophosphate pesticides (OPPs), carbamates, chloroacetamides, and pyrethroids neonicotinoids, and glyphosate. Considering their potent toxicity (acute and chronic) and easy accessible to human beings through the food chain, developing a reliable sensor for the timely detection of pesticides is of immediate requirement.

In recent years there is rapid advancement in the field of pesticide sensing by the use of NMs (e.g., metal and magnetic nanoparticles, QDs, carbon nanotubes, and graphene etc.) [120]. Reproducibility and low sensitivity limit are significantly improved, which earlier remained quite challenging task. Further, the author suggested that NMs-based pesticide sensing is low cost and easy to fabricate via user-friendly devices. Reproducibility and low LODs have proven challenging. NMs unique characteristics, such as high adsorption ability, efficient electron transfer, and the large reactive surface area, enable low LODs, which are of great need for pesticide sensing.

A significant achievement is the use of Ag nanoparticles to discriminate various pesticides (organophosphate and carbamate pesticides) based on the chemiluminescent sensor array [121]. Recently, Ag nanoparticles in their multiple forms (nanosheets, spheres, foam, core shells) were used for extraction, detection, and degradation of pesticides due to their excellent properties such as porosity, mono-dispersity, stability, and low toxicity [122]. Recently, a 3D flower-like gadolinium molybdate NMs were reported as a bifunctional catalyst for detecting and reducing organophosphate pesticide (fenitrothion) electrochemically. The catalyst acts as an efficient mediator for ultrasensitive detection (5 nM) of fenitrothion pesticide with a wide-range response range (0.02-123; 173-1823 μM). This catalyst enables the detection of fenitrothion in real water and soil samples with good recoveries. Further being an excellent photocatalytic active nanomaterial, it can degrade fenitrothion under UV irradiation with a degradation rate above 99% within 80 minutes (Fig. 7.2F) [123]. Further, Au nanorods are reported for the first time as an efficient substrate for rapid ultralow detection (0.15 ppt) of pesticide residues, its degradation, and complete removal for real-time applications. The mechanism of sensing is based on self-agglomeration [124].

7.6.2 Biological sensing

In this section, we cover the detection of various bio-components including enzymes, proteins, amino acids, different biomarkers, nucleic acids, etc. The monitoring of different bio-molecules based on most convenient and precise sensing platforms using various NMs including carbon dots, organic or inorganic NMs and aggregation-induced emission/aggregation-induced enhanced emission (AIE/AIEE) NPs are reported in this section.

7.6.2.1 Glutathione, an antioxidant detection

Glutathione (GSH), a thiol group-containing tripeptide, exists in all living cells and biological fluids. It has an essential role in clearing away intracellular free radicals, peroxides and superoxides. Therefore, oxidative stress is ascribed to the imbalance of intracellular GSH level. Under oxidative stress, GSH is oxidized to GSSG through scavenging the reactive oxygen species (ROS) and the reactive nitrogen species (RNS), and then quickly reduced to GSH by the glutathione reductase enzyme. Thus, the imbalance of GSH/GSSG ratio has been demonstrated to be related to several diseases including aging,

Table 7.2 Name of various NMswhich are used for the detection of GSH along with their limit of detection (LOD) and sensing mechanisms in some recent representative literatures.

NMs	LOD	Methods	Mechanisms	Reference
carbon dots	5.7 μM	"Turn-on"	Aggregation and dis-aggregation of self-crosslinked red emissive carbon dots	[125]
Copper nanoclusters	0.89 μM	Colorimetric	Nanomaterial exhibited peroxidase like activity	[126]
MnO_2 nanosheets	10 nM	"Turn-on"	Based on redox reaction	[127]
Nitrogen-doped carbon dots	0.226 μM	"On-off-on"	Photoinduced electron transfer (PET)	[128]
QDs-MnO_2 nanocomposites	16.3 μM	"Turn-on"	Based on oxidation-reduction redox reaction.	[129]

cancers, cardiovascular disease, Alzheimer's disease etc. Thus, numerous literatures based on different types of NMs including metal oxide, carbon dots, noble metal, molecular organic framework, AIE/AIEE nanoparticles etc. for the monitoring of intracellular GSH have been published, and are listed in Table 7.2.

CPs@MnO_2-AgNPs have been reported as multifunctional nano-sensors for the detection of GSH and cancer theranostics *in vitro* and *in vivo*. The GSH detection was performed through a fluorescent "turn-on" response. At first, the fluorescence of AgNPs was efficiently quenched by MnO_2via a static quenching along with an inner filter effect (IFE) mechanisms. The emission intensity of AgNPs was further recovered by GSH that was attributed to the unique redox reaction between manganese dioxide (MnO_2) and GSH. The LOD was obtained under optimal conditions, which was calculated to be 0.55 μM. It was shown that the reaction between MnO_2 and GSH was completed within 30 minutes. The higher oxidability of MnO_2 was performed under acidic conditions at a pH value of 5.0. It was demonstrated that these nano-sensors are very selective toward the GSH. However, homocysteine (Hcy) and cysteine (Cys) interfere in a certain degree of fluorescence intensity recovery of nano-sensors (Fig. 7.3A) [130].

Further TPE-PBP have been developed by incorporating the para-dinitrophenoxy benzylpyridinium moiety to tetraphenylethylene (TPE) for the selective and sensitive monitoring of thiol using the AIE luminogen (AIEgen) nanoparticles with ratiometric fluorescent response. Moreover, mitochondrial thiols were quantitatively mapped by these TPE-PBP NPs in vitro, in vivo and ex vivo using one-and two-photon fluorescence microscopy. The mechanistic insights displayed that the TPE-PBP exhibited red emission in PBS buffer solution in the absence of thiol. While after addition of thiol, the emission of TPE-PBP gradually blue-shifted that was attributed to the breaking of the phenyl ether bond by thiol and produced a less conjugated AIEgen which induced the self-immolation of the 4-hydroxybenzyl moiety. These excellent photostable and biocompatible TPE-PBP NPs showed low LOD, of 0.61 μM [131].

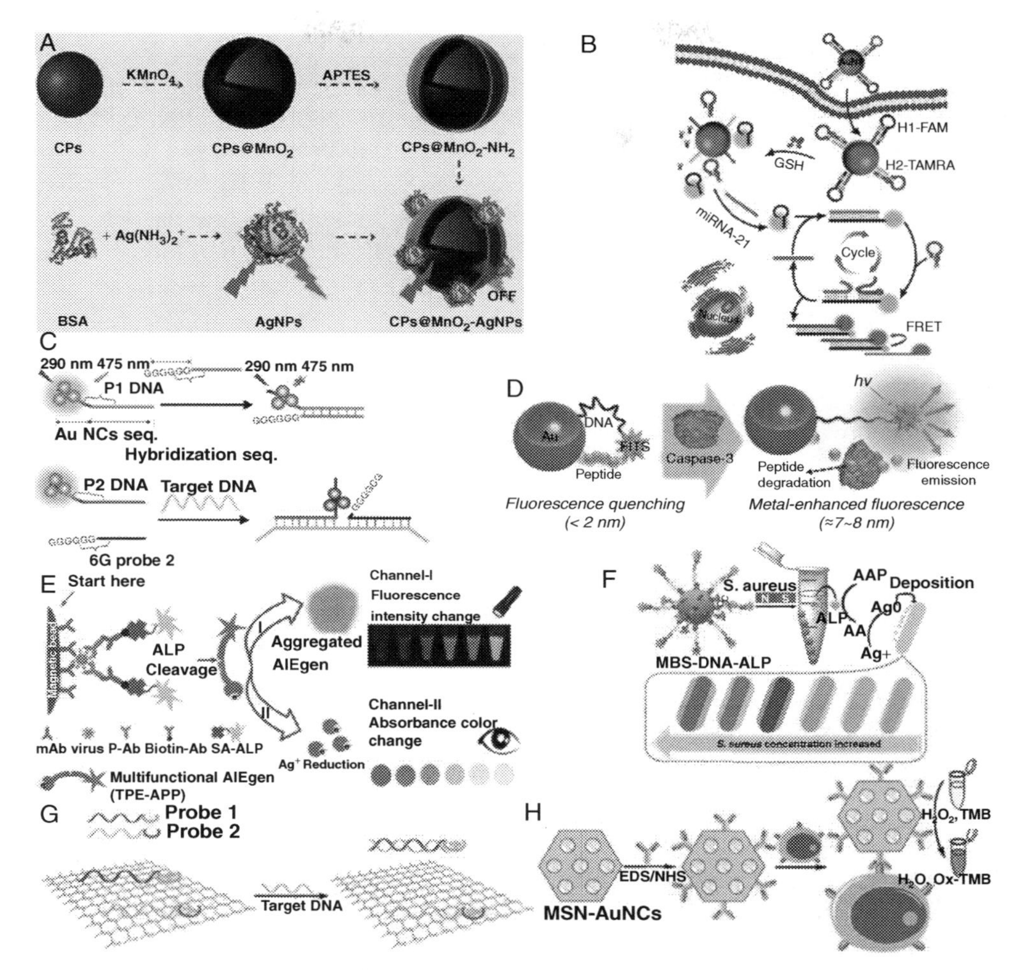

Fig. 7.3 (A) Schematic illustration for the synthesis of AgNPs, CPs@MnO_2 and the CPs@MnO_2-AgNP nanocomposite for the monitoring of GSH via fluorescent "turn-on" response. *Reproduced with permission [130]. Copyright 2019, Royal Society of Chemistry.* (B) The bio-cleavable TAMRA labeled hairpins (H1-AuNP-H2) nanosensors for the selective monitoring of miRNA-21 in living cells. *Reproduced with permission [134]. Copyright 2018, Royal Society of Chemistry.* (C) The convenient sensing platform for the monitoring of DNA based on the fluorescence quenching behavior of P1 DNA-AuNCs through proximity with Guanine- rich DNA (P1 and P2 DNA represents the different DNA sequences respectively). *Reproduced with permission [135]. Copyright 2019, Royal Society of Chemistry.* (D) The sensitive and selective monitoring of caspase-3 using AuNP through enzymatic cleavage reaction. *Reproduced with permission [136]. Copyright 2020, American Chemical Society.* (E) The virus detection platform using multifunctional AIE NPs based on plasmonic colorimetric and fluorescent dual-modality. *Reproduced with permission [146]. Copyright 2018, American Chemical Society.* (F) The sensing platform of dual enzyme-induced gold-silver NRs as a colorful substrate for the selective monitoring of *S. aureus*. *Reproduced with permission [147]. Copyright 2020, American Chemical Society.* (G) Graphical representation of the multiplexed sensing system for HAV and HBV virus using Au NPs/COF NSs (the yellow dots represents the Au NPs). *Reproduced with permission [148]. Copyright 2020, Royal Society of Chemistry.* (H) The strategy for the excellent selective and precise monitoring of $HER2^+$ breast cancerouscells via the synergistic amplified colorimetric sensing assay using the MSN-AuNC-anti-HER2 hybrid platform. *Reproduced with permission [151]. Copyright 2019, Royal Society of Chemistry.*

7.6.2.2 Nucleic acid detection

Nucleic acids have been considered to be an important information carrier of the cells. DNA is a double-stranded molecule, which is well-known as the molecule of heredity and transmits genetic information from generation to generation. RNA, is a single-stranded molecule consisting of ribose sugar, which significantly plays a crucial role in controlling the cellular functionalities, including gene regulation and transmission of genetic information. The most significant part of RNA is protein synthesis in living cells. The detection of RNA is of great significance due to its important functionalities in living cells. In 1984, the RNA-specific dye had been first reported. MicroRNAs (miRNAs) are recognized as a short single-stranded and a group of non-coding RNAs, which acts as important gene regulatory factors in enormous biological procedures, including cellular division, cellular apoptosis, cellular proliferation and gene expression. The miRNA has been proposed to be a valuable diagnostic biomarker and the abnormal expression of miRNA associated with anenormous human diseases including viral ailments, cancers, neurological and cardiovascular diseases. Therefore, the monitoring of intracellular nucleic acids by means of a most convenient and facile techniques has come into the consideration a hot topic in research field. The most commonly used methodological approaches are high-throughput sequencing, DNA microarray, polymerase chain reaction (PCR) and northern blotting. However, these traditional methods are complicated, time-consuming, and at the expense of quantitative determination. Thus, the literatures for the monitoring of DNA, RNA, messenger RNA (mRNA) and miRNA based on the different NMs are represented in references.

Recently DNA-AuNPs have been developed to detect two different mRNA in living cells with high specificity and selectively. In addition, these nanoparticles efficiently release two different anticancer drug doxorubicin and mitoxantrone [132].

Further a novel ratiometric fluorescent NM, graphene QDs and molecular fluorescent probe HVC-6@GQDs have been demonstrated for the selective detection of RNA through the FRET mechanism. Herein, HVC-6 was acted as the acceptor, while GQDs were behaved as the donor of the FRET mechanisms of the ratiometric emissive HVC-6@GQDs NMs. It was speculated that the FRET between the HVC-6 and GQDs was affected after increasing of the RNA concentrations, which could be attributed to the groove binding between RNA and HVC-6, which further increased the distance between HVC-6 and GQDs [133].

Recently bio-cleavable Au-nanoprobes functionalized with FAM and TAMRA-labeled hairpins (H1-AuNP-H2) have been strategically designed for the selective monitoring of miRNA in living cancerous cells. The mechanistic investigations showed that the glutathione cleaves the disulfide bonds in the hairpins, while target miRNAs further trigger to form many DNA duplexes via cell catalytic hairpin assembly (CHA) between the two hairpins. When the FAM dye is excited, these DNA duplexes brings the FAM donor and the TAMRA acceptor dyes into the close proximity which amplified the energy transfer (FRET process) from donor to acceptor, and are estimated for the sensitive monitoring of low amounts of under-expressed miRNA-21 in living cancrous cells. The LOD was calculated to be 3.7 pM (Fig. 7.3B) [134].

Again DNA-templated gold nanoclusters (AuNCs) were reported for the detection of guanine rich DNA (G-DNA) through the photoinduced electron transfer (PET). Photostable and highly luminescent DNA nanoclusters were employed as the indicator, while G-DNA was performed as the quencher of the fluorescence intensity. The LOD was calculated to be 200 pM. The mechanistic investigations exhibited that the emission intensity of this sensing platform in the presence of targeted DNA significantly decreased which was ascribed to the hybridization of nanoclustersand G-DNA. Herein, the 3' terminius and the 5' terminius of G-DNA and nanoclusters respectively come into close proximity which stimulate the co-operative hybridization between them (Fig. 7.3C) [135].

7.6.2.3 Enzyme and protein sensing

Proteins are considered to be a major source of energy, which plays a prominent role in maintaining the body tissue, including development and repair, and in the creation of some hormones. Additionally, it produces enzymes and helps in controlling cellular growth. The numerous diseases including Alzheimer's, prion, amyloidosis and Parkinson's are associated with the abnormal expression of proteins. The enzyme has a crucial role in maintaining the internal organ functionalities of living organisms, including metabolism, cell repair, anti-inflammatory, immunity improvement, energy production etc. Moreover, enzymes have been broadly considered as the biomarkers for distinct diseases including neurodegenerative diseases, hereditary and cancers. Therefore, the detection of different enzymes and proteins based on the facile, cost-effective and precise platforms are significantly important. Thus, most recent literatures regarding the monitoring of enzymes and proteins in cellular systems based on different NMs have been reported so far, which are mentioned in references.

Recently AuNPs have been presented for the selective and sensitive monitoring of proteolytic enzyme caspase-3 which was performed by simply enzymatically cleavage reaction and the catalytic reaction was completed within 1 hour. This simple and one step sensing platform is used for the detection of intracellular proteolytic enzymes which further can be used for the advance recogntion of neurodegenerative diseases and cancerious cell. Moreover, it displayed sensitive linear dynamic range (LDR), ranging from 0.01-10 ng/mL and detection limit was achieved less than femtomolar concentration, ranging from 10-10000 pg/mL (Fig. 7.3D) [136].

Further TPE functionalized AIE congener for the sensitive and selective monitoring of α-amylase with LOD and LDR of 0.14 U L^{-1} and 0-45.5 U L^{-1} respectively in MES buffer have been reported. The AIE probe is non-emissive when it is in totally dispersed form. While, after hydrolysis of these AIEgen by α-amylaseenzyme the enhancement in fluorescent intensity occurred and that was attributed to the restriction of intramolecular rotation (RIR) effect of the AIE congeners in the condensed state. Moreover, this system could be employed to demonstrate the applications in psychological stress analysis and acute pancreatitis diagnosis. These AIE based simple and more convenient platform exhibited promisingly valuable for the application in pharmaceutical and food industries [137].

Another peptide conjugated AIE probe TPEPy for the selective and sensitive monitoring of bacterial ALP (alkaline phosphatase) in the living cells based on "turn-on" fluorescence response has been developed. The congener showed excellent selectivity and sensitivity and LOD was found to be 6.6×10^{-3} U mL^{-1}. The fluorescent intensity of the congener was enhanced after being hydrolysed by ALP. Moreover, this wash-free proposed sensing system showed an advantageous over traditional ALP kits [138].

Recently, polyelectrolyte complex nanoparticles (PEC NPs) have been reported for the recognition of globular proteins such as bovine serum albumin, haemoglobin and human serum albumin based on FRET sensing mechanism. The emission intensity of these PEC NPs were quenched in the presence of protein and these sensing platform could be applied to detect the globular protein in the LOD concentration of 5 nM range [139].

Again colorimetric detection of protamine was performed owing to the self-assembly behavior of the AuNRs on the graphene oxide (GO) surface. The mechanistic investigations showed that the AuNRs self-assembled on the GO surface via the electrostatic interaction between the cationic AuNRs and the anionic charged of GO. The color of the aggregated AuNRs on the GO was changed from red to dark purple that was ascribed to the decreased surface plasmon resonance (SPR) absorption of AuNRs. On the contrary, the red color of well-dispersed AuNRs was maintained in the presence of protamine, which could possess tightly binding characteristics with the GO surface and prevent aggregation and adsorption of AuNRs on the GO surface. This simple sensing platform exhibited selectivity toward protamine with minimum interferences from other positively charged drugs and common biological species. This assay showed LOD of 63 ng mL^{-1} toward protamine [140].

Further a class of water-soluble distinct TPE congeners with AIE characteristics by introducing tetrazolate as novel luminiscent probes for the selective and sensitive monitoring of albumin have been reported withan LOD of 0.21 nM. These AIEgen exhibits a good LDRof 0.02-3000 mg/L for quantification of albumin. In addition, theseAIE congeners endow higher water solubility and a greater binding affinity (K_D) towards albumin due to the incorporation of tetrazole moiety into the AIE sensing system. The detection mechanism suggests that the tetrazole unit of AIE congeners specifically bind with the albumin particularly with the lysine of albumin through columbic forces and hydrogen bonds, which was proved by isothermal titration calorimetry (ITC), molecular dynamics (MD) simulation and molecular docking. The fluorescence was "turn-on" after binding with the albumin due tothe RIR effect of the TPE unit. Moreover, these

AIE NPs could be applied to evaluate the urinary albumin in analytical and medical samples [141].

7.6.2.4 Detection of different biomarkers

A biomarker or biological marker is a measurable indicator of some biological conditions or states, which shows clear variations during the diagnostics of diseases. Thus, the detection of different biomarkers is highly important for the early disease diagnosis. The detection of different biomarkers is reported in some representative literatures using different types of NMs.

Recently a novel dual-responsive colorimetric assay using GQD-Sensitized terbium/guanine monophosphate infinite coordination polymeric (GQD@Tb/GMPICP) NPs for the recognition of acetylcholinesterase (AChE) biomarker and its inhibitors organophosphorous (Ops) has been demonstrated. The AChE plays a crucial role in the regulation of activities of the neurotransmitter acetylcholine (ACh). It detects the AChE by changing the corresponding color from green to blue fluorescence, which was attributed to the "turn on and turn off" fluorescence response of GQD and Tb/GMP respectively. Moreover, the detection of Ops was performed by changing the color from greenish-blue to green fluorescence. The detection of NM is assigned to the significant changes of the coordination environment of GQD@Tb/GMPICP nanoparticles, which leads to the collapse of the ICP network [142].

Further, fluorescent metal-organic nanosheets have been strategically designed for the recognition of dopamine (DA), a biomarker of several diseases, especially Parkinson's disease. Herein, they have demonstrated ultrathin metal-organic nanosheets, ECP $[Eu(pzdc)(Hpzdc)(H_2O)]_n$ (H_2pzdc = 2,3-pyrazine dicarboxylic acid) with their higher surface area, exceptional thickness, and flexible characteristics, are employed for the detection of target analyte through the optimal surface interaction between the metal-organic nanosheets and target analytes. These ECP nanoflakes were employed to detect the DA through "turn-on" optical sensing methods with LOD of 21 nM and the association constant (Ka) of 1.25×10^5 L mol^{-1} without interfering with other similar kind of species. The mechanistic investigations displayed that the DA adsorbed at the surface of ECP nanoflakes through the electrostatic interactions and then emission intensity of europium ions was enhanced that was assigned to the efficient energy transfer of adsorbed DA to the nanoflakes [143].

Another type of AuNPs have emerged for monitoring nonenzymatic label-freeacid phosphatase (ACP) activity. This sensing method displayed selectivity toward ACP even in the presence of other similar kind of biomolecules including collagenase and dehydrogenase. The LOD was calculated to be 0.03 mU mL^{-1}. Moreover, this proposed sensing platform exhibited high sensitivity compared to the commercially available ACP kit. The ACP produceascorbic acid (AA) via the hydrolyticactivity of L-ascorbic acid 2-phosphate sesquimagnesium (AAPS), which subsequently reacts with $HAuCl_4$ and generate AuNPs. The in situ synthesized AuNPs activity toward TMB

(3,3′,5,5′-tetramethylbenzidine) and H_2O_2 gives a clear colorimetric readout. The enzymatically formed AuNPs exhibited peroxidase-like activity which produced intense blue-colored solution that could be utilized to determine the concentrations of ACP in biological media effectively [144].

Recently low-cost, label-free, visual, and facile fluorescent biosensor, F-PDA (fluorescent polydopamine) NPs have been reported for the convenient detection of ALP (alkaline phosphatase) through the FRET mechanism between F-PDA NPs and MnO_2 nanosheets. The sensing system could be applied for the selective recognition of ALP in the presence of similar kind of proteins or enzymes and the LOD was calculated to be 0.34 mU/mL. This proposed sensing platform possess excellent applicability in biological samples which could be consider to be potential applications in biomedical research and clinical diagnosis. The sensing mechanisms revealed that the emission intensity of F-PDA NMs was quenched by MnO_2. It was attributed to the formation of FPDA-MnO_2 stable complex via noncovalent bonding interaction between F-PDA NPs and MnO_2 nanosheet, which results in the FRET from F-PDA NPs to MnO_2 nanosheet. However, the emission intensity of F-PDA NMs was gradually restored with the increasing concentration of ALP and the emission intensity was almost recovered completely when the concentration of ALP increased to 150 mU/mL [145].

7.6.2.5 Bacteria and virus detection

Food poisoning or infectious diseases caused by different pathogens have become a growing threat worldwide. Thus, a rapid, portable, inexpensive and autonomous platform for the identification of pathogens is of utmost importance, which may play a important role in identifying infections and provide guidance for health monitoring. The most widely used conventional method for identifying pathogens are based on PCR, culture and colony counting and immunological techniques. However, these techniques are time-consuming or require skillful technicians and complicated procedures, which limit their broad practical applications. Hence, easy detection of pathogens is of greatest challenge. Currently, different enzymes such as catalase, lipase, urease and micrococcal nuclease etc, have been extensively used as targets for the identification of various pathogens. The most recent literatures based on bacteria and virus detection using different types of NMs are represented in references where Fig. 7.3E–G represents the sensing performances of different NMs.

Recently an expedient AIE modulator has been demonstrated for the ultrasensitive and specific monitoring of viruses by using plasmonic colorimetry and fluorescence in a single recognition platform. This immunoassay system was specific toward the EV71 virions as an example for both the robust "turn-on" fluorescence response and naked-eye colorimetric detection. The LOD was calculated to be 1.4 copies μL^{-1} under fluorescence response platform. Additionally, EV71 (Another human enterovirus 71) virions were recognized by these AIE NPs in a wide range from 1.3×10^3 to 2.5×10^6 copies μL^{-1} with the naked eye. Further, EV71 virions were successfully employed for diagnosing with 100% accuracy in 24 clinical samples. This reliable and convenient platform does not depend on the expensive instruments and complicated sample pre-treatment compared to the other conventional methodssuch as gold standard PCR. Thus, this dual-modality design strategy could be proficiently employed with the good capability for the selective diagnosis of suspected virus infected areas based on both convenient fluorescence and colorimetry preliminary screening (Fig. 7.3E) [146].

Further Au-Ag alloy NRs have been demonstrated for the rapid and sensitive recognition of *Staphylococcus aureus* (*S. aureus*) based on the colorful colorimetric sensing platform. This proposed sensing method can easily detect *S. aureus*, around 25 CFU mL^{-1} through naked-eye observations. Herein, they have exhibited oligonucleotide magnetic beads functionalized with alkaline phosphatase (MBs-DNA-ALP) for monitoring the concentration of the S. aureus. The mechanistic insight demonstrated that micrococcal nuclease was naturally secreted in the presence of S. aureus, which release the ALP followed by the spcefic cut of the AT reaches DNA sequences of MBs-DNA-ALP into fragments. Thus, the Au-Ag NRs were attained which were attributed to the catalytic activity of the most popular enzyme ALP. Finally, the rainbow color of sensing system was accomplished which was a clear identification of Au-Ag NRs. This color of sensing platform was assigned to the significant shift of the localized surface plasmon resonance (LSPR) of the Au-Ag NRs (Fig. 7.3F) [147].

Again Au NPs/COF NSs have been demonstrated as sensing platform with the green and red emissive Ag NC probe for the selective detection of HAV (hepatitis A virus) and HBV (hepatitis B virus) respectively. The demonstrated detection platform exhibited LOD for HAV and HBV of 75 pM and 150 pM respectively. Herein, green and red emissive P_1 and P_2 Ag NCs probes were equally mixed, where Au NPs/COF NSs used as aprominent quencher that was attributed to the energy or electron transfer behavior of this nanosheets. The green emission of Ag NC was displayed after introduction of HAV at 520 nm, whereas the red emission of Ag NC was shown in the presence of HBV at 615 nm. Thus this novel sensing platform could also be further used for the potential application in biomolecules detection and related biological applications (Fig. 7.3G) [148].

Recently bifunctional TriPE-NT (triphenylethylene-naphthalimide triazole) AIE congener has been reported, which exhibited excellent capability of both staining and killing of MDR (multi-drug resistant) including gram-positive and gram-negative bacteria, as an example of *E. coli* (*Escherichia coli*), MDR *E. coli*, *S. epidermidis* (*Staphylococcus epidermidis*) and MDR *S. epidermidis* with high safety along with efficacy both *in vitro* and *in vivo*. Herein, in the TriPE-NT AIE congener, the TriPE displayed the fluorescent part (AIE characteristics), whereas NT showed antibacterial efficacy. Furthermore, under light irradiation this TriPE-NT AIE NPs in water exhibited efficient ROS producing ability and drastically enhances its antibacterial ability by photodynamic therapy (PDT) toward clinically isolated MDR bacteria with a negligible toxicity toward mammalian cells. Therefore, TriPE-NT showed as an efficient antibiotic agent against MDR bacteria, which could be employed to explore the associated antibacterial mechanisms [149].

7.6.2.6 Amino acids, vitamins, and cancer cell detection

Amino acids are often referred to as the building block of proteins, which plays a significant role in living systems, including hormones, proteins and neurotransmitter synthesis. Some of the diseases including skin lesions, drowsiness, liver damage and Parkinson's are associated with the abnormal level of proteins. Vitamins are considered to be the essential nutrients, which performed hundreds of roles in the body including conversion of food into energy, repair cellular damage and boost up the immune system. The diseases such as rickets, beriberi, pellagra are associated with an abnormal level of vitamins. Therefore, the regular monitoring of amino acids and vitamins is highly required. Several analytical methods including HPLC-UV (high performance liquid chromatography-UV), microbiological assays and atomic absorption spectroscopy, electrochemistry, capillary electrophoresis etc. are applied for the selective monitoring of them. However, most of these reported approaches are high cost, require complicated sample preparation techniques and time-consuming, which limit their practical applicability. Thus, the exploration of cost-effcetive, simple and rapid detecting methods for monitoring amino acids or proteins are of utmost importance. Cancer, as an example of breast cancer, which leads to the cancer-related death in women around the world. The competent and consequent therapy of breast cancer is the advance and specific diagnosis of cancer cells, which is the identification of cancer cell at the nascent phase.

Recently 2D (two-dimensional) FeP@C (iron phosphide embedded in a carbon matrix) nanosheets have been demonstrated for the selective and sensitive monitoring of Cys (cysteine) and Cu_2^+. This novel, cost-effective sensing platform can detect cysteine even in the existence of other 19 amino acids. Thus, these simple systems are reliable and feasible for real sample analysis. The LOD was calculated to be 0.21 nM and 0.026 μM for Cu_2^+ and Cys respectively. The mechanistic investigations displayed that this iron nanosheets behaved as a robust oxidase mimics, which oxidatively catalyzed the non-emissive substrate AR (amplex red) to the highly emissive oxidized AR (AR-ox). Remarkably, Cys restrained the production of AR-ox, and resulted in a decreased emission intensity of the FeP@C/AR system. This weak fluorescence intensity again restored by Cu_2^+ that was ascribed to the strong binding characteristics of Cu_2^+ toward Cys [150].

Mesoporous silica-gold nanocluster hybrid (MSN-AuNC) has been developed for ultrasensitive and the highly selective colorimetric detection of human epidermal growth factor receptor 2-positive ($HER2^+$) breast cancerous cell. Herein, they have incorporated antibodies to MSN-AuNC hybrid system to produce MSN-AuNC-anti-HER2 sensing platform. These constructed nano-sensors exhibited excellent sensitivity with-LOD and linear detection range of down to 10 cells and 10–1000 cells respectively. The mechanistic insight showed that the high sensitivity of these biosensors was ascribed to the astonishing recognition skill of HER2 antibodies towards the HER2 receptors on the membranes expressed of SKBR3 cancerous cells and the excellent catalytic performances of Au nanoclusters. In addition, the selectivity was also conducted in the presence of eight distinct breast cancer cell including MDA-MB-231, HCC1806, Hs578T MDA-MB-468, MDA-MB-436, MDA-MB-157 and the fibrocystic, non-neoplastic breast cell MCF10A, which showed the excellent selectivity toward $HER2^+$ SKBR3 breast cancer cells. The above results imply that these simple economical sensing platforms could be considered to be expedient for the advance diagnostic of cancerous cells (Fig. 7.3H) [151].

Further, mitochondria-targeted AIEgen has been demonstrated for the identification and discrimination of various cells such as normal cells and cancerous cells, gastric cancerous cells, digestive tract cancerous cells and mixed gastric cancerous cells based on the AIE characteristics of the AIE modulators. In this prospects, five distinct TPE congeners were developed with double positive charges, which exhibited multicolor emission and identified various cells with 100% accuracy [152].

Cannabis sativa carbon dots doping with nitrogen and sulfur (N-S@CsCD) have been reported for the selective detection of vitamin B_{12}. This proposed sensing system with high quantum yield and photostability exhibited dual fluorescence characteristics toward vitamin B_{12} and temperature through the fluorescence "turn-off" behaviors. The LOD was calculated to be 7.87 mg mL^{-1}. Moreover, this N-S@CsCD system, could detect the vitamin B_{12} present in water, PBS and DMEM buffer systems. In addition, this nano-sensor exhibited potential application in the biomedical field due to its excellent biocompatibility and very high cellular uptake characteristics [153].

7.6.2.7 Adenosine and adenosine triphosphate detections

Adenosine being an endogenous nucleoside, has received an intense attention that is ascribed to its versatile functionalities inside the cellular systems. Moreover, it was consensually witnessed to be a significant regulator of tissue functionalitis and regarded as a "retaliatory metabolite" that is attributed to its equalizing the intake energy to the metabolic performances. Notably, adenosine has been witnessed as a biomarker of lung cancer. Therefore, the precisely recognition of adenosine is of utmost importance. Whereas, ATP is found in the nucleoplasm and cytoplasm of all living cells, which is witnessed as a source of energy of all living cells. It shows a most crucial role in most enzymatic activities. The concentration of ATP is an indicator of many diseases including malignant tumors, Parkinson's and Alzheimer's disease. Hence, the exploration of a convenient and reliable platforms are required to monitor adenosine and ATP. Various strategies and methods have been demonstrated to detect adenosine and ATP, including HPLC, gas/liquid chromatography-mass spectrometric method (GC/LC-MS), colorimetric analysis, fluorescent spectrometry and electrochemical assay. However, it is witnessed that the rapid development of fluorometric sensing platforms over these conventional analytical methods for monitoring the adenosine and ATP is highly desirable for its convenient, precise and simple operations.

Recently, carbon dots labelled by single-stranded DNA (ssDNA-CDs) has been developed for the selectively and sensitive monitoring of adenosine via "turn-on" fluorescent behavior through fluorescence resonance energy transfer (FRET) mechanism. The LOD was calculated to be 4.2 nM. In addition, this simple, rapid and cost-effective sensing platform were applied to detect the adenosine in biological samples also with satisfactory results, which indicates its promising future in practical applications. The mechanistic investigations demonstrated that owing to the formation of the aptamer-ssDNA duplex by aptameric-AuNPs in the absence of adenosine, the emission intensity of ssDNA was quenched. However, after incorporation of the adenosine to the sensing system, the fluorescence intensity of ssDNA-CDs was restrained, that was ascribed to the binding characteristics of the adenosine with the aptameric-AuNPs which block the energy transfer from ssDNA-CDs to aptameric-AuNPs [154].

Further, AuNPs have been designed for the label-free colorimetric detection of adenosine and ATP. The mechanistic insight displayed that the AuNPs were destabilized due to the adsorption of adenosine into the citrate-capped adenosine. At the same time, ATP stabilized the AuNPs, attributed to the negative charges from the triphosphate group of ATP [155].

7.6.2.8 Detection of cholesterols and anticoagulant

Heparin is well-known as an excellent anticoagulant both in vitro and in vivo, which can regulate numerous physiological processes and is mainly used for the prevention of thrombotic diseases. Many adverse effects including thrombocytopenia, hyperkalemia and osteoporosis could be induced and is assigned to the long-term or an overdose use of heparin. Therefore, reliable and facile techniques are required for the monitoring of accurate heparin concentration and specifically control the dose of heparin during the anticoagulant therapy and surgery. Recently, activated clotting time assay (ACT), activated partial thromboplastin time assay (aPTT) methods are applied to detect the heparin levels. However, these techniques are insufficiently reliable, high cost and time-consuming, that limited their practical application. Thus, developing a reliable and facile platform is still highly desirable for the sensitive and selective monitoring of heparin. Cholesterol, extensively exists in human blood and is witnessed as a precursor of some significant biomolecules including vitamin D, steroid hormones and bile acids. Research has indicated that various diseases including, cerebrovascular diseases, anemia, lipid metabolism disorder are associated with the imbalance of cholesterol. Thus, precise detection of cholesterol are clinically highly desirable.

Recently, Mo and S co-doped carbon QDs (Mo-CQDs) have been designed as a peroxidase mimics biosensor for the colorimetric detection of cholesterols through catalytic activity. The colorimetric Mo-CQDs biosensor unveiled high sensitivity and selectivity toward cholesterol, ranging from 0.01 to 1.0 mM and the LOD was calculated to be 7 μM. In addition, these simple sensing methods can potentially measure the total cholesterol concentration in the biological samples with satisfactory results. The cholesterol can be detected through the naked eye by these Mo-CQDs platform, which exhibits the promising technique for the portable kits and clinical diagnosis [156].

Further, mesoporous silica nanoparticle-gold nanoclusters (MSN-AuNCs) have been demonstrated for the selective and sensitive monitoring of anti-coagulant heparin (Hep). The MSN-AuNC nanoclusters in the absence of heparin exhibited enhancement of emission intensity more than 5-fold that was assigned to the electrostatic interaction between cationic MSNs and AuNC surface ligands which facilitate the self-assembly characteristics of MSN-AuNC nanoclusters. However, the emission intensity of MSN-AuNC nanoclusters in the presence of anionic charged heparin was decreased gradually that was attributed to the efficient binding characteristics of heparin on the MSN surface, which subsequently detriment the interaction between the MSNs and the AuNCs. The LOD was calculated to be 2 nM. Moreover, this proposed sensing system exhibited satisfactory application for the detection of heparin in the human serum, which shows a promising platform in clinical applications [157].

Graphene oxide (GO) and gold nanorods (AuNRs) have been developed further for the selective detection of heparin in the presence of other interfering ions and proteins using a polycationic polymer, PDADMAC (polydiallyldimethylammonium chloride), as a molecular probe. The LOD was calculated to be 10.4 nanograms. The native red color of well-dispersed AuNRs in the system was maintained and was attributed to the binding characteristics of PDADMAC polymer to the GO surface, which prevents the absorption of AuNRs on the GO surface. However, in the presence of heparin, AuNRs self-assembled and aggregated on the GO surface due to the effective binding characteristics of the negatively charged heparin with the cationic PDADMAC polymer. Lastly, the color of the AuNRs was changed from red to purple that was ascribed to the decrease of the localized surface plasmon resonance (LSPR) [158].

7.6.3 Miscellaneous sensing

7.6.3.1 Fingerprint detection

The fingerprint (FP) of each individual characterized by an intricate corrugated skin pattern is unique and remains a lifelong individual identity. The invisible latent fingerprint (LFP) formation when the finger is in contact with a surface contains both water-soluble (salts, carbohydrates, peptides, urea) and insoluble (fatty acids, sterols, wax esters, triglycerides, etc.) components reflecting various information of the individual [159]. For example, the presence of trace narcotic drugs or their metabolites in LFP is an indication that the individual uses the drug (Fig. 7.4A) [168]. Also, the detection of explosives in the LFP suggested the contraction of such chemicals by the individual. Therefore, since the 19th century, LFP has been used for the forensic investigation as physical evidence. In the past decade, researchers all over the world contributed enormously to detect this LFP. Several conventional methods are developed, such as ninhydrin test, powder dusting, cyanoacrylate, or iodine fuming and silver nitrate soaking, metal deposition and fluorescence staining. However, these methods need to improve a lot in contrast, sensitivity, selectivity. Besides, some detecting reagents are toxic and destroyed the LFP during processing. Thus, the development of a highly efficient, low-cost, user-friendly method that can give more information has attracted significant research attention.

An emerging area for LFP detection is the use of fluorescent NMs that can be CP nanoparticles, metal nanoparticles, QDs, up-conversion nanoparticles, or hybrid nanoparticles. The discriminatory identification of individuals required to give

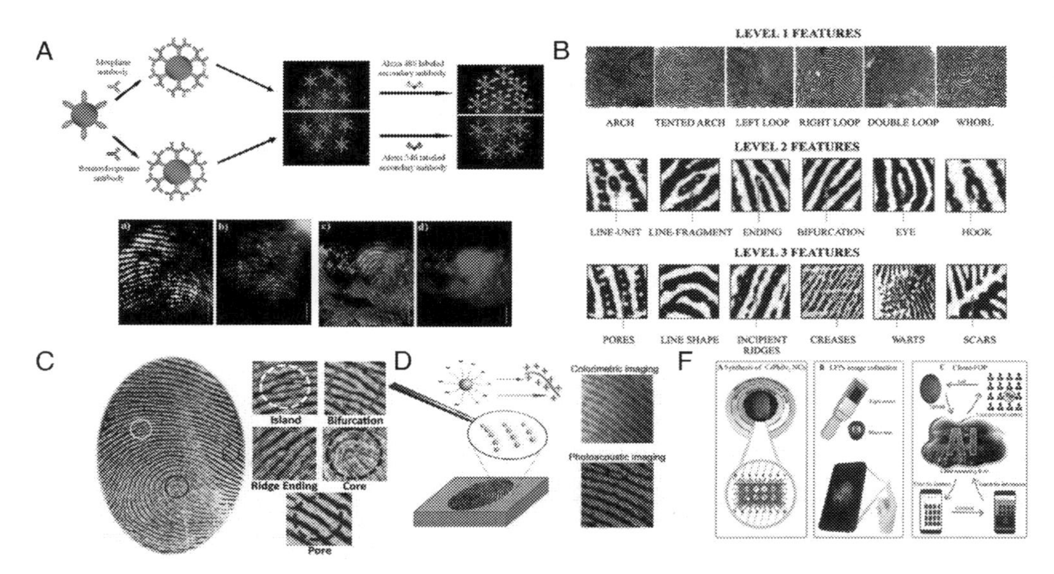

Fig. 7.4 (A) A single fingerprint used for two drug metabolites detection from a single via antibody–magnetic particle conjugates; upper portion for morphine and lower part for benzoylecgonine. Reproduced with permission [168]. American Chemical Society. (B) Various features of levels 1, 2 and 3 fingerprints. Reproduced with permission [169]. Elsevier (C) High-resolution photographs of LFP showing level 1–3 details, including island, bifurcation, ridge ending, core, and pores. Reproduced with permission [170]. Copyright 2017, American Chemical Society. (D) Photoacoustic and colorimetric imaging of LFPs. Reproduced with permission [164]. Copyright 2015, American Chemical Society. (E) Conceptual and individual fingerprint operation platform (named Cloud-FOP) for LFP identification, which integrated $CsPbBr_3$ NCs, Cloud computing,and AI deep neural network strategy. Reproduced with permission [165]. Copyright 2018, American Chemical Society.

information about levels 2 and 3. Level 1 gives macro details about ridge flow and pattern type and is not unique enough for identification. Level 2 having information about ridge endings and ridge bifurcations can adequately discriminate individuals FPs [160]. Level 3 having detailed information about deviation or shape of ridge path, shape and size of pores, the pattern of edge contours such as scars, breaks, creases and many more details are considered as permanent, unique and convey more detailed information for qualitative as well quantitative identification of an individual (Fig. 7.4B) [169]. Several reports highlighted the advancement in LFP detection by using NMs such as QDs and up-conversion NMs (UCNMs) over traditional methods [161]. The significant achievement of using NMs is high contrast due to the enhancement of signal-to-noise ratio, excellent sensitivity due to the small size, large reactive surface area, tunable shape, stickiness, and remarkable selectivity because of several surface modifications and low toxicity. The authors concluded that fluorescent NMs, especially QDs and UCNMs, could serve as a promising platform for LFP development, which supplements the traditional methods.

Recently, various background free luminescence NMs (QDs, metal-doped NMs, metal nanoclusters, dye-labeled NMs etc.) were reported for simultaneous imaging and visualization of LFPs for practical applications in forensic investigation and in-vitro diagnostic applications [162]. Further, fluorescent nano-diamond, which is biocompatible, represents a promising nanomaterial that can give information up to level 3 features with exceptional contrast, sensitivity, and selectivity (Fig. 7.4C) [163,167]. Moreover, several reports based on various classes of NMs also provide a detailed insight into LFPs.

Further, several techniques, coupled with NMs, were also reported for onsite detection. The use of PSMA-b-PS functionalized GNPs with a combined approach of photoacoustic and colorimetric imaging achieved high-resolution LFP visualization up to level 3 with hyperfine features including pores features and ridge contours (Fig. 7.4D) [164]. Remarkable improvement on the use of perovskite nanocrystals helped develop a unique cloud operation platform for a fingerprint to enhance LFP images and presenting on smartphones (Fig. 7.4E) [165].

7.7 Conclusion and future aspects

NMs are smart family of materials and are vital building blocks for numerous nanotechnology devices for translational applications. NMs can be synthesized by simple methods. The size-associated chemical and physical properties of NMs offer immeasurable opportunities for novel discoveries. To preventself-aggregation and agglomeration, NMs can be functionalised with suitable stabilizer to attain required properties for versatile applications. Owing to NMs great electrical, optical and physical properties it can be applicable in chemo-sensing, bio-sensing and this can be further employed into biomedicines, power and energy storage.

Research in NMs is growing with time and there is still space for more development since a huge diversity of NMs possible due to its vibrant chemical nature, morphology and different types of surface modification are available that makes this research area most fascinating.

References

[1] J.F.W. Herschel, On a case of superficial colour presented by a homogeneous liquid internally colourless, Philos. Trans. R. Soc. London 135 (1845) 143–145.
[2] J. Yao, M. Yang, Y. Duan, Chemistry, biology, and medicine of fluorescent nanomaterials and related systems: new insights into biosensing, bioimaging, genomics, diagnostics and therapy, Chem. Rev. 114 (12) (2014) 6130–6178.
[3] H. Kobayashi, M. Ogawa, R. Alford, P.L. Choyke, Y. Urano, New strategies for fluorescent probe design in medical diagnostic imaging, Chem. Rev. 110 (5) (2010) 2620–2640.
[4] A.P. de Silva, H.Q.N. Gunaratne, T. Gunnlaugsson, A.J.M. Huxley, C.P. McCoy, J.T. Rademacher, T.E. Rice, Signaling recognition events with fluorescent sensors and switches, Chem. Rev. 97 (5) (1997) 1515–1566.
[5] Z. Zhao, H. Su, P. Zhang, Y. Cai, R.T.K. Kwok, Y. Chen, Z. He, X. Gu, X. He, H.H.Y. Sung, I.D. Willimas, J.W.Y. Lam, Z. Zhang, B.Z. Tang, Polyyne bridged AIE luminogens with red emission: design, synthesis, properties and applications, J. Mater. Chem. B 5 (8) (2017) 1650–1657.
[6] M. Elsabahy, G.S. Heo, S.-M. Lim, G. Sun, K.L. Wooley, Polymeric nanostructures for imaging and therapy, Chem. Rev. 115 (2015) 10967–11011.
[7] C.M. Tyrakowskia, P.T. Snee, A primer on the synthesis, water-solubilization, and functionalization of quantum dots, their use as biological sensing agents, and present status, Phys. Chem. Chem. Phys. 16 (2014) 837–855.
[8] P. Makhdoumi, H. Karimi, M. Khazae, Review on metal-based nanoparticles: role of reactive oxygen species in renal toxicity, Chem. Res. Toxicol. 33 (2020) 2503–2514.
[9] J.H. Warner, A. Hoshino, K. Yamamoto, R.D. Tilley, Water-soluble photoluminescent silicon quantum dots, Angew. Chem. Int. Ed. 44 (29) (2005) 4550–4554.
[10] J.L. Heinrich, C.L. Curtis, G.M. Credo, M.J. Sailor, K.L. Kavanagh, Luminescent colloidal silicon suspensions from porous silicon, Science 255 (5040) (1992) 66.
[11] P. Li, F.-F. Cheng, W.-W. Xiong, Q. Zhang, New synthetic strategies to prepare metal-organic frameworks, Inorg. Chem. Front. 5 (2018) 2693–2708.
[12] S.B. Revin, S.A. John, Electrochemical sensor for neurotransmitters at physiological pH using a heterocyclic conducting polymer modified electrode, Analyst 137 (1) (2012) 209–215.
[13] N. Meher, P.K. Iyer, Functional group engineering in naphthalimides: a conceptual insight to fine-tune the supramolecular self-assembly and condensed state luminescence, Nanoscale 11 (28) (2019) 13233–13242.
[14] Z. Song, D. Mao, S.H.P. Sung, R.T.K. Kwok, J.W.Y. Lam, D. Kong, D. Ding, B.Z. Tang, Activatable fluorescent nanoprobe with aggregation-induced emission characteristics for selective in vivo imaging of elevated peroxynitrite generation, Adv. Mater. 28 (33) (2016) 7249–7256.
[15] R.N. Dsouza, U. Pischel, W.M. Nau, Fluorescent dyes and their supramolecular host/guest complexes with macrocycles in aqueous solution, Chem. Rev. 111 (12) (2011) 7941–7980.
[16] Z. Wang, S. Chen, J.W.Y. Lam, W. Qin, R.T.K. Kwok, N. Xie, Q. Hu, B.Z. Tang, Long-term fluorescent cellular tracing by the aggregates of AIE bioconjugates, J. Am. Chem. Soc. 135 (22) (2013) 8238–8245.
[17] R.B. Bjorklund, B. Liedberg, Electrically conducting composites of colloidal polypyrrole and methylcellulose, J. Chem. Soc. Chem. Commun. (1986) 1293–1295.
[18] B. Vincent, J. Waterson, Colloidal dispersions of electrically-conducting, spherical polyaniline particles, J. Chem. Soc. Chem. Commun. (1990) 683–684.
[19] M. Mumtaz, C. Labrugère, E. Cloutet, H. Cramail, Synthesis of polyaniline nano-objects using poly(vinyl alcohol)-, poly(ethylene oxide)- and poly[(N-vinyl pyrrolidone)-co-(vinyl alcohol)]-based reactive stabilizers, Langmuir 25 (2009) 13569–13580.

[20] H. Yabu, Self-organized precipitation: an emerging method for preparation of unique polymer particles, Polym. J. 45 (3) (2013) 261–268.
[21] Y. Zhao, C. Shi, X. Yang, B. Shen, Y. Sun, Y. Chen, X. Xu, H. Sun, K. Yu, B. Yang, Q. Lin, pH- and temperature-sensitive hydrogel nanoparticles with dual photoluminescence for bio-probes, ACS Nano 10 (6) (2016) 5856–5863.
[22] P.A. Lovell, F.J. Schork, Fundamentals of emulsion polymerization, Biomacromolecules 21 (2020) 4396–4441.
[23] J.M. Asua, Miniemulsion polymerization, Prog. Polym. Sci. 27 (2002) 1283–1346.
[24] P. Reiss, J. Bleuse, A. Pron, Highly luminescent CdSe/ZnSe core/shell nanocrystals of low size dispersion, Nano Lett. 2 (7) (2002) 781–784.
[25] P. Yang, Z. Zhu, M. Chen, W. Chen, X. Zhou, Microwave-assisted synthesis of xylan-derived carbon quantum dots for tetracycline sensing, Opt. Mater. 85 (2018) 329–336.
[26] K. Damarla, P. Bharmoria, K.S. Rao, P.S. Gehlot, A. Kumar, Illuminating microemulsions: ionic liquid–CdS quantum dots hybrid materials as potential white light harvesting systems, Chem. Commun. 52 (2016) 6320.
[27] C.M. Cobley, J. Chen, E.C. Cho, L.V. Wang, Y. Xia, Gold nanostructures: a class of multifunctional materials for biomedical applications, Chem. Soc. Rev. 40 (1) (2011) 44–56.
[28] M.-C. Daniel, D. Astruc, Gold nanoparticles: assembly, supramolecular chemistry, quantum-size-related properties, and applications toward biology, catalysis, and nanotechnology, Chem. Rev. 104 (1) (2004) 293–346.
[29] D. Liu, Z. Wang, X. Jiang, Gold nanoparticles for the colorimetric and fluorescent detection of ions and small organic molecules, Nanoscale 3 (4) (2011) 1421–1433.
[30] S.K. Ghosh, T. Pal, Interparticle coupling effect on the surface plasmon resonance of gold nanoparticles: from theory to applications, Chem. Rev. 107 (11) (2007) 4797–4862.
[31] X. Ji, X. Song, J. Li, Y. Bai, W. Yang, X. Peng, Size control of gold nanocrystals in citrate reduction: the third role of citrate, J. Am. Chem. Soc. 129 (45) (2007) 13939–13948.
[32] J. Turkevich, P.C. Stevenson, J. Hillier, A study of the nucleation and growth processes in the synthesis of colloidal gold, Discuss. Faraday Soc. 11 (0) (1951) 55–75.
[33] G. Frens, Controlled nucleation for the regulation of the particle size in monodisperse gold suspensions, Nat. Phys. Sci. 241 (105) (1973) 20–22.
[34] E.W Guo, One-Pot, High-yield synthesis of size-controlled gold particles with narrow size distribution, Inorg. Chem. 46 (16) (2007) 6740–6743.
[35] M. Brust, M. Walker, D. Bethell, D.J. Schiffrin, R. Whyman, Synthesis of thiol-derivatised gold nanoparticles in a two-phase liquid–liquid system, J. Chem. Soc. Chem. Commun. 7 (1994) 801–802.
[36] M. Brust, J. Fink, D. Bethell, D.J. Schiffrin, C. Kiely, Synthesis and reactions of functionalised gold nanoparticles, J. Chem. Soc. Chem. Commun. 16 (1995) 1655–1656.
[37] E. Skrabalak, J. Chen, Y. Sun, X. Lu, L. Au, C.M. Cobley, Y. Xia, Gold nanocages: synthesis, properties, and applications, Acc. Chem. Res. 41 (12) (2008) 1587–1595.
[38] H. Feng, Y. Yang, Y. You, G. Li, J. Guo, T. Yu, Z. Shen, T. Wu, B. Xing, Simple and rapid synthesis of ultrathin gold nanowires, their self-assembly and application in surface-enhanced Raman scattering, Chem. Commun. 15 (2009) 1984–1986.
[39] J. Zhang, J. Du, B. Han, Z. Liu, T. Jiang, Z. Zhang, Sonochemical formation of single-crystalline gold nanobelts, Angew. Chem. Int. Ed. 45 (7) (2006) 1116–1119.
[40] X. Sun, S. Dong, E.J.A.C. Wang, Large-scale synthesis of micrometer-scale single-crystalline Au plates of nanometer thickness by a wet-chemical route, Angew. Chem. 116 (46) (2004) 6520–6523.
[41] T. Wang, X. Hu, S. Dong, A renewable SERS substrate prepared by cyclic depositing and stripping of silver shells on gold nanoparticle microtubes, Small 4 (6) (2008) 781–786.
[42] P.C. Lee, D. Meisel, Adsorption and surface-enhanced Raman of dyes on silver and gold sols, J. Phys. Chem. 86 (17) (1982) 3391–3395.
[43] Y. Sun, Y. Xia, Shape-controlled synthesis of gold and silver nanoparticles, Science 298 (5601) (2002) 2176.
[44] Q. Zhang, C. Cobley, L. Au, M. McKiernan, A. Schwartz, L.-P. Wen, J. Chen, Y. Xia, Production of Ag nanocubes on a scale of 0.1 g per batch by protecting the NaHS-mediated polyol synthesis with argon, ACS Appl. Mater. Interfaces 1 (9) (2009) 2044–2048.

[45] Q. Zhang, W. Li, C. Moran, J. Zeng, J. Chen, L.-P. Wen, Y. Xia, Seed-mediated synthesis of Ag nanocubes with controllable edge lengths in the range of 30−200 nm and comparison of their optical properties, J. Am. Chem. Soc. 132 (32) (2010) 11372–11378.
[46] C.J. Orendorff, L. Gearheart, N.R. Jana, C.J. Murphy, Aspect ratio dependence on surface enhanced Raman scattering using silver and gold nanorod substrates, Phys. Chem. Chem. Phys. 8 (1) (2006) 165–170.
[47] S.D Chakraborty, S. Mondal, B. Satpati, U. Pal, S.K. De, M. Bhattacharya, S. Ray, D. Senapati, Wide range morphological transition of silver nanoprisms by selective interaction with As(III): tuning–detuning of surface plasmon offers to decode the mechanism, J. Phys. Chem. C 123 (17) (2019) 11044–11054.
[48] X. Sun, M. Hagner, Novel preparation of snowflake-like dendritic nanostructures of Ag or Au at room temperature via a wet-chemical route, Langmuir 23 (18) (2007) 9147–9150.
[49] Y. Xiong, I. Washio, J. Chen, M. Sadilek, Y.J.A.C. Xia, Trimeric clusters of silver in aqueous $AgNO_3$ solutions and their role as nuclei in forming triangular nanoplates of silver, Angew. Chem. 119 (26) (2007) 5005–5009.
[50] H. Liang, H. Yang, W. Wang, J. Li, H. Xu, High-yield uniform synthesis and microstructure-determination of rice-shaped silver nanocrystals, J. Am. Chem. Soc. 131 (17) (2009) 6068–6069.
[51] J.L. Heinrich, C.L. Curtis, G.M. Credo, M.J. Sailor, K.L. Kavanagh, Luminescent colloidal silicon suspensions from porous silicon, Science 255 (5040) (1992) 66.
[52] H. Nishimura, K. Ritchie, R.S. Kasai, M. Goto, N. Morone, H. Sugimura, K. Tanaka, I. Sase, A. Yoshimura, Y. Nakano, T.K. Fujiwara, A. Kusumi, Biocompatible fluorescent silicon nanocrystals for single-molecule tracking and fluorescence imaging, J. Cell Biol. 202 (6) (2013) 967–983.
[53] M. Dasog, G.B. De los Reyes, L.V. Titova, F.A. Hegmann, J.G.C. Veinot, Size vs surface: tuning the photoluminescence of free standing silicon nanocrystals across the visible spectrum via surface groups, ACS Nano 8 (9) (2014) 9636–9648.
[54] R.S. Wagner, W.C. Ellis, Vapor-liquid-solid mechanism of single crystal growth, Appl. Phys. Lett. 4 (1964) 89–90.
[55] C. Lieber, A. Morales, P. Sheehan, E. Wong, P. Yang, One-dimensional nanostructures: Rational synthesis, novel properties and applications, Proceedings of the Robert A. Welch Foundation 40th Conference on Chemical Research: Chemistry on the Nanometer Scale, 1997, pp. 165–187.
[56] Z. Ni, R.I. Masel, Rapid production of metal−organic frameworks via microwave-assisted solvothermal synthesis, J. Am. Chem. Soc. 128 (2006) 12394–12395.
[57] K.S. Suslick, Sonochemistry, Science 247 (1990) 1439–1445.
[58] W.J. Rieter, K.M.L. Taylor, H.Y. An, W.L. Lin, W.B. Lin, Nanoscale metal-organic frameworks as potential multimodal contrast enhancing agents, J. Am. Chem. Soc. 128 (2006) 9024.
[59] T. Tsuruoka, S. Furukawa, Y. Takashima, K. Yoshida, S. Isoda, S. Kitagawa, Nanoporous nanorods fabricated by coordination modulation and oriented attachment growth, Angew. Chem., Int. Ed. 48 (2009) 4739.
[60] R.H. Baughman, A.A. Zakhidov, W.A. Heer, Carbon nanotubes the route toward applications, Science 297 (5582) (2002) 787.
[61] E.W. Keefer, B.R. Botterman, M.I. Romero, A.F. Rossi, G.W. Gross, Carbon nanotube coating improves neuronal recordings, Nat. Nanotechnol. 3 (7) (2008) 434–439.
[62] A. Eatemadi, H. Daraee, H. Karimkhanloo, M. Kouhi, N. Zarghami, A. Akbarzadeh, M. Abasi, Y. Hanifehpour, S.W. Joo, Carbon nanotubes: properties, synthesis, purification, and medical applications, Nanoscale Res. Lett. 9 (1) (2014) 393.
[63] A. Szabó, C. Perri, A. Csató, G. Giordano, D. Vuono, J.B. Nagy, Synthesis methods of carbon nanotubes and related materials, Materials 3 (5) (2010) 3092–3140.
[64] M. José-Yacamán, M. Miki-Yoshida, L. Rendón, J.G. Santiesteban, Catalytic growth of carbon microtubules with fullerene structure, Appl. Phys. Lett. 62 (6) (1993) 657–659.
[65] A. Thess, R. Lee, P. Nikolaev, H. Dai, P. Petit, J. Robert, C. Xu, Y.H. Lee, S.G. Kim, A.G. Rinzler, D.T. Colbert, G.E. Scuseria, D. Tománek, J.E. Fischer, R.E. Smalley, Crystalline ropes of metallic carbon nanotubes, Science 273 (5274) (1996) 483.
[66] C. Berger, Z. Song, T. Li, X. Li, A.Y. Ogbazghi, R. Feng, Z. Dai, A.N. Marchenkov, E.H. Conrad, P.N. First, W.A. de Heer, Ultrathin epitaxial graphite: 2D electron gas properties and a route toward graphene-based nanoelectronics, J. Phys. Chem. B 108 (52) (2004) 19912–19916.

[67] K.S. Novoselov, A.K. Geim, S.V. Morozov, D. Jiang, Y. Zhang, S.V. Dubonos, I.V. Grigorieva, A.A. Firsov, Electric field effect in atomically thin carbon films, Science 306 (5696) (2004) 666.
[68] A.K. Geim, K.S. Novoselov, The rise of graphene, Nanosci. Technol. 6 (2009) 183–191.
[69] D. Li, M.B. Müller, S. Gilje, R.B. Kaner, G.G. Wallace, Processable aqueous dispersions of graphene nanosheets, Nat. Nanotechnol. 3 (2) (2008) 101–105.
[70] A.B. Bourlinos, D. Gournis, D. Petridis, T. Szabó, A. Szeri, I. Dékány, Graphite oxide:chemical reduction to graphite and surface modification with primary aliphatic amines and amino acids, Langmuir 19 (15) (2003) 6050–6055.
[71] S. Stankovich, R.D. Piner, X. Chen, N. Wu, S.T. Nguyen, R.S. Ruoff, Stable aqueous dispersions of graphitic nanoplatelets via the reduction of exfoliated graphite oxide in the presence of poly(sodium 4-styrenesulfonate), J. Mater. Chem. 16 (2) (2006) 155–158.
[72] A.J. Patil, J.L. Vickery, T.B. Scott, S. Mann, Aqueous stabilization and self-assembly of graphene sheets into layered bio-nanocomposites using DNA, Adv. Mater. 21 (31) (2009) 3159–3164.
[73] W. Zhang, J. Cui, C.a. Tao, Y. Wu, Z. Li, L. Ma, Y. Wen, G.J.A.C. Li, A strategy for producing pure single-layer graphene sheets based on a confined self-assembly approach, Angew. Chem. 121 (32) (2009) 5978–5982.
[74] L.R. Adil, P.K. Iyer, Effects of incorporating regioisomers and flexible rotors to direct aggregation induced emission to achieve stimuli-responsive luminogens, security inks and chemical warfare agent sensors, Chem. Commun. 56 (2020) 7633–7636.
[75] L. Moriau, M. Bele, Ž. Marinko, F.R. -Zepeda, G.K. Podboršek, M. Šala, A.K. Šurca, J. Kovač, I. Arčon, P. Jovanovič, N. Hodnik, L. Suhadolnik, Effect of the morphology of the high-surface-area support on the performance of the oxygen-evolution reaction for iridium nanoparticles, ACS Catal 11 (2021) 670–681.
[76] M.K. Rasmussen, J.N. Pedersen, R. Marie, Size and surface charge characterization of nanoparticles with a salt gradient, Nat. Commun. 11 (1) (2020) 2337–2344.
[77] H. Jalil, U. Pyell, Quantification of zeta-potential and electrokinetic surface charge density for colloidal silica nanoparticles dependent on type and concentration of the counterion: probing the outer helmholtz plane, J. Phys. Chem. C 122 (2018) 4437–4453.
[78] K. Luyts, D. Napierska, B. Nemery, P.H.M. Hoet, How physico-chemical characteristics of nanoparticles cause their toxicity: complex and unresolved interrelations, Environ. Sci.: Processes Impacts 15 (2013) 23–38.
[79] E. Korin, N. Froumin, S. Cohen, Surface analysis of nanocomplexes by X-ray photoelectron spectroscopy (XPS), ACS Biomater. Sci. Eng. 3 (2017) 882–889.
[80] S. Das, A.M. Powe, G.A. Baker, B. Valle, B.E. -Zahab, H.O. Sintim, M. Lowry, S.O. Fakayode, M.E. McCarroll, G. Patonay, M. Li, R.M. Strongin, M.L. Geng, I.M. Warner, Molecular fluorescence, phosphorescence, and chemiluminescence spectrometry, Anal. Chem. 84 (2012) 597–625.
[81] -Q. Luo, S.-C. Zhu, C. Xu, S. Zhou, C.-H. Lam, F.C.-C. Ling, Room temperature ferromagnetism in Sb doped ZnO, J. Magn. Magn. Mater. 529 (2021) 167908–167915.
[82] K. Ahmad, S.M. Mobin, Advanced Functional Nanomaterials for Explosive Sensors, in: O.V. Kharissova, L.M.T. Martínez, B.I. Kharisov (Eds.), Handbook of Nanomaterials and Nanocomposites for Energy and Environmental Applications, Springer International Publishing, Cham. (2020) 1–22.
[83] K.C. To, S. Ben-Jaber, I.P. Parkin, Recent developments in the field of explosive trace detection, ACS Nano 14 (2020) 10804–10833.
[84] L. Tang, H. Feng, J. Cheng, J. Li, Uniform and rich-wrinkled electrophoretic deposited graphene film: a robust electrochemical platform for TNT sensing, Chem. Commun. 46 (2010) 5882–5884.
[85] G.H. Shi, Z. Bin Shang, Y. Wang, W.J. Jin, T.C. Zhang, Fluorescence quenching of CdSe quantum dots by nitroaromatic explosives and their relative compounds, Spectrochim. Acta Part A Mol. Biomol. Spectrosc. 70 (2008) 247–252.
[86] W.J. Peveler, A. Roldan, N. Hollingsworth, M.J. Porter, I.P. Parkin, Multichannel detection and differentiation of explosives with a quantum dot array, ACS Nano 10 (2016) 1139–1146.
[87] X. Wu, H. Hang, H. Li, Y. Chen, H. Tong, L. Wang, Water-dispersible hyperbranched conjugated polymer nanoparticles with sulfonate terminal groups for amplified fluorescence sensing of trace TNT in aqueous solution, Mater. Chem. Front. 1 (2017) 1875–1880.

[88] A. Lichtenstein, E. Havivi, R. Shacham, E. Hahamy, R. Leibovich, A. Pevzner, V. Krivitsky, G. Davivi, I. Presman, R. Elnathan, Y. Engel, E. Flaxer, F. Patolsky, Supersensitive fingerprinting of explosives by chemically modified nanosensors arrays, Nat. Commun. 5 (1) (2014) 4195–5006.

[89] S. Pandey, G.K. Goswami, K.K. Nanda, Nanocomposite based flexible ultrasensitive resistive gas sensor for chemical reactions studies, Sci. Rep. 3 (2013) 1–6.

[90] J. García-Calvo, et al., Surface functionalized silica nanoparticles for the off–on fluorogenic detection of an improvised explosive, TATP, in a vapour flow, J. Mater. Chem. A 6 (2018) 4416–4423.

[91] M. Ahmed, S. Hameed, A. Ihsan, M.M. Naseer, Fluorescent thiazol-substituted pyrazoline nanoparticles for sensitive and highly selective sensing of explosive 2, 4, 6-trinitrophenol in aqueous medium, Sensors Actuators B Chem 248 (2017) 57–62.

[92] A.H. Malik, S. Hussain, A. Kalita, P.K. Iyer, Conjugated polymer nanoparticles for the amplified detection of nitro-explosive picric acid on multiple platforms, ACS Appl. Mater. Interfaces 7 (2015) 26968–26976.

[93] S. Ben-Jaber, et al., Photo-induced enhanced Raman spectroscopy for universal ultra-trace detection of explosives, pollutants and biomolecules, Nat. Commun. 7 (2016) 1–6.

[94] X. Sun, et al., Microwave-assisted ultrafast and facile synthesis of fluorescent carbon nanoparticles from a single precursor: preparation, characterization and their application for the highly selective detection of explosive picric acid, J. Mater. Chem. A 4 (2016) 4161–4171.

[95] Y.D. Ivanov, et al., Ultrasensitive detection of 2, 4-dinitrophenol using nanowire biosensor, J. Nanotechnol. 2018 (2018) 1–6.

[96] R. Li, Y. Yuan, L. Qiu, W. Zhang, J. Zhu, A rational self-sacrificing template route to metal-organic framework nanotubes and reversible vapor-phase detection of nitroaromatic explosives, Small 8 (2012) 225–230.

[97] M.S.S. Bharati, B. Chandu, S.V. Rao, Explosives sensing using Ag-Cu alloy nanoparticles synthesized by femtosecond laser ablation and irradiation, RSC Adv 9 (2019) 1517–1525.

[98] M. Riskin, Y. Ben-Amram, R. Tel-Vered, V. Chegel, J. Almog, I. Willner, Molecularly imprinted Au nanoparticles composites on Au surfaces for the surface plasmon resonance detection of pentaerythritol tetranitrate, nitroglycerin, and ethylene glycol dinitrate, Anal. Chem. 83 (2011) 3082–3088.

[99] Y.Y. Broza, R. Vishinkin, O. Barash, M.K. Nakhleh, H. Haick, Synergy between nanomaterials and volatile organic compounds for non-invasive medical evaluation, Chem. Soc. Rev. 47 (2018) 4781–4859.

[100] S.G.L Mirzaei, G. Neri, Detection of hazardous volatile organic compounds (VOCs) by metal oxide nanostructures-based gas sensors: a review, Ceram. Int. 42 (2016) 15119–15141.

[101] H.-W. Huang, J. Liu, G. He, Y. Peng, M. Wu, W.-H. Zheng, L.-H. Chen, Y. Li, B.-L. Su, Tunable macro–mesoporous ZnO nanostructures for highly sensitive ethanol and acetone gas sensors, RSC Adv 5 (2015) 101910–101916.

[102] W. Chen, Z. Qin, Y. Liu, Y. Zhang, Y. Li, S. Shen, Z.M. Wang, H.-Z. Song, Promotion on acetone sensing of single SnO_2 nanobelt by Eu doping, Nanoscale Res. Lett. 12 (2017) 1–7.

[103] S.-Y. Cho, H.-J. Koh, H.-W. Yoo, J.-S. Kim, H.-T. Jung, Tunable volatile-organic-compound sensor by using Au nanoparticle incorporation on MoS_2, ACS Sensors 2 (2017) 183–189.

[104] V. Postica, A. Vahl, D.S. -Carballal, T. Dankwort, L. Kienle, M. Hoppe, A.C. -Essadek, N.H. de Leeuw, M.-I. Terasa, R. Adelung, F. Faupel, O. Lupan, Tuning ZnO sensors reactivity toward volatile organic compounds via Ag doping and nanoparticle functionalization, ACS Appl. Mater. Interfaces 11 (2019) 31452–31466.

[105] H. Chen, R. Bo, A. Shrestha, B. Xin, N. Nasiri, J. Zhou, I.D. Bernardo, A. Dodd, M. Saunders, J.L. -Duffin, T. White, T. Tsuzuki, A. Tricoli, NiO–ZnO nanoheterojunction networks for room-temperature volatile organic compounds sensing, Adv. Opt. Mater. 6 (2018) 1800677.

[106] H.-C. Liao, C.-P. Hsu, M.-C. Wu, C.-F. Lu, W.-F. Su, Conjugated polymer/nanoparticles nanocomposites for high efficient and real-time volatile organic compounds sensors, Anal. Chem. 85 (2013) 9305–9311.

[107] H.-K. Chang, G.T. Chang, A.K. Thokchom, T. Kim, J. Park, Ultra-fast responsive colloidal–polymer composite-based volatile organic compounds (VOC) sensor using nanoscale easy tear process, Sci. Rep. 8 (2018) 1–11.

[108] M.E. Prévôt, A. Nemati, T.R. Cull, E. Hegmann, T. Hegmann, A zero-power optical, ppt-to ppm-level toxic gas and vapor sensor with image, text, and analytical capabilities, Adv. Mater. Technol. 5 (2020) 2000058–2000068.

[109] M.N. Khatun, A.S. Tanwar, N. Meher, P.K. Iyer, An unprecedented blueshifted naphthalimide AIEE-gen for ultrasensitive detection of 4-nitroaniline in water via "receptor-free" IFE mechanism, Chem. Asian J. 14 (2019) 4725–4731.

[110] M. Ali, N. Tit, Z.H. Yamani, Role of defects and dopants in zinc oxide nanotubes for gas sensing and energy storage applications, Int. J. Energy Res. 44 (2020) 10926–10936.

[111] Z. Li, N. Wang, Z. Lin, J. Wang, W. Liu, K. Sun, Y.Q. Fu, Z. Wang, Room-temperature high-performance H_2S sensor based on porous CuO nanosheets prepared by hydrothermal method, ACS Appl. Mater. Interfaces 8 (2016) 20962–20968.

[112] L.T. Duy, et al., Flexible transparent reduced graphene oxide sensor coupled with organic dye molecules for rapid dual-mode ammonia gas detection, Adv. Funct. Mater. 26 (2016) 4329–4338.

[113] J.A. Buledi, S. Amin, S.I. Haider, M.I. Bhanger, A.R. Solangi, A review on detection of heavy metals from aqueous media using nanomaterial-based sensors, Environ. Sci. Pollut. Res. (2020) 1–9.

[114] B. Kaur, R. Srivastava, B. Satpati, Ultratrace detection of toxic heavy metal ions found in water bodies using hydroxyapatite supported nanocrystalline ZSM-5 modified electrodes, New J. Chem. 39 (2015) 5137–5149.

[115] S. Chatterjee, X. Li, F. Liang, Y. Yang, Design of multifunctional fluorescent hybrid materials based on SiO_2 materials and core–shell Fe_3O_4@ SiO_2 nanoparticles for metal ion sensing, Small 15 (2019) 1904569–1904591.

[116] C. Feng, P. Zhao, L. Wang, T. Yang, Y. Wu, Y. Ding, A. Hu, Fluorescent electronic tongue based on soluble conjugated polymeric nanoparticles for the discrimination of heavy metal ions in aqueous solution, Polym. Chem. 10 (2019) 2256–2262.

[117] S. Venkateswarlu, A.S. Reddy, A. Panda, D. Sarkar, Y. Son, M. Yoon, Reversible fluorescence switching of metal–organic framework nanoparticles for use as security ink and detection of Pb2+ ions in aqueous media, ACS Appl. Nano Mater. 3 (2020) 3684–3692.

[118] M. Ye, J. Beach, J.W. Martin, A. Senthilselvan, Pesticide exposures and respiratory health in general populations, J. Environ. Sci. 51 (2017) 361–370.

[119] J.T. Zacharia, Identity, physical and chemical properties of pesticides, Pestic. Mod. world-trends Pestic, Anal. (2011) 1–18.

[120] G. Aragay, F. Pino, A. Merkoçi, Nanomaterials for sensing and destroying pesticides, Chem. Rev. 112 (2012) 5317–5338.

[121] Y. He, B. Xu, W. Li, H. Yu, Silver nanoparticle-based chemiluminescent sensor array for pesticide discrimination, J. Agric. Food Chem. 63 (2015) 2930–2934.

[122] G. Bapat, C. Labade, A. Chaudhari, S. Zinjarde, Silica nanoparticle based techniques for extraction, detection, and degradation of pesticides, Adv. Colloid Interface Sci. 237 (2016) 1–14.

[123] J.V Kumar, et al., 3D flower-like gadolinium molybdate catalyst for efficient detection and degradation of organophosphate pesticide (fenitrothion), ACS Appl. Mater. Interfaces 10 (2018) 15652–15664.

[124] B. MB, S.R. Manippady, M. Saxena, N.S. John, R.G. Balakrishna, A.K. Samal, Gold nanorods as an efficient substrate for the detection and degradation of pesticides, Langmuir 36 (2020) 7332–7344.

[125] J. Li, Y. Wang, S. Sun, A.M. Lv, K. Jiang, Y. Li, Z. Lia, H. Lin, Disulfide bond-based self-crosslinked carbon-dots for turn-on fluorescence imaging of GSH in living cells, Analyst 145 (2020) 2982–2987.

[126] C. Liu, Y. Cai, J. Wang, X. Liu, H. Ren, L. Yan, Y. Zhang, S. Yang, J. Guo, A. Liu, Facile preparation of homogeneous copper nanoclusters exhibiting excellent tetraenzyme mimetic activities for colorimetric glutathione sensing and fluorimetric ascorbic acid sensing, ACS Appl. Mater. Interface 12 (2020) 42521–42530.

[127] C. Yao, W. Jing, A. Zheng, L. Wu, X. Zhang, X. Liu, A fluorescence sensing platform with the MnO_2 nanosheets as an effective oxidant for glutathione detection, Sens. Actuators B 252 (2017) 30–36.

[128] Y. Liang, L. Xu, K. Tang, Y. Guan, T. Wang, H. Wang, W.W. Yu, Nitrogen-doped carbon dots used as an "on-off-on" fluorescent sensor for Fe_3^+ and glutathione detection, Dyes Pigm. 178 (2020) 108358–108366.

[129] J. Chena, Z. Huanga, H. Menga, L. Zhang, D. Jia, J. Liu, F. Yu, L. Qua, Z. Lia, A facile fluorescence lateral flow biosensor for glutathione detection based on quantum dots-MnO_2 nanocomposites, Sens. Actuators B 260 (2018) 770–777.

[130] Q. Wang, C. Wang, X. Wang, Y. Zhang, Y. Wu, C. Dong, S. Shuang, Construction of CPs@MnO_2-AgNPs as a multifunctional nanosensor for glutathione sensing and cancer theranostics, Nanoscale 11 (2019) 18845–18853.

[131] Y. Gu, Z. Zhao, G. Niu, R. Zhang, H. Zhang, G.G. Shan, H.-T. Feng, R.T.K. Kwok, J.W.Y. Lam, X. Yu, B.Z. Tang, Ratiometric detection of mitochondrial thiol with a two-photon active AIEgen, ACS Appl. Bio Mater. 2 (2019) 3120–3127.

[132] M.-E. Kyriazi, D. Giust, A.H.E. -Sagheer, P.M. Lackie, O.L. Muskens, T. Brown, A.G. Kanaras, Multiplexed mRNA sensing and combinatorial-targeted drug delivery using DNA-Gold nanoparticle dimers, ACS Nano 12 (2018) 3333–3340.

[133] G. Li, Y. Liu, J. Niu, M. Pei, W. Lin, A ratiometric fluorescent composite nanomaterial for RNA detection based on graphene quantum dots and molecular probes, J. Mater. Chem. B 6 (2018) 4380–4384.

[134] D. Li, Y. Wu, C. Gan, R. Yuana, Y. Xiang, Bio-cleavable nanoprobes for target-triggered catalytic hairpin assembly amplification detection of microRNAs in live cancer cells, Nanoscale 10 (2018) 17623–17628.

[135] H.B. Wang, H.Y. Bai, G.L. Dong, Y.M. Liu, DNA-templated Au nanoclusters coupled with proximity-dependent hybridization and guanine-rich DNA induced quenching: a sensitive fluorescent biosensing platform for DNA detection, Nanoscale Adv. 1 (2019) 1482–1488.

[136] J.-H. Choi, J.-W. Choi, Metal-enhanced fluorescence by bifunctional Au nanoparticles for highly sensitive and simple detection of proteolytic enzyme, Nano Lett. 14 (2020) 7100–7107.

[137] J. Shi, Q. Deng, Y. Li, M. Zheng, Z. Chai, C. Wan, Z. Zheng, L. Li, F. Huang, B. Tang, A rapid and ultrasensitive tetraphenylethylene-based probe with aggregation-induced emission for direct detection of α-amylase in human body fluids, Anal. Chem. 90 (2018) 13775–13782.

[138] X. Zhang, C. Ren, F. Hu, Y. Gao, Z. Wang, H. Li, J. Liu, B. Liu, C. Yang, Detection of bacterial alkaline phosphatase activity by enzymatic in situ self-assembly of the AIEgen-peptide conjugate, Anal. Chem. 92 (2020) 5185–5190.

[139] H. Talukdar, S. Kundu, Förster resonance energy transfer-mediated globular protein sensing using polyelectrolyte complex nanoparticles, ACS Omega 4 (2019) 20212–20222.

[140] N. Wiriyachaiporn, P. Srisurat, J. Cherngsuwanwong, N. Sangsing, J. Chonirat, S. Attavitaya, S. Bamrungsap, A colorimetric sensor for protamine detection based on the self-assembly of gold nanorods on graphene oxide, New J. Chem. 43 (2019) 8502–8507.

[141] Y. Tu, Y. Yu, Z. Zhou, S. Xie, B. Yao, S. Guan, B. Situ, Y. Liu, R.T.K. Kwok, J.W.Y. Lam, S. Chen, X. Huang, Z. Zeng, B.Z. Tang, Specific and quantitative detection of albumin in biological fluids by tetrazolate-functionalized water-soluble AIEgens, ACS Appl. Mater. Interfaces 11 (2019) 29619–29629.

[142] R. Ma, M. Xu, C. Liu, G. Shi, J. Deng, T. Zhou, Stimulus response of GQD-sensitized Tb/GMP ICP nanoparticles with dual-responsive ratiometric fluorescence: toward point-of-use analysis of acetylcholinesterase and organophosphorus pesticide poisoning with acetylcholinesterase as a biomarker, ACS Appl. Mater. Interfaces 12 (2020) 42119–42128.

[143] F. Moghzi, J. Soleimannejad, E.C. Sañudo, J. Janczak, Dopamine sensing based on ultrathin fluorescent metal-organic nanosheets, ACS Appl. Mater. Interfaces 12 (2020) 44499–44507.

[144] S.R. Ahmed, A. Chen, In situ enzymatic generation of Gold nanoparticles for nanozymatic label-free detection of acid phosphatase, ACS Appl. Nano Mater. 3 (2020) 9462–9469.

[145] T. Xiao, J. Sun, J. Zhao, S. Wang, G. Liu, X. Yang, FRET effect between fluorescent polydopamine nanoparticles and MnO_2 nanosheets and its application for sensitive sensing of alkaline phosphatase, ACS Appl. Mater. Interfaces 10 (2018) 6560–6569.

[146] L.-H. Xiong, X. He, Z. Zhao, R.T.K. Kwok, Y. Xiong, P.F. Gao, F. Yang, Y. Huang, H.H.-Y. Sung, I.D. Williams, J.W.Y. Lam, J. Cheng, R. Zhang, B.Z. Tang, Ultrasensitive virion immunoassay platform with dual-modality based on a multifunctional aggregation-induced emission luminogen, ACS Nano 12 (2018) 9549–9557.

[147] J. Zhou, R. Fu, F. Tian, Y. Yang, B. Jiao, Y. He, Dual enzyme-induced Au-Ag alloy nanorods as colorful chromogenic substrates for sensitive detection of staphylococcus aureus, ACS Appl. Bio Mater. 3 (2020) 6103–6109.

[148] Y. Tian, Q. Lu, X. Guo, S. Wang, Y. Gao, L. Wang, Au nanoparticles deposited on ultrathin two-dimensional covalent organic framework nanosheets for in vitro and intracellular sensing, Nanoscale 12 (2020) 7776–7781.
[149] Y. Li, Z. Zhao, J. Zhang, R.T.K. Kwok, S. Xie, R. Tang, Y. Jia, J. Yang, L. Wang, J.W.Y. Lam, W. Zheng, X. Jiang, B.Z. Tang, A bifunctional aggregation-induced emission luminogen for monitoring and killing of multidrug-resistant bacteria, Adv. Funct. Mater. 28 (2018) 1804632–1804642.
[150] C. Song, W. Zhao, H. Liu, W. Ding, L. Zhang, J. Wang, Y. Yao, C. Yao, Two-dimensional FeP@C nanosheets as a robust oxidase mimic for fluorescence detection of cysteine and Cu_2^+, J. Mater. Chem. B 8 (2020) 7494–7500.
[151] M. Li, Y.-H. Lao, R.L. Mintz, Z. Chen, D. Shao, H. Hu, H.-X. Wang, Y. Tao, K.W. Leong, A multifunctional mesoporous silica-gold nanocluster hybrid platform for selective breast cancer cell detection using a catalytic amplification-based colorimetric assay, Nanoscale 11 (2019) 2631–2636.
[152] Y. Ma, W. Ai, J. Huang, L. Ma, Y. Geng, X. Liu, X. Wang, Z. Yang, Z. Wang, Mitochondria-targeted sensor array with aggregation-induced emission luminogens for identification of various cells, Anal. Chem. 92 (2020) 14444–14451.
[153] P. Tiwari, N. Kaur, V. Sharma, H. Kang, J. Uddin, S.M. Mobin, Cannabis sativa-derived carbon dots co-doped with N-S: highly efficient nanosensors for temperature and vitamin B_{12}, New J. Chem. 43 (2019) 17058–17068.
[154] X. Shen, L. Xu, W. Zhu, B. Li, J. Hong, X. Zhou, A turn-on fluorescence aptasensor based on carbon dots for sensitive detection of adenosine, New J. Chem. 41 (2017) 9230–9235.
[155] F. Zhang, P.-J.J. Huang, J. Liu, Sensing Adenosine and ATP by aptamers and gold nanoparticles: opposite trends of color change from domination of target adsorption instead of aptamer binding, ACS Sens 5 (2020) 2885–2893.
[156] L. Zhao, Z. Wu, G. Liu, H. Lu, Y. Gao, F. Liu, C. Wang, J. Cui, G. Lu, High-activity Mo, S co-doped carbon quantum dot nanozyme-based cascade colorimetric biosensor for sensitive detection of cholesterol, J. Mater. Chem. B 7 (2019) 7042–7051.
[157] L. Ma, M. Zhang, A. Yang, Q. Wang, F. Qu, F. Qu, R.-M. Kong, Sensitive fluorescence detection of heparin based on self-assembly of mesoporous silica nanoparticle-gold nanoclusters with emission enhancement characteristics, Analyst 143 (2018) 5388–5394.
[158] S. Bamrungsap, J. Cherngsuwanwong, P. Srisurat, J. Chonirat, N. Sangsing, N. Wiriyachaiporn, Visual colorimetric sensing system based on the self-assembly of gold nanorods and graphene oxide for heparin detection using a polycationic polymer as a molecular probe, Anal. Methods 11 (2019) 1387–1392.
[159] A.H. Malik, N. Zehra, M. Ahmad, R. Parui, P.K. Iyer, Advances in conjugated polymers for visualization of latent fingerprints: a critical perspective, New J. Chem. 44 (2020) 19423–19439.
[160] S. Pankanti, S. Prabhakar, A.K. Jain, On the individuality of fingerprints, IEEE Trans. Pattern Anal. Mach. Intell. 24 (2002) 1010–1025.
[161] M. Wang, M. Li, A. Yu, Y. Zhu, M. Yang, C. Mao, Fluorescent nanomaterials for the development of latent fingerprints in forensic sciences, Adv. Funct. Mater. 27 (2017) 1606243.
[162] Y. Wang, J. Wang, Q. Ma, Z. Li, Q. Yuan, Recent progress in background-free latent fingerprint imaging, Nano Res. 11 (2018) 5499–5518.
[163] H.-S. Jung, K.-J. Cho, S.-J. Ryu, Y. Takagi, P.A. Roche, K.C. Neuman, Biocompatible fluorescent nanodiamonds as multifunctional optical probes for latent fingerprint detection, ACS Appl. Mater. Interfaces 12 (2020) 6641–6650.
[164] K. Song, P. Huang, C. Yi, B. Ning, S. Hu, L. Nie, X. Chen, Z. Nie, Photoacoustic and colorimetric visualization of latent fingerprints, ACS Nano 9 (2015) 12344–12348.
[165] M. Li, T. Tian, Y. Zeng, S. Zhu, J. Lu, J. Yang, C. Li, Y. Yin, G. Li, Individual cloud-based fingerprint operation platform for latent fingerprint identification using perovskite nanocrystals as eikonogen, ACS Appl. Mater. Interfaces 12 (2020) 13494–13502.
[166] F.E. Annanouch, Z. Haddi, M. Ling, F. Di Maggio, S. Vallejos, T. Vilic, Y. Zhu, T. Shujah, P. Umek, C. Bittencourt, C. Blackman, E. Llobet, Aerosol-Assisted CVD-Grown PdO Nanoparticle-Decorated Tungsten Oxide Nanoneedles Extremely Sensitive and Selective to Hydrogen, ACS Appl. Mater. Interfaces 8 (16) (2016) 10413–10421, doi:10.1021/acsami.6b00773 27043301.

[167] W. Xuejuan, Y. Shen, L. Haiyang, Y. Youwei, Selective fluorescence sensing of Hg^{2+} and Zn^{2+} ions through dual independent channels based on the site-specific functionalization of mesoporous silica nanoparticles, J. Mater. Chem. A 1 (2013) 10505–10512, doi:10.1039/C3TA11677H.
[168] P. Hazarika, S.M. Jickells, K. Wolff, D.A. Russell, Multiplexed detection of metabolites of narcotic drugs from a single latent fingermark, Anal. Chem. 82 (22) (2020) 9150–9154, doi:10.1021/ac1023205 20968301.
[169] K. Abhishek, A. Yogi, A minutiae count based method for fake fingerprint detection, Procedia Comput. Sci. 58 (2015) 447–452, doi:10.1016/j.procs.2015.08.061.
[170] A.H. Malik, A. Kalita, P.K. Iyer, Development of Well-Preserved, Substrate-Versatile Latent Fingerprints by Aggregation-Induced Enhanced Emission-Active Conjugated Polyelectrolyte, ACS Appl. Mater. Interfaces 9 (42) (2017) 37501–37508, doi:10.1021/acsami.7b13390 28975794.

CHAPTER 8

A comprehensive study toward the treatment of inflammatory diseases through nanoparticles

Maheswata Moharana[a], Satya Narayan Sahu[b], Subrat Kumar Pattanayak[a], Fahmida Khan[a]

[a]Department of Chemistry, National Institute of Technology Raipur, India
[b]School of Applied Sciences, Kalinga Institute of Industrial Technology (KIIT), Deemed to be University, Bhubaneswar, India

8.1 Introduction

Nanoparticles often provide advantages, including process variation time and improved delivery to targeted sites, over small molecule drugs. Immune suppression or immune stimulation, which may enhance or decrease the treatment effects of nanoparticles, may result from the interaction involving nanoparticles and immune cells [1]. The proinflammatory role of activated neutrophils is inactivated by the internalization of drug-loaded albumin nanoparticles into neutrophils, shows potential approach to the treatment of inflammatory diseases arising from insufficient sequestration and activation of neutrophils [2]. Applying of nanoparticles for selective oral drug delivery to inflamed gut tissue has been studied in inflammatory bowel disease [3]. The rolipram solution group had a high adverse effect index, while, as illustrated by a slightly decreased index; the rolipram nanoparticle groups demonstrated their ability for preserving the drug from systemic absorption. Consideration should be provided to nanoparticle deposition in inflamed tissue when developing new carrier systems for the treatment of inflammatory bowel disease [3]. Chitosan, a polymer from the chitin family, has various pharmaceutical and biomedical applications due to non-toxicity, biocompatibility, biodegradability and high drug loading capability [4]. Slow biodegradation of chitosan allows loaded moieties to be controlled and sustained; reduces the frequency of dosing; and is beneficial in infectious drug therapy to ensure better compliance [4]. Inflammation is a leading symptom in both acute and chronic inflammatory disorders. For synthesized nanoparticles have many pharmacological activities [5]. Inflammation is an important immune response which enables survival and maintains tissue homeostasis [6]. Applying of nanomedicines toward the target inflammation, shows potential approach for the treatment of inflammatory diseases, either by detecting molecules expressed on the surface of activated macrophages or endothelial cells, or by the vasculature permeability [6]. Polycaprolactone is a biodegradable polymer used to encompass certain drugs for

Advanced Nanomaterials for Point of Care Diagnosis and Therapy
DOI: https://doi.org/10.1016/B978-0-323-85725-3.00007-6

bioavailability enhancement, targeting, and continuous delivery [7]. Polycaprolactone-based nanoparticles incorporating indomethacin have topical analgesic and anti-inflammatory effects for the symptomatic treatment of inflammatory diseases [7]. Hyaluronic acid and its derivatives have been studied extensively for biological activities, such as tissue engineering, drug delivery and molecular imaging [8]. Self-assembled hyaluronic acid nanoparticles have been widely used with target-specific and long-acting nano carriers to deliver a wide range of therapeutic or diagnostic agents [8]. A polymeric antioxidant prodrug for inflammation-responsive vanillin, called poly(vanillin oxalate) which, integrates H_2O_2-reacting peroxalate ester bonds and bioactive vanillin through acid-responsive acetal linkages in its backbone [9]. It exhibit strong antioxidant activities and also reduces the expression of pro-inflammatory cytokines in activated macrophages by scavenging H_2O_2 and inhibiting the development of reactive oxygen species [9]. Effects on atorvastatin-loaded reactive oxygen lipopolysaccharide nanoparticles and macrophage membrane-coated atorvastatin nanoparticles have been investigated and macrophage membrane-coated atorvastatin nanoparticles have been found to show slightly better therapeutic efficacy in the treatment of inflammatory disease than atorvastatin nanoparticles [10]. Nanomaterials have been used for numerous biological and medical applications, including metal-based nanoparticles [11]. Various properties, such as intracellular biodistribution and cell absorption, cause different cellular responses, leading to different immune responses [11]. Earlier studies looked into the effects of nanoparticles on innate immune cells [11]. Rosmarinic acid, an important polyphenol-based compound, has gained growing attention because of its bioactive properties, including anti-inflammatory activities [12]. Preparation of PEGylated RA-derived nanoparticles and their use as a therapeutic nanomedicine in a dextran sulfate sodium-induced acute colitis mouse model for the treatment of inflammatory bowel disease was studied earlier [12]. Inflammatory bowel disease, which mainly consists of Crohn's disease and ulcerative colitis, is the current and relapsing inflammatory disorder of the gastrointestinal tract [13]. The study of nanoparticles-mediated strategies for treating inflammatory bowel disease, existing problems, and possible directions were studied [13]. One of the key spices used for medicinal purposes, and not just for cooking, is cinnamon. Plant-based silver nanoparticles have recently been investigated in various research areas [14]. Because of anti-inflammatory activity, silver nanoparticles synthesised using cinnamon oil have been used for their anti-inflammatory activity [14]. Respiratory diseases are a growing issue for the ageing population. Nanocarriers are capable of improving effectiveness and reducing systemic toxicity for a large range of drugs. The development of nanocarriers as drug delivery vehicles for the treatment of lung diseases has been studied earlier [15]. They are capable of improving efficacy and reducing systemic toxicity for a wide variety of medications. Recent research has been highlighted in rationally engineered cell membrane-based nanotherapeutics for inflammation therapy [16]. The nanodecoy, produced by the fusion of cellular membrane nanovesicles derived from human monocytes and genetically engineered cells with stable angiotensin converting enzyme II receptors, has the same antigenic appearance as the source cells [17]. The use of functionalized gold nanoparticles to track immune responses has been recorded previously [18]. The findings revealed that nanoparticles with hydrophobic zwitterionic functionality improved inflammatory outcomes while hydrophilic zwitterionic nanoparticles had minimal immunological responses[18]. In neuropathological changes due to Alzheimer's disease, persistent inflammatory behaviors appear to play a central role [19]. Glial cell activation and proliferation, as well as up-regulation of inflammatory mediators and free radicals, are all inflammatory responses seen in the Alzheimer's disease brain. The loading of quercetin into nanoparticles was intended to increase its diffusion into the brain and bioavailability across the blood-brain barrier to target potential cells [19]. Plants, vegetables, and fruits contain phytochemicals, which are natural substances derived from them. To convert metal ions to metal nanoparticles in water, phytochemicals act as reducing agents and stabilizers during nanoparticle synthesis [20]. Carvacro is a phenolic mono terpenoid. It's found in the essential oils of Origanum vulgare and thyme, and it has antimicrobial effects against a number of bacteria [21]. p-Cymene is an aromatic organic compound that occurs naturally. It's present in a variety of essential oils, the most common of which is cumin oil [22]. Lycopene, a non-toxic bright red carotenoid hydrocarbon found in tomatoes, grapefruits, and papayas [23]. Squalene is a naturally occurring unsaturated hydrocarbon. It has antioxidative properties [24]. Here in, we evaluate the interactions in the terms of binding efficacy of the carvacrol, p-cymene, (-)-linalool, lycopene, squalene molecules [25] with antilysozyme antibody protein having PDB ID:1P2C.

8.2 Computational details

The antilysozyme antibody protein's three-dimensional structure was obtained from the RCSB protein data bank (pdb ID: 1P2C). Similarly, from the Pubchem data base with Pubchem ID: 10364, 7463, 443158 and 638072, the analyzed three-dimensional structures of compounds carvacrol, p-cymene, (-)-linalool, and squalene were collected. The physiochemical and possible ADME (absorption, distribution, metabolism and excretion) properties of compounds studied by SwissADME [26] were predicted. Physicochemical characteristics such as molecular weight, partition coefficient (Log P), hydrogen bond donors, hydrogen bond acceptors, number of rotatable bonds (RB), polar surface area (PSA) and water solubility coefficient (Log S) were analyzed for each of the compounds. By using OSIRIS DataWarrior V5.2.1 [27], the toxicological and physicochemical properties of the above compounds are anticipated. In the current research, the three-dimensional protein structure was accepted to be a receptor, while carvacrol, p-cymene, (-)-linalool, and squalene were taken as ligands. Before starting the docking process, we cleaned all the bad contacts from the protein completely. We evaluate the Kollman charges and incorporate the hydrogen atoms to the polar contacts of the receptor with the help of the AutoDock tool [28]. In docking procedure, gradient optimization algorithm was used. As flexible ligands and rigid receptor for the docking phase, the grid box size of 126 × 126 × 126 in x, y and z direction and for center x, y, z are 0.394, 1.663 and 155.816 respectively with spacing of 0.514 Å were measured here. The structural visualization and outcome analysis was conducted via Discovery studio visualizing software [29].

8.3 Results and discussion

8.3.1 Physicochemical and pharmacokinetic properties of studied molecules

The physiochemical properties [30] and potential ADME properties [31–33] of all the molecules evaluated are anticipated here. Molecular weight, hydrogen bond acceptors, hydrogen bond donors, water solubility coefficient (Log S), partition coefficient (Log P), number of rotatable bonds (RB) and polar surface area (PSA) are the properties analyzed for each of the molecules. All of the molecules tested cover the needs for drug similarity properties that can be used to identify lead molecule candidates to inhibit the protein in Table 8.1.

All of the analyzed molecules have a molecular weight lower than 500 daltons. The Log P value is the absorption parameter or the hydrophobicity of the molecule is expressed, the greater the ability of the hydrophobicity to penetrate the cell plasma membrane. All the molecule here display a value of Log P greater than one. The value of the water solubility coefficient, or log S, is an important parameter for determining the pharmacokinetic activity of the lead molecule's distribution and absorption. The value of Log S was reduced to -4.5 to -1. The absorption-related polar surface area (PSA) is estimated to be less than 140 Å2. The number of rotatable bonds is always less than 10, while the minimum rotatable bond implies the best confirmation of the structure. The limits of the Lipinski rules [34–35] are taken into account since the number of acceptors of hydrogen bonds should be less than 10, the number of donors of hydrogen bonds should be less than 5, the molecular weight should be less than 500 daltons and the value of Log P should be within the range of 1 to 5. Gastrointestinal absorption (GI), blood-brain-barrier (BBB), skin permeate (Log Kp), and CYP (CYP1A2, CYP2C19,

Table 8.1 Predicted physicochemical properties of the molecules.

Molecule name	Pubchem Id	Molecular formula	Molecular weight (g/mol)	TPSA ($Å^2$)	No. of RB	Log P	Log S	HBA	HBD
Carvacrol	10364	C10H14O	150.22	20.23	1	2.82	-3.31	1	1
P-Cymene	7463	C10H14	134.22	0.00	1	3.50	-3.63	0	0
(-)-Linalool	443158	C10H18O	154.25	20.23	4	2.66	-2.40	1	1
Squalene	638072	C30H50	410.72	0.00	15	9.38	-8.69	0	0

Table 8.2 Predicted pharmacokinetic properties of studied molecules.

Molecule name	GI absorption	Log K_p in cm/s	BBB permeant	CYP1A2 inhibitor	CYP2C19 inhibitor	CYP2C9 inhibitor	CYP2D6 inhibitor	CYP3A4 inhibitor
Carvacrol	High	-4.74	Yes	Yes	No	No	No	No
P-Cymene	Low	-4.21	Yes	No	No	No	Yes	No
(-)-Linalool	High	-5.13	Yes	No	No	No	No	No
Squalene	Low	-0.58	No	No	No	No	No	No

CYP2C9, CYP2D6, CYP3A4) inhibitors are shown in Table 8.2. The pharmacokinetic properties were analyzed for each of the molecules.

Carvacrol and (-)-Linalool are highly absorbed by the gastrointestinal tract, which determines the amount of substance to be absorbed by the gastrointestinal tract. Compounds' potential to cross the blood-brain barrier is then investigated (BBB). According to the results of the BBB, all of the compounds tested can cross the blood-brain barrier, with the exception of squalene, which can be used as a lead compound. The Log Kp value specifies the capacity of the compounds to impregnate the skin. The most negative importance demonstrates the greater permeability of the skin. We are also investigating the ability of compounds to cross cytochrome P45050 through the biomolecules membranes (CYP). For the studied compound shown in the Table 8.2, the possible risk of toxicity was expected. We found from the overall analysis of the physicochemical properties of all the studied compounds, except squalene, are suitable.

8.3.2 Molecular docking study

Our main focus in this study is the evaluation of the binding energy of interaction between the anti-lysozyme antibody protein with carvacrol, p-cymene, (-)-linalool, and squalene. Formerly molecular docking has been utilized to recognize the binding affinity between the compounds. To measure the binding energy of the receptor and the ligand, molecular docking [36–51] or molecular interaction analysis is performed. Molecular docking was performed to measure the key features of binding performance and types of bonds between antilysozyme antibody protein and studied molecules. The binding performance between anti-lysozyme antibody protein with carvacrol resulting the binding energy -4.9 kcal/mol, in which total 8 numbers of hydrophobic interactions are involved. In which the residue PHE87 was binds in π- sigma and π- π stacked interaction with carbon atom and π-orbitals of carvacrol. The residue PRO44 forms π-alkyl interaction with carbon atoms of the carvacrol, whereas PHE98 was participate in both π-alkyl and π- π stacked interactions with π-orbitals and carbon atoms of carvacrol. The diagrammatic representation of interaction between anti-lysozyme antibody protein and carvacrol was shown in Fig. 8.1.

The binding performance between anti-lysozyme antibody protein with P-Cymene resulting the binding energy -5.1 kcal/mol, in which total eight numbers of hydrophobic interactions are involved. In which the residue PHE87 was binds in π- sigma and π- π stacked interaction with carbon atom and π-orbitals of p-cymene. The residue

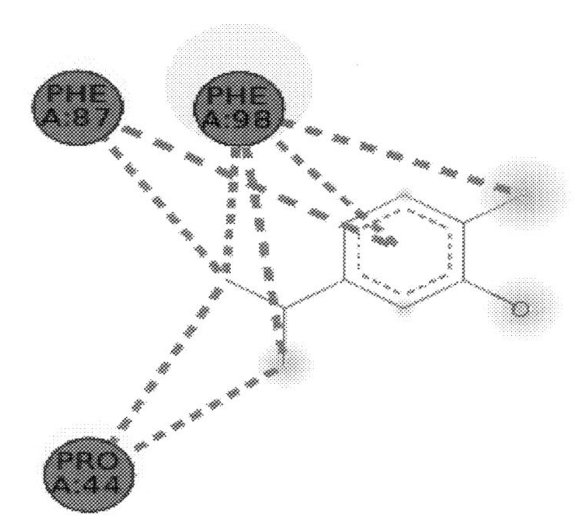

Fig. 8.1 ***The diagrammatic representation of the relationship between carvacrolol and antilysozyme antibody protein.***

PRO44 forms alkyl interaction with carbon atoms of the P-Cymene, whereas PHE98 was participate in both π-alkyl and π- π stacked interactions with pi-orbitals and carbon atoms of p-cymene. The diagrammatic representation of interaction between anti-lysozyme antibody protein and p-cymene was was shown in Fig. 8.2.

The binding performance between anti-lysozyme antibody protein with (-)-Linalool resulting the binding energy -3.9 kcal/mol, in which total four numbers of hydrophobic interactions are involved. In which the residues ARG45, LEU47 and ILE58 was binds in π-alkyl interaction with carbon atom of the (-)-Linalool. The diagrammatic representation of interaction between anti-lysozyme antibody protein and (-)-Linalool was shown in Fig. 8.3.

The binding performance between anti-lysozyme antibody protein with squalene resulting the binding energy -4.6 kcal/mol, in which total four numbers of hydrophobic interactions are involved. In which the residues VAL85, PHE87, PHE98 were binds in alkyl and π-alkyl interaction with carbon atom of the squalene. The diagrammatic representation of interaction between antilysozyme antibody protein and squalene was shown in Fig. 8.4.

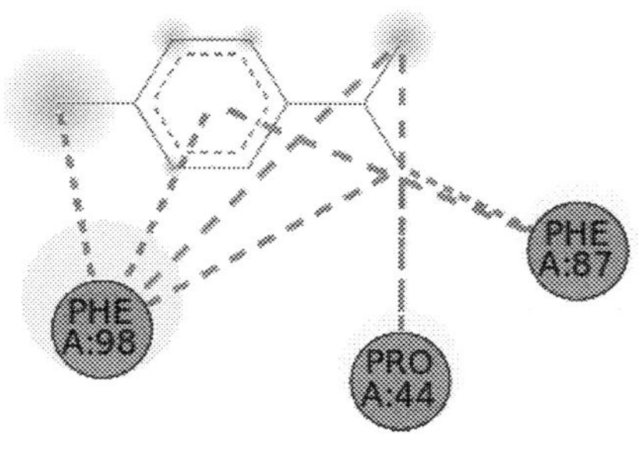

Fig. 8.2 ***The diagrammatic depiction of the interaction between the protein antilysozyme antibody and p-cymene.***

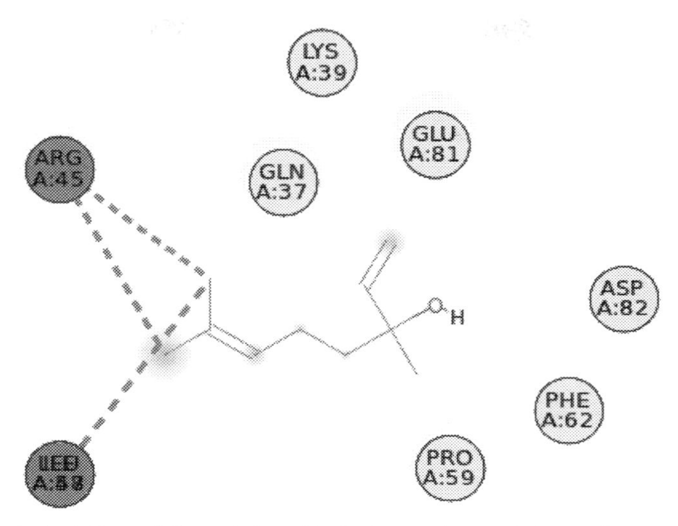

Fig. 8.3 ***Graphical depiction of the relationship between the protein antilysozyme antibody and (-)-linalool.***

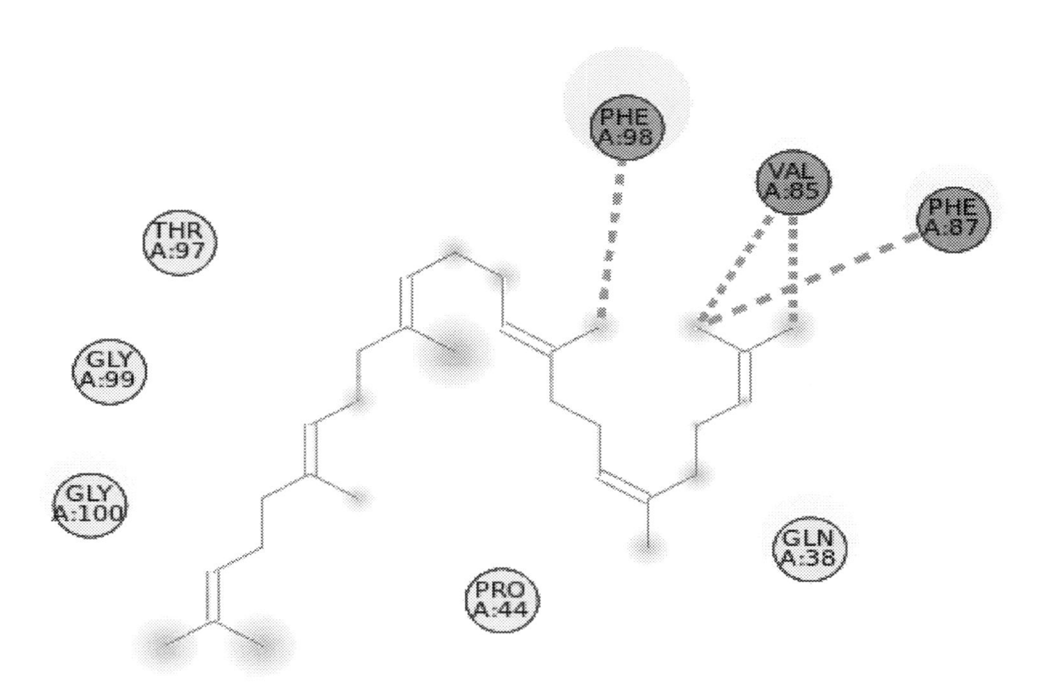

Fig. 8.4 ***Schematic diagram of the relationship between the protein of the antilysozyme antibody and squalene.***

Table 8.3 Molecular interaction of carvacrol, p-cymene. (-)-Linalool and squalene with antilysozyme antibody protein (1P2C).

Chemical compounds	Binding energy (kcal/mol)	Binding residues	Different interactions	Bond length (Å)
Carvacrol	-4.9	UNK0:C - A:PHE87	Hydrophobic	3.85608
		PHE98 -:UNK0	Hydrophobic	3.87479
		PHE87 -:UNK0	Hydrophobic	5.32655
		UNK0:C - A:PRO44	Hydrophobic	4.07016
		UNK0:C - A:PRO44	Hydrophobic	4.11072
		PHE98 -:UNK0:C	Hydrophobic	5.24172
		PHE98 -:UNK0:C	Hydrophobic	4.30681
		PHE98 -:UNK0:C	Hydrophobic	4.45591
P-Cymene	-5.1	UNK0:C - A:PHE87	Hydrophobic	3.86914
		PHE98 -:UNK0	Hydrophobic	3.85824
		PHE87 -:UNK0	Hydrophobic	5.31026
		UNK0:C - A:PRO44	Hydrophobic	4.04088
		UNK0:C - A:PRO44	Hydrophobic	4.09444
		PHE98 -:UNK0:C	Hydrophobic	5.21633
		PHE98 -:UNK0:C	Hydrophobic	4.29592
		PHE98 -:UNK0:C	Hydrophobic	4.4591
(-)-Linalool	-3.9	UNK0:C - A:ARG45	Hydrophobic	3.84061
		UNK0:C - A:ARG45	Hydrophobic	3.89873
		UNK0:C - A:LEU47	Hydrophobic	5.48004
		UNK0:C - A:ILE58	Hydrophobic	4.87815
Squalene	-4.6	UNK0:C - A:VAL85	Hydrophobic	3.98154
		UNK0:C - A:VAL85	Hydrophobic	3.94263
		PHE87 -:UNK0:C	Hydrophobic	4.9039
		PHE98 -:UNK0:C	Hydrophobic	3.96968

In overall from the results of molecular docking interaction between the anti-lysozyme antibody protein with carvacrol, p-cymene, (-)-linalool and squalene shows that the p-cymene have the best binding performance as compare to other nanoparticles because the binding affinity is high in case of p-cymene. The detail results are shown in Table 8.3.

8.4 Conclusions

Our main focus in this study is the evaluation of the binding energy of interaction between the anti-lysozyme antibody protein with carvacrol, p-cymene, (-)-linalool, and squalene. We found from the overall analysis of the physicochemical properties of molecules that all the studied molecules, except squalene, are suitable for the production of nanomedicine. The binding performance between anti-lysozyme antibody protein with carvacrol resulting the binding energy -4.9 kcal/mol, in which total eight numbers of hydrophobic interactions are involved. In which the residue PHE87 was binds in π-sigma and π- π stacked interaction with carbon atom and π-orbitals of carvacrol. The binding performance between antilysozyme antibody protein with P-Cymene resulting the binding energy -5.1 kcal/mol, in which total eight numbers of hydrophobic interactions are involved. In which the residue PHE87 was binds in π- sigma and π- π stacked interaction with carbon atom and π-orbitals of p-cymene. The residue PRO44 forms alkyl interaction with carbon atoms of the p-cymene, whereas PHE98 was participating in both π-alkyl and π- π stacked interactions with pi-orbitals and carbon atoms of p-cymene. The binding performance between anti-lysozyme antibody protein with (-)-Linalool resulting the binding energy -3.9 kcal/mol, in which total four numbers of hydrophobic interactions are involved. In which the residues ARG45, LEU47 and ILE58 was binds in π-alkyl interaction with carbon atom of the (-)-Linalool. The binding performance between antilysozyme antibody protein with squalene resulting the binding energy -4.6 kcal/mol, in which total four numbers of hydrophobic interactions are involved. In which the residues VAL85, PHE87, PHE98 were binds in alkyl and π-alkyl interaction with carbon atom of the squalene. In overall from the results of molecular docking interaction between the anti-lysozyme antibody protein with carvacrol, p-cymene, (-)-linalool, and squalene shows that the p-cymene have the best binding performance.

References

[1] G. Song, J.S. Petschauer, A.J. Madden, W.C. Zamboni, Nanoparticles and the mononuclear phagocyte system: pharmacokinetics and applications for inflammatory diseases, Curr. Rheumatol. Rev. 10 (1) (2014) 22–34.

[2] Z. Wang, J. Li, J. Cho, A.B. Malik, Prevention of vascular inflammation by nanoparticle targeting of adherent neutrophils, Nat. Nanotechnol. 9 (3) (2014) 204–210.

[3] A.L.F. Lamprecht, N. Ubrich, H. Yamamoto, U. Schäfer, H. Takeuchi, P. Maincent, C.M. Lehr, Biodegradable nanoparticles for targeted drug delivery in treatment of inflammatory bowel disease, J. Pharmacol. Exp. Ther. 299 (2) (2001) 775–781.

[4] P. Rajitha, D. Gopinath, R. Biswas, M. Sabitha, R. Jayakumar, Chitosan nanoparticles in drug therapy of infectious and inflammatory diseases, Exp. Opin. Drug Deliv. 13 (8) (2016) 1177–1194.

[5] J. Guo, D. Li, H. Tao, G. Li, R. Liu, Y. Dou, J. Zhang, Cyclodextrin-derived intrinsically bioactive nanoparticles for treatment of acute and chronic inflammatory diseases, Adv. Mater. 31 (46) (2019) 1904607.

[6] R. Brusini, M. Varna, P. Couvreur. Advanced nanomedicines for the treatment of inflammatory diseases, Adv. Drug. Deliv. Rev. 157 (2020) 161–178.

[7] W. Badri, K. Miladi, S. Robin, C. Viennet, Q.A. Nazari, G. Agusti, A. Elaissari, Polycaprolactone based nanoparticles loaded with indomethacin for anti-inflammatory therapy: from preparation to ex vivo study, Pharm. Res. 34 (9) (2017) 1773–1783.

[8] N.V. Rao, J.G. Rho, W. Um, P.K. Ek, V.Q. Nguyen, B.H. Oh, J.H. Park, Hyaluronic acid nanoparticles as nanomedicine for treatment of inflammatory diseases, Pharmaceutics 12 (10) (2020) 931.

[9] J. Kwon, J. Kim, S. Park, G. Khang, P.M. Kang, D. Lee, Inflammation-responsive antioxidant nanoparticles based on a polymeric prodrug of vanillin, Biomacromolecules 14 (5) (2013) 1618–1626.

[10] C. Gao, Q. Huang, C. Liu, C.H. Kwong, L. Yue, J.B. Wan, R. Wang, Treatment of atherosclerosis by macrophage-biomimetic nanoparticles via targeted pharmacotherapy and sequestration of proinflammatory cytokines, Nat. Commun. 11 (1) (2020) 1–14.

[11] I.J. Yu, M. Gulumian, S. Shin, T.H. Yoon, V. Murashov. https://doi.org/10.1155/2015/789312.
[12] C.H. Chung, W. Jung, H. Keum, T.W. Kim, S. Jon, Nanoparticles derived from the natural antioxidant rosmarinic acid ameliorate acute inflammatory bowel disease, ACS Nano 14 (6) (2020) 6887–6896.
[13] C. Yang, D. Merlin, Nanoparticle-mediated drug delivery systems for the treatment of IBD: current perspectives, Int. J. Nanomed. 14 (2019) 8875.
[14] M.T.A. Ka, A. Roy, S. Rajeshkumar, Anti-inflammatory activity of cinnamon oil mediated silver nanoparticles-An in vitro study, Int. J. Res. Pharm. Sci. 10 (4) (2019) 29702972.
[15] S.H. van Rijt, T. Bein, S. Meiners, Medical nanoparticles for next generation drug delivery to the lungs, 2014. doi:10.1183/09031936.00212813.
[16] H. Yan, D. Shao, Y.H. Lao, M. Li, H. Hu, K.W. Leong, Engineering cell membrane-based nanotherapeutics to target inflammation, Adv. Sci. 6 (15) (2019) 1900605.
[17] L. Rao, S. Xia, W. Xu, R. Tian, G. Yu, C. Gu, X. Chen, Decoy nanoparticles protect against COVID-19 by concurrently adsorbing viruses and inflammatory cytokines, Proc. Natl. Acad. Sci. 117 (44) (2020) 27141–27147.
[18] D.F. Moyano, Y. Liu, F. Ayaz, S. Hou, P. Puangploy, B. Duncan, V.M. Rotello, Immunomodulatory effects of coated gold nanoparticles in LPS-stimulated in vitro and in vivo murine model systems, Chem. 1 (2) (2016) 320–327.
[19] G. Testa, P. Gamba, U. Badilli, S. Gargiulo, M. Maina, T. Guina, G. Leonarduzzi, Loading into nanoparticles improves quercetin's efficacy in preventing neuroinflammation induced by oxysterols, PLoS One 9 (5) (2014) e96795.
[20] J. Lee, E.Y. Park, J. Lee, Non-toxic nanoparticles from phytochemicals: preparation and biomedical application, Bioprocess Biosyst. Eng. 37 (6) (2014) 983–989.
[21] M. Sharifi-Rad, E.M. Varoni, M. Iriti, M. Martorell, W.N. Setzer, M. del Mar Contreras, J. Sharifi-Rad, Carvacrol and human health: a comprehensive review, Phytother. Res. 32 (9) (2018) 1675–1687.
[22] M. Viuda-Martos, M.A. Mohamady, J. Fernández-López, K.A. Abd ElRazik, E.A. Omer, J.A. Pérez-Alvarez, E. Sendra, In vitro antioxidant and antibacterial activities of essentials oils obtained from Egyptian aromatic plants, Food Control 22 (11) (2011) 1715–1722.
[23] R.K. Upadhyay, Plant pigments as dietary anticancer agents, Int. J. Green Pharm. 12 (01) (2018). doi:10.22377/ijgp.v12i01.1604.
[24] S.K. Kim, F. Karadeniz, Biological importance and applications of squalene and squalane, Adv. Food. Nutr. Res. 65 (2012) 223–233.
[25] R. Conte, V. Marturano, G. Peluso, A. Calarco, P. Cerruti, Recent advances in nanoparticle-mediated delivery of anti-inflammatory phytocompounds, Int. J. Mol. Sci. 18 (4) (2017) 709.
[26] A. Daina, O. Michielin, V. Zoete, SwissADME: a free web tool to evaluate pharmacokinetics, drug-likeness and medicinal chemistry friendliness of small molecules, Sci. Rep. 7 (1) (2017) 1–13.
[27] T. Sander, J. Freyss, M. von Korff, C. Rufener, DataWarrior: an open-source program for chemistry aware data visualization and analysis, J. Chem. Inf. Model. 55 (2) (2015) 460–473.
[28] R. Huey, G.M. Morris, S. Forli, Using AutoDock 4 and AutoDock vina with AutoDockTools: a tutorial, The Scripps Research Institute Molecular Graphics Laboratory 10550 (2012) 92037 1000.
[29] D. Studio, Discovery Studio. Accelrys [2.1], 2008.
[30] O.A. Raevsky, Physicochemical descriptors in property-based drug design, Mini Rev. Med. Chem. 4 (10) (2004) 1041–1052.
[31] D. Li, L. Chen, Y. Li, S. Tian, H. Sun, T. Hou, ADMET evaluation in drug discovery. 13. Development of in silico prediction models for P-glycoprotein substrates, Mol. Pharm. 11 (3) (2014) 716–726.
[32] J. Shen, F. Cheng, Y. Xu, W. Li, Y. Tang, Estimation of ADME properties with substructure pattern recognition, J. Chem. Inf. Model. 50 (6) (2010) 1034–1041.
[33] E.L. Luzina, A.V. Popov, Synthesis of 1-aroyl (1-arylsulfonyl)-4-bis (trifluoromethyl) alkyl semicarbazides as potential physiologically active compounds, J. Fluorine Chem. 148 (2013) 41–48.
[34] M.J. Waring, Defining optimum lipophilicity and molecular weight ranges for drug candidates—molecular weight dependent lower log D limits based on permeability, Bioorg. Med. Chem. Lett. 19 (10) (2009) 2844–2851.
[35] P.R. Duchowicz, A. Talevi, C. Bellera, L.E. Bruno-Blanch, E.A. Castro, Application of descriptors based on Lipinski's rules in the QSPR study of aqueous solubilities, Bioorg. Med. Chem. 15 (11) (2007) 3711–3719.

[36] J. Panda, S.N. Sahu, J.K. Sahoo, S.P. Biswal, S.K. Pattanayak, R. Samantaray, R. Sahu, Efficient removal of two anionic dyes by a highly robust zirconium based metal organic framework from aqueous medium: Experimental findings with molecular docking study, Environ. Nanotechnol. Monit. Manag. 14 (2020) 100340.
[37] S.N. Sahu, B. Mishra, R. Sahu, S.K. Pattanayak, Molecular dynamics simulation perception study of the binding affinity performance for main protease of SARS-CoV-2, J. Biomol. Struct. Dyn. (2020). doi:10.1080/07391102.2020.1850362.
[38] S.N. Sahu, R. Sahu, S.K. Pattanayak, Molecular interaction study of phytochemicals with native and mutant protein related to nephrotic syndrome, AIP Conference Proceedings, 2270, AIP Publishing LLC, 2020 020002.
[39] R. Panda, S.N. Sahu, F. Khan, S.K. Pattanayak, Binding performance of phytochemicals with mutant threonine-protein kinase Chk2 protein: An in silico study, AIP Conference Proceedings, 2270, AIP Publishing LLC, 2020 020003.
[40] J. Panda, S.N. Sahu, R. Pati, P.K. Panda, B.C. Tripathy, S.K. Pattanayak, R. Sahu, Role of pore volume and surface area of Cu-BTC and MIL-100 (Fe) metal-organic frameworks on the loading of rifampicin: collective experimental and docking study, ChemistrySelect 5 (40) (2020) 12398–12406.
[41] S. Sahoo, S.N. Sahu, S. kumar Pattanayak, N. Misra, M. Suar, Biosensor and its implementation in diagnosis of infectious diseases. In: Smart Biosensors in Medical Care, Academic Press, 2020, pp. 29–47. https://doi.org/10.1016/B978-0-12-820781-9.00002-4.
[42] S.N. Sahu, M. Moharana, P.C. Prusti, S. Chakrabarty, F. Khan, S.K. Pattanayak, Real-time data analytics in healthcare using the Internet of Things. In: Real-Time Data Analytics for Large Scale Sensor Data, Academic Press, 2020, pp. 37–50. https://doi.org/10.1016/B978-0-12-818014-3.00002-4.
[43] M. Moharana, S.K. Pattanayak, Biosensors: a better biomarker for diseases diagnosis. In: Smart Biosensors in Medical Care, Academic Press, 2020, pp. 49–64. https://doi.org/10.1016/B978-0-12-820781-9.00003-6.
[44] S.N. Sahu, J. Panda, R. Sahu, T. Sahoo, S. Chakrabarty, S.K. Pattanayak, Healthcare information technology for rural healthcare development: insight into bioinformatics techniques. In: Internet of Things, Smart Computing and Technology: A Roadmap Ahead, Springer, Cham, 2020, pp. 151–169.
[45] S.N. Sahu, S.K. Pattanayak, Molecular docking and molecular dynamics simulation studies on PLCE1 encoded protein, J. Mol. Struct. 1198 (2019) 126936.
[46] S.N. Sahu, M. Moharana, R. Sahu, S.K. Pattanayak, Molecular docking approach study of binding performance of antifungal proteins, AIP Conference Proceedings, 2142, AIP Publishing LLC, 2019 060001.
[47] S.N. Sahu, M. Moharana, R. Sahu, S.K. Pattanayak, Impact of mutation on podocin protein involved in type 2 nephrotic syndrome: insights into docking and molecular dynamics simulation study, J. Mol. Liq. 281 (2019) 549–562.
[48] S.M. Hiremath, A. Suvitha, N.R. Patil, C.S. Hiremath, S.S. Khemalapure, S.K. Pattanayak, S. Armaković, Synthesis of 5-(5-methyl-benzofuran-3-ylmethyl)-3H-[1, 3, 4] oxadiazole-2-thione and investigation of its spectroscopic, reactivity, optoelectronic and drug likeness properties by combined computational and experimental approach, Spectrochim. Acta Part A 205 (2018) 95–110.
[49] S.N. Sahu, M. Moharana, S.R. Martha, A. Bissoyi, P.K. Maharana, S.K. Pattanayak, Computational biology approach in management of big data of healthcare sector. In: Big Data Analytics for Intelligent Healthcare Management, Academic Press, 2019, pp. 247–267. https://doi.org/10.1016/B978-0-12-818146-1.00010-6.
[50] A. Chand, P. Chettiyankandy, M. Moharana, S.N. Sahu, S.K. Pradhan, S.K. Pattanayak, S. Chowdhuri, Computational methods for developing novel antiaging interventions. In: Molecular Basis and Emerging Strategies for Anti-aging Interventions, Springer, Singapore, 2018, pp. 175–193.
[51] A. Bissoyi, S.K. Pattanayak, A. Bit, A. Patel, A.K. Singh, S.S. Behera, D. Satpathy, Alphavirus nonstructural proteases and their inhibitors. In: Viral proteases and their inhibitors, Academic Press, 2017, pp. 77–104. https://doi.org/10.1016/B978-0-12-809712-0.00004-6.

CHAPTER 9

Recent advances of nanomaterial sensor for point-of care diagnostics applications and research

Anshebo G. Alemu[a], Anshebo T. Alemu[b]
[a]Faculty of Natural & computational science, Department of Physics, Samara University, Samara, Ethiopia
[b]Faculty of Engineering, Adama Science and Technology University, Ethiopia

9.1 Introduction

Recently medical analysis has been meaningfully improved due to the advancement of novel procedures capable of carrying out recognition and quantification of explicit molecules and components, which provide valuable information about the physiological state in the absence of living being [1]. In past decades these techniques, diagnostics were based on just observable parameters in the presence of substances diagnosis represent the opportunity to perform more accurate estimates and more dynamic studies of the state of the patient. Nowadays, nanomaterial-based sensors have gotten extreme attention for healthcare applications due to their exceptional features such as elasticity, flexibility, multifunctionality, portability, simplicity, and outstanding efficiency. These developing sensors represents the incorporation of biotechnology, chemistry, material science, biomedical and molecular engineering can prominently advancement the specificity and sensitivity of biological molecule recognition, hold the ability of sensing atoms or molecules, and have countless potential in application such as living molecule biomarker, environment intensive care and pathogen diagnosis. Furthermore, the invention of unique techniques, diagnostics was based on apparent parameters. Though, solid information of other parameters to provide the diagnosis the opportunity to perform precise estimate and energetic investigation of the patient. This opportunity of assembly particular and solid evidence about the patient by computing specific molecular mechanisms needs to be supplemented by profound statistic of the philosophies associated to the physical state and the mechanism involved in the diseases. Additional way to distinguish which components should be measured to make an interpretation of the results obtained the situation of the patient. Biomarkers can be distinct constraints that can be accurately estimated in order to grow information of physiologic response to beneficial intervention [2,3].

Recently an increasing number of sensors have been explored for application on point-of-care diagnostics in health care. Their specific biochemical reactions intermediated by inaccessible enzymes, immune systems, tissues, organelles as well as entire cells to

Advanced Nanomaterials for Point of Care Diagnosis and Therapy
DOI: https://doi.org/10.1016/B978-0-323-85725-3.00009-X

detect by electrical or optical signals. The detection methods of sensors can be completely mechanized, which boosted reproducibility, permit real-time and quick analysis shows an option for cyclic application as a result of superficial regeneration [4]. Nanosensors acting critical roleof biodefense [5], food toxicity [6], fermentations [7], environmental safety [8], medical diagnostics [9,10], and plant biology [11,12]. Furthermore, cardiac sickness, lung cancerand like infectious diseases are the foremost reasons of death worldwide [10]. Nanosensor, amplifies prominence in the area of healthcare diagnostics by providing that operator-friendly, inexpensive, consistent, and quick sensing stands [9]. They have significant advantages in comparison to detection concerning chromatography. These include an exclusion of skilled operating personnel, quicker response times, portability, and higher sensitivity [13]. For instance, the time of analysis has condensed from 48-72 h to 300 sec with the assistance of current nano sensors.

Nano sensors constitute the utmost generally used group of medical sensors have many advantages over traditional analytical methods. They permit of numerous different analysts and extremely specific, sensitive and slightly in size. These nano sensors are applicable in several interdisciplinary fields such as molecular biology, chemistry, micro electromechanical arrangements, nanotechnology and micro-electronics. Recent progresses in the fields of biology, genomics and proteomics, have empowered understanding of analytic ligand interactions, the binding kinetics, dissociation-association rates, and cross reactivity [18].

9.2 Working principles of nanomaterial sensors

The working principal nanosensors have enhanced their sensitivity of analytic mixture/culture which permits innovative indicator transduction technologies in sensors. The advancement of nano sensors with enormously insignificant molecules that want to be investigated [14,15]. According to (IUPAC) description nanosensors have three authoritative elements such as receptor, transducer, and detector which convert the signal into useful data as shown in Fig. 9.1. These analytical devices for detecting an extensive

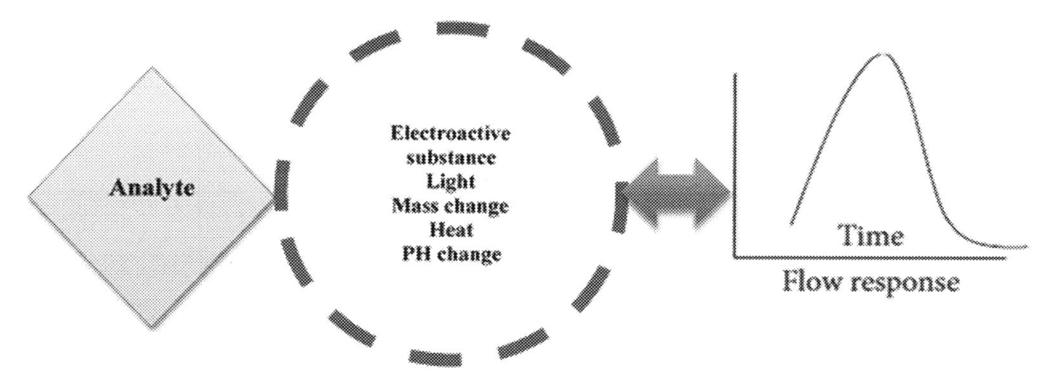

Fig. 9.1 ***A graphic diagram showing a typical nanasensor [12].***

series of analyses such as antibodies, DNA, enzymes, microorganisms, and tissues. The function transducer generates an indicator transforms the recognition into computable signal, used for the quantification of the analyst. A nanosensor senses an adjustable quantity, usually automatically translates the measurement into precise indicators. The most significant requirements of nano sensors are accuracy, diversity, sensitivity, selectivity, and stability [16,17]. Nanosensors for monitoring chemical and physical spectacles distinguishing in cellular organelles, determining nano level constituent part in the environment and industry Fig. 9.1.

9.3 Types of nanosensors

A nanosensor is an instrument work out changes in physical incitement and change them into indicators. In fact, a transducer that adapts a measurement into a signal to data. These signals can be recorded for processing such as optical, electrical, and mechanical. Nano Sensors can be grouped into by four class based on (1) the input physical number, (2) the output operating principles, (3) the type of the signal, and (4) Assembling data methods. Most research articles on nonmaterial sensors are categories based on application, energy source, and structure [20].

The purpose of this book chapter is to provide a summary of different types and classifications of non-material-based sensors, associating the characteristics and key differences among the various groups in addition to different nano materials used sensor and their application [21,22]. Other factors to group, nanomaterial sensors are their size, quantum effect, and structural differences, there are various classifications of nanomaterial sensors. As demonstrated in Table 9.1 and Fig. 9.2 nanomaterial sensors can be grouped according to its application, energy source, quantum properties, structure their applications and their properties [12,15].

9.4 Dimension of nanomaterials sensor

Recently nanotechnology has been rapidly increasing field. Through fluctuating the atomic arrangement, chemical composition, various nano materials with variations in physicochemical properties such as such, as crystalline nature, size, shape, and interaction with biological systems are produced [70]. The nanotechnology comes from our capability to gather innovative materials at the nanoscale and exercise exquisite control over their properties in the 21st century. Nowadays, nanomaterials that comprise of basic structural elements, fibers, grains, particles, sheets, or extra fundamental components. Generally, nanomaterials grouped into 0-dimensional (0D), 1-dimensional (1D), 2-dimensional (2D), and 3-dimensional (3D) in terms of dimensional characteristics. Dimensions refer to the number of dimensions [71]. In this book chapter, we will review the fundamental difference of 0-D two 3-D dimensions from the research perspective and their applications in Table 9.2.

Table 9.1 Types of nanomaterials sensors.

Type based	Sub group	Application
Application	Biosensors	For point-of-care diagnostics, for cancer detection, of various diseases, DNA detection
	Chemo sensors [15]	For pH at various ion concentrations.
	Deployable [15]	For military lightweight, portable chemical detection,
	Electron sensors [15]	for detection of electrode
Energy source [15]	Active sensors	It needs an energy source.
	Passive sensors	It does not demand energy source such as a thermocouple, and piezoelectric sensor.
Quantum properties	Carbon-based Fullerene [25,26] Nanotubes [27,28] Graphene derivatives [35–37] Carbon dots [38] Nano diamonds [40]	For various biological applications, such as cancer therapy, tissue engineering, drug delivery, bioimaging, and biosensing [23,24], used for the recognition of cysteine [28], glycaemic biomarkers [29], cholesterol [30], neurotransmitters [31], H_2O_2 [32], cancer cells [33], nucleic acids [30], pharmaceutical drugs [32], infectious bacteria [33,34] for medical diagnostics and bioimaging [39], used as biosensing, as well as bioimaging probes and contrast agents [41]
In organic	Quantum dots [42] Magnetic nanoparticles [46,47] Gold nanoparticles [48] Silver nanoparticles, [55] Nanoshells, nanowires and nanocages [57]	used for the finding of a variety of molecules, such as proteins [42], pathogens [43], lung cancer biomarkers [44] and nucleic acids [45], the food industry [46] medical diagnosis, and environmental investigations [47], used for biomarkers for cancer [49], neurological disorders [49], diabetes mellitus [50], nucleic acids [51], amino-acids [52], hemoglobin [53], and variety of pathogens [54]. used for SERS-based biosensors [56]. on electrochemical electrolysis [58]
Organic	Nanofilms [59] Nanogels [60] Dendrimers [61] Hyperbranched polymers [62] Molecular machine [95] Polymer nanocomposites [67]	For sensing glycopolymers [59], detection of nephrological diseases [59], for encapsulation vehicles; responsive materials; and sensory membrane [60], for detection of cysteine [60] sensing for medical diagnosis [61], Biosensing [62], thrombin detection [62], for medical diagnosis applications [63], used an anticancer drug [64], detect the sepsis biomarker [65,66] for glucose biosensing [100], for leukemia biomarker [67].
Structure	Electromagnetic [18,19]	Used to detect specific biomolecules enzymatic activity, and pathogens [18,68]
	Optical sensors [18,19]	Used in biotechnology, chemical industry, medicine, ecological sciences, and human guard, for pH measurement., toxins, proteins, pathogens, disease biomarkers [18,19]
	Mechanical	Applicable on monitoring the vibrational and elastic properties and microelectronic devices [18,69]

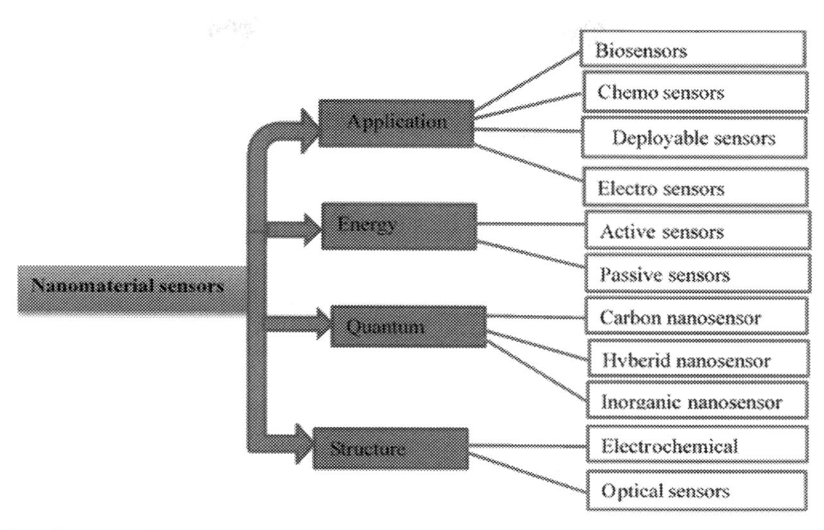

Fig. 9.2 *Classification of nanomaterial sensors [15,18].*

Table 9.2 (0D) nanomaterials sensors [75–82].

No	(0D) nanomaterials	Application	Reference
1	Carbon-based	Used for imaging, sensing and drug delivery	[75]
	Graphene Quantum Dots (GQDs)	Used for ions detection, biomarkers of ascorbic acid (AA), dopamine, DNA and amino acid and to diagnose cancer	[72]
	Carbon Quantum Dots (CQDs)	Used for bioimaging, drug delivery, biosensing, to detect intracellular biomolecules, used in cancer diagnosis.	[73]
	Fullerenes	Used for detect amino acids, ion, biomolecular recognition and disease diagnosis,	[74]
2	Inorganic Quantum dot	Used for pathogen detection, for cell detection, to detect biomarkers of cancer and brain diseases, to detection of microRNA in prostate cancer, to imaging in clinical applications, detection of influenza viruses	[76]
3	Magnetic nanoparticles	Used for biosensing and drug delivery, to detection of proteins, disease biomarkers and pathogens, to detect cancer diagnosis, for pathogen detection,	[77]
4	Noble metal nanoparticles	Used for biomolecular detection and disease diagnosis	[78]
	Gold Nanoparticles	Used for detecting cancer, neurological diseases, proteins, nucleic acids, and various pathogens, pathogen detection, to detect influenza viruses	[79]
	Silver Nanoparticles	Used for detection of cancer, markers, pathogens and drug analytes (Fig. 9.4)	[80]
5	Other types	Applicable for biosensing	[81]
	Polymer Dots	Used for evaluate anti-cancer drug efficacy	[81]
	Upconversion Nanoparticles	Used for biomolecular detection, for disease diagnosis	[82]

9.4.1 0D nanomaterials sensors

9.4.2 1D nanomaterials sensors

One-dimensional (1D) nanomaterials are being broadly employed for the restructure of sensors, and energy electronics due to 1D nanoassemblies have a fundamentally high-characteristic-fraction that empowers the structure of a conductive clarified network with slight quantity of material used while sustaining high optoelectronic efficiency. The number of energetic materials such as conductive, polymer thin films and indium tin oxide 1D nanomaterials provides exceptional advantages because of their inherent structural feature with higher, electrical, mechanical and optical properties, which permits the building of an optimal efficiency of sensors, and energy loading devices. In the context of point-of-care diagnostics 1D nanomaterials describes an optically active nanostructure with sizes between 1 nm to 100 nm. These 1D nanomaterials have inimitable properties such as good transparency, trivial, excellent sensitivity, extreme slenderness, low cost, and probable for large area construction. These have made important contributions to the sensors electronic systems by developing optoelectronic efficiency with affordable cost. 1D nanomaterials include carbon nanotubes, metallic nanowires, metal oxide nanowires, conducting polymer, nanowires, nanocages, and nanoshells [83].

9.4.3 2D nanomaterials sensors

The new innovative sensors with sensitivity, good fussiness, and fast detection present various encounters. In the previous decade, microelectronic sensors, constructed with field-effect transistors have been broadly investigated because their excellent sensitivity, swift detection, and understandable test technique. Among extraordinary sensors, 2D nano materials-based sensors have been proved with wonderful capacity for the recognition of a widespread variety of analysts which is ascribed to the exceptional structural and microelectronic characteristics of 2D nanomaterials-based sensors. The advantage of two-dimensional (2D) nanomaterials sensor can lead to conformal and close contact with metal electrodes, and they are easier to operate because of their relatively bulky adjacent sizes, simplifying control over the channel assembly in the sensor. There are numerous unique features of 2D nanometers, compared with diverse dimensionality such as electron imprisonment without exhume layer interactions, mechanical flexibility and optical transparency [93]. 2D nanomaterial devices have engrossed the prodigious attention as supportive substrates in an inclusive diversity of biosensor technologies and provide a very high density of active surface sites over a large area, making them ideal for biochemical sensing applications as shown Fig. 9.3 [94].

9.4.4 3D nanomaterials sensors

Recently the standing of 3D printing procedures has flowered in the sensing world due to their important advantages of fast fabrication, easy availability, dispensation of diverse

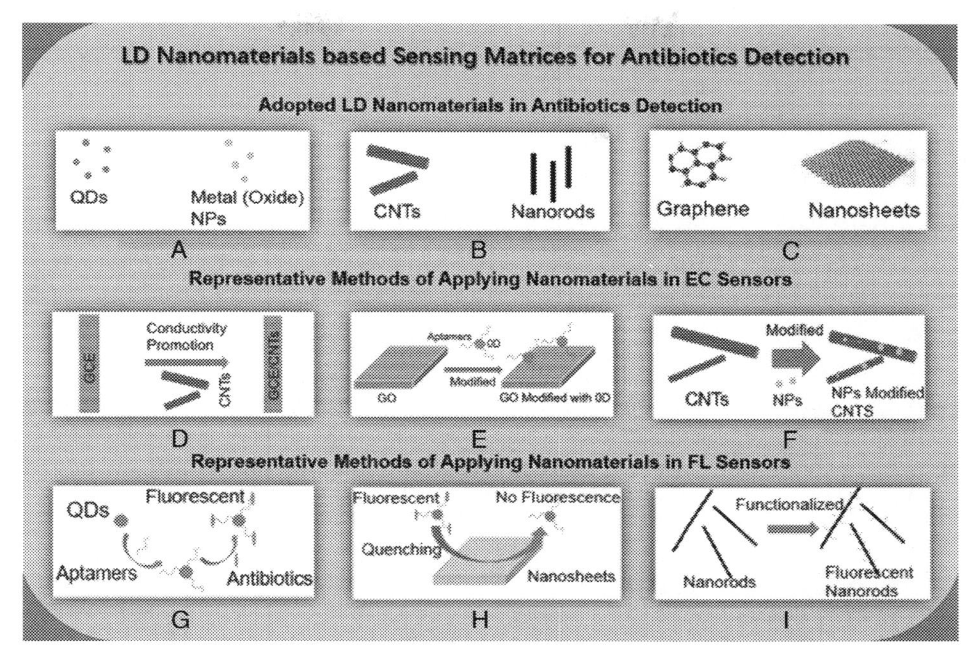

Fig. 9.3 ***Graphic diagram of LD nanomaterials for EC and FL.*** (A) 0D nanomaterials. (B) 1D nanomaterials used in the construction of sensors for antibiotic sensor. (C) 2D nanomaterials sensors for antibiotic detection. (D) A nanomaterial in EC sensors: CNTs. (E) EC sensorsmodify the electrode. (F) EC sensors: use NPs to modify carbon nanotubes. (G) A nanomaterials in FL sensors. (H) Nanomaterials in FL sensors, the 2D nanosheet. (I) A nanomaterials in FL sensors: functionalized nanorods fluoresce in response to specific antibiotics [72].

resources and sustainability. In this book chapter, we review the influence of 3D nano sensor techniques for the manufacture of sensors for diverse healthcare applications Table 9.3.

Table 9.3 (1D) nanomaterials sensors [84–92].

No	1D nanomaterials	Applications	Reference
1	Carbon nanotubes (CNTs)	Pressure sensors, Strain sensors	[84]
2	Silver Nanowires	Pressure sensors, Strain sensors	[85]
3	Copper Nanowires	Pressure sensors, Strain sensors	[86]
4	Gold Nanowires	Pressure sensors, Strain sensors	[87]
5	Zinc Oxide Nanowires	Pressure sensors, Strain sensors, including temperature and UV monitoring	[88]
6	Polymeric Nanomaterials	Strain sensors	[89]
7	Hybrid Structures	Strain sensors	[90]
8	Nanocages	Used in biosensing, for detected glucose	[91]
9	Nanos hells	Ideal components for optical sensors, functionalize biomolecules	[92]

9.5 Point-of-care applications

Nanomaterial sensors have already led to advances in the domain of sensors offer better-superiority health care, it is identical significant standards of health care by appropriate conclusions based on swift analyses, smart data investigation, and informatics. Therefore Point-of-care testing confirms that fast investigation of analysts near to the patients, enabling an improved disease diagnosis, monitoring. It also empowers speedy medical results since the diseases can be analyzed at an early stage, which leads to enhanced health conclusions for the patients empowering them to start primary treatment. In the recent decades, several potential point-of-care apparatus have been advanced and they are ways to succeed-generation point-of-care testing. Nanosensors are very significant components of point-of-care instruments are straight accountable for the bioanalytical performance of an essay. As prospective of point-of-care applications necessary to modified healthcare management typically approximation the levels of any chemical response by creating indicators mainly related with the amount of substance of an analyte and hence can notice disease causing markers of body fluids. The point-of-care diagnosis as high selectivity and sensitivity have permitted for premature diagnosis of targeted diseases; facilitating timely remedy decisions and combination with nanosensors can advance assessment of the disease onset and its progression and help to plan for treatment of many diseases. Nanosensors are considered as key devices for point-of-care diagnosis due to their automatic controllability of precise investigation for various sensing applications [105]. The field of nanotechnology renovating the scientific community, principally because of their extraordinary biological, chemical and physical, properties, has a varied diversity of applications in the spinalization of biomedical, medical imaging, catalysis, optical and electronics [106]. This book chapter explores the current progress of nanomaterial-based sensors in medical application specifically on point-of-care diagnostics.

9.5.1 Diagnosis of cancer

Among several cases, cancer is the utmost reason of death, in the developing countries as well as in the world. According to the WHO report, in the current world, there are over two hundred types of cancers, the most public types include, bladder cancer, breast cancer, colorectal cancer, esophageal cancer, kidney cancer, lung cancer, liver cancer, lymphoma, ovarian cancer, pancreatic cancer, prostate cancer, skin cancer and thyroid cancer [107]. Presently, there exist restricted devices for primary screening, diagnosis, and intensive care of cancer development. The model of point care, diagnostic instrument is movable and assures reliability. This new device would offer continuous, low cost, real time in monitoring of cancer, which provide initial analysis, drug effectiveness, and operative drug distribution [108]. According to N- Kenari et al. [107] report portable nanosensors for detection of cancer in blood vessels work focused on cancer cells placed in specific regions of the blood vessel and their exposure was based on invention and the emanation of biomarkers which imply an irregularity in the circulatory system. The authors confirmed an adjacent region of the biosensor to concentration of cancer. Furthermore, Mohanty and his coworker [37] testified the use of silicon-based biosensor (3D nanomaterial field effect transistor) devices for breast cancer analysis and screening, which permitted detection of single molecules because of they assured good sensitivities due to their outstanding electric properties and insignificant dimensions. In another group of Williams et al. [108], advanced single-walled carbon nanotube (SWCNT) biosensor for detection of human epididymis protein biomarker for ovarian cancer and to transfer antibody binding activities. Therefore, nano sensors monitor the development of the cancer comeback to medication, applicability on-site cancer analysis, price, and availability.

9.5.2 Diagnosis of diabetes

According to WHO report, diabetes is swiftly developing challenge distressing millions of individuals globally. It led to numerous serious difficulties such as blindness, cardiovascular diseases, diabetic kidney disease and lower limb amputations. Currently existing detection procedures and strategies to avoid the progression of diabetes have some limits to the approaches and would also be economically difficult for least developing countries [109]. The new non material sensors as promising method is the detection of biomarkers from reachable point-of-care biosensors can potentially improve patient care through real-time and distant health investigation. The nanomaterials for biosensor such as carbon nanotubes [109], graphene [110], electro spun nanofibers, and quantum spots have been incorporated in biosensors to enhance their sensitivity, response time, and limit of detection as shown in Table 9.4.

These nanomaterials improve efforts by growing the volume area ratio of sensors, their properties of nanosensors [109] is being used in these point-of-care devices for diagnosis of diabetes. From past reports gold nano wires were grown in the permeable anodic alumina template via an electro deposition method and restraining glucose oxides [110]. Therefore, nano material knowledge has been able to advance the sensitivity and rectilinear series of various glucose nano material sensors which is very important in point of- care diagnostic devices as illustrated Fig. 9.4.

9.5.3 Infectious disease diagnosis

One of the common key concepts for point-of-care diagnostic systems with high spatial precision, which can often induce information. There are number of infectious diseases such as cholera, dengue fever, malaria, viral hepatitis, severe respiratory syndrome, and avian influenza. And also, infectious disease is activated by pathogenic microorganisms such as bacteria, fungi, parasites and viruses, that have a deep impact on humanity due to their distinctive characteristics such as their fast multiplication and unpredictability which sets them separately from the other diseases [111]. Based on the World Health Society

Table 9.4 (2D) nanomaterials sensors [95–102].

No	2D nanomaterials sensors	Application	Reference
1	Graphene FET sensors	Point-of-care (Puce) diagnostics, optogenetics	[95]
1.1	Graphene biosensors	Detection of dopamine, ascorbic and uric acid Various metabolites, glucose monitoring, diabetes	[96]
1.2	Graphene gas sensors	The detection of gas molecules, medical diagnosis, health care, ultrasensitive biomolecule sensing, single-stranded DNA (ssDNA), DNA hybridization	[97]
1.3	Graphene water sensors	For detection heavy metals are highly toxic, which can cause long-term damage to the human body, Detect low concentrations of metal ions, used for bacterial detection. rapid detection of living bacteria sheds	[98]
2	Dichalcogenide FET sensors	Used for Mechanical exfoliation, liquid/ chemical exfoliation,	[99]
2.1	2D TMD gas sensors	For sensing volatile organic vapors, useful in high selectivity sensor,	[100]
2.2	2D TMD biosensors	Used for detecting proteins, for detecting cancer marker proteins, DNA hybridization detection	[101]
2.3	2D TMD water sensors	used in FET sensors for detecting water contaminants, used for selective heavy metal ion detection, for detecting mercury ions	[101]
3	Phosphorene FET sensors	FET-based applications	[102]
3.1	Phosphorene gas sensors	For detecting, gases, biomolecules, and heavy metal ions	[102]
3.2	Phosphorene biosensors	For human immunoglobulin G detection,	[102]
3.3	Phosphorene water sensors	To detect multiplex ions with superb selectivity and sensitivity, for detecting water contaminants, for humidity sensing	[102]

investigation the ideal diagnostic instruments for infectious diseases should have, precision, robustness, high sensitivity, specificity, user approachable, and low-cost [112]. The conventional diagnostic techniques for infectious diseases have shown numerous restrictions such as inaccuracy and slowness, and they are expensive and require skilled expertise, especially in low-in come countries. The nano material sensors have the unique properties of catalytic, electrical, magnetic, mechanical, and optical viewpoints [111], advancement in nanotechnology has seen various implementation domain especially in biomedical applications such as bioimaging, drug delivery, nanodiagnostics, and tissue engineering. Due to their exceptional characteristics their potential for point-of-care tests, diagnostics have gotten more attention for the investigation of communicable diseases due to their potential to offer transportability, robustness, and affordability Fig. 9.5.

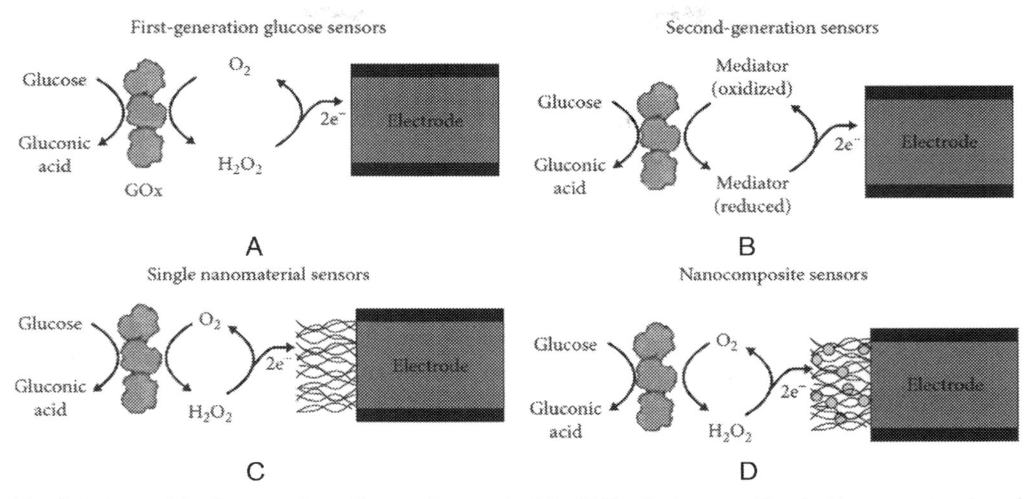

Fig. 9.4 A graphic diagram show the evolvement of (A, B) the first generation to the nanostructured materials. (C, D) shows the incorporation of nanomaterials such as CNTs or nanocomposites consisting of multiple nanomaterial sensors [39].

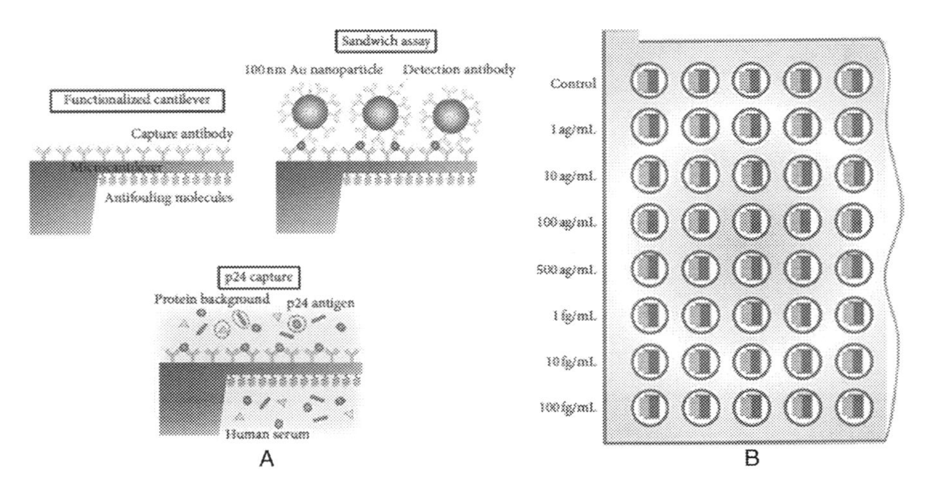

Fig. 9.5 ***A schematic picture of sandwich immunoassay:*** (A) the top view of the cantilever and (B) diagrams of the 96-well microtiter plate setup [117].

The transferable magnetic barcode assay systems had potential in point-of-care diagnosis for the efficient, swift, and low-cost detection of infectious diseases [112], Table 9.5.

9.5.4 Diagnosis of bilharzia

The common systems with high spatial precision, which can often induce information of point-of-care diagnosis of Bilharzia. It is a very devastating disease which disturbs above two hundred million people and whose highest problem of morbidity and mortality is

Table 9.5 Categories of 3D printing sensor for biomedical application [103–104].

No	3D sensor	Application	Reference
1	Fused Deposition Modelling	For biomedical applications, to detect the three cancer biomarker proteins, to control living cells	[103]
2	Stereolithography	For rapid detection of enzymatic activity, for malaria diagnosis, α-fetoprotein (AFP) biomarker detection, to detect the AFP antigen.	[103]
3	Polyjet Process	For examine of drug transport, for analyzing adenosine triphosphate (ATP) and dopamine sensing, facilitating an accurate surgical procedure	[103]
4	Selective laser sintering	To monitor PH, for controlling enzymatic assays	[104]
5	3D Inkjet Printing	To improve the hearing sensing, to receive information, by therapies and diseases	[104]
6	Digital Light Processing	To measure the cell number,	[104]

found in low-income African countries. Recent diagnostic procedures for bilharziaare the use of the microscopic determination of parasite such as eggs, in urine, stool by immunological methods antigen detection) depends on the severity, sensitivity of the infecting [113]. The current nanomaterial sensor takes stayed used to improve the sensitivity, has the potential to offer only improvement to current approaches, but also unexpectedly delivers many new tools and capabilities [114]. As recent reports Odundo et al. designates the use nano material sensor develop a simple and highly sensitive non-stop, comprising of gold nanoparticles associated with antibody and demonstrated its potential for diagnosis of soluble egg antigen (SEA) a bilharzia antigen. As they concluded that conjugating the gold nano material sensors the sensitivity and specificity for detection of light infections might be simply notable. They further determined that the compulsory could also decline the quantity of monoclonal antibody spent in the evaluate and hence lesser the price. From the different research report, it is very clear that nanomaterial sensor enhances the sensitivity and specificity of the present sensors for the exposure of bilharzias, which can be very important in the advance of point-of care instrument for bilharzias [114].

9.5.5 Diagnosis of malaria

Based on the world health association description around partial of the world's people lives in malaria-endemic areas, and more than half a million deaths resulting from malaria and its complications are reported each year, making it a significant global health problem [115]. Analysis of malaria involves documentation and quantification of mark metabolites in biological fluids, mainly blood, urine, and saliva. A variety of biomarkers for malaria exist include which paramagnetic nanoparticle byproduct of the malaria parasite, also called malaria dye. The point-of-care diagnostic device attracting much researchattention, but a few scholars have anticipated potential utensils for malaria diagnostics [116].

9.5.6 The human immunodeficiency virus diagnosis

According to worldwide public health report viruses are the most communal root cases of mankind disease, and there is growing capacity to be utilized as a medical instrument [116]. Among different types perilous disease HIV is a main universal public health problem which calls for sophisticated clinical organization [117]. The capacity of distinguishing such viruses further successfully will lead to enhanced health and improved security.

In generally HIV is investigated with severe care of the pathological load and premature detection of HIV infection is the greatest technique to avoid its spread and to progress the effectiveness of the antiretroviral rehabilitation. The old-style approaches for the recognition of the HIV epidemiologic load such as culturing, enzyme-linked immunosorbent assay (ELISA), polymerase chain reaction (PCR) have several encounters to point-of care application [117] in terms of overpriced, strictly require trained technicians. Nowadays Si based nanowire sensors effectively distinguish various hazardous viruses, such as Dengue, Influenza A H3N2, H1N1, and HIV [118], their antibodies explicit to the target virus. Furthermore, plasmonic-based biosensor has been reported for the recognition of HIV at clinically appropriate concentrations. According to Shen et al. Report the detection of target viruses captured by the antibodies of the Silicon-based biosensor for the rapid analysis of exhaled breath condensate samples. And also based on Islam et al. [117] report graphene nano sensor used for exposure of HIV, related diseases such as cardiovascular sicknesses. Recently another report by Ng et al. [117], demonstrated the implementation of magneto nanosensor and attractive nanoparticles for the verdict of HIV in saliva and leukocytosis in plasma and entire blood. Therefore, we concluded that the portability, high sensitivity, and affluence of practice of their nano material sensor has the potential device for HIV and hence enable premature detection of the diseases.

9.5.7 Diagnosis of COVID-19

According to Zhou et al. report in Wuhan, China in December 2019, severe respiratory disaster, with pneumonia like indications, its RNA sequencing from the bronchoalveolar lavage fluid of the infected patients identified a new RNA virus [118,119]. These rapid community-spread of new human coronavirus (COVID19) and morbidity statistics has put Forth an extraordinary impulse for swift diagnostics for fast, sensitive detection followed by contact tracing, containment strategies, especially when no vaccine or therapeutics are known. Currently, quantifiable actual-period polymerase cable response (qrt-PCR) is being used broadly to detect COVID-19 from several types of living specimens, which is time-uncontrollable, labor-intensive and may not be quickly deployable in resource-restricted settings. This might lead to hindrance in gaining realistic figures of communal east of SARS-CoV-2 in the population. This book chapter reviews the recent station of analytic methods, possible limitations, the recompenses of biosensor-based investigation over the conventional detection of SARS- COVID-19. Recently nanosensors have comprised of biochemical receptors specifically with a target and the transducer convert

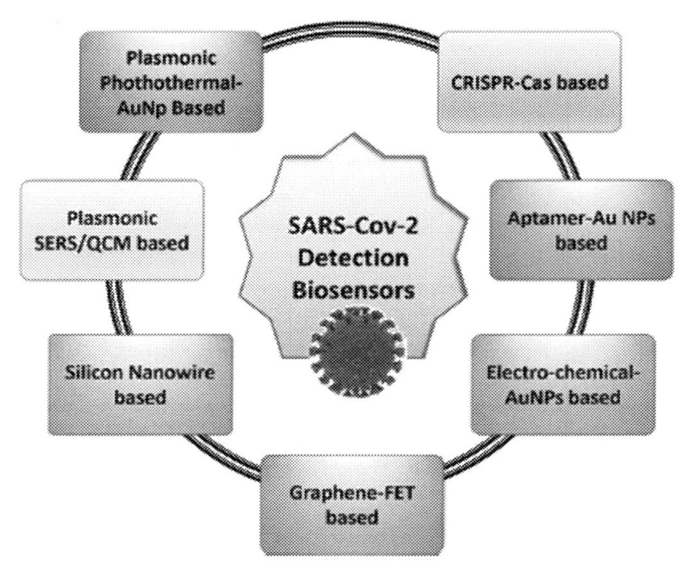

Fig. 9.6 ***The graphic diagram for detection COVID-19 virus [120].***

the recognition process into a quantitative signal. Novel nanodevices used to perceive COVID 19, created paper strip, chemical biosensor, photosensitive biosensor, nanoparticles sensor and surface plasmon resonance [120]. Although a substantial number of procedures are available for detecting virus particles, but there are several difficulties, that limit the practical use of these methods as illustrated in Fig. 9.6.

These limitations such as, lower accuracy and sensitivity, the necessity for sample and purification, time-consuming, higher instrument, accessories, maintenance cost, large-scale availability, composite operation of the devices and an obligation of extremely competent technical staffs and not suitable for rapid, on-site analysis. There is recently invented nanosensor such as piezoelectric immunosensor, thermal biosensor available for detection research of SARS-CoV-2 [121], and prospects of biosensors in rapid diagnosis of the mass population to contain the spread of this virus. Therefore, nanosensors are mostly based on detecting virus surface proteins and internal genetic material. In the near coming, developing new machineries such as graphene-FET, Au/Ag nanoparticles-based nanosensor, optical sensor, and surface plasmon (SPR) based pioneering platforms could overlay the effective customs of rapid, highly sensitive and more promising biomedical sensing diagnostic devices for COVID-19 and other unique pandemics.

9.5.8 Diagnosis of biomarker detection

The description procedure of nanomaterial sensors applications is highly multipurpose and multifunctional as from the estimation diagnosis in the health-linked for toxicants and physical aspects like humidity, heavy metal toxicity. In generally, the benefit of nanomaterial sensors are speedy reply, trivial size, high sensitivity, and transportability associated to present bulkyelectrode, sensors. Incorporation of nanosensors, microfluidics, unconsciousoperators, and transduction devices on a single chip provides numerous advantages for point of attention devices. Nanosensors have main applications in medicine and molecular level research, drug discovery, environmental intensive care, bio-security, agriculture, food, and in the dispensation industry. Newly proteomics and genomics research have exposed numerous novel biomarkers that have excessive potential for use in disease diagnosis of point care detection [122].The heterogeneity of difficult diseases, such as, HIV, cancer, prohibits the single indicator test from provided that satisfactory diagnosis results and requests the increased desires for various biomarkers [122]. In recent years, numerous studies have fixated on building comprehensive platforms for direct finding of diseases from full blood [123].The type of distinguished molecules has also prolonged to contain a widespread diversity of biomarkers. Nanomaterials have been broadly practical as label free electrical detection procedures for many proteins, such as autoinflammatory diseases, and atherothrombosis, cancer, diabetes, Parkinson's disease, pregnancy [123]. Shehada et al. report thatnanowire-based cancer diagnosis system using breath volatolome as the input of modified surface that can selectively perceive volatile biological compounds which is linked with gastric cancer conditions.The potential of biomarker sensor become, transferable, painless, simple and rapid method for the premature detection of disease. The detection flow is given in Fig. 9.7A, and the typical I-V vs. curves of the sensor are presented in Fig. 9.7B and C.Therefore, the inclusive biomarker detection system also has

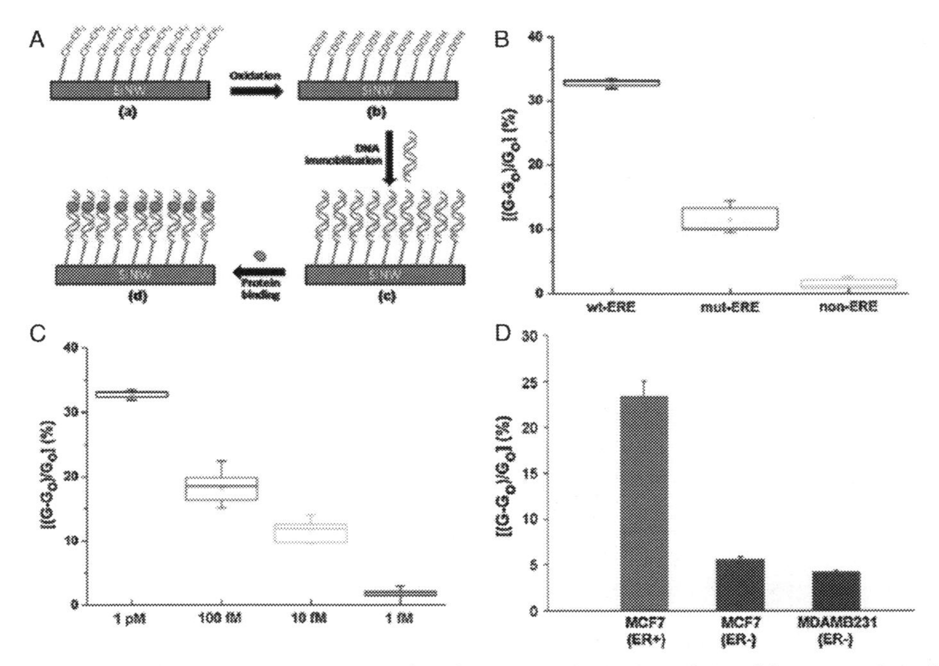

Fig. 9.7 ***Nanowire biosensors.*** (A) Diagram of working principle. (B) Specificity of the interaction of ERα with three diverse EREs as well as wild-type, scrambled ERE and mutant. (C) Response of the wt-ERE-functionalized biosensor to various concentrations of specific ERα [124].

the possibility to be further advanced as a nanowire-on-a-chip system done by nanowire medical sensor, provided that an express and practical resolution for medical applications.

9.5.9 DNA and RNA detection diagnosis

A wide verity range of biomolecules, such as antigens, antibodies, deoxyribonucleic acid (DNA), enzymes, microRNA can be immobilized, peptide nucleic acid (PNA), ribonucleic acid (RNA), and viruses, operation of carbon-based sensor for the translation of biorecognition inputs to highly sensitive and computable yields [124]. Furthermore, nanowire sensors are accomplished of detecting precise sequences of DNA. Balancing single-stranded orders of PNAs are working as acceptor for DNA on silicon nanowire surfaces [125]. In addition silicon nanowire sensor was shown to have the ability to detect DNA, quartz-crystal microbalance [125], nanoparticle-enhanced SPR for DNA detection and superficial plasmon reverberation (SPR) [126]. Furthermore, to the detection of individual strands of DNA, nano material sensor also be utilized to detect the bonding between protein and DNA. According to Zhang et al. report developed self-assembled monolayer Si nanowire sensor to detect the interface of estrogen receptor (ERα) and DNA, in which ER is an important protein for breast cancer proliferation and invasion. By distinguishing the DNA or RNA stated by the cells as biomarkers [127]. Furthermore, nanowire sensors are able to monitor numerous DNA level cancer biomarkers, such as telomerase, opening a door for using nanowire-based DNA sensor for cancer diagnosis and treatment.

9.6 Future perspectives and research

Nanosensors were possessing extraordinarily large superficial area, with feature of their slight size exhibit fascinating optical, electromagnetic, and piezoelectric properties, which grasp enormous potential for investigation in the disciplines of bioimaging, healthcare diagnostics and medical diagnostics. Moreover, nanosensors own exceptional attraction for biomolecules, enzymes, enabling the governor of antibodies, nucleic acids, proteins, and various extra medically applicable substances, introductory potentials to progress an inclusive variability of sensing platforms, such as enzymatic sensors, environmental sensors, immunosensors, sandwich analyzes, and various others. Current progresses in medical sensing podiums have active several different systems of nanomaterials, ranging from 0D to comparatively 3DLevel. Through the advancement of investigative knowledge, health treatment can be modified to patients grounded on the individual's explicit medical characteristics, including age, diet, environment gender, height, and weight. The swift, profitable, and facetious operative procedures presented by nanomaterial sensors are predictable to renovation unadventurous luxurious distinguishing schemes in the nearby coming. The real success of adapted medication depends diagnostic techniques such as medical-sensors applicable for actual monitoring, link with drug injection instrumentation, and measure medication release frequency. Nanomaterial-based sensor characterize novel source of alternatives for medical instruments, they can advance the performance of the nanosensor progress to achieve better-quality signal to noise ratio, better time of response and increase stability, shelf life, sensitivity, to develop sensors and other medical appliances.

9.7 Conclusion

The paper presents a substantial review on nanosensors for medicinal applications. These biosensors employ the exceptional properties to diagnose a target molecule and effect transduction of microelectronic signal. The overall, advantages of nanomaterial-based biosensors are the fast response, small size, high sensitivity, and transportability compared to existing traditional electrode sensors. Incorporation of nanomaterials, micro fluidics, automatic testers, and transduction devices on an individual chip delivers numerous advantages for fact of care devices due to advanced sensitivity intensely. Finally, will review their applications for the detection, biomarkers, and DNA, as well as for and recent developments in their performance and functionality of sensors are also surveyed to highlight recent progress and future direction in healthcare Application.

Conflicts of interest

The, authors have no conflicts of interest concerning the publication of this book chapter.

Acknowledgments

Anshebo Getachew Alemu would like thanks *Dr. Sushma Dave*, for invitation, constructive comments and guidance.

References

[1] C. Choi, M. Kee Choi, T. Hyeon, D. Hyeong Kim, Nanomaterial-based soft electronics for healthcare applications. ChemNanoMat 2 (2016) 1006–1017.
[2] Alejandro Chamorro-Garcia, Merkoçi Arben, Nanobiosensors in diagnostics, Nanobiomedicine 3 (2016) doi:10.1177/1849543516663574.
[3] S. K., A. Mohan, R. Guleria, Biomarkers in cancer screening, research and detection: present and future: a review., Biomarkers 11 (5) (2006) 385–405.
[4] M. Pirzada, A. Zeynep, Nanomaterials for healthcare biosensing applications, Sensors 19 (23) (2019) 5311.
[5] S.A. Hashsham, L.M. Wick, J.M. Rouillard, E. Gulari, J.M. Tiedje, Potential of DNA microarrays for developing parallel detection tools (PDTs) for microorganisms relevant to biodefense and related research needs, Biosens. Bioelectron. 20 (2004) 668.
[6] V. Scognamiglio, F. Arduini, G. Palleschi, G. Rea, Biosensing technology for sustainable food safety, Trends Anal. Chem. 62 (2014) 1–10.
[7] Y. Chen, D. Feng, C.Y. Bi, S.R. Zhu, J.G. Shi, Recent progress of commercially available biosensors in china and their applications in fermentation processes development of serial SBA biosen, Northeast Agric. Univ. 21 (2014) 73–85.

[8] B.M. Woolston, S. Edgar, G. Stephanopoulos, Metabolic engineering: past and future, Annu Rev. Chem. Biomol. Eng. 4 (2013) 259–288.
[9] M. Pirzada, A. Zeynep, Nanomaterials for healthcare biosensing applications, Sensors 19 (23) (2019) 5311.
[10] P. Mehrotra, Biosensors and their applications–a review, J. Oral Biol. Craniofacial Res. 6 (2016) 153–159.
[11] A. Walia, R. Waadt, J.A.M., Genetically encoded biosensors in plants: pathways to discovery, Annu. Rev. Plant Biol. 69 (2018) 497–524.
[12] M. Pirzada, Z. Altintas, Can quantum-mechanical description of physical reality be considered complete? Sensors 19 (2019) 5311, doi:10.3390/s19235311.
[13] M. Saito, N. Uchida, S. Furutani, M. Murahashi, W. Espulgar, N. Nagatani, H. Nagai, Y. Inoue, T. Ikeuchi, S. Kondo, Field-deployable rapid multiple biosensing system for detection of chemical and biological warfare agents, Microsyst. Nanoeng 4 (2018) 1–11.
[14] N.M. Noah, P.M. Ndangili, Current trends of nanobiosensors for point-of-care diagnostics, J. Anal. Methods. Chem. (2019) ID 2179718.
[15] R.A. Karim, Y. Reda, A.A. Fattah, Tetrabutylammonium chloride modified carbon paste electrode for rapid and highly sensitive voltammetric determination of carbendazim, J. Electrochem. Soc. 167 (2020) 037554.
[16] M. Pirzada, Z. Altintas, Recent progress in optical sensors for biomedical diagnostics, Micromachines 11 (2020) 356, doi:10.3390/mi11040356.
[17] E. Primiceri, M.S. Chiriacò, F.M. Notarangelo, A. Crocamo, D. Ardissino, M. Cereda, A.P. Bramanti, M.A. Bianchessi, G. Giannelli, G. Maruccio, Key enabling technologies for point-of-care diagnostics, Sensors 18 (2018) 3607.
[18] A.J. Saleh, J. Joon, M. Aminur, Electrochemical sensors based on carbon nanotubes, Sensors 9 (2009) 2289.
[19] P. Damborský, J. Švitel, J. Katrlík, Optical biosensors, Essays Biochem 60 (2016) 91–100.
[20] M.H. Shamsi, S. Chen, Biosensors-on-chip: A topical review, J. Micromech. Microeng. 27 (2017) (2017) 083001.
[21] A.K. Pulikkathodi, I. Sarangadharan, Y.-H. Chen, G.-Y. Lee, J.I. Chyi, G.B. Lee, Y.L. Wang, A comprehensive model for whole cell sensing and transmembrane potential measurement using FET biosensors, J. Solid State Sci. Technol. 7 (2018) Q3001.
[22] Y.C. Syu, W.-E. Hsu, C.T. Lin, Review—field-effect transistor biosensing: devices and clinical applications, J. Solid State Sci. Technol. 7 (2018) Q3196.
[23] K. Bhattacharya, S.P. Mukherjee, A. Gallud, S.C. Burkert, S. Bistarelli, S. Bellucci, M. Bottini, A. Star, B. Fadeel, Biological interactions of carbon-based nanomaterials: From coronation to degradation, Nanomed. Nanotechnol. Biol. Med. 12 (2016) 333–351.
[24] G. Hong, S. Diao, A.L. Antaris, H. Dai, Carbon nanomaterials for biological imaging and nanomedicinal therapy, H. Chem. Rev. 115 (2015) 10816–10906.
[25] S. Emelyantsev, E. Prazdnova, V. Chistyakov, I. Alperovich, Biological effects of C60 fullerene revealed with bacterial biosensor—toxic or rather antioxidant? Biosensors 9 (2019) 81.
[26] H.H. Nguyen, S.H. Lee, U.J. Lee, C.D. Fermin, M. Kim, Immobilized enzymes in biosensor applications, Materials 12 (2019) 121.
[27] (a) S. Fortunati, A. Rozzi, F. Curti, M. Giannetto, R. Corradini, M. Careri, Nanomaterials for healthcare biosensing applications, Sensors 19 (2019) 588. (b) Y. Zhou, Y. Fang, R.P. Ramasamy, Sensors 19 (2019) 392.
[28] C. Chaicham, T. Tuntulani, V. Promarak, B. Tomapatanaget, Nanomaterials for healthcare biosensing applications, Sens. Actuators B Chem. 282 (2019) 936. D.R. Kumar, M.L. Baynosa, J. Shim, J. Sens. Actuators B Chem. 293 (2019) 107–114.
[29] N.J. Hoboken, NJ, USA (2017); ISBN 978-1-119-06501-2.
[30] J.A. Rather, E.A. Khudaish, A. Munam, A. Qurashi, P. Kannan, Electrochemically reduced fullerene–graphene oxide interface for swift detection of Parkinsons disease biomarkers, Sens. Actuators B Chem. (2016) 672.
[31] B. Thirumalraj, S. Palanisamy, S.M. Chen, B.S. Lou, Preparation of highly stable fullerene C60 decorated graphene oxide nanocomposite and its sensitive electrochemical detection of dopamine in rat brain and pharmaceutical samples, J. Colloid Interface Sci 462 (2016) 375–381.
[32] J. Diao, T. Wang, L. Li, Graphene quantum dots as nanoprobes for fluorescent detection of propofol in emulsions, R. Soc. Open Sci. 6 (2019) 181753.
[33] S. Savas, Z. Altintas, Graphene quantum dots as nanozymes for electrochemical sensing of Yersinia enterocolitica in milk and human serum, Materials 12 (2019) 2189.

[34] W. Dong, Y. Ren, Z. Bai, Y. Yang, Q. Chen, Fabrication of hexahedral Au-Pd/graphene nanocomposites biosensor and its application in cancer cell H2O2 detection, Bioelectrochemistry 128 (2019) 274–282.
[35] S. Shahrokhian, R. Salimian, Ultrasensitive detection of cancer biomarkers using conducting polymer/electrochemically reduced graphene oxide-based biosensor: Application toward BRCA1 sensing, Actuators B Chem. 266 (2018) 160–169.
[36] P. Suvarnaphaet, S.A. Pechprasarn, Graphene-based materials for biosensors: a review, Review. Sensors 17 (2017) 2161.
[37] S. Savas, Z. Altintas, Graphene quantum dots as nanozymes for electrochemical sensing of Yersinia enterocolitica in milk and human serum, Materials 12 (2019) 2189.
[38] S.Y. Lim, W. Shen, Z. Gao, Carbon quantum dots and their applications, Chem. Soc. Rev. 44 (2015) 362–381.
[39] D. Bhattacharya, M.K. Mishra, G.J. De, Electrochemical synthesis of carbon dots with a Stokes shift of 309 nm for sensing of Fe3+ and ascorbic acid, Phys. Chem. C 121 (2017) (2017) 28106–28116.
[40] J.C. Arnault, Elsevier: Amsterdam, Nanomaterials for healthcare biosensing applications, The Netherlands (2017), ISBN 978-0-323-43029-6.
[41] P. Karami, S.S. Khasraghi, M. Hashemi, S. Rabiei, A. Shojaei, Polymer/nanodiamond composites-a comprehensive review from synthesis and fabrication to properties and applications, Adv. Colloid Interface Sci. 269 (2019) 122.
[42] K. Wang, Y. Dong, B. Li, D. Li, S. Zhang, Y. Wu, Differentiation of proteins and cancer cells using metal oxide and metal nanoparticles-quantum dots sensor array, Sen. Actuators B Chem. 250 (2017) 69–75. W.H. Zhang, W. Ma, Y.T. Long, Anal. Chem. 88 (2016) 5131–5136.
[43] E.Ü. Kasap, D. Cetin, Z. Suludere, H.I. Boyaci, C. Türkyilmaz, N. Ertas, U. Tamer, Rapid detection of bacteria based on homogenous immunoassay using chitosan modified quantum dots, Sens. Actuators B Chem. 233 (2016) 369–378.
[44] S. Wu, L. Liu, G. Li, F. Jing, H. Mao, Q. Jin, W. Zhai, H. Zhang, J. Zhao, C. Jia, Multiplexed detection of lung cancer biomarkers based on quantum dots and microbeads, Talanta 156–157 (2016) 48.
[45] H. Deng, Q. Liu, X. Wang, R. Huang, H. Liu, Q. Lin, X. Zhou, D. Xing, A review of one-dimensional tio2 nanostructured materials for environmental and energy applications, Bioelectron 87 (2017) 931–940. S. Lv, F. Chen, C. Chen, X. Chen, H. Gong, C. Cai, Talanta 165 (2017) 659–663.
[46] A. Zeynep (Ed.), Biosensors and nanotechnology: applications in health care diagnostics. John Wiley & Sons, 2017.
[47] Z. Altintas, S.S. Kallempudi, U. Sezerman, Y. Gurbuz, A novel magnetic particle-modified electrochemical sensor for immunosensor applications, Sens. Actuators B Chem. 174 (2012) 187–194.
[48] Q. Yuan, J. He, Y. Niu, J. Chen, Zhou, Y. Zhang, C. Yu, Sandwich-type biosensor for the detection of α2, 3-sialylated glycans based on fullerene-palladium-platinum alloy and 4-mercaptophenylboronic acid nanoparticle hybrids coupled with Au-methylene blue-MAL signal amplification, Biosens. Bioelectron 102 (2018) 321–327.
[49] D.T. Tran, V.H. Hoa, L.H. Tuan, N.H. Kim, J.H. Lee, Worm-like gold nanowires assembled carbon nanofibers-CVD graphene hybrid as sensitive and selective sensor for nitrite detection, Bioelectron. 119 (2018) 134–140.
[50] D. Ji, N. Xu, Z. Liu, Z. Shi, S.S. Low, J. Liu, C. Cheng, J. Zhu, T. Zhang, H. Xu, Smartphone-based differential pulse amperometry system for real-time monitoring of levodopa with carbon nanotubes and gold nanoparticles modified screen-printing electrodes, Biosens. Bioelectron. 129 (2019) 216–223.
[51] D.L. Escosura, A. Muñiz, L.Ba. Pires, L. Serrano, L. Altet, O. Francino, A. Sánchez, A. Merkoçi, Magnetic Bead/Gold Nanoparticle Double-Labeled Primers for Electrochemical Detection of Isothermal Amplified *Leishmania* DNA, Small 12 (2016) 205–213.
[52] B. Thirumalraj, N. Dhenadhayalan, S.M. Chen, Y.J. Liu, T.W. Chen, P.H. Liang, K.C. Lin, Highly sensitive fluorogenic sensing of L-Cysteine in live cells using gelatin-stabilized gold nanoparticles decorated graphene nanosheets, Sens. Actuators B Chem. 259 (2018) 339–346.
[53] J.G. Egan, N. Drossis, I. LEbralidze, H.M. Fruehwald, N.O. Laschuk, J. Poisson, H.W. DeHaan, O.V. Zenkina, Hemoglobin-driven iron-directed assembly of gold nanoparticles, RSC Adv. 8 (2018) 15675–15686.
[54] Z. Altintas, M. Akgun, G. Kokturk, Y. Uludag, A fully automated microfluidic-based electrochemical sensor for real-time bacteria detection, Y Bioelectron. 100 (2018) 541–554.
[55] H. Malekzad, Z. Sahandi, P. Zangabad, M. Karimi, M.R. Hamblin, Noble metal nanoparticles in biosensors: recent studies and applications, Nanotechnol. Rev. 6 (2017) 301.

[56] W. Ombati, A. Setiono, M. Bertke, H. Bosse, Cantilever-droplet-based sensing of magnetic particle concentrations in liquids, Sensors 19 (2019) 4758.
[57] H. Yang, J. Hou, Z. Wang, T. Zhang, C. Xu, An ultrasensitive biosensor for superoxide anion based on hollow porous PtAg nanospheres, C. Biosens. Bioelectron. 117 (2018) 429–435.
[58] X. Zhu, T. Liu, H. Zhao, L. Shi, X. Li, M. Lan, Ultrasensitive detection of superoxide anion released from living cells using a porous Pt–Pd decorated enzymatic sensor, Biosens. Bioelectron. 79 (2016) 449–456.
[59] J.-S. Do, Y.-H. Chang, T. Ming-Liao, Highly sensitive amperometric creatinine biosensor based on creatinine deiminase/Nafion®-nanostructured polyaniline composite sensing film prepared with cyclic voltammetry, Mater. Chem. Phys. 219 (2018) 1–12.
[60] M. Razavi, A. Thakor, Nanobiomaterials Science, Nanobiomaterials Science, Development and Evaluation, Woodhead publishers, Cambridge, UK, 2017.
[61] D.P. Nikolelis, G.P. Nikoleli, Nanotechnology and Biosensors, Elsevier, Amsterdam, 2018.
[62] Y. Niu, M. Chu, P. Xu, S. Meng, Q. Zhou, W. Zhao, B. Zhao, An aptasensor based on heparin-mimicking hyperbranched polyester with anti-biofouling interface for sensitive thrombin detection, Biosens. Bioelectron. 101 (2018) 174.
[63] Z. Altintas, A. Takiden, T. Utesch, M.A. Mroginski, B. Schmid, F.W. Scheller, R.D. Süssmuth, Integrated approaches toward high-affinity artificial protein binders obtained via computationally simulated epitopes for protein recognition, Funct. Mater 29 (2019) 1–11.
[64] Z. Bagheryan, J.B. Raoof, R.A. Ojani, A switchable Gquadruplex device with the potential of a nanomachine for anticancer drug detection, Int. J. Biol. Macromol. 83 (2016) 97–102.
[65] S.M. Russell, A. Alba-patiño, M. Borges, R. De, Multifunctional motion-to-color janus transducers for the rapid detection of sepsis biomarkers in whole blood, R. Biosens. Bioelectron. 140 (2019) 111346.
[66] F.B. Emre, M. Kesik, F.E. Kanik, H.Z. Akpinar, E. Aslan Gurel, R.M. Rossi, LA. Toppare, A benzimidazole-based conducting polymer and a PMMA–clay nanocomposite containing biosensor platform for glucose sensing, Synth. Met. 207 (2015) 102–109.
[67] A. Soni, C.M. Pandey, M.K. Pandey, G. Sumana, Highly efficient Polyaniline-MoS2 hybrid nanostructures based biosensor for cancer biomarker detection, Anal. Chim. Acta 1055 (2019) 26–35.
[68] J.M. Perez, L. Josephson, R. Weisleder, Use of magnetic nanoparticles as nanosensors to probe for molecular interactions, Chem Bio Chem 5 (2004) 261.
[69] Z. Altintas, A. Guerreiro, S.A. Piletsky, I.E. Tothill, NanoMIP based optical sensor for pharmaceuticals monitoring, Nano MIP Actuators B Chem 213 (2015) 305–313.
[70] P. Carneiro, S. Morais, M.C. Pereira, Application of zero-dimensional nanomaterials in biosensing, Nanomaterials 9 (2019) 1663. K. Vikrant, N. Bhardwaj, S.K. Bhardwaj, K., Kim, H. Deep, A. Biomaterials 214 (2019) 119215.
[71] M.A. Raza, Z. Kanwal, A. Rauf, A.N. Sabri, S. Riaz, S. Naseem, Toxicity of zero- and one-dimensional carbon nanomaterials, Nanomaterials 6 (2016) 74. S. Zhu, L. Gong, J. Xie, Z. Gu, Y. Zhao, Small Methods 1 (2017) 1700220.
[72] Y. Dong, F. Li, Y. Wang, Strategies to improve photodynamic therapy efficacy of metal-free semiconducting conjugated polymers, Front. Chem. 8 (2020) 551. Y. Yan, J. Gong, J. Chen, Z. Zeng, W. Huang, K. Pu, Adv. Mater. 31 (2019) 1808283.
[73] S. Pandit, P. Behera, J. Sahoo, M. De, In situ synthesis of amino acid functionalized carbon dots with tunable properties and their biological applications, ACS Appl. Bio. Mater. 2 (2019) 3393–3403.
[74] N. Dhenadhayalan, K.C. Lin, T.A. Saleh, Application of zero-dimensional nanomaterials in biosensing, Small 16 (2020) 1905767. I.S. Raja, S.J. Song, M.S. Kang, Y.B. Lee, B. Kim, S.W. Hong, Nanomaterials 9 (2019) 1214.
[75] Q. Xu, W. Li, L. Ding, W. Yang, H. Xiao, W. Ong, Function-driven engineering of 1D carbon nanotubes and 0D carbon dots: mechanism, properties and applications, J. Nanoscale 11 (2019) 1475–1504.
[76] C. Li, H.L W.Li, Y. Zhang, G. Chen, Artificial Light-Harvesting Systems Based on AIEgen-branched Rotaxane Dendrimers for Efficient Photocatalysis, Chem. Int. Ed. 59 (2020) 247–252.
[77] N.Ž. Knežević, I. Gadjanski, J.O. Durand, Magnetic nanoarchitectures for cancer sensing, imaging and therapy, J. Mater. Chem. B 7 (2019) 9–23.
[78] Z. Wang, T. Hu, R. Liang, R, M. Wei, Application of Zero-Dimensional Nanomaterials in Biosensing, Front. Chem 8 (2020) 320.
[79] X. Zhang, G. Xie, D. Gou, P. Luo, Y. Yao, H. Chen, A novel enzyme-free electrochemical biosensor for rapid detection of Pseudomonas aeruginosa based on high catalytic Cu-ZrMOF and conductive Super P, Biosens. Bioelectron. 142 (2019) 111486.

[80] Q. Wang, C. Wang, Y. Zhang, Y. Wu, C. Dong, Construction of CPs@ MnO 2–AgNPs as a multifunctional nanosensor for glutathione sensing and cancer theranostics, Nanoscale (11) (2019) 18845.
[81] L. Liu, H. Zhang, Z. Wang, D. Song, Application of zero-dimensional nanomaterials in biosensing, Biosens. Bioelectron 141 (2019) 111403. J.H. Luo, Q. Li, S.H. Chen, R. Yuan, ACS. App. Mater. Interfaces 11 (2019) 27363.
[82] H. Sun, J. Deng, L. Qiu, X. Fang, H. Peng, Recent progress in solar cells based on one-dimensional nanomaterials, Energy Environ. Sci. 8 (2015) 1139.
[83] W. Zeng, L. Shu, Q. Li, S. Chen, F. Wang, X.M. Tao, Creation of additional electrical pathways for the robust stretchable electrode by using UV irradiated CNT-elastomer composite, Adv. Mater 26 (2014) 5310. T. Sekitani, Y. Noguchi, K. Hata, T. Fukushima, T. Aida, T. Someya, Science 321 (2008) 1468.
[84] D. Langley, G. Giusti, C. Mayousse, C. Celle, D. Bellet, J.P. Simonato, Flexible transparent conductive materials based on silver nanowire networks: a review, Nanotechno 24 (2013) 452001.
[85] A.R. Rathmell, S.M. Bergin, Y.L. Hua, Z.Y. Li, B.J. Wiley, Cu–Ag core–shell nanowires for electronic skin with a petal molded microstructure, Adv. Mater. 22 (2010) 3558. Y. Tang, S. Gong, Y. Chen, L.W. Yap, W. Cheng, ACS Nano 8 (2014) 5707.
[86] F.Y. Yang, Y. You, G. Li, J. Guo, T. Yu, Z. Shen, T. Wu, B. Xing, Nanoformulation of metal complexes: Intelligent stimuli-responsive platforms for precision therapeutics, Chem. Commun. 1984 (2009).
[87] J. Zhou, Y. Gu, P. Fei, W. Mai, Y. Gao, R. Yang, G. Bao, Z.L. Wang, Nano Lett. 8 (2008) 3035.
[88] B. Sun, Y.Z. Long, Z.J. Chen, S.L. Liu, H.D. Zhang, J.C. Zhang, W.P. Han, Recent advances in flexible and stretchable electronic devices via electrospinning, J. Mater. Chem. C. *2* (2014) 1209.
[89] H.C. Chang, C.L. Liu, W.C. Chen, Flexible nonvolatile transistor memory devices based on One-Dimensional electrospun P3HT:Au hybrid nanofibers, Adv. Funct. Mater. 23 (2013) 4960.
[90] J. Liang, L. Li, D. Chen, T. Hajagos, Z. Ren, S.Y. Chou, W. Hu, Q. Pei, Intrinsically stretchable and transparent thin-film transistors based on printable silver nanowires, carbon nanotubes and an elastomeric dielectric, Nat. Commun. 6 (2015) 7647.
[91] V. Ghini, S. Chevance, P. Turano, About the use of 13C-13C NOESY in bioinorganic chemistry, J. Inorg. Biochem. 192 (2019) 25–32.
[92] I.G. Jadach, D. Kalinowska, M. Drozd, M. Pietrzak, Controllable fabrication and characterization of gold nanoparticles: enhanced therapeutic approaches in cancer, Biomed. Pharmacother. 111 (2019) 1147–1155.
[93] H. Zhang, Ultrathin two-dimensional nanomaterials, ACS Nano (10) (2015) 9451–9469.
[94] C. Zhu, D. Du, Y. Lin, Graphene and graphene-like 2D materials for optical biosensing and bioimaging: A review, Mater 2 (2015) 032004.
[95] T. Yang, H. Chen, C. Jing, S. Luo, W. Li, K. Jiao, Using poly (m-aminobenzenesulfonic acid)-reduced MoS2 nanocomposite synergistic electrocatalysis for determination of dopamine, Sens. Actuators B249 (2017) 451–457.
[96] A. Facchinetti, G. Sparacino, S. Guerra, Y.M. Luijf, J.H. De Vries, J.K. Mader, M. Ellmerer, C. Benesch, D. Bruttomesso, A. Avogaro, C. Cobelli, Real-time improvement of continuous glucose monitoring accuracy: the smart sensor concept, Diabetes Care. 36 (2013) 793.
[97] T.Y. Chen, L. Phan Thi Kimchi, L. Hsu, Y.H. Lee, J.T.W. Wang, K.H. Wei, C.T. Lin, L.-J. Li, Label-free detection of DNA hybridization using transistors based on CVD grown graphene, Biosens. Bioelectron. 41 (2013) 103–109.
[98] S. Mao, J. Chang, G. Zhou, J. Chen, Nanomaterial-enabled rapid detection of water contaminants, Small 11 (2015) 5336–5359.
[99] Z. Yin, H. Li, H. Li, L. Jiang, Y. Shi, Y. Sun, G. Lu, Q. Zhang, X. Chen, H. Zhang, Single-Layer MoS_2 Phototransistors, ACS Nano 6 (2012) 74.
[100] S.L. Zhang, H. Yue, X. Liang, W.C. Yang, Liquid-phase Co-exfoliated graphene/MoS_2 nanocomposite for methanol gas sensing, J. Nanosci. Nanotechnol 15 (2015) 8004–8009.
[101] L. Wang, Y. Wang, J.I. Wong, T. Palacios, J. Kong, H.Y. Yang, Functionalized MoS_2 Nanosheet-Based Field-Effect Biosensor for Label-Free Sensitive Detection of Cancer Marker Proteins in Solution, Small 10 (2014) 1101–1105.
[102] A.N. Abbas, B. Liu, L. Chen, Y. Ma, S. Cong, N. Aroonyadet, M. Koepf, T. Nilges, C. Zhou, Black phosphorus gas sensors, ACS. Nano 9 (2015) 5618–5624.
[103] K. Kadimisetty, L.M. Mosa, S. Malla, J.E.S. Warden, T.M. Kuhns, R.C. Faria, N.H. Lee, J.F Rusling, 3D-printed supercapacitor-powered electrochemiluminescent protein immunoarray, Biosens. Bioelectron 77 (2016) 188–193.

[104] C. Ude, T. Hentrop, P. Lindner, T. Scheper, S. Beutel, New perspectives in shake flask pH control using a 3D-printed control unit based on pH online measurement, Sens. Actuators B Chem 221 (2015) 1035.

[105] A. Kaushik, M. Mujawar, Nano-enabled biosensing systems for intelligent healthcare: towards COVID-19 management, Sensors 18 (2018) 4303.

[106] G. Bagherzade, M.M. Tavakoli, M.H. Namaei, The antimicrobial and immunomodulatory effects of Cotyledon Orbiculata extracts, A.P.J. of T. Biomedicine 7 (2017) 227–233. A.I. Usman, A.A. Aziz, O.A. Noqta, J. Sci. Eng. 8 (2019) 171182.

[107] P. Paul, A.K. Malakar, S. Chakraborty, The significance of gene mutations across eight major cancer types, Reviews in Mutation Research 781 (2019) 88–99.

[108] M. Pritiraj, Y. Chen, X. Wang, M. K. Hong, C. L. Rosenberg, D.T. Weaver, S. Erramilli, Field effect transistor nanosensor for breast cancer diagnostics. arXiv preprint arXiv:1401.1168 (2014).

[109] E.M. Freer, O. Grachev, X. Duan, S. Martin, D.P. Stumbo, High-yield self-limiting single-nanowire assembly with dielectrophoresis, Nat. Nanotechnol. 5 (2010) 525.

[110] J.H. Chua, R.E. Chee, A. Agarwal, M.W. She, G.J. Zhang, Label-free electrical detection of cardiac biomarker with complementary metal-oxide semiconductor-compatible silicon nanowire sensor arrays, Anal. Chem. 81 (2009) 6266.

[111] C.C. Wu, F.H. Ko, Y.S. Yang, D.L. Hsia, B.S. Lee, T.S. Su, Label-free biosensing of a gene mutation using a silicon nanowire field-effect transistor, Biosens. Bioelectron. 4 (2009) 820.

[112] G.J. Zhang, M.J. Huang, J.J. Ang, Q. Yao, Y. Ning, Label-free detection of carbohydrate–protein interactions using nanoscale field-effect transistor biosensors, Anal. Chem. 85 (2013) 4392–4397.

[113] T.T. Nielsen, L. Duroux, M. Hinge, K. Shimizu, L. Gurevich, P.K. Kristensen, C. Wingren, K.L. Larsen, Nonfouling tunable βCD dextran polymer films for protein applications, ACS. Appl. Mater. Interfaces (2015).

[114] C.B. Hutson, J.W. Nichol, H. Aubin, H. Bae, S. Yamanlar, S. Al-Haque, S.T. Koshy, A. Khademhosseini, Synthesis and characterization of tunable poly (ethylene glycol): gelatin methacrylate composite hydrogels, Tissue Eng. A 17 (2011) 1713–1723.

[115] J. Schmitt, H. Hess, G. S. Hendrik, Affinity purification of histidine-tagged proteins, Mol. Biol. Rep. 18 (3) (1993) 223–230.

[116] E. Stern, J.F. Klemic, D. Routenberg, P.N. Wyrembak, D.B. Turner- Evans, A.D. Hamilton, D. a LaVan, T.M. Fahmy, M.a Reed, Label-free immunodetection with CMOS-compatible semiconducting nanowires, Nature (2007) 519–522.

[117] M.A. Lifson, M.O. Ozen, F. Inci, et al., Advances in biosensing strategies for HIV-1 detection, diagnosis, and therapeutic monitoring, Adv. Drug Deliv. Rev. 103 (2016) 90–104.

[118] E. Ng, C. Yao, T.O. Shultz, S. Ross-Howe, S.X. Wang, An automated and mobile magnetoresistive biosensor system for early hepatocellular carcinoma diagnosis, Nanomed.: Nanotechnol. Biol. Med. 16 (2019) 10–19.

[119] P. Zhou, X.-.L. Yang, X.-.G. Wang, et al., Addendum: A pneumonia outbreak associated with a new coronavirus of probable bat origin, Nature (2020).

[120] L. Yu, S. Wu, X. Hao, et al., Rapid detection of COVID-19 coronavirus using a reverse transcriptional loop-mediated isothermal amplification (RT-LAMP) diagnostic platform, Clin. Chem. (2020).

[121] T. Lee, J.H. Ahn, S.Y. Park, et al., Recent advances in AIV biosensors composed of nanobio hybrid material, Micromachines 9 (2018) 651.

[122] R. Etzioni, N. Urban, S. Ramsey, M. McIntosh, S. Schwartz, B. Reid, J. Radich, G. Anderson, L. Hartwell, The case for early detection, Nat. Rev. Cancer 3 (2003) 243–252.

[123] A. Kim, C.S. Ah, H.Y. Yu, J.H. Yang, I.B. Baek, C.G. Ahn, C.W. Park, M.S. Jun, S. Lee, Nanowire-based sensors for biological and medical applications, Appl. Phys. Lett. 91 (2007) 29–32.

[124] G.J. Zhang, M.J. Huang, J.J. Ang, E.T. Liu, K.V. Desai, Nanowire-based sensors for biological and medical applications, Biosens. Bioelectron. 26 (2011) 3233.

[125] J. Hahm, C.M. Lieber, Direct ultrasensitive electrical detection of DNA and DNA sequence variations using nanowire nanosensors, Nano Lett. 4 (2004) 51–54.

[126] L. He, M.D. Musick, S.R. Nice warner, F.G. Salinas, S.J. Benkovic, M.J. Natan, C.D. Keating, Colloidal Au-enhanced surface plasmon resonance for ultrasensitive detection of DNA hybridization, J. Amer. Chem. Soc. 122 (2000) 9071–9077.

[127] G.J. Zhang, M.J. Huang, J.J. Ang, Q. Yao, Y. Ning, Label-free detection of carbohydrate–protein interactions using nanoscale field-effect transistor biosensors, Anal. Chem. 85 (2013) 4392.

CHAPTER 10

Hybrid organic or inorganic nanomaterials for healthcare diagnostics

Indu Sharma, Satabdi Bhattacharjee
Department of Microbiology, Assam University, Silchar, Assam, India

10.1 Introduction

Hybrid nanomaterials are emerging as a very powerful material due to their diverse properties. These materials consists of both the inorganic and organic components that has a synergistic effects [1,2]. The area of hybrid materials is constantly expanding. Hybrid organic and inorganic nanomaterials allows an excellent motive force for the expeditious development of several research areas in nanomedicine. These hybrid materials are potent imitator of natural structures and attain the properties that are very difficult to produce a copy.

The organic molecule of hybrid material ranges from simpler to advanced form and are modifiable. The inorganic hybrid materials includes various inorganic moieties that constructs polymers with metals that are directly integrated in the polymeric chains. The hybrid organic–inorganic systems filled the space in between the combination of molecules, nanoparticles and their function. Researches in the area of hybrid molecule lead to the growth of modern techniques to commercialize in future.

10.2 Classification of organic and inorganic materials

Organic materials plays a vital role in the development of hybrid materials [3]. The inorganic materials exist in different sizes and shapes ranging from nano to large. Organic and inorganic materials are divided into the following:

(i) Chemical matrices
(ii) Biological matrices

(i) Chemical matrices
 (a) Hydrogels: Hydrogels consist of very large molecules [4] made up of cross-linked polymers. The 3D structures permits the supply of nutrients, gases, wastes, and biomolecules that are required in biomedical engineering involving tissues and regenerative medicine [5]. The hydrogel stiffness provides a several dimensional cell culture platform to reproduce tissue [6]. The initiation

Advanced Nanomaterials for Point of Care Diagnosis and Therapy
DOI: https://doi.org/10.1016/B978-0-323-85725-3.00024-6

of inorganic material into hydrogel permits in making the catalytically active interfaces that increases the rate of chemical reaction [7].

(b) Layer-by-layer: Layer-by-Layer (LbL) assembly has appeared as an adaptable process in coating biological and non-biological surfaces. Capsule and planar films are the mostly used [8,9]. Various carbon based nano materials and gold nano particles are applied in polyelectrolyte polymers and brushes. The enhancement of thermal properties is also one such function [10,11].

(c) Polymer brushes: Polymer brushes is a type of coating constructed by long chain polymeric molecule. The polymer density is high. The di-block polymers is used for attachment of chain and are combined with a chain growth polymerization [12]. Few examples are bacterial adherence prevention, attachment of cells, and making of colloid particles [13].

(d) Block copolymers: Block copolymers consist of least two polymeric subunits. The changes of the polymer shell with magnetic nanoparticles manage the discharge of hybrid vesicles. Other property of magnetic nanoparticles includes the magnetic resonance imaging (MRI) and drug delivery.

(ii) Biological Matrices

The biological matrices are designed by using biological molecules and microorganisms [14]. These are:

(a) Lipids: Lipid bilayers are consist two layers of lipid and [15] the functional property is its permeability to molecules [16]. It act as a barrier surrounding the cells and provides protection [17]. The two thick molecules of lipid with inorganic nanoparticles of nanometer size have been used to functionalize the lipid membrane.

(b) Proteins and enzymes, carbohydrates, and nucleic acids: These are the cell components [14]. The enzyme coating act as a drug delivery vehicles [18] and the DNA molecule is used for assembling gold nanoparticles [19]. Magnetic particles act as a device to measure the rheoogical properties [20].

(c) Cells, bacteria, and other microorganisms: The bacteria, yeast, actinimycetes, and fungus are capable of synthesizing nanomaterials. The micro organisms grab the target ion from the environment and convert it into metal ion using microbial enzymes, thereby synthesizing nanoparticles.

10.3 Different types of inorganic and organic nanomaterial in treatment

Lipid nanoparticles such as liposome is the most commonly used nanomaterial for cancer therapy [21,22]. Polymeric nanomaterial are used in treatment due to their unique size, degradation rate, and their capacity to deliver different therapeutic agents and imaging probes [23]. Dendrimer-based nanoparticles act as a imaging agents to identify the targeted cancer cells [24,25].

Gold, silver, and platinum are used as theranostics since many years [26]. These noble metal nanoparticles (NPs) have optical properties that can be efficiently adjusted to acceptable wavelengths based on their size, shape, and composition [27]. They can be organized to make functional with antibodies, peptides, DNA or RNA in achieving different cells with biocompatible polymers [28].

Gold nanomaterials can be made into nano rods, nano shells and nano cages by adjusting the composition and making it a therapeutic agent (Choi et al., 2011). Moreover, silver nanoparticles (Ag NPs) can detect the drug molecule by targeting the drug delivery and fluorescence imaging inside the cancer cells. Magnetic NP are used as a probe for MRI. Among the magnetic NPs, superparamagnetic iron oxide nanoparticles (SPIONPs) are the most commonly used nanomaterials due to their functional properties such as targeting, imaging and therapeutic features [29].

Silica based nanoparticles (SiNPs) are also useful in diagnostic and therapeutic purposes [30]. There are two types of silica nanoparticles: solid silica nanoparticles and mesoporous silica nanoparticles. Sol-gel synthesis and microemulsion silica based nanoparticles are useful in diagnostic imaging like MRI and therapeutic applications such as use of chemotherapeutic agents like doxorubicin, camptothecin, and paclitaxel [31]. These are safe and expected to satisfy the clinical requirements to finally find their applications in clinical practices [27,32].

10.4 Applications in healthcare

Nanomaterials act as therapeutic agents due to its low toxicity, high anticancer effects and are required in low concentration. One of the major advantage is these type of nano drugs doesnot accumulate in cardiac muscle. Doxorubicin was approved by the Food and Drug Administration (FDA) for the treatment of refractory ovarian carcinoma and Kaposi sarcoma [33]. Another widely used drug Paclitaxel is also approved by FDA as antineoplastic agent and is useful for the prevention of restenosis in coronary vessels and the occlusion of the vascular access for hemodialysis [34]. Liposomal amphotericin B are useful in treating visceral leishmaniasis and fungal infections in immunocompromised patients [35].

The nanoparticle drugs are used to prevent stented site restenosis in cardiovascular surgery, age-related macular degeneration (AMD), diabetic retinopathy, AIDS-related cytomegalovirus retinitis and several types of uveitis and chorioretinitis [36]. Nanobiosensors for molecular diagnostics is another application of nanotechnology in medicine. Due to the advancement in omics and molecular biology the identification of molecule in the pathogenesis of disease have become easier. There are currently several nano-bio sensors for cardiovascular diseases using gold NPs (AuNPs) or silicon nano-wires [37,38]. Varieties of nanodrug have been used in diagnosis and therapy due to improved drug bioavailability and better synergistic outcomes compared to conventional chemotherapeutic agents [39] as shown in Fig. 10.1.

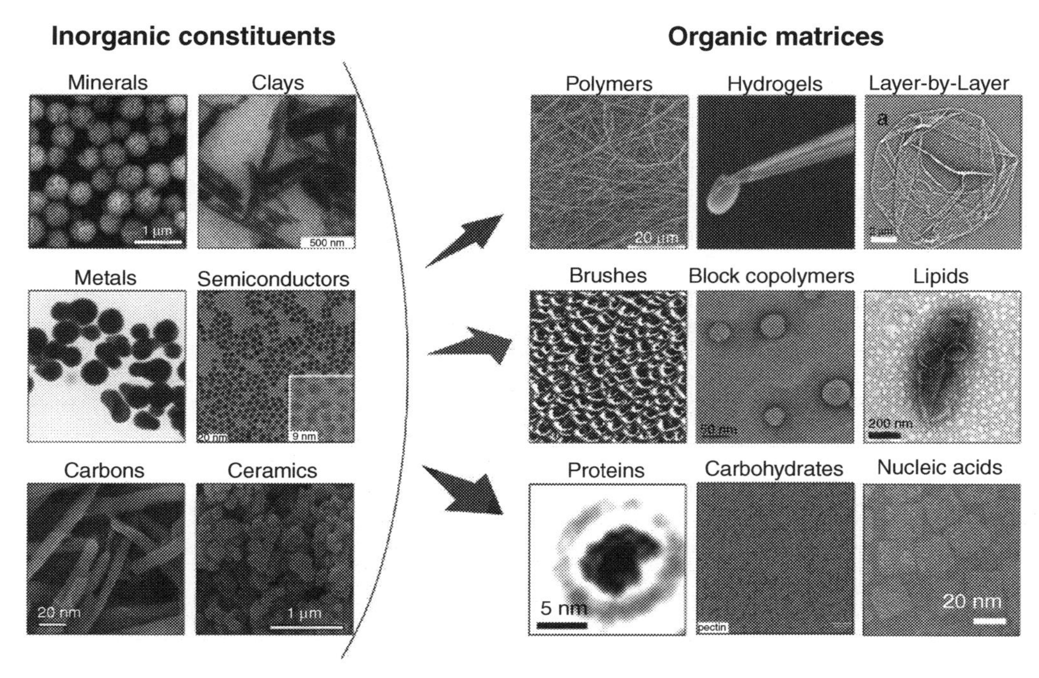

Fig. 10.1 Schematic representation showing the classification of inorganic and organic matrices [49].

10.5 Prevailing scenario of nanomaterials in diagnostics and therapeutics

Nanotheranostics has entered the phase of therapeutics for cancer due to its ability to deliver therapeutic effects. Nanotubes can restore electrical function to damaged hearts. Nanoparticles of per fluorohydrocarbons combined with a lipid layer are used as an ultrasonic contrast agent [40]. Superparamagnetic nanoparticles of iron oxide are used clinically as an MRI contrast agent [41,42]. The metal-containing nanoparticles can be heated using near-infrared radiation [43] or a rapidly oscillating magnetic field to kill the tumor cells [44,45]. Thin layer of nanocrystalline structure can reduce problems with artificial joints [46]. Titanium dioxide nanoparticles also have a bactericidal effect. Silver nanoparticles have antiplatelet properties that may act as future antithrombotic drugs [41]. Silver NP is also used as antiseptic and disinfectant. Silver nanoparticles are effective as they can be integrated with other materials like globular or fibrous proteins and polymers [47]. The nanoparticles then act as depots that continually release new silver ions (Shrivastava et al., 2008). When applied to medical instruments or implants, antimicrobial layers of this help reduce the number of infections. Antimicrobial wound dressings containing nanocrystalline silver are already on the market [48]. The application of nanotheranostics might help in the future for discovering new molecules and manipulating those already available to improve health care as shown in Fig. 10.2.

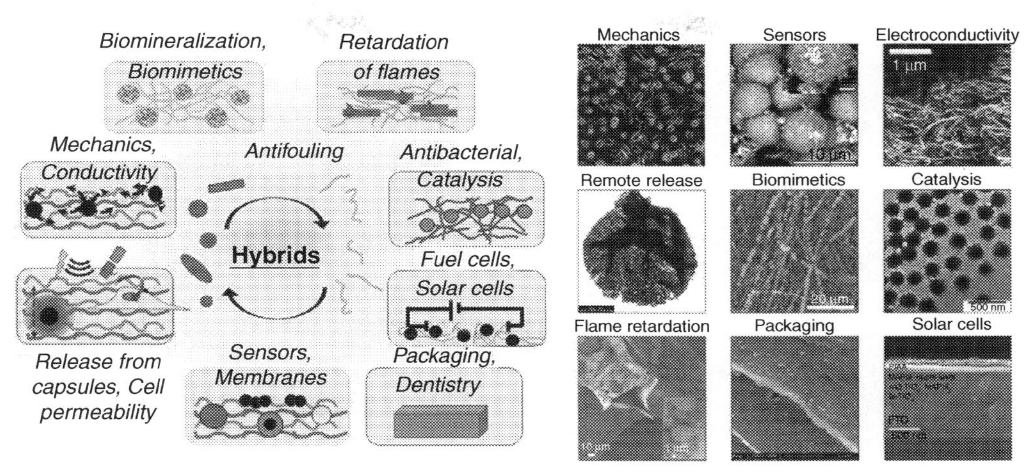

Fig. 10.2 Schematic representation of different types of hybrid nanomaterial [49].

10.6 Future perspectives and conclusions

In recent years, there has been tremendous growth in the field of nanotechnology in diagnosis, treatment, and prevention of diseases. The use of nanodevices will identify the disease at the cellular and molecular level at an early stage. Genomics and proteomics research is rapidly illuminating the molecular basis of many diseases. This has led to the development of a powerful diagnostic tool to identify the genetic pre-deposition of the diseases. Nanotechnology plays a vital role in developing cost-effective diagnostic tools. In the future, nanodevices could provide better medical facilities that could monitor human health. Nanotechnology is a part of a predicted future in which therapy and diagnostic may become more effective. Though the introduction of nanotheranostics as a routine health care system has a long way to go and evaluations concerning cytotoxicity, immunotoxicity, and genotoxicity, their cost-effectiveness and availability of testing systems still need to be developed, monitored and extensively studied further. Therefore, future research and concepts in this field are required to explore nanomedicines toward developing theranostic applications that will act as potential carriers to address these issues in the future.

References

[1] G. Kickelbick, Concepts for the incorporation of inorganic building blocks into organic polymers on a nanoscale, Prog. Polym. Sci. 28 (2003) 83–114.
[2] C. Sanchez, G.J.A.A. Soler-Illia, Hybrid materials, Encyclop. Chem. Proc (2006).
[3] R. Mastria, A. Rizzo, C. Giansante, D. Ballarini, L. Dominici, O. Inganas, et al., Role of polymer in hybrid polymer/PbS quantum dot solar cells, J. Phys. Chem. C 119 (2015) 14972–14979.
[4] J.L. Drury, D.J. Mooney, Hydrogels for tissue engineering: scaffold design variables and applications, Biomaterials 24 (2003) 4337–4351.
[5] R.S. Stowers, S.C. Allen, L.J. Suggs, Dynamic phototuning of 3D hydrogel stiffness, Proc. Natl. Acad. Sci. U.S.A. 112 (2015) 1953–1958.

[6] M. Robitaille, J.Y. Shi, S. Mcbride, K.T. Wan, Mechanical performance of hydrogel contact lenses with a range of power under parallel plate compression and central load, J. Mech. Behav. Biomed. Mater. 22 (2013) 59–64.
[7] G. Agrawal, M.P. Schurings, P. Van Rijn, A. Pich, Formation of catalytically active gold-polymer microgel hybrids via a controlled in situ reductive process, J. Mater. Chem. A 1 (2013) 13244–13251.
[8] V. Selin, V. Albright, J.F. Ankner, A. Marin, A.K. Andrianov, S.A. Sukhishvili, Biocompatible nanocoatings of fluorinated polyphosphazenes through aqueous assembly, ACS Appl. Mater. Interfaces 10 (2018) 9756–9764.
[9] R. Von Klitzing, Internal structure of polyelectrolyte multilayer assemblies, Phys. Chem. Chem. Phys. 8 (2006) 5012–5033.
[10] J. Banjare, Y.K. Sahu, A. Agrawal, A. Satapathy, Physical and thermal characterization of red mud reinforced epoxy composites: an experimental investigation, in: S. Narendranath, M.R. Ramesh, D. Chakradhar, M. Doddamani, S. Bontha (Eds.), International Conference on Advances in Manufacturing and Materials Engineering, Mangalore, 2014, pp. 755–763.
[11] D. Boyaciyan, P. Krause, R. Von Klitzing, Making strong polyelectrolyte brushes pH-sensitive by incorporation of gold nanoparticles, Soft Matter 14 (2018) 4029–4039.
[12] A. Chremos, J.F. Douglas, A comparative study of thermodynamic, conformational, and structural properties of bottlebrush with star and ring polymer melts, J. Chem. Phys. (2018) 149.
[13] N. Ayres, Polymer brushes: applications in biomaterials and nanotechnology, Polym. Chem. 1 (2010) 769–777.
[14] G.M. Cooper, The Cell: A Molecular Approach, Sinauer Associates, Sunderland, MA, 2000.
[15] J.F. Nagle, S. Tristram-Nagle, Structure of lipid bilayers, Biochim. Biophys. Acta 1469 (2000) 159–195.
[16] O.S. Andersen, R.E. Koeppe, Bilayer thickness and membrane protein function: an energetic perspective, Annu. Rev. Biophys. Biomol. Struct. 36 (2007) 107–130.
[17] H. Hauser, M. Stubbs, M.C. Phillips, Ion permeability of phospholipid bilayers, Nature 239 (1972) 342–344.
[18] L.V. Sigolaeva, D.V. Pergushov, M. Oelmann, S. Schwarz, M. Brugnoni, I.N. Kurochkin, et al., Surface functionalization by stimuli-sensitive microgels for effective enzyme uptake and rational design of biosensor setups, Polymers 10 (2018) 791.
[19] A. Kuzyk, R. Schreiber, Z.Y. Fan, G. Pardatscher, E.M. Roller, A. Hogele, et al., DNA-based self-assembly of chiral plasmonic nanostructures with tailored optical response, Nature 483 (2012) 311–314.
[20] F. Ziemann, J. Radler, E. Sackmann, Local measurements of viscoelastic moduli of entangled actin networks using an oscillating magnetic bead micro-rheometer, Biophys. J. 66 (1994) 2210–2216.
[21] R. Kumar, A. Kulkarni, D.K. Nagesha, S. Sridhar, In vitro evaluation of theranostic polymeric micelles for imaging and drug delivery in cancer, Theranostics 2 (7) (2012) 714.
[22] K. Shashi, K. Satinder, P. Bhart, A complete review on liposomes, Int. Res. J. Pharm 3 (2012) 10–16.
[23] UK. Sukumar, B. Bhushan, P. Dubey, I. Matai., A. Sachdev, G. Packirisam, Emerging applications of nanoparticles for lung cancer diagnosis and therapy, Int. Nano Lett 3 (2013) 1–17.
[24] A. Fernandez-Fernandez, R. Manchanda, A.J. McGoron, Theranostic applications of nanomaterials in cancer: drug delivery, image-guided therapy, and multifunctional platforms, Appl. Biochem. Biotechnol. 165 (7-8) (2011) 1628–1651.
[25] J.B. Wolinsky, M.W. Grinstaff, Therapeutic and diagnostic applications of dendrimers for cancer treatment, Adv. Drug Deliv. Rev. 60 (9) (2008) 1037–1055.
[26] G. Doria, J. Conde, B. Veigas, L. Giestas, C. Almeida, M. Assunção, J. Rosa, P.V. Baptista, Noble metal nanoparticles for biosensing applications, Sensors 12 (2) (2012) 1657–1687.
[27] B. Sisay, S. Abrha, Z. Yilma, A. Assen, F. Molla, E. Tadese, A. Wondimu, N. Gebre-Samuel, G. Pattnaik, Cancer nanotheranostics: a new paradigm of simultaneous diagnosis and therapy, J. Drug Deliv. Therap. 4 (5) (2014) 79–86.
[28] J. Conde, G. Doria, P. Baptista, Noble metal nanoparticles applications in cancer, J. Drug Deliv. 2012 (2012).
[29] J. Xie, G. Liu, H.S. Eden, H. Ai, X. Chen, Surface-engineered magnetic nanoparticle platforms for cancer imaging and therapy, Acc. Chem. Res. 44 (10) (2011) 883–892.

[30] L.S. Wang, M.C. Chuang, J.A.A. Ho, Nanotheranostics–a review of recent publications, Int. J. Nanomedicine 7 (2012) 4679.
[31] J.L.Vivero-Escoto,Y.T. Huang, Inorganic-organic hybrid nanomaterials for therapeutic and diagnostic imaging applications, Int. J. Mol. Sci. 12 (6) (2011) 3888–3927.
[32] Q. He, M. Ma, C. Wei, J. Shi, Mesoporous carbon@ silicon-silica nanotheranostics for synchronous delivery of insoluble drugs and luminescence imaging, Biomaterials 33 (17) (2012) 4392–4402.
[33] A.G. Cattaneo., R. Gornati., E. Sabbioni., M. Chiriva-Internati., E. Cobos., R.J. Marjorie, G. Bernardinia, Nanotechnology and human health: risks and benefits, J. Appl. Toxicol. 30 (2010) 730–744.
[34] J. Margolis, J. McDonald, R. Heuser, P. Klinke, R. Waksman, R. Virmani, N. Desai, D. Hilton, Systemic nanoparticle paclitaxel (nab-paclitaxel) for in-stent restenosis I (SNAPIST-I): a first-in-human safety and dose-finding study, Clin. Cardiol. 30 (4) (2007) 165–170.
[35] R.J. Hay, Liposomal amphotericin B. AmBisome. J. Infect. 28 (1994) 35–43.
[36] SA. Wickline, AM. Neubauer, P. Winter, S. Caruthers, Applications of nanotechnology to atherosclerosis, thrombosis, and vascular biology, Arterioscler. Thromb. Vasc. Biol. 26 (2006) 435–441.
[37] J.H. Chua, R.E. Chee, A. Agarwal, S.M. Wong, G.J. Zhang, Label-free electrical detection of cardiac biomarker with complementary metal-oxide semiconductor-compatible silicon nanowire sensor arrays, Analyt. Chem. 81 (15) (2009) 6266–6271.
[38] V. Pavlov, Y. Xiao, B. Shlyahovsky, I. Willner, Aptamer-functionalized Au nanoparticles for the amplified optical detection of thrombin, J. Am. Chem. Soc. 126 (38) (2004) 11768–11769.
[39] H. Peng, X. Liu, G. Wang, M. Li, K.M. Bratlie, E. Cochran, Q. Wang, Polymeric multifunctional nanomaterials for theranostics, J. Mat. Chem. B. 3 (34) (2015) 6856–6870.
[40] P.A. Dayton, K.W. Ferrara, Targeted imaging using ultrasound, J. Magn. Reson. Imaging 16 (4) (2002) 362–377.
[41] S. Shrivastava, T. Bera, S.K. Singh, G. Singh, P. Ramachandrarao, D. Dash, Characterization of antiplatelet properties of silver nanoparticles, ACS Nano. 3 (6) (2009) 1357–1364.
[42] R. Weissleder, G. Elizondo, J. Wittenberg, C.A. Rabito, H.H. Bengele, L. Josephson, Ultrasmall superparamagnetic iron oxide: characterization of a new class of contrast agents for MR imaging, Radiology 175 (2) (1990) 489–493.
[43] L.R. Hirsch, R.J. Stafford, J.A. Bankson, S.R. Sershen, B. Rivera, R.E. Price, J.D. Hazle, N.J. Halas, J.L. West, Nanoshell-mediated near-infrared thermal therapy of tumors under magnetic resonance guidance, Proc. Natl. Aca. Sci. 100 (23) (2003) 13549–13554.
[44] M. Johannsen, B. Thiesen, A. Jordan, K. Taymoorian, U. Gneveckow, N. Waldöfner, R. Scholz, M. Koch, M. Lein, K. Jung, S.A. Loening, Magnetic fluid hyperthermia (MFH) reduces prostate cancer growth in the orthotopic Dunning R3327 rat model, Prostate 64 (3) (2005) 283–292.
[45] D.P. O'Neal, L.R. Hirsch, N.J. Halas, J.D. Payne, J.L. West, Photo-thermal tumor ablation in mice using near infrared-absorbing nanoparticles, Cancer Lett. 209 (2) (2004) 171–176.
[46] G.E. Park, T.J. Webster, A review of nanotechnology for the development of better orthopedic implants, J. Biomed. Nanotech. 1 (1) (2005) 18–29.
[47] F. Furno, K.S. Morley, B. Wong, B.L. Sharp, P.L. Arnold, S.M. Howdle, R. Bayston, P.D. Brown, P.D. Winship, H.J. Reid, Silver nanoparticles and polymeric medical devices: a new approach to prevention of infection? J. Antimicrob. Chemother. 54 (6) (2004) 1019–1024.
[48] A.B. Lansdown, A guide to the properties and uses of silver dressings in wound care, Prof. Nurse (London, England) 20 (5) (2005) 41–43.
[49] M.S. Saveleva, K. Eftekhari, A. Abalymov, T.E. Douglas, D. Volodkin, B.V. Parakhonskiy, A.G. Skirtach, Hierarchy of hybrid materials—the place of inorganics-in-organics in it, their composition and applications, Front. Chem. 7 (2019) 179.

CHAPTER 11

Carbon nanomaterials: Application as sensors for diagnostics

Naveen K. Dandu[a], Ch. G. Chandaluri[b], Kola Ramesh[c], D. Saritha[c], N. Mahender Reddy[c], Gubbala V. Ramesh[c]
[a]Material Science Division, Argonne National Lab, Argonne, IL
[b]Faculty of Chemistry, Humanities and Sciences Division Indian Institute of Petroleum and Energy, Visakhapatnam, India
[c]Department of Chemistry, Chaitanya Bharathi Institute of Technology (A), Gandipet, Hyderabad, Telangana, India

11.1 Introduction

The sensors devised utilizing nanomaterials for the diagnostic applications is of primary objective of research around the globe to exploit the intrinsic properties of nanomaterials. Carbon nanomaterials were rigorously investigated among the numerous available materials for the design and improvement of various sensors in medical diagnosis fields. Apart from the extraordinary electronic, magnetic and optical properties, the chemical versatility, various easily adaptable synthetic methods, substantial biocompatibility and superior chemical stability making carbon nanomaterials as potential candidates for their wide usage in the design of sensors. Carbon nanomaterials based on their shape and dimensions can be grouped as carbon nanotubes, carbon quantum dots, graphene, carbon nano horns, graphene oxide, and carbon nanodiamonds. All these carbon allotropes are employed for the creation of sensors to exploit their inherent properties.

The prospective of carbon nanomaterials is perfect in sensing applications owing to its huge surface/volume proportion, superior conductivity, and electron flexibility. The execution of diverse carbon nanomaterials in the growth of biosensors has seized the concerns of researcher's globally since the finding of carbon-centered nanomaterial's in the 1990s [1]. Enormously adaptable, functioning with diverse segments are the key benefits of Carbon nanomaterials [2]. Carbon nanoparticles have been investigated for their long-term use in the exploitation of bio sensing devices [3]. These materials have been employed as electrochemical probe sowing to its exclusive electrochemical assets viz. an enormous potential window, low price, and minor background current. Carbon nanomaterial's have modernized the electrochemical finding of numerous analytes due to its biocompatibility and also utilized for the qualitative and measurable of electro active analytes. These approaches are precise, consistent and economical.

The carbon atom has valency of four and has the capability to form covalent bonds among themselves or with other essentials and also undergo polymerization. The carbon nanotubes find immense practical applications in electronics, storage of gas, plastic

Advanced Nanomaterials for Point of Care Diagnosis and Therapy
DOI: https://doi.org/10.1016/B978-0-323-85725-3.00015-5

production, composites, decorates, fabrics, batteries and biosensors, etc. accredited to their little contaminated nature and massive fabrication for consumption [4]. Carbon nanotubes are employed as imperative transducer resources in biosensors owing to its great aspect ratios, extraordinary mechanical property, great surface area, brilliant thermal/chemical steadiness, and electronic/optical properties [5]. Graphene is a possible aspirant electrochemical sensor. Graphene has vast utilization in sensors owing of huge surface area and/or great charge carrier movement. Fullerenes have explored in therapeutic sectors. But, the applied employment of fullerenes is restricted owing to their production price and small yields of the products. The Carbon dots displayed a biocompatibility and it causes the progress of biosensors. The carbon dots founded biosensors creating a promising atmosphere for the finding of biologically momentous ions. The carbon dots, carbon quantum dots have showed enhanced productivity and choosiness for the connection of proteins, antibodies and enzymes through adsorption, chemical responses. Chemical steadiness, solubility, brilliant electro catalytic to the finding of H_2O_2 and O_2 are found in these materials. The carbon dots have been working to sense an enormous quantity of biomolecules with great sensitivity and selectivity.

Biosensors are influential and pioneering diagnostic devices that associate biological constituents and transducers for the finding of biological constituents. A biosensor is a distinctive device and it has accurate identification ability, employing diverse biomolecules viz. enzymes, antibodies as identification components for the finding of samples including biomarkers, drugs, metabolites, contaminants, poison etc. and changing biological signals into physicochemical indications that ultimately mention the quantity of illustration as showed in Fig. 11.1 [6]. Biosensors changes biochemical responses into assessable signals and it is proportionate to the concentration of analyte [7]. Biosensors performance depends on Choosiness, Sensitivity, reproducibility, Steadiness, reaction time.

Biosensors have comprehensively explored for the finding of biological fragments, pathogens and further illness producing mediators in the medical segment. The biosensors are essentially chemical sensors that create practice of the finding the properties of the biomolecules. A distinctive biosensor comprises of three measures (1) a recognition fragment namely, protein, enzyme, DNA, etc., (2) a transducer component registers the interface as an indication among the analyte and the identifying fragment, and (3) a signal processor. Biosensors have emerging extensive exploitation in the nutrition production, medical checkup, biomedicine, and ecological valuation. Biological sensors have the various benefits and are extensively explored in numerous fields including ecological review [8], biomedical study [9], nutrition analysis [10], and fermentation manufacturing [11].

Biosensors have employed in food, ecological, and human samples [12]. It is imperative to discover materials and ability to associate with detecting components and a decent signal-transducing property to enable the advanced progress of biosensors.

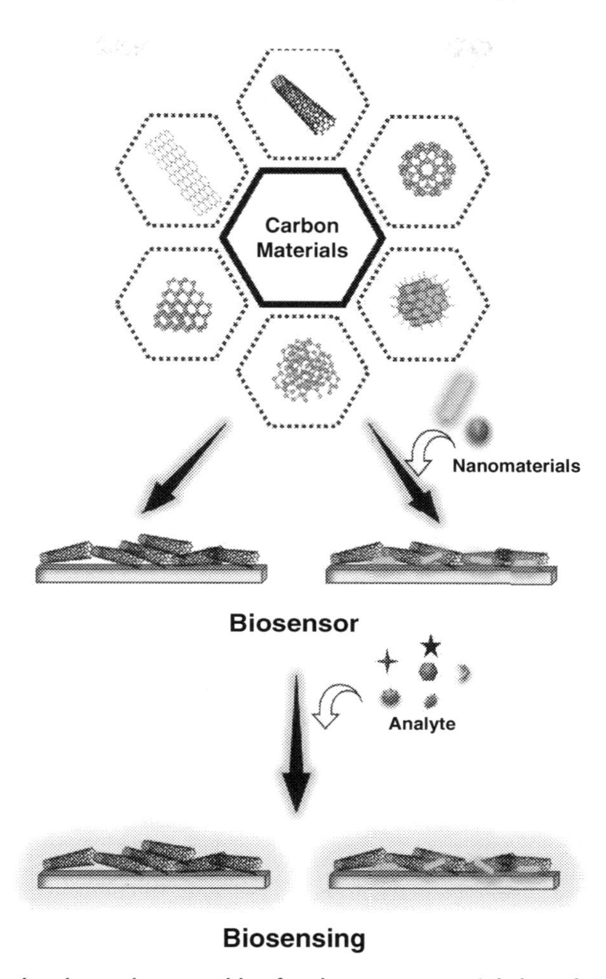

Fig. 11.1 *Schematic plot shows the assembly of carbon nanomaterials-based sensors for diagnostic applications.*

Carbon based nanomaterial's exposed brilliant integration with identifying components because of their alteration ability and precise conjugated structure [13]. Carbon nanomaterials are extensively utilized in biosensors for the recognition of fragments. This chapter discusses current advancements of carbon-based nanomaterials as biosensors.

11.2 Carbon-based nanomaterials

Carbon-supported nanomaterials including graphene, nanodiamonds, carbon nanotubes, graphene oxides, carbon dots, and so on, are explored intensively in various applications (Fig. 11.2). Research in these nanomaterials was awarded two Nobel prizes. One for the discovery and study of Fullerenes and other for the discovery of graphene

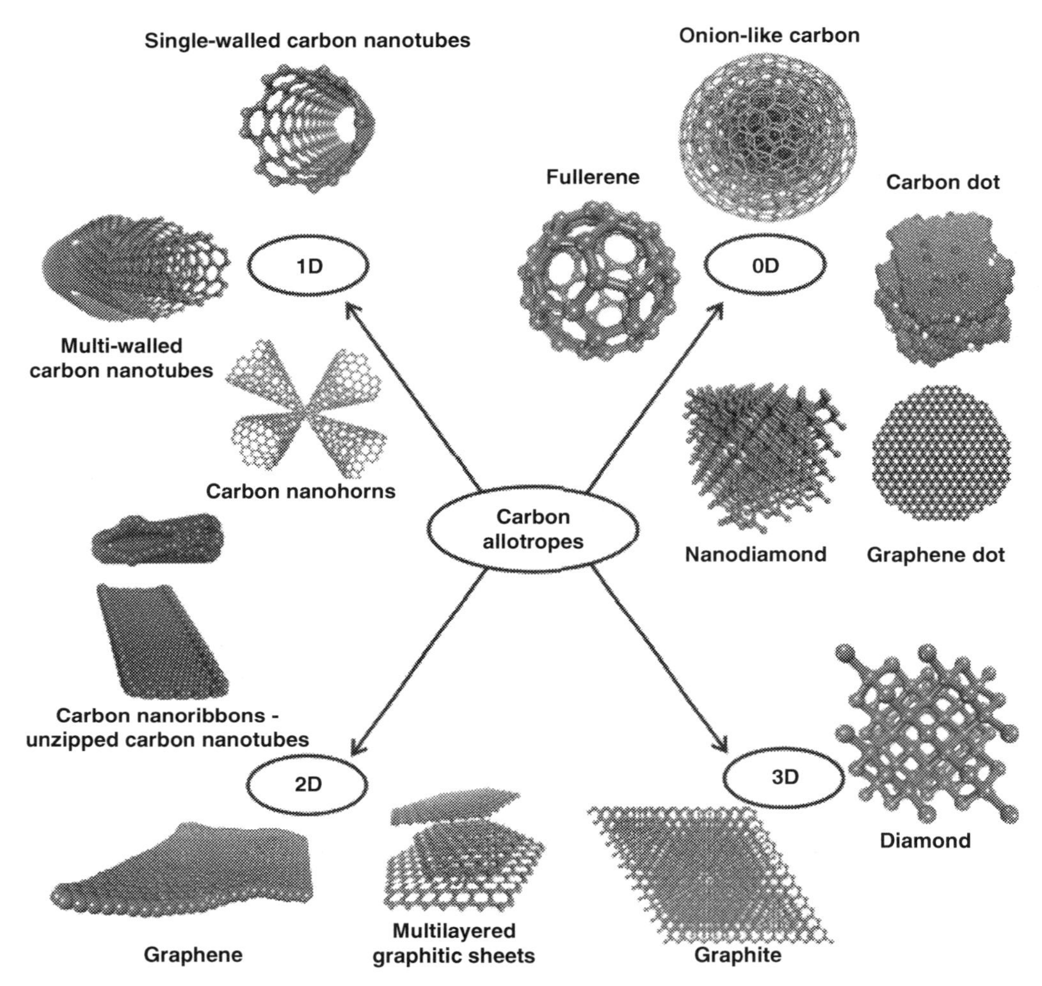

Fig. 11.2 ***Schematic structures of nanoscale carbonaceous materials [14].*** *(Copyright © 2010, AAAS).*

[14–15]. In the following sections of the chapter, we have discussed the details of each of these nanomaterials, including their synthesis and applications.

11.2.1 Fullerenes

It was discovered by Harold et al. in 1985 [15]. They comprise of sp^2 hybridized carbon atoms, with an empirical formula of C_n (n>20), that form closed or partially closed spheres, with 5 to 7 atoms fused rings. Among many fullerenes, C60, that can be found in soot, has been studied extensively. Their structure resembles to that of a soccer ball [16], containing five and six membered fused rings as shown in Fig 11.1. They are made by evaporating graphitic electrodes in helium [17–18]. Chemical method is another route, but the yields are very low that these methods are used only for research purposes.

One of the examples of the chemical methods used in synthesizing fullerenes is the synthesis of C_{60} from pyrolytic hydrogenation of naphthalene, which requires a high energy. Dopants can be introduced into fullerene structures to change their properties. Endohedral fullerenes are the ones where atoms are enclosed inside the fullerenes. There are two forms of endohedral fullerenes: endohedral metallofullerenes, which utilize metals as dopants, and endohedral non-metal doped fullerenes, which use non-metals as dopants [19–20]. Different fullerenes are used for the applications based on the desired properties. They are used in cancer therapies and MRI [17].

11.2.2 Carbon nanotubes

CNTs were exposed by S. Ijima in 1991 [18]. CNTs have pseudo one dimensional structures are produced by rolling a graphene sheet into a cylindrical shape. The CNTs form in variable lengths and diameters. Moreover, CNTs can form single walled CNTs (SWCNTs) and multi-walled CNTs (MWCNTs), depends on their structures. SWCNTs range with diameters between 1-3 nm and MWCNTs range between 5–25 nm.

The layout of carbon atoms in x and y directions of a graphene sheet are not equivalent due to the lack of axis of symmetry. For the path traversing in the x-direction leads in the "armchair" shape, while in the y-direction, it follows a "zigzag" path. Rolling a graphene in to a CNT involves the superposition of two points on the plane. The variance in superposition at different points lead to large diversity of chemical structures. A chiral vector (Ch), formed by drawing a vector between the points, characterizes the formation of CNT. Electronic and optical properties of the formed CNTs significantly depend on their chirality. CNTs are metallic or semiconducting based on their chirality [19]. Armchair SWCNTs have the tendency to exhibit electric conductivity whereas zigzag SWCNTs show semiconducting properties, applicable to sensor fabrications [20]. On the other hand, MWCNTs exhibit extraordinary mechanical properties. CNTs have been synthesized mainly using one of the following three methods. Each of these processes has its own set of benefits and drawbacks when it comes to generating high-quality CNTs.

- **Chemical vapor deposition (CVD):** This method produces good quality CNTs, but with low yields. In a tube furnace, hydrocarbon gas is transported through a catalyst. Usually, a transition metal nanoparticle supported on alumina is used as a catalyst. The material is cleaned after the furnace has cooled to room temperature. Depending on the type of catalyst, hydrocarbon and operating temperature, various CNTs are produced [21].
- **Arc discharge:** When compared to CVD, it produces excellent yields. The carbon is evaporated by helium plasma and employing cobalt as metal catalyst in this process. The formed CNTs bind together due to van der Waals interactions.
- **Laser vaporization:** Laser vaporization produces high yields compared to CVD. In a 1200°C tube furnace, lasers are used to ablate a carbon target. CNTs form as aggregates of carbon-saturated catalyst nanoparticles over the metal in this process [22].

The length of synthesized CNT is determined by the amount of time it takes for them to grow. CNTs can be employed in medication delivery applications. Various medications have been shown to load successfully on the inner and exterior surfaces of CNTs in tests.

11.2.3 Graphene

Graphene is a 2D nanomaterial comprising of a single layer of carbon atoms bond through sp^2 hybridization joined by single and double bonds as exposed in Fig. 11.3 [23–24]. It has 0.34 nm thickness and 0.14 nm C-C bond distance. It has analogous properties such as mechanical, thermal, and electrical properties those of carbon nanotubes. Furthermore, graphene is more easier to synthesise than carbon nanotubes, with higher yields and higher quality.Graphene is used as a building block for producing other carbon allotropes such as 3-D graphite (by stacking graphene layers), 1-D nanotubes (by rolling) and 0-D fullerenes (by wrapping) as shown in Fig. 11.3.

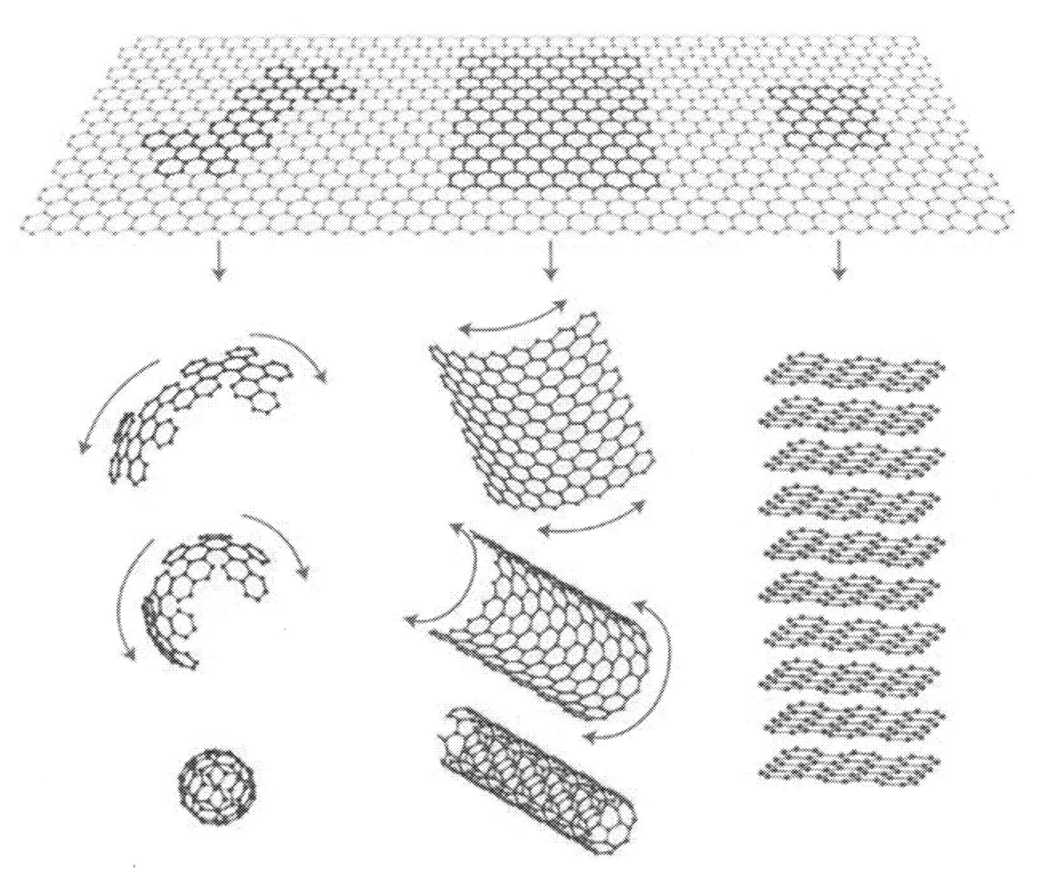

Fig. 11.3 ***Graphene build up into different structures [24].*** *(Copyright © 2007, Springer Nature).*

Graphene exhibits remarkable properties including advanced inherent mobility at 27°C (250,000 $cm^2V^{-1}s^{-1}$) [25], Young's modulus equal to 1 TPa [26], thermal conductivity equal to 3000 W/Mk [27], and exceptional optical transmittance (≈97.7%) [28]. Other distinct properties of graphene are they possess zero band gap semiconducting properties that can be tuned for various electronic applications [29], exhibits anomalous quantum Hall effect (QHE) [30], has an ability to carry high current densities [31], and are impermeable to gases [32].

Following five methods are considered the best in synthesizing graphene with high quality and scalability (Fig. 11.4) [33]:

- **Mechanical exfoliation:** Using Scotch tape, bulk graphite is exfoliated into graphene sheets, which are then pressed on Si, SiO2, or Ni substrates [38]. It is a low-cost approach, however it performs poorly in terms of scalability.

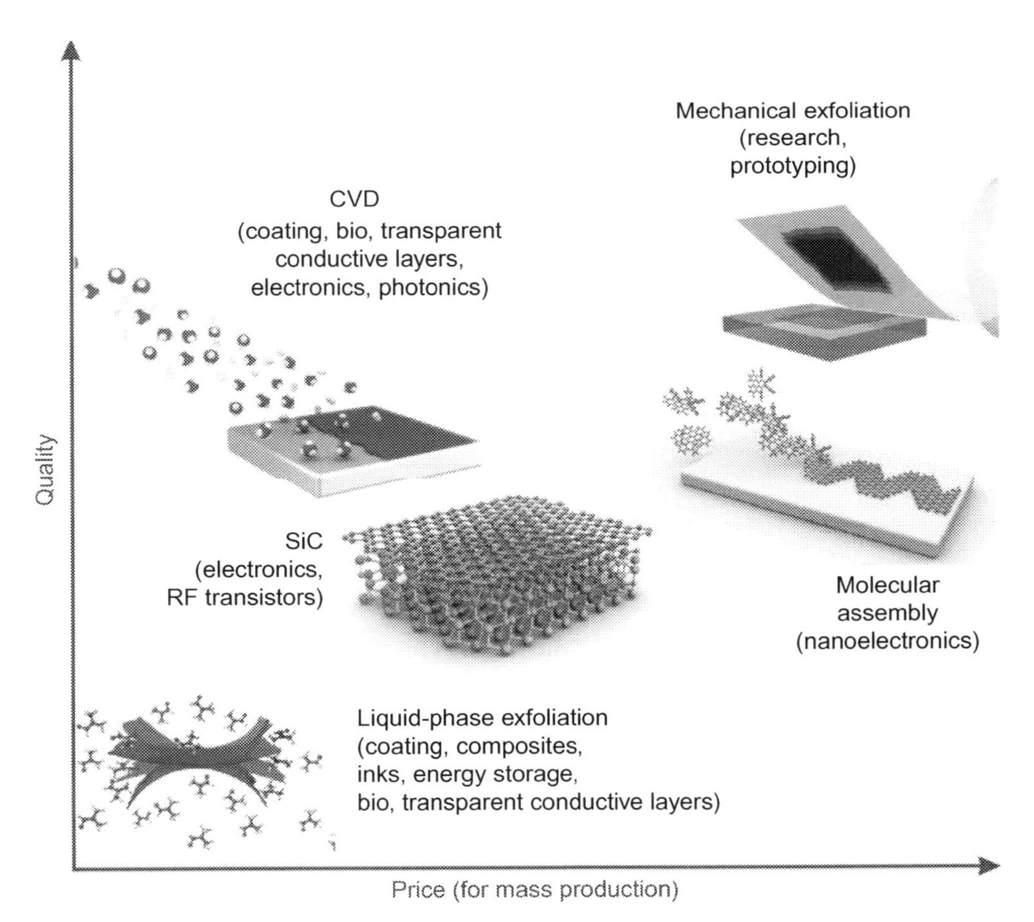

Fig. 11.4 ***Several methods for the production of graphene [34].*** *(Copyright © 2012, Springer Nature).*

- **Chemical exfoliation:** In this method, graphite is exfoliated to graphene by chemical method in presence of a solvent. This can happen either by enthalpy of mixing or due to charge transfer. In case of enthalpy of mixing, if the surface energy of graphene matches with the solvent, enthalpy of mixing is minimal and hence exfoliation is possible under sonication. Examples of such solvents that match the surface energy of graphene are dimethyl formamide, benzyl benzoate, etc. [34]. In the instance of charge transfer, interactions with ionic liquids result in charge transfer with graphite, resulting in exfoliation [35]. A disadvantage with this method is that due to sonication, graphene sheet may get damaged. However, with electrochemical exfoliation, problem with sonication was avoided, that produced better quality graphene [36].
- **Chemical exfoliation via graphene oxide:** For the manufacture of graphene, graphene oxide, which is also a good semiconductor, has been employed as a precursor. Reduced graphene oxide (rGO) can be prepared from graphene oxide (GO) *via* thermal or chemical procedure. Both GO and rGO can be differentiated easily. When GO is immersed in DMF, the solution becomes brown, whereas rGO solution becomes black. While C to O ratio in GO is typically 4 to 1, after reduction the formed rGO can have C to O ratio of 12 to 1. In the process of thermal reduction, GO undergoes thermal annealing to form GO [37]. Wu et al. reduced GO to rGO using arc discharge method [38]. Although rGO was produced in high quantity, high energy is consumed while using thermal reduction process. Chemical method is cheaper and easier process in reducing GO to rGO. Chemical reagents such as hydrazine [39] hydroiodic acid and ascorbic acid [40] were used to convert GO to rGO. Although conversion of GO to rGO was a feasible method, quality of produced rGO was poor.
- **CVD:** Graphene was generated from C containing gases via catalytic metal surfaces including Fe [41], Ni [42], Co, Pt, and Pd [43]. CVD and surface segregation happen simultaneously during the CVD process, causing carbon atoms from the gas source to permeate into the metal [44]. However, the CVD technique has three major drawbacks: (1) it is expensive because to the high energy required, (2) substrate transfer is difficult, and (3) crystallographic growth orientation is crucial for electronic applications [49].
- **Synthesis on SiC:** SiC is a fantastic material for high-efficiency transistors and other electronic components. Graphene synthesis on SiC is known as graphitization. C or Si can be used to grow graphene. It has been found that growth on Si has a control over the amount of graphene layers [45]. The high cost and temperature of SiC wafers are two major drawbacks of this approach.

11.2.4 Carbon nanodiamonds

Carbon nanodiamonds (CNDs) have the crystal structure of diamond and exhibits brilliant properties. CNDs have N-V defect sites (N stands for nitrogen and V stands for vacancy), due to which they display excellent fluorescence. Due to their fluorescence nature, CNDs found their applications as biomarkers [46–47]. CNDs are synthesized by explosives detonations [48]. Doping N vacancies with electron irradiation and heat annealing produces fluorescent CNDs [49]. For extra stability, CNDs are functionalized using covalent or non-covalent methods [50]. Using covalent methods, stable drugs are produced, while using noncovalent methods drugs are easily attached but the lesser stability. These methods are especially useful in the drug delivery applications [51]. CND functionalization with biomolecules is also very easy [52].

11.2.5 Carbon dots

These carbonaceous nanomaterials are described as 0D quasi-spherical nanoparticle sizes varies from 2 to 10 nm [53]. They are amorphous to nanocrystalline core with graphene or graphene oxide flakes (sp^2 carbon domains) surrounded by amorphous carbon frames (sp^3 carbon) with a typical interlayer distance of ca. 0.34 nm [54]. Due to notable properties such as water solubility, tunable functionalization; cheap and easy synthetic routes [55], biocompatibility and quantum confinement effect, these CQDs have potential in replacing toxic metals and semiconductor quantum dots that are presently in use. They are classified into two types, carbon and graphene quantum dots. Carbon quantum dots exhibit little poisonousness. Due to their biocompatibility nature, they can be used as electrochemical biosensors.

11.3 Biosensor applications

11.3.1 Carbon nanotubes

11.3.1.1 Carbon nanotubes in diagnosis of cancer

Electrochemical biosensors are one of prominent techniques explored for the recognition of various kinds of cancer diseases. Electrochemical biosensors are two- or three-electrode electrochemical cells that may convert a physicochemical change into an electrochemical signal. They frequently have a sensing element on the working electrode of the sensor that reacts electrochemically with the analyte and generates an electrochemical signal, which can be a biological material. The apparent challenge is sensitivity which is influenced by electron transfer toward the electrode and selectivity affected by specific reaction between electrode and analyte or target molecule. This challenge can be addressed by increasing surface area of the electrode thereby increasing reactive sites on the surface of the electrode, integrating decent conducting constituents with the electrode for the effective transfer of electrons and catalyzing the electrochemical redox reaction to increase the selectivity and sensitivity. CNTs play crucial role in CNT-based biosensors due to their inherent rewards including high surface area, sensitivity, quick response, simple process, and auspicious transportability.

The early diagnosis of cancer can be accomplished by detecting the biomarkers present in body fluids, which confirms the cancer in the body. Commonly known cancer biomarkers are proteins, glucose, peptides, polysaccharides, uric acid, lipids, DNAs, RNAs, dopamine. Another vital way to diagnose the cancer is the detection of antigens present body fluids.

The unique chemical, electrical and mechanical properties of CNTs tender novel approach for the fabrication new biosensors. For the designing of the better sensors the CNTs should be functionalized or hybridized with organic and inorganic materials/nanomaterials. The functionalization will increase the application prospects. General biomarker capture probes are ssDNAs, (aptamers) and proteins (antigen). Immunoassays depending on antibody antigens, aptamer-RNA, aptamer protein and binding proteins are perhaps the most important analytical techniques for the quantitative detection of biomarkers.

Prostate cancer can be diagnosed early by detecting the presence of biomarkers such as prostate-specific membrane antigen (FOLH1), osteopontin (OPN), prostate cancer antigen 3 (PCA3), prostate stem cell antigen (PSCA), prostate specific antigen (PSA), etc. CNTs were used to design the sensing electrodes to detect the PSA in the body fluids using antibodies as sensing elements on the surface of sensor.

James F. Rusling and his team reported electrochemical recognition of PSA in serum and tissue lysate sample employing SWNT forest platforms. It was reported that the high HRP/Ab2 bioconjugates increased sensitivity to 4 pg mL-1 (100 amol mL^{-1}) for PSA in the 10 μL undiluted of calf blood significantly by the horseradish peroxydase (HRP) labels, as well as secondary antibodies (Ab2) linked to carbon nanotubes Fig. 11.5 [56].

Another MWNT based labeled sensor was reported where polypyrrole (Pp) coated MWNT were decorated with gold nanoparticles (AuNP) and used for the construction of sensing electrode to detect the PSA using primary antibody and HRP label [57]. The label free electrochemical sensing electrodes were designed using SWNT [58] and AuNP supported cross linked starch functionalized MWNT [59] for the recognition of PSA using monoclonal antibody-antigen interaction with very good sensitivities. Carboxylic acid functionalized SWNTs were used to fabricate transparent electrical immunosensor for the recognition of prostate cancer biomarker osteopontin with detection limit of 0.3 pg mL^{-1} [60]. The prostate cancer antigen 3 (PCA3) is a prostate cancer biomarker and for its detection an impedance biosensor was assembled using MWNT. The layer of single-stranded DNA (ssDNA) was immobilized on a layer-by-layer (LbL) film of chitosan (CHT) and MWCNT [61] for the efficient recognition of PAC3 with detection limit of 0.128 nmol mL^{-1}. Oligonucleotide probes were attached on to a nanostructured electroactive polymer/MWCNT composite for the electrochemical recognition of miR-141, a miRNA biomarker. The presence of MWCNT increased the electroactivity of the copolymer. Increment in current owing

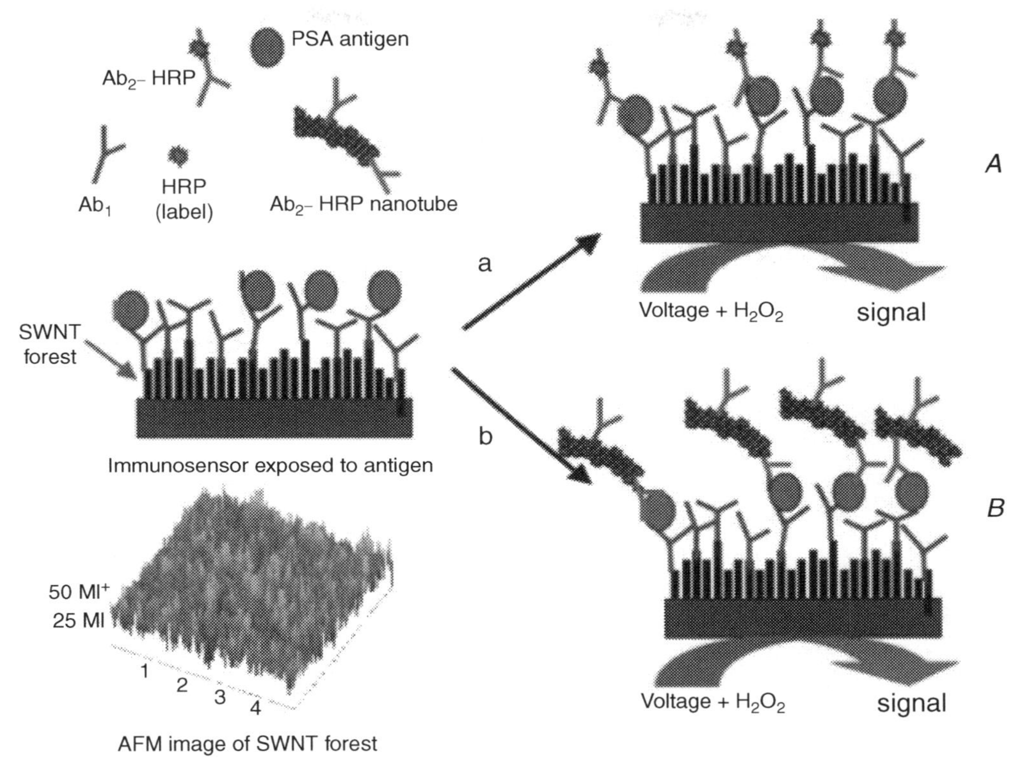

Fig. 11.5 ***Illustration of detection principles of SWNT immunosensors [57].*** *(Copyright © 2006, American Chemical Society).*

to improvement of the polymer electroactivity in the presence of micro-RNA miR-141 and detection limit observed is 8 fM [62]. The field-effect transistors are one of the elegant techniques used for the detection of various biomarkers. In the course, CNTs were modified with the Carnation Italian ringspot virus p19 protein [63] and hereditarily caused single-chain flexible portion (scFv) protein [64] to design carbon nanotubes field-effect transistor for the recognition of prostate biomarkers microRNA miR-141 and osteopontin respectively with good selectivity and sensitivity.

MicroRNA miR-21 is a biomarker for the finding of breast cancer. A glass carbon electrode was altered with MWNT-COOH and immobilized with ss-DNA to get an electrochemical biosensor (ss-DNA/MWCNT-COOH/GCE) for the recognition of miRNA-21 in methylene blue redox indicator presence [65]. The study states the MWNT modification increased conductance. The sensor can detect the miRNA varies from 0.1 to 500.0 pM with a relatively low recognition limit of 84.3 fM.

Another fascinating analysis was conducted by Benvidi et al. to detect the gene marker BRCA1 for the finding of breast cancer [66]. The GCE electrode was modified with reduced graphene oxide (RGO) and multi-walled carbon nanotubes (MWCNTs).

Then the BRCA1 DNA probe was immobilized using 1-pyrenebutyric acid-N-hydroxy succinimide ester as a scaffold molecule. The resulting MWCNT-modified electrode is exhibiting good sensitivity with finding limit of 3.1×10^{-18} M and RGO-modified electrode with finding limit of 3.5×10^{-19} M.

The gene marker MUC1 is very significant biomarker for the primary diagnosis of breast cancer. The electrochemical aptasensors were developed by modifying the carbon electrodes with SWCNT [67] and core-shell nanofibers, MWCNT and gold nanoparticles [68]. Then the MUC1 aptamer was immobilized on these electrodes and utilized for the recognition of MUC1 protein thereby early diagnosis of breast cancer can be realized. The methodology offers great potential not only for the recognition of the MUC1 biomarker, but also to develop various biosensors by modifying the electrodes with other enzymes or any other sensing elements.

Human epidermal growth factor receptor-2 (HER2) another biomarker for detection of breast cancer. In the pursuit and electrochemical aptamer-based biosensor was developed by altering the glassy carbon electrode with AuNP deposited on a reduced graphene oxide and SWCNT composite, subsequently an aptamer directed against HER2 was immobilized on the electrode [69]. The ensuing electrode was used for the recognition of HER2 in the serum sample with recognition limit of 50 $fg.mL^{-1}$. An impedimetric sensing electrode (AuNPs/MW-CILE) [70] was fabricated using MWCNT/gold nanoparticles (GNPs) in a carbon ionic liquid electrode (CILE) for the recognition of HER2 in serum samples with enhanced electrochemical signal. Another exciting analysis was conducted by Freitas et al., where screen-printed carbon electrodes (SPCEs) were altered by gold nanoparticles (AuNPs) and MWCNT combined with AuNPs [71]. Then these electrodes are modified using HER2 antibody for the recognition of HER2-ECD in serum samples and attained the detection limit of 0.16 ng/mL for SPCE-MWCNT/AuNP and 8.5 ng/mL for SPCE-AuNP.

Carcinoembryonic Antigen (CEA) is a glycoprotein is an imperative biomarker and its detection can aid early diagnosis of various cancers such as breast, ovarian, gastric, lung, and rectal pancreatic cancer. Various interesting results were reported, where, glassy carbon electrodes were modified by uniform nano multilayer film assembly of MWNTs wrapped by poly(diallyldimethylammoniumchloride) & poly(sodium-p-styrene- sulfonate) ($Au/PDCNT/(PSS/PDCNT)_2$ [72], multilayer films produced from Prussian Blue (PB), graphene and carbon nanotubes ($Au\text{-}NPs/(PB\text{-}NPs/rGO\text{-}MWCNTs)_5$) [73]. Then the AuNPs were deposited on these layers and CEA antibodies were immobilized. These sensors were utilized for the detection CEA with very good sensitivity.

A modest method to alter microelectrode array with chitosan–MWCNT–thionine (CS–MWCNTs–THI) hybrid film using single step electrochemical deposition method [74]. Then the anti-CEA was immobilized on this hybrid film employed for the finding CEA with detection limit of 0.5 pg mL^{-1}.

An electrochemical sensor (AuNPs/CNOs/SWCNTs/CS) [75] fabricated by modifying the GCE by means of a nanocomposite made-up of AuNPs, carbon nano-onions (CNOs), SWCNTs and chitosan (CS). The modification rendered enhanced electronic conductivity due to the increased surface area of the electrode and electronic properties of the AuNPs, SWCNTs and CNOs. The sensor showed the low finding limit of 100 fg mL^{-1} for the CEA recognition.

Recently, a novel electrochemical immunosensor was developed using nanofibers decorated with AuNPs and MWCNTs [76]. The core-shell nanofibers were prepared by electrospinning on the electrode surface directly and then modified with AuNPs and MWCNTs. CEA antibody was immobilized on the electrode and used for the recognition of CEA with finding limit of 0.09 ng mL^{-1}.

Recognition of volatile organic compounds in the human breath for the finding of cancer and other diseases is a fascinating approach. For the effective detection of these VOCs many biosensors were developed using carbon nanotubes. SWCNTs were functionalized with organic molecules tricosane (C23H48) or pentadecane (C15H32) to construct a device for the detection of VOCs exposed to representative VOCs 1,2,4-trimethybenzene (TMB) and decane separately [77]. The observations reveal that a change in the resistance of the SWCNT biosensor when exposed to the VOCs. Another sensor was designed using a composite comprising of molecularly imprinted polymer nanoparticles and MWCNT for finding of hexanol a lung cancer biomarker [78]. An electronic nose (e-nose) was developed using conductive polymer nanocomposite (CPC) and carbon nanotubes [79]. The sensor was able to detect the polar and nonpolar VOCs for finding lung cancer with decent selectivity and sensitivity. The initial finding of lung cancer can be realized by the detection of biomarkers anti-MAGE A2 and anti-MAGE A11. The electrochemical sensors were fabricated by modifying the graphite surface with SWCNT and Chitosan (CS) composite. Then the antigens (MAGE A2 and MAGE A11) were immobilized on these electrodes (MAGE A2/CNT-CHI/graphite and MAGE A11/CNT-CHI/graphite) and successfully employed for recognition of the aforementioned lung cancer biomarkers [80].

Biomedical imaging:

Among the imaging techniques fluorescence is an imperative methodology for finding of many cancer diseases. The method suffers from the limited penetration depth of light, contrast probes should be accumulated in the tumor region and probes should uptake lower healthy tissue than malignant tissue. Hence, an enormous demand is there to develop contrast probes to address the said limitations and to offer better protocols for early diagnosis of cancer. Carbon nanotubes offer better solution owing to their admirable properties such as small size, high surface area, conductivity, ease of functionalization there by good biocompatibility, etc.

Robinson et al. studied the fluorescence imaging by means of functionalized SWCNTs with synthetic polymers for the detection of breast tumors in Balb/c mice [81]. Phospholipid-PEG (C18-PMH-mPEG) solubilized SWNTs were prepared for video-rate imaging of cancer tumors depends on the intrinsic fluorescence of SWNTs in the second near-infrared (NIR-II, 1.1−1.4 μm) region after intravenous injection. The significance of the study is ultrahigh accumulation in tumors and high tumor uptake of SWCNTs. The same group elegantly explored the application of SWCNTs for the simultaneous tumor imaging and thermal therapy [82]. The SWCNTs with la large diameters were used for fluorescence imaging in vivo in the long-wavelength NIR region (1500–1700 nm, NIR-IIb) [83]. Recently, He Li et al. reported another interesting application of SWCNTs in real time, non-invasive imaging of tumors using SWCNTs bearing tetrazine (TZ@SWCNTs) and trans-cyclooctene-caged (TCO-caged) molecules by means of bioorthogonal chemistry [84].

Raman imaging is another interesting technique for the imaging and diagnosis of cancer. CNTs play an imperative role in this phenomenon owing to their inherent optical properties. Lamprecht et al. studied the detection of urinary bladder carcinoma (T24) cells by Raman imaging using the folic acid (FA) functionalized double-walled CNTs (DWCNTs) by means of targeted delivery. The exploration indicates the accumulation of majority of DWCNTs around the nucleus after 8-10 h of incubation and the technique is also useful for the location mapping and tracking of cancer cells [85]. Another interesting Raman imaging study using SWCNTs for multi color optical imaging and detection cancer was reported by Liu et al. The SWCNTs were synthesized with five different 13C/12C isotope compositions and conjugated to five targeting ligands to convey molecular specificity. Then, the multiplexed live-cell Raman imaging by staining the cells with a five-color mixture of SWCNTs was carried out. The imaging of cancer cells revealed the momentous up-regulation of epidermal growth factor receptor on LS174T colon cancer cells and the fine resolution can be achieved [86]. Wang et al. prepared nanocomposites by coating the DNA-functionalized SWCNTs with noble metals (Ag or Au) and then the surface modification with polyethylene glycol (PEG). These nanocomposites were conjugated to folic acid (FA) and employed for the targeted Raman imaging of cancer cells [87].

Photo acoustic imaging (PAI) a fascinating imaging technique developed recently, where contrast agents will expand when they absorb incident light and expansion causes in wide-band ultrasonic emission. The emission can be sensed by an ultrasound microphone and the data can be exploited to construct 2D or 3D images. Since the technique depends on detection of sound it has rewards such as good tissue penetration and fine spatial resolution. Various nanomaterials were explored as contrast agents due to their small size; they can penetrate better in to the tissue and can get the better resolution.

The carbon nanotubes explored as contrast agents PAI. Zerda et al. described the preparation of SWCNT based contrast agents utilizing diverse optical dyes. The SWCNTs were coated with optical dyes and conjugated to cyclic Arg-Gly-Asp peptides to molecularly target the $\alpha_v\beta_3$ integrin, that is, coupled with tumor angiogenesis and obtained improved photoacoustic signal [88]. Another fascinating analysis was conducted by Wang et al. for the AI imaging of the gastric cancer cells using SWCNTs. The silica coated gold nanorods were attached on to the surface of MWCNTs and subsequently conjugated with RGD peptide to yield RGD-conjugated sGNR/MWNT probes. These probes were used to target in vivo gastric cancer cells and achieved strong PA imaging in the nude mice [89]. Further the carbon nanotubes not only used in diagnosis but also used for real time imaging and cancer therapies [90,91].

Magnetic Resonance Imaging (MRI) is an influential imaging technique employed for diagnostic and therapeutic procedures because it commends high spatial resolution and penetration capacity coupled with the potential for whole-body imaging. Often MRI procedures require contrast agents and CNTs modified with metals are explored as contrast agents. Wu et al. synthesized the magnetic hybrids of MWCNT and cobalt ferrite ($CoFe_2O_4$) using solvothermal method and successfully applied as contrast agent for the MR imaging of cancer tumors. The application can be extended to chemotherapy by loading with the anti-cancer drug doxorubicin (DOX) and subsequently releasing in a sustainable way [92]. An another study reported by surface functionalizing the iron filled MWCNTs with Gd^{+3} ions for the MR imaging and cancer therapy [93]. In a slightly different study Marangon et al. attached Gd^{+3} ions to the functionalized MWCNTs with chelating ligand diethylenetriaminepentaacetic dianhydride (DTPA). Because of the covalent bond between MWCNT and the ligand possess high association constant for Gd^{3+} and thus dropping the risk of Gd^{3+} release. The intravenous administration of Gd-CNTs in mice was succeeded in strong T1 contrast MR signal enhancement on cellular internalization and the uptake into organs for instance the liver and spleen were magnificently displayed using MRI [94].

11.3.1.2 Carbon nanotubes in glucose detection

Detection of glucose is the most significant and repeatedly used medical test for the diagnosis and manage diabetes, correspondingly there is a humongous demand for the biosensors with great accuracy. Numerous enzymatic and non-enzymatic biosensors were established via carbon nanotubes.

The glucose oxidase (GOx) was immobilized on MWCNTs functionalized with Chitosan (CS) to get MWNT-CS-NW/GOx biosensor. So fabricated biosensor was effectively applied for the glucose electrochemical recognition [95]. The conduction of electrons between GOx and target molecules was aided by MWCNTs-CS-NW. The hybrid films of reduced graphene oxide–multiwalled carbon nanotube (ERGO–MWCNT) [96] and graphene-MWCNTs (G-CNT) [97] were also used for the immobilization of glucose oxidase (GOx) to fabricate an electrochemical biosensor to detect glucose and achieved better sensitivity. The methodology offers vast potential for the immobilization of other enzymes also. Another way to fabricate sensor was reported by using polymer-SWCNT composite. SWCNT was modified with poly(4-vinylpyridine) (P4VP) and glucose oxidase (GOx) was immobilzed to get chemiresistive sensor for the recognition of glucose with high selectivity and quick response time (3s) [98].

The CNTs were also explored to develop composite materials and elegantly used for the construction of biosensors. A composite made of multi-wall carbon nanotubes (MWCNTs), polyaniline (PANI), and gold nanoparticles (AuNPs) used for alteration of the electrode surface and then glucose oxidase enzyme (GOx) carrier was fixed on it to offer a glucose sensor (PTFE/GOx/AuNPs/PANI/MWCNTs/GCE) with high sensitivity of sensitivity of 29.17 mA mM^{-1} cm^{-2} and a detection limit of 0.19 mM [99].

The enzyme-based sensors fabrication is associated with certain draw backs such as multi-step immobilization procedures are often complicated and instability of the sensors on the long run. The performance of these sensors is influenced by the ecological situations such as temperature, pH, and humidity and enzyme-poisoning molecules in the sample. Hence, the sensors may lose their activity and to confront this problem the alternate solutions are required. Non-enzymatic biosensors offer such solutions and CNTs decorated with various metal nanoparticles were used for the creation of biosensors for the glucose detection.

In progress, a biosensor was developed using the vertically aligned array of MWCNTs were modified with cupric oxide (CuO) nanoparticles in two step two-step electrodeposition technique. The Cu nanoparticles (Cu Nps) were deposited onto MWCNTs in the first step, then CuNps were oxidized to CuO in second step. The study claims resulting sensor was presenting higher sensitivity than that of many enzymatic sensors with sensitivity of 2190 μA mM^{-1} cm^{-2} and recognition limit of 800 nM [100]. Another interesting biosensor was developed by depositing the Ni and Cu nanoparticles on MWCNTs. First the MWCNTs were functionalized with carboxylic group and were drop casted on GC electrode. Then Cu and Ni nanoparticles were electro deposited on MWCNT/GC electrode to get a biosensor. The sensor was exhibiting high sensitivity with recognition limit of 0.025 μM [101]. The pulsed laser deposition method was used for the deposition of gold nanostructures on CNTs to develop biosensor for the recognition of glucose. The obtained biosensor was able to detect with greater sensitivity of 25 μA cm^{-2} mM^{-1} and a detection limit of 0.1 mM [102]. A facile strategy to develop a sensor used, where MWCNT were functionalized with metformin (MH). The silver nanoparticles (AgNps) were deposited electro chemically on metformin functionalized MWCNTs. The sensor Ag@MH/MWCNT was used in enzyme-free, amperometric detection of glucose with low recognition limit of 0.0003 μM [103].

11.3.1.3 Carbon nanotubes in the detection of other diseases

The detection of dopamine can lead to diagnose neurological diseases such as Alzheimer's and Parkinson's. Carbon nanotubes were explored for the detection of dopamine. For instance, MWCNTs were used to modify the mesoporous silica (SiO_2) by the sol-gel process utilizing HF as the catalyst to prepare a hybrid material. The hybrid material was coated on a GC electrode for the construction of sensing electrode SiO_2/MWCNT/G and employed for the concurrent detection of dopamine, uric acid and paracetamol with detection limit of 0.014, 0.068 and 0.098 μmol L^{-1}, correspondingly Fig. 11.6 [104].

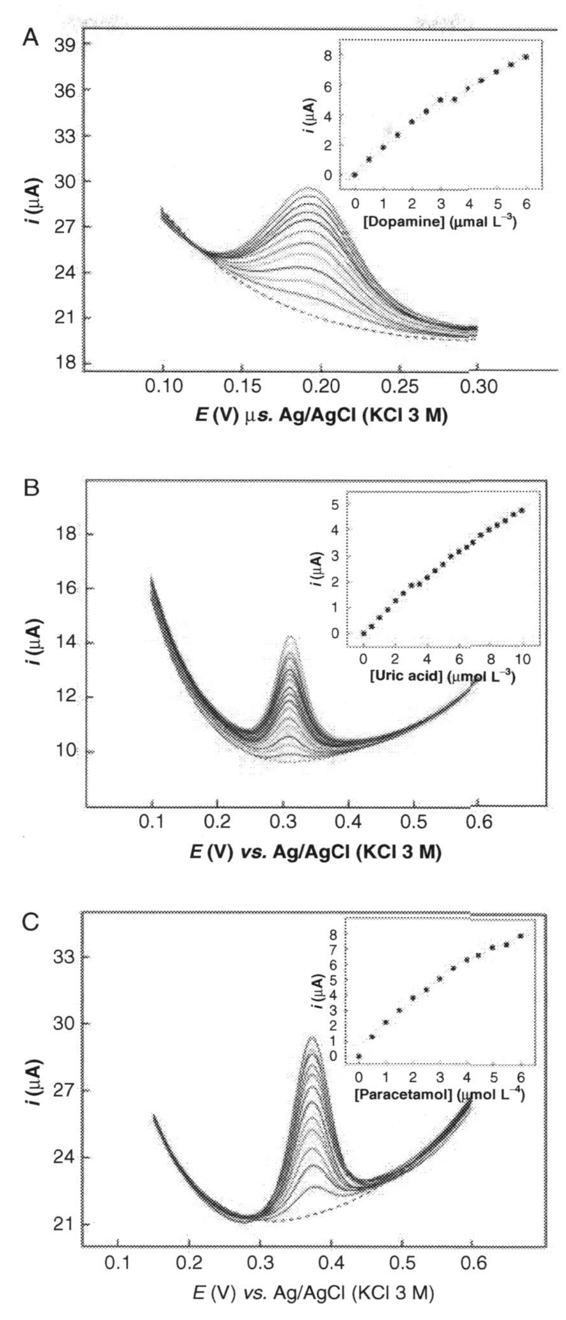

Fig. 11.6 Separate discrepancy pulse voltammograms to SiO2/MWCNT/func/GCE of (A) dopamine, (B) uric acid, and (C) paracetamol with dopamine concentrations extending from 5.0×10^{-7} to 6.0×10^{-6} mol L^{-1}, and uric acid and paracetamol concentrations varies from 5.0×10^{-7}–1.0×10^{-5} mol L^{-1} and 5.0×10^{-7}–6.0×10^{-6} mol L^{-1}, separately, in PBS, pH 7.0. Inset figures: Current intensities plotted against concentration of dopamine, uric acid and paracetamol. [105] *(Copyright © 2013, Elsevier).*

Another endeavor where the nanocomposites were prepared using CNTs and coated on the carbon electrodes for the construction of biosensors. In the course MWCNTs were modified with biofunctional polydopamine and Pt Nps [105]. amphiphilic copolymer poly(vinylbenzyl-thymine-co-styrene-co-maleic anhydride) (PSVM) and Au Nps [106]. platinum nanochains (PtNCs) and graphene nanoparticles (GNPs), [107] to prepare composites used to develop the biosensors for the detection of dopamine with remarkable sensitivities. Recently, a hybrid material was prepared using carboxylated MCNTs and hydroxylated SWCNTs to build a biosensor for the detection of dopamine with very good sensitivity [108]. The netrin-1 is a cardiovascular disease biomarker and MWCNTs, nafion, thionine-coated gold nanoparticles (Thi@AuNPs), and monoclonalantibodies against netrin 1 used for the develop sensing electrode for its detection with detection limit of 30 fg mL^{-1} [109].

11.3.2 Graphene

Graphene centered nanomaterial's have been explored for the electrochemical bio sensing applications such as recognition of bio related molecules and proficient nano transporters for the drug carriage systems. Graphene is appropriate to employ in biosensing procedures owing to its non-toxic nature. Graphenes, including graphene oxide, reduced graphene oxide have exclusive electronic, adsorption, and fluorescence assets. They have been explored as dominant crucial essentials of biosensors for sensing biomarkers [110].

11.3.2.1 Cancer biomarker detection

Cancer is one of the foremost reasons of death in the ecosphere. The most notorious cancers are bladder, ovarian, kidney, pancreatic, lung, skin, breast, lymphoma, esophageal, thyroid, prostate, colorectal cancer [111]. Blood comprises an extensive diversity of protein biomarkers which aids in initial cancer diagnostics and recognition.

The RNA biomarker PCA3 is connected to prostate cancer. PCA3 biosensor fabricated depended on NaYF4:Yb, Er up conversion Nano particles as emitters connected to DNAs as capture evaluates and graphene oxide as the fluorescence quencher. Up conversion nanoparticles are not network with Graphene oxide owing to goal-capture DNA and preserve fluorescence sign at the presence of target PCA3. FET biosensors made-up build on LbL collected graphene complexes for prostate precise antigen (PSA) detection by Zhang et al. [112]. The substrate was dipped into poly(diallyldiamine chloride), poly(styrene sulfonate), and graphene suspension. The target was captured by restraining the PSA antibodies on uppermost graphene layer.

The enhanced performance of sensor was witnessed by hanging the multilayer of the graphene among gold electrode sensor [113]. Lung cancer growth indication (ANXA2, ENO1, and VEGF) is done by fabricating FET biosensor using single crystalline graphene to detect. Antibodies were restrained on graphene via poly-L-lysine.

Sensor performance was amplified because of the lack of grain boundary and substrate scattering. ANXA2, ENO1, and VEGF graphene composite biosensor was also fabricated via self-assembly procedure. PDDA, graphene, and TiO2 layers were deposited on a polymer. The exterior of the composite was deposited by lung cancer antibodies for biomarker capture. Carcinoembryonic antigen (CEA) sensor was fabricated by modifying the graphene foam via polydopamine (pDA) linker, concanavalin A (conA), and HRP-branded anti-CEA antibodies through the lectin-mediated strategy [114]. The assembly and recognition procedure of the sensor is exposed in Fig. 11.7. Xu et al. established tumor marker for recognition of CEA, PSA, and AFP simultaneously [115]. Reduced graphene oxide, PSS altered GCE and nano composite labels of carbon-AuNP were utilized to made this biosensor. Crucial antibodies were restrained on the altered electrode surface. Tags of biocomposites were ready independently for the object biomarkers through adsorbing thionin (Thi), 2,3-diaminophenazine (DAP), Cd2+ on Au nano particles and restraining anti-CEA, anti-PSA, and anti-AFP, correspondingly. Sandwich sort biosensor was assembled via hybridization sequence response and biotin-streptavidin indication intensification approach for detection of CEA, AFP, CA125, and PSA simultaneously [116]. The sensor stage was ready through simultaneous immobilizing of the the four antibodies on uniform GR-Au multilayer film altered GCE. Joining of secondary biotinylated antibodies, SA, oligonucleotides for HCR and redox probes labelled SA on Au/SiO_2-Fe_3O_4 NPs produces Bioconjugate signal tags. The sensor displayed superior sensitivity for the recognition of the four biomarkers simultaneously.

The anthropoid epidermal tumor aspect receptors ErbB2, HER2, and HER3 and carbohydrate antigen 15-3 (CA 15-3) are breast cancer biomarkers. Microfluidic ErbB2

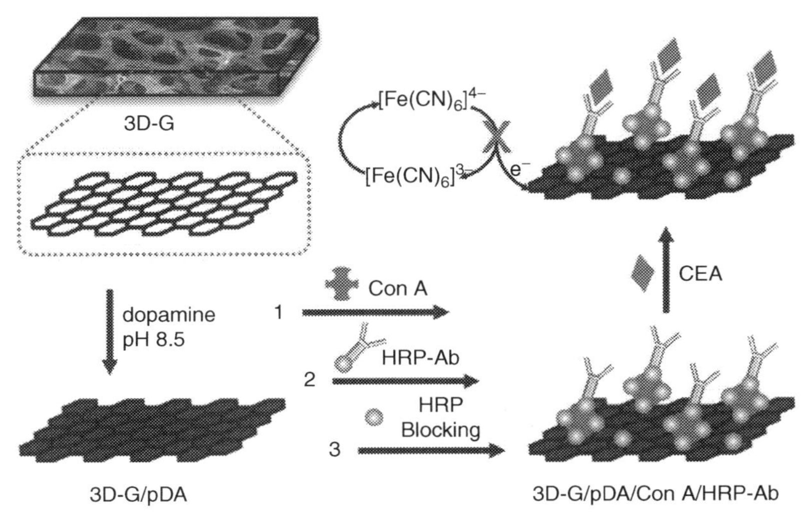

Fig. 11.7 ***Image of the assembly stages and recognition procedure of the graphene foam electrode-founded immunosensor [114].*** *(Copyright © 2015, Elsevier).*

immunosensor assembled founded on permeable GF electrode altered via carbon-doped TiO2 nanofibers and ErbB2 antibodies [117]. Rajesh et al. [118] made-up a delicate HER3 biosensor built on graphene FET decorated with Pt nanoparticles functionalized with antibody. Pt nanoparticles were connected to graphene *via* the 1-methyl pyrene amine linker and thiol-comprising antibodies were restrained on Pt nano particles.

Tumor indication for cervical cancer is Squamous cell carcinoma antigen (SCCA). Sandwich-type immunosensors fabricated using N-doped graphene (N-GS)–chitosan composite film N-GS film on GCE correspondingly by anti-SCCA. The trace tags were Pt-Fe3O4 and Pd-Au/carbon NPs. Later sensor showed improved performance of sensor per LOD of 1.7 pg/mL [119–122]. The biomarker for bladder cancer is the nuclear matrix protein 22 (NMP22). Arresting NMP22 antibodies on AuPdPt nano particles and coating bio conjugates on rGO-TEPA modified GCE produces NMP22 immunosensor [120].

11.3.2.2 Diagnosis of diabetes

Diabetes is an emerging concern presently disturbing billions of publics globally. Deficiency of insulin and hyperglycemia results the metabolic disorder of diabetes mellitus. The diabetes is detected by lesser or upper than the glucose concentration range is 80–120 mg dL^{-1}. Heart disease, kidney failure, or blindness complications are high as a consequence of diabetes. Diabetes is expected to affect over 9% of adults worldwide in 2014, which also lists the disease as a foremost root of mortality and disability. A restricted examination of glucose levels in blood samples is required for the analysis and detection of diabetes mellitus [121].

Graphene based biosensors have utilized for glucose sensing in diabetic diagnostics to improve their sensitivity, time of response, and boundary of recognition. The detection of glucose was affirmed in 2009 via grapheme build electrochemical biosensor. The device exhibited a detection limit in the sort of 2–14 nM and produces steady output. A permeable Graphene biosensor has made via the addition of catalase enzyme [122]. Modified graphene surface using the chitosan-rGO complex can be utilized for the recognition of Glucose and urea with in the sample. Enhancement of the pH value measures the detection of urea via the oxidation of urea such as urease and the reduction of electron transfer of $[Fe(CN)_6]^{3-}$. However the glucose can be detected by the reduction of pH value via the Glucose Oxidase and allows the transfer of electrons in $[Fe(CN)_6]^{3-}$. The glucose in food samples was detected via glucose oxidase enzyme biosensor *via* multiwalled carbon nanotubes functionalized with 4-(pyrrole-1-yl) benzoic acid through reduced graphene oxide [123].

Kang et al. described the graphene founded glucose biosensor made by coating the blend of graphene and chitosan on a carbon electrode has generated gifted outcomes. The detection limit of glucose is 0.3603 mg/dl and range of sensing is from 1.4412 to 216.1871 mg/dl [124]. Polypyrrole centered graphene biosensor was established via the polypyrrole to capture and encapsulate graphene and glucose oxidase on a glass carbon electrode. The detection limit of glucose is 5.4×10^{-2} mg/dL and linear finding range is from $3.60 \times 10^{-2} - 0.7206$ mg/dl [125]. Additional graphene-based glucose sensor was established by chemical vapor deposition. Multilayered graphene petal was grown on a silicon founded exterior. Pt nanoparticles deposited on a three-dimensional graphene petal *via* electrochemical deposition method and further depositied with poly(3,4-ethylenedioxythiophene)-poly(styrenesulfonate) doped through Graphe oxide. The detection limit of glucose is 5.40×10^{-3} mg/dl and linear sensing range is 0.1801–900.7795 mg/dl [126].

Numerous graphene-centered glucose biosensors have been stated [127]. The initial graphene founded glucose biosensor described with graphene/polyethylenimin electrode displayed extensive linear glucose response 2 to 14 mM, decent reproducibility and great steadiness. Chemically reduced graphene oxide (CR-GO) described as glucose sensor [128]. CR-GO biosensor displayed significantly improved amperometric signals for identifying glucose. The linear range is 0.01 – 10 mM, great sensitivity and detection limit is 2.00 mM (S/N ¼ 3). GOD/CR-GO/GC electrode shows quick response to glucose and extremely stable which creates this electrode as a prominent biosensor to detect the plasma glucose level for the analysis of diabetes. Biocompatible chitosan was hired to diffuse graphene and build glucose biosensors [129]. Chitosan facilitated to produce highly spread graphene suspension and arrest the enzyme fragments. It shows outstanding sensitivity and durable steadiness for evaluating glucose. Metal nanoparticles/graphene biosensors have established for the detection of diabetes. Graphene combined with Au nanoparticles to produce a composite film via chitosan which displayed decent electrocatalytical action to H_2O_2 and O_2. Wu et al. explored the GOD/graphene electrode with Pt nano particles via chitosan. The limit of glucose detection is 0.6 mM. Huge surface area and worthy graphene conductivity, synergistic outcome of graphene and metal nanoparticles enhanced the performance [130]. The ADH-graphene-GC electrode reveals quicker response, broader linear range, and lesser detection limit for ethanol detection compared to graphite electrode. The improved performance can be elucidated by the active transfer of substrate and graphene matrixes comprising enzymes in addition to the intrinsic biocompatibility of graphene.

Sophisticated results and steadiness are the attractive features for detection of glucose via a non-enzymatic amperometric sensor. Non-enzymatic electrodes attempt to straightly oxidise glucose in the tester. The capable electron transfer speed and brilliant catalytic material are the two aspects which stimuli the performance of non-enzymatic glucose sensors. Enzyme-free glucose sensors fabricating via graphene gained attention.

11.3.2.3 Detection of genetic complaints

The finding of DNA acts a crucial part in the analysis of numerous genetic complaints in the body. Low price, remarkable sensitivity and amazing choosiness of the electrochemical biosensor explored for the recognition of DNA arrangements for the several human sicknesses. The electrochemical biosensor functioned on the oxidation of DNA principle using the graphene. Graphene electrode has showed an advanced electro catalytic activity for the oxidation of DNA. Oxidation recent indications of permitted bases such as guanine, adenine, thymine and cytosine on DPV are well exhibited by the graphene electrode. The reduced graphene oxide nanowalls has active for the recognition of four bases of DNA via the extremely resolved oxidation signals [131]. The DNA bases have been completely noticed by the anodized epitaxial Graphene with Oxygen defects. The limit of detection is 1 μg mL^{-1}. The Graphene centered sensors have engaged to sense a comprehensive variety of DNA molecules. Reported limit of detection is 40 pM [132]. Enhanced redox properties has been reported by graphene with gold nanoparticles, Fe_3O_4, nickel hydroxide, and cobalt phthalocyanine due to the improved electron affinity for the analyte. Human diseases connected to DNA can be detected by graphene based electrochemical sensors. Chemically reduced graphene oxide (CR-GO) based DNA electrochemical sensor described by Zhou et al. [133]. Four free bases of the DNA such as guanine (G), adenine (A), thymine (T), and cytosine (C) are detected by the signals on the CR-GO/GC electrode.

11.3.2.4 Diagnosis of human immunodeficiency virus

HIV is a foremost international community fitness concern which demands for sophisticated medical supervision. Multisystem situation connecting the cardiovascular disorders (CVDs) and rheumatoid arthritis (RA) has developed [134]. HIV is identified by watching the virus-related load comprise culturing, enzyme-connected immunosorbent assay, and polymerase sequence response [135]. The initial recognition of HIV contamination is the greatest approach to inhibit its extent and to progress the productivity of the antiretroviral therapy. Nucleic acid amplification assessments are common methods for sensing low virus level in blood [136]. Time consuming, expensive, and technically expert technicians are the drawbacks linked with this procedure.

Scientists have dedicated their determinations to progress diagnostic strategies to observer the HIV viral through superior sensitivity via nanoscale skills and employing them to observe antiretroviral treatment and initial infant recognition of HIV. Islam et al. developed the nano sensor using graphene-built sensor for the finding of HIV. Graphene functionalized with amines and conjugated with antibodies such as anti-p24, anti-cardiac troponin, anti-cyclic citrullinated peptide respectively for HIV, CVDs, RA through carbodiimide stimulation to sense numerous biomarkers. These graphene conjugated antibodies are characterized via numerous methods such as UV-Vis, Raman spectroscopy. Superior sensitivity with a decent linear reaction from 1 fg/mL to 1 μg/mL was observed for Anti p24, anti-cardiac troponin, anti-cyclic citrullinated peptide. The graphene-nano based sensor confirmed brilliant performance for the recognition of HIV, CVD, and RA biomarkers in actual samples. Hence graphene is promising to employ for diagnosis of HIV and therefore permit initial recognition of the infections.

11.3.2.5 Salivary detection

Graphene printed on water soluble silk permits to achieve the bio transfer graphene nanosensors on biomaterials. Bio-interfaced sensing platform is formed as product and is explored for the detection of proposed analytes. Bacteria is detected bio selectively at single-cell levels by self-assembling of antimicrobial peptides on graphene. This combined nano senor incorporating on a tooth to distantly observe the breathing and sense the bacteria present in saliva [137]. Graphene/carbon nanofiber electrode is explored to detect glucose concentrations in real saliva samples [138]. Recognized biomarker is cortisol in saliva for checking mental strain. Khan et al. makings a sensing methodology for sensing ways of Prostate-specific antigen in anthropoid saliva through a graphene nanocomposite through co-polymers and gold electrodes. The sensor hires the connection of an anti-PSA antibody by an antigen-PSA to aid as a resistor in a circuit and offers an impedance variation which permits to sense and enumerate PSA in saliva samples. Impedance analyzer was located in the edge of a sensor chip and the statistics was recorded through bluetooth-enabled module. This can detect PSA broadly extending among 0.1 pg mL^{-1} and 100 ng mL^{-1} through a finding limit of 40 fg mL^{-1} [139]. Graphene nanosensing scheme was employed to sense cytokine biomarkers in saliva. The emersion of numerous organizations deliver mono and bilayered graphene nanosheets on various interfaces has delivered an additional motivation to apply graphene in salivary-founded biosensor utilizations [140].

11.3.4 Carbon dots

Carbon dots (CDs) and their fluorescence activities have opened up a novel insight for cancer detection due to their little cytotoxic effects, small size and excellent biocompatibility. Dual emission CDs (430 and 642 nm) were synthesized by a single step facile hydrothermal process via alizarin carmine and used as a ratiometric sensor for glutathione (GSH). The fluorescence intensity ratio at 430 nm and 642 nm (I430 nm/I642 nm) varies with the change in the concentration of GSH. The fluorescence intensity at 430 nm improved with the growing concentration of GSH, while the intensity decreased marginally at 642 nm. Detection limit of these dual emission CDs are 0.26 μM and sensitivity in the linear range of 1−10 to 25−150 μM. Confocal fluorescent imaging of Cancer cells and normal cells showed that CDs synthesized could be used as an efficient method to image cancer cells [141]. The level of expression of miRNA-21 is closely linked to cancer, in particular gastrointestinal cancer. Monitoring miRNA-21 has therapeutic use in the analysis and assessment of gastrointestinal cancer. Wang et al. was developed miRNA-21 fluorescence bioassay sensor based on the T7 exonuclease-mediated cyclic enzymatic amplification method by using CDs and the FAM-labeled ssDNA as signal source. The F_{FAM} F_{CD} value suggests a clear linear relation with miRNA-21 concentration within a range of 0.05–10 nM, and the detection limit for miRNA-21 was 1 pM with excellent reproducibility and selectivity. Moreover, in clinical blood samples of healthy people and gastrointestinal cancer patients, the sensor was successfully testing miRNA-21 expression levels, and the findings were compatible with those of qRT-PCR, showing a high clinical significance for diagnosing miRNA-21 expression levels. In this technique, CDs have shown the three roles of built-in internal fluorescence, quencher and probe carrier [142]. Bladder cancer (BC) in the male urinary system is one of the common diseases. Nuclear matrix protein 22 (NMP22) is a probable and modern biomarker of BC in many of the studies in recent years [143–145]. Recently, Othman and co-workes synthesized nitrogen doped carbon dots (NCDs) by hydrothermal method with high quantum yield and demonstrated a selective, eco-friendly, and rapid sensor of NMP22 antigen on the basis of fluorescence immunoassy technology. A linear connection between fluorescence intensity changes and the concentration of NMP22 within the choice of 1.3-16.3 ng/mL and a finding limit of 0.047 ng/mL (47 pg/mL) was observed under optimum conditions [146].

Fluorescence resonance energy transfer (FRET) is a spectacle used in biosensors. FRET depends on energy transfer based on the distance between the donor and acceptor molecules. Sidhu et al. have identified a FRET phenomenon between CDs and naphthalimide to detect the activity of thioredoxin reductase (TrxR), which is frequently overexpressed in several cancer cells. The covalent disulfide bond was established between CDs and Naphthalimide, which is responsible for the FRET mechanism. In the absence of TrxR, CDs act as a donor and Naphthalimide as an acceptor, and the emission observed at 565 nm when excited at 360 nm. In comparison, disulfide bond breaks down at the higher TrxR level thus destroys the FRET pair which results florescence emission intensity change at 565 nm and 440 nm (Fig. 11.8) [147]. Another FRET-based sensor consists of CDs and GO, which was combined with catalytic hairpin assembly (CHA) was used for the recognition of prostate-specific antigen (PSA) with a detection limit of 0.22 ng mL^{-1}. CDs fluorescence was quenched in the absence of target since hydrophobic and π-π stacking connections between GO and absorbed hairpin DNA, whereas fluorescence intensity recovered in the vicinity of target due to CHA circuits to produce Y-shaped double-stranded DNA (dsDNA). This sensor also used for the recognition of carcinoembryonic antigen (CEA) (LOD of 0.56 ng mL^{-1}) and adenosine triphosphate (ATP) (80 nM) [148].

The Electrochemiluminescence resonance energy transfer (ECL-RET) assay is another effective diagnostic tool that has attracted significantly in the detection and biosensing. The sandwich-type Nafion/GO-HBP biosensor based on ECL-RET assay was intended by Bahari et al. for the recognition of the CA 19-9 antigen whereas Ru-complex as ECL emitter. Working range and finding limit of this sensor were 2 mU mL^{-1}- 50 U mL^{-1} and 0.25 mU mL^{-1}, respectively [149].

Numerous methods for glucose detection have been developed and only a few cases can currently be used for clinical diagnosis. FRET-based glucose sensor made of NCDs and silver nanoprisms recently reported by Wang and coworkers. NCDs acted

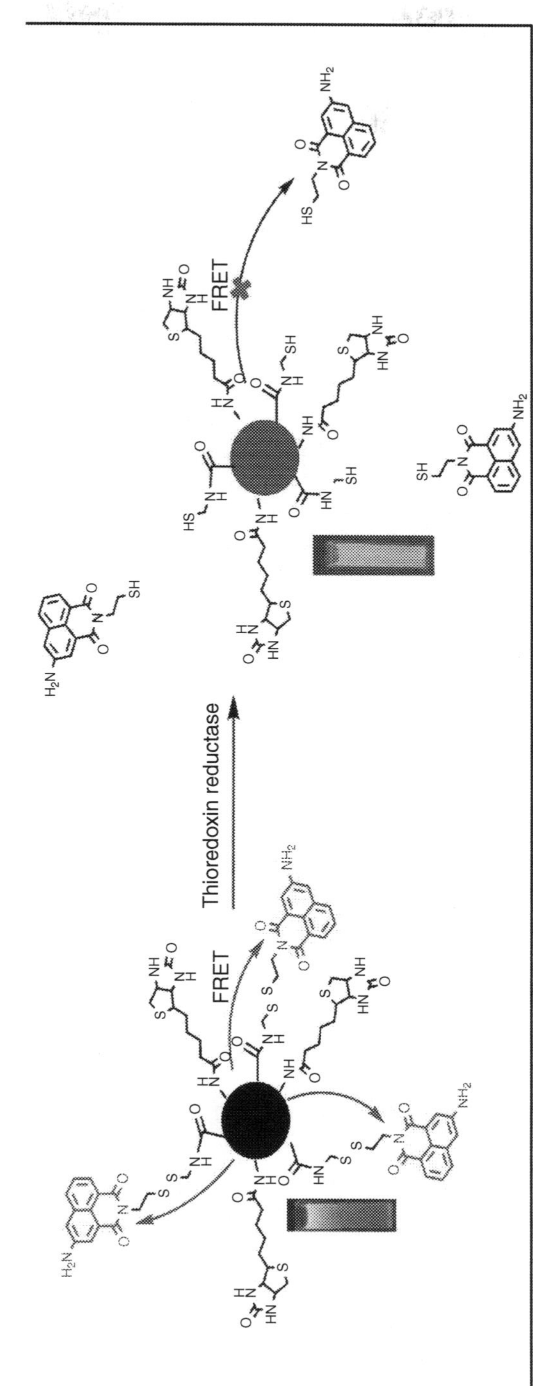

Fig. 11.8 ***Schematic representation of sensing mechanism of TrxR [147].*** *(Copyright © 2017, American Chemical Society).*

as an energy donor and Ag nanoprisms (NPRs) acted as energy acceptor and established good spectral overlap. In the occurrence of glucose and glucose oxidase FRET mechanism was blocked by the oxidative etching of the acceptor and H_2O_2 produced to convert Ag NPRs into Ag^+ ions. This new technique is a simple, robust, precise and profitable approach for quantitative finding of glucose which is built on the basis of a turn-on fluorescent signal versus glucose concentration [150]. Another turn-on FRET-based glucose sensor was reported by Li et al. which consist of CDs and manganese dioxide nanosheet (CDs/MnO_2) assembly. Initially, MnO_2 nanosheets quench CD fluorescence in the CD/MnO_2 FRET system. As H_2O_2 was added to the system, the MnO_2 nanosheets would be destroyed by the redox reaction and the CD fluorescence restored. The fluorescent intensity of CDs was increased linearly under ideal conditions with increased glucose levels ranging from 2 to 200 μM, and the glucose recognition limit was determined to be 0.83 μM [151].

The orange-red emissive CDs used as a fluorimetric detection of glucose. The emission peak was greatly quenced by the H_2O_2 produced due to the oxidation of glucose by glucose oxidase. The response was linear in the range between 0.5 to 100 μM concentrations of glucose with a finding limit of 0.33 μM. The system was selective to glucose and used to assess blood glucose in diluted human serums and in urine from diabetic people and healthy people [152]. Several other fluorescence-based biosensors are reported such as titanium carbide combined with red emitting CDs [153], gadolinium-doped CDs and carbon microparticles [154].

Few biosensors founded on colorimetry have been reported for glucose recognition using doped CDs. Li et al. developed N, Fe doped CDs as a mimic peroxidase for detecting hydrogen peroxide produced by glucose oxidation. This is a vastly effective and sensitive colorimetric recognition technique for glucose. A finding limit for H_2O_2 and glucose was 0.52 and 3.0 μM respectively [155]. Multielement doped CDs (ME-CDs) were prepared by hydrothermal treatment of animal blood. The prepared ME-CDs comprise mostly of carbon, oxygen, and nitrogen elements with a trace quantity of iron and sulfur elements, which show a good fluorescence emission with a quantum yield of 32.6 %. The ME-CDs demonstrated admirable peroxidase-like catalytic activity due to the presence of doped iron [156].

11.3.5 Fullerene

Among carbon materials, fullerene has drawn considerable attention as biosensors for diagnosis and phototheranostics due to biocompatibility, favorable electrical conductivity, mechanical strength and price-effectiveness. Significant attention has been paid to fullerene-C60 owing to its favorable electrochemical properties. Latest analysis demonstrated that partially reduce fullerene-C60 films show curiously improved electrochemical performances in the aqueous solution hence it has been utilized for alterating electrodes for different electroactive molecules [157]. In the electrochemical immunoassay of prostate-specific antigen (PSA), Suresh et al. developed an efficient immune-sensing device, integrating immobilized hydroquinone (HQ), fullerene-C60, and copper nanoparticles nanocomposite films on glassy carbon electrode (HQ@CuNPs-reduced fullerene-C60/GCE). Nano-composite HQ@CuNP-reduced-fullerene-C60 film demonstrated high sensitivity owing to the huge surface area and brilliant electro-catalytic activity of Cu nanoparticles and fullerene-C60. The immunosensor demonstrated a varied linearity from 0.005 ng/mL to 20 ng/mL with LOD of 0.002 ng/mL. In the clinical study, the anticipated immunosensor demonstrated decent biocompatibility, admirable selectiveness, low matrix, reasonable reproductivity and PSA steadiness [158]. Fullerene is also used for phototheranostic applications integrating photo-imagery and phototherapy, which demonstrate tremendous promise in today's accurate tumor therapy. Shi and coworkers reported near-infrared light-harvesting fullerene-based nanoparticles for cancer treatment [159].

Fullerene based electrochemiluminescence biosensor for glucose detection developed by Ye et al. The catalytic film consists of immobilized glucose oxidase on a C60 modified glass carbon electrode embedded in tetraocylammonium bromide ($TOAB^+$). The established glucose biosensor displayed a linear response varies from 500 nM to 13 mM under the ideal setting. The biosensor showed excellent selectivity, long-term durability and exceptional reproducibility [160] miR-3675-3p is a hopeful biomarker for Idiopathic pulmonary fibrosis (IPF) and it is an interstitial lung illness. Fullerene based nanohybrid electrochemical miRNA biosensor was developed to sense miR-3675-3p using multiple signal amplification strategies. This biosensor demonstrated an extensive linear range for miR-3675-3p from 10 fm to 10 nM under optimum conditions, with a finding limit of 2.99 fM [161].

11.3.6 Carbon nanohorns

The single walled carbon nanohorns (SWCNH) were explored for the fabrication of biosensors by immobilization of biorecognition/sensing elements on them. For example, carbon nanohorns (CNHs) were immobilized with Fibrinogen (Fib) and used for the development tof disposable electrochemical biosensor to detect Fibrinogen, that is useful for the diagnosis of various diseases like liver, cardiovascular and cancers. The immunosensor employed for the outcome of fibrinogen (Fib) in conjunction with anti-Fib branded with horseradish peroxidase (HRP) and the redox media to rhydroquinone (HQ). The report claims the sensor is exhibiting better performance than the commercial ELISA kits and detection limit is 58 ng/mL [162]. Another interesting result was reported, where CNHs were used for the construction of enzyme free biosensor for the determination of microRNA-21 a cancer biomarker. The GC electrode was decorated with SWCNTHs and the electrode surface was modified with thiosemicarbazide (TSC) and AuNps to get sensing electrode. The sensor was employed for the finding of microRNA 21 using electrochemiluminescence (ECL) method with wide range and detection limit of 0.03 fM [163]. Thrombinaserine protease and its concentration levels play an important physiological and pathological role. The detection of thrombin can lead to the diagnosis of cardiovascular and Alzheimer's diseases. Single walled carbon nanohorns (SWCNHs) were modified with dye-labeledthrombin aptamer (dye–TA) to design fluorescent aptasensor for thrombin detection. Initially SWCNHs were oxidized and dye-labeledthrombin aptamer (dye–TA) was adsorbed on to it to get a sensor that is able to sense the thrombin with finding limit of 100 pM [164]. Glucose detection is another diagnostically important task and CNHs were used to design and develop biosensors for the same. SWCNTHs were immobilized with phosphomolybdic acid and to get sensor that offered high sensitivity for the recognition of glucose with finding limit of 2.4 μM [165]. Another report published where Cu-MOF and carbon nanohorns (CNHs) were used to develop a biosensing interface for the glucose detection with detection limit of 78nM [166]. DNA detection is greater importance in medical diagnosis and SWCNHs were used as effective fluorescence platforms. A simple two-step, combination sense approach employed for fluorescent finding of a DNA target. The study illustrates that the SWCNHs have high fluorescence quenching capacity on top of diverse affinities toward ss- and dsDNA, thereby DNA detection can be achieved easily [167].

11.3.7 Nanodiamonds

Nanodiamonds play very an imperative role in the construction of biosensors for diverse application such as medical diagnosis, detection of various drugs & their intermediates, organic chemicals present in the human body fluids and in the water bodies [168]. The nanodiamonds can also be modified like other carbon nanomaterials and used to design the biosensors. Nanodiamonds were used in the imaging methods for the recognition of cancer. The polyethylene glycol (PEG) modified nanodiamonds (NDs) were conjugated to an anti-Human epidermal growth factor receptor-2 (HER2) peptide and used as molecularly targeted contrast agent for the recognition of breast cancer with high resolution photoacoustic (PA) imaging [169]. Another interesting application was reported where the nanodiamods with europium-chelating tags used for the background-free fluorescent imaging with increased intensity [170]. Nanocrystalline nanodiamonds were doped with boron for construction of enzyme free biosensors for the recognition of glucose. The sensor was exhibiting good sensitivity and choosiness toward glucose in ascorbic acid and uric acid presence [171]. Then in the other investigations these boron doped nanodiamond electrodes were modified with nickel (Ni) [172] and carbon coated Ni particles [173] and used for fabrication of biosensors to detect glucose with high sensitivity and selectivity. Later another research group further modified the boron doped nanodiamond electrode with a porous layer of Au/Ni and used for the detection of glucose [174]. Detection and monitoring of neurochemicals is another important task in the neurosciences and in the diagnosis of neurological diseases. Dopamine is on of the vital neurochemical and nanodiamonds were used to fabricate biosensors for its detection. The boron-doped ultra-nanocrystalline diamond (UNCD) used for coating on metal microwires and employed as microsensors for the in vivo recognition of dopamine. The UNCD coated on paryleneinsulatedtantalum metal microwire and the resulting microelectrode array was proficient of measuring dopamine with detection limit of 27nM [175]. Another microelectrode devised using the hybrid of multiwall carbon nanotube (MWCNT) and boron-doped ultra-nanocrystalline diamond (UNCD) was used for dopamine detection with improved sensitivity and recognition limit of 9.5±1.2% nM [176]. Hydrogen peroxidea toxic substance in the cellular system and for its recognition a biosensor was established using the horseradish peroxidase (HRP) enzyme decorated nanocrystalline diamond thin films with a finding limit of 35 nM [177].

11.4 Conclusion

During the last two decades extensive research on carbon nanomaterials, including graphene, CNTs, fullerenes, carbon nanohorns, carbon dots, nanodiamonds have demonstrated their possible applications in numerous fields of research, including bio-sensing, bio-imaging, catalysis, drug delivery and electronic devices owing to their chemical stability, unique dimensionality, electrical and optical properties. This chapter illustrated the introduction of various carbon nanomaterials and its applications in diagnostics as sensors. Both optical and electrochemical based biosensors are discussed for cancer biomarker detection, diagnosis of diabetes, detection of genetic complaints, human immunodeficiency virus (HIV), and glucose.

References

[1] M. Rizwan, D. Koh, M.A. Booth, M.U. Ahmed, Combining a gold nanoparticle-polyethylene glycol nanocomposite and carbon nanofiber electrodes to develop a highly sensitive salivary secretory immunoglobulin A immunosensor, Sens. Actuators B. Chem. 255 (2018) 557.

[2] A.C. Power, B. Gorey, S. Chandra, J. Chapman, Carbon nanomaterials and their application to electrochemical sensors: a review, Nanotechnol. Rev. 7 (2018) 9.

[3] W.R. Yang, P. Thordarson, J.J. Gooding, S.P. Ringer, F. Braet, Carbon nanotubes for biological and biomedical applications, Nanotechnology 18 (2007) 412001.

[4] R.G. Mendes, A. Bachmatiuk, B. Buechner, G. Cuniberti, M.H. Ruemmeli, Carbon nanostructures as multi-functional drug delivery platforms, J. Mater. Chem. B. 1 (2013) 401.

[5] P.M. Ajayan, Nanotubes from Carbon, Chem. Rev. 99 (1999) 1787.

[6] J. Adhikari, M. Rizwan, N.A. Keasberry, M.U. Ahmed, Current progresses and trends in carbon nanomaterials-based electrochemical and electrochemiluminescence biosensors, J. Chin. Chem. Soc. (2020) 1–24.

[7] S.A. Lim, M.U. Ahmed, Electrochemical immunosensors and their recent nanomaterial-based signal amplification strategies: a review, RSC Adv. 6 (2016) 24995.

[8] P. Ramnani, N.M. Saucedo, A. Mulchandani, Carbon nanomaterial-based electrochemical biosensors for label-free sensing of environmental pollutants, Chemosphere 143 (2016) 85–98.

[9] D. Chimene, D.L. Alge, A.K. Gaharwar, Two-Dimensional Nanomaterials for Biomedical Applications: Emerging Trends and Future Prospects, Adv. Mater. 27 (2015) 7261–7284.

[10] K.M. Lee, M. Runyon, T.J. Herrman, R. Phillips, J. Hsieh, Review of Salmonella detection and identification methods: Aspects of rapid emergency response and food safety, Food Control 47 (2015) 264–276.
[11] P. Giménez-Gómez, M. Gutiérrez-Capitán, F. Capdevila, A. Puig-Pujol, C. Fernández-Sánchez, C. Jiménez-Jorquera, Monitoring of malolactic fermentation in wine using an electrochemical bienzymatic biosensor for l-lactate with long term stability, Anal. Chim. Acta 905 (206) (2016) 126–133.
[12] R.P. Singh, Prospects of Nanobiomaterials for Biosensing, Int. J. Electrochemical. Sci. (2011) 1–30.
[13] V. Georgakilas, et al., Functionalization of Graphene: Covalent and Non-Covalent Approaches, Derivatives and Applications. Chem. Rev. 112 (2012) 6156–6214.
[14] V. Georgakilas, J.A. Perman, J Tucek, R Zboril, Broad Family of Carbon Nanoallotropes: Classification, Chemistry, and Applications of Fullerenes, Carbon Dots, Nanotubes, Graphene, Nanodiamonds, and Combined Superstructures, 115, Chem. Rev., 2015, pp. 4744–4822.
[15] H.W. Kroto, J.R. Heath, S.C. O'Brien, R.F. Curl, R.E. Smalley, C60: Buckminsterfullerene, Nature 318 (1985) 162–163.
[16] J.I. Peredes, S.V. Rodil, A.M. Alonso, J.M.D. Tascon, Graphene Oxide Dispersions in Organic Solvents, Langmuir 24 (2008) 10560.
[17] J. Shi, L. Wang, J. Gao, Y. Liu, J. Zhang, R. Ma, R. Liu, Z. Zhang, A fullerene-based multi-functional nanoplatform for cancer theranostic applications, Biomaterials 35 (2014) 5771–5784.
[18] SS. Iijima, Helical microtubules of graphitic carbon, Nature 354 (1991) 56–58.
[19] R. Saito, M. Fujita, G. Dresselhaus, M.S. Dresselhaus, Electronic structure of chiral graphene Tubules, Appl. Phys. Lett. 60 (1992) 2204.
[20] M.S. Dresselhaus, G. Dresselhaus, A. Jorio, Unusual Properties and Structure of Carbon Nanotubes, Annu. Rev. Mater. Res. 34 (2004) 247–278.
[21] S.B. Sinnott, R. Andrews, D. Qian, A.M. Rao, Z. Mao, E.C. Dickey, F. Derbyshire, Model of carbon nanotube growth through chemical vapor deposition, Chem. Phys. Lett. 315 (1999) 25–30.
[22] C.E. Baddour, C. Briens, Carbon Nanotube Synthesis: A Review, Int. J. Chem. React. Eng. 3 (2005).
[23] B.S. Wong, S.L. Yoong, A. Jagusiak, T. Panczyk, H.K. Ho, W.H. Ang, G. Pastorin, Carbon nanotubes for delivery of small molecule drugs, Adv. Drug Delivery Rev. 65 (2013) 1964–2015.
[24] A.K. Geim, K.S. Novoselov, The rise of graphene, Nat. Mater. 6 (2007) 183–191.
[25] A.S. Mayorov, R.V. Gorbachev, S.V. Morozov, L. Britnell, R. Jalil, L.A. Ponomarenko, P. Blake, K.S. Novoselov, K. Watanabe, T. Taniguchi, A.K. Geim, Micrometer-Scale Ballistic Transport in Encapsulated Graphene at Room Temperature. Nano. Lett. 11 (2011) 2396–2399.
[26] C. Lee, X. Wei, J.W. Kysar, J. Hone, Measurement of the elastic properties and intrinsic strength of monolayer graphene, Science 321 (2008) 385–388.
[27] A.A. Balandin, Thermal properties of graphene and nanostructured carbon materials, Nat. Mater. 10 (2011) 569–581.
[28] R.R. Nair, P. Blake, A.N. Grigorenko, K.S. Novoselov, T.J. Booth, T. Stauber, N.M.R. Peres, A.K. Geim, Fine Structure Constant Defines Visual Transparency of Graphene, Science 320 (2008) 1308.
[29] K. Novoselov, Graphene Mind the Gap, Nat. Mater. 6 (2007) 720–721.
[30] K.S. Novoselov, E. McCann, S.V. Morozov, V.I. Fal'ko, M.I. Katsnelson, U. Zeitler, D. Jiang, F. Schedin, A.K. Geim, Unconventional quantum Hall effect and Berry's phase of 2π inbilayer graphene, Nat. Phys. 2 (2006) 177–180.
[31] J. Moser, A. Barreiro, A. Bachtold, Current-induced cleaning of graphene, Appl. Phys. Lett. 91 (2007) 163513.
[32] J.S. Bunch, S.S. Verbridge, J.S. Alden, A.M. van der Zande, J.M. Parpia, H.G. Craighead, P.L. McEuen, Impermeable Atomic Membranes from Graphene Sheets, Nano. Lett. 8 (2008) 2458–2462.
[33] K.S. Novoselov, V.I. Fal'ko, L. Colombo, G.e.l.l.e.r.t. P.R, M.G. Schwab, K. Kim, A roadmap for Graphene. Nature 490 (2012) 192–200.
[34] Y. Hernandez, V. Nicolosi, M. Lotya, F.M. Blighe, Z. Sun, S. De, I.T. McGovern, I.B. Holland, M. Byrne, Y.K. Gun'Ko, J.J. Boland, P. Niraj, G. Duesberg, S. Krishnamurthy, R.R. Goodhue, J. Hutchison, V. Scardaci, A.C. Ferrari, J.N. Coleman, High-Yield Production of Graphene by Liquid-Phase Exfoliation of Graphite, Nat. Nanotechnol. 3 (2008) 3563–3568.
[35] D. Nuvoli, L. Valentini, V. Alzari, S. Scognamillo, S.B. Bon, M. Piccinini, J. Illescas, A.J. Mariani, High concentration few-layer graphene sheets obtained by liquid phase exfoliation of graphite in ionic liquid. Mater. Chem. 21 (2011) 3428–3431.

[36] C.Y Su, A.Y. Lu, Y. Xu, F.R. Chen, A.N. Khlobystov, L.J. Li, High-Quality Thin Graphene Films from Fast Electrochemical Exfoliation, ACS Nano. 5 (2011) 2332–2339.
[37] S. Pei, H.M. Cheng, The reduction of graphene oxide, Carbon 50 (2012) 3210–3228.
[38] Z.S. Wu, W. Ren, L. Gao, J. Zhao, Z. Chen, B. Liu, D. Tang, B. Yu, C. Jiang, H.M. Cheng, Synthesis of graphene sheets with high electrical conductivity and good thermal stability by hydrogen arc discharge exfoliation. ACS Nano. 3 (2009) 411–417.
[39] Q. He, H.G. Sudibya, Z. Yin, W.u. S, H. Li, F. Boey, W. Huang, P. Chen, P.H. Zhang, Graphene-based electronic sensors. ACS Nano. 4 (2010) 3201–3208.
[40] M.J. Fernández-Merino, G.u.a.r.d.i.a. L, J.I. Paredes, S. Villar-Rodil, P. Solís-Fernández, A. Martínez-Alonso, J.M.D. Tascón, Vitamin C Is an Ideal Substitute for Hydrazine in the Reduction of Graphene Oxide Suspensions, J. Phys. Chem. C. 114 (2010) 6426–6432.
[41] H.J. Grabke, W. Paulitschke, G. Tauber, H. Viefhaus, Equilibrium surface segregation of dissolved nonmetal atoms on iron (100) faces, Surf. Sci. 63 (1977) 377–389.
[42] J.C. Shelton, H.R. Patil, J.M. Blakely, Equilibrium segregation of carbon to a nickel (111) surface: A surface phase transition. Surf. Sci. 43 (1974) 493–520.
[43] J.C. Hamilton, J.M. Blakely, Carbon segregation to single crystal surfaces of Pt, Pd and Co, Surf. Sci. 91 (1980) 199–217.
[44] Y. Zhang, L. Gomez, F.N. Ishikawa, A. Madaria, K. Ryu, C. Wang, A. Badmaev, C. Zhou, Comparison of Graphene Growth on Single-Crystalline and Polycrystalline Ni by Chemical Vapor Deposition, J. Phys. Chem. Lett. 1 (2010) 3101–3107.
[45] W. Norimatsu, M. Kusunoki, Epitaxial graphene on SiC{0001}: advances and perspectives, Phys. Chem. Chem. Phys. 16 (2014) 3501–3511.
[46] E. Perevedentseva, Y.C. Lin, M. Jani, C.L. Cheng, Biomedical applications of nanodiamonds in imaging and therapy, Nanomed. 8 (2013) 2041–2060.
[47] R. Lam, D. Ho, Nanodiamonds as vehicles for systemic and localized drug delivery. Expert opinion on drug delivery, Exp. Opin. Drug Deliv. 6 (2009) 883–895.
[48] V.V. Danilenko, On the history of the discovery of nano diamond synthesis, Phys. Solid State 46 (2004) 595–599.
[49] R. Schirhagl, K. Chang, M. Loretz, C.L. Degen, Nitrogen-Vacancy Centers in Diamond: Nanoscale Sensors for Physics and Biology, Annu. Rev. Phys. Chem. 65 (2014) 83–105.
[50] V.N. Mochalin, P. Amanda, L.X. Mei, Y. Gogotsi, Adsorption of Drugs on Nanodiamond: Toward Development of a Drug Delivery Platform, Mol. Pharm. 10 (2013) 3728.
[51] G. Reina, S. Orlanducci, C. Cairone, E. Tamburri, S. Lenti, I. Cianchetta, M. Rossi, M.L. Terranova, Rhodamine/Nanodiamond as a System Model for Drug Carrier, J. Nanosci. Nanotechnol. 1 (2015) 1022.
[52] R. Kaur, I. Badea, Nanodiamonds as novel nanomaterials for biomedical applications: drug delivery and imaging systems, Int. J. Nanomed. 8 (2013) 203.
[53] S.Y. Lim, W. Shen, Z. Gao, Carbon quantum dots and their applications, Chem. Soc. Rev. 44 (2015) 362–381.
[54] S. Zhu, Y. Song, X. Zhao, J. Shao, J. Zhang, B. Yang, The photoluminescence mechanism in carbon Dots (graphene quantum dots, carbon nanodots, and polymer dots): Current state and future Perspective, Nano. Res. 8 (2015) 355–381.
[55] J.C.G. Esteves da Silva, H.M.R. Gonçalves, Analytical and bioanalytical applications of carbon dots, Trends Anal. Chem. 30 (2011) 1327–1336.
[56] X. Yu, B. Munge, V. Patel, G. Jensen, A. Bhirde, J.D. Gong, S.N. Kim, J. Gillespie, J.S. Gutkind, F. Papadimitrakopoulos, F. James, Carbon Nanotube Amplification Strategies for Highly Sensitive Immunodetection of Cancer Biomarkers, J. Am. Chem. Soc. 128 (2006) 11199–11205.
[57] Y.J. Zou, C.L. Xiang, L.X. Sun, F. Xu, H.Y. Zhou, Ultrasensitive Prostate Specific Antigen Immunosensor Based on Gold Nanoparticles Functionalized Polypyrrole@Carbon Nanotubes, Asian J. Chem 26 (2014) 8002–8006.
[58] J. Okuno, K. Maehashi, K. Kerman, Y. Takamura, K. Matsumoto, E. Tamiya, Label-free immunosensor for prostate-specific antigen based on single-walled carbon nanotube array-modified microelectrodes. Biosens. Bioelectron. 22 (2007) 2377–2381.
[59] J. Tian, J. Huang, Y. Zhao, S. Zhao, Electrochemical immunosensor for prostate-specific antigen using a glassy carbon electrode modified with a nanocomposite containing gold nanoparticles supported with starch-functionalized multi-walled carbon nanotubes, Microchim. Acta. 178 (2012) 81–88.

[60] A. Sharma, S. Hong, R. Singh, J. Jang, Single-walled carbon nanotube based transparent immunosensor for detection of a prostate cancer biomarker osteopontin, Analytica. Chimica. Acta. 869 (2015) 68–73.
[61] J.C. Soares, A.C Soares, V.C. Rodrigues, M.E. Melendez, A.C. Antos, E.F. Faria, R.M. Reis, A.L. Carvalho, O.N. Oliveira Jr., Biosensors, Detection of the Prostate Cancer Biomarker PCA3 with Electrochemical and Impedance-Based Biosensors, ACS Appl. Mater. Interfaces 11 (2019) 46645−46650.
[62] H.V. Tran, B. Piro, S. Reisberg, L.D. Tran, H.T. Duc, M.C. Pha, Label-free and reagentless electrochemical detection of microRNAs using a conducting polymer nanostructured by carbon nanotubes: Application to prostate cancer biomarker miR-141, Biosens. Bioelectron. 49 (2013) 164–169.
[63] P. Ramnani, Y. Gao, M. Ozsoz, A. Mulchandani, Mulchandani, Electronic Detection of MicroRNA at Attomolar Level with High Specificity. Anal. Chem. 85 (2013) 8061–8064.
[64] B. Mitchell, J.D. Souza Lerner, T. Pazina, J. Dailey, B.R. Goldsmith, M.K. Robinson, A.T.C. Johnson, Hybrids of a Genetically Engineered Antibody and a Carbon Nanotube Transistor for Detection of Prostate Cancer Biomarkers, ACS Nano. 6 (2012) 5143–5149.
[65] H.A.R. Pour, M. Behpour, M. Keshavarz, A novel label-free electrochemical miRNA biosensor using methylene blue as redox indicator: application to breast cancer biomarker miRNA-21, Biosens. Bioelectron. 77 (2016) 202–207.
[66] A. Benvidi, M.D. Tezerjani, S. Jahanbani, M.M. Ardakani, S.M. Moshtaghioun, Comparison of impedimetric detection of DNA hybridization on the various biosensors based on modified glassy carbon electrodes with PANHS and nanomaterials of RGO and MWCNTs, Talanta. 147 (2016) 621–627.
[67] M.A.H. Nawaz, S. Rauf, G. Catanante, M.H. Nawaz, G. Nunes, J.L. Marty, A. Hayat, One Step Assembly of Thin Films of Carbon Nanotubes on Screen Printed Interface for Electrochemical Aptasensing of Breast Cancer Biomarker, Biomarker. Sensors, 16 (2016) 1651.
[68] G. Paimard, M. Shahlaei, P. Moradipour, V. Karamali, E. Arkan, Impedimetric aptamer based determination of the tumor marker MUC1 by using electrospun core-shell nanofibers, Microchim. Acta. 187 (2019).
[69] P. Rostamabadi, E.H. Bafrooei, Impedimetric aptasensing of the breast cancer biomarker HER2 using a glassy carbon electrode modified with gold nanoparticles in a composite consisting of electrochemically reduced graphene oxide and single-walled carbon nanotubes. Microchim. Acta. 186 (2019).
[70] E. Arkan, R. Saber, Z. Karimi, M.A. Shamsipur, A novel antibody-antigen based impedimetric immunosensor for low level detection of HER2 in serum samples of breast cancer patients via modification of a gold nanoparticles decorated multiwall carbon nanotube-ionic liquid electrode, Anal. Chim. Acta. 874 (2015) 66–74.
[71] M. Freitas, H.P.A. Nouws, C.D. Matos, Electrochemical Sensing Platforms for HER2-ECD Breast Cancer Biomarker Detection. Electroanalysis, Electroanalysis 31 (2019) 121–128.
[72] X. Gao, Y. Zhang, H. Chen, Z. Chen, Xi. Lin, Amperometric immunosensor for carcinoembryonic antigen detection with carbon nanotube-based film decorated with gold nanoclusters, Anal. Biochem. 414 (2011) 70–76.
[73] D. Feng, Xi. Lu, Xi. Dong, Y. Ling, Y. Zhang, Label-free electrochemical immunosensor for the carcinoembryonic antigen using a glassy carbon electrode modified with electrodeposited Prussian Blue, a graphene and carbon nanotube assembly and an antibody immobilized on gold Nanoparticles, Microchim Acta 180 (2013) 767–774.
[74] H. Xu, Y. Wang, L. Wang, Yi. Ong, J. Luo, Xi. Cai, A Label-Free Microelectrode Array Based on One-Step Synthesis of Chitosan–Multi-Walled Carbon Nanotube–Thionine for Ultrasensitive Detection of Carcinoembryonic Antigen, Nanomaterials 6 (2016) 132.
[75] M. Rizwan, S. Elma, S.A. Lim, M.U. Ahmed, AuNPs/CNOs/SWCNTs/chitosan-nanocomposite modified electrochemical sensor for the label-free detection of carcinoembryonic antigen, Biosens. Bioelectron. 107 (2018) 211–217.
[76] G. Paimard, M. Shahlaei, P. Madaripur, H. Akbari, M. Jafaria, E. Arkan, An Impedimetric Immunosensor modified with electrospun core-shell nanofibers for determination of the carcinoma embryonic antigen, Sens. Actuators B. Chem. 311 (2020) 127928.
[77] F.L. Liu, P. Xiao, H.L. Fang, H.F. Dai, L. Qiao, Y.H. Zhang, Single-walled carbon nanotube-based biosensors for the detection of volatile organic compounds of lung cancer, Physica E. 44 (2011) 367–372.

[78] S. Sajjad, M.B.Nojavani Janfaza, M. Nikkhah, T. Alizadeh, A. Esfandiar, M.R. Ganjali, A selective chemiresistive sensor for the cancer-related volatile organic compound hexanal by using molecularly imprinted polymers and multiwalled carbon nanotubes. Microchim. Acta. 186 (2019).

[79] S. Chatterjee, M. Castro, J.F. Feller, An e-nose made of carbon nanotube based quantum resistive sensors for the detection of eighteen polar/nonpolarVOC biomarkers of lung cancer, J. Mater. Chem. B. 1 (2013) 4563–4575.

[80] M. Choudhary, A. Singh, S. Kaur, A. Kavita, Enhancing Lung Cancer Diagnosis: Electrochemical Simultaneous Bianalyte Immunosensing Using Carbon Nanotubes–Chitosan Nanocomposite, Appl. Biochem. Biotechnol. 174 (2014) 1188–1200.

[81] J.T. Robinson, G. Hong, Y. Liang, B. Zhang, O.K. Yaghi, H. Dai, In Vivo Fluorescence Imaging in the Second Near-Infrared Window with Long Circulating Carbon Nanotubes Capable of Ultrahigh Tumor Uptake. J. Am. Chem. Soc. 134 (2012) 10664−10669.

[82] A.L. Antaris, J.T. Robinson, O.K. Yaghi, G. Hong, S. Diao, R. Luong, D. Hongjie, Ultra-Low Doses of Chirality Sorted (6,5) Carbon Nanotubes for Simultaneous Tumor Imaging and Photothermal Therapy, ACS Nano. 7 (2013) 3644–3652.

[83] S. Diao, L.J. Blackburn, G. Hong, A.L. Antaris, J. Chang, J.Z. Wu, B. Zhang, K. Cheng, C.J. Kuo, H. Dai, Fluorescence imaging in vivo at wavelengths beyond 1500 nm, Angew. Chem. Int. Ed. 54 (2015) 1–6.

[84] H. Li, J. Conde, A. Guerreiro, J.L. Gonçalo, Tetrazine Carbon Nanotubesfor Pretargeted In Vivo "Click-to-Release" Bioorthogonal Tumour Imaging, Angew. Chem. Int. Ed. 59 (2020) 16023–16032.

[85] C. Lamprecht, N. Gierlinger, E. Heister, B. Unterauer, B. Plochberger, M. Brameshuber, P. Hinterdorfer, S. Hild, A. Ebner, Mapping the intracellular distribution of carbon nanotubes after targeted delivery to carcinoma cells using confocal Raman imaging as a label-free technique, J. Phys. Condens. Matter 24 (2012) 164206.

[86] Z. Liu, S. Tabakman, S. Sherlock, Xi. Li, Z. Chen, K. Jiang, S. Fan, H. Dai, Multiplexed five-color molecular imaging of cancer cells and tumor tissues with carbon nanotube Raman tags in the near-infrared, Nano. Res. 3 (2010) 222–233.

[87] X. Wang, C. Wang, L. Cheng, S.T. Lee, Z. Liu, Noble Metal Coated Single-Walled Carbon Nanotubes for Applications in Surface Enhanced Raman Scattering Imaging and Photothermal Therapy, J. Am. Chem. Soc. 134 (2012) 7414–7422.

[88] A.D.L. Zerda, S. Boyapati, R. Teed, Y.M. Salomon, S.M. Tabakman, Z. Liu, B.T.K. Yakub, Xi. Chen, H. Dai, S. Sanjiv, Family of Enhanced Photoacoustic Imaging Agents for High-Sensitivity and Multiplexing Studies in Living Mice, ACS Nano 6 (2012) 4694–4701.

[89] C. Wang, C. Bao, S. Liang, H. Fu, K. Wang, M. Deng, Q. Liao, D. Cui, RGD-conjugated silica-coated gold nanorods on the surface of carbon nanotubes for targeted photoacoustic imaging of gastric cancer, Nanoscale Res. Lett. 9 (2014) 264.

[90] Li. Xie, G. Wang, H. Zhou, F. Zhang, Z. Guo, C. Liu, Xi. Zhang, L. Zhu, Functional long circulating single walled carbon nanotubes for fluorescent/photoacoustic imaging-guided enhanced phototherapy, Biomaterials 103 (2016) 219–228.

[91] J. Zhang, L. Song, S. Zhou, M. Hu, Y. Jiao, Y. Teng, Y. Wang, Xi. Zhang, Enhanced ultrasound imaging and anti-tumor in vivo properties of Span–polyethylene glycol with folic acid–carbon nanotube–paclitaxel multifunctional microbubbles, RSC Adv. 9 (2019) 35345.

[92] H. Wu, G. Liu, X. Wang, J. Zhang, Y. Chen, J. Shi, H. Yang, He. Hu, S. Yang, Solvothermal synthesis of cobalt ferrite nanoparticles loaded on multiwalled carbon nanotubes for magnetic resonance imaging and drug delivery, Acta Biomaterialia 7 (2011) 3496–3504.

[93] T. TazePeci, S.J. Dennis, M. Baxendale, Iron-filled multiwalled carbon nanotubes surface-functionalized with paramagnetic Gd (III): A candidate dual-functioning MRI contrast agent and magnetic hyperthermia structure. Carbon 87 (2015) 226–232.

[94] I. Marangon, C.M. Moyon, JK. Tabi, M.L. Béoutis, L. Lartigue, D. Lloyeau, E. Pach, B. Ballesteros, G. Autret, T. Ninjbadgar, D.F. Brougham, A. Bianco, F. Gazeau, Covalent Functionalization of Multi-walled Carbon Nanotubes with a Gadolinium Chelate for Efficient T1-Weighted Magnetic Resonance Imaging, Adv. Funct. Mater. 24 (45) (2014) 7173–7186.

[95] P. Gomathi, M.K. Kim, J.J. Park, D. Ragupathy, A. Rajendran, S. Lee, J.C. Kim, S.H Lee, H.D. Ghim, Multiwalled carbon nanotubes grafted chitosan nanobiocomposite: A prosperous functional nanomaterials for glucose biosensor application, Sens. Actuators B 155 (2011) 897–902.

[96] V. Mani, B. Devadas, S.M. Chen, Direct electrochemistry of glucose oxidase at electrochemically reduced graphene oxide-multiwalled carbon nanotubes hybrid material modified electrode for glucose biosensor, Biosens. Bioelectron. 41 (2013) 309–315.
[97] T.T. Thakoor, K. Komori, P. Ramnani, I. Lee, A. Mulchandani, Mulchandani, Electrochemically Functionalized Seamless Three-Dimensional Graphene-Carbon Nanotube Hybrid for Direct Electron Transfer of Glucose Oxidase and Bio electrocatalysis, Langmuir 31 (2015) 13054–13061.
[98] S. oylemez, B.Yoon, L.Toppare, T.M. Swager, Quaternized Polymer–Single-Walled Carbon Nanotube Scaffolds for a Chemiresistive Glucose Sensor, Langmuir 31 (2015) 13054–13061.
[99] X. Zeng, Y. Zhang, Xi. Du, Y. Li, W.Tang, A highly sensitive glucose sensor based on a gold nanoparticles/polyaniline/multi-walled carbon nanotubes composite modified glassy carbon electrode, New J. Chem. 42 (2018) 11944–11953.
[100] J. Yang, L.C. Jiang, W.D. Zhang, S. Gunasekaran, A highly sensitive non-enzymatic glucose sensor based on a simple two-step electrodeposition of cupric oxide (CuO) nanoparticles onto multi-walled carbon nanotube arrays, Talanta 82 (2010) 25–33.
[101] K.C. Lin, Y.C. Lin, S.M. Chen, A highly sensitive nonenzymatic glucose sensor based on multi-walled carbon nanotubes decorated with nickel and copper nanoparticles, Electrochem. Acta. 96 (2013) 164–172.
[102] M. Gougis, A.T. Aoul, D. Ma, M. Mohamedi, Laser synthesis and tailor-design of nanosized gold onto carbon nanotubes for non-enzymatic electrochemical glucose sensor, Sens. Actuators B. 193 (2014) 363–369.
[103] M. Baghayeri, A. Amiri, S. Farhadi, Development of non-enzymatic glucose sensor based on efficient loading Ag nanoparticles on functionalized carbon nanotubes, Sens. Actuators B: Chem. 225 (2016) 354–362.
[104] C.T. Canevari, A.P.R. Pereira, R. Landers, V.E. Benvenutti, S.A.S. Machado, Sol–gel thin-film based mesoporous silica and carbon nanotubes for the determination of dopamine, uric acid and paracetamol in urine, Talanta 116 (2013) 726–735.
[105] M. Lin, H. Huang, Y. Liu, C. Liang, S. Fei, Xi.C. Chunlin, High loading of uniformly dispersed Pt nanoparticles on polydopamine coated carbon nanotubes and its application in simultaneous determination of dopamine and uric acid. Nanotechnology 24 (2013) 065501.
[106] J. Liu, Y. Xie, K. Wang, Q. Zeng, R. Liu, Xi. Liu, A nanocomposite consisting of carbon nanotubes and gold nanoparticles in an amphiphilic copolymer for voltammetric determination of dopamine, paracetamol and uric acid, Microchim. Acta. 184 (2017) 1739–1745.
[107] Z.N. Huang, J. Zou, J. Teng, Q. Liu, M.M. Yuan, F.P. Jiao, Xi. Yu. Jiang, J.G. Yu, A novel electrochemical sensor based on self-assembled platinum nanochains - Multi-walled carbon nanotubes-graphene nanoparticles composite for simultaneous determination of dopamine and ascorbic acid. Ecotoxicol. Environ. Saf. 172 (2019) 167–175.
[108] J.F. Guan, J. Zou, Y.P. Liu, Xi.Y. Jiang, J.G. Yu, Hybrid carbon nanotubes modified glassy carbon electrode for selective, sensitive and simultaneous detection of dopamine and uric acid, Ecotoxicol. Environ. Saf. 201 (2020) 110872.
[109] W. Xu, J. He, L. Gao, J. Zhang, C. Yu, Immunoassay for netrin 1 via a glassy carbon electrode modified with multi-walled carbon nanotubes, thionine and gold nanoparticles, Microchim. Acta. 182 (2015) 2115–2122.
[110] C.I.L. Justino, A.R. Gomes, A.C. Freitas, T.A.P. Rocha-Santos Duarte, Graphene based sensors and biosensors, TrAC Trends Anal. Chem. 91 (2017) 53–66.
[111] P. Paul, A.K. Malakar, S. Chakraborty, The significance of gene mutations across eight major cancer types, Mutation Research/Reviews in Mutation Research 781 (2019) 88–99.
[112] B. Zhang, T. Cui, An ultrasensitive and low-cost graphene sensor based on layer-by-layer nano self-assembly, Appl. Phys. Lett. 98 (2011) 073116.
[113] B. Zhang, Q. Li, T. Cui, Ultra-sensitive suspended graphene nanocomposite cancer sensors with strong suppression of electrical noise, Biosens. Bioelectron. 31 (2012) 105–109.
[114] J. Liu, J. Wang, T. Wang, D. Li, F. Xi, J. Wang, E. Wang, Three-dimensional electrochemical immunosensor for sensitive detection of carcinoembryonic antigen based on monolithic and macroporous graphene foam, Biosens. Bioelectron. 65 (2015) 281–286.

[115] T. Xu, N. Liu, J. Yuan, Z. Ma, Triple tumor markers assay based on carbon–gold nanocomposite. Biosens. Bioelectron. 70 (2015) 161–166.
[116] Q. Zhu, Y. Chai, Y. Zhuo, R. Yuan, Ultrasensitive simultaneous detection of four biomarkers based on hybridization chain reaction and biotin–streptavidin signal amplification strategy, Biosens. Bioelectron. 68 (2015) 42–48.
[117] A. Md Ali, K. Mondal, Y. Jiao, S. Oren, Z. Xu, A. Sharma, L. Dong, Microfluidic Immuno-Biochip for Detection of Breast Cancer Biomarkers Using Hierarchical Composite of Porous Graphene and Titanium Dioxide Nanofibers, ACS Appl. Mater. Interfaces 8 (2016) 20570–20582.
[118] Z.Gao Rajesh, R. Vishnubhotla, P. Ducos, M.D. Serrano, J. Ping, M.K. Robinson, A.T.C. Johnson, Genetically Engineered Antibody Functionalized Platinum Nanoparticles Modified CVD-Graphene Nanohybrid Transistor for the Detection of Breast Cancer Biomarker, HER3, Adv. Mater. Interfaces 3 (2016) 1600124.
[119] J. Gao, B. Du, X. Zhang, A. Guo, Y. Zhang, D. Wu, H. Ma, Q. Wei, Ultrasensitive enzyme-free immunoassay for squamous cell carcinoma antigen using carbon supported Pd–Au as electrocatalytic labels, Anal. Chim. Acta. 833 (2014) 9–14.
[120] H. Ma, X. Zhang, X. Li, R. Li, B. Du, Q. Wei, Electrochemical immunosensor for detecting typical bladder cancer biomarker based on reduced graphene oxide–tetraethylene pentamine and trimetallic AuPdPt nanoparticles, Talanta 143 (2015) 77–82.
[121] K.J. CashH.A. Clark, Nanosensors and nanomaterials for monitoring glucose in diabetes, Trends Mol. Med. 16 (2010) 584–593.
[122] F. Liu, Q. Xu, W. Huang, Z. Zhang, G. Xiang, C. Zhang, C. Liang, H. Lian, J. Peng, Green synthesis of porous graphene and its application for sensitive detection of hydrogen peroxide and 2,4-dichlorophenoxyacetic acid, Electrochim. Acta. 295 (2019) 615.
[123] Y.H. Song, H.Y. Liu, H.L. Tan, F.G. Xu, J.B. Jia, L.X. Zhang, Z. Li, L. Wang, pH-Switchable Electrochemical Sensing Platform based on Chitosan-Reduced Graphene Oxide/Concanavalin A Layer for Assay of Glucose and Urea, Anal. Chem. 86 (2014) 1980–1987.
[124] X. Kang, J. Wang, H. Wu, I.A. Aksay, J. Liu, Y. Lin, Glucose Oxidase–graphene–chitosan modified electrode for direct electrochemistry and glucose sensing, Biosens. Bioelectron. 25 (2009) 901–905.
[125] S. Alwarappan, C. Liu, A. Kumar, C.-Z. Li, Enzyme-Doped Graphene Nanosheets for Enhanced Glucose Biosensing, J. Phys. Chem. C. 114 (2010) 12920–12924.
[126] J.C. Claussen, A. Kumar, D.B. Jaroch, Nanostructuring Platinum Nanoparticles on Multilayered Graphene Petal Nanosheets for Electrochemical Biosensing, Adv. Funct. Mater. 22 (2012) 3399–3405.
[127] C.S. Shan, H.F. Yang, J.F. Song, D.X. Han, A. Ivaska, L. Niu, Direct Electrochemistry of Glucose Oxidase and Biosensing for Glucose Based on Graphene, Anal. Chem. 81 (2009) 2378–2382.
[128] M. Zhou, Y.M. Zhai, S.J. Dong, Electrochemical sensing and biosensing platform based on chemically reduced graphene oxide, Anal. Chem. 81 (2009) 5603–5613.
[129] X.H. Kang, J. Wang, H. Wu, A.I. Aksay, J. Liu, Y.H. Lin, Glucose oxidase-graphene-chitosan modified electrode for direct electrochemistry and glucose sensing, Biosens. Bioelectron. 25 (2009) 901–905.
[130] H. Wu, J. Wang, X.H. Kang, C.M. Wang, D.H. Wang, J. Liu, I.A. Aksay, Y.H. Lin, Glucose biosensor based on immobilization of glucose oxidase in platinum nanoparticles/graphene/chitosan nanocomposite film, Talanta 80 (2009) 403–406.
[131] H. Sim, J.H. Kim, S.K. Lee, M.J. Song, D.H. Yoon, D.S. Lim, S.I. Hong, High sensitivity non-enzymatic glucose biosensor based on Cu (OH) 2 nanoflower electrode covered with boron-doped nanocrystalline diamond layer, Thin Solid Films 520 (2012) 7219–7223.
[132] X. Xing, X. Liu, Y. He, Y. Lin, C. Zhang, H. Tang, D. Pang, Amplified Fluorescent Sensing of DNA Using Graphene Oxide and a Conjugated Cationic Polymer, Biomacromolecules 14 (2013) 117.
[133] M. Zhou, Y.M. Zhai, S.J. Dong, Electrochemical Sensing and Biosensing Platform Based on Chemically Reduced Graphene Oxide, Anal. Chem. 81 (2009) 5603–5613.
[134] S. Islam, S. Shukla, V.K. Bajpai, A smart nanosensor for the detection of human immunodeficiency virus and associated cardiovascular and arthritis diseases using functionalized graphene-based transistors, Biosens. Bioelectron. 126 (2019) 792–799.
[135] F. Inci, O. Tokel, S. Wang, Nanoplasmonic quantitative detection of intact viruses from unprocessed whole blood, ACS Nano. 7 (2013) 4733–4745.

[136] P.M. Kosaka, V. Pini, M. Calleja, J. Tamayo, Ultrasensitive detection of HIV-1 p24 antigen by a hybrid nanomechanical-optoplasmonic platform with potential for detecting HIV-1 at first week after infection, PLoS One 12 (2017) Article ID e0171899.

[137] M.S. Mannoor, H. Tao, J.D. Clayton, A. Sengupta, D.L. Kaplan, R.R. Naik, F.G. Omenetto N. Verma, M.C. McAlpine, Graphene-based wireless bacteria detection on tooth enamel, Nat. Commun. 3 (2012) 763.

[138] D. Ye, G. Liang, H. Li, J. Luo, S. Zhang, H. Chen, J. Kong, A novel nonenzymatic sensor based on CuO nanoneedle/graphene/carbon nanofiber modified electrode for probing glucose in saliva, Talanta 116 (2013) 223–230.

[139] M.S. Khan, K. Dighe, Z. Wang, I. Srivastava, E. Daza, A.S. Schwartz-Dual, J. Ghannam, S.K. Misra, D. Pan, Detection of prostate specific antigen (PSA) in human saliva using an ultra-sensitive nanocomposite of graphene nanoplatelets with diblock-co-polymers and Au electrodes, Analyst 143 (2018) 1094–1103.

[140] M. Coros¸ S. Pruneanu, R.I. Stefan-van Staden, Review—Recent Progress in the Graphene-Based Electrochemical Sensors and Biosensors, J. Electrochem. Soc. 167 (2019) 037528.

[141] L. Lin, L. Shi, J. Jia, O. Eltayeb, W. Lu, Y. Tang, C. Dong, Dual Photoluminescence Emission Carbon Dots for Ratiometric Fluorescent GSH Sensing and Cancer Cell Recognition, S. Shuang. 12 (2020) 18250–18257.

[142] Z. Wang, Z. Xue, Xi. Hao, C. Miao, J. Zhang, Y. Zheng, Z. Zheng, X. Lin, S. Weng, Ratiometric fluorescence sensor based on carbon dots as internal reference signal and T7 exonuclease-assisted signal amplification strategy for microRNA-21 detection, Anal. Chim. Acta. 1103 (2020) 212–219.

[143] P. Moonen, L. Kiemeney, J.A. Witjes, Urinary NMP22® BladderChek® Test in the Diagnosis of Superficial Bladder Cancer, Eur. Urol. 48 (2006) 951–956.

[144] S. Goodison, C.J. Rosser, V. Urquidi, Bladder Cancer Detection and Monitoring: Assessment of Urine- and Blood-Based Marker Tests, Mol. Diagn. Ther. 17 (2013) 71–84.

[145] D.S. Stampfer, G.A. Carpinito, J.R. Villanueva, L.W. Willsey, C.P. Dinney, H.B. Grossman, H.A. Fritsche, W.S. McDougal, Evaluation Of Nmp22 In The Detection Of Transitional Cell Carcinoma Of The Bladder, J. Urol. 159 (1998) 394–398.

[146] H.O. Thmana, F. Salehniab, M. Hosseinic, R. Hassana, A. Faizullaha, M.R. Ganjalib, Fluorescence immunoassay based on nitrogen doped carbon dots for the detection of human nuclear matrix protein NMP22 as biomarker for early stage diagnosis of bladder cancer, Microchem. J. 157 (2020) 104966.

[147] J.S. Sidhu, A. Singh, N. Garg, N. Singh, Naphthalimide Coupled FRET Pair for Highly Selective Ratiometric Detection of Thioredoxin Reductase and Cancer Screening, ACS Appl. Mater. Interfaces 9 (2017) 25847–25856.

[148] J.H. He, Y.Y. cheng, Q.Q. Zhang, H. Liu, C.H. Huang, Carbon dots-based fluorescence resonance energy transfer for the prostate specific antigen (PSA) with high sensitivity, Talanta 219 (2020) 121276.

[149] D. Bahari, B. Babamiri, A. Salimi, R. Hallaj, S.M. Amininasa, A self-enhanced ECL-RET immunosensor for the detection of CA19-9 antigen based on Ru(bpy)2(phen-NH2)2+ - Amine-rich nitrogen-doped carbon nanodots as probe and graphene oxide grafted hyperbranched aromatic polyamide as platform, Anal. Chim. Acta. 1132 (2020) 55–65.

[150] Y. Wang, C. Han, L. Yu, Ji. Wu, Y. Min, J. Tan, Y. Zhao, P. Zhang, Etching-controlled suppression of fluorescence resonance energy transfer between nitrogen-doped carbon dots and Ag nanoprisms for glucose assay and diabetes diagnosis, Spectrochim. Acta. A. Mol. Biomol. Spectrosc. 242 (2020) 118713.

[151] L. Yong, L. Xuan, T. Hongliang, Z.Z. Huang, A turn-on fluorescent assay for glucose detection based on carbon dots/manganese dioxide assembly, Microchem. J. 158 (2020) 105266.

[152] F.P. Mutuyimana, J. Liu1, M. Na, S. Nsanzamahoro, Z.R.H. Chen, Xi. Chen, Synthesis of orange-red emissive carbon dots for fluorometric enzymatic determination of glucose, Microchim. Acta. 185 (2018) 518.

[153] Xi. Zhu, Xi. Pang, Y. Zhang, S. Yao, Titanium carbide MXenes combined with red-emitting carbon dots as a unique turn-on fluorescent nanosensor for label-free determination of glucose, J. Mater. Chem. B. 7 (2019) 7729.

[154] H. Meixin, Q. Jianrong, J. Ruan, G. Shen, Highly Sensitive Detection of Glucose by a "Turn-Off-On" Fluorescent Probe Using Gadolinium-Doped Carbon Dots and Carbon Micro particles, J. Biomed. Nanotechnol. 14 (2018) 1117–1124.

[155] Y. Li, Y. Weng, S. Lu, M. Xue, B. Yao, W. Weng, T. Zheng, One-Step Hydrothermal Synthesis of N, Fe-Codoped Carbon Dots as Mimic Peroxidase and Application on Hydrogen Peroxide and Glucose Detection. J. Nanomaterials, Volume 2020, Article ID 1363212, https://doi.org/10.1155/2020/1363212.
[156] B. Wang, F. Liu, Y. Wu, Y. Chen, B. Weng, C.M. Li, Synthesis of catalytically active multielement-doped carbon dots and application for colorimetric detection of glucose, Sens. Actuators B: Chem. 255 (2018) 2601–2607.
[157] R.N. Goyal, V.K. Gupta, M. Oyama, N. Bachheti, Voltammetric determination of adenosine and guanosine using fullerene-C60-modified glassy carbon electrode, Talanta 71 (2007) 1110–1117.
[158] L. Suresh, J.S. Bondili, P.K. Brahman, Development of proof of concept for prostate cancer detection: an electrochemical immunosensor based on fullerene-C60 and copper nanoparticles composite film as diagnostic tool, Mater. Today Chem. 16 (2020) 100257.
[159] S. Huaxia, G. Rui, X. Wenjing, H. Huang, L. Xue, W. Wang, Y. Zhang, S. Weili, D. Xiaochen, Near-Infrared Light-Harvesting Fullerene-Based Nanoparticles for Promoted Synergetic Tumor Phototheranostics, ACS Appl. Mater. Interfaces 11 (2019) 44970−44977.
[160] Y. Cui, X.i. Zhong, R. Yuan, Y. Chai, A novel ECL biosensor based on C60 embedded in tetraoctylammonium bromide for the determination of glucose, Sens. Actuators B. 199 (2014) 101–107.
[161] Z. Jianli, Z.Y. Yuan, M. Zhao, J. Wang, Y. Chen, Q. Zhu, L. Bai, An efficient electrochemical assay for miR-3675-3p in human serum based on the nanohybrid of functionalized fullerene and metal-organic framework, Anal. Chim. Acta 1140 (2020) 78–88.
[162] I. Ojeda, B. Garcinuño, M.M. Guzmán, A.G. Cortés, M. Yudasaka, S. Iijima, F. Langa, P.Y. Sedeno, J.M. Pingarron, S. Iijima Yudasaka, F. Langa, P.Y. Sedeno, J.M. Pingarron, Carbon Nanohorns as a Scaffold for the Construction of Disposable Electrochemical Immunosensing Platforms. Application to the Determination of Fibrinogen in Human Plasma and Urine, Anal. Chem. 86 (2014) 7749–7756.
[163] Y.Q. Yu, J.P. Wang, M. Zhao, L.R. Hong, Y.Q. Chai, R. Yuan, Y. Zhuo, Target-catalyzed hairpin assembly and intramolecular/intermolecular co-reaction for signal amplifiedelectrochemiluminescent detection of microRNA, Biosens. Bioelectron. 77 (2016) 442–450.
[164] S. Zhu, S. Han, L. Zhang, S. Parveena, G. Xu, A novel fluorescent aptasensor based on single-walled carbon nanohorns, Nanoscale 3 (2011) 4589–4592.
[165] J. Chen, H. Ping, H. Bai, H. Lei, K. Liu, F. Dong, X. Zhang, Novel phosphomolybdic acid/single-walled carbon nanohorn-based modified electrode for non-enzyme glucose sensing, J. Electroanal. Chem. 784 (2017) 41–46.
[166] W. Zheng, Y. Liu, P. Yang, Y. Chen, J. Tao, J. Hu, P. Zhao, Carbon nanohorns enhanced electrochemical properties of Cu-based metal organic framework for ultrasensitive serum glucose sensing, J. Electroanal. Chem. 862 (2020) 114018.
[167] S. Zhu, Z. Liu, W. Zhang, S. Han, L. Hua, G. Xu, Nucleic acid detection using single-walled carbon nanohorns as a fluorescent sensing platform, Chem. Commun. 47 (2011) 6099–6101.
[168] S. Baluchova, A. Danhel, H. Dejmkova, V. Ostatna, M. Fojta, K.S. Peckov, Recent progress in the applications of boron doped diamond electrodes in electroanalysis of organic compounds and biomolecules – A review, Anal. Chim. Acta. 1077 (2019) 30–66.
[169] T. Zhang, H. Cui, C.Y. Fang, K. Cheng, X. Yang, H.C. Chang, M. Laird, Targeted nanodiamonds as phenotype-specific photoacoustic contrast agents for breast cancer, Nanomedicine (Lond) 10 (4) (2015) 573–587.
[170] M.N. Cordina, N.S.L.M. Parker, A.E. Dass, L.J. Brown, N.H. Packer, Reduced background autofluorescence for cell imaging using nanodiamonds and lanthanide chelates, Sci. Rep. 8 (2018) 4521.
[171] Y.S. Zou, L.L. He, Y.C. Zhang, X.Q. Shi, Z.X. Li, Y.L. Zhou, C.J. Tu, L. Gu, H.B. Zeng, The microstructure and electrochemical properties of boron-doped nanocrystalline diamond film electrodes and their application in non-enzymatic glucose detection, J. Appl. Electrochem. 43 (2013) 911–917.
[172] H. Longa, X. Liuc, Y. Xieb, N. Hub, Z. Dengb, Y. Jiangb, Q. Weib, Z. Yub, S. Zhang, Thickness effects of Ni on the modified boron doped diamond by thermal catalytic etching for non-enzymatic glucose sensing, J. Electroanal. Chem. 832 (2019) 353–360.

[173] C. Li, T. Zhao, Q. Wei, Z. Deng, H. Long, K. Zheng, H. Li, Y. Guo, Z. Yu, L.i. Ma, K. Zhou, N. Huang, C.T. Lin, The effect of heat treatment time on the carbon-coated nickel nanoparticles modified boron-doped diamond composite electrode for non-enzymatic glucose sensing, J. Electroanal. Chem. 841 (2019) 148–157.

[174] K. Yao, B. Dai, X. Tan, V. Ralchenko, L. Yang, B. Liu, S. Zhenhua, J. Zhao, K. Liu, J. Han, J. Zhu, Fabrication of Au/Ni/boron-doped diamond electrodes via hydrogen plasma etching graphite and amorphous boron for efficient non-enzymatic sensing of glucose, J. Electroanal. Chem. 871 (2020) 114264.

[175] P.U. Arumugam, H. Zeng, S. Siddiqui, D.P. Covey, A.J. Carlisle, A.P. Garris, Characterization of ultrananocrystalline diamond microsensors for in vivo dopamine detection, Appl. Phys. Lett. 102 (2013) 253107.

[176] G.Dutta C.Tan, H. Yin, S. Siddiqui, P.U. Arumugam, Detection of neurochemicals with enhanced sensitivity and selectivity via hybrid multiwall carbon nanotube-ultrananocrystalline diamond microelectrodes, Sens. Actuators B: Chem. 258 (2018) 193–203.

[177] Q. Wang, A. Kromka, J. Houdkova, O. Babchenko, B. Rezek, M. Li, R. Boukherroub, S. Szunerits, Langmuir 28 (2012) 587–592.

CHAPTER 12

Modification of screen printed biosensors using nanomaterials

Silpa P A
Department of ECE, Sahrdaya College of Engineering & Technology, Thrissur, Kerala, India

12.1 Introduction

Recently many analytical instruments are using for environmental, food, pharmaceutical, and clinical applications to determine harmful components, compositions, and certain medical conditions. Among these instruments, biosensors play a major role [1]. Biosensors works based on the chemical species or biological components in the samples by recognizing a target which is detected by a transducer. The main recognition molecules are affinity and catalytic based. Signaling receptors, antibodies, and nucleic acids which belongs to affinity-based recognition. At the same time, enzymes and biological cells are recognized using catalytic biosensors. A transducer is in cooperation with this to do the final detection. In olden days such bulky electrochemical sensors were widely used, but one of the challenges faced was to generate results at the point of care itself. Screen-printed electrodes (SPE) can overcome this challenge. It can take the electrochemical sensors to next level. The main advantage is miniaturization and size reduction which in turn requires a very small sample. Different substrates such as Alumina, ceramics, and plastic (polyvinylchloride, polycarbonate) gold, iron, silver, and fiberglass are possible.

The material using for making SPE decides the selectivity and sensitivity of the sensor. Usually, silver and carbon inks are using for printing SPE, but it is possible to use other materials such as gold and platinum. As shown in Fig. 12.1 in the construction procedure the viscous fluid (ink or paste) applied through a mesh screen onto the substrate. The mesh screen is having a pattern and the paste connects the recognition receptor to the substrate. Adhesives are added to attach paste to the substrate. Again powdered silver, nanoparticles, gold, or platinum can be added to this to enhance the characteristics. Recognition Receptors are attached to the ink over immobilization techniques such as physical absorption or chemical cross-linking [2]. The contents of paste is decides the overall analytical performance and commercial value of resulting sensors. Low resistivity inks are preferred and due to this reason Carbon inks are unpopular. Best alternatives is conductive crystalline substances with low residual currents. Adhesives are used along with inks. Ethylene Glycol, Cellulose acetate, Resin, cyclohexanone, are the commonly using adhesives. To enhance the electrochemical transduction, ink can be modified using

Advanced Nanomaterials for Point of Care Diagnosis and Therapy
DOI: https://doi.org/10.1016/B978-0-323-85725-3.00018-0

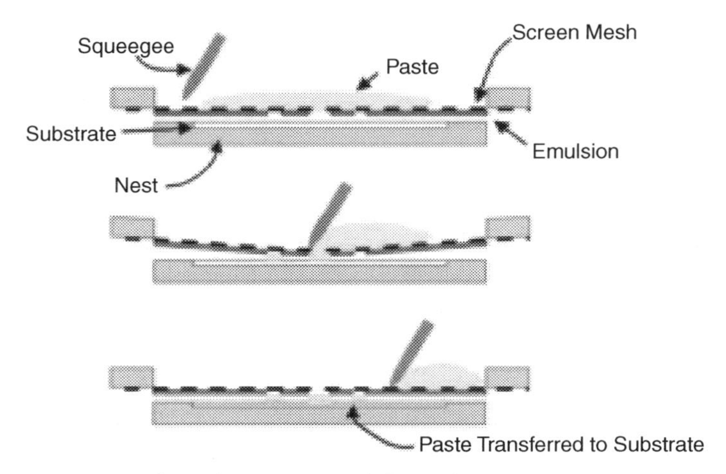

Fig. 12.1 Construction procedure of screen printed electrodes.

silver, gold and platinum nanoparticles. The common forms of SPE include a disc, ring, and a band. Some novel designs of SPE are as shown in Fig. 12.2.

The nanomaterials can alter the features of sensors. It is found that morphological features like size, shape, and surface properties of nanomaterial can determine the characteristics of sensors. The energy band gaps of the nanomaterials are correspond to energy levels of biological reactions. So the nanomaterials are very important in biosensors.

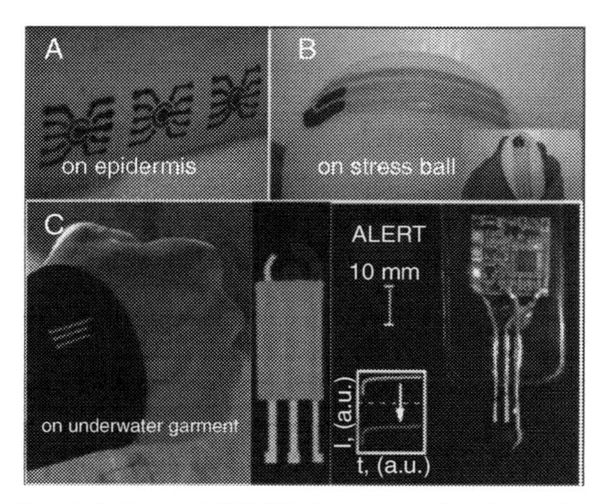

Fig. 12.2 Some of the Novel designs of SPE (A) three to eight electrode carbon ink-based arrays stamped on the epidermis, (B) electrode configuration on stress ball, and (C) on underwater garment to check the seawater indicated with red led.

The coming sections will discuss the nanomaterials using in biosensors and the modification of screen printed electrodes with Carbon nanotubes, Golden nanoparticle, Prussian blue magnetic nanoparticles, and quantum dots. For each section different possible sensors are discussed with cyclic volumetric graphs.

12.2 Classification of nanomaterials for biosensors

Nanomaterials act as amplifying components in biosensors since they enhance the performance. They will increase the sensitivity of detection up to one molecule. They can immobilize a large number of bioreceptor units at reduced volumes. Numerous nanomaterials are available including carbon allotropes and polymers. Among them for the biosensing applications commonly used nanomaterials are Gold nanoparticles, Quantum dots, Magnetic nanoparticles, Prussian blue [3]. The coming sections will discuss the properties of each nanoparticle toward the biosensing application in detail.

a) Carbon nanotube

Carbon nanotubes are popular in terms of nanowire morphology, biocompatibility, and electronic properties [4]. Carbon nanotube films exhibit high areas of interfaces with enhanced capabilities. They are showing specific moorage sites or redox mediation of bioelectrochemical reactions. It shows highly electrochemical reactions. CNT films having high electroactive surface areas due to the natural development of highly porous 3-D networks. The fast and efficient electron transfers are enabled by CNT with the driving force of hydrophobic reactions. In cases where the situation of electron transfers is not available, the mediated electron transfer is also possible with anchor molecules/ or with redox-active species. The important purpose of effective enzyme cabling on nanostructured carbon is nano energy conversion.

b) Golden nanoparticles

It is the most used one due to the biocompatibility of golden nanoparticles. The golden nanoparticles' properties can be tuned and adjusted. Since the golden nanoparticles are having surface plasmons the color change after a particular reaction is visible to the naked eye. This property has been used in a lot of sensors. The optical behavior of the gold surface opens enormous opportunities for sensing. The irradiation of light causes an oscillation of electron in the conduction band called surface plasmons. This was explained by mie theory [5] and is strongly depends on the size, shape, and environmental factors. The environmental dependency provides recognition through a change in oscillation frequency and hence to a color change recognition using the naked eye. Bioanalysis using SPR transduction is also possible. It detects the alteration of dielectric constant of propagating surface plasmons and the detection of the analyte can be recorded like changes of angle, intensity, or phase. Golden nanoparticles also show the refractive index sensing by coupling with ELISA. Gold nanoparticles can transfer electrons between an electroactive biological species. They can act as electron shuttle by approaching the redox center of enzyme and regenerating biocatalyst. A clear increase in the signal can be observed. Besides it is used as a probe in surface-enhanced Raman spectroscopy (SERS) [6]. It can be cut in any size and shape. Different synthesis methods are available for making golden nanoparticles. The phase synthesis method is the most popular. Other than this seed-based synthesis [7], one-step synthesis methods [8] also there.

c) Prussian blue (PB)

Firstly Prussian blu was used toward electrochemical behavior by depositing on platinum foil [9] followed by cyclic voltammetry. After this successful use, the electrochemistry of Prussian blu was completely explored. The most important feature of Prussian blue was the catalytic effect for the reduction of O_2 and H_2O_2 [10]. The reason behind this was a zeolitic structure with a cubic cell of 10.2 A^0, which allows the diffusion of low molecular weight molecules. So the PB is referred to as a 3D catalyst. PB can be used as an optical transducer too. In an experimental session, PB was used as a hydrogen peroxide detector where the color change of PB can be used as a detection purpose. Other than this PB can be used as a mediator for other substances. The features of PB is very promising for the oxidase based biosensors.

d) Magnetic nanoparticles

Magnetic nanoparticles are used as alternating materials of fluorescent labeling. Nanosized magnetic nanoparticles show diverse magnetic behaviors due to the reduction in magnetic domains. This causes superparamagnetisation. So the magnetization can be flipped within neel's reduction time and the magnetization will be present even if the external magnetic field is zero [11]. Iron oxide is the widely used nanomaterial for bioanalytical applications. When combining with other techniques, the magnetic nanoparticles offer further high sensitive transduction techniques. The magnetic labels offer no noise or sensing interference, as the biological samples are not having any magnetic behavior [12].

e) Quantum dots (QD)

A promising candidate for the biosensor purpose is quantum dots or luminescent semiconducting nanocrystals. They have shown their appropriateness either as a transducer or as optical labels. The most studied ones are cadmium chalcogenides. The bandwidth of quantum dots will vary with size and they emit a different wavelength of light [13]. It can be used in optical transduction analyzers since they enable efficient multiplexed analysis. One major issue with the quantum dots was sometimes the structural defects may cause non-radiative relaxation, but can be suppressed by a wide bandgap semiconductor which yields photo-stability [14]. It is possible to make QD's with inert and biocompatible coatings which will resist the toxicity. The phenomenon of transferring energy (nonradiative) between excited QD (Donor) and a quencher (acceptor) is described by Forster Resonance energy transfer (FRET). This can be used for optical transduction. The same way Bioluminescence resonant energy transfer (BRET) can be used. A light-emitting protein label handovers the energy to QDs and removes the need for an external excitation light source.

12.3 Modification of screen printed electrodes using nanomaterials

A. SPE modified with carbon nanotubes

A single-walled carbon nanotube can act as a semiconductor or metal depending on its chirality. Semiconducting nanotubes can be used for sensing purposes. It was found that carbon nanotubes can improve the sensing capability of screen printed electrodes. The performance of CNT modified screen-printed electrode is much superior compared to bare carbon electrodes as it is having a high surface area which increases the possibility of the reaction of molecules.

In a study by [15] multiwalled carbon nanotube was deposited on the graphene electrode through evaporation of carbon nanotube in dimethyl formaldehyde. The process includes immobilization of OPH (organophosphorus hydrolase) on the working electrode by dropping OPH solution on the sensor surface and drying at room temperature for a few hours followed by washing. The properties were explored using cyclic volumetry measurements (CV), it is found that there is a substantial amount of current in the modified one. The cyclic volumetric measurements are as shown in Fig. 12.3. The current showing is on the order of 100 μA.

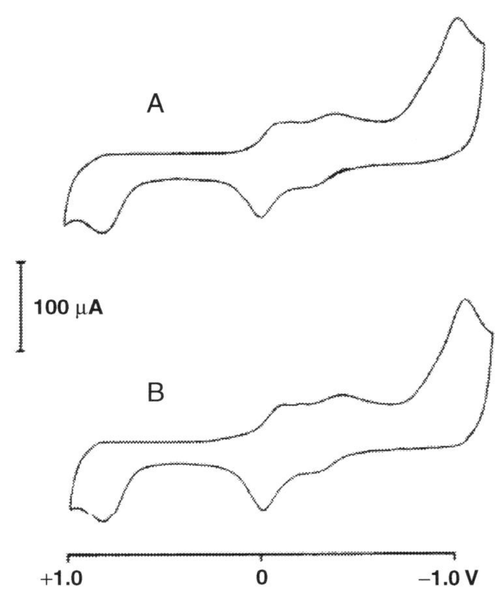

Fig. 12.3 CV measurements for the modified screen-printed electrode (A) 2mM solution and (B) in 50μl solution [15].

L lactate is an important component in the biochemical process. Reduced lactate can cause the death of the patient through organ. Most lactate amperometric biosensors are based on lactate dehydrogenase (LDH) or lactate oxidase (Lox). In an article [16] invented L-lactate biosensors to detect lactate dehydrogenase and lactate oxidase. They have used Multi wall carbon nanotube (MWCNT) which is purified by stirring on nitric acid followed by a drying procedure. The membrane was prepared by sonication of MWCNT. The sonicated solution was deposited onto the working electrode. The CV measurements are as shown in Fig. 12.4. The current is on the order of 10^{-4}. The sensitivity reported was 7334-8300 μA. The repeatability of the sensor was good with an RSD of 4.3%. The constructed one showed linearity from 1×10^{-6} to 2×10^{-5} with a lower limit detection of 3.7×10^{-7}.

The use of antidepressants in humans is a topic of current interest. Antidepressants are eliminated with urine. So the sensor was developed to detect antidepressants from urine. In a work [17] amperometric biosensors were developed based on multiwall carbon nanotube and oxide modified electrode for the determination of components such as imipramine, bifonazole and phenazepam(common antidepressants). The MWCNT was deposited on the electrode surface using the drop evaporation method. The amount of MWCNT varies from 2.5 to 0.5 μL at a concentration of 1 mg/mL in DMF or chitosan. The modified one showed a lot of advantages such as a higher value of the analytical signal better operational characteristics, more clear voltammograms over the unmodified sensors. The correlation coefficient and sensitivity were improved so much. The CV curve and calibration curve is as shown in Fig. 12.5.

In [18] a CNT modified biosensor for monitoring glucose dehydrogenase (GDH) was presented. GDH determination is very important in diabetic patients. CNT is used to modify the surface of the material. CNT material was dispersed in a solution of SDBS and Triton X-100 followed by a 2-hour sonication. From this, a part of the solution was coated on the reaction area and dried at room temperature. The sensor needs

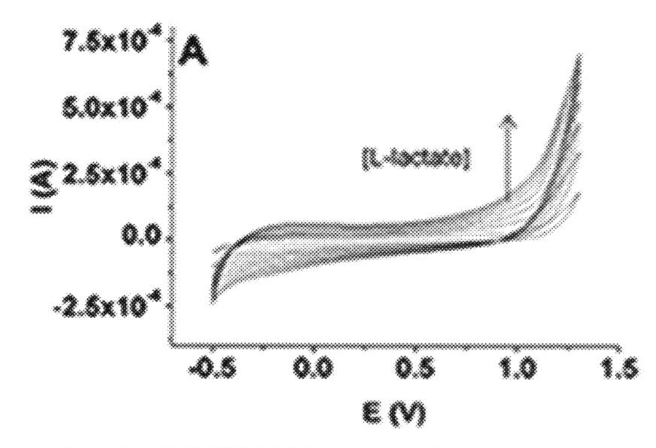

Fig. 12.4 CV Measurements by a Lox/MWCNT/PS biosensor [16].

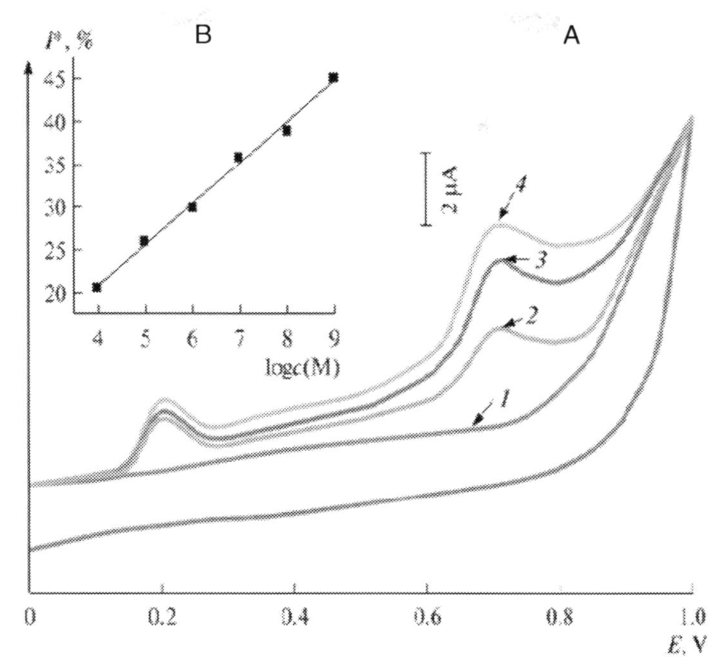

Fig. 12.5 CV obtained with a biosensor (A) in the presence of adrenalin and phenazepam and (B) a Calibration curve for determining phenazepam [17].

only a 2 μL solution for sensing. The CNT modified electrode showed a larger current response. The presence of CNT increases the background current. From Fig. 12.6 it can be understood that there is a large rise in current. An oxidation peak (around 600 mv) and a reduction peak (around -275 mv) were observed at the same time no peak was observed on the unmodified one. This points to the fact that an obvious increase in electron transfer between the enzyme and electrode. At the same time, it can be seen that no redox peaks are observed for merely CNT modified electrodes.

In [19], presented a sensor modified with silicon sol-gel conducting matrix containing carbon nanotubes. Sol gels can be effectively used for synthesis because of their high biocompatibility, protectiveness, heat, and biological stability. To modify electrode tetraethoxysilane and dimethyl diethoxy silane were mixed in a particular ratio. 5 μl of 1.2 M NaF was added to each mixture. The whole mixture was blended thoroughly and transferred to the electrode. The sensor exhibited a sensitivity of 31.0 μA mm^{-1} to glucose with a range of 1-35 mM and response time was 20 s. The cyclic voltammogram is as shown in Fig. 12.7. The cyclic voltammogram reveals that the process is limited by diffusion if the enzyme is entrapped in a sol-gel matrix modified with CNT. But still getting high heterogenic electron transfer constant which ensures the high efficiency of the electrode. It can detect the glucose within the concentration range 0.045-1.045 Mm.

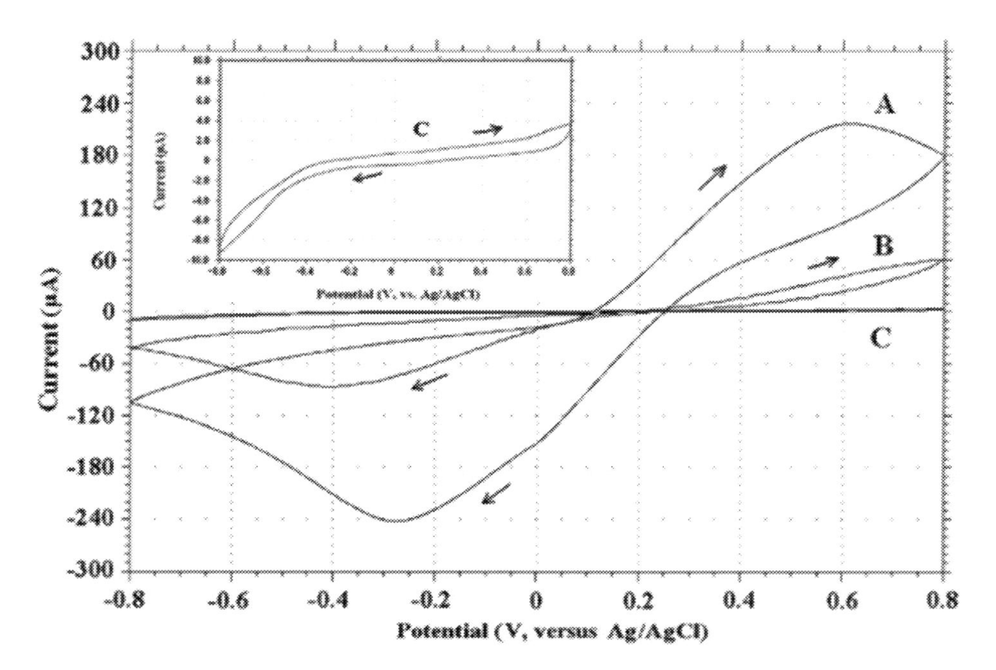

Fig. 12.6 CV of SPE modified with (A) PQQ-GDH/ferricyanide/CNT, (B) PQQ-GDH/ferricyanide, and (C) PQQ-GDH/CNT [18].

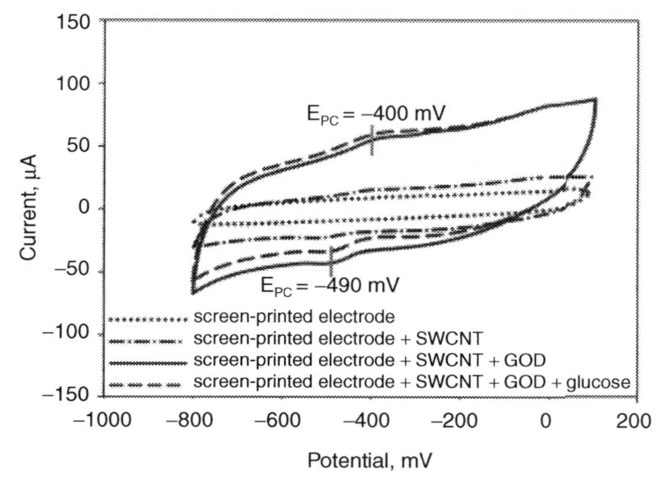

Fig. 12.7 The cyclic voltammogram for the glucose oxidase graphite electrode is coated immobilized in organosilicon (Kamanina et al.).

In a cholesterol sensor, rhodium graphite screen-printed electrodes have been used modified by multiwalled carbon nanotubes and by cytochromes p450cc as a catalytic enzyme [20]. MWCNT were drop casted rhodium-graphite electrodes. Before drop-casting CNT's were prepared in a solution of chloroform and sonicated for 2 hours. The sensors showed very much improvement in insensitivity. The cyclic voltammograms are as shown in Fig. 12.8. It shows the oxidation peak corresponding to the electrochemical reduction of the cytochrome with the oxygen. An enhancement in the current peak is observed compared to the previous works without modification. So this is a clear indication that MWCNT's can mediate electron transfer.

B. SPE modified with golden nanoparticle

In a work [21] a disposable biosensor was proposed to sense ethanol in beer samples by depositing PEDOT (polyethylene dioxythiophene) on to gold nanoparticle modified screen-printed electrodes. The SPE was modified by depositing 15 microliters of volume colloidal gold suspension and it was dried under darkness at room temperature and the electrode was rinsed by mechanically stirring at 250 rpm for the 30s. The sensitivity showed was 31.1 μA/mM. The various cyclic voltammograms are as shown in Fig. 12.9. Higher peak currents and smaller peak to peak potential separation (ΔE) was observed at voltammogram b) when compared with voltammogram a), which indicates the higher electrochemical activity of golden nanoparticle modified electrodes. The final voltammogram was recorded at a PEDOT/SPCE. The presence of polymer produced slight

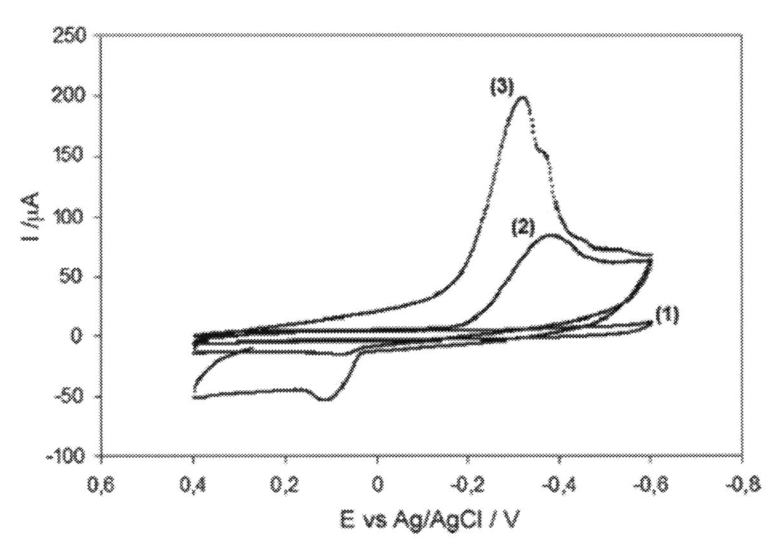

Fig. 12.8 ***Cyclic voltammograms of screen-printed bare rhodium–graphite electrode.*** (A) Electrodes modified with golden nanoparticles, (B) modified with multi-walled carbon nanotube, and (C) modified with Prussian blu [20].

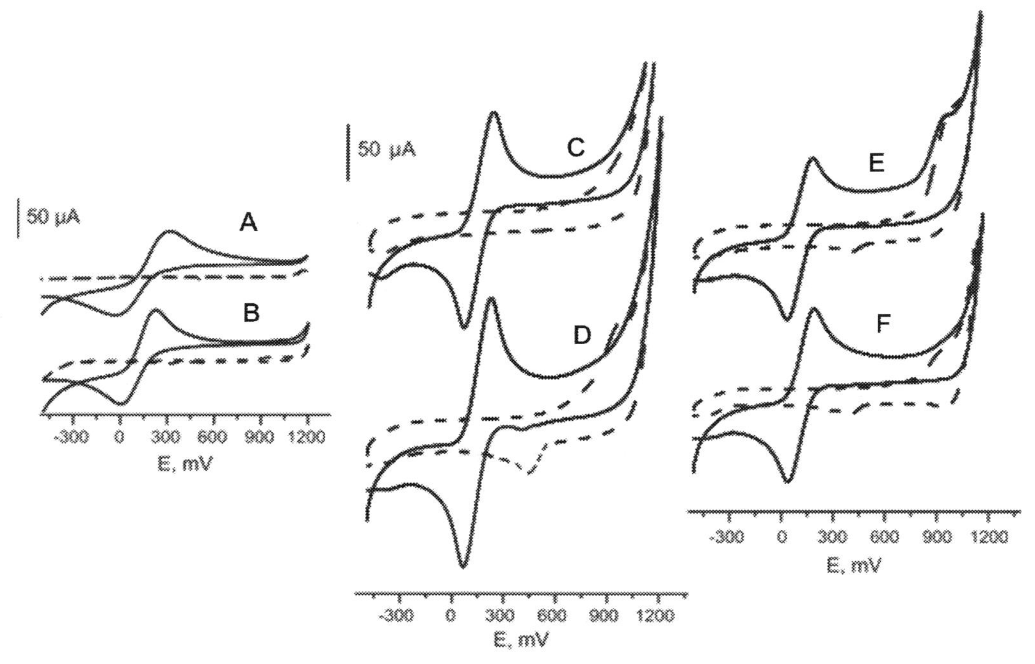

Fig. 12.9 CV of (A) screen printed carbon electrode, (B) nAu/SPE, (C) PEDOT/SPE, (D) PEDOT/nAU/SPE, (E) ADH/PEDOT/nAU/SPE, and (F) Tyr/PEDOT/nAU/SPE [21].

shifts of potentials and more values. Giving rise to a smaller peak to peak separation the peak current increased notably. The voltammogram d) gave rise to voltammograms with a further increase in peak current. These results suggest that there is active surface area enhancement due to the presence of nanoparticles.

A senor was developed for detecting ethanol in wine [22]. It is found that the inclusion of gold nanoparticles will increase the sensitivity of SPE. Its sensitivity is found to be as high as 0.490 μA m M^{-1} as shown in Fig. 12.10. The same way a sensor has been developed for selecting malic acid in wine [23] with golden nanoparticles. The viability has been checked for white, rose and red wines, the result showed great improvements in the performance of sensors. It showed a sensitivity of 850 -1700 μA/M

Laccase enzymes belong to the family of multicopper oxidases (MCO) found in plants, fungi, and bacteria. A laccase biosensor was developed by modifying with gold nanoparticle by integrating into carbon nanotube [24]. The MWCNT modified screen-printed electrodes were modified by depositing golden nanoparticle solution and after that photopolymer was spread onto the electrode surface. Then the electrode was dried using a UV-vis lamp. This step ensures the trapping of golden nanoparticle onto the electrode surface. The CV curve is shown in Fig. 12.11.

In [25] glucose sensor has been developed by using golden nanowires. The deposition golden nanowire was electroless using nanoporous polycarbonate. The residues

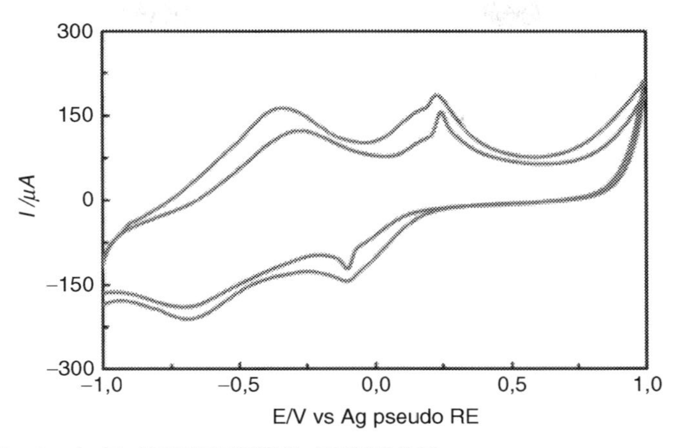

Fig. 12.10 CVs obtained with SPCE/MWCNT/AuNP/PNR [22].

were removed by washing the surface with distilled water, followed by HNO_3. The sensitivity showed was 18.2 $nAmM^{-1}$. The Cv diagram for two sensors for comparison purposes is as shown in Fig. 12.12. It is the voltammetric response recorded at 50 mVs^{-1} at different concentration values of the redox probe. The main parameter improved was stability and linearity.

C. SPE modified with Prusian blue

Prussian blue is a dark blue pigment used widely in biosensors. Prussian blue consists of alternating iron (II) and iron (III) located on a face-centered cubic lattice, in which iron (III) ions are surrounded octahedrally by nitrogen atoms and iron (II) are surrounded by carbon atoms.

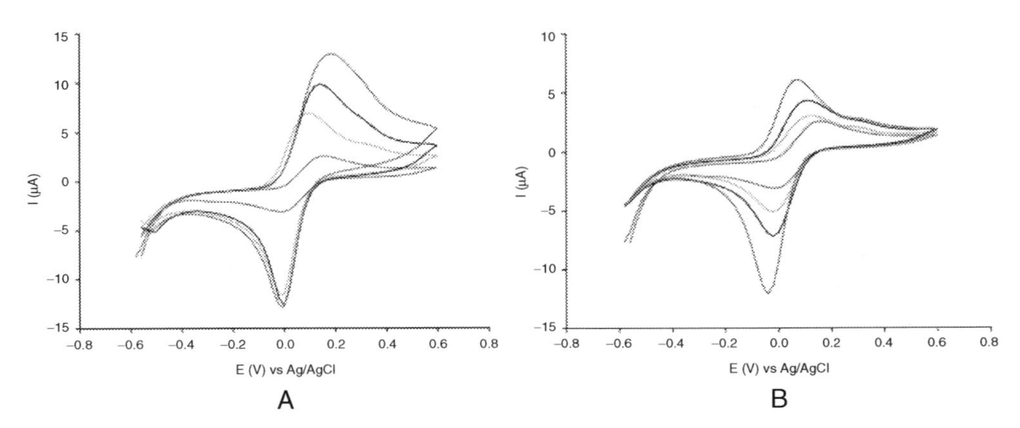

Fig. 12.11 CV of potassium ferricynaide solution at unmodified (MWCNT) (red curve) and at MWCNT modified with AuNP (Fig. 12.10A) and 10 nm at the following concentrations (Fig. 12.10B) 7×10^{-4} (green curve) 1×10^{-3} (black curve) 1.4×10^{-3} (blue curve) [24].

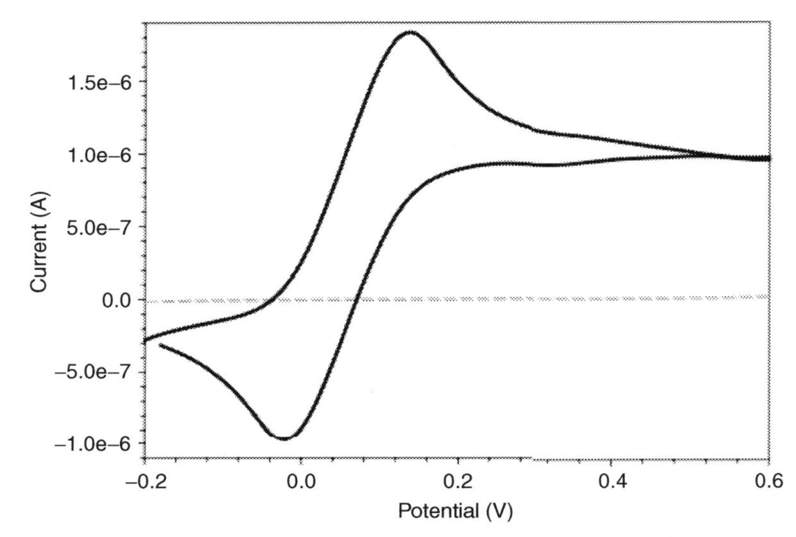

Fig. 12.12 ***Cyclic voltammograms using a typical NEE/SPS in 10^{-3} mol L^{-1} $NaNO_3$ as supporting electrolyte [25].***

In an H_2O_2 sensor Prussian blue modified sensor was used for detection [26]. Prussian blue film was found to be a good catalyst for H_2O_2. It was prepared by mixing $K_4[Fe(CN)_6]$ with $FeCl_3$ after that crystallization was induced by adding acetone into the mixture. The such obtained crystals were powdered and mixed with carbon ink. The CV is as shown in Fig. 12.13. The obtained sensitivity was 137 μAmM^{-1} CM^{-2}. On the CV two electrolytes were examined. Both showed Fe2+/Fe3+ redox couple.

In a glucose sensor [27] the deposition of PB was conducted on the screen-printed electrode after an electrochemical treatment at Ag/AgCl in a 0 .05 M buffer phosphate solution. The chemical deposition was carried out by dropping a solution of $K_3Fe(CN)_6$ in 10 mM HCl onto the working electrode area followed by a washing and drying procedure. The Cyclic voltammogram is as shown in Fig. 12.14. Both PB depositions presented the typical two pairs of redox peaks.

In an electrochromic glucose biosensor [28] Prussian Blue electrochromic pigment was used. It was equipped by Chemically growing a thin PB layer on the surface of tin oxide –based conducting particles covered by a thin antimony tin oxide (ATO) shell. The PB modified pigments combined with a binder system yields good screen printing pastes. The cyclic voltammogram is as shown in Fig. 12.15. The most striking feature in the graph is much higher currents observed for the blue pasts.

Carbamic and organophosphorus pesticides in water samples can be detected by Prussian blue modified sensors. They have used both co-phthalocyanine and Prussian blue. A precursor solution obtained by adding 5 ul of 0.1 M potassium ferricyanide in 10 mM HCL to 5 ul of 0.1 M ferric chloride in 10 mM HCL applied to the electrode's

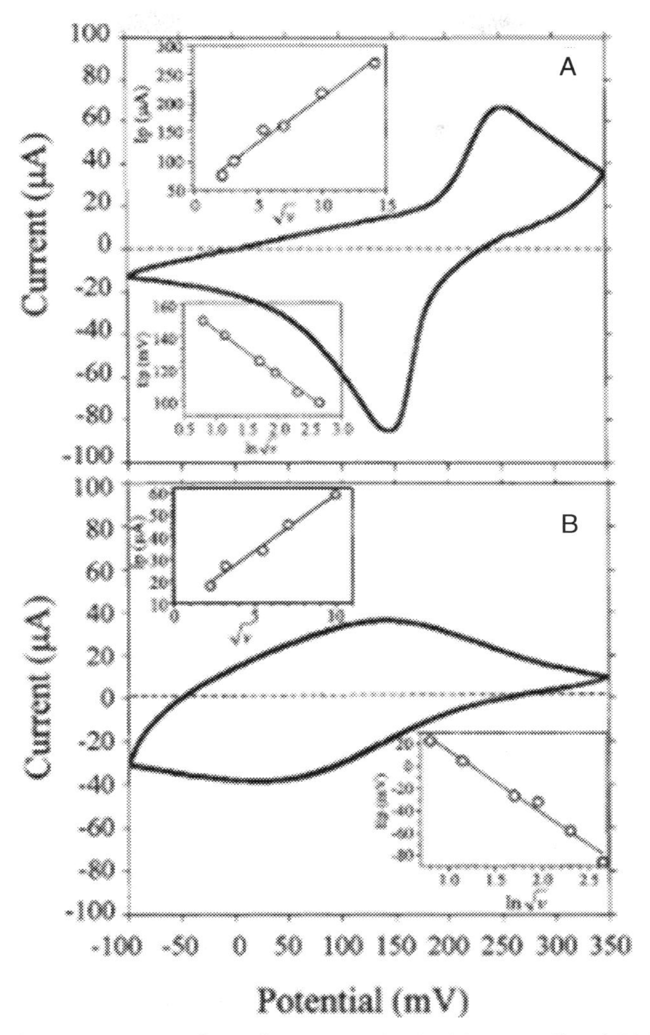

Fig. 12.13 ***Cyclic voltammograms performed at 10% PB-SPE in:*** (A) 0.1 M KCl and (B): PBS, pH: 7.40 [26].

working area. The electrodes were shaken and rinsed with HCl. The probes were left 90 min in the oven and dried. Prussian blue electrodes are having good sensitivity compared to co-phthalocyanine. The sensitivity reported was 143 mAM^{-1} CM^{-2}.

In [29] a sensor was developed to detect hydrogen peroxide. The Prussian blue nanoparticles(PBNP) were deposited using piezoelectric inkjet printing. The PBNP dispersion was prepared by mixing potassium ferrocyanide and iron (III) chloride. The sensor showed excellent detection. The CV is as shown in Fig. 12.16. At the bare SPE, there was a little or no visible reduction of H_2O_2 at the electrode. For PNB-SPE the

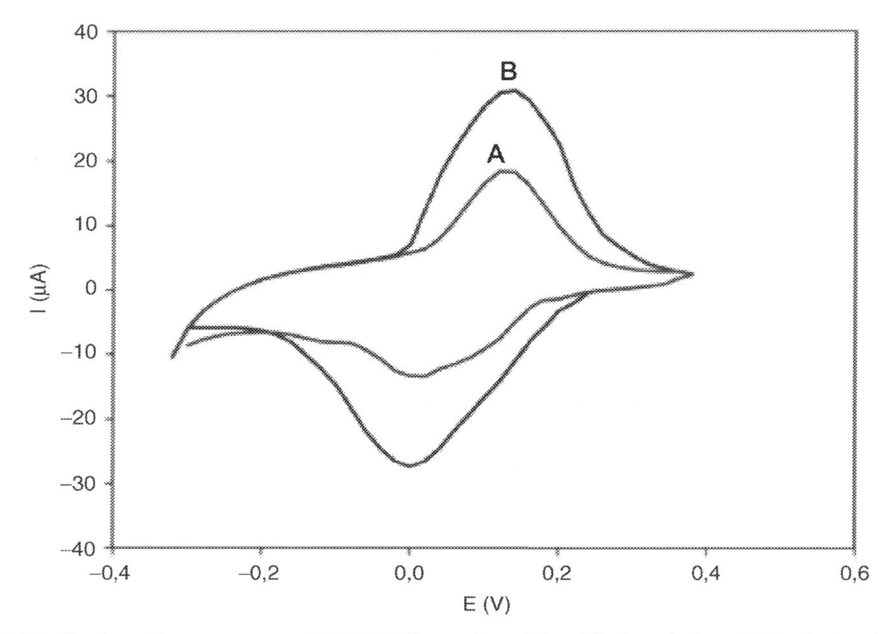

Fig. 12.14 Cyclic voltammograms of PB–SPCEs produced by (A) chemical and (B) electrochemical deposition in [27].

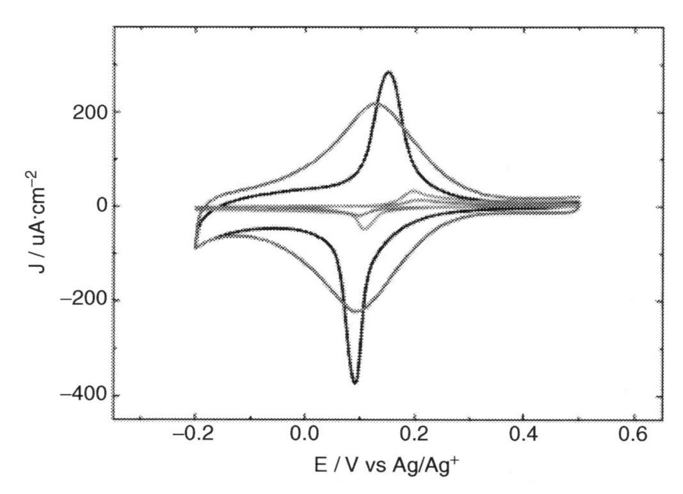

Fig. 12.15 ***Cyclic voltammograms were obtained in supporting electrolyte for the different types of PB-SPEs.*** SiO2-ATO/PB (*black*), ITO/PB (*blue*), DropSend C/PB (*orange*), Gwent C/PB (*green*) [28].

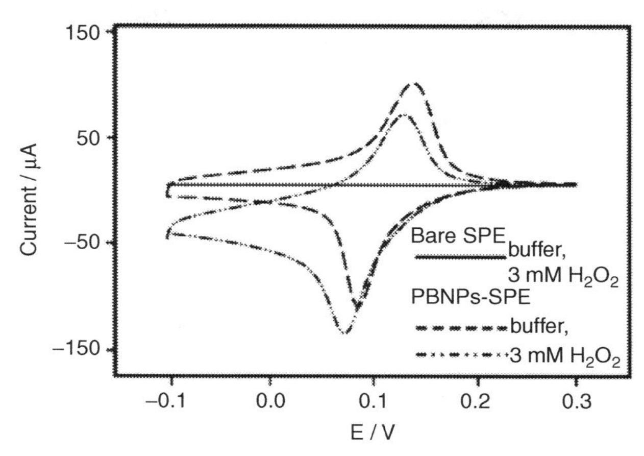

Fig. 12.16 ***Cyclic voltammograms of bare SPE in 3 mM H_2O_2 (solid line), SPE modified with 20 layers of PBNP dispersion in a buffer (dashed line) and 3 mM H_2O_2 (dashed-dotted line).*** All in 0.05 M phosphate buffer, 0.1 M KCl, pH 7.4 at 50 mV•s^{-1} vs Ag/AgC [29].

redox couple for PB could be seen. At the same time, the presence of H_2O_2 the cathode peak indicates the formation of PW was shifted to more negative potential accompanied by an increase in cathode current. The anode peak goes down into lower potentials.

In [30] modified gold and platinum printed electrodes with Prussian blue for hydrogen peroxide detection were discussed. The Au and pt screen printed electrodes were mechanically polished with alumina powder and subsequent rinsing with ultrapure water. The deposition of PB film was deposited galvanostatically. The cyclic voltammogram is as shown in Figs. 12.17 and 12.18. It was found that the PB modified electrode as more appropriate for hydrogen peroxide monitoring than the equivalently prepared PB modified paltinum electrode (PB/Pt). The current response was found to be about 700 nA; a value twice higher than for PB/Pt. The PB/Au electrodes shown a sensitivity for H_2O_2 equal to 2A M^{-1} Cm^{-2}.

In [31] amperometric biosensors were developed to improve NADH detection. The modification of screen printed electrodes was accomplished by precipitation of 0.1 M $k_3Fe_9CN_6$ and 0.1 M $FeCl_3$ solutions, prepared in 0.01 M HCl. The pretreatment was done by using a phosphate buffer. The cyclic voltammograms of with and without the modified electrode are as shown in Fig. 12.19. In the case of pretreated electrodes, a dramatic decrease of cathodic and anodic waves was observed.

In [32] glucose biosensor was developed using Prussian blue. The PB was prepared by an aqueous solution of $K_4[Fe(CN)_6]$ and $FeCl_3$ was mixed and the crystallization was induced by adding acetone. The crystals were separated by using vacuum filtration and it is powdered. The powder is blended in the mortar and mixed with carbon ink. Got sensitivity up, to 200 μAm^{-1} CM^{-2}. The cyclic voltammogram of the optimized

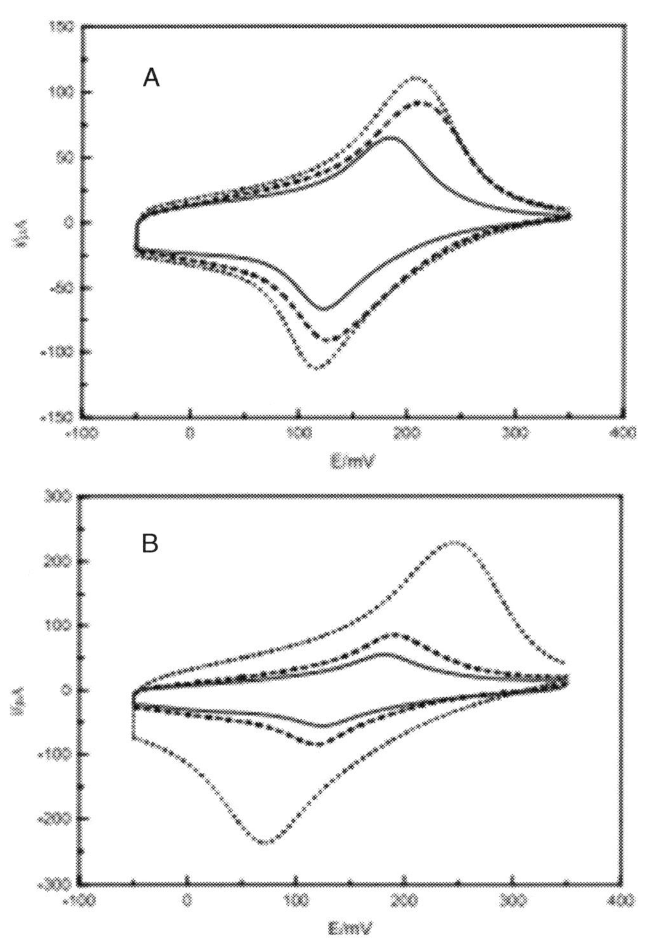

Fig. 12.17 ***Cyclic voltammograms of PB modified screen-printed electrodes in R1.*** **(A) Au and (B) Pt: solid, dashed, and dotted lines refer to 100, 160, and 240 s of electrodeposition, respectively. In [30].**

electrode is as shown in Fig. 12.20. The CV in the absence and presence of glucose is similar to that of PB without GOX. The background-subtracted CV shows the catalytic reaction peak of glucose with an maximum between 0 and -50 mv.

In [33] made a hydrogen peroxide and glucose sensor with modified electrodes. The glassy graphite powder was suspended in 10ml of a solution of $K_3Fe(CN)_6$ and 0.1 mol/1 in 10 mmol/1 HCl. Next 10 ml of a 0.1 mol/l solution of $FeCl_3$ in 10 mmol/I HCl were added and the resulting mixture was stirred for 10 min. The powder was washed and dried. The dependence of the peak current toward the scan rate has been studied as shown in Fig. 12.21. The straight relation among peak current and scan is in the range between 2 and 200 mv/s reveals that the electrodic behavior is not influenced by diffusion problems.

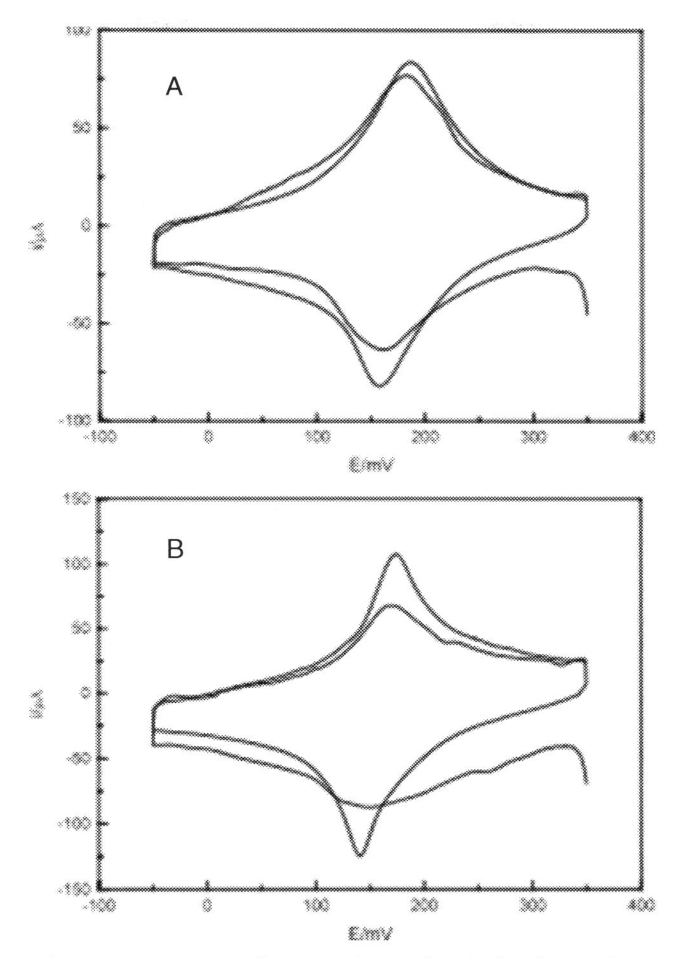

Fig. 12.18 ***Cyclic voltammograms revealing the electrochemical redox activation of the PB films.*** (A) Au and (B) Pt. Activation in R2 reagent in [30].

In [34] presented Prussian blue H_2O_2 sensor as a stable and long life sensor. They prepared Prussian blue modified SPE by placing a loco solution consists of potassium ferricyanide, HCl, and ferric chloride. The drop was placed on the working electrode area and dried. The effectiveness of the PB deposition procedure onto the SPE was verified by cyclic modified electrodes between the voltage -0.5 to 1.2v as shown in Fig. 12.22. Two typical pairs of redox waves showing the oxidation as well as reduction of PB are existing. Along with this increase of the cathodic wave at around 0.1V in presence of H_2O_2 can be noticed, proving the activity of the catalyst on the electrode.

D. Magnetic nanoparticles

In [35] a glucose biosensor has been designed using magnetic nanoparticle modified screen-printed electrodes. The Nano Fe_3O_4 was prepared through the

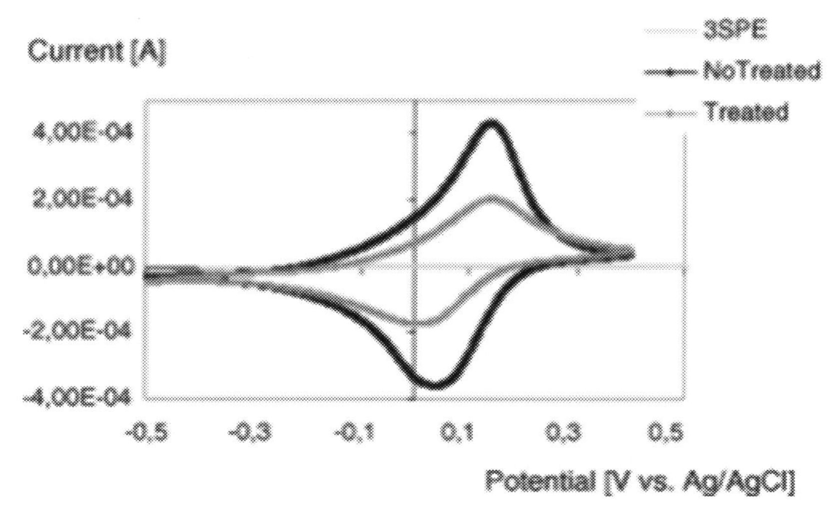

Fig. 12.19 ***Cyclic voltammogram of unmodified and Prussian blue modified electrodes [31].***

coprecipitation method. The sensor was prepared by drop coating a mixture of ferricyanide (Ferri)-nano-Fe_2O_4 onto the surface of screen-printed carbon electrodes and glucose oxidase was then layered on them. The cyclic voltammogram was as shown in Fig. 12.23. A remarkable increase in response current was observed. The result might suggest that nano-Fe_3O_4 provides the necessary conduction pathway to enhance the electron transfer but also act like nanoscaled electrodes for the molecules. The reason behind this is nano-Fe_3O_4 can absorb the iron cyanide complexes which increases the surface concentration redox probe and as a result the current will increase.

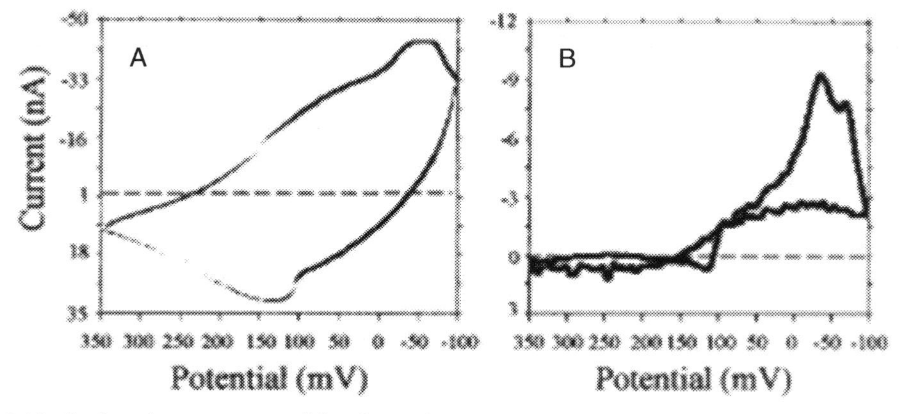

Fig. 12.20 ***Cyclic voltammogram of the electrode containing 5% of PB and 8% GOx.*** (A) 0 and 40 mM glucose, (B) background-subtracted CV of 40 mM glucose in PBS (Pravda et al.).

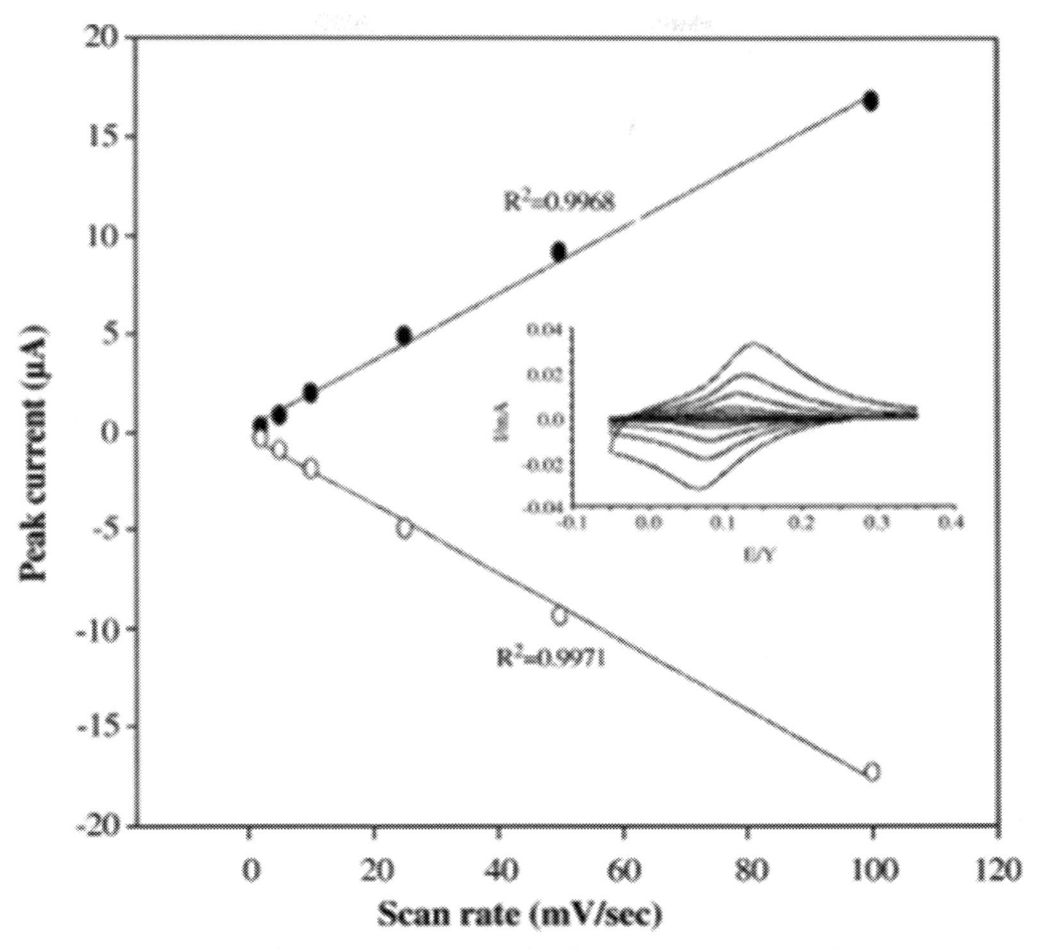

Fig. 12.21 ***Dependance of peak current (μA) toward cyclic volumetry scan rate [34].***

In [36] a biosensor for detecting chlorpromazine was discussed based on Magnetic core-shell manganese ferrite nanoparticles (MCMNP) The bare screen printed electrodes were coated with MCSNP. A solution of MCSNP in a 1 ml aqueous solution was prepared by dispersing 1 mg MCSNP. And after that, The solution was ultrasonicated for 1 hr and a 2 microliter aliquots of the MCSNP/H_2O suspension solution was cast on the carbon working electrodes. The CV curve is as shown in Fig. 12.24. The peak potential was observed at 500 mv, which is about 130 mv more negative compared to unmodified SPCE. The modified one shows a much high anodic peak current for the oxidation of chloropromazine compared to the unmodified one. This indicates that modification significantly improved the performance of electrodes toward chlorpromazine oxidation.

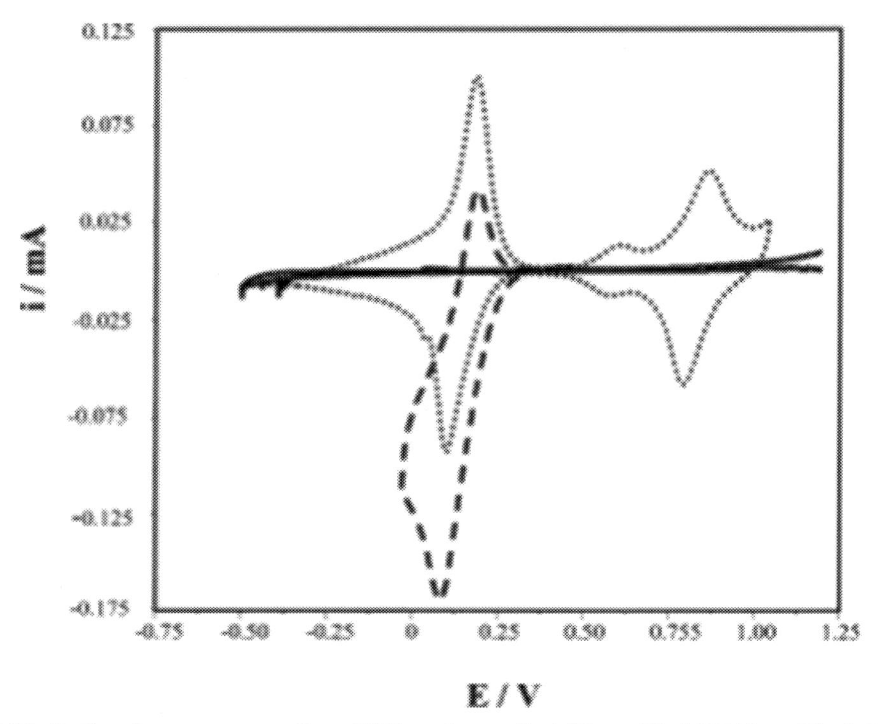

Fig. 12.22 *Cyclic voltammograms of bare SPE (continuous line); PB modified SPE (dotted line); PB modified SPE in presence of H_2O_2.*

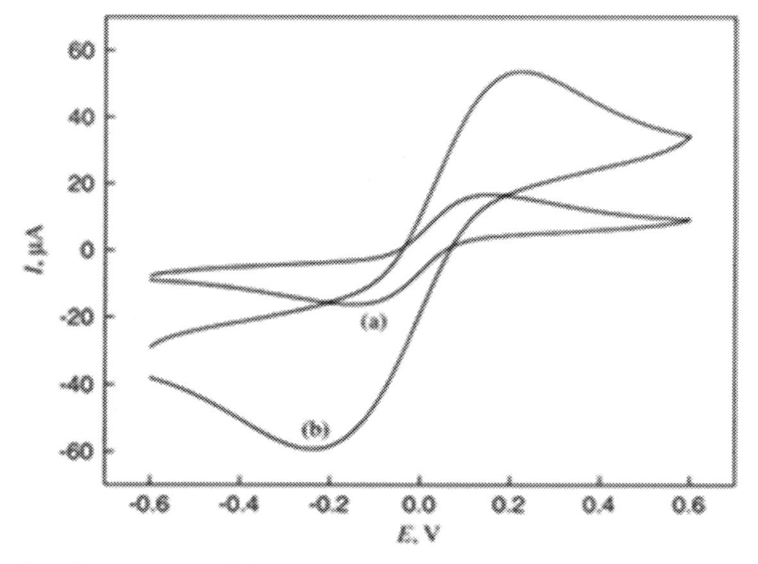

Fig. 12.23 Cyclic voltammogram of glucose at (A) GOD/Ferri-coated SPCEs and (B) GOD/Ferri-Nano-Fe3o4 coated SPCE [35].

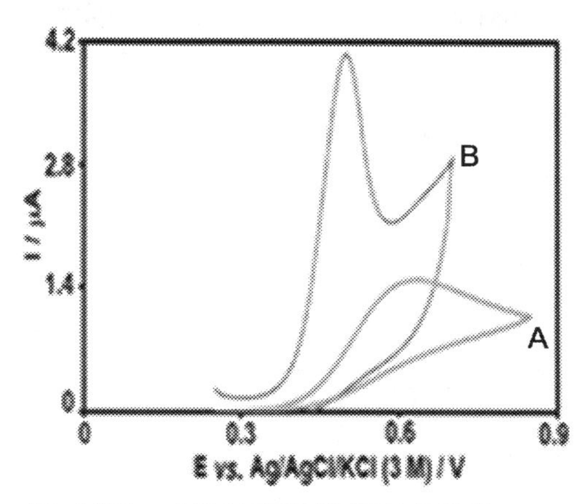

Fig. 12.24 CV of (A) unmodified SPE and (B) MCSNP/SPCE (S. Z. [36]).

In [37] presented amlodipine based screen printed electrodes, they have prepared the electrode by dispersing 1mg MCSNP with ultrasonication for 1h and a microliter aliquots of the MCSNP/H_2O suspension was cast on the carbon working electrodes. It is found that the sensor can detect the amlodipine (AML) concentration in tablets and urine.

In core-shell magnetic nanoparticles based sensor was developed for amperometric immunosensing. They have modified the surface with 1.0 g of core-shell ($CdFe_2O_4$-SiO_2). Magnetic nanoparticles were reacted with 10 ml of a fresh ethanol solution of 3-aminopropyltriethoxysilane (APTES) for 10 hr. These precipitates were washed with water and ethanol several times in an ultrasonic bath. Good reproducibility for four consecutive measurements of the same concentration of under the investigational conditions on the redeveloped surface. The comparative standard deviation of the measured current was 5%.

E. Quantum dots

[38] designed a sensor for sensing diethylstilbestrol (DCS) by means of graphene quantum dots (GQD) surface modified screen-printed electrode. The GQD was prepared using the citric acid pyrolysis method, The SPE's were modified with GQD using electrodeposition. The cyclic voltammogram is as shown in Fig. 12.25. It shows a linear

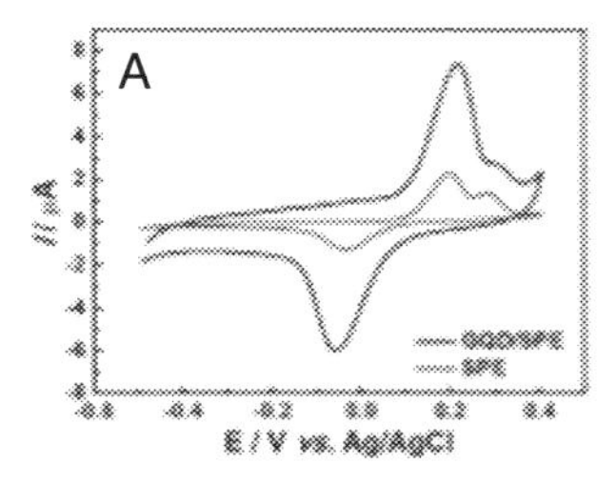

Fig. 12.25 ***Cyclic voltammograms obtained for SPE (red) and GQD/SPE (black) [38].***

dependence between $v^{1/2}$ for both anode and cathode current peaks. This indicates that electric reactions are incomplete by the diffusion of DES molecules from the bulk solution to the electrode surface.

In [39] detail a sensor to sense dopamine and uric acid together. Dopamine is one of the key molecules that play a significant role in renal, cardiovascular, central nervous, and hormonal systems and mammals. The major inhibition of the detection of DA is the coexistence of uric acid in high relative concentration. The sensor was made using disposable graphene nanosheets (GR) and NiO nanoparticles (NiO/GR/SPE). A solution of GR was prepared through the ultrasonication process. The GR/H2O solution was cast onto carbon working electrodes and allowed to evaporate. Then the GR/SPE was immersed into 0.5 mm $Ni(NO_3)$ at a potenstatic potential of -1.1v for 7s. Fig. 12.26 depicts the CV response for electro-oxidation of 75.0 μMA at an unmodified SPE(curve a) GR/SPE (curve b) and Nio/GR/SPE (curve c). NiO/Gr/SPE shows a much higher anodic peak current for the oxidation of DA compared with anodic SPE indicating that the combination of GR and

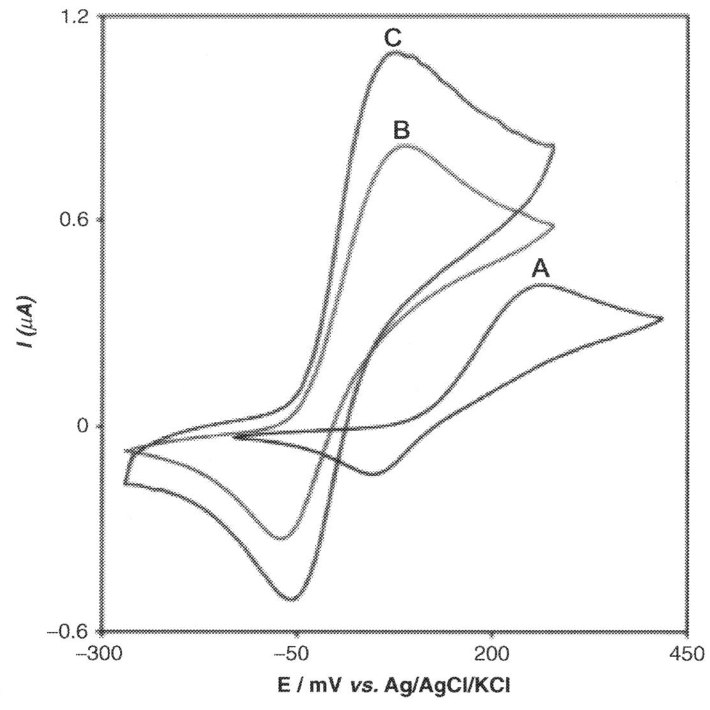

Fig. 12.26 CVs of (A) unmodified SPE, (B) GR/SPEs, and (C) NiO/ GR/SPE in the presence of 75.0 mM DA at a pH 7.0, respectively. In all cases, the scan rate was 50 m Vs^{-1} [40].

Table 12.1 Comparison of different nanomaterial.

Sl no	Publication	Modified material	Linear range	LOD	Sensitivity
1	[18]	Carbon nanotube	1.35×10^{-3} M	0.15×10^{-3} M	31.0 μA
2	[25]	Golden nanotube	2×10^{-2} M	1.5×10^{-4} M	18.2 nA
3	(Pravda et al.)	Prussian blue	4×10^{-3} M	0.2×10^{-3} M	140 nA
4	[35]	Magnetic nanoparticle	33.3×10^{-3} M	Not mentioned	1.74 μA
5	[41]	Unmodified	5 × 10-3	0.18×1^{-3} M	1.025 μA

Nio nanoparticles has pointedly improved the performance of electrode toward the oxidation.

Punrat et al. [40] presented a sensor using polyaniline/graphene quantum dot modified SPE for the rapid determination of Cr(VI). Cr(VI) is a toxic agent that has more mutagenic properties. The golden quantum dot was prepared using pyrolysis method. Aliquot aniline monomer was added into aGQD solution. The mixture was then sonicated before adding $4MH_2SO_4$ and diluting. The solution was aspirated into the holding coil and subsequently injected through the flowcell resulting entire surface covered with solution. The sensor has high sensitivity with a detection limit as low as 0.097 mgl^{-1}.

12.4 Advantages of modification

A lot of examples are discussed in the previous sessions. To understand the advantages of modification with nanomaterials, the glucose sensor is selected as an example. Table 12.1 summarizes the glucose sensors with different nanoparticles. When considering things the linear range magnetic nanoparticles are the best material. At the same time, golden nanotube provides the good limits of detection. In terms of sensitivity carbon nanotube is a good choice.

12.5 Conclusion

In this review, we have demonstrated some of the nanomaterial modified screen-printed electrodes. Electrochemical sensors are a good analytical tool as a demand for sensitive, rapid, and selective determination of analytes. Recently SPE has been replaced by conventional electrodes as it is having attractive features like low cost, reproducibility, disability besides this it can be incorporated into miniaturized devices. The use of nanomaterials along with SPE has increased signal sensitivity and stability. Nanoparticles modified SPE's are already available commercially. So in the coming years screen printed electrodes will find the new heights of application.

References

[1] Z. Taleat, A. Khoshroo, M. Mazloum-Ardakani, Screen-printed electrodes for biosensing: a review (2008- 2013), Microchim. Acta 181 (9–10) (2014) 865–891, https://doi.org/10.1007/s00604-014-1181-1.

[2] M.U. Ahmed, M.M. Hossain, M. Safavieh, Y.L. Wong, I.A. Rahman, M. Zourob, E. Tamiya, Toward the development of smart and low cost point-of-care biosensors based on screen printed electrodes, Crit. Rev. Biotechnol. 36 (3) (2016) 495–505, https://doi.org/10.3109/07388551.2014.992387.

[3] M. Holzinger, A. Le Goff, S. Cosnier, Nanomaterials for biosensing applications: a review, Front. Chem. 2 (2014) 1–10, https://doi.org/10.3389/fchem.2014.00063.

[4] F. Valentini, M. Carbone, G. Palleschi, Carbon nanostructured materials for applications in nano-medicine, cultural heritage, and electrochemical biosensors, Anal. Bioanalyt. Chem. 405 (2–3) (2013) 451–465, https://doi.org/10.1007/s00216-012-6351-6.

[5] P. Mulvaney, Surface plasmon spectroscopy of nanosized metal particles, Langmuir 12 (3) (1996) 788–800, https://doi.org/10.1021/la9502711.

[6] S. Zeng, K.T. Yong, I. Roy, X.Q. Dinh, X. Yu, F. Luan, A review on functionalized gold nanoparticles for biosensing applications, Plasmonics 6 (3) (2011) 491–506, https://doi.org/10.1007/s11468-011-9228-1.

[7] N.R. Jana, L. Gearheart, C.J. Murphy, Seeding growth for size control of 5-40 nm diameter gold nanoparticles, Langmuir 17 (22) (2001) 6782–6786, https://doi.org/10.1021/la0104323.

[8] M. Aslam, L. Fu, M. Su, K. Vijayamohanan, V.P. Dravid, Novel one-step synthesis of amine-stabilized aqueous colloidal gold nanoparticles, J. Mater. Chem. 14 (12) (2004) 1795–1797, https://doi.org/10.1039/b402823f.

[9] V.D. Neff, Electrochemical oxidation and reduction of thin films of Prussian blue, J. Electrochem. Soc. 125 (6) (1978) 886–887, https://doi.org/10.1149/1.2131575.

[10] K. Itaya, N. Shoji, I. Uchida, Catalysis of the reduction of molecular oxygen to water at Prussian blue modified electrodes, J. Am. Chem. Soc. 106 (12) (1984) 3423–3429, https://doi.org/10.1021/ja00324a007.

[11] K.J.M. Bishop, C.E. Wilmer, S. Soh, B.A. Grzybowski, Nanoscale forces and their uses in self-assembly, Small 5 (14) (2009) 1600–1630, https://doi.org/10.1002/smll.200900358.

[12] C.R. Tamanaha, S.P. Mulvaney, J.C. Rife, L.J. Whitman, Magnetic labeling, detection, and system integration, Biosens. Bioelectron. 24 (1) (2008) 1–13, https://doi.org/10.1016/j.bios.2008.02.009.

[13] H. Weller, Colloidal semiconductor Q-particles: chemistry in the transition region between solid state and molecules, Angew. Chem. Int. Ed. Engl. 32 (1) (1993) 41–53, https://doi.org/10.1002/anie.199300411.

[14] B.O. Dabbousi, J. Rodriguez-Viejo, F.V. Mikulec, J.R. Heine, H. Mattoussi, R. Ober, K.F. Jensen, M.G. Bawendi, (CdSe)ZnS core-shell quantum dots: Synthesis and characterization of a size series of highly luminescent nanocrystallites, J. Phys. Chem. B 101 (46) (1997) 9463–9475, https://doi.org/10.1021/jp971091y.

[15] M. Trojanowicz, A. Mulchandani, M. Mascini, Carbon nanotubes-modified screen-printed electrodes for chemical sensors and biosensors, Anal. Let. 37 (15) (2004) 3185–3204, https://doi.org/10.1081/AL-200040320.

[16] S. Pérez, S. Sánchez, E. Fàbregas, Enzymatic Strategies to construct L-mlactate bosensors based on polysulfone/carbon nanotubes Membranes, Electroanalysis 24 (4) (2012) 967–974, https://doi.org/10.1002/elan.201100628.

[17] E.P. Medyantseva, D.V. Brusnitsyn, R.M. Varlamova, M.A. Beshevets, H.C. Budnikov, A.N. Fattakhova, Capabilities of amperometric monoamine oxidase biosensors based on screen-printed graphite electrodes modified with multiwall carbon nanotubes in the determination of some antidepressants, J. Anal. Chem. 70 (5) (2015) 535–539, https://doi.org/10.1134/S106193481505010X.

[18] G. Li, H. Xu, W. Huang, Y. Wang, Y. Wu, R. Parajuli, A pyrrole quinoline quinone glucose dehydrogenase biosensor based on screen-printed carbon paste electrodes modified by carbon nanotubes, Meas. Sci. Technol. (6) (2008) 19, https://doi.org/10.1088/0957-0233/19/6/065203.

[19] O.A. Kamanina, S.S. Kamanin, A.S. Kharkova, V.A. Arlyapov, Glucose biosensor based on screen-printed electrode modified with silicone sol–gel conducting matrix containing carbon nanotubes, 3 Biotech. 9 (7) (2019) 1–7, https://doi.org/10.1007/s13205-019-1818-1.

[20] S. Carrara, V.V. Shumyantseva, A.I. Archakov, B. Samorì, Screen-printed electrodes based on carbon nanotubes and cytochrome P450scc for highly sensitive cholesterol biosensors, Biosens. Bioelectron. 24 (1) (2008) 148–150, https://doi.org/10.1016/j.bios.2008.03.008.
[21] V. Serafín, L. Agüí, P. Yáñez-Sedeño, J.M. Pingarrón, A novel hybrid platform for the preparation of disposable enzyme biosensors based on poly(3,4-ethylenedioxythiophene) electrodeposition in an ionic liquid medium onto gold nanoparticles-modified screen-printed electrodes, J. Electroanal. Chemi. 656 (1–2) (2011) 152–158, https://doi.org/10.1016/j.jelechem.2010.11.038.
[22] M. Bilgi, E. Ayranci, Biosensor application of screen-printed carbon electrodes modified with nanomaterials and a conducting polymer: Ethanol biosensors based on alcohol dehydrogenase, Sens. Actuators, B: Chem. 237 (2016) 849–855, https://doi.org/10.1016/j.snb.2016.06.164.
[23] B. Molinero-Abad, M.A. Alonso-Lomillo, O. Domínguez-Renedo, M.J. Arcos-Martínez, Malate quinone oxidoreductase biosensors based on tetrathiafulvalene and gold nanoparticles modified screen-printed carbon electrodes for malic acid determination in wine, Sens. Actuators, B: Chem. 202 (2014) 971–975, https://doi.org/10.1016/j.snb.2014.06.057.
[24] G. Favero, G. Fusco, F. Mazzei, F. Tasca, R. Antiochia, Electrochemical characterization of graphene and MWCNT screen-printed electrodes modified with AuNPs for laccase biosensor development, Nanomaterials 5 (4) (2015) 1995–2006, https://doi.org/10.3390/nano5041995.
[25] W. Vastarella, L. Della Seta, A. Masci, J. Maly, M. De Leo, L.M. Moretto, R. Pilloton, Biosensors based on gold nanoelectrode ensembles and screen printed electrodes, Int. J. Environ. Anal. Chem. 87 (10–11) (2007) 701–714, https://doi.org/10.1080/03067310701332626.
[26] M.P. O'Halloran, M. Pravda, G.G. Guilbault, Prussian Blue bulk modified screen-printed electrodes for H2O2 detection and for biosensors, Talanta 55 (3) (2001) 605–611, https://doi.org/10.1016/S0039-9140(01)00469-6.
[27] D. Albanese, A. Sannini, F. Malvano, R. Pilloton, M. Di Matteo, Optimisation of glucose biosensors based on sol-gel entrapment and Prussian blue-modified screen-printed electrodes for real food analysis, Food Anal. Methods 7 (5) (2014) 1002–1008, https://doi.org/10.1007/s12161-013-9705-6.
[28] M. Aller-Pellitero, J. Fremeau, R. Villa, G. Guirado, B. Lakard, J.Y. Hihn, F.J. del Campo, Electrochromic biosensors based on screen-printed Prussian Blue electrodes, Sens. Actuators, B: Chem. 290 (2019) 591–597, https://doi.org/10.1016/j.snb.2019.03.100.
[29] S. Cinti, F. Arduini, D. Moscone, G. Palleschi, A.J. Killard, Development of a hydrogen peroxide sensor based on screen-printed electrodes modified with inkjet-printed prussian blue nanoparticles, Sensors (Switzerland) 14 (8) (2014) 14222–14234, https://doi.org/10.3390/s140814222.
[30] I.L. De Mattos, L. Gorton, T. Ruzgas, Sensor and biosensor based on Prussian Blue modified gold and platinum screen printed electrodes, Biosens. Bioelectron. 18 (2–3) (2002) 193–200, https://doi.org/10.1016/S0956-5663(02)00185-9.
[31] A.M. Gurban, T. Noguer, C. Bala, L. Rotariu, Improvement of NADH detection using Prussian blue modified screen-printed electrodes and different strategies of immobilisation, Sens. Actuators, B: Chem. 128 (2) (2008) 536–544, https://doi.org/10.1016/j.snb.2007.07.067.
[32] M. Pravda, M.P. O'Halloran, M.P. Kreuzer, G.G. Guilbault, Composite glucose biosensor based on screen-printed electrodes bulk modified with Prussian Blue and glucose oxidase, Anal. Lett. 35 (6) (2002) 959–970, https://doi.org/10.1081/AL-120004548.
[33] F. Ricci, A. Amine, C.S. Tuta, A.A. Ciucu, F. Lucarelli, G. Palleschi, D. Moscone, Prussian Blue and enzyme bulk-modified screen-printed electrodes for hydrogen peroxide and glucose determination with improved storage and operational stability, Anal. Chim. Acta 485 (1) (2003) 111–120, https://doi.org/10.1016/S0003-2670(03)00403-3.
[34] F. Ricci, A. Amine, G. Palleschi, D. Moscone, Prussian Blue based screen printed biosensors with improved characteristics of long-term lifetime and pH stability, Biosens. Bioelectron. 18 (2–3) (2002) 165–174, https://doi.org/10.1016/S0956-5663(02)00169-0.
[35] B.W. Lu, W.C. Chen, A disposable glucose biosensor based on drop-coating of screen-printed carbon electrodes with magnetic nanoparticles, J. Magn. Magn. Mater. 304 (1) (2006) 400–402, https://doi.org/10.1016/j.jmmm.2006.01.222.
[36] S.Z. Mohammadi, A.H. Sarhadi, F. Mosazadeh, Screen-printed electrode modified with magnetic core-shell nanoparticles for detection of chlorpromazine, Anal. Bioanal. Chem. Res. 5 (2) (2018) 363–372, https://doi.org/10.22036/abcr.2018.128511.1205.

[37] S.Z.I.A. Mohammadi, S.Tajik, H. Beitollahi, Screen printed carbon electrode modified with magnetic core shell manganese ferrite nanoparticles for electrochemical detection of amlodipine, J. Serbian Chem. Soc. 84 (9) (2019) 1005–1016, https://doi.org/10.2298/JSC1810056036M.
[38] A. Gevaerd, C.E. Banks, M.F. Bergamini, L.H. Marcolino-Junior, Graphene quantum dots modified screen-printed electrodes as electroanalytical sensing platform for diethylstilbestrol, Electroanalysis 31 (5) (2019) 838–843, https://doi.org/10.1002/elan.201800838.
[39] S. Jahani, H. Beitollahi, Selective detection of dopamine in the presence of uric acid using NiO nanoparticles decorated on graphene nanosheets modified screen-printed electrodes, Electroanalysis 28 (9) (2016) 2022–2028, https://doi.org/10.1002/elan.201501136.
[40] E. Punrat, C. Maksuk, S. Chuanuwatanakul, Talanta Polyaniline /graphene quantum dot-modi fi ed screen-printed carbon electrode for the rapid determination of Cr (VI) using stopped- fl ow analysis coupled with voltammetric technique, Talanta 150 (2016) 198–205, https://doi.org/10.1016/j.talanta.2015.12.016.
[41] N. Chandra Sekar, S.A. Mousavi Shaegh, S.H. Ng, L. Ge, S.N Tan, A paper-based amperometric glucose biosensor developed with Prussian Blue-modified screen-printed electrodes, Sens. Actuators, B: Chem. 204 (2014) 414–420, https://doi.org/10.1016/j.snb.2014.07.103.

CHAPTER 13

Hybrid organic or inorganic nanomaterials for healthcare diagnostics

Pallab K. Bairagi[a], Pravat Rajbanshi[b], Prateek Khare[c]
[a]Ordinance Factory Nalanda, Department of Defence Production, Ministry of Defence, Rajgir, Bihar, India
[b]Department of Chemical Engineering, Indian Institute of Technology Kanpur, Kanpur, Uttar Pradesh, India
[c]Department of Chemical Engineering, Madan Mohan Malaviya University of Technology, Gorakhpur, Uttar Pradesh, India

13.1 Introduction

The advancement in the field of molecular diagnostics allows detections of the diseases prior to its dangerous clinical symptoms, which ultimate helps to apply predictive and preventive medicines at early stage of diagnosis tools like medical imagery, electrocardiography, and X-ray facility or by analytical methods developed in clinical laboratories. It also leads to treat disease at the earliest which not only reduced the treatment costs, and also improves better quality of patient life, without prone to the side effects of treatment [1]. Recently, a very famous invention of prognosis tool comes into limelight (antigen test) for COVID-19 detection, which is relatively inexpensive and less complex, fast and easy to interpret result. Other than this, a very commonly used diagnostic practice in the case of pregnancy test. However, in both the cases, the result is qualitative, i.e., conformance/non-conformance is affirmed. Generally, in this class of diagnostic's tool, target biomolecules, are identified as biomarkers, which played predominant role in determining efficient correlation between biosensors and their respective diagnostics. The presence of the specific biomolecules is characteristic for a specific disease (molecular biomarkers, some example of biomarkers are; human chorionic gonadotropin is a molecular biomarker for pregnancy prostate-specific antigen (PSA) enzyme in blood are molecular biomarker for prostate cancer, elevated concentration of these biomarkers, are indicative clinical-symptoms of the onset types of disease [2]. Although, there are many other examples apart from these two typeset of diagnostic tool, for example, in the case of hyperglycemia/hypoglycemia, kidney dysfunction, infection, etc. The presence of glucose, creatinine, and many other different infecting agents (including bacteria, virus, parasite, etc.) or other biomolecules, show an elevated change in their concentration from the permissible values which clearly indicates from its highly intensive values. In the case of different complex diseases, including different kinds of cancers, the biomarkers are made of different larger biomolecules such as enzymes, antibodies (proteins), micro-Ribonucleic acids (RNAs), damaged deoxyribonucleic acid (DNAs) etc. These

Advanced Nanomaterials for Point of Care Diagnosis and Therapy
DOI: https://doi.org/10.1016/B978-0-323-85725-3.00014-3

enzymes are generally detected using antibody-based immune assays as each of these antibodies which are need to be selective to a single biomolecule, termed as biomarkers, and this characteristics of any types of diagnostic tool make more peculiar and efficient for diagnosis purposes [3].

The environmental, industrial and social challenges demand the development of less expensive diagnostic tools like sensors with high selectivity and sensitivity, nontoxicity along with the portability, so that, it may directly apply in the different fields like agricultural, biomedical, environmental and industrial applications. As this type of diagnostics tools are now frequently attached to the robotics and drones for the exploration for survey purposes.

In light of these upcoming demands from different fields of application, electrochemical type of sensors have potential applicability and randomly fulfilled the most criteria. The noteworthy aspect about electrochemical sensor is the automation and adaptation (i.e useful in biomedical applications, and feasibility), allowing low limits of detection (LOD), fast analytical response even in flow analysis and alert systems. Offering an unlimited choice of electrode materials, geometrical-configurations, that can be fabricated at large-scale production and directly apply for end use purposes like pharmaceutical analysis [4], cancer cell detection [5], and many other application like food quality and safety aspects [6,7].

Apart from these advantageous, electrochemical sensors also have some concerns issues like; there may have a problem in the validation of analytical methods when prepared with different synthetic and non-biomolecules-based recognition elements especially metal and metal oxide nanoparticles (NPs). Lack of selectivity may be due to presence of the complex sample matrices in electrochemical interface, biofouling in vivo measurements. Limited lifetime of biosensors has impeded huge demands of newer class of engineering materials. Introduction of nanomaterials whose at least one of the dimensions falls in the range of 1–100 nm are termed nanomaterials, owing to their small size, that is, nano-meter, this class of materials have exhibited remarkable distinction in their chemical and physical properties from the bulk materials [8]. In recent times, quantum effect, as an another factor causing significant differences in the characteristics of nanomaterials which may be exist in such nanomaterial due to the discontinuous quantum confinement of delocalized electrons [2,9,10]. The shape of these particles is crucial to their properties as this would offer high surface area per unit mass for enchasing the surface catalytic activity of nanomaterial by an approximately 1000-fold [8]. Based on chemical constitution, nanomaterials can mainly be classified into three groups: (1) carbon allotrope-based nanomaterials, (2) inorganic nanomaterials (consist of metallic or non-metallic constituents such as Au, Ag, SiO_2, and (3) organic nanomaterials composed of polymeric nanomaterials, which are discussed later in details. Based on geometrical classification, there are several examples of these types of nanomaterials, for example, nanoparticles, nanorods, nanowires, nanofibers, etc., have been extensively in different field of engineering, biomedical, and other application, due to their salient physical and chemical properties [11,12].

In recent times, (micro-)nanotechnology, to discover and develop new and efficient nanomaterials to be directly attached or used in smart, flexible and wearable sensing devices, optoelectronics, antimicrobial and anti-corrosive coatings, nanofibers, energy devices, etc. Development, fabrication and advancement of different biosensing devices, smart nanomaterials, wearable and nonwearable biosensors for diagnosis, sepsic detection, and healthcare monitoring have also been found in different studies showing high potential and practical impact. Most of these sensors use enzymes or other biomolecules as the recognition element, which shows promising electrochemical characteristics—high selectivity and sensitivity over a wide concentration range. However, these sensors are prone to denaturation and exhibit poor reproducibility and reusability, as they are temperature and pH sensitive [13]. To overcome these shortcomings, numerous non-enzymatic recognition elements have been developed using different organic and inorganic compounds, including polymers, organometallic compounds, and metal or metal oxide NPs. Among these, electropolymerization is a facile method for control over the thickness of the polymeric layer on a suitable electrode and does not require an extensive experimental set-up.

Graphitic carbon-based materials, such as carbon nanofibers (CNFs), carbon nanotubes (CNTs), carbon nanoparticles, graphene/graphene oxide (GO), and reduced graphene oxide (rGO) dispersed with metal nanoparticles (NPs) have been used in different biochemical applications, such as adsorption, pollutant and heavy metal removal, waste water treatments, and drug delivery [14]. The substantial electrical conductivity, mechanical strength, and biocompatibility of these materials make them attractive electrode materials for use in different energy and sensing applications [15].

13.2 Different classes of nanomaterials and their applications

13.2.1 Shape- and size-based classification

In nanotechnology, nano-scale dimension (at least lies below 100 nm range at least in one dimension) of the material plays a considered role in tailoring the unique properties of nanomaterials, which are usually based on basis of their size. The physical and chemical properties associate with the nanomaterials has shown distinguished characteristics than the bulk materials for the same elements or compounds owing to their shape and size. Generally, the size and shape of nanomaterials have tremendous influence on its characteristic's properties specifically in the different environment [14]. Shapes are further classified based on their nanoscale morphology including nano-rod, nanowire, nano-particle, films, layers, sphere or in other geometry etc. Broadly it was classified into four categories: (i) Zero-dimensional (0-D): nanomaterials of these shapes have all three dimensions in nanoscale (i.e., $10^{-9} - 10^{-7}$ m). Nanoparticles, carbon dots etc.

Table 13.1 Classification of nanomaterials based on the size and dimension.

Categories	Number of dimension(s) fall within nanoscale range (<100 nm)	Example
0-D	Three	Nanodiamonds, metal NPs - Au, Ag, Pt, Cu, Ni, Co, Fe, Pd, carbon dots, quantum dots, etc.
1-D	Two	CNFs, CNTs, metal nanowires, Zn nanorods, nano graphite etc.
2-D	One	Graphene, rGO, few layered-GO, electro-polymeric film (poly methylene blue (PMB), poly pyrrole(PPy), poly indole(PIn), poly-aniline (PAni) etc. with thickness within nanoscale range) etc.
3-D	Zero	Coreshells (Ag@ZrO_2 core-shell nanoparticles, Pt-maghemite core-shell NPs, ZrO_2 coated Ag NPs, silica coated gold NPs etc.), ZnO/GO, MnO_2/graphene, NiO/graphene, MnO_2 nanoflowers etc.

fall under this category, (ii) *One-dimensional (1-D)*: Nanomaterials of these shapes have two dimensions in nanoscale (i.e., 10^{-9} m), (iii) *Two-dimensional (2-D)*: Nanomaterials of these shapes have one dimension in nanoscale (i.e., 10^{-9} m). The rest two dimensions of 2-D nanomaterials are out of the nanoscale. Nanolayers, nanofilms, nano coatings etc. fall under this category, and (iv) *Three-dimensional (3-D)*: Nanomaterials with 0-D, 1-D and 2-D shapes merge of form a composite with complex geometry and all the three dimensions are looking like greater than 100 nm, i.e., out of nanoscale (i.e., 10^{-9} m) are commonly categorized as a 3-D nanomaterial. Core@shells, nanocomposites, bundle of nanofibers/nanowires/nano tubes, multiple nanolayers etc. fall under this category. Summary of all categories of NPs with examples are presented in Table 13.1 [16–23].

13.2.2 Classification based on chemical composition

Nanomaterials are also broadly classified into various categories such as: based on their chemical composition, properties and sometimes morphology. Different noble metals and metal oxide NPs are successfully developed and used as the sensing elements because of their high purity; sensing-suitable physical and chemical properties, mainly chemical inertness, biocompatibility and nontoxicity; and electrochemical compatibility especially because of their excellent electrocatalytic characteristics. But nonbiological component-based sensor based on these NPs have low selectivity and sensitivity issue. These nanoparticles are also commonly used in different enzymatic and biological components-based sensing systems which are developed using commercial electrodes such as glassy carbon electrodes (GCE), carbon screen printed electrode (SPE) and others.

Transition metals and metal oxides are also developed and tested in various studies for their different biomedical applications such as biochemical sensing, drug delivery etc. These NPs offer: high sensitivity, high electrocatalytic characteristics in different potential-driven biochemical redox reactions involving biomolecules, efficiency in drug delivery etc., and therefore, are found to be effective in different biomedical application especially in biosensing and electrochemical sensing of biomolecules. These are used to develop both enzymatic and non-enzymatic sensing systems; however, low selectivity is the main drawbacks of these sensors especially when applied for measuring different biomolecules electrochemically. This is because, these nanomaterials have a tendency to reduce or oxidize a number of biomolecules available in human blood over a narrow range of potential. Several metal chalcogenides and pnictogenides are also developed and used for the sensing applications of the biological components in some studies. But these nanomaterials are commonly used as the semiconductor materials based on their unique chemical properties.

In last few decades, carbon-based nanomaterials such as CNTs, CNFs, graphene, GO, rGO, and nanographite are also been developed and studied for their biomedical application. These biocompatible materials showed excellent electrochemical and catalytic properties in different biomedical application especially in biomolecule sensing and drug delivery. These materials have high thermal and mechanical strength, high electrical and thermal conductivity. These materials also showed superior antibacterial activity along with their numerous applicability in different fields such as energy applications, environmental remediation, etc. Different functionalized polymeric nanomaterials, which are highly biocompatible and electroactive are also tested for different biomedical and found to be promising. But low sensitivity is the main drawback of these polymeric nanomaterials when applied for sensing applications.

Biocompatible polymers/macromolecules and carbonaceous materials, because of having multiple active sites and functional groups can easily be incorporated /anchored with redox sensitive metal and metal oxide nanoparticles. because of the above-mentioned properties, pave new platforms to develop different hybrid nanomaterials by incorporating/anchoring with for different biomedical applications. The idea behind developing hybrid nanomaterials is to use the sensitivity and superior electrocatalytic characteristics of the metal and/or metal nanoparticles as well as selectivity and biocompatibility of the carbonaceous and/or polymeric electrode materials and substrate. In this context, several sensing systems have been fabricated and tested efficiently for different biomolecule and biological components sensing and measurements. The commonly developed hybrid inorganic and organic nanomaterials can be divided into different categories based on the type of the substrate or the electrode materials and the sensing elements, such as: Core@Shell nanomaterials; metal and metal oxide NPs-dispersed 1-D carbon nanomaterials (CNT and CNFs); metal and metal oxide NPs-doped 2-D carbon nanomaterials (graphene, rGO, GO etc.); polymers and macromolecules functionalized with different metal and/or respective metal oxide NPs, other organic components, different biological components etc. These materials are found to measure Development of these biocompatible and non-toxic nanomaterials is simple and less expensive also. Some biomolecule-specific polymers are also developed electrochemically (electro-polymers) such as PMB, PPy, poly(methyl orange) etc. and tested for different analyte-specific sensing of biomolecules showing excellent electrochemical outcomes including high selectivity, repeatability and reproducibility. In some studies, these electro-polymers are developed on the surface of different metal and metal oxides NPs dispersed carbon or polymeric electrode materials and substrates. However, commercialization of any of these hybrid and non-hybrid sensing system is scarce. Several studies are undergoing in different laboratories for the commercialization of these class of sensor. All of the above-mentioned hybrid and non-hybrid nanomaterials are broadly classified based on their composition and physicochemical properties and are summarized in Table 13.2 [27–57].

Table 13.2 shows that due to their distinct chemical composition and properties, their applicability are changed depending upon prerequisite properties required in many applications. It can be observed that metal-oxide NPs show better stability for catalytic and redox type applications, whereas pure metal-based NPs are purely preferred sensing-based application due to high sensitivity and catalytic properties. In contrast to above two categories, carbon-based nanomaterials and polymeric nanocomposites materials provides stable matrix as precursor for such small dimensional NPs, owing to this such materials are potentially tested for process engineering applications.

21ST century belongs to nanomaterials, specially metal and carbon based quantum nanodots, exhibit fluorescent characteristics, this class of fluorescent-materials have shown wide spread applications in field of biomedical for bioimaging, biosensing, drug delivery and photodynamic therapy [add]. Organic fluorophores type's materials are used in bioimaging, but impeded its wide applicability due to various drawbacks, as organic based florophores are prone to photobleaching and undergone fast degradation while irradiating with light during the imaging process, and exhibited narrow excitation and emission wavelength range (no possibility of tuning the optical properties). Owing to its quick drip out characteristics from the cytoplasm to media, therefore, organic fluorophores are restricted to very limited biomedical applications.

Over the last one decades, In context of limited applications of organic-fluorophores materials, metal based chalcogenide inorganic semiconductor nanocrystals such as quantum dots (QDs) of (Se, Te) are capped with long-chain alkyl thiols were used as fluorescent probe. Semiconductor QDs have shown unique properties like high photo-stability is ascribed to core-shell type capping of ligands, quantum confinement (optical properties changes with the change in particle of QDs due to increase in particle size that shift the absorption in the longer wavelength). However, metal base QDs have shown, poor biocompatibility, due to their toxic nature of heavy metals like Cd, Pb and Hg, and released toxic metals ions into the medium that may causes threat to the live

Table 13.2 Classification of nanomaterials (NPs) based on its constituents, morphology, and properties.

Class of nanomaterials	Properties	Example
Metal NPs [30–34]	High surface area, excellent adsorption capability, some metals NPs are biocompatible, offering high sensitivity, catalytic and electrocatalytic activity	Different transition metal and group of two/three different metal particles: Au, Ag, Pt, Cu, Ni, Co, Fe, Pd, Pt/Pd, Au/Ag, Cu/Ni etc.
Metal oxide NPs [35–38], metal chalcogenides NPs [39,40], core@shell nanomaterials [41]	Good catalytic and redox properties, ability to form complex with biomolecules, shows semiconductor properties also, low conductivity	AgO, CuO, Cu_2O, Ni_xO_y, MnO_2, ZnO, FeO, Fe_2O_3 etc.; metal NPs and metal oxide NPs composite: Au/AgO, Pt/AgO, Pt/NiO, Cu/Cu_2O; metal chalcogenides: Ag_2S, Cu_2S (chalcocite) and CuS (covellite) etc.
Semiconductor nanomaterials [42–47]	Metallic and nonmetallic properties, exhibit large band gap. Graphene-based materials also show piezoelectric properties and gas sensing properties. Some metal chalcogenides and pnictogenides have also been used as the semiconductor nanomaterials.	Graphene and graphene composites, ZnO, ZnS, CdSe, CdTe, GaN, GaP, GaAs, etc.
Carbon-based nanomaterials [48–57]	High electrical conductivity, large surface area, biocompatible, high mechanical strength, strong electron affinity etc.	Carbon nanofibers [42,43]; carbon nano tubes [44,45]; nanographite, graphene and its derivatives [46–49]; cellulose nanofibers and nitrate fibers [50,51], etc.
Polymeric nanomaterials and electro-polymers (nano films, fibers and layers) [58–64]	Catalytic and electrocatalytic activity, biocompatibility, controllable thickness, high stability, easy surface functionalization and rafting facility	Polyacrylamide, poly pyrrole, poly methylene blue, polystyrene, poly indole, poly (β-cyclodextrin), etc.
Quantum dots [65–70]		Graphene oxide-based quantum dots [54], graphene-based quantum dots [54–56], carbon quantum dots [56,57], etc.
Nanocomposites (NCs) [71–74]	High surface to volume ratio, catalytic activity, some nanocomposites show magnetic properties and super magnetism, ability to change refractive index in some cases	Ceramic/metal NCs, Polymer matrix-based NCs, metal matrix-based NCs, metal/polymer matrix NCs, etc.

cells. In this context, carbon quantum dots (CQDs), offers, better potential candidates to overcome the current challenges of biodegradability and toxicity as QDs possess. CQDs is a novel fluorescent types materials exhibited better optical and biological properties applied for bioimaging of live cells (Sumit Sonkar). CQDs in present century is considered as a new invented member of nanocarbon family, emerged superior in terms of low cytotoxicity, high aqueous solubility and almost equally in emission quantum yield for in vivo and in vitro bioimaging. Discovery of fluorescent carbon nanoparticles was claimed by two different processes, first method was reported by Xu et al. (2004) [24] during separation and purification of single-walled carbon nanotubes via electrophoresis, while in second method, CQDs has been derived from graphite powder using laser ablation technique by Sun et al 2006 [25]. In last two decades, CQDs exhibit distinct properties, such as high surface area, flexible functionalization, nontoxicity, photostability, high emission quantum yield, high water soluble nature, and most recommendable its synthesis is relatively facile and inexpensiveness, derived from natural extract of food and plants [25–28].

13.3 Selection criteria of nanodevices for disease diagnosis

13.3.1 Analytical performances

Nano-devices are commonly used in disease diagnosis and their performance is primarily relied on the interactions which occurred between recognition element attached to the device and the biomarkers responsible to cause disease. Analytical performance in biomedical devices, for decease diagnosis and in some cases therapy, are generally count on the following different criteria including sensitivity, selectivity, limit of detection, limit of quantification, signal to noise ratio, analyte measurement range (concentration), repeatability, recovery, reproducibility etc. Among these criteria some of them are either belong to nanomaterials or biomarker. A brief detail of these parameters is described below.

13.3.2 Sensitivity

This term is allied with the slope of the calibration data, determined from experimental data measured the known analyte concentration, determined in a sensing device/sensor. This term has a significant importance, especially when one prefers to measure quantitatively. This term has significance that shows the change in response measured (detectable properties) per unit change in concentration of analyte. Thus, a higher value of sensitivity may indicate the sensor/device to be superior one as compared to similar devices for low sensitivity values. This term is commonly evaluated in form of empirical formula which is determined from real samples specifically for a known detectable concentration range.

Specificity-selectivity: This is the second most important criteria after sensitivity used for analysing the performance of any sensing devices. This term is generally related to the different metal, metal oxide, polymeric, hybrid, organometallic and other inorganic and organic nanomaterials, especially when these materials are to be used as the recognition elements. When the recognition elements of different sensing devices are enzymes, antibodies, DNAs, RNAs, peptides or other biomaterials which are highly specific toward a fixed biomarkers or biological components, the measurements (both qualitative and quantitative) become easier in a complex body fluid like blood, blood serum, cerebral-spinal fluid or others. This property defines the selectivity of the devices to be used. However, these biomaterials are highly sensitive to the temperature, pH and can be damaged in presence toxic ingredients. It has repeatability issue also. In this context, sensing devices developed using different synthetic nanomaterials especially metal and metal nanoparticles, polymers, macromolecules, organometallics as the recognition elements. However, for these devices, selective measurement is the prime focus.

Detection Limit: It is the lowest concentration/quantity of the biomarkers which can be measured/detected using the measurement devices for the pathogen or biomarker detection and measurements. The value depends on the different physical, chemical, optical, electrochemical and other properties of the nanoparticles to be used for diagnosis purposes; and their different interaction/measurement phenomenon occurring between biomarkers and the recognition elements. This parameter can be measured by "Hit and Trial" approach by starting the measurement with zero concentration or can be determined using available empirical formula and measured calibration data for the developed sensing device.

Quantification limit: It is another measurable analytical performance for the different nanomaterials-incorporated disease diagnosis devices. It indicates the minimum change in the concentration or quantity of the biomarker that can be measured using a nanodevice during the measurement. The data can be evaluated by "hit and trial" method or from the sensitivity value and standard deviation in the measured calibration data.

Signal to noise ratio: At the time of the biomarker measurement using different analytical methods, some noise in the response may be observed which may interfere the response to be measured for the measurable analytes. According to the different statistical correlation and instrument's internal characteristics, the signal to noise ratio must be 3:1 or more.

Working range of the nanodevice: This is an important criterion for the disease diagnosis as most of the biomolecules either in their elevated or lower magnitudes than the permissible limits (for example, 80-120 mg dL^{-1} for glucose, 40-200 mg/dL for cholesterol, 13-17.5 g dL^{-1} (men) and 12-15.5 g dL^{-1} (women) for hemoglobin, 4.35 to 5.65 million per microliter (mcL) (men) and 3.92 to 5.13 million per mcL (women) for red blood cells, 4,500 to 11,000 μL^{-1} for white blood cells, etc., may indicate a disease in living body [75,76]. Sometimes existence of different pathogen, antigen, virus, bacteria, microbes or raptured biological components and living cell also indicate some critical diseases. In some cases, a very low concentration of some of these species may not be harmful. All of the above clearly mention how important the working range for a nanodevice to be developed for disease diagnosis is. The device should be functional at the required biomarker range which indicate the present human health condition and diagnosis whether the human is suffering from any diseases due to the undesired concentration of that biomarker.

Recovery: Recovery of the nanodevice associated with disease diagnosis and therapy is directly related to the accuracy of the measurement which is highly critical in confirming the occurrence of the unusual biological phenomena in the human body. A high recovery value is always preferable for a developed device.

Reproducibility of the measurement data: Reproducibility of the measuring device indicates the precession in the measurement. A high reproducibility always acceptable as it confirms the measurement data. Most of the commercial sensors, incorporated with different biomaterials and biological elements, show poor reproducibility due to their less uniform distribution and degrading tendency with the change in working environment. The standard deviation in the measured data using different devices fabricated/developed in one or more production batches should be as low as possible which indicate a high reproducibility value and confirms the commercial and practical applicability of the device.

13.3.3 Challenges in development and implementation

Apart from the different types of analytical performances, there some special challenges associated to the development/fabrication nanodevices for the disease diagnosis applications. These are summarized below in brief [77].

Real sample preparation: One of the main challenges for developing a biosensor or similar sensing devices is to test the developed senor, sensing system or the sensing materials in the real atmosphere, i.e., in real body fluids. For the common biomolecules like glucose cholesterol, creatinine, ascorbic acid and others, blood samples collected in different pathological laboratory for different routing check-up can be used or reused for the testing purposes (as the requirements of clinical samples for testing is very small when tested using a biosensor) following proper ethical guidelines as mentioned by the Government regulatory and others as well as institute protocol. However, when the measurement is to be performed for the biomolecules or physiological components (like tissue and others) which are not commonly found in the body fluids collected for the routine check-up or when the samples are not available, artificial/Synthetic clinical samples based on the literature need to be prepared, if possible, which is a very challenging task. In each of the above cases, preparation of the real samples is highly challenging.

Multi-analyte detection: Most of the commercial sensors, incorporated with different biological components, inorganic and organic hybrid nanomaterials, organometallics and others as the recognition elements, have a major limitation. These devices are being fabricated/developed to measure single analytes in most of the cases as this makes the signal processing simple during the data analysis. However, a facility to measure multi-analyte simultaneously is complex in nature, but highly acceptable as it facilitates to confirm the other physiological phenomena happening in the human biological systems as these phenomena are associated and influence with each other [77].

Stability: Lifetime of a nanodevice depends on the physicochemical properties of the materials used in it, storage condition, exposure time, reusability. Stability of the sensors and biomarkers measuring devices represents their behavior when used for determining the presence of the biomarkers qualitatively and quantitatively over the period of time. Most of the devices, using different biomaterials, have short lifetime because of the proneness of the biomaterials in presence of different pathogens and toxic elements, and because of their temperature and pH sensitiveness. When incorporated with different inorganic and organic nanomaterial, macromolecules or organometallics, the lifetime of the devices is relatively higher as compared to when incorporated with different biological components/bodies. However, the lifetime of these nanomaterials-based disease diagnosing devices is highly depends on the physicochemical properties of the nanomaterials, storage and working condition [62,77].

Reusability: It justifies the practical applicability of the nanodevices on the basis economic as well as social background. Higher the reusability of the device with high accuracy and precession, especially for the biomarkers of common but chronic and severe diseases, higher is the chance to be purchased and used by the patients belong to rural and remote area and with low economic background [61,62].

13.3.4 Other requirements

There are some other requirements as well as limitations for the commonly developed/used nanomaterials-based diagnosing devices.

Invasive approaches: Most of the synthetic and artificial nanomaterials-based devices fabricated/developed in different studies use non-invasive methods for measuring/detecting biomarkers or pathogens responsible for different diseases. Some devices have semi-invasive approaches also for measuring biomarker concentration in sweat and sputum, some use using pressure sensor for measuring blood pressure and heart condition. But most of these studies measures the physiological conditions based on some empirical corelations of biomarkers present in the internal and external body fluids, and are not accurate. Also, the number of devices developed with semi-invasive and invasive approach is scarce. Invasive approach is also required for the continuous and point of care monitoring of the patients suffering from critical and chronic diseases.

Nontoxicity and biocompatibility: For the invasive approach it is necessary to be the device to be non-toxic to avoid any unnecessary and undesired physiological reactions and phenomena during and after the use of the devices. Different metal nanoparticles-based synthetic nanomaterials can interact with different biomolecules and biological components other than the target biomarker(s) and affect the occurring desired physiological phenomena. Therefore, it is a hard challenge to develop devices with invasive approach and non-toxic behavior. However, for the case of non-invasive approaches, the toxicity of the nanomaterials has less effect in the measurement and disease diagnosis except when it has some selectivity problems. The nanomaterials with less toxic nature toward the biological components and biomarkers, and has capability to interact with the required biological system can be considered as the biocompatible species.

Sterility: It is one of the most common practices, pre- and post-use of any nanodevices, to make the device free from any pathogens, bacteria, viruses or other toxic agents. It helps to protect the devices from unnecessary damages and to useful over a period of time repeatedly. Commonly, the devices and solutions to be used for the biomarker measurement have been autoclaved under specific conditions. It has been observed that this safety process has negligible effect in the chemical, analytical, optical or electrochemical performances of the fabricated devices.

Ethical consideration: With the emergence of the sensing applications and the sensory materials development for the disease diagnosis across the world and the extensive research in the same, it is highly required to establish and follow clear and uniform rules and regulations to protect experimental findings, analysis and applications [78]. Transmission of the findings and third-party use must also be regulated properly. For the confidentiality of patients consent and data, encryption and security methods need to be followed also. Research permission related regulatory and real-field application trials need to be monitored also based on the biomedical advancements, social and global requirements.

13.4 Synthesis of smart nanomaterials

Synthesis of nanomaterials for their direct exposure toward the physiological components in different biomedical applications must facilitate high purity, negligible presence of toxic particles, and preferably a green synthesis procedure [79]. However, the nanomaterials, to be integrated and used as the electrode and for developing sensory materials over it, some of the above requirements may be ignored during its synthesis. Different properties especially colloidal stability and biocompatibility facilitate different nanoparticles and other nanomaterials to be potentially suitable for different biomedical applications [80]. These properties commonly depend on different tunable properties, for example, size of the nanoparticles and its surface to volume ratio, stability, sensitivity, selectivity etc. which make the nanoparticles to be optimal substrates for developing therapeutic devices. These properties can be improved via surface modification (commonly by surfactant-treatment or polymer protection which helps to protect from agglomeration, corrosion, surface rupture; increase the capability of functionalization etc.). In this context, several synthesis procedures have been developed which can be categorized into biological, physical, chemical, electrochemical, photochemical methods etc. (depending on environmental condition, working facilities and requirements).

Table 13.3 Methods involving synthesis of different metal nanoparticles.

Name of the metal NPs	Methods of synthesis
Gold [81–87]	Chemical reduction from its salt, photochemical reduction, biological synthesis, laser ablation technique, electrochemical methods, ultraviolet irradiation lithographic technique, ultrasonication, aerosol technologies, etc.
Silver [86–92]	Evaporation/condensation, laser ablation technique, chemical and photochemical reduction, electrochemical methods, biological synthesis, gamma and electron irradiation, microemulsion, thermal decomposition of silver oxalate, etc.
Platinum [87,93–95]	Chemical and sonochemical reduction, thermally induced reduction, biological synthesis, etc.
Palladium [96–98]	Chemical and sonochemical reduction, thermally induced reduction, biological synthesis, etc.
Iron [99–102]	Co-precipitation, biological synthesis, thermal decomposition, microemulsion route hydrothermal synthesis, etc.
Cu [103–106]	Chemical and thermal reduction, biological synthesis, metal vapor synthesis, radiation, microemulsion, laser ablation method, mechanical attrition, etc.
Nickel [107–109]	
Zn [110–114]	Sol-gel process, biological synthesis, microemulsion, vapor phase oxidation, thermal vapor transport and condensation, sonochemical reduction, precipitation, hydrothermal and polyol methods, etc.
In [115,116]	Sol-gel process, pulse-laser deposition, mechanical chemical processing, thermal decomposition and hydrolysis, biological synthesis, microemulsion, spray pyrolysis, hybrid induction and laser heating method, hydrothermal synthesis, etc.

Ways of different metal nanoparticle synthesis using various methods are described in Table 13.3. A brief detail about these methods described below.

13.4.1 Physical methods

In these methods, synthesis of nanoparticles is performed via two different approach: Top-down approach (in which the larger particles are commonly converted into small and desired sized nanoparticles) and bottom-up approach (in this method small molecules are combined to form large particles) [117]. Different methods applied via Top-down approach to prepare nanoparticles are: mechanical milling/grinding for TiO_2 NPs [118]; electro-explosion MoO_2 and ZnS NPs [119,120]; physical vapor deposition for Fe/MgO, Pb/Cu, Al/Pb, Al/SiC, Cu/Al_2O_3, Al/Mo, and Ag/Au etc. [121]; sputtering [122]; laser ablation method for Au, Ag, CoPt, Ni/NiO, FeNi, etc. [123,124]. Development of nanoparticles following Top-down approach, the main challenge is to maintain uniform size of the particles. RF plasma [125,126]; spinning for ZnO, $Na_3V_2(PO_4)_3/C$ etc. [127,128]; atomic/molecular condensation [129] methods; supercritical fluid synthesis [130,131]; laser evaporation etc. are commonly used for developing nanomaterials following Bottom-up approach. These methods can produce nanoparticles with uniform and desired size efficiently; however, to form stable nanoparticles, addition of different stabilizers may be required during the preparation stage(s).

13.4.2 Biological methods

These methods of developing nanomaterials are simple, requires minimal steps, easy to operate, and can be considered as green synthesis. The use of different microorganisms and part of different plants in metallic salt solutions is the common practice to develop different metal nanoparticles using these methods.

Nanoparticle synthesis using microorganisms: Synthesis of nanoparticles involving different microorganisms by using its protein parts is commonly defined as Biomineralization process [132]. Different bacteria, fungi, algae etc. used in different studies for the synthesis of different metal and metal oxides are presented in Table 13.4. Fungal cells secret higher amount of protein as compared to bacterial cells. However, the microorganisms are pathogenic in most of cases, and therefore, safety is a major concern during its use in NPs synthesis.

Preparation by using biotemplate: Different biological methods have been used for preparing different intracellular, unique nanostructured and sophisticated nanomaterials using biological templates. Proteins are the main constituents in the nanocomposites developed using these methods. DNA templates have also been found very useful for the nanomaterials' synthesis. Different biosensors, bioelectronic devices, bioNEMS can be developed using these templates. A brief summary of these templates is presented in Table 13.4.

Preparation using different plants their extract: Different phytochemicals, for example, organic acids, flavones, quinonesare, different sugars, terpenoids etc present in different plants have been found as the good reducing agents. Using these materials, in last few decades, several metal NPs has been synthesized for their numerous therapeutic and drug delivery applications. A brief about some useful plants for the synthesis of different NPs are summarized in Table 13.4.

13.4.3 Chemical methods

Most of the methods used for developing different metal nanoparticles follow one or more chemical route (i.e., via different types of chemical reactions) especially when the synthesis is carried out in gas and liquid phase. Based on the requirement of different properties such as size and shape of the nanomaterials, feasibility of the process, ease of synthesis and the physical-chemical properties of nanomaterials different chemical methods for example, co-precipitation, chemical vapor deposition, hydrothermal

Table 13.4 List of microorganisms used for different nanoparticles synthesis.

Class of micro-organism	Name of the micro-organism/plant(s)	Class name according to the properties	Name of the nanoparticles synthesized	Features of the microorganism
Bacteria	Magneto tactic bacteria	Magneto tactic bacteria	Magnetosomes – protein-coated nanosized magnetic iron oxide crystals [133]	Magnetosomes are useful in medical applications such as hyperthermia
	Rhodopseudomonas capsulata	Photosynthetic bacteria	Gold NPs [134]	NADH-dependent reductase helps to reduce Au ions to Au NPs
	Pseudomonas cell (found in alpine site)	Photosynthetic bacteria	Palladium NPs [135]	
Algae	Sargassum wightii algae		Gold NPs [136]	With 12 h incubation, 95% production was achieved as shown in the study
Fungi	T. reesei		Cas9 protein [137]	It has wide application in the field of medicine, paper, textile and food industries
	Fusarium oxysporum fungus		Silver NPs [138]	Enzymatic activity in the NADH-based reductase provide long term stability in the developed NPs
Biological templates	Ferritin	Intracellular iron storage-protein	Intracellular iron oxide [139]	Present in prokaryotic and eukaryotic cells. It controls the iron deficiency/overloading in human physiological system by acting as a buffer.

(Continued)

Table 13.4 (Cont'd)

Class of micro-organism	Name of the micro-organism/plant(s)	Class name according to the properties	Name of the nanoparticles synthesized	Features of the microorganism
	Horse spleen ferritin	Intra cellular NPs storage	Gold NPs [140]	Same as above
	Apoferritin	Intra cellular NPs storage	Biotin-conjugated Yttrium phosphate NPs [141]	Same as above
	Plasmid	DNA template	CdS-DNA NP conjugate [142]	Closed circular molecules; found in different bacteria; NPs synthesis was commonly performed using spin-coating.
Plant(s) and it's parts	Alfaalfa plant (*Medicago Sativa*)	Plant	Gold NPs [143]	Sugars and/or Terpenoids present in this plant show reducing behavior.
	Neem (*Azadirachta indica*)	Plant leaves	Bimetallic gold [144], silver [145], Bimetallic gold core-silver shell [146], etc.	Same as above
	Geranium (*Pelargonium graveolens*)	Plant leaves	Gold NPs [147]	Same as above
	Sunflower (*Helianthus annuus*)	Flower, seed, plants stem	Silver, cobalt, zinc, nickel, and copper NPs [148]	Same as above
	Indian mustard (*Brassica juncea*)	Seeds	Silver NPs [149]	
	Aloe vera	Plant leaf extract	Gold NPs [150]	Same as above

method, sonochemical method, pyrolysis, microemulsion, sol-gel method, ion exchange, solvothermal method, reflux, etc., have been applied for the synthesis of different nanomaterials. A brief detail of these methods, with a view of their successful applications in nanomaterials synthesis, are presented in Table 13.5.

13.4.4 Electrochemical and photochemical methods

In recent times, a type of chemical method based on electrochemical phenomena has been extensively used for the synthesis of different nanomaterials from different electrosensitive materials. The method is governed by electrochemical potential-driven simple redox reactions is highly useful in the synthesis metal and metal oxides nanoparticles [191,192], graphene and graphene oxide nanofilms [193], various organic and inorganic hybrid nanomaterials including different electro-polymers [194], organometallic compounds [195,196] etc.

Different nanomaterials have been extensively developed using the UV-ray or sunlight assisted chemical reactions following the photochemical reaction method, however, their application is mainly limiting in the energy related studies [197].

13.5 Nanomaterials characterization techniques

13.5.1 Physicochemical methods

Nanomaterials of both the classes (inorganic and organic) mostly characterized by several physicochemical techniques for their efficient syntheses, purity, texture, morphology, lattice structure and others. For example, X-ray diffraction spectroscopy, governed with several establish laws (like Bragg's law, Debye-Scherrer formula), has been used to determine lattice structure, particle size, purity etc, for the metal nanoparticles and other crystalline nanomaterials. Raman spectroscopy and Fourier Transform Infra-Red Spectroscopy have been used to determine the presence of different functional groups in case of hybrid inorganic and organic nanomaterials, including organometallics, nanostructured polymer nanofilms, etc., which confirm the efficient and expected materials synthesis. The data also help to determine graphitic contents and other properties for the carbon-based graphitic nanomaterials. Scanning electron microscopy, transmission electron microscopy and other microscopic techniques have been commonly used to determine surface morphology. Energy-dispersive X-ray spectroscopy, X-ray photoelectron spectroscopy, X-ray fluorescence and other similar methods are being used to determine elemental composition, surface rupture and others especially for measuring the metal nanoparticles and other critical elements composition. Concentration of metal nanoparticles has also been measured using atomic emission spectroscopy, inductively coupled plasma atomic emission spectroscopy, inductively coupled plasma optical emission spectroscopy, inductively coupled plasma mass spectroscopy and similar techniques. Several chromatographic methods such as HPLC and UHPLC, GC, GCMS

Table 13.5 A brief details of various chemical methods used for the synthesis of nanomaterial.

Name of the method	Details of the method	Examples of nanomaterials synthesized following this method
co-precipitation	In this method, during the mixing of two or more multivalent water-soluble metallic salts in an alkaline aqueous medium, sometimes the metal ion with higher valency gets reduced to form one or more water-insoluble salt(s) and is/are precipitated out by the influence of other. The precipitate is then filtered out to collect metal or metal oxide nanoparticles. The method is also called reactive crystallization. Key steps of this method are: nucleation, growth, Ostwald ripening, agglomeration, stabilization [main paper]. The reaction may or may not be influenced by stirring rate or heating. Several factors, such as pH, salt ration, type of base used and temperature control the size of the nanoparticles to be formed.	Fe_3O_4@SiO_2@PPh_3@$Cr_2O_7^-$) [151], Cobalt Ferrite NPS [152], Fe_3O_4@SiO_2–OSO_3H nanocatalyst [153], Fe_3O_4/Au [154], Fe_2O_3/SiO_2 and Gd/SiO_2 [155], N-doped TiO_2 [156], cobalt aluminate [157], Magnetic nanoparticles [158], Fe_3O_4 [159], C- and N-doped cobalt ferrite [160].
chemical vapor deposition		Metal NPs-dispersed CNF [161], Metal doped CNT [162], $SnSe_2$ [163] and other metal dichalcogenides [164], carbon/metal oxide hybrid materials [165], TiO_2 [166].
hydrothermal method	In this process, a solution (commonly prepared using organic solvents) containing the mixture of one or more multivariant metallic salts is subjected to high pressure and temperature (commonly in an autoclave). To obtain desired size and surface properties for the nanomaterials, pH of the solution, temperature, pressure, aging time and reaction time has crucial role.	The method is highly effective to produce small-sized nanomaterials which have large applications in biological fields. $NiFe_2O_4$/MWCNT [167], $Co_{1-x}Zn_xFe_2O_4$ [168], MnZn Ferrites [169], $ZnFe_2O_4$ [170], Copper ferrites [171], Zeolites [172], selenides, metal oxides, metal doped-graphene oxide-based nanomaterials [173] etc. are commonly prepared using this method.

sono-chemical method	The method involves the use of ultrasonic irradiation which creates bubble/cavities inside the liquid layers which collapses by the influence of ultrasonic energy, leading to sono-chemical extraction of the materials.	Metal oxides; metal sulfides such as CoS_2 [174]; different alloys such as Mo, Pd dispersed in MWCNT [175]; selenides such as CdSe, ZnSe etc. [176].
pyrolysis		N- and S-doped carbon nanofiber [177,178]
microemulsion		Pt [179], Pd [180], silica-doped TiO_2 [181], Fe–Co/MgO [182].
sol-gel method	In presence of different precursor, metal oxides undergo acid- or base-catalysed hydrolysis in its aqueous or alkaline solution. The hydrolysed product mixture is then converted to gel phase via polycondensation, followed by to the powder state. The powdered materials can get some additional crystalline nature via heat treatment.	The method is useful to prepare different metal oxides, $Bi_{0.5}(Na_{1-x}K_x)_{0.5}TiO_3$, SiO_2, composite nanomaterials, inorganic, and organic hybrid compounds etc. [183–187].
solvothermal method		Functionalized CNT [188], silver/cobalt oxide-doped graphene oxide [189], copper/zirconia-doped CNF [190].

etc. are used to determine the purity of hybrid organic nanomaterials. Some of these methods have also been used for performance evaluation sometimes.

13.5.2 Electrochemical methods

A number of application-specific electrochemical methods have been developed for determining different electrochemical properties, such as different impedance components, electrochemical surface area, or potentials, corresponding to oxidation and reduction reactions. A brief description of the common electrochemical methods used to measure different organic and inorganic analytes including biomarkers and their concentration, are described below [198].

Electrochemical impedance spectroscopy is operated with an open circuit potential of an electrochemical system over a wide frequency range with a sinusoidal input. Resistance and phase angle are used to determine the electrical conductivity, predict the type of circuit, and the nature of the working electrode. Resistance can also be used to measure the concentration of different analytes, as resistance changes with varying analyte concentration. However, changes in concentration result in minimal changes in resistance except when the solution is highly viscous, for example, in different petroleum products. Therefore, this method is not applicable for measuring small molecule chemicals, biomarkers, or other biomolecules.

Cyclic voltammetry (CV) is the most common electrochemical method and has been used for numerous electrochemical applications. CV operates within a potential window, which may be predefined, under a constant or variable scan rate, scan cycles, and other operating conditions. It is used to measure redox peak potentials and the peak currents, which are used to calculate or determine the nature of the reaction, redox behavior, reversibility of the reaction, electro-catalytic effect of the electrode or sensor used electrochemical surface area, diffusion and kinetic rate constants, nature of electron transfer, and calibration data for measuring the analyte concentration. The data can also be used to interpret the capacitive behavior of the electrode, number of electrons involved in a reaction, and plausible reaction kinetics. CV has been recently applied to grow an electro-polymer over an electrode surface in a controlled manner by controlling the number of scan cycles and scan rate. This can be used to impart different non-enzymatic recognition elements onto an electrochemical sensor or different materials for corrosion-resistance.

CV experiments, operating for a half cycle at very low scan rate (0.1 mv s^{-1}), can be used to determine the open circuit potential, current, and power density for electrodes or different kinds of electrochemical system and devices. This method is termed linear sweep voltammetry (LSV) and is highly applicable in all energy applications, such as developing supercapacitors, microbial fuel cells, and microbial energy storage devices.

Amperometric electrochemistry, that is, chronoamperometry, has been used to measure analyte concentrations by measuring the changes in current with an electrochemical sensor or sensing system. It can also be used to determine the effect of interfering molecules/compounds during the measurement of a target analyte. It is most commonly used to determine electrokinetic parameters or conduct electrochemical functions, such as determination of diffusion and kinetic rate constants, reaction type and successive reaction kinetics, stability and response time of a senor or electrode, and electrode charging and discharging to determine specific conductivity with various current and energy densities.

Differential pulse voltammetry (DPV) is a recently developed electrochemical method, which is an extension of CV, used to electrochemically measure the concentration of an analyte. The method records the change in current (Δi) during potential sweeping over a potential window. The slope of the calibration data, determined from the peak current (Δi_p) from Δi_p versus V plots, is higher than the calibration data measured using CV, LSV, or chronoamperometry, resulting in highly sensitive measurements. However, additional advantages of this method have not been fully explored.

13.6 Application of hybrid inorganic and organic nanomaterials in healthcare diagnostics

The recent advances in nanomaterials are primarily focused on enhancing their biocompatibility and surface functionalization capabilities so that these nanomaterials can be adopted in different healthcare applications, for example, developing biosensing devices, tissue engineering, and drug delivery. Here in this section, different applications of nanomaterials in developing different diagnosing devices are reviewed.

13.6.1 Application as electrode materials

13.6.1.1 Carbon-based nanomaterials

Carbon nanofibers (CNFs): CNFs are cylindrical nanomaterials with graphitic layers of sp^2 hybridized, arranged as stacked cones, cups, or plates [199]. The material possesses high mechanical strength, greater surface area, thermal conductivities and good electrical conducticity due to the presence of sp^2-hybridized carbon and the delocalized π-electron cloud. CNFs also display high stability in acidic and basic environments. Several metal-CNFs hybrid materials have been developed for various applications, including Cu- and Cu/Zn-CNFs have also been used as drug delivery and antibiotic materials [200,201], demarcated its biocompatibility aspects. Defect sites with electronic charges that are present within the structure of CNFs hold different metal and metal oxide NPs, providing superior electrochemical properties compared to naive CNFs. However, CNFs doped with metal NPs are rarely used in electrochemical sensors [202,203], despite their electrochemical compatibility. Therefore, development of single and bimetal NPs-dispersed in CNF-based electrodes for electrochemical sensing applications is one of the prime focuses of the present study.

Carbon Nanotubes: Over the last decade, single- and multi-walled carbon nanotubes (SWCNT and MWCNT, respectively) [204] have been developed for use in numerous electrochemical applications. CNTs are promising electrode materials as they have a large surface area, a high mechanical strength, and a high electro-conductivity with superior electron transport capabilities. Both SWCNT and MWCNT have been extensively used as efficient electrode materials in different electrochemistry-based diagnosis: CNTs have also been successfully employed as the electrode in electrochemical sensors for detection of biomolecules and chemical compounds such as glucose [205,206], putrescine [207], cholesterol [208,209], dopamine [210], rutin [211]. However, these electrodes are applicable only to high analyte concentrations, resulting in low sensitivity. DNA has been deposited onto MWCNTs for use as supercapacitors and artificial muscles [212]. The compact cylindrical structure with minimal corner defects of the material limits its dispersion in different solvents and lowers its binding efficiency with metal NPs, limiting the electrochemical response. In addition, CNT growth requires a metallic or commercial electrode support.

Graphene and graphene-based nanomaterials: Graphene has hexagonal, honeycomb-like two-dimensional geometry with thin sheet-like structure [213]. It exhibits good thermal and chemical stability with high mechanical strength because of phononic vibration; large electrical conductivity because of the presence of large number of conjugated sp^2-C atoms and aromatic structure in it [214]; high electrochemical and BET surface area [215]; and biocompatibility [216,217]. Functionalizing graphene with different compounds makes it a suitable material for various biomedical applications [218–220], such as brain stimulators [221], tissue engineering [222], drug delivery [223], and fabrication of bio-electronic materials [224]. However, graphene commonly exists as nanoparticles and is hard to disperse in any solvent. Thus, it has been deposited onto GCEs or other commercial electrodes, such as Au and other noble metal rods/wires/meshes, polymers, and metal oxides. It has been successfully applied as the electrode material for the qualitative and quantitative determination of different biomaterials, including biomarkers of chronic and life-threatening diseases, such as CA 15-3 and other cancer biomarkers [225,226], ascorbic acid, dopamine, uric acid [227], cholesterol [228], and glucose [229].

However, the low electrical conductivity of GCE or the noble metals results in low sensitivity and a high detection limit. Thus, different biocompatible graphene derivatives, such as graphene oxide (GO), reduced graphene oxide (rGO), and nano-graphite rods with or without dispersed metal and metal oxide NPs have been used as the electrode materials in different electrochemical studies [230]. GO and rGO are easy to disperse in any suitable solvent, including different polymeric suspensions [231], due to the inclusion of different oxygen-based functional groups with a reasonably high oxygen to carbon ratio compared to graphene or graphite. The presence of oxygen-containing functional groups and π-electron clouds also helps these materials to bind and trap metal

NPs. GO or rGO, modified with or without different materials, have been efficiently used in different bio-sensing methods, including electrochemical sensing for biochemically and biomedically relevant targets of thrombin [232], proteins [233], and ATP [234]. However, rGO has a higher electro-conductivity compared to GO due to its higher graphitic content. Surprisingly, these materials have not been explored for the development of electrodes for electrochemical sensing.

13.6.1.2 Polymeric films (nanosized and nanostructured) and metal-polymer hybrids

Polymeric materials, from the time of its development, are of massive interest to the scientific community for their applicability in different interdisciplinary domains including the development of different biomedical devices. Several conductive polymeric materials, either coated on some substrate or in hybrid form with other compounds, have been widely used as the electrode materials in different biochemical and biomedical studies based on their biocompatibility in the respective applications. Among these, poly(aniline), poly(pyrrole), poly(thiophene), poly(methylene blue) have been extensively used as the electrode materials of different electrochemical and other biosensors because of their biocompatibility and similarity to the different biopolymers. Some of these materials is developed in nanosized and nanostructured geometrical shape by coating on different commercial and synthetic substrates for several biomedical applications including sensing.

Several polymeric materials, doped with different metal and metal oxide nanoparticles, have been found to exhibit excellent biomedical properties such as sensitivity, selectivity, detection limit etc. during their application in the biosensing devices due to the enhancement in biocompatibility, active surface area and surface functionalization post association. The commonly used metal nanoparticles in these materials are Au, Ag, Pt, and Pd [77]. Incorporation of these tiny sized particles in the polymeric matrix helps to absorb biomolecules in its surface via Vander Waal force of attraction and/other electrostatic forces. Also, association of different conductive carbon-based biocompatible materials with different conductive polymeric materials amplified the electrochemical properties in the prepared carbon-polymer hybrid materials form that in their base materials. Common examples of these hybrids are – poly(pyrrole), poly(dopamine) etc. associated with different carboxylic acid, chitin and cellulose with their promising applications in developing different flexible electronics including different wearable and nonwearable sensing devices.

All of these facilitates the entrapment of different biological components and biomolecules, for example, protein, antibody, DNA, RNA or other low molecular components inside the matrix for acting as the biomarker recognition elements and different metabolites. These also facilitates the development of molecular imprinted polymer-based sensors for different biomedical applications (MIP sensors). Au NPs- and graphene- associated carboxylic-pyrrole hybrid for pyoverdine to diagnosis and therapy management of nosocomial infections [235] and carcinoembryonic antigen to diagnosis gastrointestinal cancer [236], Au NPs-functionalized sulfonic acid-modified Poly(aniline) nanowires to detect dopamine and serotonin in presence of uric acid and ascorbic acid [237,238].

13.6.1.3 Electrochemically synthesized metal- or graphitic carbon-polymer hybrids and electropolymers

Most of the conducting polymers and hybrid polymeric materials (associated with metal/metal oxide nanoparticles, conductive carbon-based materials and others) discussed in the section 6.1.3 have been developed using several chemical methods. Among these, the materials developed using different electrochemical methods is easy, provide almost green synthesis as the main reagent is electron. This facilitates to develop electrode and sensor materials of high purity. Thickness and density of these materials during their growth over any substrate is uniform and can easily be controlled, and have large influence in their diagnostic applications. Most of these electro-polymers, especially, poly(methylene blue), poly(indole), poly(thiophene), poly(3,4-ethylenedioxythiophene): poly(styrenesulfonate) (PEDOT:PSS), PEDOT:polystyrene, poly(pyridine) etc. have been grown onto either commercial glassy carbon electrode (GCE), carbon screen-printed electrode (SPE), Pt or Au rod/wire, or any other substrate, and were used in developing different biosensing devices.

13.6.2 Application as substrate for biosensor development

Different electroconducting organic and inorganic materials have been used as the support material, or substrate, to fabricate the electrode of an electrochemical sensor due to the low mechanical stability of different carbonaceous materials in the working electrolyte solution. Different noble and transition metal rods/plates/wires can serve as the support for the electrode materials by depositing different carbonaceous electrode materials, with or without dispersed metal or metal oxide NPs, onto it, including, Pt and Pt-Ru rods for CNTs, stainless steel mesh of different mesh size for carbon NPs, and Cu foil for CuO NPs [239–242]. However, the materials are expensive. Different precursor-based activated carbon micro-fibers have been used as the support materials for growing CNFs and other conducting materials, but the material suffers from low electro-conductivity [243–245]. Several organic polymers have also been developed as the support for electrode materials, such as PET for TiO_x/Ag/AZO for optoelectronic applications [246] and poly-(4,3-ethylenedioxythiophene): poly(styrene-sulfonate) [247], which can be used to fabricate flexible electrochemical sensors. The materials are chemically stable at different temperatures, pH, and humidity, but exhibit low conductivities, which limits their sensitivity and other electrochemical characteristics. However, studies using different conducting polymeric films as the support materials are scarce.

13.6.3 Application as biomarker recognition element

13.6.3.1 Metal and metal oxide nanoparticles as recognition elements

Noble metal and metal oxide NPs, some of which incorporate transition metal oxides, deposited onto different electrodes have also been used as recognition elements, especially as non-enzymatic recognition elements for determining the concentrations of different chemical- and bio-markers. For example, Au, Au/CuO, Pt/NiO, and Ag/CuO NPs have been used to sense glucose [248–251]; Ag and Pt NPs to measure cholesterol [252,253]; Au rods for determination of different biogenic amines [254]; and have been studied. The materials have exhibited good electrochemical compatibility in measuring different chemical and biochemical compounds (especially biomolecules); however, low sensitivities due to the low electrical conductivity and the lack of selectivity are significant challenges. To overcome these challenges, various transition metals and their oxide NPs have been analyzed as the recognition element, such as NiO for cholesterol [255]; Cu/Ni, Ni/Co, and Co_3O_4 for glucose [256–258]; These materials have demonstrated promising experimental outcomes; however, the materials also exhibit low selectivity.

13.6.3.2 Hybrid organic nanomaterials as recognition elements

To overcome the above shortcomings, several synthetic organic compounds, including different organometallic compounds, polymers, and macromolecules, have been developed and tested as recognition elements for various analytes, especially biomarkers. The materials are either drop-casted or grown over the electrode surface before electrochemical measurement. For example, imidazole-derivative/rGO/GCE has been used to detect dopamine, uric acid, and L-cysteine [259]; Ag NPs dispersed in ((pyridin-2-yl)methyleneamino)benzenethiol-modified styrene has been used to sense cholesterol [260]; which have exhibited excellent electrochemical responses during the measurements. These are novel methods and materials for developing different non-biomaterial based electrochemical sensors for selective measurement. However, low selectivity and complex preparation processes to maintain purity are significant drawbacks of these materials. The layer thickness of the polymers and macromolecules developed in these studies are also difficult to control.

13.6.3.3 Polymeric and electropolymeric recognition elements

Similar to many other compounds, several polymeric compounds have been developed using different electrochemical methods to serve as the recognition element for different analytes. The methods are simple, easy to control, and require limited reagents as compared to other conventional polymerization techniques. The thickness of the electro-polymeric layer is critical for implementation as the recognition element and is easy to control when prepared using CV by optimizing the scan cycle (N), scan rate (ν), and working potential. Therefore, most of the reported electro-polymers have been developed using CV under optimum N and ν over a predefined potential window based on the redox behavior of the monomer molecule(s). For example, poly(methylene blue) for oxidation of NADH [261], poly(methyl orange) for detection of dopamine [262], poly(β-cyclodextrin) for sensing of uric acid [263], and poly(aniline) for detection of H_2O_2 [264–266] have been used due to their electro-catalytic activity in the specific redox reactions involving the analyte molecule(s) and exhibit good selectivity and sensitivity with high repeatability. This methodology is new, and therefore, the materials and the methodologies have been used in a limited number of studies.

13.7 Conclusion

Hybrid nanomaterials consist of inorganic, organic or organometallic nanoparticles as their main constituents provided a new platform especially sensing of physiological components, bioimaging, drug delivery and tissue engineering. Hybrid materials provides efficient redox-sensitive, due to the synergetic effects of active noble or transition metal and metal oxide nanoparticles, incorporated with biocompatible or on electro-conductive carbonaceous electrode materials, showed good sensitivity and low detection limit for developing electrochemical biosensor Hybrid nanomaterials developed using natural and synthetic polymers (which showed excellent biomedical applicability especially when applied in tissue engineering and bioimaging), applied for sensing of different biomolecules including different biomarkers and critical components. Presence of the polymeric component in the electro-polymers helps a crucial role in the selectivity of the sensor and its flexibility. These materials also pave a new platform for developing wearable sensors. Hybrid nanomaterials showed very good selectivity and sensitivity, with high thermal and mechanical strength, flexibility and reproducibility in the measurement data. These materials confirmed their applicability as the new and promising sensing materials for electrochemical and optical sensing of biomarkers and physiological components. In spite of showing such promising electrochemical characteristics and sensing performances, hybrid nanomaterials-based and metal/metal oxide nanoparticles-based electrochemical and optical biosensors for various biological components and biomarkers are yet to be commercialized. Studies are being done in different research laboratories and institutes for the same, which is a prime interest to the scientific community in recent days.

References

[1] J. Castillo, et al., Biosensors for life quality: design, development and applications, Sens. Actuators B: Chem. 102 (2) (2004) 179–194.
[2] A. Gallotta, E. Orzes, G. Fassina, Biomarkers quantification with antibody arrays in cancer early detection, Clin. Lab. Med. 32 (1) (2012) 33–45.
[3] T. Prasada Rao, R. Kala, Potentiometric transducer based biomimetic sensors for priority envirotoxic markers—An overview, Talanta 76 (3) (2008) 485–496.
[4] D. Sharma, C.M. Hussain, Smart nanomaterials in pharmaceutical analysis, Arabian J. Chem. 13 (1) (2020) 3319–3343.

[5] K. Bhattacharyya, et al., Gold nanoparticle–mediated detection of circulating cancer cells, Clin. Lab. Med. 32 (1) (2012) 89–101.
[6] S. Kumar, et al., Recent advances in biosensors for diagnosis and detection of sepsis: a comprehensive review, Biosens. Bioelectron. 124-125 (2019) 205–215.
[7] F. Arduini, et al., Electrochemical biosensors based on nanomodified screen-printed electrodes: recent applications in clinical analysis, TrAC Trends Anal. Chem. 79 (2016) 114–126.
[8] E. Roduner, Size matters: why nanomaterials are different, Chem. Soc. Rev. 35 (7) (2006) 583–592.
[9] S.N. Baker, G.A. Baker, Luminescent carbon nanodots: emergent nanolights, Angew. Chem. Int. Ed. 49 (38) (2010) 6726–6744.
[10] B.R. Cuenya, Synthesis and catalytic properties of metal nanoparticles: Size, shape, support, composition, and oxidation state effects, Thin Solid Films 518 (12) (2010) 3127–3150.
[11] J. Gao, et al., Multifunctional yolk–shell nanoparticles: a potential MRI contrast and anticancer agent, J. Am. Chem. Soc. 130 (35) (2008) 11828–11833.
[12] C.R. Patra, et al., Application of gold nanoparticles for targeted therapy in cancer, J. Biomed. Nanotechnol. 4 (2) (2008) 99–132.
[13] A. Sharma, et al., Wearable biosensors: an alternative and practical approach in healthcare and disease monitoring, Molecules 26 (3) (2021) 748.
[14] D. Maiti, et al., Carbon-based nanomaterials for biomedical applications: a recent study, Front. Pharmacol. 9 (1401) (2019).
[15] K. Bhattacharya, et al., Biological interactions of carbon-based nanomaterials: from coronation to degradation, Nanomedicine: Nanotechnology, Biology and Medicine, 12 (2) (2016) 333–351.
[16] M. Sireesha, et al., A review on carbon nanotubes in biosensor devices and their applications in medicine, Nanocomposites 4 (2) (2018) 36–57.
[17] D. Ji, et al., Smartphone-based differential pulse amperometry system for real-time monitoring of levodopa with carbon nanotubes and gold nanoparticles modified screen-printing electrodes, Biosens. Bioelectron. 129 (2019) 216–223.
[18] S.K. Krishnan, et al., A review on graphene-based nanocomposites for electrochemical and fluorescent biosensors, RSC Advances, 9 (16) (2019) 8778–8881.
[19] F. Lin, Y.-W. Bao, F.-G. Wu, Carbon dots for sensing and killing microorganisms, C 5 (2) (2019) 33.
[20] S. Kumar, et al., Electrochemical sensors and biosensors based on graphene functionalized with metal oxide nanostructures for healthcare applications, ChemistrySelect 4 (18) (2019) 5322–5337.
[21] V.D.N. Bezzon, et al., Carbon nanostructure-based sensors: a brief review on recent advances, Adv. Mater. Sci. Eng. 2019 (2019) 4293073.
[22] M. Baccarin, et al., Nanodiamond based surface modified screen-printed electrodes for the simultaneous voltammetric determination of dopamine and uric acid, Microchim. Acta 186 (3) (2019) 200.
[23] V.N. Mochalin, et al., The properties and applications of nanodiamonds, Nat. Nanotechnol. 7 (1) (2012) 11–23.
[24] X. Xu, et al., Electrophoretic analysis and purification of fluorescent single-walled carbon nanotube fragments, J. Am. Chem. Soc. 126 (40) (2004) 12736–12737.
[25] Y.-P. Sun, et al., Quantum-sized carbon dots for bright and colorful photoluminescence, J. Am. Chem. Soc. 128 (24) (2006) 7756–7757.
[26] P. Khare, et al., Brightly fluorescent zinc-doped red-emitting carbon dots for the sunlight-induced photoreduction of Cr(VI) to Cr(III), ACS Omega 3 (5) (2018) 5187–5194.
[27] A. Bhati, et al., Sunlight-induced photocatalytic degradation of pollutant dye by highly fluorescent red-emitting Mg-N-embedded carbon dots, ACS Sustain. Chem. Eng. 6 (7) (2018) 9246–9256.
[28] P.G. Luo, et al., Carbon "quantum" dots for optical bioimaging, J. Mater. Chem. B 1 (16) (2013) 2116–2127.
[29] V.D.N. Bezzon, et al., Carbon nanostructure-based sensors: a brief review on recent advances, Adv. Mater. Sci. Eng. 2019 (2019) 4293073.
[30] M. Baccarin, et al., Nanodiamond based surface modified screen-printed electrodes for the simultaneous voltammetric determination of dopamine and uric acid, Microchim. Acta 186 (3) (2019) 200.
[31] V.N. Mochalin, et al., The properties and applications of nanodiamonds, Nat. Nanotechnol. 7 (1) (2012) 11–23.

[32] A.A. Nyle, et al., The nanomaterials and recent progress in biosensing systems: a review, Trends Environ. Anal. Chem. 26 (2020) e00087.
[33] M. Pohanka, Overview of piezoelectric biosensors, immunosensors and DNA sensors and their applications, Materials 11 (3) (2018) 448–460.
[34] Y. Ma, et al., Review on porous nanomaterials for adsorption and photocatalytic conversion of CO2, Chin. J. Catal. 38 (2017) 1956–1969.
[35] S. Chen, R. Yuan, Y. Chai, et al., Electrochemical sensing of hydrogen peroxide using metal nanoparticles: a review, Microchim. Acta 180 (2013) 15–32.
[36] G. Doria, et al., Noble metal nanoparticles for biosensing applications, Sensors 12 (2) (2012) 1657–1687.
[37] S.F. Himmelsto, T. Hirsch, A critical comparison of lanthanide based up conversion nanoparticles to fluorescent proteins, semiconductor quantum dots, and carbon dots for use in optical sensing and imaging, Methods Appl. Fluoresc. 7 (2) (2019) 022002.
[38] L. Lamon, et al., Grouping of nanomaterials to read-across hazard endpoints: a review, Nanotoxicology 13 (1) (2019) 100–118.
[39] T. Mazhar, et al., Green synthesis of bimetallic nanoparticles and its applications: a review, J. Pharm. Sci. Res. 9 (2) (2017) 102–110.
[40] J.M. George, et al., Metal oxide nanoparticles in electrochemical sensing and biosensing: a review, Microchim. Acta 185 (358) (2018).
[41] A.P. Dral, J.E. Elshof, 2D metal oxide nanoflakes for sensing applications: review and perspective, Sens. Actuators B 272 (2018) 369–392.
[42] A.S. Agnihotri, et al., Transition metal oxides in electrochemical and bio sensing: a state-of-art review, Appl. Surf. Sci. 4 (2021) 100072.
[43] N. Tripathy, D.H. Kim, Metal oxide modified ZnO nanomaterials for biosensor applications, Nano Convergence 5 (1) (2018) 27.
[44] Y.-H. Wang, et al., Recent advances in transition-metal dichalcogenides based electrochemical biosensors: a review, Biosens. Bioelectron. 97 (2017) 305–316.
[45] L. Argueta-Figueroa, Nanomaterials made of non-toxic metallic sulfides: a systematic review of their potential biomedical applications, Mater. Sci. Eng: C, 76 (2017) 1305–1315.
[46] P.K. Kalambate, et al., Core@shell nanomaterials based sensing devices: a review, Trends Anal. Chem. 115 (2019) 147–161.
[47] M.H. Naveen, et al., Applications of conducting polymer composites to electrochemical sensors: a review, Appl. Mater. Today 9 (2017) 419–433.
[48] T.R. Pavase, et al., Recent advances of conjugated polymer (CP) nanocomposite-based chemical sensors and their applications in food spoilage detection: a comprehensive review, Sens. Actuators B 273 (2018) 1113–1118.
[49] P.L. Suárez, et al., Functionalized phosphorescent nanoparticles in (bio)chemical sensing and imaging –a review, Analytical Chimica Acta 1046 (2019) 16–31.
[50] L. Clarizia, et al., Hydrogen generation through solar photocatalytic processes: a review of the configuration and the properties of effective metal-based semiconductor nanomaterials, Energies 10 (10) (2017) 1624.
[51] Y. Zhang, et al., Current advances in semiconductor nanomaterial-based photoelectrochemical biosensing, Chem. Eur. J. 24 (53) (2018) 14010–14027.
[52] J. Joy, et al., Nanomaterials for photoelectrochemical water splitting – review, Int. J. Hydrogen Energy 43 (10) (2018) 4804–4817.
[53] C.M. Das, Multifaceted hybrid carbon fibers: applications in renewables, sensing and tissue engineering, J. Compos. Sci. 4 (3) (2020) 117–139.
[54] Z. Wang, et al., Carbon nanofiber-based functional nanomaterials for sensor applications, Nanomaterials 9 (7) (2019) 1045.
[55] S. Gupta, et al., Recent advances in carbon nanotube based electrochemical biosensors, Int. J. Biol. Macromol. 108 (2018) 687–703.
[56] Z. Zhu, An overview of carbon nanotubes and graphene for biosensing applications. Nano-Micro Letters, 2017. 9(25).
[57] I. Prattis, et al., Graphene for biosensing applications in point-of-care testing, Trends Biotechnol. (2021), https://doi.org/10.1016/j.tibtech.2021.01.005.

[58] W. Wang, et al., Review—Biosensing and biomedical applications of graphene: a review of current progress and future prospect, J. Electrochem. Soc. 166 (6) (2019) B505–B520.
[59] S. Joshi, et al., A review on peptide functionalized graphene derivatives as nanotools for biosensing, Microchim. Acta 187 (27) (2020).
[60] S. Priyadarsini, et al., Graphene and graphene oxide as nanomaterials for medicine and biology application, J. Nanostructure. Chem. 8 (2018) 123–137.
[61] M. Chang, et al., Cellulose-based biosensor for bio-molecules detection in medical diagnosis: a mini-review, Curr. Med. Chem. 27 (28) (2020) 4593–4612.
[62] T. Park, et al., An organic/inorganic nanocomposite of cellulose nanofibers and zno nanorods for highly sensitive, reliable, wireless, and wearable multifunctional sensor applications, ACS Appl. Mater. Interfaces 11 (51) (2019) 48239–48248.
[63] M.A. Beluomini, Electrochemical sensors based on molecularly imprinted polymer on nanostructured carbon materials: a review, J. Electroanal. Chem. 840 (2019) 343–366.
[64] J. Zhao, et al., Responsive polymers as smart nanomaterials enable diverse applications, Annu. Rev. Chem. Biomol. Eng. 10 (1) (2019) 361–382.
[65] I. Pandey, et al., Electrochemically grown polymethylene blue nanofilm on copper-carbon nanofiber nanocomposite: an electrochemical sensor for creatinine, Sens. Actuators B 277 (2018) 562–570.
[66] P.K. Bairagi, N. Verma, Electrochemically deposited dendritic poly (methyl orange) nanofilm on metal-carbon-polymer nanocomposite: a novel non-enzymatic electrochemical biosensor for cholesterol, J. Electroanal. Chem. 814 (2018) 134–143.
[67] P.K. Bairagi, N. Verma, Electro-polymerized polyacrylamide nano film grown on a Ni-reduced graphene oxide- polymer composite: a highly selective non-enzymatic electrochemical recognition element for glucose, Sens. Actuators B 289 (216) (2019).
[68] S. Shrivastava, et al., Spermine biomarker of cancerous cells voltammetrically detected on a poly(β-cyclodextrin) - electropolymerized carbon film dispersed with Cu - CNFs, Sens. Actuators B 313 (128055) (2020).
[69] J. Wang, et al., Nanomaterial-doped conducting polymers for electrochemical sensors and biosensors, J. Mater. Chem. B 6 (2018) 4173–4190.
[70] V. Galstyan, Quantum dots: Perspectives in next-generation chemical gas sensors - a review, Anal. Chim. Acta 1152 (238192) (2021).
[71] Y. Park, et al., Medically translatable quantum dots for biosensing and imaging, J. Photochem. Photobiol. C 30 (2017) 51–70.
[72] P. Zheng, N. Wu, Fluorescence and sensing applications of graphene oxide and graphene quantum dots: a review, Chem. Asian J. 12 (18) (2017) 2343–2353.
[73] S. Chung, et al., Graphene quantum dots and their applications in bioimaging, biosensing, and therapy, Adv. Mater. 33 (22) (2021) 1904362.
[74] S. Iravani, R.S. Varma, Green synthesis, biomedical and biotechnological applications of carbon and graphene quantum dots. A review, Environ. Chem. Lett. 18 (2020) 703–727.
[75] M.J. Molaei, Principles, mechanisms, and application of carbon quantum dots in sensors: a review, Anal. Methods 12 (2020) 1266–1287, doi:10.1039/C9AY02696G.
[76] M.M. Shameem, et al., A brief review on polymer nanocomposites and its applications, Materials Today Proceedings 45 (2) (2021) 2536–2539.
[77] N. Karak, Chapter 1 - Fundamentals of Nanomaterials and Polymer Nanocomposites, in: N. Karak (Ed.), Nanomaterials and Polymer Nanocomposites: Raw Materials to Applications, Elsevier, Amsterdam, 2019, pp. 1–45.
[78] S.K. Krishnan, et al., A review on graphene-based nanocomposites for electrochemical and fluorescent biosensors, RSC Advances, 9 (2019) 8778–8881.
[79] D.W. Silva, et al., Nanostructured Poly(Phenazine)/Fe2O3 nanoparticle film modified electrodes formed by electropolymerization in ethaline – Deep eutectic solvent. Microscopic and electrochemical characterization, Electrochimca Acta 347 (2020) 136284.
[80] H.A.C. Krebs, Chemical composition of blood plasma and serum, Annu. Rev. Biochem. 19 (1950) 409–443.
[81] S.H. Holm, et al., Separation of parasites from human blood using deterministic lateral displacement, Lab. Chip. 11 (7) (2011) 1326–1332.

[82] O. Hosu, et al., 1 - Recent approaches to the synthesis of smart nanomaterials for nanodevices in disease diagnosis, in: S. Kanchi, D. Sharma (Eds.), Nanomaterials in Diagnostic Tools and Devices, Elsevier, Amsterdam, 2020.
[83] N. Sultan, Reflective thoughts on the potential and challenges of wearable technology for healthcare provision and medical education, Int. J. Inf. Manage. 35 (5) (2015) 521–526.
[84] R.G. Saratale, et al., A comprehensive review on green nanomaterials using biological systems: recent perception and their future applications, Colloids Surf. B 170 (2018) 20–35.
[85] Q.X. Mu, et al., Chemical basis interactions between engineered nanoparticles and biological systems, Chem. Rev. 114 (15) (2014) 7740–7781.
[86] L.A. Kolahalam, et al., Review on nanomaterials: synthesis and applications, Materials Today Proceedings 18 (2019) 2182–2190.
[87] K. Kalimuthu, et al., Eco-friendly synthesis and biomedical applications of gold nanoparticles: a review, Microchem. J. 152 (104296) (2020).
[88] J. Santhoshkumar, et al., Phyto-assisted synthesis, characterization and applications of gold nanoparticles – a review, Biochem. Biophys. Rep. 11 (2017) 46–57.
[89] C.D. De Souza, et al., Review of the methodologies used in the synthesis gold nanoparticles by chemical reduction, J. Alloys Compd. 798 (2019) 714–740.
[90] M. Sengani, et al., Recent trends and methodologies in gold nanoparticle synthesis – a prospective review on drug delivery aspect, OpenNano 2 (2017) 37–46.
[91] K. Banerjee, V.A. Ravishankar Rai, Review on mycosynthesis, mechanism, and characterization of silver and gold nanoparticles, BioNanoScience 8 (2018) 17–31.
[92] R. Nishanthi, et al., Green synthesis and characterization of bioinspired silver, gold and platinum nanoparticles and evaluation of their synergistic antibacterial activity after combining with different classes of antibiotics, J. Mater. Chem. C 96 (2019) 693–707.
[93] A.A. Yaqoob, Silver nanoparticles: various methods of synthesis, size affecting factors and their potential applications–a review, Appl. Nanosci. 10 (2020) 1369–1378.
[94] S. Ahmed, et al., Green nanotechnology: A review on green synthesis of silver nanoparticles — an eco-friendly approach, Int. J. Nanomed. 14 (2019) 5087–5107.
[95] J.R. Koduru, Phytochemical-assisted synthetic approaches for silver nanoparticles antimicrobial applications: a review, Adv. Colloid Interface Sci. 256 (2018) 326–339.
[96] S. Rajeshkumar, L.V. Bharath, Mechanism of plant-mediated synthesis of silver nanoparticles – a review on biomolecules involved, characterisation and antibacterial activity, Chem. Biol. Interact. 273 (2017) 219–227.
[97] R. Shanmuganathan, et al., Synthesis of silver nanoparticles and their biomedical applications - a comprehensive review, Curr. Pharm. Des. 25 (2019) 2650–2660.
[98] C. Dong, et al., Size-dependent activity and selectivity of carbon dioxide photocatalytic reduction over platinum nanoparticles, Nat. Commun. 9 (2018) 1252–1262.
[99] K. Gupta, T.S. Chundawat, Bio-inspired synthesis of platinum nanoparticles from fungus Fusarium oxysporum: its characteristics, potential antimicrobial, antioxidant and photocatalytic activities, Mater. Res. Express 6 (10) (2019) 101050d6.
[100] V.S. Ramkumar, et al., Synthesis of platinum nanoparticles using seaweed Padina gymnospora and their catalytic activity as PVP/PtNPs nanocomposite towards biological applications, Biomed. Pharmacother. 92 (2017) 479–490.
[101] S. Luo, et al., Metal organic frameworks as robust host of palladium nanoparticles in heterogeneous catalysis: synthesis, application, and prospect, ACS Appl. Mater. Interfaces 11 (36) (2019) 32579–32598.
[102] T.T.V. Phan, et al., An up-to-date review on biomedical applications of palladium nanoparticles, Nanomaterials 10 (1) (2020) 66–84.
[103] M. Hazarika, et al., Biogenic synthesis of palladium nanoparticles and their applications as catalyst and antimicrobial agent, PlosOne (2017), https://doi.org/10.1371/journal.pone.0184936.
[104] A. Ebrahiminezhad, et al., Plant-mediated synthesis and applications of iron nanoparticles, Mol. Biotechnol. 60 (2018) 154–168.
[105] Green Synthesis of Iron Oxide Nanoparticles and Its Biomedical ApplicationsM. Thakur, S. Poojary, N. Swain, Nanotechnology applications in health and environmental sciences, in: N. Saglam, F. Korkusuz, R. Prasad (Eds.), Nanotechnology in the Life Sciences, Springer, Cham, 2021.

[106] Y.P. Yew, et al., Green biosynthesis of superparamagnetic magnetite Fe3O4 nanoparticles and biomedical applications in targeted anticancer drug delivery system: a review, Arabian J. Chem. 13 (2020) 2287–2308.
[107] N.V.S. Vallabani, S. Singh, Recent advances and future prospects of iron oxide nanoparticles in biomedicine and diagnostics, 3 Biotech 8 (6) (2018) 279.
[108] M. Rafique, A review on synthesis, characterization and applications of copper nanoparticles using green method, Nano 12 (4) (2017) 1750043.
[109] A. Waris, et al., A comprehensive review of green synthesis of copper oxide nanoparticles and their diverse biomedical applications. Inorg. Chem. Commun., 2021. 123(108369).
[110] Z. Qing, et al., Progress in biosensor based on DNA-templated copper nanoparticles, Biosenssors Bioelectronics 137 (2019) 96–109.
[111] S. Pourbeyram, et al., Green synthesis of copper oxide nanoparticles decorated reduced graphene oxide for high sensitive detection of glucose, J. Mater. Chem. C 94 (2019) 850–857.
[112] M.R. Ahghari, et al., Synthesis of nickel nanoparticles by a green and convenient method as a magnetic mirror with antibacterial activities, Sci. Rep. 10 (2020) 12627.
[113] R. Rawal, et al., A contrivance based on electrochemical integration of graphene oxide nanoparticles/nickel nanoparticles for bilirubin biosensing, Biochem. Eng. J. 125 (2017) 238–245.
[114] A.K. Vivekanandan, et al., Sonochemical synthesis of nickel-manganous oxide nanocrumbs decorated partially reduced graphene oxide for efficient electrochemical reduction of metronidazole, Ultrason. Sonochem. 68 (2020) 105–176.
[115] D.M. Cruz, et al., Green nanotechnology-based zinc oxide (ZnO) nanomaterials for biomedical applications: a review. J. Phys.: Mater. 3 (3) (2020) 034005.
[116] W. Chen, C. Wu, Synthesis, functionalization, and applications of metal–organic frameworks in biomedicine, Dalton Trans. 47 (2018) 2114–2133.
[117] V.N. Kalpana, V.D. Rajeswari, A review on green synthesis, biomedical applications, and toxicity studies of ZnO NPs, Bioinorg. Chem. Appl. 3569758 (2018), https://doi.org/10.1155/2018/3569758.
[118] M.L.M. Napi, et al., Electrochemical-based biosensors on different zinc oxide nanostructures: a review, Materials 12 (18) (2019) 2985.
[119] N.R. Shanmugam., et al., A review on ZnO-based electrical biosensors for cardiac biomarker detection, Future Science OA 3 (4) (2017), https://doi.org/10.4155/fsoa-2017-0006.
[120] C. Kaçar, et al., Amperometric biogenic amine biosensors based on Prussian blue, indium tin oxide nanoparticles and diamine oxidase– or monoamine oxidase–modified electrodes, Anal. Bioanal. Chem. 412 (8) (2020) 1933–1946.
[121] E.B. Aydın, M.K. Sezgintürk, Indium tin oxide (ITO): a promising material in biosensing technology, Trends Anal. Chem. 97 (2017) 309–315.
[122] S. Iravani, et al., Synthesis of silver nanoparticles: chemical, physical and biological methods, Res. Pharm. Sci. 9 (6) (2014) 385–406.
[123] M. Singh, et al., A novel ball milling technique for room temperature processing of TiO2 nanoparticles employed as the electron transport layer in perovskite solar cells and modules, J. Mater. Chem. A 6 (2018) 7114–7122.
[124] A. Hashemzadeh, et al., Synthesis of MoO2 nanoparticles via the electro-explosion of wire (EEW) method, Mater. Res. Express, 6 (2019) 1250d3.
[125] N. Goswami, P. Sen, Photoluminescent properties of ZnS nanoparticles prepared by electro-explosion of Zn wires, J. Nanopart. Res. 9 (2007) 513–517.
[126] A.V. Rane, et al., Methods for synthesis of nanoparticles and fabrication of nanocomposites, in: S.M. Bhagyaraj et al (Ed.), Micro and Nano Technologies, Synthesis of Inorganic Nanomaterials, Woodhead Publishing, Duxford, 2018, pp. 121–139.
[127] O.K. Alexeeva, V.N. Fateev, Application of the magnetron sputtering for nanostructured electrocatalysts synthesis, Int. J. Hydrogen Energy 41 (5) (2016) 3373–3386.
[128] J. Zhang, et al., Colloidal metal nanoparticles prepared by laser ablation and their applications, Chem. Phys. Chem 18 (9) (2017) 986–1006.
[128] S. Dou, et al., Plasma-assisted synthesis and surface modification of electrode materials for renewable energy, Adv. Mater. 30 (21) (2018) 1705850.

[130] H. Çolak, E. Karaköse, Green synthesis and characterization of nanostructured ZnO thin films using Citrus aurantifolia (lemon) peel extract by spin-coating method, J. Alloys Compd. 690 (2017) 658–662.

[131] L. Wu, et al., Room-temperature pre-reduction of spinning solution for the synthesis of Na3V2(PO4)3/C nanofibers as high-performance cathode materials for Na-ion batteries, Electrochimca Acta 274 (2018) 233–241.

[132] Z. Samiei, et al., Fe3O4@C@OSO3H as an efficient, recyclable magnetic nanocatalyst in Pechmann condensation: green synthesis, characterization, and theoretical study, Mol. Divers. 25 (2021) 67–86.

[133] M.K.M. Lane, J. Zimmerman, Controlling metal oxide nanoparticle size and shape with supercritical fluid synthesis, Green Chem. 21 (2019) 3769–3781.

[134] M. Türka, C. Erkey, Synthesis of supported nanoparticles in supercritical fluids by supercritical fluid reactive deposition: current state, further perspectives and needs, J. Supercrit. Fluids 134 (2018) 176–183.

[135] G. Gahlawat, A.R. Choudhury, A review on the biosynthesis of metal and metal salt nanoparticles by microbes, RSC Advances, 9 (23) (2019) 12944–12967.

[136] L. Yan, et al., Magnetotactic bacteria, magnetosomes and their application, Microbiol. Res. 167 (9) (2012) 507–519.

[137] S. He, et al., Biosynthesis of gold nanoparticles using the bacteria Rhodopseudomonas capsulate, Mater. Lett. 61 (18) (2007) 3984–3987.

[138] M. Schlüter, et al., Synthesis of novel palladium(0) nanocatalysts by microorganisms from heavy-metal-influenced high-alpine sites for dehalogenation of polychlorinated dioxins, Chemosphere 117 (2014) 462–470.

[139] G. Singaravelu, et al., A novel extracellular synthesis of monodisperse gold nanoparticles using marine alga, Sargassum wightii Greville, Colloids Surf. B 57 (1) (2007) 97–101.

[140] A. Rantasalo, et al., Novel genetic tools that enable highly pure protein production in Trichoderma reesei, Sci. Rep. 9 (5032) (2019).

[141] A. Ahmad, et al., Extracellular biosynthesis of silver nanoparticles using the fungus Fusarium oxysporum, Colloids Surf. B 28 (4) (2003) 313–318.

[142] B. Jiang, et al., Ferritins as natural and artificial nanozymes for theranostics, Theranostics 10 (2) (2020) 687–706.

[143] M.S. Ardejeni, et al., Complete shift of ferritin oligomerization toward nanocageassemblyvia engineered protein–protein interactions, Chem. Commun. 49 (2013) 3528–3530.

[144] H. Wu, et al., Apoferritin-templated yttrium phosphate nanoparticle conjugates for radioimmunotherapy of cancers, J. Nanosci. Nanotechnol. 8 (5) (2008) 2316–2322.

[145] P. Rajapaksha, et al., A review of methods for the detection of pathogenic microorganisms, Analyst 144 (2019) 396–411.

[146] J. Gardea-Torresdey, et al., Alfalfa sprouts:a natural source for the synthesis of silver nanoparticles, Langmuir 19 (4) (2003) 1357–1361.

[147] F. Zamarchi, I.C. Vieira, Determination of paracetamol using a sensor based on green synthesis of silver nanoparticles in plant extract, J. Pharm. Biomed. Anal. 196 (2021) 113912.

[148] S.Z. Ghazali, et al., Anti-plasmodial activity of aqueous neem leaf extract mediated green synthesis-based silver nitrate nanoparticles, Prep. Biochem. Biotechnol. (2021), https://doi.org/10.1080/10826068.2021.1913602.

[149] S.S. Shankar, et al., Rapid synthesis of Au, Ag, and bimetallic Au core–Ag shell nanoparticles using Neem (Azadirachta indica) leaf broth, J. Colloid Interface Sci. 275 (2004) 496–502.

[150] M. Pandian, Development of biogenic silver nano particle from Pelargonium graveolens leaf extract and their antibacterial activity, J. Nanosci. Nanotechnol. 1 (2) (2013) 57–64.

[151] P.A. Raymundo-Pereira, et al., Polyphenol oxidase-based electrochemical biosensors: a review, Analytical Chimica, Acta, 1139 (2020) 198–221.

[152] G.S. Shekhawat, V. Arya, Biological synthesis of Ag nanoparticles through in vitro cultures of Brassica juncea C. zern, Adv. Mater. Res. 67 (2009) 295–299.

[153] S.P. Chandran, et al., Synthesis of gold nanotriangles and silver nanoparticles using aloevera plant extract, Biotechnol. Prog. 22 (2) (2006) 577–583.

[154] R. Rahimi, et al., Synthesis and characterization of magnetic dichromate hybrid nanomaterials with triphenylphosphine surface modified iron oxide nanoparticles (Fe3O4@SiO2@PPh3@Cr2O7-), Solid State Science 28 (2014) 9–13.

[155] S. Amiri, et al., The role of cobalt ferrite magnetic nanoparticles in medical science, Mater. Sci. Eng. C 345 (2013) 18–23.
[156] A. Maleki, et al., One-pot three-component synthesis of pyrido [2′,1′:2,3]imidazo [4,5-c]isoquinolines using Fe3O4@SiO2–OSO3H as an efficient heterogeneous nanocatalyst, RSC Adv. 4 (2014) 64169–64173.
[157] X. Yan, et al., Controllable synthesis and characterization of Fe3O4/Au composite nanoparticles, J. Magn. Magn. Mater. 380 (2015) 150–156.
[158] D. Jana, et al., Preparation and properties of various magnetic nanoparticles, Sensors 9 (4) (2009) 2352–2362.
[159] A. Sanchez-Martinez, et al., N-doped TiO2 nanoparticles obtained by a facile coprecipitation method at low temperature, Ceram. Int. 44 (5) (2018) 5273–5283.
[160] Z. Shuyan, et al., Fabrication of cobalt aluminate nanopigments by coprecipitation method in threonine waterborne solution, Dyes Pigm. 151 (2018) 130–139.
[161] B. Philip, et al., Synthesis, characterization, and applications of magnetic nanoparticles featuring polyzwitterionic coatings, Polymers 10 (1) (2018) 91.
[162] H. El Ghandoor, et al., Synthesis and some physical properties of magnetite (Fe3O4) nanoparticles, Int. J. Electrochem. Sci. 7 (2012) 5734–5745.
[163] D. Cao, et al., Investigation on the structures and magnetic properties of carbon or nitrogen doped cobalt ferrite nanoparticles, J. Mater. Chem. C 4 (5) (2016) 951–957.
[164] A. Yahyazadeh, B. Khoshandam, Carbon nanotube synthesis via the catalytic chemical vapor deposition of methane in the presence of iron, molybdenum, and iron–molybdenumalloy thin layer catalysts, Results Phys. 7 (2017) 3826–3837.
[165] Y. Zhang, et al., Synthesis and surface-enhanced raman scattering of ultrathin SnSe2 nanoflakes by chemical vapor deposition, Nanomaterials 8 (7) (2018) 515.
[166] J. You, et al., Synthesis of 2D transition metal dichalcogenides by chemical vapor deposition with controlled layer number and morphology, Nano Convergence 5 (1) (2018) 26.
[167] C. Young, et al., Controlled chemical vapor deposition for synthesis of nanowire arrays of metal–organic frameworks and their thermal conversion to carbon/metal oxide hybrid materials, Chem. Mater. 30 (10) (2018) 3379–3386, doi:10.1021/acs.chemmater.7b04944.
[168] A.M. Alotaibi, et al., Chemical Vapor Deposition of Photocatalytically Active Pure Brookite TiO_2 Thin Films, Chem. Mater. 30 (4) (2018) 1353–1361.
[169] Z. Zhenzhen, et al., Development of a validated HPLC method for the determination of tenofovir disoproxil fumarate using a green enrichment process, Anal. Methods 7 (15) (2015) 6290–6298.
[170] W. Unchista, et al., EXAFS study of cations distribution dependence of magnetic properties in Co1−xZnxFe2O4 nanoparticles prepared by hydrothermal method, Microelectron. Eng. 146 (2015) 68–75.
[171] L. Mingling, et al., Synthesis and characterization of nanosized MnZn ferrites via a modified hydrothermal method, J. Magn. Magn. Mater. 439 (2017) 228–235.
[172] Z. Jie, et al., ZnFe2O4 nanoparticles: synthesis, characterization, and enhanced gas sensing property for acetone, Sens. Actuators B 221 (2015) 55–62.
[173] A. Maqusood, et al., Copper ferrite nanoparticle-induced cytotoxicity and oxidative stress in human breast cancer MCF-7 cells, Colloids Surf. B 142 (2016) 46–54.
[174] X. Siqi, et al., Synthesis EMT-type zeolite by microwave and hydrothermal heating, Microporous Mesoporous Mater. 278 (2019) 54–63.
[175] W. Tao, et al., Hydrothermal preparation of Ag-TiO2-reduced graphene oxide ternary microspheres structure composite for enhancing photocatalytic activity, Physica E 112 (2018) 128–136.
[176] M.B. Muradov, et al., Synthesis and characterization of cobalt sulfide nanoparticles by sonochemical method, Infrared Phys. Technol. 89 (2018) 255–262.
[177] C.U. Okoli, et al., Solvent effect in sonochemical synthesis of metal-alloy nanoparticles for use as electrocatalysts, Ultrason. Sonochem. 41 (2018) 427–434.
[178] P.-K. Mokhtar, et al., A simple sonochemical approach for synthesis of selenium nanostructures and investigation of its light harvesting application, Ultrason. Sonochem. 23 (2015) 246–256.
[179] E. Azwar, et al., Transformation of biomass into carbon nanofiber for supercapacitor application – a review, Int. J. Hydrogen Energy 43 (45) (2018) 20811–20821.
[180] C. Lv, et al., 3D sulfur and nitrogen codoped carbon nanofiber aerogels with optimized electronic structure and enlarged interlayer spacing boost potassium-ion storage, Small 15 (23) (2019) 1900816.

[181] E.T. Adesuji, et al., From nano to macro: hierarchical platinum superstructures synthesized using bicontinuous microemulsion for hydrogen evolution reaction, Electrochimca Acta 354 (2020) 136608.
[182] A.M. Perez-Coronado, et al., Catalytic reduction of bromate over catalysts based on Pd nanoparticles synthesized via water-in-oil microemulsion, Appl. Catal. B 237 (2018) 206–213.
[183] M. Karbasi, et al., Microemulsion-based synthesis of a visible-light-responsive Si-doped TiO2 photocatalyst and its photodegradation efficiency potential, Mater. Chem. Phys. 220 (2018) 374–382.
[184] M. Akbari, et al., Microemulsion based synthesis of promoted Fe–Co/MgO nanocatalyst: Influence of calcination atmosphere on the physicochemical properties, activity and light olefins selectivity for hydrogenation of carbon monoxide, Mater. Chem. Phys. 249 (2020) 123003.
[185] M.Vladimir, et al., Development of magnetic nanoparticles for cancer gene therapy: a comprehensive review, ISRN Nanomater (2013) 646284, https://doi.org/10.1155/2013/646284.
[186] S. Manlika, et al., Synthesis and characterization of Bi0.5(Na1- xKx)0.5TiO3 powders by sol–gel combustion method with glycine fuel, Ceram. Int. 44 (2018) S168–S171.
[187] S.Yu, et al., Fabrication of Nd:YAG transparent ceramics using powders synthesized by citrate sol-gel method, J. Alloys Compd. 772 (2019) 751–759.
[188] H. Shasha, et al., Research on cracking of SiO2 nanofilms prepared by the sol-gel method, Mater. Sci. Semicond. Process. 91 (2019) 181–187.
[189] Xu Heju, et al., Large area MoS2/Si heterojunction-based solar cell through sol-gel method, Mater. Lett. 238 (2019) 13–16.
[190] L.Y. Jun, et al., Comparative study of acid functionization of carbon nanotube via ultrasonic and reflux mechanism, J. Environ. Chem. Eng. 6 (5) (2018) 5889–5896.
[191] A. Deepi, et al., One pot reflux synthesis of reduced graphene oxide decorated with silver/cobalt oxide: a novel nano composite material for high capacitance applications, Ceram. Int. 44 (16) (2018) 20524–20530.
[192] I.U. Din, et al., Carbon nanofiber-based copper/zirconia catalyst for hydrogenation of CO2 to methanol, J. CO_2 Util. 21 (2017) 145–155.
[193] N.K. Awad, et al., A review of TiO2 NTs on Ti metal: Electrochemical synthesis, functionalization and potential use as bone implants, Mater. Sci. Eng. C 76 (2017) 1401–1412.
[194] A. Serrà, E. Vallés, Advanced electrochemical synthesis of multicomponent metallic nanorods and nanowires: Fundamentals and applications, Appl. Mater. Today 12 (2018) 207–234.
[195] A.G. Olabi, et al., Recent progress of graphene based nanomaterials in bioelectrochemical systems, Sci. Total Environ. 749 (141225) (2020).
[196] A.A. Lahcen, A. Amine, Recent advances in electrochemical sensors based on molecularly imprinted polymers and nanomaterials, Electroanalysis 31 (2) (2018) 188–201.
[197] B.A. AlMashrea, et al., Polyaniline coated gold-aryl nanoparticles: electrochemical synthesis and efficiency in methylene blue dye removal, Synth. Met. 269 (2020) 116528.
[198] X. Zhang, et al., Advances in organometallic/organic nanozymes and their applications, Coord. Chem. Rev. 429 (2020) 213652.
[199] H.R. Heydarnezhad, et al., Conducting electroactive polymers via photopolymerization: a review on synthesis and applications, Polym. Plast. Technol. Eng. 57 (11) (2017) 1093–1109.
[200] A.J. Bard, L.R. Faulkner, Electrochemical Methods – Fundamentals and Applications, second ed., Wiley, New York, USA, 2001.
[201] K.L. Klein, et al., Surface characterization and functionalization of carbon nanofibers, J. Appl. Phys. 103 (2008) 061301.
[202] M. Asfaq, et al., Synthesis of PVA-CAP-based biomaterial in situ dispersed with Cu NP and carbon micro-nanofibers for antibiotic drug delivery applications, Biochem. Eng. J. 90 (2014) 79–89.
[203] Asfaq, et al., Copper/zinc bimetal nanoparticles-dispersed carbon nanofibers: a novel potential antibiotic material, Mater. Sci. Eng. C 59 (2016) 938–947.
[204] E. Rand, et al., A carbon nanofiber based biosensor for simultaneous detection of dopamine and serotonin in the presence of ascorbic acid, Biosens. Bioelectron. 42 (2013) 434–438.
[205] I. Tapsoba, et al., SWV determination of glyphosate in Burkina Faso soils using carbon fiber microelectrode, Int. J. Biol. Chem. Sci. 6 (2012) 2211–2220.
[206] C. Gao, et al., The new age of carbon nanotubes: an updated review of functionalized carbon nanotubes in electrochemical sensors, Nanoscale 4 (2012) 1948–1963.

[207] H. Nie, et al., Nonenzymatic electrochemical detection of glucose using well-distributed nickel nanoparticles on straight multi-walled carbon nanotubes, Biosens. Bioelectron. 30 (2011) 28–34.
[208] Y. Tang, et al., Nonenzymatic glucose sensor based on icosahedron AuPd@CuO core shell nanoparticles and MWCNT, Sens. Actuators B 251 (2017) 1096–1103.
[209] J.H.T. Luong, et al., Multiwall carbon nanotube (MWCNT) based electrochemical biosensors for mediatorless detection of putrescine, Electroanalysis 17 (2005) 47–53.
[210] A. Wisitsoraat, et al., Fast cholesterol detection using flow injection microfluidic device with functionalized carbon nanotubes based electrochemical sensor, Biosens. Bioelectron. 26 (2010) 1514–1520.
[211] X. Tan, et al., An amperometric cholesterol biosensor based on multiwalled carbon nanotubes and organically modified sol-gel/chitosan hybrid composite film, Anal. Biochem. 337 (2005) 111–120.
[212] X. Kan, et al., Imprinted electrochemical sensor for dopamine recognition and determination based on a carbon nanotube/polypyrrole film, Electrochimca Acta 63 (2012) 69–75.
[213] J.-L. He, et al., β-Cyclodextrin incorporated carbon nanotube-modified electrode as an electrochemical sensor for rutin, Sens. Actuators B 114 (1) (2006) 94–100.
[214] S.R. Shin, et al., DNA-wrapped single-walled carbon nanotube hybrid fibers for supercapacitors and artificial muscles, Adv. Mater. 20 (3) (2008) 466–470.
[215] K.S. Novoselov, et al., Electric field effect in atomically thin carbon films, Science 306 (2004) 666–669.
[216] T. Torres, Graphene chemistry, Chem. Soc. Rev. 46 (2017) 4385–4386.
[217] X. Bo, et al., Electrochemical sensors and biosensors based on less aggregated graphene, Biosens. Bioelectron. 89 (2017) 167–186.
[218] A.M. Pinto, et al., Graphene-based materials biocompatibility: A review, Colloids Surf. B 111 (2013) 188–202.
[219] K. Wang, et al., Biocompatibility of graphene oxide, Nanoscale Res. Lett. 6 (2011) 8.
[220] Y. Zhang, et al., Graphene: a versatile nanoplatform for biomedical applications, Nanoscale 4 (2012) 3833–3842.
[221] M.E. Foo, S.C.B. Gopinath, Feasibility of graphene in biomedical applications, Biomed. Pharmacother. 94 (2017) 354–361.
[222] K. Kostarelos, K.S. Novoselov, Graphene devices for life, Nat. Nanotechnol. 9 (2014) 744–745.
[223] Y. Lu, et al., Flexible neural electrode array based-on porous graphene for cortical microstimulation and sensing, Sci. Rep. 6 (2016) 33526.
[224] S. Sayyar, et al., Covalently linked biocompatible graphene/polycaprolactone composites for tissue engineering, Carbon 52 (2013) 296–304.
[225] K. Liu, et al., Green and facile synthesis of highly biocompatible graphene nanosheets and its application for cellular imaging and drug delivery, J. Mater. Chem. B 21 (32) (2011) 12034–12040.
[226] J. Choi, et al., Graphene bioelectronics, Biomed. Eng. 3 (4) (2013) 201–208.
[227] H. Li, et al., Electrochemical immunosensor with N-doped graphene-modified electrode for label-free detection of the breast cancer biomarker CA 15-3, Biosens. Bioelectron. 43 (2013) 25–29.
[228] F. Shahzad, et al., Highly sensitive electrochemical sensor based on environmentally friendly biomass-derived sulfur-doped graphene for cancer biomarker detection, Sens. Actuators B 241 (2017) 716–724.
[229] Z.H. Sheng, et al., Electrochemical sensor based on nitrogen doped graphene: simultaneous determination of ascorbic acid, dopamine and uric acid, Biosens. Bioelectron. 34 (1) (2012) 125–131.
[230] A.K. Rengaraj, et al., Electrodeposition of flower-like nickel oxide on CVD-grown graphene to develop an electrochemical non-enzymatic biosensor, J. Mater. Chem. B 3 (2015) 6301–6309.
[231] S. Lin, et al., A flexible and highly sensitive nonenzymatic glucose sensor based on DVD-laser scribed graphene substrate, Biosens. Bioelectron. 110 (2018) 89–96.
[232] C. Chung, Biomedical applications of graphene and graphene oxide, Acc. Chem. Res. 46 (10) (2013) 2211–2224.
[233] L. Tang, et al., Uniform and rich-wrinkled electrophoretic deposited graphene film: a robust electrochemical platform for TNT sensing, Chem. Commun. 46 (2010) 5882–5884.
[234] H. Chang, et al., Graphene fluorescence resonance energy transfer aptasensor for the thrombin detection, Anal. Chem. 82 (2010) 2341–2346.
[235] Y. He, et al., Low background signal platform for the detection of ATP: when a molecular aptamer beacon meets graphene oxide, Biosens. Bioelectron. 29 (2011) 76–81.

[236] J. Wang, Electrochemical glucose biosensors, Chemical Review 108 (2) (2008) 814–825.
[237] I. Gandouzi, et al., A nanocomposite based on reduced graphene and gold nanoparticles for highly sensitive electrochemical detection of Pseudomonas aeruginosa through its virulence factors, Materials 12 (7) (2019) 1180.
[238] S. Rauf, et al., Carboxylic group riched graphene oxide based disposable electrochemical immunosensor for cancer biomarker detection, Anal. Biochem. 545 (2018) 13–19.
[239] K. Khoshnevisan, et al., Nanomaterial based electrochemical sensing of the biomarker serotonin: a comprehensive review, Microchimca Acta 186 (1) (2019) 49.
[240] M. Stoytcheva, et al., Advances in the electrochemical analysis of dopamine, Curr. Anal. Chem. 13 (2) (2017) 89–103.
[241] M. Welsch, M. Perchthaler, Catalyst Support Material and Electrode Fabrication, in: Q. Li, D. Aili, H. Hjuler, J. Jensen. (Eds.), High Temperature Polymer Electrolyte Membrane Fuel Cells, Springer, Cham, 2016.
[242] N. Jha, et al., Pt–Ru/multi-walled carbon nanotubes as electrocatalysts for direct methanol fuel cell, Int. J. Hydrogen Energy 33 (2008) 427–433.
[243] S. Singh, et al., Candle soot-derived carbon nanoparticles: An inexpensive and efficient electrode for microbial fuel cells, Electrochimca Acta 264 (2018) 119–127.
[244] Z. Wang, et al., CuO nanostructures supported on Cu substrate as integrated electrodes for highly reversible lithium storage, Nanoscale 3 (2011) 1618–1623.
[245] A. Yadav, et al., Nickel nanoparticle-doped and steam-modified multiscale structure of carbon micro-nanofibers for hydrogen storage: effects of metal, surface texture and operating conditions, Int. J. Hydrogen Energy 42 (9) (2017) 6104–6117.
[246] A. Yadav, N. Verma, Enhanced hydrogen storage in graphitic carbon micro-nanofibers at moderate temperature and pressure: synergistic interaction of asymmetrically-dispersed nickel-ceria nanoparticles, Int. J. Hydrogen Energy 42 (44) (2017) 27139–27153.
[247] A. Modi, et al., In situ nitrogen-doping of nickel nanoparticle-dispersed carbon nanofiber-based electrodes: its positive effects on the performance of a microbial fuel cell, Electrochimca Acta 190 (2016) 620–627.
[248] L. Kinner, et al., Polymer interlayers on flexible PET substrates enabling ultra-high performance, ITO-free dielectric/metal/dielectric transparent electrode, Material Design 168 (2019) 107663.
[249] E.C. Webb, Enzyme Nomenclature 1992: Recommendations of the Nomenclature Committee of the International Union of Biochemistry and Molecular Biology on the Nomenclature and Classification of Enzymes, sixth edition, Academic Press, San Diego, California, 1992.
[250] S. Karra, et al., Morphology of gold nanoparticles and electrocatalysis of glucose oxidation, Electrochemica Acta 218 (2016) 8–14.
[251] K. Dhara, et al., Single step synthesis of Au–CuO nanoparticles decorated reduced graphene oxide for high performance disposable nonenzymatic glucose sensor, J. Electroanal. Chem. 743 (2015) 1–9.
[252] M. Li, et al., Electrodeposition of nickel oxide and platinum nanoparticles on electrochemically reduced graphene oxide film as a nonenzymatic glucose sensor, Sens. Actuators B 192 (2014) 261–268.
[253] D. Xu, et al., Design and fabrication of Ag-CuO nanoparticles on reduced graphene oxide for nonenzymatic detection of glucose, Sens. Actuators B 265 (2018) 435–442.
[254] Y. Li, et al., A nonenzymatic cholesterol sensor constructed by using porous tubular silver nanoparticles, Biosens. Bioelectron. 25 (2010) 2356–2360.
[255] J. Yang, et al., Non-enzymatic cholesterol sensor based on spontaneous deposition of platinum NP on layer-by-layer assembled CNT thin film, Sens. Actuators B 171–172 (2012) 374–379.
[256] G. Favaro, Determination of biogenic amines in fresh and processed meat by ion chromatography and integrated pulsed amperometric detection on Au electrode, Food Chem. 105 (2007) 1652–1658.
[257] A.K. Rengaraj, Electrodeposition of flower-like nickel oxide on CVD-grown graphene to develop an electrochemical non-enzymatic biosensor, J. Mater. Chem. B 3 (2015) 6301–6309.
[258] L. Shabnam, et al., Nonenzymatic multispecies sensor based on Cu-Ni nanoparticle dispersion on doped graphene, Electrochimca Acta 224 (2017) 295–305.
[259] A. Arvinte, et al., Comparative electrodeposition of Ni–Co nanoparticles on carbon materials and their efficiency in electrochemical oxidation of glucose, J. Appl. Electrochem. 46 (2016) 425–439.
[260] M. Li, et al., Facile synthesis of ultrafine Co3O4 nanocrystals embedded carbon matrices with specific skeletal structures as efficient non-enzymatic glucose sensors, Anal. Chim. Acta 861 (2015) 25–35.

[261] A. Benvidi, et al., Development of a carbon paste electrode modified with reduced graphene oxide and an imidazole derivative for simultaneous determination of biological species of N-acetyl-L-cysteine, uric acid and dopamine, Electroanalysis 28 (7) (2016) 1625–1633.
[262] P. Goswami, et al., A novel non-enzymatic sensing probe for detection of cholesterol in solution, J. Chem. Pharm. Res. 6 (1) (2014) 603–610.
[263] A. Silber, et al., Poly(methylene blue)-modified thick-film gold electrodes for the electrocatalytic oxidation of NADH and their application in glucose biosensors, Biosens. Bioelectron. 11 (1996) 215–223.
[264] K. Reddaiah, et al., Electrochemical investigation of L-dopa and simultaneous resolution in the presence of uric acid and ascorbic acid at a poly (methyl orange) film coated electrode: a voltammetric study, J. Electroanal. Chem. 682 (2012) 164–171.
[265] M.B. Wayu, et al., Electropolymerization of β-cyclodextrin onto multi-walled carbon nanotube composite films for enhanced selective detection of uric acid, J. Electroanal. Chem. 783 (2016) 192–200.
[266] Q. Wang, et al., Nonenzymatic hydrogen peroxide sensor based on a polyaniline-single walled carbon nanotubes composite in a room temperature ionic liquid, Microchim. Acta 167 (2009) 153.

CHAPTER 14

Emerging strategies in nanotheranostics: A paradigm shift

Jnana R. Sahu[a], Tejaswini Sahoo[a], Anulipsa Priyadarshini[a], Deepak Senapati[a], Debasis Bisoi[b], Sabyasachi Parida[b], Goutam Ghosh[c], Rojalin Sahu[a]
[a]School of Applied Sciences, Kalinga Institute of Industrial Technology, Deemed to be University, Bhubaneswar, Odisha, India
[b]Pradyumna Bal Memorial Hospital, Kalinga Institute of Medical Sciences, Bhubaneswar, Odisha, India
[c]UGC-DAE Consortium for Scientific Research, Mumbai Centre, CFB Building, BARC Campus, Mumbai, India

14.1 Introduction

Theranostics implies the combined treatment modalities, therapeutic and diagnosis [1,2]. This is a synergistic integration of therapy and diagnosis that helps in superior prognosis and enhanced treatment methods for various illnesses including cancer. The main aim of this treatment method is to develop a precise therapeutic strategy toward precision medicine. In this method, the combination of diagnostic and therapeutic properties in single units resulted in a new treatment protocol. A precise and efficient method is developed for the management of diseases based on the diagnosis results. These agents are the materials which combine the diagnostic and therapeutic properties so that the detection and treatment of the disease detection can be achieved in a single platform. Nanotheranostic agents are now developed based on the advancement in nanotechnology [3]. These combine agents primarily depend on the various intrinsic characteristics of nanoparticles. Various types of molecules can be combined on the surface of the nanotheranostic agents that can accomplish multiple purposes. This could be achieved because the surface area of nanoparticles is very high. Again, the surfaces of nanoparticles can be functionalized by anchoring agents to tether the delivery agents for targeted delivery. Here the anchoring agents are basically the stabilizing agents like polyethylene glycol, PEG, and other ligands. The site specific drug delivery is achieved by using the nanoparticles-based theranostics agents because the localization of these materials is significantly easy [4]. Hence, the side effects due to the drug are minimized. Further, these co-delivery agents or nanoparticles based theranostics agents are not easily removed by the kidney owing to their nanometer size so that the circulation time of these materials is extended. Moreover, blood vessels of tumor cells are abnormal and normally leaky. Owing to this characteristic property of tumor cells, these nanotheranostic materials able to escape from the blood flow to this cell and remained there for a longer time due to the poor lymphatic drainage. This is how the tumor cells selectively absorb the nanotheranostic materials or co-delivery agents and it is termed as enhanced

Advanced Nanomaterials for Point of Care Diagnosis and Therapy
DOI: https://doi.org/10.1016/B978-0-323-85725-3.00013-1

retention effect. The key challenges in the drug delivery are to reducing toxicity and competitive targeting the tissues affected by the disease. A proper targeting molecule enhances the agglomeration of the therapeutic molecule in the diseased tissues. Hence, it increases the concentration of the therapeutic molecule locally and also it raises the imaging contrast [5]. The beauty of this co-delivery agent, i.e, the nanoparticle based imaging and therapy treatment method adds to the earlier paradigm for enabling diagnosis to be done neither after nor before, rather during the treatment method. It is well known that various nanoparticles, which are already used as imaging agents, can be used as nanotheranostics by encapsulating therapeutic agents on their surfaces. The essential requirement of this combined therapy necessitates adequate accumulation of these nanotheranostics in the infected areas. Here, two research domains are fused, diagnosis and therapy. Design of the targeting agent can be modified as per the requirement. For example, in case of cancer, a suitable biomarker is identified this is expressed on the cancer cells and is loaded on the surface of the nanoparticle to recognize the cancer cells.

14.2 Nanotheranostics

Nanotheranostics are in the budding stage and struggling for the clinical trials. Moreover, progress in the field of nanotechnology and personalized medicines have made this a burning research area. In this chapter, we have made sincere efforts to organize the theranostic materials based on their core nonmaterial. The nanomaterials already used for diagnostic i.e. imaging and therapeutic purpose are listed here with applications in Tables 14.1 and 14.2, respectively. The nanoparticles used for diagnosis (imaging) with applications is illustrated in Table 14.1. The application of nanotheranostics is presented in the Fig. 14.1.

The recent developments in this research are compiled in this chapter. From the previous reports, it is found that optical and MRI is the preferred diagnosis modalities in these treatment methods. These diagnosis modalities are integrated with the therapeutic agents on the surface of the nanoparticles to form the nanotheranostics. The schematic representation of theranostics function is illustrated in Fig. 14.2.

14.3 Metal-based nanomaterials

Metal based nanoparticles, especially the noble metal nanoparticles have been utilized for biomedical applications. These nanoparticles are used for the diagnosis and therapy of cancer, radiotherapy, drug delivery, bacterial treatment, etc [42].

14.3.1 Gold-based nanomaterials

Gold based nanoparticles are extensively studied as multifunctional probes owing to their extraordinary biocompatibility nature. These nanoparticles are also well studied for

Table 14.1 Nanoparticles used for diagnosis (imaging) with applications.

Types of imaging	Particle size	NPs use	Applications	References
Magnetic resonance imaging (MRI)	Milli to micromolar	Iron oxide, gadolinium	Inflammation or infection in an organ, breast cancer	[6,7]
Computed tomography (CT)	Millimolar	Barium-based nanoparticles, gold based nanoparticles, iodine based micelles and liposomes	In vivo PET-CT imaging	[8–13]
Ultrasound (US)	Single MB detectable	Nanobubbles, air-releasing polymers	Visualization of subcutaneous body structures including tendons, muscles, joints, vessels and internal organs	[14–16]
Optical imaging (OI)	Nanomolar	Quantum dots, fluorescent nanoparticle probes	In vivo multimodal imaging, In vivo fluorescence imaging (FLI) of the reticuloendothelial system	[17–20]
Photoacoustic imaging (PAI)	Milli to nanomolar	Gold nanoparticles, gold nanorods, carbon nanotubes	Breast imaging, dermatologic imaging, vascular imaging, carotid artery imaging, musculoskeletal imaging, gastrointestinal imaging	[21–23]
Positron emission tomography (PET)	Picomolar	Liposomes	Targeted delivery and imaging with bioconjugated ^{64}Cu-BATPEG-liposome	[24]
Single photon emission computed tomography (SPECT)	Picomolar	Liposomes	Clinical biodistribution and also imaging of breast, head and neck, glioma and lung cancers in C-26 tumor-bearing BALB/cByJ mice	[25–27]
γ-Camera	Millimolar	Iron oxide, liposomes, immunoliposome	Tumor angiogenesis	[28,29]

Table 14.2 The nanoparticles used for therapeutic applications.

Types of therapy	Nanoparticles	Applications	References
Photothermal therapy (PTT)	CdTe and CdSe QDs/silica shell	Skin cancer treatment	[30]
Photodynamic therapy (PDT)	Gold NPs, Human serum Albumin (HAS)	Colon cancer treatment	[31,32]
NIR therapy	Pluronic NPs	Liver cancer treatment	[33]
Radiochemotherapy	Liposomes	Imaging, biodistribution, pharmacokinetics, therapeutic efficacy	[34–37]
RNA interference (RNAi) therapy	Gold NPS	Cancer treatment	[38]
Radio therapy	Liposomes	Radionuclide therapy in a head and neck squamous cell carcinoma Xenograft positive surgical margin mode	[39,40]
Photoacoustic therapy	PL-PEG	Breast tumor	[41]

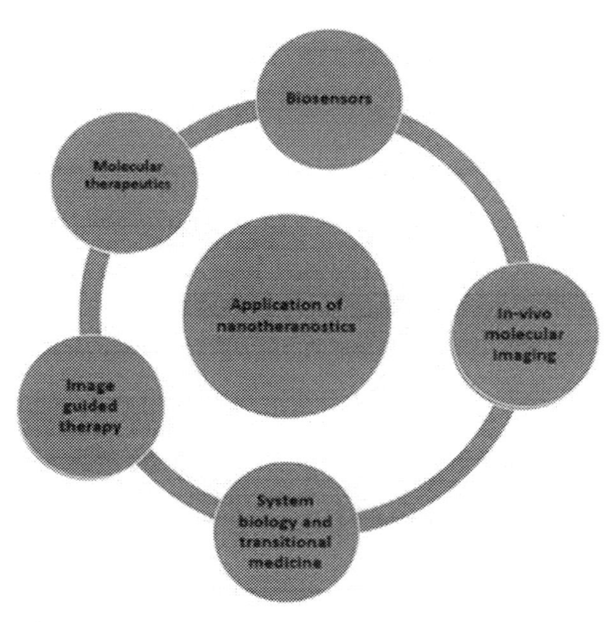

Fig. 14.1 ***Different applications nanotheranostics.***

Fig. 14.2 ***Schematic representation of theranostics function.***

strategic modification of their surface. The exclusive optical and photothermal properties of gold nanoparticles allow them to be used as sensing materials and the photothermal property enables them to be used for therapeutic purposes. Gold nanoparticles are well known for its excellent Local Surface Plasmon Resonance (LSPR) [43]. Further, this LSPR can be adjusted by tuning the morphologies of gold based nanoparticles such as nanorod, nanocage, nanoparticles, and nanoshell and these nanoparticles with different morphologies show different optical and thermal characteristics [44]. These are the unique properties of gold based nanoparticles to make gold based nanoparticles as the prospective theranostics materials. Gold based nonmaterial offers a wide platform for codelivery of therapeutic and diagnostic agents on its surface by utilizing covalent bonding between gold and sulfur. The delivery of the drugs can be released intracellular from the gold based theranostics agents by exchanging reactions with glutathione present in the cytosolic. Glutathione is plentily available in the cell. From the reported literature, it is known that the nanoparticles surface can be customized with various ligands like PEG, β-cyclodextrin, biotin and paclitaxel to be useful for theranostic application [45]. These theranostic agents are used in the treatment in cancer. Regular gold nanoparticles were rarely utilized as imaging materials. The high atomic number of gold and X-ray attenuation coefficient as compared to Iodine. There are some reports on the study of gold nanoparticles used as contract agents in computed tomography (CT) imaging technology [46]. A novel gold nanoparticle is used for targeted CT imaging and therapy of prostate cancer. Here, prostate specific membrane antigen (PSMA) is functionalized on the surface of the gold nanoparticle. Lymph Node Carcinoma of the Prostate (LNCap) are the human cells used in the study of oncology [47]. It is also confirmed from the previous studies that, PSMA conjugated gold nanoparticles have shown greater CT intensity toward Lymph Node Carcinoma of the Prostate (LNCap) cells when compared with non targeted prostate cancer cells (PC3). The anti cancer drug Doxorubicin (DOX) has shown more efficacy against LNCap as compared to PC3 cells. The gold nanoparticle is a very good contrast agent in CT imaging due to its high X-ray absorption coefficient. Gold nanomaterials are good radiotherapy sensitizer due to their high X-ray absorption coefficient. The big challenge for radiation therapy is to cure the disease without major cause of morbidity. Precise radiation treatment can be achieved by targeted release of radiotherapy sensitizer to the cancer cells [48]. By using this targeted drug delivery treatment method, the radiation therapy treatment will be more successful without affecting the nearby healthy tissues. Gold nanorod modified with silica and conjugated with folic acid has been used as an efficient agent for X-ray radiotherapy and photothermal therapy [49]. Further, these multifunctional nanoprobes were developed to acquire many functions for targeting tumor cells. These probes are very helpful in targeting, imaging and for selective therapy of tumor tissues without harming the healthy tissue.

Gold nanoparticles based on transient photothermal vapor nanobubbles, known as plasmonic nanobubbles (PNB) are used as contrast agents in imaging [50]. Intracellular PNB can be optically formed and tuned via laser application [51]. Further, some researchers have reported that PNB is an efficient in vivo theranostic agent and is confirmed in case prostate cancer of zebrafish. A unique nanotheranostics agent based on gold nanomaterials has been developed which was sensitive to metalloprotease, (MMP-AuNR) for the diagnosis and treatment of cancer [52]. Matrix metallopeptidases (MMP) are also known as matrix metallo proteins are metal based, Zinc dependent proteins and are associated with the metabolism of cancer. Imaging by using MMP could give important information about the diagnosis of cancer. Detection and monitoring of cancer can be achieved by conjugating Cy5.5 peptide complex and a fluorescent probe on the surface of the gold nanorod. The photothermal property induced by the MMP conjugated gold nanorods has raised the temperature up to 60°C resulting in the destruction of the tumor tissues. The encapsulated drug molecules released from the gold nanocages by applying NIR laser. The temperature is raised when the NIR laser is focused on the tumor cells. Further, high intensity ultrasound can be used to deliver the encapsulated drugs enabling a novel theranostic systems utilizing gold nanomaterials.

14.3.2 Magnetic nanomaterials

Theranostics based on magnetic nanoparticles is superior because magnetic nanoparticles are biocompatible and economical. The magnetic characteristic helps as contrast in MRI imaging. Iron oxide nanoparticles are the important magnetic material which exhibit super magnetic properties even at room temperature. It is the well studied magnetic nanoparticle among all the magnetic nanoparticles. These iron oxide nanoparticles are generally called super paramagnetic iron oxide nanoparticles (SPIONs). These super paramagnetic iron oxide nanoparticles are immensely studied nanoparticles for their biomedical applications [53]. When these nanoparticles are of less than or greater than 50 nm are called Ultra Small Iron Oxides Nanoparticles (USPIONs). Polymers containing polar groups in its polymer chain, hydrophilic are added to passivate the surface of the nanoparticles which restrict the aggregation of the nanoparticles. These polymers which are used to increase the stability of the nanoparticles are dendrimer, PVP, polyaniline and dextran. Dextran and the derivatives of dextran are the most studied polymers in this regard [54]. Nanoparticles coated with polymers facilitate the drug encapsulation and help in covalent conjugation of other moieties. Poly acrylic acid is used to coat the surface of the nanoparticles to encapsulate anticancer drug taxol and a NIR dye to be used as nanotheranostic [55]. Here it helps in monitoring the delivery of the drug on the basis of magnetic resonance imaging and fluorescence. Folate conjugated Poly acrylic acid iron oxide and again conjugated with folate nanoparticles has been used for killing the cancer cells and diagnosed by optical or magnetic resonance based imaging. Thermosensitive nano vehicles were reported to be coated with different polymers with a number of functionalities. A temperature sensitive polymer, *N*-isopropylacrylamide (NIPAAM) has been used to facilitate the conjugation on the surface of the iron oxide magnetic nano material [56]. This polymer also helps to tune the temperature due to its hydrophilic character. Polyethylene glycol methacrylate has been used as a coating material on the surface of the magnetic nano materials. The OH groups present in the polymer helps in the coupling with folic acid and also increases the circulation time. The polymeric nanocomposites with the size of approximately 200 nm have shown darker T2 relaxation than the control. Precise release of the anticancer drug, Doxorubicin (DOX) has been achieved by using *N*-isopropylacrylamide (NIPAAM) owing to its thermoresponsive character of the polymer, NIPAAM. DOX is released at a temperature of 72% ± 5.25% in 48 h which is higher than the normal physiological condition. The hydrophobic and hydrophilic characters are present in the amphiphilic polymers. Amphiphilic polymers are used as coating materials on the surface of magnetic nanomaterials so that different polymers can avoid getting the hydrophobic and hydrophilic properties which are required during drug delivery. The anti cancer drug currcumin has been delivered by using a polymer, Pluronic F127 [57]. This polymer is composed of a polyoxypropylene and joined by polyoxyethylene further coated with β-cyclodextrin. This copolymer, Pluronic F127 coated nanoparticle is a very efficient material for the encapsulation of the drug curcumin. The optimized formula of the polymer, F127250 has the benefits like, its smaller size, high drug loading ability, lower binding affinity toward proteins and excellent uptake capacity in the carcinoma cells. This polymer does not affect the magnetic property of the nanoparticles. Delayed contrast property in MRI with effective loading of the anti cancer drugs has been achieved using the polymer, Pluronic F127. Near Infrared Dyes has been studied to localize the tumor in mice model without using any magnetic field. Further, these magnetic nanoparticles are slowly accumulated on the tumor tissues and reach the maximum at 48 h and then gradually decrease over 11 days after the injection of these dyes to the mice. pH-sensitive anti cancer drug delivery has been achieved by using a nano carrier modified with HER-2, the receptors present in breast cells. Further, this helped in effective management of cancer by simultaneous delivery of the drugs with imaging. α-pyrenyl-ω-carboxyl poly ethylene glycol is used for the encapsulation of $MnFe_2O_4$ magnetic nano material and the anticancer drug Doxorubicin. The drug, Doxorubicin has been released quickly due to the weak non covalent π–π interaction between the drug and the pyrene groups resulting in the suppression of the growth of tumor. The growth of the tumor is controlled by the therapeutic properties of the anticancer drug Doxorubicin as well as the antibody. The weak van der Waals interactions such as electrostatic interaction have been used to conjugate the drug molecule onto the surface of the magnetic nano materials for its theranostic uses. β-galactosidase has been conjugated as the loading material onto the heparin coated magnetic nano material, iron oxide. This nanocomposite is very efficient for delivering the loaded β-galactosidase into the brain tumor in a very selective manner which is studied in a mice model. This method has been stated to be applicable in various protein therapeutics. There are some reports on the codelivery of various drugs using the same nano carrier. Magnetic nano materials conjugated with polylactide-co-glycolide matrices (PLGA-MNPs) have been used as twofold delivery systems for drugs as well as the imaging material [58]. This nanocomposite has been proven one of the better contrast agents as compared to the commercially available contrast agents. This nanocomposite is very efficient in encapsulating both the hydrophobic and hydrophilic anticancer drugs. The hydrophilic drug, carboplatin and the hydrophobic drug, paclitaxel are successfully delivered by using this nano composite. Magnetic nano materials functionalized with grafted agents making available for coupling with covalent bonding are reported. These grafted moieties help in conjugation with drug molecules as well as the targeting molecules for theranostic purposes. Superparamagnetic iron oxide nanoparticles (SPIONs) modified with *N*-phosphonomethyl iminodiacetic acid has been used for site specific drug delivery [59]. This nanocomposite is hydrophilic, biodegradable and biocompatible. This nanocomposite is very efficient in delivering drugs in the intracellular region. Here the drug release is based on the pH. The ester linkages facilitate the release of the anticancer drug in the acidic compartment of the cancer cells. The delivery of the drugs is controlled through the magnetically guided magnetic nanoparticles. A nano composite is synthesized by loading the anticancer drug paclitaxel on the surface of the magnetic nanoparticles which are functionalized with ethyl(dimethylaminopropyl) carbodiimide/*N*-hydroxysuccinimide. This is proven as an active nanotherapy method with lesser IC50 value. Further, these nanocomposites have shown a very long period of circulation in the blood serum resulting in the higher therapeutic value as well as the bioavailability of the drugs. A SPION conjugated with quantum dots of CdS:Mn/ZnS has been designed as nanotheranostics [60]. These nano composites are functionalized with STAT3 inhibitor, PEG, vitamins, anticancer drugs to be used as nano vehicles to deliver therapeutics as well as the imaging materials. The drug uptake and the delivery of the drugs are monitored by the luminescent quantum dots. The cytosol which contains the glutathione where the nanocomposite is exposed and it is the "on" state and the "off" state is the quenching of the luminescent of the quantum dots when conjugated with ligands rich in electrons. This is further validated by MRI imaging.

Photodynamic therapy (PDT) combines treatment modality which contains a drug which destroys the cancer cells after exposure of laser light of a particular wavelength. These therapeutics are the photosensitizers which are activated upon exposure to light of certain wavelengths. The cancer cells are destroyed by the reactive oxygen species (ROS). The oxygen atoms get energy from the activated photosensitizers, i.e. the therapeutic molecules. The main drawbacks of these photosensitizers are their poor solubility in aqueous medium and poor selectivity. These drawbacks were taken care by designing second generation sensitisers. These second generation sensitisers, the photosensitizers are combined with the nanomaterial. This photosensitizers are combined with the magnetic nanoparticle using silane as the coupling agent, which results the nanotheranostics. These nanocomposites have shown good solubility in water, noncytotoxic and biocompatible which make these efficient in photodynamic therapy. Further, these nanocomposites were very effective in treating cancer in vivo. Effective treatment of cancer has been demonstrated by using photosensitizer nanocomposite. The efficacy of hyperthermia is highly dependent on the formulation of the nano composite. Iron cobalt magnetic nanoparticles and graphite for the site specific delivery to tumor cells based on hyperthermia therapy. Fe_3O_4 magnetic nanoparticles have been utilized for hyperthermia therapy. The temperature has been raised by 47°C by applying radiofrequency at 25 kW. In Ehrlich tumors, monitoring of the destruction of cancer cells are done by the presence of the intensity in ($T1$)-weighted images. Iron oxide magnetic nanoparticles coated with sugar have been synthesized to be used as negative contrast agents for MRI [61]. The average size of the magnetic nanoparticles was 4 to 35 nm. The surface of the nanoparticles was coated with organic molecules with phosphate groups and sugars. The sugar groups were having ribose, mannose and rhamnose moieties. These nanocomposites were hydrophilic in nature resulting the formations of colloids. These colloids were very stable for a long time. These nanocomposites have shown better T2 relaxation as compared to the commercially available products. Further, these composites release very large amount of heat upon exposure to radiofrequency in electromagnetic radiation. The applied radiofrequency was having frequency and amplitude very close to the tolerance limit of human being. Iron oxide nanoparticles are the most studied magnetic nano materials in clinical diagnosis and for therapeutic application. The supermagnetism displayed by the magnetic iron oxide nanoparticles enables these nanoparticles to be very useful for biomedical application.

14.4 Polymeric nanomaterials

Polymeric nanomaterials are always well known for their excellent properties like biodegradability, biocompatibility, and structural versatility [62,63]. Polymeric nanoparticles synthesized using synthetic or natural polymers provide a wide scope in drug delivery with ideal encapsulation property. Numerous methods are used for the preparation of polymeric nanoparticles like polymerization, solvent diffusion, spontaneous emulsification, or solvent evaporation. Several formulations and process parameters that affect release profile, stability and size of the polymeric nanoparticles are composition of the polymer, molecular weight, drug-to-polymer ratio, solubility of drug, etc. Polymer nanoparticles have emerged as a prefect theranostic platform for the diagnosis and treatment of cancer disease. Polymers such as poly (ε-caprolactone), poly(D,L-lactic acid), polyethylene glycol (PEG) and poly (D,L-glycolic acid), are already being used clinically. In general, polymer based nanomaterial which is used for the theranostic purpose have the following components:

(i) Stability and biocompatibility
(ii) Therapeutic molecule such as drugs
(iii) Imaging material which will be used for MRI, PET, CT etc

Sometimes, the therapeutic molecules simultaneously used as therapeutic as well as imaging agent as in the case of anticancer drug, Doxrubicin [64]. Clinically used imaging materials are integrated with therapeutic molecules in the polymeric nanomaterials for theranostics uses [65]. Presently, radiopharmaceuticals are used for theranostics purposes. Here, the imaging techniques are mainly PET or SPECT, magnetic resonance imaging (MRI) contrast agents, nano/microbubbles for ultrasound imaging and fluorescent agents for fluorescent imaging. These modalities have their boon and bane and are used for desired outcome to be obtained. Polymer nanoparticles as nano-carriers of various types, including dendrimers [66-68], micelles [69-71] and polymer conjugates [72,73], have been developed to deliver the drugs to cancerous sites and have shown excellent efficacy in multiple types of cancers. Polymeric nanoparticle cores are loaded with different types of imaging or therapeutic agents. They provide continuous and controlled release of these agents resulting from surface or bulk erosion, stimulation by the local environment, or diffusion through the polymer matrix [74]. Contrast agents like gadolinium (Gd) and superparamagnetic iron oxide (SPIO) are most often used for MRI, and polymeric nanoparticles are found to be very useful carriers for both of these contrast agents [75-77]. A successfully developed prototype on a multifunctional polymeric micelle system has been reported by Guthi and coworkers [78]. In this system, SPIOs and DOX were encapsulated in the micelle core for MR imaging and therapeutic delivery for lung cancer, respectively. Methoxy-terminated PEG-co-poly(d-, l-lactic acid), maleimide-terminated PEG-co-poly(D-L-lactic acid) and amphiphilic block copolymers were synthesized as the micelle core and incorporated with SPIOs for ultra-sensitive MR detection. Gadolinium compounds are also used to formulate polymeric nanoparticles based theranostics for tumor accumulation and real-time visualization of blood circulation. From the different studies by Ye et al., It was found that conjugates of gadolinium- (1,4,7,10-tetraazacyclododecane-1,4,7-triacetic acid) (Gd-DO3A) and poly(L-glutamic acid) (PGA) with higher molecular weights shown strong MR signal [79]. These conjugates also exhibited higher tumor accumulation and prolonged blood circulation. Imaging based on radionuclides are very sensitive and efficient. Important radionuclides which are used in this imaging are ^{99m}Tc, ^{177}Lu, ^{111}In, etc with multiple copolymers to formulate robust nanosized delivery system [80-82]. Mitra et al. have conjugated ^{90}Y and ^{99m}Tc to N-(2- hydroxypropyl) methacrylamide and it was further integrated with a peptide. Polymeric nanoparticles integrated with fluorescence imaging techniques allow researchers to monitor drug pharmacokinetics, real time drug tumor accumulation, and intratumoral drug distribution. Peng and coworkers have developed multi-functional polymeric nanoparticles composed of PEG-polycaprolactone (PCL) di-block copolymer to load a NIR fluorescent dye named IR-780 for both NIR imaging and photodynamic therapy [83]. Gao et al. has designed a model for the in vivo study of breast cancer by utilizing anticancer drug, Doxorubicin conjugated micelles (composed of PEG-poly(L-lactic acid) and PEG-polycaprolactone block copolymers and encapsulated in DOX) and DOX-loaded, PFP encapsulating nanobubbles [84]. Generally, nanobubbles and microbubbles are used as contrast agents in ultrasound imaging. Presently, an imaging technique known as Photoacoustic has been used in the theranostic approach. This method facilitates the high-performance molecular imaging studies to be visualized. Pu et al. have recently reported the use of near-infrared light absorbing semiconducting polymer nanoparticles as new types of contrast agents for photo acoustic imaging [85]. These nanoparticles provide real-time in vivo imaging and produce a stronger signal. Natural polymeric material chitosan has also been widely used as nanovehicles of thepeutic and imaging agents.

Theranostic chitosan nanoparticles have been developed which carry Cy5.5 for imaging during the treatment of cancer. Here the anticancer drug is paclitaxel [86]. Further, glycol chitosan nanoparticles have been modified with hydrophobic 5β-cholanic acid. This modified nanoplatform is delivered for the encapsulation of hydrophobic paclitaxel.

14.5 Silica-based nonmaterial

Silica based nano materials are well known for their high capacity for high drug loading capacity owing to their very large surface area, high mechanical, thermal and chemical stability, easier surface modification and excellent biocompatibility. Honeycomb like pores are present in these nanoparticles. This unique pores present in the silica based nanoparticles allows an excellent control on the drug delivery. It has the capacity to close the pore outlets with various gatekeepers which are sensitive to internal as well as external stimuli. The drug delivery can be controlled by designing different gatekeepers anchored on the pore outlets [87]. These can be developed through the responsive polymer shells so that the drug release can be controlled. The release can be allowed or avoided with respect to a particular internal or external stimulus. The internal stimulus are enzymes, pH and external stimulus are temperature, magnetic fields and light etc.

Various materials as gatekeepers can be designed and developed for this purpose. The surface of the silica nanoparticles can be functionalized with different molecules having different functional groups like carboxylic groups, thiols and amino acids. The surface fictionalization enhances the drug loading capacity of these nano materials. Different synthetic methods have been used to produce different silica based nanoparticles with different morphologies. The pore diameter and volumes are also controlled by the synthetic methods. The silica based nanoparticles with diversified morphology can load various types of therapeutic molecules in the pores of the nanomaterials. The pore network can be tuned according to the size of the therapeutic molecules. These silica based nano platforms are excellent nano vehicles for delivery of drug to the cancer cells. These materials have shown potential in precise and selective drug delivery. Development of a nano composite using phosphonate containing silica matrix encapsulating methylene blue dye has been used for photodynamic therapy [88]. Here the dye molecule has been used as the photosensitizer molecule. The methylene blue dye encapsulated phosphonate containing silica nanomaterial has been exposed to a light of wavelength, 635 nm generated reactive oxygen resulting in the destruction of Hela cells. The methylene blue dye encapsulated phosphonate containing silica matrix has been used for site specific drug delivery. The luminescence emitted by this nanocomposite provides the imaging part for this drug delivery approach by using this nano formulation. Mesoporous silica nanoparticles encapsulating dye molecules in its pores has been synthesized and are called hematoporphyrin. These mesoporous silica based nano particles has been used as the vehicle for the photosensitizing molecules and helps in the photo oxidation-reduction reaction. Development of trifunctionalized mesoporous silica nanoparticles has been used successfully for theranostic applications [89]. In this single unit, imaging, therapy and targeting has been achieved for the drug delivery application. This nano vehicle conjugated with cRGDyK peptides to modify the surface of the mesoporous silica nanoparticles. This nano platform has been proved as an excellent material to target $\alpha v\beta 3$ integrins of U87-MG of human glioblastoma cells. Further, this nanocomposite has shown high therapeutic properties. These silica based nonmaterials provide space to store and release of the drugs because these materials are having very high surface area. This nanoplatform has been proven for adequate supply of the therapeutic molecules. Biocompatibility of fluorescent mesoporous silica nanoparticles has been studied in mice model [90]. This is also used to study the human cancer xenograft using mass spectroscopy and fluorescent microscopy for in vivo imaging. Suppress of tumor growth has been confirmed from the study. Further, the researchers also reported that the fluorescent mesoporous silica nanoparticles has been well established in mice model and these nanocomposites are highly concentrated in the tumor tissues and cancer cells are destroyed with folic acid bonded fluorescent mesoporous silica nanoparticles. A highly versatile silica based nanocomposite has been developed by modifying the surface of the dye encapsulated nanoparticle with magnetic nano materials. These magnetic nano materials help in diagnosis or in magnetic resonance imaging. This resulted in the improvement of theranostics. A new treatment modality has been developed to overcome the multiple drug resistance in the tumor cells. Accurate delivery of the anticancer drug has been achieved in the drug-resistant cancer cells. Polyethyleneimine-modified mesoporous silica nanoparticles loaded with the anticancer drug, Doxorubicin and siRNA have been used as an effective drug delivery nano vehicle. Here, the cell death has been achieved in a synergistic fashion.

14.6 Carbon nanomaterials

The carbon based nanomaterials are categorized into three different dimesions. These are zero dimensional, one dimensional and two dimensional based on their bonding structures. Example of zero dimensional carbon nanomaterial is carbon dots; one dimensional is carbon naotubes and two dimensional is graphene. These nanomaterials have extraordinary properties like physical, optical, chemical, electrical, mechanical and biological. Carbon dots synthesized by green approach have good aqueous stability, low cytotoxicity and bright luminescence [91]. Thus have great potential in biomedical applications such as imaging and detection. Du et al. have studied about properties of carbon quantum dots synthesized from bagasse using hydrothermal carbonization [92]. They are useful in bioimaging and biolabelling of cancer cells. 1D carbon nanotubes (CNTs) have single (SWNTs) and multilayered (MWNTs) cylindrical structures. CNTs are used for imaging, gene therapy and drug delivery. Ligand-functionalized SWNTs are good carriers for the continuous delivery of anticancer drugs, genes, and proteins to the target. Robinson and coworkers were the first to demonstrate the dual use of PEGylated phospholipid functional single-walled carbon nanotube in NIR imaging and photothermal therapy. The in vivo and in vitro detection and drug delivery were investigated by Liu et al. by loading DOX onto CNTs [93]. As a theranostic agent, SWNT is a viable candidate because of its strong optical absorption in the NIR region for imaging of tumor cells. Further, SWNT behave as an aborsber in the NIR region of the electromagnetic spectrum for photo thermal ablation of tumors with low injected dose. Elimination of tumor in an effective way was obtained using this system as compared to Au nanoparticles. Fullerene (C60) has the antioxidant property which makes it suitable for biomedical applications. However, functionalized C60 is used as an effective nanotheranostic agent [94]. It is used in neuroprotection, apoptosis, photodynamic therapy, drug and gene delivery, and MRI. Graphene in the NIR region has been revealed as a photovoltaic transducer because of its high optical absorption. According to a study conducted by Yang et al., PEGylated nanographene sheets (NGS) showed greater proficient for targeting tumor cells [95]. This PEGylated NGS was used further in vivo photo-thermal therapy showing better results. From the study, it was found that PEGylated NGS has no side effects.

14.7 Composite nonmaterial

Design and the development of nanomaterials with multiple functions like diagnosis, therapy and targeting are very crucial. These nanomaterials with multiple functionalities are estimated to beat the limitations related with conventional treatment of cancer. But, most of the nanomaterials do not contain all the characteristics to be used as the multifunctional material for the treatment of cancer. Sometimes they have one of the three functionalities which can be used either for diagnosis or therapy. A new category of nanomaterials known as hybrid materials which consists of different nanomaterials have shown as an excellent nano platform in theranostics. These are the potential nano materials for diagnosis, therapeutic and imaging applications. Gold coated iron oxide magnetic nanoparticles (nanoroses) of 20-50 nm sizes have been utilized in photothermal therapy. Further, the authors claimed that the uptake capacity of the macrophages has been improved when the nano formulation was coated with dextran. The dextran coated nanorods have produced intense near infrared contrast in hyperspectral microscopy. This is studied in vivo and in vitro rabbit models of atherosclerosis. They have also mentioned that these nanoroses with multiple functionalities in imaging, magnetic and therapeutic applications provide a huge opportunity in site specific drug delivery. Superparamagnetic iron oxide has been coated with silica and a layer of gold to form the composite nanomaterial, SPIO@AuNS and used to combine photodynamic therapy and MRI imaging [96]. This nano formulation has shown an absorption in the near infrared region which is very important for photothermal ablation of utilizing a laser. This nanocomposite has been studied in a mice model. A novel photothermal nano composite was synthesized by using gold nanorods conjugated with magnetic nanomaterials and the surface was modified with mesoporous silica. This nano formulation offered multiple functions such as MRI for imaging. Photothermal therapy and the high surface area provided enough space for drug loading. This unique nano formulation has the MRI, thermal, photothermal therapy, chemotherapy, optical imaging in one system. The drug delivery has been studied by this nanocomposite using the encapsulated anticancer drug Doxorubicin. This shows drug delivery in acidic condition, i.e. pH dependent drug release within 5 h. Rates of inhibition of tumor cells were observed by the synergistic effect of photothermal therapy and chemotherapy using the anticancer drug, Doxorubicin. The drug loading capacity of this formulation was very high which is the basic requirement for an ideal nano vehicle for nanotheranostic application. Most of the hybrid nanocomposites which are used in theranostics have utilized mesoporous silica owing to its larger surface area, pore network, controllable pore sizes and easy surface modification. Further, these mesoporous silica based nanostructures are biocompatible. These are used as the drug reservoir for hybrid nanomaterials. There are two main drawbacks in using silica based nano material for hybrid nanocomposite formulation for theranostics application. A hybrid nanocomposite has been synthesized by using phospholipid and mesoporous silica nanoparticles. This nano vehicle was found to be effective in preventing self-aggregation. Further, this nanoshuttle was found to be capable of preventing nonspecific binding to proteins in physiological conditions as well as the environment containing salt. A mesoporous silica nanocomposite was synthesized by conjugating with phospholipid anchored with folic acid [97]. This nanoshuttle has shown targeting function toward the folate receptor-overexpression Hela cells. The authors have believed that the modification on the surface of the nano formulation with phospholipids resulted in the biocompatible nano platform for the development in theranostics. In a single platform, imaging and therapeutic are combined.

14.8 Other nanomaterials

Beside common nanomaterials like metal, oxides, carbon, polymer etc. there are also different other types of nanomaterials which possess amazing features like other commonly used nanomaterials. Upconversion nanoparticles have been widely employed in theranostic applications. Upconversion is a process in which higher energy photons are obtained from lower energy photons of near-infrared radiation through a non-linear optical method. Some examples of upconversion nanoparticles are lanthanide-doped nanoparticles (rare earth doped material) and fluorophores. These nanoparticles have the ability to penetrate deeply into biological tissues. The features like high photostability, long half-life, sharp absorption and emission lines, higher stokes shifts and high signal to noise ratio make the upconversion nanoparticles suitable for bioimaging and detection in vivo and in vitro. Combining upconversion nanoparticles with other nanomaterials also make it multifunctional for imaging and therapy. Conjugation of these nanoparticles with polymer modified photosynthesizer can be executed under near-infrared radiation for photodynamic therapy [98,99]. A nanoprobe has been developed for the photo thermal therapy and diagnosis in the near IR and mid IR regions of the electromagnetic spectrum. This nanoprobe was prepared by the adsorption of IONPs onto the surface of NaYF4 modified upconversion nanoparticles [100]. Quantum dots are the semiconductor nanoparticles that possess bright and stable fluorescence. ZnSe, InAs, CdTe, and CdSe have been utilized to synthesize quantum dots in different sizes. These synthesized quantum dots extends the full visible and NIR spectrum. There are less investigations on the quantum dots as theranostic carriers because of its instability and toxicity. However, for obtaining highly stable quantum dots, thiolated ligands were employed through disulfide linkage. Optical properties of quantum dots likely of NIR quantum dot loaded micelles have been investigated for imaging purposes. 10, 12-pentacosadiynoic acid (PCDA) was selected as the hydrophobic moiety-forming component for encapsulation of drugs and because of its responsiveness toward the UV irradiation, a strong outer shell was formed through cross-linking. This type of conjugated quantum dots reduces the toxicity and improve tumor cell uptake and retention. Also the conjugation of PCDA-herceptin and PEG-PCDA with quantum dots increases the stability and application in the bioimaging and therapy of cancer cells. Liposomes are composed of bilayered thin lipid films made up of either synthetically or naturally. Biodegradability, non immunogenicity, biocompatibility, amphiphilic nature, structural versatility, make the liposomes attractive for the delivery of drugs to specific cellular targets. Some liposome-based therapies such as Doxil® have been already accepted for the treatment of certain cancers. However, liposomes undergo a major drawback of rapid eradication from blood by reticuloendothelial system. Nanostructured lipid carriers (NLC) and Solid lipid nanoparticles (SLN) are the lipid based carriers initially developed as the alternative to liposomes. SLN have rigid cores of solid lipids which are biodegradable whereas NLC have the mixture of both liquid and solid lipids. These lipid based carriers enable the passive targeting of lipophilic drugs and are explored for antiparasitic therapy.

14.9 Conclusion and future scope

This chapter provides an insight into the current publications in nanotheranostics. We have summarized the publications on theranostics based on noble metal nanoparticles, gold, magnetic nanomaterials, polymer, silica based nano materials, nanohybrids, carbon based nano materials and quantum dots. Although there is remarkable progress in nanotheranostics, still development is required to design novel nano platforms to meet clinical standards. The nanotheranostics discussed here have their unique advantages as well as individual limitations. For example, in case of gold nanoparticles the cost is the key limitation. In the case of silica based nano materials, the main limitations are its larger size. In the case of iron oxide nanoparticles, inadequate sensitivity toward the contrast agent in MRI is the biggest drawback. Non biodegradability is the challenge faced in the case of carbon based nano platforms. These limitations should be overcome to develop a novel nano platform to be used in theranostics. Several perspectives can be found to achieve the ultimate goal in theranostics. Nanotheranostics is an excellent platform for therapy and diagnosis, i,e. killing two birds using a single stone. Substantial in vivo studies are very much required for the clinical application of these nanotheranostics. Quantum dots have been used for imaging at the cellular level and it suffers difficulty for imaging in the body. Dyes absorbing in the near infrared regions can be used as an alternative for biomedical imaging. The development in nanotechnology research has paved a great opportunity for the development of nanotheranostics. The beautiful part of this treatment modality is that the diagnosis is not performed before or after of the treatment but during the treatment process. From the above discussion on nanotheranostics, it has been confirmed that after a thorough evaluation on the cytotoxicity, immunotoxicity and genotoxicity, the nanotheranostics can be applied in a routine health care system. The introduction of nanotheranostics in the health care system may be an important treatment modality in personalized and predictive medicine.

References

[1] S.S. Kelkar, T.M. Reineke, Theranostics: combining imaging and therapy, Bioconjug. Chem. 22 (10) (2011) 1879–1903 no.

[2] M.E. Caldorera-Moore, W.B. Liechty, N.A. Peppas, Responsive theranostic systems: integration of diagnostic imaging agents and responsive controlled release drug delivery carriers, Acc. Chem. Res. 44 (10) (2011) 1061–1070.

[3] J.K. Patra, G. Das, L.F. Fraceto, E.V.R. Campos, M.P. Rodriguez-Torres, L.S. Acosta-Torres, L.A. Diaz-Torres, et al., Nano based drug delivery systems: recent developments and future prospects, J. Nanobiotechnology 16 (1) (2018) 71.

[4] C. Bharti, U. Nagaich, A.K. Pal, N. Gulati, Mesoporous silica nanoparticles in target drug delivery system: a review, International journal of pharmaceutical investigation 5 (3) (2015) 124.
[5] Li-S. Wang, M.-.C. Chuang, Ja-A Ho, Nanotheranostics–a review of recent publications, Int. J. Nanomed. 7 (2012) 4679.
[6] C. Zhang, M. Jugold, E.C. Woenne, T. Lammers, B. Morgenstern, M.M. Mueller, H. Zentgraf, et al., Specific targeting of tumor angiogenesis by RGD-conjugated ultrasmall superparamagnetic iron oxide particles using a clinical 1.5-T magnetic resonance scanner, Cancer Res. 67 (4) (2007) 1555–1562.
[7] S.G. Crich, B. Bussolati, L. Tei, C. Grange, G. Esposito, S. Lanzardo, G. Camussi, S. Aime, Magnetic resonance visualization of tumor angiogenesis by targeting neural cell adhesion molecules with the highly sensitive gadolinium-loaded apoferritin probe, Cancer Res. 66 (18) (2006) 9196–9201.
[8] Y. Liu, Hybrid BaYbF 5 nanoparticles: novel binary contrast agent for high-resolution in vivo X-ray computed tomography angiography. In: Multifunctional Nanoprobes, Springer, Singapore, 2018, pp. 105–120.
[9] A. Jakhmola, N. Anton, T.F. Vandamme, Inorganic nanoparticles based contrast agents for X-ray computed tomography, Adv. Healthc. Mater. 1 (4) (2012) 413–431.
[10] Y.-H. Kim, J. Jeon, SuH Hong, W.-K. Rhim, Y.-S. Lee, H. Youn, J.-.K. Chung, et al., Tumor targeting and imaging using cyclic RGD-PEGylated gold nanoparticle probes with directly conjugated iodine-125, Small 7 (14) (2011) 2052–2060.
[11] J. Leike, A. Sachse, C. Ehritt, Biodistribution and CT-imaging characteristics of iopromide-carrying liposomes in rats, J. Liposome Res. 6 (4) (1996) 665–680.
[12] V.P. Torchilin, M.D. Frank-Kamenetsky, G.L. Wolf, CT visualization of blood pool in rats by using long-circulating, iodine-containing micelles, Acad. Radiol. 6 (1) (1999) 61–65.
[13] N.K. Devaraj, E.J. Keliher, G.M. Thurber, M. Nahrendorf, R. Weissleder, 18F labeled nanoparticles for in vivo PET-CT imaging, Bioconjug. Chem. 20 (2) (2009) 397–401.
[14] Z. Gao, A.M. Kennedy, D.A. Christensen, N.Y. Rapoport, Drug-loaded nano/microbubbles for combining ultrasonography and targeted chemotherapy, Ultrasonics 48 (4) (2008) 260–270.
[15] E. Kang, HSu Min, J. Lee, M.H. Han, H.J. Ahn, In-C Yoon, K. Choi, K. Kim, K. Park, I.C. Kwon, Nanobubbles from gas-generating polymeric nanoparticles: ultrasound imaging of living subjects, Angew. Chem. Int. Ed. 49 (3) (2010) 524–528.
[16] A. Carovac, F. Smajlovic, D. Junuzovic, Application of ultrasound in medicine, Acta. Informatica. Medica. 19 (3) (2011) 168.
[17] X. Gao, Y. Cui, R.M. Levenson, L.W.K. Chung, S. Nie, In vivo cancer targeting and imaging with semiconductor quantum dots, Nat. Biotechnol. 22 (8) (2004) 969–976.
[18] S. Santra, D. Dutta, G.A. Walter, B.M. Moudgil, Fluorescent nanoparticle probes for cancer imaging, Technol. Cancer Res. Treat. 4 (6) (2005) 593–602.
[19] F. Ducongé, T. Pons, C. Pestourie, L. Hérin, B. Thézé, K. Gombert, B. Mahler, et al., Fluorine-18-labeled phospholipid quantum dot micelles for in vivo multimodal imaging from whole body to cellular scales, Bioconjug. Chem. 19 (9) (2008) 1921–1926.
[20] Y. Inoue, K. Izawa, K. Yoshikawa, H. Yamada, A. Tojo, K. Ohtomo, In vivo fluorescence imaging of the reticuloendothelial system using quantum dots in combination with bioluminescent tumour monitoring, Eur. J. Nucl. Med. Mol. Imaging 34 (12) (2007) 2048–2056.
[21] X. Cai, W. Li, C.-H. Kim, Y. Yuan, L.V. Wang, Y. Xia, In vivo quantitative evaluation of the transport kinetics of gold nanocages in a lymphatic system by noninvasive photoacoustic tomography, ACS Nano 5 (12) (2011) 9658–9667.
[22] La De, A. Zerda, C. Zavaleta, S. Keren, S. Vaithilingam, S. Bodapati, Z. Liu, J. Levi, et al., Carbon nanotubes as photoacoustic molecular imaging agents in living mice, Nat. Nanotechnol. 3 (9) (2008) 557–562.
[23] A.B.E. Attia, G. Balasundaram, M. Moothanchery, U.S. Dinish, R. Bi, V. Ntziachristos, M. Olivo, A review of clinical photoacoustic imaging: current and future trends, Photoacoustics 16 (2019) 100144.
[24] J.W. Seo, H. Zhang, D.L. Kukis, C.F. Meares, K.W. Ferrara, A novel method to label preformed liposomes with 64Cu for positron emission tomography (PET) imaging, Bioconjug. Chem. 19 (12) (2008) 2577–2584.
[25] K.J. Harrington, S. Mohammadtaghi, P.S. Uster, D. Glass, A. Michael Peters, R.G. Vile, J. Simon W. Stewart, Effective targeting of solid tumors in patients with locally advanced cancers by radiolabeled pegylated liposomes, Clin. Cancer Res. 7 (2) (2001) 243–254.

[26] K.J. Harrington, C.R. Lewanski, J.S.W. Stewart, Liposomes as vehicles for targeted therapy of cancer. Part 1: preclinical development, Clin. Oncol. 12 (1) (2000) 2–15.
[27] M. Hamoudeh, M.A. Kamleh, R. Diab, H. Fessi, Radionuclides delivery systems for nuclear imaging and radiotherapy of cancer, Adv. Drug. Deliv. Rev. 60 (12) (2008) 1329–1346.
[28] B.R. Line, A. Mitra, A. Nan, H. Ghandehari, Targeting tumor angiogenesis: comparison of peptide and polymer-peptide conjugates, J. Nucl. Med. 46 (9) (2005) 1552–1560.
[29] M. Assadi, K. Afrasiabi, I. Nabipour, M. Seyedabadi, Nanotechnology and nuclear medicine; research and preclinical applications, Hell. J. Nucl. Med. 14 (2) (2011) 149.
[30] M. Chu, X. Pan, D. Zhang, Q. Wu, J. Peng, W. Hai, The therapeutic efficacy of CdTe and CdSe quantum dots for photothermal cancer therapy, Biomaterials 33 (29) (2012) 7071–7083.
[31] H. Jeong, M.S. Huh, SoJ Lee, H. Koo, I.C. Kwon, S.Y. Jeong, K. Kim, Photosensitizer-conjugated human serum albumin nanoparticles for effective photodynamic therapy, Theranostics 1 (2011) 230.
[32] R. Al-Majmaie, N. Alattar, D. Zerulla, M. Al-Rubeai, Toluidine blue O-conjugated gold nanoparticles for photodynamic therapy of cultured colon cancer, Proc. SPIE 8427, Biophotonics: Photonic Solutions for Better Health Care III, 842722, (8 May 2012). https://doi.org/10.1117/12.921813.
[33] C.-K. Lim, J. Shin, Y.-D. Lee, J. Kim, SOh Keun, H.Y. Soon, Y.J. Seo, C.K. Ick, K. Sehoon, Phthalocyanine-aggregated polymeric nanoparticles as tumor-homing near-infrared absorbers for photothermal therapy of cancer, Theranostics 2 (9) (2012) 871.
[34] L.-C. Chen, C.-H. Chang, C.-Yu Yu, Ya-J. Chang, Yu-H. Wu, W.-C. Lee, C.-H. Yeh, Te-W. Lee, G. Ting, Pharmacokinetics, micro-SPECT/CT imaging and therapeutic efficacy of 188Re-DXR-liposome in C26 colon carcinoma ascites mice model, Nucl. Med. Biol. 35 (8) (2008) 883–893.
[35] Yi-Yu Lin, J.-Je Li, C.-H. Chang, Yi-C. Lu, J.-J. Hwang, Y.-L. Tseng, W.-J. Lin, G. Ting, H.-E. Wang, Evaluation of pharmacokinetics of 111In-labeled VNB-PEGylated liposomes after intraperitoneal and intravenous administration in a tumor/ascites mouse model, Cancer Biother. Radiopharm. 24 (4) (2009) 453–460.
[36] T.-H. Chow, Yi-Yu Lin, J.-J. Hwang, H.-E. Wang, Y.-L. Tseng, S.-J. Wang, R.-S. Liu, W.-J. Lin, C.-S. Yang, G. Ting, Improvement of biodistribution and therapeutic index via increase of polyethylene glycol on drug-carrying liposomes in an HT-29/luc xenografted mouse model, Anticancer Res. 29 (6) (2009) 2111–2120.
[37] Ya-J. Chang, C.-H. Chang, C.-Yu Yu, T.-J. Chang, L.-C. Chen, M.-H. Chen, Te-W. Lee, G. Ting, Therapeutic efficacy and microSPECT/CT imaging of 188Re-DXR-liposome in a C26 murine colon carcinoma solid tumor model, Nucl. Med. Biol. 37 (1) (2010) 95–104.
[38] Y. Matsushita-Ishiodori, T. Ohtsuki, Photoinduced RNA interference, Acc. Chem. Res. 45 (7) (2012) 1039–1047.
[39] S.X. Wang, A. Bao, S.J. Herrera, W.T. Phillips, B. Goins, C. Santoyo, F.R. Miller, R.A. Otto, Intraoperative 186Re-liposome radionuclide therapy in a head and neck squamous cell carcinoma xenograft positive surgical margin model, Clin. Cancer Res. 14 (12) (2008) 3975–3983.
[40] C.L. Zavaleta, B.A. Goins, A. Bao, L.M. Mcmanus, C. Alex Mcmahan, W.T. Phillips, Imaging of 186Re-liposome therapy in ovarian cancer xenograft model of peritoneal carcinomatosis, J. Drug Target. 16 (7-8) (2008) 626–637.
[41] J. Zhong, S. Yang, X. Zheng, T. Zhou, Da Xing, In vivo photoacoustic therapy with cancer-targeted indocyanine green-containing nanoparticles, Nanomedicine 8 (6) (2013) 903–919.
[42] X. Huang, PK. Jain, IH. El-Sayed, MA. El-Sayed. Gold nanoparticles: interesting optical properties and recent applications in cancer diagnostics and therapy, Nanomedicine 2 (5) (2007) 681-693.
[43] J.B. Vines, J.-H. Yoon, Na-E. Ryu, D.-J. Lim, H. Park, Gold nanoparticles for photothermal cancer therapy, Front. Chem. 7 (2019) 167.
[44] M. Kim, J.-H. Lee, J.-M. Nam, Plasmonic photothermal nanoparticles for biomedical applications, Adv. Sci. 6 (17) (2019) 1900471.
[45] D.N. Heo, H.Y. Dae, M. Ho-Jin, B.L. Jung, S.B. Min, C.L. Sang, J.L. Won, S. In-Cheol, K.K. Il, Gold nanoparticles surface-functionalized with paclitaxel drug and biotin receptor as theranostic agents for cancer therapy, Biomaterials 33 (3) (2012) 856–866.
[46] B. Aydogan, Ji Li, T. Rajh, A. Chaudhary, S.J. Chmura, C. Pelizzari, C. Wietholt, M. Kurtoglu, P. Redmond, AuNP-DG: deoxyglucose-labeled gold nanoparticles as X-ray computed tomography contrast agents for cancer imaging, Mol. Imaging Biol. 12 (5) (2010) 463–467.

[47] S.S. Lee, P.J.R. Roche, P.N. Giannopoulos, E.J. Mitmaker, M. Tamilia, M. Paliouras, M.A. Trifiro, Prostate-specific membrane antigen–directed nanoparticle targeting for extreme nearfield ablation of prostate cancer cells, Tumor Biol. 39 (3) (2017) 1010428317695943.
[48] G. Song, L. Cheng, Yu Chao, K. Yang, Z. Liu, Emerging nanotechnology and advanced materials for cancer radiation therapy, Adv. Mater. 29 (32) (2017) 1700996.
[49] P. Huang, C.Z. Le Bao, J. Lin, T. Luo, D. Yang, M. He, et al., Folic acid-conjugated silica-modified gold nanorods for X-ray/CT imaging-guided dual-mode radiation and photo-thermal therapy, Biomaterials 32 (36) (2011) 9796–9809.
[50] E.Y. Lukianova-Hleb, X. Ren, R.R. Sawant, X. Wu, V.P. Torchilin, D.O. Lapotko, On-demand intracellular amplification of chemoradiation with cancer-specific plasmonic nanobubbles, Nat. Med. 20 (7) (2014) 778.
[51] D. Lapotko, Plasmonic nanobubbles as tunable cellular probes for cancer theranostics, Cancers 3 (1) (2011) 802–840.
[52] D.K. Yi, S. In-Cheol, R. Ju Hee, K. Heebeom, W.P. Chul, Y. In-Chan, C. Kuiwon, C.K. Ick, K. Kwangmeyung, A. Cheol-Hee, Matrix metalloproteinase sensitive gold nanorod for simultaneous bioimaging and photothermal therapy of cancer, Bioconjug. Chem. 21 (12) (2010) 2173–2177.
[53] S. Laurent, A. Ata Saei, S. Behzadi, P. Arash, M. Morteza, Superparamagnetic iron oxide nanoparticles for delivery of therapeutic agents: opportunities and challenges, Expert Opin. Drug Deliv. 11 (9) (2014) 1449–1470.
[54] H. Su, Y. Liu, D. Wang, C. Wu, C. Xia, G. Qiyong, S. Bin, Ai Hua, Amphiphilic starlike dextran wrapped superparamagnetic iron oxide nanoparticle clsuters as effective magnetic resonance imaging probes, Biomaterials 34 (4) (2013) 1193–1203.
[55] J. Zhao, D. Zhong, S. Zhou, NIR-I-to-NIR-II fluorescent nanomaterials for biomedical imaging and cancer therapy, J. Mater. Chem. B 6 (3) (2018) 349–365.
[56] R. Rastogi, N. Gulati, R.K. Kotnala, U. Sharma, R. Jayasundar, V. Koul, Evaluation of folate conjugated pegylated thermosensitive magnetic nanocomposites for tumor imaging and therapy, Colloids Surf. B 82 (1) (2011) 160–167.
[57] A. Sahu, N. Kasoju, P. Goswami, U. Bora, Encapsulation of curcumin in Pluronic block copolymer micelles for drug delivery applications, J. Biomater. Appl. 25 (6) (2011) 619–639.
[58] F. Xiong, S. Huang, N. Gu, Magnetic nanoparticles: recent developments in the drug delivery system, Drug Dev. Ind. Pharm. 44 (5) (2018) 697–706.
[59] J. Xu, J. Sun, Y. Wang, J. Sheng, F. Wang, Mi Sun, Application of iron magnetic nanoparticles in protein immobilization, Molecules 19 (8) (2014) 11465–11486.
[60] S.Y. Lee, SIk Jeon, S. Jung, InJ Chung, C.-H. Ahn, Targeted multimodal imaging modalities, Adv. Drug. Deliv. Rev. 76 (2014) 60–78.
[61] L. Lartigue, C. Innocenti, T. Kalaivani, A. Awwad, M.M.S. Duque, Y. Guari, J. Larionova, et al., Water-dispersible sugar-coated iron oxide nanoparticles. An evaluation of their relaxometric and magnetic hyperthermia properties, J. Am. Chem. Soc. 133 (27) (2011) 10459–10472.
[62] C. Clawson, L. Ton, S. Aryal, V. Fu, S. Esener, L. Zhang, Synthesis and characterization of lipid–polymer hybrid nanoparticles with pH-triggered poly (ethylene glycol) shedding, Langmuir 27 (17) (2011) 10556–10561.
[63] C.-M.J. Hu, S. Kaushal, H.S.T Cao, S. Aryal, M. Sartor, S. Esener, M. Bouvet, L. Zhang, Half-antibody functionalized lipid− polymer hybrid nanoparticles for targeted drug delivery to carcinoembryonic antigen presenting pancreatic cancer cells, Mol. Pharm. 7 (3) (2010) 914–920.
[64] P. Mohan, N. Rapoport, Doxorubicin as a molecular nanotheranostic agent: effect of doxorubicin encapsulation in micelles or nanoemulsions on the ultrasound-mediated intracellular delivery and nuclear trafficking, Mol. Pharm. 7 (6) (2010) 1959–1973.
[65] T. Krasia-Christoforou, T.K. Georgiou, Polymeric theranostics: using polymer-based systems for simultaneous imaging and therapy, J. Mater. Chem. B 1 (24) (2013) 3002–3025.
[66] K.T. Al-Jamal, S. Akerman, J.E. Podesta, A. Yilmazer, J.A. Turton, A. Bianco, N. Vargesson, et al., Systemic antiangiogenic activity of cationic poly-L-lysine dendrimer delays tumor growth, Proc. Natl. Acad. Sci. 107 (9) (2010) 3966–3971.
[67] I.J. Majoros, A. Myc, T. Thomas, C.B. Mehta, J.R. Baker, PAMAM dendrimer-based multifunctional conjugate for cancer therapy: synthesis, characterization, and functionality, Biomacromolecules 7 (2) (2006) 572–579.

[68] R. Rupp, S.L. Rosenthal, L.R. Stanberry, VivaGel™(SPL7013 Gel): a candidate dendrimer–microbicide for the prevention of HIV and HSV infection, Int. J. Nanomed. 2 (4) (2007) 561.
[69] D.-W. Kim, S.-Y. Kim, H.-K. Kim, S.-W. Kim, S.W. Shin, J.S. Kim, K. Park, M.Y. Lee, D.S. Heo, Multicenter phase II trial of Genexol-PM, a novel Cremophor-free, polymeric micelle formulation of paclitaxel, with cisplatin in patients with advanced non-small-cell lung cancer, Ann. Oncol. 18 (12) (2007) 2009–2014.
[70] R. Trivedi, U.B. Kompella, Nanomicellar formulations for sustained drug delivery: strategies and underlying principles, Nanomedicine 5 (3) (2010) 485–505.
[71] H. Xin, L. Chen, J. Gu, X. Ren, J. Luo, Y. Chen, X. Jiang, S. Xianyi, F. Xiaoling, Enhanced anti-glioblastoma efficacy by PTX-loaded PEGylated poly (ε-caprolactone) nanoparticles: in vitro and in vivo evaluation, Int. J. Pharm. 402 (1-2) (2010) 238–247.
[72] B. Jr, A. A., M. Mazzanti, G. Castillo, E. Bolotin, Protected graft copolymer (PGC) in imaging and therapy: a platform for the delivery of covalently and non-covalently bound drugs, Theranostics 2 (6) (2012) 553.
[73] R. Duncan, Drug-polymer conjugates: potential for improved chemotherapy, Anticancer Drugs 3 (3) (1992) 175–210.
[74] D. Peer, J.M. Karp, S. Hong, O.C. Farokhzad, R. Margalit, R. Langer, Nanocarriers as an emerging platform for cancer therapy, Nat. Nanotechnol. 2 (12) (2007) 751–760.
[75] S. Balasubramaniam, S. Kayandan, Y.-N. Lin, D.F. Kelly, M.J. House, R.C. Woodward, T.G.St. Pierre, JS. Riffle, R.M. Davis, Toward design of magnetic nanoparticle clusters stabilized by biocompatible diblock copolymers for T 2-weighted MRI contrast, Langmuir 30 (6) (2014) 1580–1587.
[76] P. Mi, D. Kokuryo, H. Cabral, M. Kumagai, T. Nomoto, I. Aoki, Y. Terada, A. Kishimura, N. Nishiyama, K. Kataoka, Hydrothermally synthesized PEGylated calcium phosphate nanoparticles incorporating Gd-DTPA for contrast enhanced MRI diagnosis of solid tumors, J. Control. Release 174 (2014) 63–71.
[77] K.S. Kim, W. Park, J. Hu, Y.H. Bae, K. Na, A cancer-recognizable MRI contrast agent using pH-responsive polymeric micelle, Biomaterials 35 (1) (2014) 337–343.
[78] J.S. Guthi, Su-G. Yang, G. Huang, S. Li, C. Khemtong, C.W. Kessinger, M. Peyton, J.D. Minna, K.C. Brown, J. Gao, MRI-visible micellar nanomedicine for targeted drug delivery to lung cancer cells, Mol. Pharm. 7 (1) (2010) 32–40.
[79] F. Ye, T. Ke, E.-K. Jeong, X. Wang, Y. Sun, M. Johnson, Z.-R. Lu, Noninvasive visualization of in vivo drug delivery of poly (L-glutamic acid) using contrast-enhanced MRI, Mol. Pharm. 3 (5) (2006) 507–515.
[80] F. Yamamoto, R. Yamahara, A. Makino, K. Kurihara, H. Tsukada, E. Hara, I. Hara, et al., Radiosynthesis and initial evaluation of 18F labeled nanocarrier composed of poly (L-lactic acid)-block-poly (sarcosine) amphiphilic poly depsipeptide, Nucl. Med. Biol. 40 (3) (2013) 387–394.
[81] Y.W. Cho, SAh Park, T.H. Han, JiS Park, S.J. Oh, D.H. Moon, K.-Ja Cho, et al., In vivo tumor targeting and radionuclide imaging with self-assembled nanoparticles: mechanisms, key factors, and their implications, Biomaterials 28 (6) (2007) 1236–1247.
[82] A. Mitra, A. Nan, B.R. Line, H. Ghandehari, Nanocarriers for nuclear imaging and radiotherapy of cancer, Curr. Pharm. Des. 12 (36) (2006) 4729–4749.
[83] C.-L. Peng, Y.-H. Shih, P.-C. Lee, T.M.-H. Hsieh, T.-Y. Luo, M.-J. Shieh, Multimodal image-guided photothermal therapy mediated by 188Re-labeled micelles containing a cyanine-type photosensitizer, ACS nano 5 (7) (2011) 5594–5607.
[84] Z. Gao, A.M. Kennedy, D.A. Christensen, N.Y. Rapoport, Drug-loaded nano/microbubbles for combining ultrasonography and targeted chemotherapy, Ultrasonics 48 (4) (2008) 260–270.
[85] K. Pu, A.J. Shuhendler, J.V. Jokerst, J. Mei, S.S. Gambhir, Z. Bao, J. Rao, Semiconducting polymer nanoparticles as photoacoustic molecular imaging probes in living mice, Nat. Nanotechnol. 9 (3) (2014) 233–239.
[86] J.H. Na, H. Koo, S. Lee, K.H. Min, K. Park, H. Yoo, S.H. Lee, et al., Real-time and non-invasive optical imaging of tumor-targeting glycol chitosan nanoparticles in various tumor models, Biomaterials 32 (22) (2011) 5252–5261.
[87] Y. KyungáJung, Sweet nanodot for biomedical imaging: carbon dot derived from xylitol, RSC Adv. 4 (44) (2014) 23210–23213.

[88] F. Du, M. Zhang, X. Li, J. Li, X. Jiang, Z. Li, Ye Hua, et al., Economical and green synthesis of bagasse-derived fluorescent carbon dots for biomedical applications, Nanotechnology 25 (31) (2014) 315702.
[89] Z. Liu, S. Tabakman, K. Welsher, H. Dai, Carbon nanotubes in biology and medicine: in vitro and in vivo detection, imaging and drug delivery, Nano Res. 2 (2) (2009) 85–120.
[90] R. Partha, J.L. Conyers, Biomedical applications of functionalized fullerene-based nanomaterials, Int. J. Nanomed. 4 (2009) 261.
[91] K. Yang, S. Zhang, G. Zhang, X. Sun, S.-T. Lee, Z. Liu, Graphene in mice: ultrahigh in vivo tumor uptake and efficient photothermal therapy, Nano Lett. 10 (9) (2010) 3318–3323.
[92] C. Wang, H. Tao, L. Cheng, Z. Liu, Near-infrared light induced in vivo photodynamic therapy of cancer based on upconversion nanoparticles, Biomaterials 32 (26) (2011) 6145–6154.
[93] J. Shan, S.J. Budijono, G. Hu, N. Yao, Y. Kang, Y. Ju, R.K. Prud'homme, Pegylated composite nanoparticles containing upconverting phosphors and meso-tetraphenyl porphine (TPP) for photodynamic therapy, Adv. Funct. Mater. 21 (13) (2011) 2488–2495.
[94] L. Cheng, K. Yang, Y. Li, J. Chen, C. Wang, M. Shao, S.-T. Lee, Z. Liu, Facile preparation of multifunctional upconversion nanoprobes for multimodal imaging and dual-targeted photothermal therapy, Angew. Chem. 123 (32) (2011) 7523–7528.
[95] Y. Wang, Q. Zhao, N. Han, L. Bai, J. Li, J. Liu, E. Che, et al., Mesoporous silica nanoparticles in drug delivery and biomedical applications, Nanomed. Nanotechnol. Biol. Med. 11 (2) (2015) 313–327.
[96] X. He, Xu Wu, K. Wang, B. Shi, L. Hai, Methylene blue-encapsulated phosphonate-terminated silica nanoparticles for simultaneous in vivo imaging and photodynamic therapy, Biomaterials 30 (29) (2009) 5601–5609.
[97] S.-H. Cheng, C.-H. Lee, M.-C. Chen, J.S. Souris, F.-G. Tseng, C.-S. Yang, C.-Y. Mou, C.-Tu Chen, L.-W. Lo, Tri-functionalization of mesoporous silica nanoparticles for comprehensive cancer theranostics—the trio of imaging, targeting and therapy, J. Mater. Chem. 20 (29) (2010) 6149–6157.
[98] J. Lu, M. Liong, Z. Li, J.I. Zink, F. Tamanoi, Biocompatibility, biodistribution, and drug-delivery efficiency of mesoporous silica nanoparticles for cancer therapy in animals, Small 6 (16) (2010) 1794–1805.
[99] X. Ji, R. Shao, A.M. Elliott, R. Jason Stafford, E. Esparza-Coss, J.A. Bankson, G. Liang, et al., Bifunctional gold nanoshells with a superparamagnetic iron oxide– silica core suitable for both MR imaging and photothermal therapy, J. Phys. Chem. C 111 (17) (2007) 6245–6251.
[100] Li-S. Wang, Li-C. Wu, S.-Yi Lu, Li-L. Chang, I.-T. Teng, C.-M. Yang, Ja-A. Ho, Biofunctionalized phospholipid-capped mesoporous silica nanoshuttles for targeted drug delivery: improved water suspensibility and decreased nonspecific protein binding, ACS Nano 4 (8) (2010) 4371–4379.

CHAPTER 15

Nanoparticles in dentistry

Debarchita Sarangi[a], **Snigdha Pattanaik**[b]
[a]Department of Prosthodontics, Crown & Bridge, Institute of Dental Sciences, Siksha O Anusandhan (Deemed to be University), Bhubaneshwar, Odisha, India
[b]Department of Orthodontics & Dentofacial Orthopaedics, Institute of Dental Sciences, Siksha O Anusandhan (Deemed to be University), Bhubaneshwar, Odisha, India

15.1 Introduction

15.1.1 What is nanoscale?

Standard units are those which we regularly measure in daily life. The examples are weight, length, volume, etc. These standard units are internationally approved so that they are accurate and reliable throughout the world. The Systeme Internationale (SI) approved metric system is universally accepted. There are various units that are used in day to day life. The SI has two such groups of units. They are base units and derived units. The seven standard base units are as follows in Table 15.1.

Derived units are those which are derived from the above base units. Some examples of derived units are: force and weight with SI unit Newton, energy/work/heat with SI unit Joule, electric charge with SI unit Coulomb, etc.

The description to denote larger and smaller measurements use base unit which is multiplied by powers of ten or minus ten. There are prefixes used to represent the powers of tens multiplied with the base unit, as given in the following table. The base unit used in the table is meter (m).

"Nano" mathematically means ten to the power minus nine of any measurement in a metric system, that is, 10^{-9}. This word has been used as a prefix in metric systems to describe a measurement which is one-billionth that of the base unit (Table 15.2). The word "nano" is a Greek word, meaning dwarf. This explains the use of the word as a prefix to describe measurements, which are smaller than the base unit. According to ISO, the term nanoscale is defined as the length range approximately from 1 nm to 100 nm [1–5].

15.1.2 Nanotechnology

The study of phenomena at a small scale or minute scale is called "nanoscience." According to the Royal Society/Royal Academy of Engineering Working Group, the definition of nanoscience is "the study of phenomena and manipulation of materials at atomic, molecular and macromolecular scales, where properties differ significantly from those at larger scale." Nanotechnology is the application of nanoscience in technology. The Royal Society/Royal Academy of Engineering Working Group in 1994, also gave the definition

Advanced Nanomaterials for Point of Care Diagnosis and Therapy
DOI: https://doi.org/10.1016/B978-0-323-85725-3.00008-8

Table 15.1 SI base units and measures.

	SI base units	Measure
1.	Meter (m)	Length
2.	Kilogram (kg)	Mass
3.	Second (s)	Time
4.	Ampere (A)	Electric current
5.	Kelvin (K)	Thermodynamic temperature
6.	Mole (mol)	Amount of substance
7.	Candela (cd)	Luminous intensity

Table 15.2 Powers of 10 and prefixes.

Powers of 10	Prefixes used	Measurement using base unit
$1*10^3$	Kilo-	1000 meter
$1*10^2$	Hecta-	100 meter
$1*10^1$	Deca-	10 meter
$1*10^0$	Meter	1 meter
$1*10^{-1}$	Deci-	0.1
$1*10^{-2}$	Centi-	0.01
$1*10^{-3}$	Milli-	0.001
$1*10^{-6}$	Micro-	0.000001
$1*10^{-9}$	Nano-	0.000000001
$1*10^{-12}$	Pico-	0.000000000001
$1*10^{-15}$	Femto-	0.000000000000001

of nanotechnology, which is as follows: nanotechnologies are the design, characterization, production, and application of structures, devices and systems by controlling shape and size at nanometer scale. Nanotechnology application is for a scale of less than 100 nanometers [6,7].

15.1.3 What are nanomaterials and their classification?

Nanomaterials are cornerstones of nanoscience and nanotechnology. Nanostructure science and technology is a broad and interdisciplinary area of research and development activity that has been growing explosively worldwide in the past few years. It has the potential for revolutionizing the ways in which materials and products are created and the range and nature of functionalities that can be accessed. It is already having a significant commercial impact, which will assuredly increase in the future. Nanoscale materials are defined as a set of substances where at least one dimension is less thanapproximately 100 nanometers. A nanometer is one millionth of a millimeter – approximately100,000 times smaller than the diameter of a human hair. Nanomaterials are of interest because at this scale unique optical, magnetic, electrical, and other properties emerge. These emergent properties have the potential for great impacts in electronics, medicine, and other fields. Richard W. Siegel has classified nanomaterials into the following, as shown in Table 15.3 [8].

Table 15.3 Classification of nanomaterials by Richard W. Siegel.

	No. of dimensions	As present in
1.	Zero dimension	Atomic clusters, filaments, and cluster assemblies
2.	One dimension	Multilayers
3.	Two dimensions	Ultrafine-grained overlayers or buried layers
4.	Three dimensions	Nanophase materials consisting of equiaxed nanometer sized grains

Table 15.4 Classification according to morphology.

	Morphology	As present in
1.	Nanoclusters	
2.	Nanofibers	Electrospun polymers/silk
3.	Nanowires/tubes	Carbon-based nanomaterials
4.	Nanoparticles	Silica, HA, gold, silver

Table 15.5 Classification according to chemistry.

	Material chemistry	As present in
1.	Carbon-based nanomaterials	Fullerenes nanotubes
2.	Polymer nanomaterials	Electrospun silk nanofibers
3.	Metallic nanomaterials	Nanogold and nanosilver
4.	Nanoceramics	Silica, titanium oxide, and HA nanoparticles
5.	Nanocomposites	Ceramic/polymer nanocomposites Silica/resin nanocomposites

The other two major classifications of nanomaterials are based on morphology and chemistry is shown in Tables 15.4 and 15.5, respectively [8,9].

15.1.4 What are nanoparticles and classification of nanoparticles?

The European Commission defines nanoparticles as nano-objects, which have all the external dimensions, measured in nanoscale. These have one or more external dimensions, ranging from 1 nanometer to 100 nanometers [10].

International Standard Organization describes nanoparticles as a subgroup of nano-objects. They have one, two or may be three external dimensions. Nanoparticles have the second and third external dimensions, orthogonal to each other. Other nano-objects are nanofibers, which have one large external dimension and two similar external dimensions; nanoplates, which have only one of its external dimension in the nanoscale and nanorods, which are solid nanofibers [11,12]. Types of nanoparticles have been summarized in Tables 15.6 and 15.7.

Table 15.6 Summary of type of nanoparticles and their description.

	Types of Nano-Particles	Description [13]
1.	Single nanoparticles	Which cannot be dissociated (deagglomerated or deaggregated) into smaller constituent particles under the dispersion conditions
2.	Aggregates	Aggregates are particles comprising strongly bonded or fused particles where the resulting specific external surface area is significantly smaller than the sum of calculated surface areas of the individual components.
3.	Agglomerates	Agglomerates are collections of weakly bound particles or aggregates or mixtures of the two where the resulting external surface area is similar to the sum of the surface areas of the individual components.

Table 15.7 Classification according to binding of particles.

	Based on how the particles are bound [14]	Examples
1.	Free	As in uncured resin pastes
2.	Fixed	As on implant surfaces
3.	Embedded	As on resin-based composites

15.1.5 Basic proerties of nanomaterials

The nano-materials can be said to useful, based on their applications. Thus, the following properties have been found useful:

1. When used in composites, they allow for better biodegradability, as compared to that of conventional [15,16].
2. The surface finish is better when nanofillers are used.
3. There is improved translucency and other optical properties with the use of nanofillers [17].
4. Use of nanomaterials increases the surface area, thus enabling better mechanical interlocking between the polymer and nanomaterials [18].
5. Resistance to crack propagation gets improved because of reduced areas of stress concentration [19].
6. Improved mechanical properties are seen with the use of inorganic ceramic nanoparticles which are hard and brittle [20].

15.1.6 Synthetic and natural nanomaterials

Nanomaterials on the Earth spheres are either formed by anthropogenic activites or produced by biological species. Naturally formed nanoparticles are found in the troposphere, hydrosphere, lithosphere, and also the biosphere. Synthetic nanomaterials are those which are engineered by machines. The increased production of nano-materials is due to increased industrial applications. This also comes with the outcome increased environmental hazards in future [21–24].

15.1.7 Sources of nanomaterials

The sources may be: incidental, engineered, and naturally produced. Incidental nanomaterials are those which are byproducts of other major processes in the industries or may be a byproduct of natural happenings like fire in the forest. Industrial activities, burning of carbon compounds, dust storms, volcanic activities, fires in the forest, skin and shedding of biological species, both plants and animals contribute to the incidental nanomaterial composition about 90% of atmospheric aerosol are generated naturally and remaining through humans. Engineered ones are those which are intentionally manufactured by humans or may be a released during these processes. Carbon nanoparticles, titanium oxide, hydroxyapatite present in cosmetic products, sports goods, etc., are some examples of engineered products. But, the manufacturing of these products may be detrimental to the environment. Naturally occurring nanomaterials are those which can be found in the biological species, plants and animals both [25–30].

Plant-based nanomaterials are a result of physiological activities where in the nutrients absorbed by plants from soil and water is converted into nanobiominerals. Nanowax protective coatings are present on the insects. The DNA and RNA, which are the genetic materials are nanostructured. Enymes, antibodies, and other bodily secretions are also nanostructures [21].

15.2 Nanoparticles in dental materials

Nanoparticles used in dentistry are:

1. Carbon nanotubes: These are carbon-based nanomaterials used in restorative materials and also other materials. They have been chosen because of their mechanical properties such as high strength and low density. They have good electrical and thermal properties such as heat stability and transmission efficiency [31,32].
2. Graphene: Graphene is another carbon-based material. It is an allotrope of carbon. Acrylic teeth are layered with graphene particles are used for their fracture resistant property. Graphene incorporated bio-films are used as a nanocomposite material also. The use of graphene in dental materials, intraorally reduces the formation of biofilm, which eventually leads to caries formation [33,34].
3. Hydroxy apatite: Nanohydroxy-apatite crystals can bind to the protein, bacteria and plaque due to their surface area. These are used to seal the dentinal tubules opening, thus preventing the nerve exposure to external stimuli. These can easily adhere to caries surface, present on the enamel and inhibit demineralization. Thus, dentifrices, mouthwashes now-a-days have hydroxyapatite nanocrystals incorporated in them. Hydroxy-apatite creates an enamel-like layer around the tooth due to its biomimetic

nature. These particles ensure immediate bonding between tissue surface and titanium implant [35,36].

4. Zirconia: Nanoparticles of zirconia and alumina have the properties of ceramic. Due to the presence of tetragonal zirconia particles in the alumina matrix, toughness and longevity is retained. Zirconia nanoparticles are also used in polishing agents due to their biofilm inhibiting activity [37–39].
5. Iron oxide: These particles also help in inhibiting biofilm formation on implant surfaces [40].
6. Silica: Polishing agents with silica nanoparticles polish the enamel, leaving behind a very less roughened surface. This leads to less cariogenic activity on the surface [41–43].
7. Titanium: The surface modification of implant with titanium nanoparticles reduces the microbial adhesion. This also promotes the adhesion of bone to the implant at nanolevel [44]
8. Silver: The antibacterial property of silver particles has made the utility of these particles ranging from domestic products to medical products. They have been used in prosthetic materials, restorative materials, etc., and also in disinfectants. The only disadvantage is that it leads to toxicity when there is exposure of high doses [32,45–49].

15.3 Nanoparticles in diagnostic dentistry

The information about and from the patient is first collected when s/he arrives the clinic. The reason for visit is asked to the patient and then a detailed history is taken. A detailed examination is also done, following which an initial diagnosis is reached. Thus, various scientific methods are applied clinically to find out the reason for the complaint and correlate it with existing symptoms. These are further confirmed by getting certain tests and investigations done. The identification of the nature of disease is the first step toward treating the patient [50].

There are various scientific principles and methods utilized in reaching to a diagnosis of any disease. Nanodiagnostics have been used in identifying disease at a primary stage. Nanodevices are inserted into the body for early detection of a disease. These devices work selectively and make several analysis at a subcellular level [51,52].

"Dento-facial medicine and diagnosis" is a branch of dentistry which deals with diagnosis and management of diseases in the oral cavity and their surrounding structures. The specialist in dentofacial medicine and diagnosis diagnoses and treats the oral diseases, their systemic manifestations, interprets the symptoms and investigations, plans the due course of treatment that includes referral of patient to other specialists [53–55].

The earlier the diagnosis, the more are the chances of better treatment and treatment response by the body. Identifiication of a disease can happen at the minute cellular or molecular levels. This helps in early diagnosis of the disease. Here, nanodiagnostic devices come into play.

Even the efficiency of diagnostics can be enhanced with the use of nanomedicines as they are selective in nature. Nanodevices can be used to collect tissue or fluid samples for various analysis at the nano-level; to detect a disease and also presence of toxic substances or tumor [51,52].

Cancer treatments include chemotherapy, radiation therapy and surgical management. Delayed detection of disease is a major challenge faced by cancer therapists. The inability and inadequacy of drugs reaching the target tumor cells, lac of optimum distribution, etc., are some the major setbacks faced in cancer detection and therapy [53–55].

Invasive biopsy is one of the common clinical diagnostic procedure. Histopathological changes in the tissues are examined and reported. These changes may not be evident at a preliminary stage of tumor [56–58].

In case of increased level of malignancy (cancer), there is an elevated level of a proteomic and genomic marker. These are present in a secretory vesicle called exosome, which is membrane bound. Nanoparticles are used in the study of these markers under atomic-force microscope. Other devices used for cancer diagnosis are optical nanobiosensor, oral fluid nanosensors, and the nanoelectromechanical system or multiplexing modalities [59,60].

The major and one of the first applications of nano-particles are in the diagnostic procedure. Conventional fluorescent markers have been used. But due to their disadvantages like single use before fading, dye bleeding, etc., nanoparticles are now taking their place. Examples are quantum dots, probes encapsulated by biologically localized embedding, per fluorocarbon, etc. [61].

One of the major advantages of using nanoparticles are their ability to be tagged inside and outside the cells for studying more than one biomolecule during the progress of the disease. Nanoparticles are also used as contrast agents for magnetic Resonance imaging. Polymeric nanospheres are used to target the required tissues before imaging. Real-time imaging of these nanospheres and selective local drug delivery have been studied by Lanza et al. [62–64].

Genotypic analysis of DNA microarrays, gold nanoparticles in DNA diagnostics, transdermal monitoring of changes in interstitial fluids and utilization of solid nanoparticles coated with fluorescent enzyme, hollow nanocapsules with fluorescent indicators in glucose concentration monitoring are few areas of application of nanoparticles in diagnosis [61].

15.4 Nanotechnology in preventive dentistry

Prevention is a major aspect of dentistry. It is an altruist concept. Prevention of initiation or spread of a lesion can save the patient from invasive treatments, intense medications and also avoids use of excessive medications, thus saving time and brings about a sense of wellness in the individual. Similarly, preventive dental procedures can save the patient as well as the clinician from several rounds of extensive procedures, chairside time, laboratory time, dental pain, etc.

Modern dentistry aims at prevention. This can be achieved by early prevention. The use of advanced materials for regular oral hygiene maintenance like mouthwashes and dentifrices. Besides, the use of biomimetic materials can also limit certain lesions. Controlling bacterial biofilms, preventive restorations to limit carious activity, remineralization products are some products which may be used for prevention and nano-modified products have been developed for the same [65].

Prevention of demineralization defects and other lesions have always laid a challenge to clinicians. Nanomaterials have laid an arena to overcome these challenges by their utility in a number of dental care products challenges in dentistry. Following are few such uses.

15.4.1 In periodontal problems or gum related problems

Doxycycline gel has been nano-structured for local application. This has been used in experimental disease model for bone loss. Dentifrobots are nanorobots in dental mouthwashes and dentifrices. These can rapidly move to supragingival and subgingival surfaces when left on the occlusal surface and removes the organic residues there. This prevents calculus deposition [66,67].

15.4.2 In carious lesions and demineralized tooth defects

1. As nanotherapy for prevention of bacterial growth:
   Nanoparticles incorporated in dental materials, such as dental composites and bonding agents prevent the growth of bacteria by disrupting bacterial cell membrane, displacing magnesium ions for enzymatic action, preventing DNA replication, inhibiting active transport and sugar break down, generating reactive oxygen, etc. Antibacterial nanoparticle incorporated coating on tooth surfaces not only kills bacteria on the surface, but also inhibits bacterial adhesion to tooth surface [68–83].

15.4.3 In remineralization of small demineralized defects

Calcium carbonate nanoparticles act as delivery vehicle. They retain well on oral tissue surfaces and deliver slow and continuous release of calcium ions into oral fluids. They make the oral pH alkaline. In experimental studies, it has been proved they reverse the demineralization on incipient caries when used in dentifrice.

Nanoparticles of calcium fluorides have high solubility. When used in a mouth rinse, they can act as potentially good anticaries agents as their concentration in oral fluids is high. This process helps in remineralization of the calcified tooth defects.

Biomimetic nanoparticles of carbonate hydroxyl apatite have been used in-vitro to repair micro defects on tooth surfaces. These have been incorporated in dentifrices and mouthwashes to promote repair of demineralized surface defects. Dentifrices with nanohydroxyl apatite, either spheroidal or needle shaped enhance the remineralization potential of etched enamel. These agents have been successful on enamel surfaces. However, when tested in-vitro with bio-active nanoglass particles and beta-tricalcium phosphate, the mechanical properties of dentine could not be reproduced due to its complex organic and inorganic structure.

Calcium and phosphate ions are important for remineralization. Nanocomposites having remineralizing action by continuous release of calcium and phosphate ions, prevented recurrent or secondary caries even under restorations. These materials could sustain a large amount of compressive stress, even with the calcium and phosphate ions releasing properties. They also help in increasing the pH of the oral environment when it is increased due to cariogenic activity. These nanoparticles chemically bind to the tooth surface in areas of cariogenic erosion. The remineralization action takes place by forming a protective layer.

Lots of researches are going on to find out a suitable nanomaterial to be used as a restorative material, having biomimetic properties and also required physical properties for long term application [84–89].

15.4.4 In caries vaccine

Nanodelivery vehicles have been used for increasing the immunogenicity of DNA vaccine for anti-caries action [90–99].

15.5 Nanoparticles in therapeutic dentistry

Therapeutics means the science of treating a disease through medicines. Nanotechnology has been utilized in therapeutics for treating oral cancer, dentinal hypersensitivity, root canal disinfection, tissue engineering, etc.

15.5.1 Cancer therapy

Nano-modified drugs for cancer therapeutics are being researched, besides the diagnosis part. They are being utilized as drug delivery systems, targeted cancer therapy [100,101].

15.5.2 Gold particles in cancer therapeutics

Gold nanoparticles are available in as nano-spheres, nano-cages, nano-rods, surface enhanced Raman spectroscopy particles. These particles scatter light resonantly when activated at their surface Plasmon resonance frequency. The radiation sensitivity of gold particles are dependent on their sizes. Due to overexpression of epidermal growth factor biomarker in epithelial cancer cells, there is very high binding capacity of the anti-epidermal growth factor receptor antibody conjugated nano-particles with the cancer cells. The addition of 10% common salt confirms the conjugation of anti-bodies by change in color. Gold nanoshells have been used to improve contrast of blood vessels during angiography through magnetic resonance imaging [102–106].

Gold nanoprobes increase absorption in the localized tissue, decrease toxicity and lead to better elimination from the body. These have been used for photothermal therapy. They are used as radiosensitisers in radiation therapy. These gold nanostructures can be used to improve apoptotic activity and production of reactive oxygen to kill cancerous cells. Mitochondria-mediated autophagy mediated selective inhibition of cancer cell growth can be seen with use of iron-core gold nanosell particles.

In addition to treating cancer, nanotechnology has been also extended to offer an effective management of the breakthrough pain associated with cancer. A nanodelivery transbuccal system was developed to rapidly and efficiently deliver the opioid analgesia at a consistent and controlled diffusion into the target tissue. Accordingly, it minimizes the risk of overdosing in patients and also protects the patients from needle injection. It also avoids the enzymatic and spontaneous degradation of the drug associated with oral administration [107–112].

15.5.3 Nanocapsule drug delivery

Decreased dosage requirement, increased protection of drugs, increased drug stability are the advantages of using nanocapsules. The intended effect is ensured and the side effects are reduced.

Nanoparticles can efficiently reach the target areas as they can pass through small blood vessels. Once reaching the target area, they can undergo sustained and controlled release. Nanoassemblers are microscopic devices that can be controlled by computers for various activities.

15.5.4 Carbon nanotubes

These have a special tendency of being internalized by healthy somatic cells and target malignant cells. After these single-walled carbon nanotubes have entered the tumor site, the treatment can begin. The patient has to be taken to a region with radio-frequency and infra-red radiation to see the effects. These SWNTs convert the radio-frequency into heat. Vibrational movement also gets converted to heat. Heat denatures the protein in the cancer cells and hence death of malignant cell.

15.5.5 Liposomes

Drug delivery can be done by nanocarriers. They have to go across numerous barriers. To overcome this, the carriers are bound to specific cells through ligand molecules. Some drugs cannot diffuse properly within the tumor. Hence, it is not a feasible option. This may be the reason for failure due to drug resistance and cancer cell resistance. Liposome molecules offer greater advantage at diffusing into cells and also improving the drug uptake process [113–122].

15.5.6 Oncolysis through silver nanoparticles

Silver particles have the ability to absorb the bacteria and destroy it in affected tissues. Oligodynamic silver nanoparticles are effective against various cancer related infections such as humanherpes virus, human immunodeficiency virus, human papillomavirus, etc. The presence of oligodynamic silver particles in a silver-based drugs improve their antimicrobial properties. Various studies support the fact that they can be safe. Even on administering a dose of 500cc of oligodynamic silver for 79 days or 1000 cc of the same for 39 days, the risk threshold of developing argyria is not reached [123–135].

15.5.7 Nanoshells and nanovectors

Nanoshells which are miniscale beads are specific tools in cancer therapeutics. Nanoshells have outer metallic layers that selectively destroy cancer cells while leaving normal cells intact. Undergoing trial are nanoparticle-coated, radioactive sources placed close to or within the tumor to destroy it [136].

Nanovectors have been used for genetherapy in nonviral gene systems. Nanomaterials like Brachysil have been used for brachytherapy in cancer patients [137].

15.5.8 Treatment for dentinal hypersensitivity

Nanorobots are being employed to selectively target for occluding the open tubules. The loss of enamel layer from the crown portion of the tooth and loss of cementum layer from the root portion of the tooth exposes the dentin layer. A change in fluid dynamics in the dentinal tubules generates a sensation called dentinal hypersensitivity. Gold nanoparticles were easily adsorbed by the inner dentinal tubules. Silver staining occlude the open tubules, hence reducing the sensitivity. Laser radiation therapy after brushing the opened tubules with concentrated gold nanoparticles also blocks the dentinal tubules [138,139].

15.5.9 Disinfecting the root canal

Zinc oxide and chitosan are used in root canal sealants for disinfection of root canals. A reduction in *Enterococcus faecalis* was seen in the canal. Magnesium oxide nanoparticles have been proved to have promising anti-bacterial activity, when used as root canal irrigant. They have shown to have long term elimination of *E. faecalis* from the root canal [140,141].

15.5.10 Tissue engineering

Scaffolding matrices need to have specific properties to promote attachment to cell, their proliferation and differentiation. They should also disintegrate or degrade to match with the regeneration speed of tissues. Nanoengineered scaffolds have been developed that present unique 3D matrix [142–155].

The nanofibrous scaffolds have been used for regeneration of various dento-facial tissues like enamel, pulp-dentin complex, temporomandibular joint, periodontal fibers, alveolarbone, etc. Nanofibrous scaffolds are produced by electro-spinning, phase separation or peptide synthesis. These have also been employed to mimic enamel. The electrospun nanofibers have been used to culture human periodontal cells. Phase separation method has been used to prepare nanohydroxyapatite nanocomposite scaffolds, combining with marrow stromal cells and implanted in rabbit jaw bone. The study showed that there was increased stimulation of osteogenesis. But there was no quantitative improvement in new bone formation there. Nano-beta-tricalcium phosphate was also used for a similar kind of study. The result showed that there was improved bone regeneration when used with mesenchymal stem cells [156–173].

15.5.11 Drug delivery

Scaffolds or nano-particles are used as carriers to deliver drugs at the required site. Signaling molecules are used in combination with the scaffold surface or 3-D structure when these 3-D scaffolds are the carriers. These molecules are loaded in such a way that they allow controlled and sustained release of drug. Nanoparticles of glycydilmethacrylate, derivative dextran, and gelatin were used to encapsulate bone morphogenic protein to get a sustained release of the growth factor. To augment alveolar ridge, nanodiamonds have been used [174–176].

15.6 Nanoparticles in implant dentistry

One of the applications of implants is in dentistry. Dental implants are used to replace missing teeth. The success of implants depends upon how the implant surface integrates with the bone around it [177–179].

Branemark introduced implants in late 1960s. He studied a piece of titanium in rabbit's bone which integrated to the bone strongly. This concept of binding of bone to titanium surface was called osseointegration and was introduced into dentistry in 1971. The use of implants has more of success stories to tell [180–183].

15.6.1 Nanoscale topography modifications

To improve surface attachment, between the bone-implant interface, certain modification of topography are done. Application of thin layer of ceramic shall help bond the bone to implant. Bioactive materials like calcium phosphates, hydroxyapatite, tricalcium phosphate, and bioactive glasses have been used as cell attachment, differentiation and bone-formation promoters. A carbonated apatite layer forms on the surfaces by dissolution and precipitation process, which is similar to mineral phase. Collagen fibers can be introduced into these apatites [184–189].

15.6.2 Surface functionalization

Different methods have been used to functionalize titanium surface. This shall promote protein adhesion and subsequent cell attachment. Physical or chemical adsorption, covalent bonding, self-organized layers are being used for functionalisation. The idea behind functionalization is improvisation of osseointegration. This can be achieved by adapting techniques to the surfaces of specified alloys, each with different surface chemistries. This shall allow display of different molecules with limited spatial distribution and adhesion. It is yet to be proved through in-vivo studies [190–192].

15.6.3 Surface topography

Surface topography is vital in cases of bioinert materials. Rough surfaces allow mechanical interlocking. They help in more extracellular matrix apposition as they may affect differentiation of cells. The topography can be modified at a nanoscale, which shall promote osseointegration [193–196].

15.7 Nanoparticles in sterilization and disinfection

Sterilization is a process of destroying or eliminating all forms of microbial life. Disinfection is a process of eliminating many or all pathogenic micro-organisms, excluding the bacteria in spore form, from inanimate objects. Cleaning is the removal of visible soil from objects and surfaces with the use of enzymes or detergents [197].

15.7.1 Disinfection using nanotechnology

Use of physical and chemical factors to reduce the chances of contamination through use of personal protective equipments and disinfecting the surfaces has been recommended by World Health Organization. There are various types of disinfectant agents, such as sodium hypochlorite, hydrogen peroxide, alcohols, soaps, etc. Some of these agents may not work properly. Thus, to improve properties and longevity of action, new disinfectants have been developed. Nanotechnology has been employed to develop improved disinfectants. Antimicrobial activity in improved products may allow slow release of disinfectants with increased duration of action. Metal nanoparticles incorporated disinfectants have broader spectrum of antiviral and antimicrobial action. For example, silver nanoparticles are used as potent, broad spectrum antiviral agent with or without surface modification. Disinfectants with titanium dioxide and silver nanoparticles have been developed for self-sterilization. Though various studies have to be done to eliminate the scalability and production, which may be their disadvantages, and also to produce improved varieties of disinfectants [198–209].

15.7.2 Antimicrobial spray nanocoating

This method has the advantages of volume reduction, restricted and defined particle size, changing chemical and physical properties, biological stability, etc. They also allow controlled way of dosage administering and handling. The choice of nanoparticle spray is dependent on the field of application. For example, polymer incorporated materials have been used for public places.

Chemical disinfectants like chlorines, peroxides, quaternary amines, and alcohols have been used for disinfection, but they have been used in high concentration with limited efficiency over time with risk to health of public. Thus, nanoparticles of metals have been proposed to be used as alternatives. The most common metal used as nanoparticle for sterilization and disinfection is silver [210–211].

15.8 Nanotechnology in different branches in dentistry

	Nanomaterial derivative	Branch in dentistry
1.	• Bioactive multifunctional composite (BMC) [6,212]	Restorative dentistry, orthodontics
2.	• Fullerene-like tungsten disulfide nanoparticles coated orthodontic wire to reduce frictional force. • Nitrogen-doped titanium oxide coated brackets for enamel remineralization and in gingivitis cases • Orthodontic nanorobots used for manipulating periodontal tissues [213–216].	Orthodontics, pedodontics
3.	• Local anesthesia with nanorobots that block the pulpal and dental sensation till the completion of the process. These nanorobots are reinstructed to withdraw the blocking of tooth to restore normal sensation [60].	All branches
4.	• Nanoneedles and nanotweezers [60]	All branches, especially oral surgery, cell surgeries
5.	• Bone replacement Materials [60]	Surgical procedures in periodontics, Oral surgery
6.	• Toothbrushes, mouth rinses, dentifrices [217,218]	Preventive dentistry
7.	• Nano-modified ceramics and impression materials [219,220]	Restorative dentistry and prosthodontics
8.	• Modified root canal irrigant [221,222]	Endodontics
9.	• Disinfecting agents [198–209]	In all branches
10.	• Nanoimplant coatings [223]	Implant dentistry

15.9 Conclusion

Nanotechnology has definitely grabbed into various fields. Its application in is widespread and is still continuing to grow. The issue is, even if there have been so many developments recently, the actual number of materials currently available in market for utility are very less. The applications in diagnostic, therapeutic, treatment, surgical, prosthetic, and restorative procedures have been found out. But their utility is presently, not at its expected peak.

Besides, its utility, the safety and efficacy of the materials should not be overlooked. Long-term research back-up is needed to support the clinical performance, safety, efficacy for utilization of this technology in full flow.

References

[1] Nano.gov. 2021. *Standards for Nanotechnology | National Nanotechnology Initiative*. [online] Available at: https://www.nano.gov/you/standards.

[2] Scale W. What is the Nanoscale – Size and Scale [Internet]. MRSEC Education Group. 2021. Available from: https://education.mrsec.wisc.edu/what-is-the-nanoscale-size-and-scale/.

[3] ISO/TR 10993-22, Biological Evaluation of Medical Devices—Part 22: Guidance on Nanomaterials, International Organization for Standardization, 2016.

[4] ISO/TS 80004-1, Nanotechnologies –Vocabulary – Part 1: Core Terms, International Organization for Standardization, 2010.

[5] G. Schmalz, R. Hickel, K. van Landuyt, F. Reichl, Nanoparticles in dentistry, Dent. Mater. 33 (11) (2017) 1298–1314.

[6] Royal Society and Royal Academy of Engineering, Nanoscience and Nanotechnologies: Opportunities and Uncertainties, Royal Society, London, 2004, http://www.royalsoc.ac.uk.

[7] Nanotechnology—what is it? Should we be worried? Occup Med (Lond) 56 (2006) 295–299, doi:10.1093/occmed/kql050.

[8] (PDF) Chapter - INTRODUCTION TO NANOMATERIALS [Internet]. ResearchGate. 2021 [cited 18 October 2021]. Available from: https://www.researchgate.net/publication/259118068_Chapter_-_INTRODUCTION_TO_NANOMATERIALS.

[9] Z. Mok, G. Proctor, M. Thanou, Emerging nanomaterials for dental treatments, Emerging Topics in Life Sciences 4 (6) (2020) 613–625.

[10] Definition - Nanomaterials - Environment - European Commission [Internet]. Ec.europa.eu. 2021. Available from: https://ec.europa.eu/environment/chemicals/nanotech/faq/definition_en.htm.

[11] ISO/TR 10993-22, Biological Evaluation of Medical Devices—Part 22: Guidance on Nanomaterials, International Organization for Standardization, 2016.

[12] ISO/TS 80004-1, Nanotechnologies –Vocabulary – Part 1: Core Terms, International Organization for Standardization, 2010.

[13] H. Rauscher, G. Roebben, A.B. Sanfeliu, H. Emons, N. Gibson, R. Koeber, et al., Towards a Review of the EC Recommendation for a De Nition of the Termnanomaterial—Part 3 Scienti c-Technical Evaluation of Options to Clarify the De Nition and to Facilitate its Implementation, European Commission, 2015. [Internet]. Publications.jrc.ec.europa.eu. 2021 [cited]. Available from: https://publications.jrc.ec.europa.eu/repository/bitstream/JRC91377/jrc_nm-def_report2_eur26744.pdf.

[14] Scientific Committee on Emerging and Newly IdentifiedHealth Risks (SCENIHR), Opinion on the Guidance on the Determination of Potential Health Effects of Nanomaterials used in Medical Devices, European Commission, 2015. [Internet]. Toxicology.org. 2021. Available from: https://www.toxicology.org/groups/ss/MDCPSS/docs/20160121Webinar_S.

[15] S.S. Ray, M. Okamoto, Biodegradable polylactide and its nanocomposites: opening a new dimension for plastics and composites, Macromol Rapid Commun. 24 (2003) 815–840.

[16] A. Mohanty, L. Drzal, M. Misra, Nano reinforcements of bio-based polymers—the hope and the reality, J. Am. Chem. Soc. 225 (2003) 33.

[17] E.G. Mota, H. Oshima, L.H. Burnett Jr., L. Pires, R.S. Rosa, Evaluation of diametral tensile strength and Knoop microhardness of five nanofilled composites in dentin and enamel shades, Stomatologija 8 (2006) 67–69.

[18] R.W. Arcís, A. López-Macipe, M. Toledano, E. Osorio, R. Rodríguez-Clemente, J. Murtra, et al., Mechanical properties of visible light-cured resins reinforced with hydroxyapatite for dental restoration, Dent. Mater. 18 (2002) 49–57.
[19] C.P. Turssi, J.L. Ferracane, K. Vogel, Filler features and their effects on wear and degree of conversion of particulate dental resin composites, Biomaterials 26 (2005) 4932–4937.
[20] E.G. Mota, H. Oshima, L.H. Burnett Jr., L. Pires, R.S. Rosa, Evaluation of diametral tensile strength and Knoop microhardness of five nanofilled composites in dentin and enamel shades, Stomatologija 8 (2006) 67–69.
[21] J. Jeevanandam, A. Barhoum, Y. Chan, A. Dufresne, M. Danquah. Review on nanoparticles and nanostructured materials: history, sources, toxicity and regulations, Beilstein J. Nanotechnol. 9 (2018) 1050–1074.
[22] M. Hochella, M. Spencer, K. Jones. Nanotechnology: nature's gift or scientists' brainchild? Environ Sci: Nano 2 (2015) 114–119.
[23] V. Sharma, J. Filip, R. Zboril, R. Varma. Natural inorganic nanoparticles – formation, fate, and toxicity in the environment, Chem Soc Rev. 44 (23) (2015) 8410–8423.
[24] S. Wagner, A. Gondikas, E. Neubauer, T. Hofmann, F. von der Kammer. Spot the Difference: Engineered and Natural Nanoparticles in the Environment-Release, Behavior, and Fate, Angew Chem Int Ed. 53 (2014) 12398–12419.
[25] D. Taylor. Dust in the wind, Environ Health Perspect. 110 (2) (2002) A80.
[26] W.P. Linak, C.A. Miller, J.O.L. Wendt. Comparison of particle size distributions and elemental partitioning from the combustion of pulverized coal and residual fuel oil, J. Air Waste Manage Assoc. 50 (2000) 1532–1544.
[27] F. Rogers, P. Arnott, B. Zielinska, J. Sagebiel, K. Kelly, D. Wagner, et al., Real-Time Measurements of Jet Aircraft Engine Exhaust, J. Air Waste Manage Assoc. 55 (5) (2005) 583–593.
[28] M. De Volder, S. Tawfick, R. Baughman, A. Hart. Carbon Nanotubes: Present and Future Commercial Applications, J. Science 339 (6119) (2013) 535–539, doi:10.1126/science.1222453.
[29] A. Weir, P. Westerhoff, L. Fabricius, K. Hristovski, N. Von Goetz. Titanium Dioxide Nanoparticles in Food and Personal Care Products, Environ Sci. Technol. 46 (4) (2012) 2242–2250.
[30] M. Sadat-Shojai, M. Khorasani, A. Jamshidi, S. Irani. Nano-hydroxyapatite reinforced polyhydroxybutyrate composites: A comprehensive study on the structural and in vitro biological properties, Mater. Sci. Eng. 33 (5) (2013) 2776–2787.
[31] P. Ajayan, O. Zhou, Applications of Carbon Nanotubes, Topics in Applied Physics, (2001), 391–425.
[32] S. Priyadarsini, S. Mukherjee, M. Mishra, Nanoparticles used in dentistry: a review, J. Oral. Biol. Craniofac Res. 8 (1) (2018) 58-67.
[33] S. Kulshrestha, S. Khan, R. Meena, B.R. Singh, A.U. Khan, A graphene/zinc oxide nanocomposite film protects dental implant surfaces against cariogenic Streptococcus mutans, Biofouling 30 (2014) 1281–1294.
[34] P. Bartolo, J.-P. Kruth, J. Silva, et al., Biomedical production of implants by additive electro-chemical and physical processes, Cirp. Ann. Manuf. Technol. 61 (2012) 635–655.
[35] S.Lodha Khetawat, Nanotechnology (nanohydroxyapatite crystals): recent advancement in treatment of dentinal hypersensitivity, J. Interdiscipl Med. Dent. Sci. 3 (2015) 181.
[36] Besharat, L.K. Pepla, G. Palaia, G. Tenore, G. Migliau, Nano-hydroxyapatite and its applications in preventive, restorative and regenerative dentistry: a review of literature, Ann. Stomatol. 5 (2014) 108.
[37] S. Deville, J. Chevalier, G. Fantozzi, et al., Low-temperature ageing of zirconia-toughened alumina ceramics and its implication in biomedical implants, J. Eur. Ceram. Soc. 23 (2003) 2975–2982.
[38] C. Pecharroman, J.F. Bartolome, J. Requena, et al., Percolative mechanism of aging in zirconia-containing ceramics for medical applications, J. Eur. Ceram. Soc. 15 (2003) 507–511.
[39] J. Guerreiro-Tanomaru, A. Trindade-Junior, B. Cesar Costa, G. da Silva, L. Drullis Cifali, M. Basso Bernardi, et al., Effect of zirconium oxide and zinc oxide nanoparticles on physicochemical properties and antibiofilm activity of a calcium silicate-based material, Scientific World J. 2014 (2014) 1-6.
[40] M. Sathyanarayanan, R. Balachandranath, Y. Genji Srinivasulu, S. Kannaiyan, G. Subbiahdoss, The effect of gold and iron-oxide nanoparticles on biofilm-forming pathogens, ISRN Microbiol. 2013 (2013) 1-5.
[41] J. Roulet, T. Roulet-Mehrens, The surface roughness of restorative materials and dental tissues after polishing with prophylaxis and polishing pastes, J. Periodontol. 53 (1982) 257–266.

[42] S. Guorgan, S. Bolay, R. Alacam, In vitro adherence of bacteria to bleached or unbleached enamel surfaces, J. Oral Rehab 24 (1997) 624–627.
[43] A. Banerjee, T. Watson, Dentine caries excavation: a review of current clinicaltechniques, Br. Dent. J. 188 (2000) 476.
[44] B. Grosner-Schreiber, M. Griepentrog, I. Haustein, et al., Plaque formation on surface modified dental implants, Clin. Implant Dent. Relat. Res. 12 (2001) 543–551.
[45] M. Samiei, M. Aghazadeh, M. Lotfi, S. Shakoei, Z. Aghazadeh, S.M.V. Pakdel, Antimicrobial efficacy of mineral trioxide aggregate with and without silver nanoparticles, Iran Endod J. 8 (2013) 166.
[46] H. Jia, W. Hou, L. Wei, B. Xu, X. Liu, The structures and antibacterial properties of nano-SiO 2 supported silver/zinc–silver materials, Dent. Mater. J. 24 (2008) 244–249.
[47] K.-Y. Nam, In vitro antimicrobial effect of the tissue conditioner containing silver nanoparticles, J. Adv. Prosthodont 3 (2011) 20–24.
[48] F.A. Sheikh, N.A. Barakat, M.A. Kanjwal, et al., Electrospun titanium dioxide nanofibers containing hydroxyapatite and silver nanoparticles as future implant materials, J. Mater. Sci. Mater. Med. 21 (2010) 2551–2559.
[49] S. Sadhasivam, P. Shanmugam, K. Yun, Biosynthesis of silver nanoparticles by Streptomyces hygroscopicus and antimicrobial activity against medically important pathogenic microorganisms, Colloids Surf. B 81 (2010) 358–362.
[50] BRITISH DENTAL JOURNAL VOLUME 213 NO. 1 JUL 14 2012 Oral diagnosis and treatment planning: part 1. Introduction P. Newsome,1 R. Smales2 and K. Yip3.
[51] R.A. Freitas Jr., Nanodentistry, J. Am. Dent. Assoc. 131 (2000) 1559–1565.
[52] C. Lampton, Nanotechnology promises to revolutionize the diagnosis and treatment of diseases, Genet Eng News 15 (1995) 23–25.
[53] D. Hanahan, R.A. Weinberg, The hallmarks of cancer, Cell 100 (2000) 57–70.
[54] B. Ehdaie, Application of nanotechnology in cancer research: Review of progress in the National Cancer Institute's Alliance for Nanotechnology, Int. J. Biol. Sci. 3 (2007) 108–110.
[55] Alok, et al., Nanotechnology: A boon in oral cancer diagnosis and therapeutics, SRM Journal of Research in Dental Sciences |4| |(4) |(2013).
[56] W. Arap, R. Pasqualini, E. Ruoslahti, Cancer treatment by targeted drug delivery to tumor vasculature in a mouse model, Science 279 (1998) 377–380.
[57] D. Peer, J.M. Karp, S. Hong, O.C. Farokhzad, R. Margalit, R. Langer, Nanocarriers as an emerging platform for cancer therapy, Nat. Nanotechnol. 2 (2007) 751–760.
[58] I. Brigger, C. Dubernet, P. Couvreur, Nanoparticles in cancer therapy and diagnosis, Adv. Drug. Deliv. Rev. 54 (2002) 631–651.
[59] I. Abiodun-Solanke, D. Ajayi, A. Arigbede, Nanotechnology and its application in dentistry, Ann Med Health Sci. Res. 4 (Suppl 3) (2014) S171–S177, doi:10.4103/2141-9248.141951.
[60] K.R. Saravana, R. Vijayalakshmi, Nanotechnology in dentistry, Ind. Dent. Res. 17 (2006) 62–65.
[61] Nanomedicine – prospective therapeutic and diagnostic applications, F. Dwaine, L.C.T. Emerich, World J. Pharm. Pharm. Sci. Volume 7, Issue 11, 1611-1633. Review Article ISSN 2278.
[62] W. Zhang, Y. Wang, B.T. Lee, C. Liu, G. Wei, et al., A novel nanoscale-dispersed eye ointment for the treatment of dry eye disease, Nanotechnology 25 (2014) 125101.
[63] G.M. lanza, R.L. Trousil, K.D. Wallace, et al., In vitro characterization of a novel, tissue-targeted ultrasonic contrast system with acoustic microscopy, J. Acoust. Soc. Am. 104 (1998) 3665–3672.
[64] M. Nishizawa, V.P. Menon, C.R. Martin, Metal nanotubule membranes with electrochemically switchable ion-transport selectivity, Science 268 (1995) 700–702.
[65] M. Hannig, C. Hannig, Nanomaterials in preventive dentistry, Nature Nanotechnology 5 (8) (2010) 565–569.
[66] M.A. Botelho, J.G. Martins, R.S. Ruela, D.B. Queiroz, W.S. Ruela, Nanotechnology in ligature-induced periodontitis: protective effect of a doxycycline gel with nanoparticules, J. Appl. Oral. Sci. 18 (4) (2010) 335–342.
[67] K.R. Saravana, R. Vijayalakshmi, Nanotechnology in dentistry, Ind. J. Dent. Res. 17 (2006) 62–65.
[68] H. Gu, D. Fan, J. Gao, et al., Effect of ZnCl2 on plaque growth and biofilm vitality, Arch. Oral. Biol. 57 (4) (2012) 369–375.
[69] Y. Xie, Y. He, P.L. Irwin, T. Jin, X. Shi, Antibacterial activity and mechanism of action of zinc oxide nanoparticles against Campylobacter jejuni, Appl. Environ. Microbiol. 77 (7) (2011) 2325–2331.

[70] K. Blecher, A. Nasir, A. Friedman, The growing role of nanotechnology in combating infectious disease, Virulence 2 (5) (2011) 395–401.
[71] Q.M.S. Li, D.Y. Lyon, L. Brunet, M.V. Liga, D. Li, PJ. Alvarez, Antimicrobial nanomaterials for water disinfection and microbial control: potential applications and implications, Water Res. 42 (18) (2008) 4591–4602.
[72] R.P. Allaker, The use of nanoparticles to control oral biofilm formation, J. Dent. Res. 89 (11) (2010) 1175–1186.
[73] D.K. Shvero, M.P. Davidi, E.I. Weiss, N. Srerer, N. Beyth, Antibacterial effect of polyethyleneimine nanoparticles incorporated in provisional cements against Streptococcus mutans, J. Biomed. Mater. Res. B Appl. Biomater. 94 (2) (2010) 367–371.
[74] S. Kasraei, L. Sami, S. Hendi, M.Y. Alikhani, L. Rezaei-Soufi, Z. Khamverdi, Antibacterial properties of composite resins incorporating silver and zinc oxide nanoparticles on Streptococcus mutans and Lactobacillus, Restor. Dent. Endod. 39 (2) (2014) 109–114.
[75] C. Chen, M.D. Weir, L. Cheng, et al., Antibacterial activity and ion release of bonding agent containing amorphous calcium phosphate nanoparticles, Dent. Mater. 30 (8) (2014) 891–901.
[76] N. Kawabata, M. Nishiguchi, Antibacterial activity of soluble pyridinium-type polymers, Appl. Environ. Microbiol. 54 (1998) 2532–2535.
[77] R. Foldbjerg, P. Olesen, M. Hougaard, D.A. Dang, H.J. Hoffmann, H. Autrup, PVP-coated silver nanoparticles and silver ions induce reactive oxygen species, apoptosis and necrosis in THP-1 monocytes, Toxicol. Lett. 190 (2009) 156–162.
[78] B. Song, L.G. Leff, Influence of magnesium ions on biofilm formation by Pseudomonas fluorescens, Microbiol. Res. 161 (4) (2006) 355–361.
[79] M. Rai, A. Yadav, A. Gade, Silver nanoparticles as a new generation of antimicrobials, Biotechnol. Adv. 27 (2009) 76–83.
[80] P.B. das Neves, J.A. Agnelli, C. Kurachi, C.W. de Souza, Addition of silver nanoparticles to composite resin: effect on physical and bactericidal properties in vitro, Braz. Dent. J. 25 (2) (2014) 141–145.
[81] F. Li, M.D. Weir, A.F. Fouad, H.H.K. Xu, Effect of salivary pellicle on antibacterial activity of novel antibacterial dental adhesives using a dental plaque microcosm biofilm model, Dent. Mater. 30 (2) (2014) 182–191.
[82] A. Besinis, T. De Peralta, R.D. Handy, Inhibition of biofilm formation and antibacterial properties of a silver nano-coating on human dentine, Nanotoxicology 8 (7) (2014) 745–754.
[83] Z. Lu, K. Rong, J. Li, H. Yang, R. Chen, Size-dependent antibacterial activities of silver nanoparticles against oral anaerobic pathogenic bacteria, J. Mater. Sci. Mater. Med. 24 (6) (2013) 1465–1471.
[84] S. Nakashima, M. Yoshie, H. Sano, A. Bahar, Effect of a test dentifrice containing nano-sized calcium carbonate on remineralization of enamel lesions in vitro, J. Oral Sci. 51 (1) (2009) 69–77.
[85] H.H. Xu, L. Sun, M.D. Weir, et al., Nano DCPA-whisker composites with high strength and Ca and PO4 release, J. Dent. Res. 85 (8) (2006) 722–727.
[86] H.H. Xu, M.D. Weir, L. Sun, S. Takagi, LC. Chow, Effects of calcium phosphate nanoparticles on CaPO4 composite, J. Dent. Res. 86 (4) (2007) 378–383.
[87] H.H. Xu, M.D. Weir, L. Sun, Calcium and phosphate ion releasing composite: effect of pH on release and mechanical properties, Dent. Mater. 25 (4) (2009) 535–542.
[88] H.H. Xu, L. Sun, M.D. Weir, S. Takagi, L.C. Chow, B. Hockey, Effects of incorporating nanosized calcium phosphate particles on properties of whisker-reinforced dental composites, J. Biomed. Mater. Res. B Appl. Biomater. 81 (1) (2007) 116–125.
[89] N. Roveri, E. Foresti, M. Lelli, et al., Synthetic biomimetic carbonate hydroxyapatite nanocrystals for enamel remineralization, Adv. Mater. Res. 4 (7–50) (2008) 821–824.
[90] G.P. Talwar, M. Diwan, F. Razvi, R. Malhotra, The impact of new technologies on vaccines, Natl. Med. J. India 12 (6) (1999) 274–280.
[91] J. Wang, J.G. Liu, J. Jiang, W. Wang, G.H. Bai, XY. Guan, Study on gene vaccine pcDNA3-PAc against dental caries by intranasal immunization in rabbits, Shanghai Kou Qiang Yi Xue 23 (2) (2014) 133–137.
[92] L.K. Su, F. Yu, Z.F. Li, C. Zeng, Q.A. Xu, M.W. Fan, Intranasal co-delivery of IL-6 gene enhances the immunogenicity of anti-caries DNA vaccine, Acta Pharmacol. Sin. 35 (5) (2014) 592–598.
[93] G. Liu, M. Fan, J. Guo, Efficacy of immune responses induced by anti-caries DNA vaccine-loaded bacterial ghost in mice, Zhonghua Kou Qiang Yi Xue Za Zhi 49 (1) (2014) 37–41.

[94] J.H. Guo, R. Jia, M.W. Fan, Z. Bian, Z. Chen, B. Peng, Construction and immunogenic characterization of a fusion anti-caries DNA vaccine against PAc and glucosyltransferase I of Streptococcus mutans, J. Dent. Res. 83 (3) (2004) 260–270.
[95] S. Kt, M. Kmk, B. N, S. Jimson, S. R, Dental caries vaccine – a possible option? J. Clin. Diagn. Res. 7 (6) (2013) 1250–1253.
[96] T. Koga, N. Okahashi, I. Takahashi, T. Kanamoto, H. Asakawa, M. Iwaki, Surface hydrophobicity, adherence, and aggregation of cell surface protein antigen mutants of Streptococcus mutans Serotype c, Infect. Immun. 58 (2) (1990) 289–296.
[97] Y. Yamashita, W.H. Bowen, R.A. Burne, HK. Kuramitsu, Role of the Streptococcus mutans gtf genes in caries induction in the specific-pathogen-free rat model, Infect. Immun. 61 (9) (1993) 3811–3817.
[98] L. Chen, J. Zhu, Y. Li, et al., Enhanced nasal mucosal delivery and immunogenicity of anti-caries DNA vaccine through incorporation of anionic liposomes in chitosan/DNA complexes, PLoS One 8 (8) (2013) e71953 71951–71913.
[99] M. Hannig, C. Hannig, Nanotechnology and its role in caries therapy, Adv. Dent. Res. 24 (2) (2012) 53–57.
[100] Nanomedicine Taxonomy - Nanowerk [Internet]. Nanowerk.com. 2003. Available from: https://www.nanowerk.com/nanotechnology/reports/reportpdf/report31.pdf.
[101] A. Aliosmanoglu, I. Basaran, Nanotechnology in cancer treatment, J. Nanomedicine Biotherapeutic Discov 2 (2012) 107.
[102] A. Alok, P. Sunil, A. Aggarwal, N. Upadhyay, N. Agarwal, M. Kishore, Nanotechnology: a boon in oral cancer diagnosis and therapeutics, SRM J. Res. Dent. Sci. 4 (2013) 154-6024.
[103] W. Cai, K. Chen, Z.B. Li, S.S. Gambhir, X. Chen, Dual-function probe for PET and near-infrared fluorescence imaging of tumor vasculature, J. Nucl. Med. 48 (2007) 1862–1870.
[104] P.S. Reddy, P. Ramaswamy, C. Sunanda, Role of gold nanoparticles in early detection of oral cancer, J. Indian Acad Oral Med. Radiol. 22 (2010) 30–33.
[105] I.H. El-Sayed, X. Huang, M.A. El-Sayed, Surface plasmon resonance scattering and absorption of anti-EGFR antibody conjugated gold nanoparticles in cancer diagnostics: Applications in oral cancer, Nano Lett. 5 (2005) 829–834.
[106] J.C. Kah, K.W. Kho, C.G. Lee, C. James, R. Sheppard, Z.X. Shen, et al., Early diagnosis of oral cancer based on the surface plasmon resonance of gold nanoparticles, Int. J. Nanomedicine 2 (2007) 785–798.
[107] X.D. Zhang, D. Wu, X. Shen, et al., Size-dependent radiosensitization of PEG-coated gold nanoparticles for cancer radiation therapy, Biomaterials 33 (27) (2012) 6408–6419.
[108] N. Sulfikkarali, N. Krishnakumar, S. Manoharan, R.M. Nirmal, Chemopreventive efficacy of naringenin-loaded nanoparticles in 7,12-dimethylbenz(a)anthracene induced experimental oral carcinogenesis, Pathol. Oncol. Res. 19 (2013) 287–296.
[109] B. Kang, M.A. Mackey, M.A. El-Sayed, Nuclear targeting of gold nanoparticles in cancer cells induces DNA damage, causing cytokinesis arrest and apoptosis, J. Am. Chem. Soc. 132 (5) (2010) 1517–1519.
[110] M.A. Mackey, F. Saira, M.A. Mahmoud, M.A. El-Sayed, Inducing cancer cell death by targeting its nucleus: solid gold nanospheres versus hollow gold nanocages, Bioconjug. Chem. 24 (6) (2013) 897–906.
[111] Y.N. Wu, L.X. Yang, X.Y. Shi, et al., The selective growth inhibition of oral cancer by iron core-gold shell nanoparticles through mitochondria-mediated autophagy, Biomaterials 32 (20) (2011) 4565–4573.
[112] M. Sprintz, C. Benedetti, M. Ferrari, FERRARI applied nanotechnology for the management of breakthrough cancer pain, Minerva Anestesiol. 71 (2005) 419–423.
[113] K. Ramanujan, Like burrs on your clothes, virus-size capsules stick to cells to target drug delivery, Chronicle Online (2009).
[114] X. Gao, L. Yang, J.A. Petros, F.F. Marshall, J.W. Simons, S. Nie, In vivo molecular and cellular imaging with quantum dots, Curr. Opin. Biotechnol. 16 (2005) 63–72.
[115] I.L. Medintz, H.T. Uyeda, E.R. Goldman, H. Mattoussi, Quantum dot bioconjugates for imaging, labelling and sensing, Nat. Mater. 4 (2005) 435–446.
[116] X. Gao, Y. Cui, R.M. Levenson, L.W. Chung, S. Nie, In vivo cancer targeting and imaging with semiconductor quantum dots, Nat. Biotechnol. 22 (2004) 969–976.
[117] T.L. Schleyer, Nanodentistry. Fact or fiction? J. Am. Dent. Assoc. 131 (2000) 1567–1568.
[118] P. Chakravarty, R. Marches, N.S. Zimmerman, A.D. Swafford, P. Bajaj, I.H. Musselman, et al., Thermal ablation of tumor cells with antibodyfunctionalized single-walled carbon nanotubes, Proc. Natl. Acad. Sci. U. S. A. 105 (2008) 8697–8702.

[119] N.W. Kam, M. O'Connell, J.A. Wisdom, H. Dai, Carbon nanotubes as multifunctional biological transporters and near-infrared agents for selective cancer cell destruction, Proc. Natl. Acad. Sci. U. S. A. 102 (2005) 11600–11605.
[120] Broadwith P. Are nanotubes the future for radiotherapy? [Internet]. Chemistry World. 2021. Available from: https://www.chemistryworld.com/news/are-nanotubes-the-future-for-radiotherapy/3000389.article.
[121] M.M. Gottesman, T. Fojo, SE. Bates, Multidrug resistance in cancer: role of ATP-dependent transporters, Nat. Rev. Cancer 2 (2002) 48–58.
[122] D. Peer, R. Margalit, Fluoxetine and reversal of multidrug resistance, Cancer Lett. 237 (2006) 180–187.
[123] A.B. Lansdown, Microbial multidrug resistance (mdr) and Oligodynamic silver, J. Wound Care 11 (2002) 125–130.
[124] J.P. Spano, G. Carcelain, C. Katlama, D. Costagliola, Non-AIDS-defining malignancies in HIV patients: clinical features and perspectives, Bull. Cancer 93 (2006) 37–42.
[125] CB. Fields, Method for treating blood borne viral pathogens such as immunodeficiency virus. United States Patent No. 6,066,489. [2000 May 23].
[126] P. Sola, R. Bedin, F. Casoni, P. Barozzi, J. Mandrioli, E. Merelli, New insights into the viral theory of amyotrophic lateral sclerosis: study on the possible role of Kaposi's sarcoma-associated virus/human herpesvirus 8, Eur. Neurol. 47 (2002) 108–112.
[127] D. Kuck, T. Lau, B. Leuchs, A. Kern, M. Müller, L. Gissmann, et al., Intranasal vaccination with recombinant adeno-associated virus type 5 against human papillomavirus type 16 L1, J. Virol. 80 (2006) 2621–2630.
[128] D. Serraino, P. Piselli, C.l. Angeletti, M. Scuderi, G. Ippolito, M.R. Capobianchi, Infection with Epstein-Barr virus and cancer: an epidemiological review, J. Biol. Regul. Homeost. Agents 19 (2005) 63–70.
[129] J.A. Abelson, T. Moore, D. Bruckner, J. Deville, K. Nielsen, Frequency of fungemia in hospitalized pediatric inpatients over 11 years at a tertiary care institution, Pediatrics 116 (2005) 61–67.
[130] K.M. Knapp, P.M. Flynn, Newer treatments for fungal infections, J. Support. Oncol. 3 (2005) 290–298.
[131] E. Liakopoulou, K. Mutton, D. Carrington, S. Robinson, C.G. Steward, N.J. Goulden, et al., Rotavirus as a significant cause of prolonged diarrhoeal illness and morbidity following allogeneic bone marrow transplantation, Bone Marrow Transplant. 36 (2005) 691–694.
[132] R.F. Chemaly, H.A. Torres, R.Y. Hachem, G.M. Nogueras, E.A. Aguilera, A. Younes, et al., Cytomegalovirus pneumonia in patients with lymphoma, Cancer 104 (2005) 1213–1220.
[133] A.A. Marino, T.J. Berger, R.O. Becker, JA. Spadaro, The effect of selected metals on marrow cells in culture, Chem. Biol. Interact. 9 (1974) 217–223.
[134] M. Bruchez Jr., M. Moronne, P. Gin, S. Weiss, A.P. Alivisatos, Semiconductor nanocrystals as fluorescent biological labels, Science 281 (1998) 2013–2016.
[135] C. Mah, I. Zolotukhin, T.J. Fraites, J. Dobson, C. Batich, B.J. Byrne, Microsphere-mediated delivery of recombinant AAV vectors in vitro and in vivo, Mol. Ther. 1 (2000) S239.
[136] N.J. Shetty, P. Swati, K. David, Nanorobots: future in dentistry, Saudi Dent. J. 25 (2013) 49–52.
[137] D. Pradhan, L. Pratik, A. Sharma, Chaudri Nanotechnology, Future of Dentistry, Int. J. Oral Health Med. Res. 3 (6) (2017) 134–136.
[138] C.H.C. Liu, Filling in dentinal tubules, Nanotechnology 18 (2007) 475104.
[139] R.A. Freitas Jr., Nanodentistry, J. Am. Dent. Assoc. 131 (11) (2000) 1559–1566.
[140] A. Kishen, Z. Shi, A. Shrestha, K.G. Neoh, An investigation on the antibacterial and antibiofilm efficacy of cationic nanoparticulates for root canal disinfection, J. Endod. 34 (12) (2008) 1515–1520.
[141] A. Monzavi, S. Eshraghi, R. Hashemian, F. Momen-Heravi, In vitro and ex vivo antimicrobial efficacy of nano-MgO in the elimination of endodontic pathogens, Clin. Oral Investig. 19 (2) (2015) 349–356.
[142] R. Langer, J.P. Vacanti, Tissue engineering, Science 260 (1993) 920–926.
[143] B.P. Chan, K.W. Leong, Scaffolding in tissue engineering: general approaches and tissue-specific considerations, Eur. Spine. J. 17 (suppl 4) (2008) 467–479.
[144] G. Balasundaram, T.J. Webster, Nanotechnology and biomaterials for orthopedic medical applications, Nanomedicine (Lond) 1 (2006) 169–176.
[145] J. Shi, A.R. Votruba, O.C. Farokhzad, R. Langer, Nanotechnology in drug delivery and tissue engineering: from discovery to applications, Nano Lett. 10 (2010) 3223–3230.
[146] J.D. Hartgerink, E. Beniash, S.I. Stupp, Self-assembly and mineralization of peptide-amphiphile nanofibers, Science 294 (2001) 1684–1688.

[147] J. Song, V. Malathong, C.R. Bertozzi, Mineralization of synthetic polymer scaffolds: a bottom-up approach for the development of artificial bone, J. Am. Chem. Soc. 127 (2005) 3366–3372.
[148] C. Du, G. Falini, S. Fermani, C. Abbott, J. Moradian-Oldak, Supramolecular assembly of amelogenin nanospheres into birefringent microribbons, Science 307 (2005) 1450–1454.
[149] G. He, T. Dahl, A. Veis, A. George, Nucleation of apatite crystals in vitro by self-assembled dentin matrix protein 1, Nat. Mater. 2 (2003) 552–558.
[150] B. Inanç, Y.E. Arslan, S. Seker, A.E. Elçin, Y.M. Elçin, Periodontal ligament cellular structures engineered with electrospun poly(DL-lactide-co-glycolide) nanofibrous membrane scaffolds, J. Biomed. Mater. Res. A 90 (2009) 186–195.
[151] T. Qu, X. Liu, Nano-structured gelatin/bioactive glass hybrid scaffolds for the enhancement of odontogenic differentiation of human dental pulp stem cells, J. Mater Chem. B Mater Biol. Med 1 (2013) 4764–4772.
[152] X. Yang, F. Yang, X.F. Walboomers, Z. Bian, M. Fan, J.A. Jansen, The performance of dental pulp stem cells on nanofibrous PCL/gelatin/nHA scaffolds, J. Biomed Mater Res A 93 (2010) 247–257.
[153] Z. Huang, C.J. Newcomb, P. Bringas Jr., S.I. Stupp, M.L. Snead, Biological synthesis of tooth enamel instructed by an artificial matrix, Biomaterials 31 (2010) 9202–9211.
[154] Y. Kumada, S. Zhang, Significant type I and type III collagen production from human periodontal ligament fibroblasts in 3D peptide scaffolds without extra growth factors, PLoS One 5 (2010) e10305.
[155] J. Wang, X. Liu, X. Jin, et al., The odontogenic differentiation of human dental pulp stem cells on nanofibrous poly(L-lactic acid) scaffolds in vitro and in vivo, Acta Biomater. 6 (2010) 3856–3863.
[156] J. Wang, H. Ma, X. Jin, et al., The effect of scaffold architecture on odontogenic differentiation of human dental pulp stem cells, Biomaterials 32 (2011) 7822–7830.
[157] N. Ganesh, R. Jayakumar, M. Koyakutty, U. Mony, S.V. Nair, Embedded silica nanoparticles in poly(caprolactone) nanofibrous scaffolds enhanced osteogenic potential for bone tissue engineering, Tissue Eng. Part A 18 (2012) 1867–1881.
[158] K.M. Woo, V.J. Chen, PX. Ma, Nano-fibrous scaffolding architecture selectively enhances protein adsorption contributing to cell attachment, J. Biomed. Mater. Res. A 67 (2003) 531–537.
[159] G.H. Kim, Y.D. Park, S.Y. Lee, et al., Odontogenic stimulation of human dental pulp cells with bioactive nanocomposite fiber, J. Biomater. Appl 29 (2015) 854–866.
[160] X.L. Deng, M.M. Xu, D. Li, G. Sui, X.Y. Hu, X.P. Yang, Electrospun PLLA/MWNTs/HA hybrid nanofiber scaffolds and their potential in dental tissue engineering, Key Eng. Mater. 33 (0–332) (2007) 393–396.
[161] M.M. Xu, F. Mei, D. Li, et al., Electrospun poly(L-lacticacid)/nano-hydroxyapatite hybrid nanofibers and their potential in dental tissue engineering, Key Eng. Mater. 33 (0–332) (2007) 377–380.
[162] A. Besinis, R. van Noort, N. Martin, Infiltration of demineralized dentin with silica and hydroxyapatite nanoparticles, Dent. Mater. 28 (2012) 1012–1023.
[163] T. Qu, J. Jing, Y. Jiang, et al., Magnesium-containing nanostructured hybrid scaffolds for enhanced dentin regeneration, Tissue Eng. Part A 20 (2014) 2422–2433.
[164] K.M. Galler, A. Cavender, V. Yuwono, et al., Self-assembling peptide amphiphile nanofibers as a scaffold for dental stem cells, Tissue Eng Part A 14 (2008) 2051–2058.
[165] A. Vallés-Lluch, E. Novella-Maestre, M. Sancho-Tello, M.M. Pradas, G.G. Ferrer, CC. Batalla, Mimicking natural dentin using bioactive nanohybrid scaffolds for dentinal tissue engineering, Tissue Eng. Part A 16 (9) (2010) 2783–2793.
[166] J. Fletcher, W. Dominic, C. Emma Fowler, S. Mann, Electrospun mats of PVP/ACP nanofibres for remineralization of enamel tooth surfaces, Cryst. Eng. Comm. 13 (2011) 3692.
[167] S. Zhang, Y. Huang, X. Yang, et al., Gelatin nanofibrous membrane fabricated by electrospinning of aqueous gelatin solution for guided tissue regeneration, J. Biomed. Mater. Res. A 90 (2009) 671–679.
[168] K.T. Shalumon, S. Sowmya, D. Sathish, K.P. Chennazhi, S.V. Nair, R. Jayakumar, Effect of incorporation of nanoscale bioactive glass and hydroxyapatite in PCL/chitosan nanofibers for bone and periodontal tissue engineering, J. Biomed. Nanotechnol 9 (2013) 430–440.
[169] H. He, J. Yu, J.E.L. Cao, D. Wang, H. Zhang, et al., Biocompatibility and osteogenic capacity of periodontal ligament stem cells on nHAC/PLA and HA/TCP scaffolds, J. Biomater Sci. Polym. Ed. 22 (1-3) (2011) 179–194.
[170] C.L. Yang, J.S. Lee, U.W. Jung, Y.K. Seo, J.K. Park, S.H. Choi, Periodontal regeneration with nano-hyroxyapatite-coated silk scaffolds in dogs, J. Periodontal Implant Sci. 43 (2013) 315–322.

[171] S.J.R. Srinivasan, K.P. Chennazhi, S.V. Nair, R. Jayakumar, Biocompatible alginate/nano bioactive glass ceramic composite scaffolds for periodontal tissue regeneration, Carbohydr. Polym. 87 (2012) 274–283.
[172] J. Guo, Z. Meng, G. Chen, et al., Restoration of critical-size defects in the rabbit mandible using porous nanohydroxyapatite-polyamide scaffolds, Tissue Eng. Part A 18 (2012) 1239–1252.
[173] X. Zhang, M. Xu, X. Liu, et al., Restoration of critical-sized defects in the rabbit mandible using autologous bone marrow stromal cells hybridized with nano-β-tricalcium phosphate/collagen scaffolds, J. Nanomater 2013 (2013) 1–8.
[174] K. Kettler, K. Veltman, A. van Wezel, A. Jan Hendriks, Cellular uptake of nanoparticles as determined by particle properties, experimental conditions, and cell type, Environ. Toxicol. Chem. 33 (2014) 481–492.
[175] L. Treuel, X. Jiang, G.U. Nienhaus, New views on cellular uptake and trafficking of manufactured nanoparticles, J. R. Soc. Interface 10 (2013) 20120939.
[176] F.M. Chen, Z.W. Ma, G.Y. Dong, Z.F. Wu, Composite glycidyl methacrylated dextran (Dex-GMA)/gelatin nanoparticles for localized protein delivery, Acta Pharmacol. Sin. 30 (2009) 485–493.
[177] B. Gokcen-Rohlig, M. Yaltirik, S. Ozer, E.D. Tuncer, G. Evlioglu, Survival and success of ITI implants and prostheses: retrospective study of cases with 5-year follow-up, Eur. J. Dent. 3 (2009) 42−49.
[178] M.R. Baig, M. Rajan, Effects of smoking on the outcome of implant treatment: a literature review, Indian J. Dent. Res. 18 (2007) 190−195.
[179] J. Zupnik, S.-.W. Kim, D. Ravens, N. Karimbux, K. Guze, Factors associated with dental implant survival: a 4-year retrospective analysis, J. Periodontol. 82 (2011) 1390−1395.
[180] P. Worthington, Introduction: history of implants, in: R.JE. Wplbr (Ed.), Osseointegration in Dentistry: An Overview, Quintessence Publishing, Illinois, 2003, p. 2.
[181] P.I. Branemark, R. Adell, U. Breine, B.O. Hansson, J. Lindstrom, A. Ohlsson, Intraosseous anchorage of dental prostheses. I. Experimental studies, Scand J. Plast Reconstr Surg 3 (1969) 81–100.
[182] L. Linder, T. Albrektsson, P.I. Branemark, H.A. Hansson, B. Ivarsson, U. Jonsson, I. Lundstrom, Electron-microscopic analysis of the bone titanium interface, Acta Orthop. Scand. 54 (1) (1983) 45–52.
[183] S. Hobo, E. Ichida, L. Garcia, Osseointegration and occlusal rehabilitation. In: Osseointegration Implant Systems, Quintessence Publishing, Tokyo, 1990, pp. 3–4.
[184] P. Tomsia, M.E. Launey, J.S. Lee, M.H. Mankani, U.G.K. Wegst, E. Saiz, Nanotechnology approaches for better dental implants, Int. J. Oral Maxillofac. Implants. 26 (2011) 25–49.
[185] B.D. Boyan, C.H. Lohmann, D.D. Dean, Sylvia V.L., D.L. Cochran, Z. Schwartz, Mechanisms involved in osteoblast response to implant surface morphology, Annu. Rev. Mater. Res. 31 (2001) 357–371.
[186] B. Kasemo, Biological surface science, Surf. Sci. 500 (2002) 656–677.
[187] L.L. Hench, Bioceramics, J. Am. Ceram. Soc. 81 (7) (1998) 1705–1728 [Review].
[188] L.L. Hench, O. Andersson, Bioactive glasses, in: L.L. Hench, J. Wilson (Eds.), An Introduction to Bioceramics, World Scientific, Singapore, 1993, pp. 41–62.
[189] L.L. Hench, I.D. Xynos, J.M. Polak, Bioactive glasses for in situ tissue regeneration, J. Biomater Sci. (Polymer) 15 (4) (2004) 543–562 [Review].
[190] R. Beutner, J. Michael, B. Schwenzer, D. Scharnweber, Biological nano-functionalization of titanium-based biomaterial surfaces: a flexible toolbox, J. R. Soc., Interface 7 (2010) S93–S105.
[191] H. Schliephake, D. Scharnweber, Chemical and biological functionalization of titanium for dental implants, J. Mater. Chem. 18 (21) (2008) 2404–2414.
[192] H. Schliephake, D. Scharnweber, M. Dard, A. Sewing, A. Aref, S. Roessler, Functionalization of dental implant surfaces using adhesion molecules, J. Biomed Mater Res B-Appl Biomater 73B (1) (2005) 88–96.
[193] K. Kieswetter, Z. Schwartz, T.W. Hummert, D.L. Cochran, J. Simpson, D.D. Dean, B.D. Boyan, Surface roughness modulates the local production of growth factors and cytokines by osteoblast-like MG-63 cells, J. Biomed. Mater. Res. 32 (1) (1996) 55–63.
[194] Z. Schwartz, J.Y. Martin, D.D. Dean, J. Simpson, D.L. Cochran, B.D. Boyan, Effect of titanium surface roughness on chondrocyte proliferation, matrix production, and differentiation depends on the state of cell maturation, J. Biomed. Mater. Res. 30 (2) (1996) 145–155.
[195] O. Zinger, G. Zhao, Z. Schwartz, J. Simpson, M. Wieland, D. Landolt, B. Boyan, Differential regulation of osteoblasts by substrate microstructural features, Biomaterials 26 (14) (2005) 1837–1847.

[196] D. Deporter, Dental implant design and optimal treatment outcomes, Int. J. Periodontics Restorative Dent 29 (6) (2009) 625–633.
[197] IntroductionDisinfection & Sterilization Guidelines | Guidelines Library | Infection Control | CDC [Internet]. Cdc.gov. 2021. Available from: https://www.cdc.gov/infectioncontrol/guidelines/disinfection/introduction.html.
[198] A. Chang, A.H. Schnall, R. Law, A.C. Bronstein, J.M. Marraffa, H.A. Spiller, et al., Cleaning and disinfectant chemical exposures and temporal associations with COVID-19—National Poison Data System, United States, January 1, 2020–March 31, 2020, MMWR Morb. Mortal. Wkly. Rep. 69 (2020) 496–498.
[199] A. Kapoor, R. Saha, Hand washing agents and surface disinfectants in times of Coronavirus (COVID-19) outbreak, Indian J. Community Health 32 (2020) 225–227.
[200] H. Huang, C. Fan, M. Li, H.-.L. Nie, F.-.B. Wang, H. Wang, et al., COVID-19: a call for physical scientists and engineers, ACS Nano 14 (4) (2020) 3747–3754.
[201] M.M. Querido, L. Aguiar, P. Neves, C.C. Pereira, JP. Teixeira, Self-disinfecting surfaces and infection control, Colloids Surf. B Biointerfaces 178 (2019) 8–21.
[202] F. Geyer, M. D'Acunzi, A. Sharifi-Aghili, A. Saal, N. Gao, A. Kaltbeitzel, et al., When and how self-cleaning of superhydrophobic surfaces works, Sci. Adv. 6 (2020) eaaw9727.
[203] S.P. Dalawai, M.A. Saad Aly, S.S. Latthe, R. Xing, R.S. Sutar, S. Nagappan, et al., Recent advances in durability of superhydrophobic self-cleaning technology: a critical review, Prog. Org. Coat. 138 (2020) 105381.
[204] L. Dyshlyuk, O. Babich, S. Ivanova, N. Vasilchenco, A. Prosekov, S. Sukhikh, Suspensions of metal nanoparticles as a basis for protection of internal surfaces of building structures from biodegradation, Case Stud. Constr. Mater. 12 (2020) e00319.
[205] M. Rai, S.D. Deshmukh, A.P. Ingle, I.R. Gupta, M. Galdiero, S. Galdiero, Metal nanoparticles: the protective nanoshield against virus infection, Crit. Rev. Microbiol. 42 (2016) 46–56.
[206] N. Vaze, G. Pyrgiotakis, J. McDevitt, L. Mena, A. Melo, A. Bedugnis, et al., Inactivation of common hospital acquired pathogens on surfaces and in air utilizing engineered water nanostructures (EWNS) based nanosanitizers, Nanomed. Nanotechnol. Biol. Med. 18 (2019) 234–242.
[207] Coronavirus: Nanotech Surface Sanitizes Milan with Nanomaterials Remaining Self-sterilized for Years | STATNANO [Internet]. Statnano.com. 2021. Available from: https://statnano.com/news/67531/Coronavirus-Nanotech-Surface-Sanitizes-Milan-with-Nanomaterials-Remaining-Self-sterilized-for-Years.
[208] Mineral Nanocrystal-based Coating Activated by Light Kills Coronavirus | STATNANO [Internet]. Statnano.com. 2021. Available from: https://statnano.com/news/67583/Mineral-Nanocrystal-based-Coating-Activated-by-Light-Kills-Coronavirus.
[209] WCW. Chan, Nano research for COVID-19, ACS Nano 14 (2020) 3719–3720.
[210] M. Saccucci, E. Bruni, D. Uccelletti, A. Bregnocchi, M. Sarto, M. Bossù, et al., Surface disinfections: present and future, J. Nanomater 2018 (2018) 1–9.
[211] M. Kchaou, K. Abuhasel, M. Khadr, F. Hosni, M. Alquraish, Surface disinfection to protect against microorganisms: overview of traditional methods and issues of emergent nanotechnologies, Appl. Sci. 10 (17) (2020) 6040.
[212] Z.D. Yesil, S. Alapati, W. Johnston, RR. Seghi, Evaluation of the wear resistance of new nanocompositeresin restorative materials, J. Prosthet. Dent. 99 (2008) 435–443.
[213] M. Redlich, A. Katz, L. Rapoport, H.D. Wagner, Y. Feldman, R. Tenne, Improved orthodontic stainlesssteel wires coated with inorganic fullerene-like nanoparticles of WS(2) impregnated in electroless nickel-phosphorous film, Dent. Mater. 24 (2008) 1640–1646.
[214] B. Cao, Y. Wang, N. Li, B. Liu, Y. Zhang, Preparation of an orthodontic bracket coated with an nitrogen-doped TiO (2-x) N (y) thin film and examination of its antimicrobial performance, Dent. Mater. J. 32 (2013) 311–316.
[215] L. Eslamian, A. Borzabadi-Farahani, N. Mousavi, A. Ghasemi, The effects of various surface treatmentson the shear bond strengths of stainless steel brackets to artificially-aged composite restorations, Aust Orthod J 27 (2011) 28–32.
[216] J.P. Alcock, M.E. Barbour, J.R. Sandy, AJ. Ireland, Nanoindentation of orthodontic archwires: the effect ofdecontamination and clinical use on hardness, elastic modulus and surface roughness, Dent. Mater. 25 (2009) 1039–1043.

[217] B. Gibbins, L. Warner, The role of antimicrobial silver nanotechnology, MDDI; (2005).
[218] E. Giertsen, Effects of mouth-rinses with tri-closan, zinc ions, copolymer, and sodium lauryl sulphate combined with fluoride on acid formation by dental plaque in vivo, Caries Res. 38 (2004) 430–443.
[219] S. Deville, J. Chevalier, G. Fantozzi, et al., Low-temperature ageing of zirconia-toughened alumina ceramics and its implication in biomedical implants, J. Eur. Ceram. Soc. 23 (2003) 2975–2982.
[220] Pecharroman, J.F. Bartolome, J. Requena, et al., Percolative mechanism of aging in zirconia-containing ceramics for medical applications, J. Eur. Ceram. Soc. 15 (2003) 507–511.
[221] A. Kishen, Z. Shi, A. Shrestha, K.G. Neoh, An investigation on the antibacterial and antibiofilm efficacy of cationic nanoparticulates for root canal disinfection, J. Endod. 34 (12) (2008) 1515–1520.
[222] A. Monzavi, S. Eshraghi, R. Hashemian, F. Momen-Heravi, In vitro and ex vivo antimicrobial efficacy of nano-MgO in the elimination of endodontic pathogens, Clin. Oral Investig. 19 (2) (2015) 349–356.
[223] R. Smeets, B. Stadlinger, F. Schwarz, B. Beck-Broichsitter, O. Jung, C. Precht, et al., Impact of dental implant surface modifications on osseointegration, BioMed Res. Int. 2016 (2016).

CHAPTER 16

Advanced nanomaterial for point-of-care chemotherapy

Soumya S. Panda, Hemlata Das, Saroj Prasad Panda, Spoorthy Kolluri, Sindhu Kilaru
Department of Medical Oncology and Department of Pathology, IMS and Sum Hospital, Bhubaneswar, India

16.1 Introduction

Cancer is the second most common cause of mortality worldwide, accounting for 13% of total deaths [1]. International Agency for Research on Cancer (GLOBOCAN 2018) estimated 18.1 million new cancer cases, and about 9.6 million cancer deaths which may reach an estimated 19.3 million new cancer cases per year by 2025 [2]. Low- and middle-income countries share around half of the overall cancer cases and related mortality [3]. The current therapeutic armamentarium for cancer includes surgery, radiotherapy, and systemic treatments comprising hormone therapy, chemotherapy, immunotherapy and targeted therapie [4]. Chemotherapy remains the main modality for cancer treatment despite significant developments in other modalities [5].

16.2 Challenges with conventional chemotherapy agents

Numerous challenges hamper effective cancer treatment and development of effective drugs. Many chemotherapeutic agents are hydrophobic requiring solvents or carriers as solubilization strategies which may cause unwarranted toxicities, adding to the already innate toxicity of the chemotherapeutic agents [6,7]. The highly non-specific targeting of cancer cells is another particular concern with the conventional chemotherapy preparations, which exposes normal cells to the undesirable effects of drugs, thereby limiting the required dose within the tumor resulting in suboptimal treatment [8,9]. High doses are required to compensate the rapid elimination and non-specific tissue distribution, which may result in toxicity leading to a low quality of life and economic burden [10,11]. Furthermore, resistance may develop, which may lead to cross-resistance to a wide range of different drugs hampering further mangement [10]. Drug resistance is of particular concern with conventional chemotherapy, with majority of the patients with metastatic tumor eventually affected [12]. Decreased drug absorption in the tumor cells due to activated efflux pumps, such as P-glycoprotein, activated glutathione detoxification system, or altered apoptosis control are the underlying cellular mechanisms for multidrug resistance (MDR) [13,14].

Advanced Nanomaterials for Point of Care Diagnosis and Therapy
DOI: https://doi.org/10.1016/B978-0-323-85725-3.00006-4

To address these issues, molecularly targeted therapies have emerged. Conventional chemotherapy kills most of the cancer cells initially, reducing the tumor size. However, the spared cancer cells result in relapse and metastasis. Whereas, with targeted chemotherapy, cancer cells may eventually get eliminated because of the loss of self-renewal and proliferation capacities [15]. Targeted drug delivery and considerations for the channelization of active drug moiety to the site of action has led to the development of novel drug delivery systems in the past few decades. The efficacy, safety and treatment compliance can be significantly improved with the evolution of a conventional drug molecule to a novel delivery system [16]. The major components of a novel drug delivery systems are the transporting vehicle (for example lipid), the loaded drug, the "programmable" targeting agent (for example receptor specific ligand) that enables the appropriate delivery routes to ensure the required concentration of drug reaches the tumor site avoiding toxicity on healthy tissues [10]. Nanoparticles are considered as appropriate carriers for personalized cancer treatment due to their ability of targeted delivery, and thereby, overcoming limitations of conventional chemotherapy [8,10,11].

16.3 Nanotherapeutics to overcome conventional chemotherapy limitations

Nanotechnology comprises the use of nano size range (0.1–100 nm) drugs or devices. The use of nanotechnology based drug delivery has revolutionized cancer management [17]. The medical applications of nanotechnology – nanomedicine – facilitate high specificity of drugs to interact at subcellular levels [18]. Nanocarrier drug delivery systems can increase the solubility of drugs, cross biological barriers and act against multiple types of cancer cells through targeting signaling pathways associated with tumor development and specific tumor markers [4,6,19].

16.3.1 Ideal drug delivery

The characteristics of an ideal chemotherapeutic drug delivery system includes the use of multifunctional nanocarriers, which combine the active (pH-, light-, and thermosensitivity) and passive targeting (enhanced permeability and retention [EPR] effects) aspects, and magnetic properties [20]. The multifunctional nanocarriers combine with pH-, light-, and thermosensitive and magnetic particles and deliver the chemotherapeutic agents or monoclonal antibodies at the cancer target site (Fig. 16.1) [21-24].

The ideal drug delivery system should allow the solubilzation of the drug in an aqueous media decreasing the infusion time, and achieve stable drug concentrations at tumor sites with minimal drug exposure to healthy cells (Table 16.1) [25].

The nanotherapeutics provides a longer shelf-life, can incorporate both hydrophilic and hydrophobic substances, improve the biodistribution and therapeutic index of drugs, facilitate sustained release of the drug both during the transportation and at

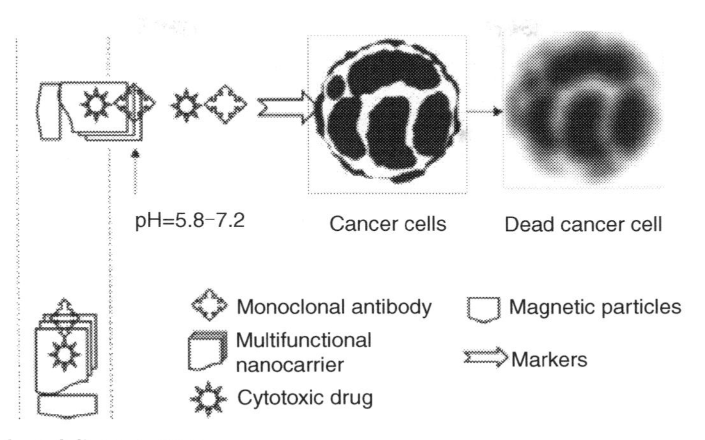

Fig. 16.1 ***Ideal drug delivery system.***

Table 16.1 Characteristics of an ideal drug delivery system.

Low cost
Shorter infusion time
Less frequent dosing schedule
Easy to prepare at clinics for administration
High therapeutic index
• High accumulation in tumors – active targeting
• Low accumulation in health tissues
• Passive targeting
• Enhanced permeability and retention effects
• High drug concentration – long circulation, slow drug release
Simple, scalable, and reproducible manufacturing

the site of action, and lead to the increased intercellular drug concentration through enhanced permeability retention (EPR) or endocytosis [6,26].

The opsonisation (protein covering) of nanotherapeutics occur in the intravascular circulation and improves the targeting and treatment of tumors along with a marked decrease in the drug concentration, thus, minimizing the toxicities of anticancer drugs [27]. Fig. 16.2 enlists the characteristics of various drug delivery nanosystems [27,28].

16.4 Nanocarriers in drug delivery systems

Nanocarriers used in drug delivery systems consists of structures with sizes in the nm or μm ranges, made of polymers (polymeric carriers, micelles or dendrimers), lipids (liposomes), solid lipid carriers, gold carriers, nanotubes and magnetic carriers (Fig. 16.3) [9,11,29].

16.4.1 Key properties of nanomaterials

The key physicochemical properties of nanomaterials including their ultra-small size, large surface to volume ratio, ability to load diverse drugs, presence of biochemical

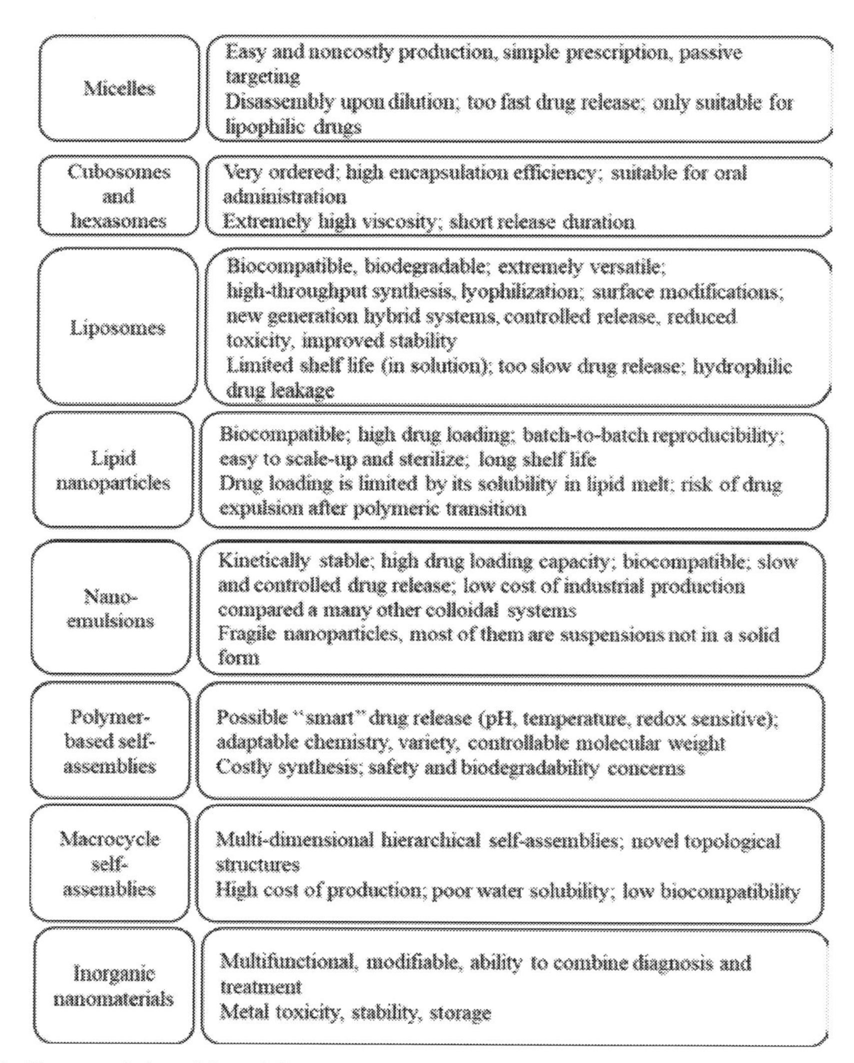

Micelles	Easy and noncostly production, simple prescription, passive targeting Disassembly upon dilution; too fast drug release; only suitable for lipophilic drugs
Cubosomes and hexasomes	Very ordered; high encapsulation efficiency; suitable for oral administration Extremely high viscosity; short release duration
Liposomes	Biocompatible, biodegradable; extremely versatile; high-throughput synthesis, lyophilization; surface modifications; new generation hybrid systems, controlled release, reduced toxicity, improved stability Limited shelf life (in solution); too slow drug release; hydrophilic drug leakage
Lipid nanoparticles	Biocompatible; high drug loading; batch-to-batch reproducibility; easy to scale-up and sterilize; long shelf life Drug loading is limited by its solubility in lipid melt; risk of drug expulsion after polymeric transition
Nano-emulsions	Kinetically stable; high drug loading capacity; biocompatible; slow and controlled drug release; low cost of industrial production compared a many other colloidal systems Fragile nanoparticles, most of them are suspensions not in a solid form
Polymer-based self-assemblies	Possible "smart" drug release (pH, temperature, redox sensitive); adaptable chemistry, variety, controllable molecular weight Costly synthesis; safety and biodegradability concerns
Macrocycle self-assemblies	Multi-dimensional hierarchical self-assemblies; novel topological structures High cost of production; poor water solubility; low biocompatibility
Inorganic nanomaterials	Multifunctional, modifiable, ability to combine diagnosis and treatment Metal toxicity, stability, storage

Fig. 16.2 ***Characteristics of drug delivery nanosystems.***

moieties on surface, hydrophilic or hydrophobic nature, and physical appearance (shape or morphology) facilitate better functionality and efficiency in therapeutic applications [30–32].

16.4.1.1 Shape and size

The shape and size of the nanoparticles affect drug stability and transport behavior in blood circulation, especially in the tumor vasculature, drug-tissue interactions, loading and release of the drug, and accumulation in the tumor for therapeutic effects [33–35]. Furthermore, their nanoparticle size, shape and geometry also affect fluid dynamics

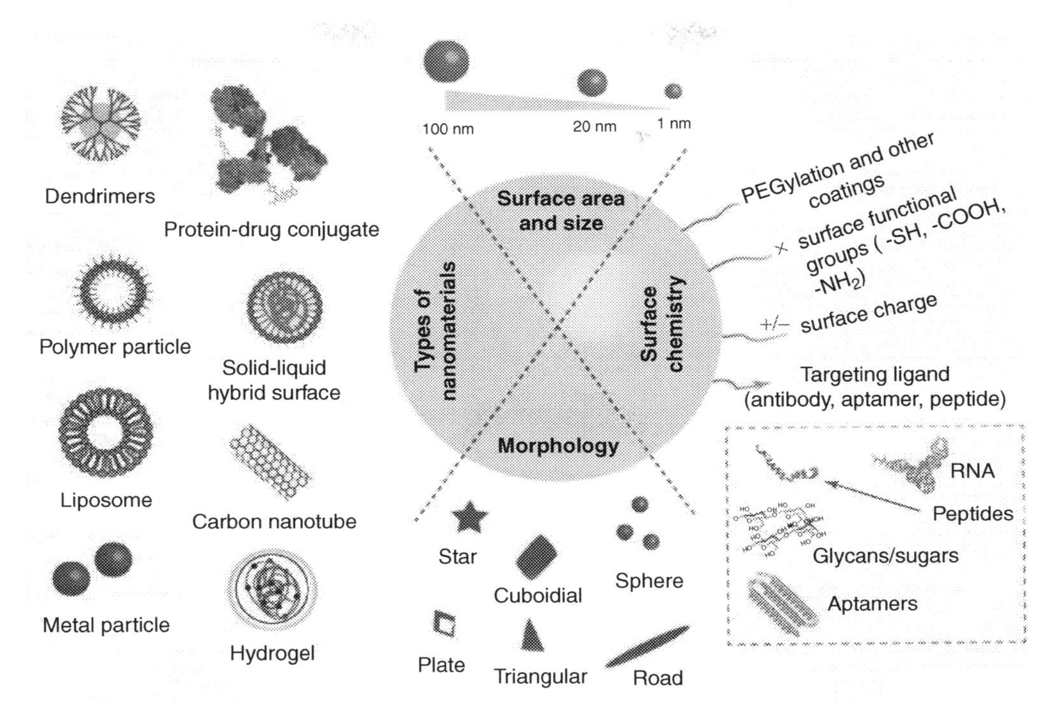

Fig. 16.3 ***Nanomaterials in chemoterhapy.[29]*** Adopted from Navya et al., Nano Converg. 2019;6:23[29] under the terms of the Creative Commons Attribution 4.0 International License (http://creativecommons.org/licenses/by/4.0/). *Credits: Navya PN, Kaphle A, Srinivas SP, Bhargava SK, Rotello VM, Daima HK. Current trends and challenges in cancer management and therapy using designer nanomaterials. Nano Converg. 2019;6(1):23. Published 2019 Jul 15. doi:10.1186/s40580-019-0193-2.*

including the hemodynamic, van der Waals, buoyancy and electrostatic forces in the tumor environment, and in the intracellular uptake of the drugs [29,33]. The development of nano-electromechanical system has led to the production of uniformly sized and shaped nanoparticles [36]. Nanoparticles smaller than 100 nm remain in blood vessels within fenestrae of the endothelial lining [33].

16.4.1.2 Surface properties

The high surface charges on the nanoparticles can dominate their material properties. The high surface area to volume ratio of nanoparticles facilitates the availability of a greater interface with their surrounding environment [29,35,37]. The physicochemical properties of nanoparticles are dynamic and the size and shape of the particles can change over time and with the environment. Nanoparticle interaction with biological media leads to the formation of a protein coating (corona), which increases the size of the nanoparticles. Also, nanoparticles adsorb material from the environment to reduce the surface energy, which further increases their size. The environmental

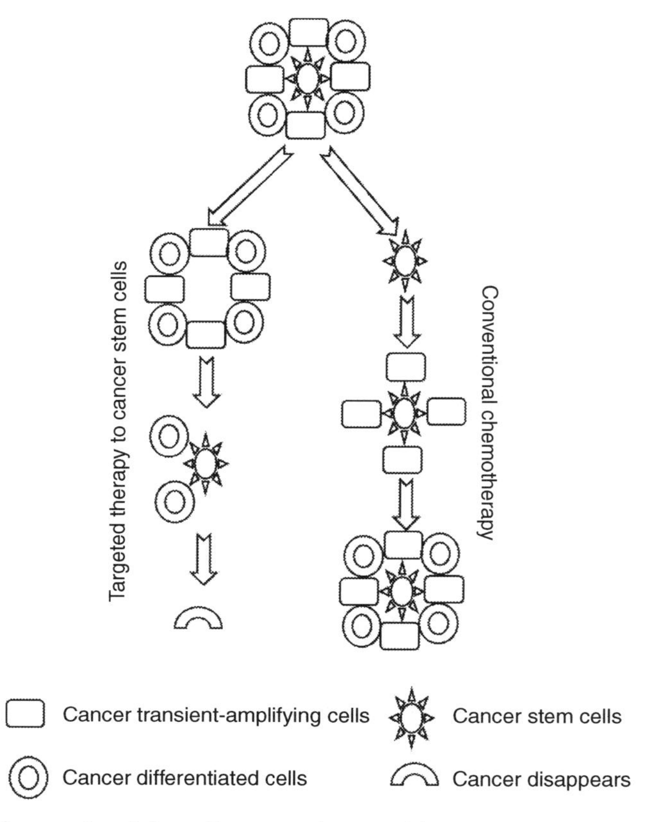

Fig. 16.4 *Effects of conventional chemotherapy and targeted therapy.*

reactions such as oxidation or corrosion or adsorption of material on nanoparticle surface can transform the nanoparticles into core-shell particles [38], and may hamper their efficiency.

16.4.1.3 Active and passive targeting

Due to issues like relapse, resistance and metastasis with conventional chemotherapy, the use of targeted therapies has increased. The targeted therapy prevents self-renewal and proliferation of the cells, and thus minimizes the aforementioned challenges with conventional chemotherapy (Fig. 16.4) [20,39].

16.4.1.3.1 Passive targeting of nanoparticles

The effectiveness of chemotherapeutic agents depend on the time required to reach the tumor and the proper dosage to exert maximum antitumor activity [29]. The leaky vasculature and compromised lymphatic drainage of the tumor allows the accumulation of nanoparticles through passive targeting through the EPR effects [35,40]. The retention of nanoparticles in the tumor environment is facilitated by the absence of normal lymphatic drainage in the tumor vasculature. Hence, the small molecule drugs encapsulated in nanoparticles allows increased systemic circulation, provides tumor selectivity and minimizes the toxic effects [27]. The passive targeting through EPR can be enhanced through chemical (bradykinin, nitric oxide, peroxynitrite, prostaglandins, vascular permeability factor/vascular endothelial growth factor) or mechanical (ultrasound, radiation, hyperthermia or photo-immunotherapy) stimulations.

16.4.1.3.2 Active targeting

The active targeting significantly increases the drugs delivery to the tumors as compared with passive targeting. Active targeting utilizes the ligands showing preferential binding toward receptors overexpressed onto the cancer cells. This phenomenon increases the nanocarrier affinity to the cancer cell surface and improves drug penetration. The common ligands used for active targeting of chemotherapeutics are epidermal growth factor receptor, transferrin, death receptor complexes and folate ligand [27,41].

16.4.1.4 Minimizing reticuloendothelial system uptake

The reticuloendothelial system (RES) uptake led clearance of nanoparticles decreases the accumulation of nanoparticles in the tumors. The RES uptake can be minimized through decreased affinity of proteins responsible for the opsonization of nanoparticles through coating with polyethylene glycol (PEG) or other amphipathic agents [41,42].

16.4.1.5 Overcome multi-drug resistance

Multiple drug resistance (MDR) of chemotherapeutics or novel targeted drugs causes ~90% cancer mortality rate [5]. The mechanism of MDR includes increased drug efflux, accelerated metabolism of xenobiotics, and genetic factors [5]. Nanoparticles increase the drug efflux altered metabolism, activate DNA repair and change apoptotic pathways thus leading to activation of MDR inhibitors [6,43].

16.4.2 Tumor uptake of nanoparticles

Tumor microenvironment and vasculature, RES, blood-brain barrier, and kidney filtration are the biological barriers that influence drug delivery [44]. Several strategies have been investigated to improve the tumor penetration of nanomedicines through nanoparticle design [45]. Table 16.2 enlists the different strategies for tumor uptake of nanoparticles [44].

16.4.3 Typical nanocarrier systems

Several nanoparticles, including inorganic nanomaterials like carbon nanotubes, silica nanoparticles, gold nanoparticles, magnetic nanoparticles and quantum dots, and organic nanomaterials, such as polymeric micelles, liposomes, dendrimers have been investigated for cancer chemotherapy (Fig. 16.5) [46].

Table 16.2 Strategies for tumor uptake of nanoparticles.

Nanoparticle design
Tumor infiltration • Personalized nanoparticle design • Smaller size nanoparticles for deeper penetration • Stimuli-responsiveness modifications of nanoparticle physicochemical properties • Decrease the uptake by macrophages through surface functionalization Tumor retention • Use of nanoparticles that respond to external stimuli • Specific biomarker use • Self-assembly and self-expansion of nanoparticles in the tumors • Nanoparticle conjugation with ligands
Tumor environment modification
• Vasoconstrictor use to adjust the blood pressure • Use of radio and immunotherapies to enhance nanoparticle accumulation in the tumor • Use of photodynamic therapy for increased endothelial gaps • Use of mechanical stimulation for enhanced capillary permeability • Use of antineogenesis agents to manage interstitial fluid pressure

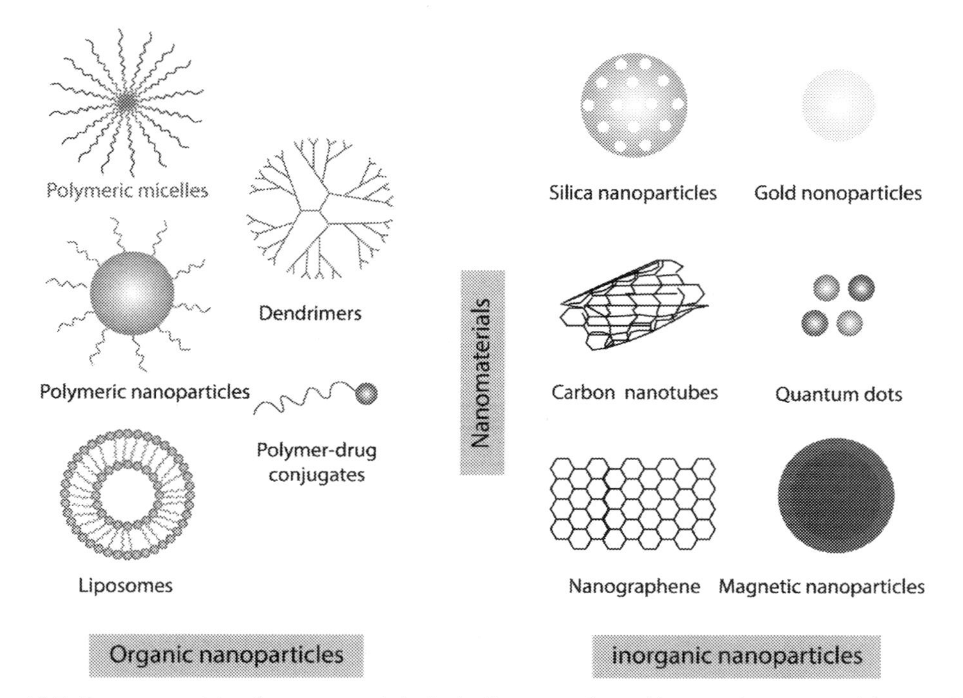

Fig. 16.5 ***Representation of nanomaterials, including organic and inorganic nanoparticles, applied for cancer therapy.*** *Adapted from Zhou et al., Nanotechnol Rev 2017; 6(5): 473–496 [46].*

16.4.4 Organic nanoparticles

Organic nanoparticles such as liposomes, polymeric nanoparticles, dendrimers and micelles have been investigated as drug deliver carrier due to their attractive properties and low toxicity profile [29,47].

16.4.4.1 Polymeric micelles

Polymeric micelles are spherical, nano-sized (20–200 nm) colloidal nanocarriers developed with a hydrophilic shell and hydrophiobic core architecture that enable loading of hydrophobic drugs to the core and improves the solubility [46,48,49]. The hydrophobic polymer form the micelle core whereas hydrophilic part forms the corona (shell) [49]. Polymeric micelles can increase the bioavailability of poorly water soluble drugs. Polysaccharides, poly(ε-caprolactone) (PCL), poly(lactide) (PLA), and poly(lactic-co-glycolic acid) (PLGA) are the most commonly used polymers in hydrophobic core construction whereas PEG is commonly used for hydrophilic shell of the polymeric micelles. The hydrophobic core allows entrapment of the hydrophobic drugs thus increasing drug solubilization. In addition, the small particle size can facilitate polymeric micelles to circumvent a faster renal excretion and subsequently increase their circulation time [31,46].

16.4.4.2 Polymer nanoparticles

Polymeric nanoparticles (10–1000 nm) are colloidal particles that are comprised of drug encapsulated within or adsorbed to the polymers [31,50]. Polymeric nanoparticles have demonstrated stability, biocompatibility, biodegradability, tailorability and associated with low cost for cancer treatment [51]. Polymeric nanoparticles can improve the effectiveness of chemotherapeutic agents through the EPR effects [46].

16.4.4.3 Dendrimers

Dendrimers are nanosized (1–15 nm) radially symmetric monodisperse molecules with a highly branched 3D architecture including a symmetric core, and inner and outer shells [31,46,52]. The varied characteristics of dendrimers including polyvalency, self-assembling, electrostatic interactions, chemical stability, low cytotoxicity and solubility. The presence of a higher density of surface functional group as well as empty internal cavities makes them a good choice for anticancer nanomedicines [46,52]. Nanoparticle dendrimers can improve the drugs' therapeutic index through surface derivatization with PEGylation, acetylation, glycosylation, and various amino acids [53].

16.4.4.4 Liposomes

Liposomes (10–1000 nm) are closed bilayer spherical structures comprised of an aqueous compartment which is covered by ≥1 lipid bilayer. Liposomes are formulated through dispersion of phospholipids with hydrophobic long chain tails and hydrophilic heads [46,54]. Liposomes enable entrapment of the water soluble drugs in their hydrophilic core and lipophilic drugs in the lipid bilayer. The limited tissue uptake of the chemotherapeutic agent due to encapsulation within liposomes leads to increased therapeutic index [55].

16.4.4.5 Polymer-drug conjugates

Polymer-drug conjugates (5–50 nm) are characterized by the conjugation of therapeutic agents to the water soluble polymeric carriers, such as PEG [46,56]. Polymer-drug conjugates provide improved drug solubilization, prolonged circulation, reduced immunogenicity, controlled release and enhanced safety [56]. Drug conjugates are the most successful nanotherapeutics used in clinical cancer treatment [46].

16.4.4.6 Inorganic nanoparticles [31,47]

In cancer management, inorganic nanoparticles with unique physical and chemical characteristics have been exploited as therapeutic agents.

16.4.4.7 Silica nanoparticles

In the past few decades, several silica-based nanostructures have been synthesized. Silica based nanoparticles (20–100) can be used to encapsulate various chemotherapeutic agents [46]. Silica nanoparticles can be solid silica or mesoporus silica nanoparticles. The mesoporus silica nanoparticles have emerged as a drug delivery system with potential for many anticancer drugs. The mesoporus silica nanoparticles can exert specific targeting of the cells with minimal toxicity, preserve cargo as it escapes the endo-lysosomal membrane and can be biodegraded or cleared from the organism to minimize toxicity [57].

16.4.4.8 Gold nanoparticles

Gold nanoparticles (1–100 nm) are promising novel agents for anti-cancer therapy [46]. Gold particles has been used since the 1900s' but recently gold nanoparticles are increasingly used [58]. The unique physicochemical properties of gold nanoparticles include the surface plasmon resonance exerting strong optical properties, and surface modification through binding with amine and thiol groups. Gold nanoparticles can be targeted to cancer cells overexpressing cell surface receptors [59]. Furthermore, the size and shape of gold nanoparticles can be easily controlled [60]. Owing to the advantage of being fully multifunctional, and possibility of combining different desired functionalities in one molecular-sized package, gold nanoparticles are preferred [61].

16.4.4.9 Carbon nanotubes

Carbon nanotubes (0.4–2 nm) are thin sheets of benzene ring carbons rolled up into a seamless tubular structure [62]. Carbon nanotubes possess characteristics such as biocompatibility, surface-to-volume ratio, and thermal conductivity [46]. Carbon nanotubes have been used in cancer therapy due to their favorable properties like electrics, thermotics, optics, mechanics and biology characteristics [46]. Discharge, laser ablation and chemical vapor deposition are the common synthesis techniques for carbon nanotubes [62].

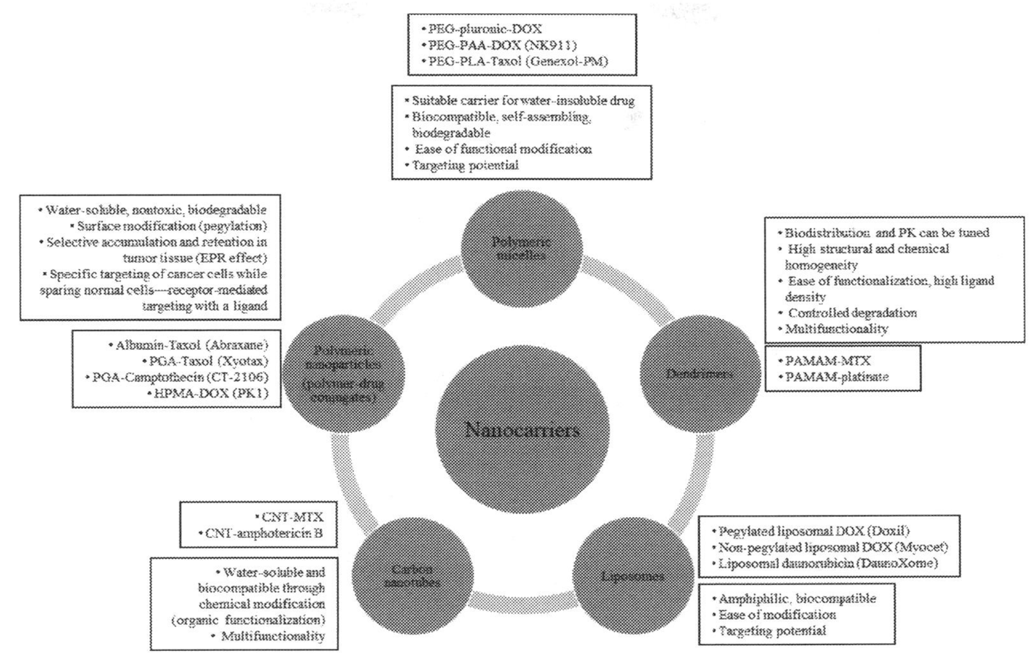

Fig. 16.6 ***Examples of common nanocarriers used in cancer drug delivery.*** CNT, carbon nanotube; DOX, doxorubicin; EPR, enhanced permeability and retention; MTX, methotrexate; PAMAM, poly(amidoamine); PEG, polyethylene glycol; PGA, poly(L-glutamic acid).

16.4.4.10 Quantum dots

Quantum dots are 2–100 nm sized metal compounds that exert tunable optical and EPR effects. Most of the quantum dots are semiconductors, and their optical properties during production can be regulated by temperature and choice of surfactants [46]. In quantum dots, the core consisting of elements from II–VI or III–V of the periodic table is covered by a semiconductor shell. Indium phosphide, indium arsenide, gallium arsenate, gallium nitride, zinc sulfide, zinc selenium, cadmium selenium, and cadmium tellurium are the common quantum dots. Targeted polymer coating on the surface of quantum dots facilitates quantum dots to exert targeted action for the treatment of cancers [63].

16.4.4.11 Magnetic nanoparticles

Magnetic nanoparticles can be synthesized into various sizes and can be modified with different functional groups in order to load a number of molecules, and have been established as a promising drug carrier [64]. Magnetic nanoparticles are generally coated with surfactants or polymers to increase their biocompatibility [46]. Fig. 16.6 illustrates characteristics and examples of various nanocarrier system commonly used in the cancer treatment [9,11].

16.4.5 Nanomedicine drug release strategies

The nanoparticle drugs are either encapsulated or conjugated with nanocarriers, and the delivery of the drug depends on the characteristics of the nanocarriers [29]. Drug release from nanomedicine depends on several factors including drug solubility, pH, temperature, desorption of the surface-bound or adsorbed drug, drug diffusion, nanoparticle matrix swelling and erosion, and diffusion processes [53]. There are open-loop and closed-loop control systems to control the drug release from nanosystems. In open-loop control system, the drug release is controlled by external stimuli such as magnetic pulses, thermal, acoustic pulses or electric fields, whereas in closed-loop systems, the presence and intensity of internal stimuli in the tumor environment controls the drug release [46]. The drug release from nanocarriers can be based on the diffusion-controlled, solvent-controlled, chemical reaction-controlled, and stimuli-controlled mechanisms (Fig. 16.7) [65].

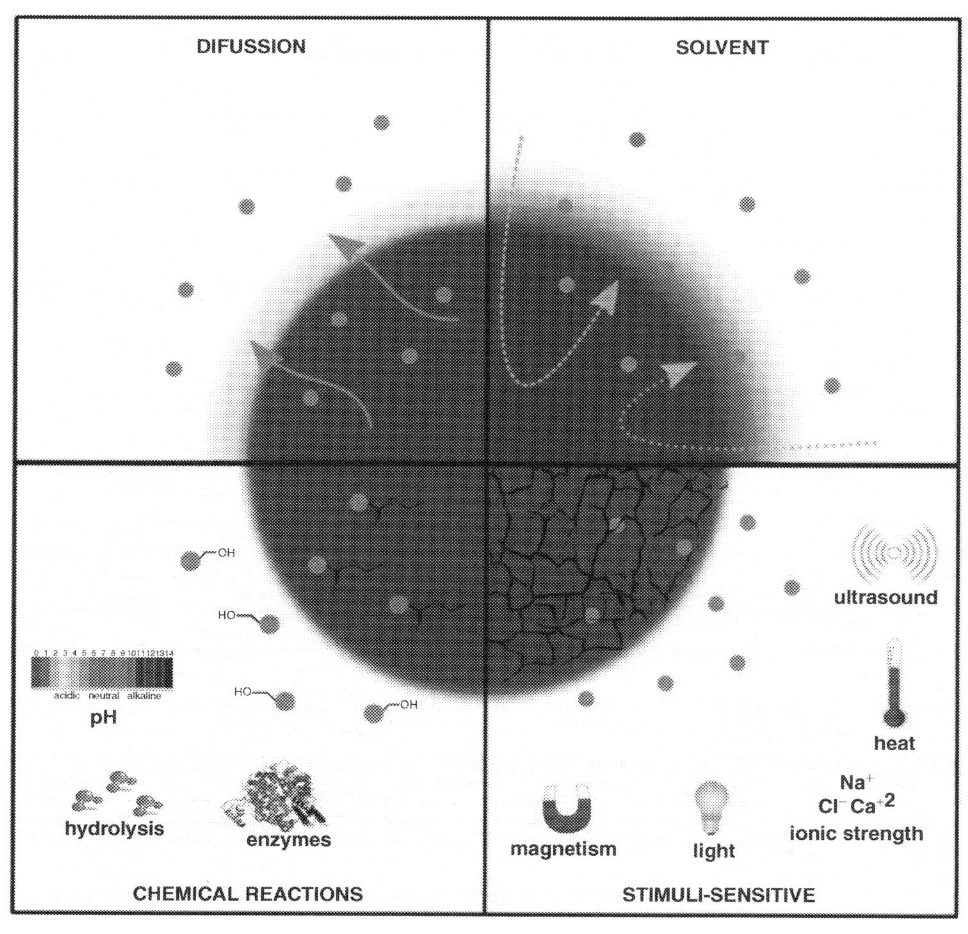

Fig. 16.7 ***Mechanisms for controlled release of drugs using different types of nanocarriers.*** *Adopted from Patra et al. J Nanobiotechnology. 2018; 16: 71 under the terms of the Creative Commons Attribution 4.0 International License (http://creativecommons.org/licenses/by/4.0/).*

In diffusion-controlled systems, the drug release is based on the difference in the concentration across the membrane with initial faster release and subsequently decreasing release rate [66]. The osmosis-controlled release and swelling-controlled release are the solvent-controlled release systems. In osmosis-controlled release, there is a flow of water from outside of the carrier with a low drug concentration to inside the carrier into the drug-loaded core with a high drug concentration [67]. The water uptake from the aqueous systems to the hydrophilic polymeric systems results in swelling followed by drug-release in the swelling-controlled release system [66]. Internal or external stimuli such as temperature, pH, ionic strength, chemicals and electric or magnetic fields controls the drug release from chemical and stimuli-responsive nanocarriers [29,55].

16.4.6 Nanoparticles and anticancer response modalities

The cytotoxic effects of nanoparticles are exerted through alteration of the chemical and/or physical environment surrounding the tumor tissues. Nanoparticle localization to the tumor occurs via passive (EPR) or active targeting (ligands) [41]. The anticancer effects of nanoparticles can be exerted through individual or multimodality approach including drug release and thermal ablation [41,68–70].

16.4.6.1 Nanoparticle-mediated drug release

Limited drug delivery due to lack of drug penetration or sufficient concentration has been a significant concern for the conventional chemotherapeutic agents as well as new targeted drug therapies [68,71]. The perivascular accumulation and slow release hinders the drug delivery of nanoparticles too [68]. To increase the drug penetration, hyperthermia can be applied but it may also result in subpar penetration [63]. The nanoparticle mediated drug release controls the biodistribution profile of the drug and thus controls the cytotoxicity profile. Furthermore, the reduced excretion rate of low molecular weight drugs in the nanoparticle mediated delivery results in the accumulation of the drugs and greater yield at tumor site. For liposome nanomedicines, sustained drug release could be obtained through electrostatic interactions between a cationic drug, anionic gel and a barrier function of the lipid bilayer [72]. pH sensitive liposomes are developed which exert pH-dependent release of the drug based on tumor environment [66]. In polymeric micelles, the interactions between drugs and micellar cores controls the drug release, and the drug release can be extended through drug conjugation to the hydrophobic segment of polymeric micelle [73].

Liposome nanoparticle mediated delivery of doxorubicin is the successful example of nanoparticle-mediated drug delivery as it demonstrates a substantial reduction in the cardiotoxicity as compared with the free doxorubicin [74]. An example for the improved efficacy with nanoparticle-mediated drug release include the albumin-conjugated paclitaxel (Abraxane), which has been approved by the US Food and Drug Administration (FDA) [75]. The use of lipid based nanoparticle formulations to increase the drug penetration via EPR effects has been researched. Nanosomal lipid suspension formulation of docetaxel is developed, which has shown greater systemic availability than conventional docetaxel [76], and has demonstrated improved efficacy via EPR effects [77]. Several newer agents are being developed with nanoparticle based strategies with active or passive targeting and the ability to circumvent macrophage phagocyte systems [41].

16.4.6.2 Hyperthermia or thermal ablation

Thermal ablative approaches including radiofrequency ablation, laser-induced thermotherapy and microwave ablation, currently used in the treatment of cancers, especially metastatic diseases, do not use nanoparticles. The use of nanoparticles in the ablative approaches has led to molecular recognition of metastatic lesions. The use of nanoparticles for eradicating macrometastatic diseases is the most important objectives of the nanomedicines for cancer treatment. Nanoparticle based thermal ablation has been evidenced as cancer therapeutics with direct thermal effects, immunomodulation, and overall anticancer effects [41,78].

16.4.6.3 Challenges in nanoparticle drug delivery

The unique characteristics of nanoparticles, including the size, surface charges, drug loading, and targeting potential are the key parameters to be evaluated for physicochemical characterization of nanoparticles [29]. The development of nanoparticles is very technical due to the unique physicochemical properties, which may have batch-to-batch inconsistencies making large-scale production of nanomedicines very challenging [29]. The scale-up of nanomedicines include several components such as the material status as generally regarded as safe (GRAS), size and shape of nanoparticle, associated toxicity profiles, biodegradability, etc [79]. Furthermore, the high cost of raw materials and complicated multistage production process makes the nanoparticles an expensive option [69]. Several methods have been proposed to facilitate the scale-up of nanomedicine developments including life cycle assessment models to guide production process choice, use of green chemistry to provide safe and environment friendly synthetic routes, and implementation of continuous flow methods to obtain reproducible nanoparticle formulations [80].

Regulatory approval of nanomedicines also poses a significant challenge as there are no specific guidelines. A delay in commercialization of nanomedicines occurs due to time-consuming regulatory evaluations as nanomedicines are developed based on multifunctional nanoplatforms [29].

The toxicity profile of nanomedicine is a potential risk as anti-cancer therapy. The toxicity profile of the nanomedicine cannot be generalized as they differ in size, shape, and other physicochemical properties [29]. The nanoparticles can increase the reactive-oxygen species (ROS) leading to DNA mutations and can be carcinogenic [81]. Nanoparticles possess oxidative potential themselves, and the physicochemical properties of nanoparticles including surface charge, size, chemical composition, and presence of transition metals attribute to prooxidant effects [82]. Also, the redox imbalance due to nanoparticles may lead to inflammation, genotoxicity, fibrosis, and carcinogenesis [82].

16.4.7 Common chemotherapeutic agents as nanotherapeutics

With the advent of nanoparticle drug delivery systems, newer generations of nanoparticles have emerged over few decades, and have reached to clinical trials, of which few nanomedicines are approved by drug regulatory agencies [83,84], while many are still in the development stages [85]. Antimetabolites, alkylating agents, mitotic spindle inhibitors, topoisomerase inhibitors are the commonly used anticancer drug classes based on the mechanism of action [5]. Anthracyclines, taxanes, and topoisomerase inhibitors are one the most widely used chemotherapeutic agents. Although these agents have demonstrated clinical benefits, they are associated with numerous challenges with regarding to safety and efficacy [86], and hence, several nanotechnology based formulations of these agents have been developed to improve their tolerability, toxicity, and efficacy profiles.

16.4.7.1 Doxorubicin

Doxorubicin has a relatively low therapeutic index, and is cardiotoxic, risk of which increases with higher cumulative doses of doxorubicin [87]. Furthermore, it may cause rechallenge on relapse or resistance to the drug owing to hampered dosing due to the narrow therapeutic index [88].

Liposomal encapsulated formulation of doxorubicin has been developed to address the cardiotoxicity concerns of conventional doxorubicin formulation by altering the pharmacokinetics or tissue distribution of the drug [87,89]. Liposomes range from 25 to 100 nM size, and the EPR effect allows the below threshold size liposomes to accumulate in tumor environment. It is well established that RES can uptake the bigger drug molecules leading to a faster metabolism and thus there are decreased effects due to subpar levels of the drug in the circulation. A decrease in the size of nanotransporter is an option but it allows lesser amount of drug to be transported. Hence, new methods that overcome this issue are researched in which coating with polymers has been widely used, especially with PEG [89].

The PEG-coated liposomes are stable and long-circulating drug carriers. Pegylated-liposome encapsulated doxorubicin (PLD) contains the drug encapsulated in liposomes with surface-bound methoxypolyethylene glycol with a diameter of <100 nm, which exerts an extremely long circulating half-life, reduced plasma clearance and volume of distribution compared with liposomal doxorubicin or free doxorubicin formulations [90]. Doxorubicin-containing PEG liposomes circulate for 2-3 weeks after injection of PLD. Pegylated liposomes extravagate through leaky tumor vasculature into the tissue compartments and doxorubicin penetrates into the tumor and exerts its chemotherapeutic effect. PLD remains intact in the circulation, which is responsible for the reduced toxicity without sacrificing efficacy [91,92]. Several pegylated and nonpegylated liposomal formulations of doxorubicin are available such as Doxil, Pegadria, Myocet, etc.

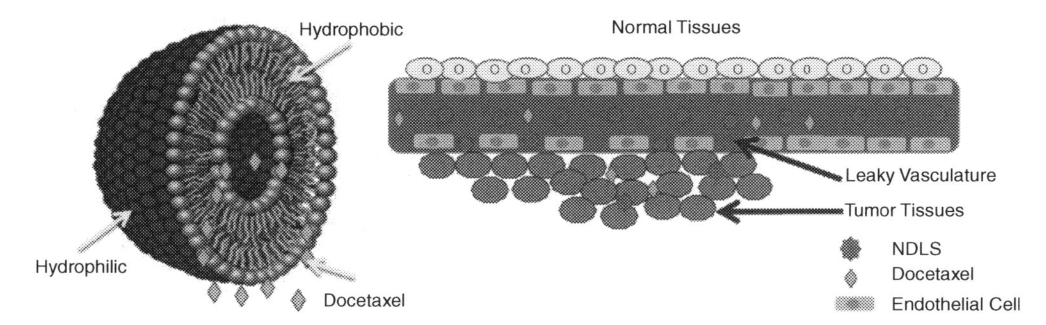

Fig. 16.8 ***Novel nanosomal docetaxel lipid suspension.***

16.4.7.2 Docetaxel

Taxanes (docetaxel and paclitaxel) are one of the most common active moieties used as chemotherapy for several cancers. The conventional docetaxel is formulated with surfactant polysorbate 80 (Tween 80), the major component of which is polyoxyethylene-20-sorbitan monooleate, structurally similar to polyethylene glycols (PEGs) [93,94]. The presence of polysorbate 80 and ethanol in the conventional docetaxel formulation is reported to cause numerous adverse effects, including acute hypersensitivity reactions, cumulative fluid retention, and neurotoxicity [95]. Several novel formulations such as taxane analogues and prodrugs, docetaxel-encapsulated nanoparticle-aptamer bioconjugates albumin nanoparticles, polyglutamates, emulsions, liposomes, docetaxel fibrinogen-coated olive oil droplets and submicronic dispersion have been developed to avoid the toxicities of carrier used in docetaxel formulations [93]. So far, a nanosomal docetaxel lipid suspension (NDLS; DoceAqualip) formulation is the only approved formulation. The NDLS was developed with 'NanoAqualip' technology, that has nanocarriers in suspension form composed of lipids Generally Regarded As Safe (GRAS) by the US Food and Drug Administration (FDA). NDLS is developed with addition of docetaxel to high-pressure homogenized soy phosphatidylcholine and sodium cholesteryl sulfate in sodium citrate buffer to obtain docetaxel lipid suspension, which is mixed with sucrose solution and filtered through sterile 0.45/0.22-μm polyvinylidene fluoride filters [96]. The nanosized particles of NDLS (<100 nm) can increase the delivery of docetaxel to the tumor sites through the EPR effect, which is further facilitated through the leaky tumor vasculature (Fig. 16.8) [76,77,97,98]. The efficacy and safety of NDLS formulation have also been demonstrated in the treatment of breast, ovarian, cervical, penile, gastric, hormone refractory prostate, non-small cell lung, head and neck cancers, and sarcoma in several published reports [95,99–105].

16.4.7.3 Paclitaxel

Taxol(paclitaxel), the conventional paclitaxel formulation, is indicated for the treatment of ovarian and breast cancers. Paclitaxel is formulated with ethanol and cremophor

EL (CrEL), a polyoxyethylated castor oil vehicle to enhance the solubility. Cremophor EL and ethanol in the conventional formulation may result in important clinical implications with reported adverse events such as severe anaphylactic hypersensitivity reactions, infusion related toxicities, hyperlipidemia, abnormal lipoprotein patterns, aggregation of erythrocytes and peripheral neuropathy (PN) [106,107]. To control the undesirable side effects due to the solvent, patients are pre-medicated with corticosteroids, however, minor reactions were still observed despite premedication [108–110]. Cremophor EL also leaches the plasticizer diethylhexyl phthalate from polyvinyl chloride containing solution bags and administration sets [111], which necessities the use of special infusion sets. In order to circumvent the toxicities related to the vehicles, several paclitaxel formulations have been developed using liposomes, polymeric micelles, protein and nanospheres to avoid or minimize the use of solvents.

The nanoparticle albumin-bound paclitaxel (nab-paclitaxel) formulation uses albumin as a carrier to facilitate targeted delivery of paclitaxel to tumors. This carrier results in a lower distribution in healthy tissues, thus resulting in a higher uptake in tumor and improved clinical efficacy [60]. Paclitaxel injection concentrate for nanodispersion (PICN) is a polyoxyethylated castor oil- and albumin-free formulation of paclitaxel. PICN is under investigation as a polyoxyethylated castor oil- and albumin-free self-assembly nanoparticle formulation of paclitaxel (100–150 nm) stabilized with a polymer (polyvinylpyrrolidone) and a lipid (cholesteryl sulfate and caprylic acid) [112]. The nanoparticle polymer-based paclitaxel (Nanoxel) formulation is water soluble and, unlike conventional paclitaxel, does not need CrEL, leading to little or no infusion reactions [113]. A novel nanosomal paclitaxel Lipid Suspension (NPLS; PacliAqualip) formulation was developed, which is free from CrEL and ethanol [114].

16.4.7.4 Irinotecan

Camptosar, the conventional irinotecan formulation, is indicated for the treatment of colon and small cell lung cancers. The utility of irinotecan has been limited partly attributed to the rapid elimination and dose-limiting toxicities (severe diarrhea and cytopenias) [115,116]. The liposomal formulation of irinotecan (Onivyde) was developed to overcome these limitations. The liposomal formulation also prolongs the systemic circulation of irinotecan, improving its distribution with preferential tumor tissue targeting through EPR effects [117]. The prolonged exposure of liposomal irinotecan in tumor tissue has led to increased antitumor activity when compared with conventional formulation along with the decreased incidence of drug-related systemic toxicities [117].

16.4.8 Nanomaterials as drug carriers: Advantages and disadvantages

Table 16.3 enlists the advantages and disadvantages of the nanomaterials as drug carriers [28,41].

Table 16.3 Nanomaterials as drug carriers.

Advantages	Disadvantages
• Increases aqueous solubility • Protect drugs dissolved in the bloodstream • Improves the pharmacokinetic and pharmacological properties • Targeted delivery • Limiting the drug accumulation in the kidneys, liver, spleen, and other non-targeted organs • Enhancing therapeutic efficacy • Real-time monitoring of therapeutic efficacy	• May increase ROS production and DNA mutations • Carcinogenesis • Toxicity: associated with asthma, bronchitis, Alzheimer's disease, and Parkinson's disease • Vascular-related events such as blood clots • Difficulty in regulatory approvals

16.5 Summary

The introduction of nanotherapeutics has overcome the challenges associated with conventional chemotherapeutic agents. Nanotherapeutics provide a longer shelf-life, improve the biodistribution; both hydrophilic and hydrophobic substances can be incorporated, thus eliminating the need of solvents. They provide control with sustained release both during the membrane transportation and at the site of action thereby increasing their intercellular concentration. Nanocarrier drug delivery systems can increase the solubility of drugs, cross biological barriers and inhibit multiple types of cancer cells by targeting specific markers, and/or specific signaling pathways linked to tumor development. Subsequently, nanoparticles are considered the appropriate vehicles for personalized cancer treatment due to their ability of targeted delivery. Despite several advantages, only a few numbers of nano-drugs are approved for the treatment of cancer. It is anticipated that improved cancer therapies using nanotechnology will result in improved therapeutic outcomes.

References

[1] D.K. Yadav, H. Pawar, S. Wankhade, S. Suresh, Development of novel docetaxel phospholipid nanoparticles for intravenous administration: quality by design approach, AAPS PharmSciTech. 16 (2015) 855–864.

[2] F. Bray, J. Ferlay, I. Soerjomataram, R.L. Siegel, L.A. Torre, A. Jemal, Global cancer statistics 2018: GLOBOCAN estimates of incidence and mortality worldwide for 36 cancers in 185 countries, CA Cancer J. Clin. 68 (2018) 394–424.

[3] K.I. Block, C. Gyllenhaal, L. Lowe, A. Amedei, A. Amin, A. Amin, K. Aquilano, J. Arbiser, A. Arreola, A. Arzumanyan, et al., Designing a broad-spectrum integrative approach for cancer prevention and treatment, Semin. Cancer Biol. 35(Suppl.) (2015) S276–S304.

[4] E.R. Arakelova, S.G. Grigoryan, A.M. Khachatryan, K. Avjyan, M. LiliaSavchenko, G. Arsenyan, Flora, New Drug Delivery System for Cancer Therapy, World Acad. Sci., Eng. Technol., Int. J. Med., Health, Biomed., Bioeng. Pharm. Eng. 7 (2013) 927–932

[5] K. Bukowski, M. Kciuk, R. Kontek, Mechanisms of multidrug resistance in cancer chemotherapy, Int. J. Mol. Sci. 21 (2020).

[6] M. Chidambaram, R. Manavalan, K. Kathiresan, Nanotherapeutics to overcome conventional cancer chemotherapy limitations, J. Pharm. Pharm. Sci. 14 (2011) 67–77.
[7] M. Narvekar, H.Y. Xue, J.Y. Eoh, H.L. Wong, Nanocarrier for poorly water-soluble anticancer drugs–barriers of translation and solutions, AAPS Pharm Sci. Tech. 15 (2014) 822–833.
[8] R.H. Prabhu, P.VB, M.D. Joshi, Polymeric nanoparticles for targeted treatment in oncology: current insights, Int. J. Nanomedicine 10 (2015) 1001–1018.
[9] D. Khan, The use of nanocarriers for drug delivery in cancer therapy, J. Cancer Sci. Ther. 02 (2010) 058–062.
[10] A.G. Tzakos, E. Briasoulis, T. Thalhammer, W. Jäger, V. Apostolopoulos, Novel oncology therapeutics: targeted drug delivery for cancer, J. Drug Deliv. 2013 (2013) 918304.
[11] K. Cho, X. Wang, S. Nie, C. ZG, D.M. Shin, Therapeutic nanoparticles for drug delivery in cancer, Clin. Cancer Res. 14 (2008) 1310–1316.
[12] G. Housman, S. Byler, S. Heerboth, K. Lapinska, M. Longacre, N. Snyder, S. Sarkar, Drug resistance in cancer: an overview, Cancers 6 (2014) 1769–1792.
[13] A.A. Stavrovskaya, Cellular mechanisms of multidrug resistance of tumor cells, Biochemistry (Mosc) 65 (2000) 95–106.
[14] B. Mansoori, A. Mohammadi, S. Davudian, S. Shirjang, B. Baradaran, The different mechanisms of cancer drug resistance: a brief review, Adv. Pharm. Bull. 7 (2017) 339–348.
[15] H. Liu, L. Lv, K. Yang, Chemotherapy targeting cancer stem cells, Am. J. Cancer Res. 5 (2015) 880–893.
[16] A.D. Seidman, D. Berry, C. Cirrincione, L. Harris, H. Muss, P.K. Marcom, G. Gipson, H. Burstein, D. Lake, C.L. Shapiro, Randomized phase III trial of weekly compared with every-3-weeks paclitaxel for metastatic breast cancer, with trastuzumab for all HER-2 overexpressors and random assignment to trastuzumab or not in HER-2 nonoverexpressors: final results of cancer and leukemia group B protocol 9840, J. Clin. Oncol. 26 (2008) 1642–1649.
[17] M. Prasad, U.P. Lambe, B. Brar, I. Shah, M. J, K. Ranjan, R. Rao, S. Kumar, S. Mahant, S.K. Khurana, et al., Nanotherapeutics: an insight into healthcare and multi-dimensional applications in medical sector of the modern world, Biomed. Pharmacother. 97 (2018) 1521–1537.
[18] M. Saha, Nanomedicine: promising tiny machine for the healthcare in future-a review, Oman Med. J. 24 (2009) 242–247.
[19] R.L. Hong, C.J. Huang, Y.L. Tseng, V.F. Pang, S.T. Chen, J.J. Liu, F.H. Chang, Direct comparison of liposomal doxorubicin with or without polyethylene glycol coating in C-26 tumor-bearing mice: is surface coating with polyethylene glycol beneficial? Clin. Cancer Res. 5 (1999) 3645–3652.
[20] H. Liu, L. Lv, K. Yang, Chemotherapy targeting cancer stem cells, Am. J. Cancer Res. 5 (2015) 880–893.
[21] S. Senapati, A.K. Mahanta, S. Kumar, P. Maiti, Controlled drug delivery vehicles for cancer treatment and their performance, Signal Transduct. Target. Ther. 3 (7) (2018).
[22] M.D.K. Glasgow, M.B. Chougule, Recent developments in active tumor targeted multifunctional nanoparticles for combination chemotherapy in cancer treatment and imaging, J. Biomed. Nanotechnol. 11 (2015) 1859–1898.
[23] S. Gurunathan, M.-.H. Kang, M. Qasim, J.-.H. Kim, Nanoparticle-mediated combination therapy: two-in-one approach for cancer, Int. J. Mol. Sci. 19 (3264) (2018).
[24] R.R. Sawant, V.P. Torchilin, Multifunctional nanocarriers and intracellular drug delivery, Curr. Opin. Solid State Mater. Sci. 16 (2012) 269–275.
[25] L. Feng, R.J. Mumper, A critical review of lipid-based nanoparticles for taxane delivery, Cancer Lett. 334 (2013) 157–175.
[26] A. Hafner, J. Lovrić, G.P. Lakoš, I. Pepić, Nanotherapeutics in the EU: an overview on current state and future directions, Int. J. Nanomed. 9 (2014) 1005–1023.
[27] M.F. Attia, N. Anton, J. Wallyn, Z. Omran, T.F. Vandamme, An overview of active and passive targeting strategies to improve the nanocarriers efficiency to tumour sites, J. Pharm. Pharmacol. 71 (2019) 1185–1198.
[28] Z. Li, S. Tan, S. Li, Q. Shen, K. Wang, Cancer drug delivery in the nano era: an overview and perspectives (Review), Oncol. Rep. 38 (2017) 611–624.
[29] P.N. Navya, A. Kaphle, S.P. Srinivas, S.K. Bhargava, V.M. Rotello, H.K. Daima, Current trends and challenges in cancer management and therapy using designer nanomaterials, Nano Converg. 6 (23) (2019).

[30] P.N. Navya, H.K. Daima, Rational engineering of physicochemical properties of nanomaterials for biomedical applications with nanotoxicological perspectives, Nano Converg. 3 (1) (2016).
[31] C.-Y. Zhao, R. Cheng, Z. Yang, Z.-.M. Tian, Nanotechnology for cancer therapy based on chemotherapy, Molecules. 23 (2018) 826.
[32] S.M. Moghimi, A.C. Hunter, J.C. Murray, Nanomedicine: current status and future prospects, FASEB J. 19 (2005) 311–330.
[33] M. Caldorera-Moore, N. Guimard, L. Shi, K. Roy, Designer nanoparticles: incorporating size, shape and triggered release into nanoscale drug carriers, Exp. Opin. Drug Deliv. 7 (2010) 479–495.
[34] T. Sun, Y.S. Zhang, B. Pang, D.C. Hyun, M. Yang, Y. Xia, Engineered nanoparticles for drug delivery in cancer therapy, Angew. Chem. Int. Ed. Engl. 53 (2014) 12320–12364.
[35] F. Kratz, A. Warnecke, Finding the optimal balance: Challenges of improving conventional cancer chemotherapy using suitable combinations with nano-sized drug delivery systems, J. Control. Release 164 (2012) 221–235.
[36] B.D. Gates, Q. Xu, M. Stewart, D. Ryan, C.G. Willson, G.M. Whitesides, New approaches to nanofabrication: molding, printing, and other techniques, Chem. Rev. 105 (2005) 1171–1196.
[37] C. Bantz, O. Koshkina, T. Lang, H.J. Galla, C.J. Kirkpatrick, R.H. Stauber, M. Maskos, The surface properties of nanoparticles determine the agglomeration state and the size of the particles under physiological conditions, Beilstein J. Nanotechnol. 5 (2014) 1774–1786.
[38] D.R. Baer, M.H. Engelhard, G.E. Johnson, J. Laskin, J. Lai, K. Mueller, P. Munusamy, S. Thevuthasan, H. Wang, N. Washton, et al., Surface characterization of nanomaterials and nanoparticles: Important needs and challenging opportunities, J. Vac. Sci. Technol. 31 (50820) (2013).
[39] D.R. Camidge, Targeted therapy vs chemotherapy: which has had more impact on survival in lung cancer? Does targeted therapy make patients live longer? Hard to prove, but impossible to ignore, Clin Adv. Hematol Oncol. 12 (2014) 763–766.
[40] J.K. Patel, A.P. Patel, Passive Targeting of Nanoparticles to Cancer, in: Y.V. Pathak (Ed.), Modification of Nanoparticles for Targeted Drug Delivery, Springer International Publishing, Cham, 2019, pp. 125–143.
[41] W.H. Gmeiner, S. Ghosh, Nanotechnology for cancer treatment, Nanotechnol. Rev. 3 (2015) 111–122.
[42] J.V. Jokerst, T. Lobovkina, R.N. Zare, S.S. Gambhir, Nanoparticle PEGylation for imaging and therapy, Nanomedicine (Lond) 6 (2011) 715–728.
[43] J.L. Markman, A. Rekechenetskiy, E. Holler, J.Y. Ljubimova, Nanomedicine therapeutic approaches to overcome cancer drug resistance, Adv. Drug Deliv. Rev. 65 (2013) 1866–1879.
[44] I.I. Lungu, A.M. Grumezescu, A. Volceanov, E. Andronescu, Nanobiomaterials used in cancer therapy: an up-to-date overview, Molecules 24 (2019).
[45] Y.-R. Zhang, R. Lin, H.-J. Li, He W-l, J.-Z. Du, J. Wang, Strategies to improve tumor penetration of nanomedicines through nanoparticle design, Wiley Interdiscip. Rev. Nanomed. Nanobiotechnol. 11 (2019) e1519.
[46] Q. Zhou, L. Zhang, H. Wu, Nanomaterials for cancer therapies, Nanotechnol. Rev. 6 (2017) 473.
[47] Z. Qing, Z. Li, W. Hong, Nanomaterials for cancer therapies, Nanotechnol. Rev. 6 (2017) 473–496.
[48] M. Amin, A.M. Butt, M.W. Amjad, P. Kesharwani, Chapter 5 - Polymeric Micelles for Drug Targeting and Delivery, in: V. Mishra, P. Kesharwani, M.C.I. MohdAmin, A. Iyer (Eds.), Nanotechnology-Based Approaches for Targeting and Delivery of Drugs and Genes, Academic Press, Cambridge, Massachusetts, 2017, pp. 167–202.
[49] A.M. Jhaveri, V.P. Torchilin, Multifunctional polymeric micelles for delivery of drugs and siRNA, Front. Pharmacol. 5 (2014).
[50] B.L. Banik, P. Fattahi, J.L. Brown, Polymeric nanoparticles: the future of nanomedicine, Wiley Interdiscip. Rev. Nanomed. Nanobiotechnol. 8 (2016) 271–299.
[51] E. Espinosa-Cano, R. Palao-Suay, M.R. Aguilar, B. Vazquez, J. San Roman, Polymeric nanoparticles for cancer therapy and bioimaging. 2018. doi:10.1007/978-3-319-89878-0_4.
[52] E. Abbasi, S.F. Aval, A. Akbarzadeh, M. Milani, H.T. Nasrabadi, S.W. Joo, Y. Hanifehpour, K. Nejati-Koshki, R. Pashaei-Asl, Dendrimers: synthesis, applications, and properties, Nanoscale Res. Lett. 9 (247) (2014).
[53] S.A.A. Rizvi, A.M. Saleh, Applications of nanoparticle systems in drug delivery technology, Saudi Pharm. J. 26 (2018) 64–70.
[54] S.C.A. Lopes, C.S. Giuberti, T.G.R. Rocha, D.S. Ferreira, E.A. Leite and M.C. Oliveira (May 9th 2013). Liposomes as carriers of anticancer drugs, cancer treatment - conventional and innovative

approaches, IntechOpen, London, UK. DOI: 10.5772/55290. Available from: https://www.intechopen.com/books/cancer-treatment-conventional-and-innovative-approaches/liposomes-as-carriers-of-anticancer-drugs.
[55] T.O.B. Olusanya, R.R. Haj Ahmad, D.M. Ibegbu, J.R. Smith, A.A. Elkordy, Liposomal drug delivery systems and anticancer drugs, Molecules 23 (2018) 907.
[56] I. Ekladious, Y.L. Colson, M.W. Grinstaff, Polymer–drug conjugate therapeutics: advances, insights and prospects, Nat. Rev. Drug Discov. 18 (2019) 273–294.
[57] N. Iturrioz-Rodríguez, M.A. Correa-Duarte, M.L. Fanarraga, Controlled drug delivery systems for cancer based on mesoporous silica nanoparticles, Int. J. Nanomed. 14 (2019) 3389–3401.
[58] S. Jain, D.G. Hirst, J.M. O'Sullivan, Gold nanoparticles as novel agents for cancer therapy, Brit. J. Radiol. 85 (2012) 101–113.
[59] J.H. Park, G. von Maltzahn, E. Ruoslahti, S.N. Bhatia, M.J. Sailor, Micellar hybrid nanoparticles for simultaneous magnetofluorescent imaging and drug delivery, Angew. Chem. Int. Ed. Engl. 47 (2008) 7284–7288.
[60] Abraxane Prescribing Information, Abraxis BioScience, LLC (2013) Revised September.
[61] Z.-Z.J. Lim, J.-E.J. Li, C.-T. Ng, L.-Y.L. Yung, B.-H. Bay, Gold nanoparticles in cancer therapy, Acta Pharmacol. Sin. 32 (2011) 983–990.
[62] S. Iijima, Helical microtubules of graphitic carbon, Nature 354 (1991) 56–58.
[63] S. Nikazar, V.S. Sivasankarapillai, A. Rahdar, S. Gasmi, P.S. Anumol, M.S. Shanavas, Revisiting the cytotoxicity of quantum dots: an in-depth overview, Biophys. Rev. 12 (2020) 703–718.
[64] R. Tietze, J. Zaloga, H. Unterweger, S. Lyer, R.P. Friedrich, C. Janko, M. Pöttler, S. Dürr, C. Alexiou, Magnetic nanoparticle-based drug delivery for cancer therapy, Biochem. Biophys. Res. Commun. 468 (2015) 463–470.
[65] J.K. Patra, G. Das, L.F. Fraceto, E.V.R. Campos, M.D.P. Rodriguez-Torres, L.S. Acosta-Torres, L.A. Diaz-Torres, R. Grillo, M.K. Swamy, S. Sharma, et al., Nano based drug delivery systems: recent developments and future prospects, J. Nanobiotechnol. 16 (71) (2018).
[66] J.H. Lee, Y. Yeo, Controlled drug release from pharmaceutical nanocarriers, Chem. Eng. Sci. 125 (2015) 75–84.
[67] S. Herrlich, S. Spieth, S. Messner, R. Zengerle, Osmotic micropumps for drug delivery, Adv. Drug. Deliv. Rev. 64 (2012) 1617–1627.
[68] A.A. Manzoor, L.H. Lindner, C.D. Landon, J.-Y. Park, A.J. Simnick, M.R. Dreher, S. Das, G. Hanna, W. Park, A. Chilkoti, et al., Overcoming limitations in nanoparticle drug delivery: triggered, intravascular release to improve drug penetration into tumors, Cancer Res. 72 (2012) 5566–5575.
[69] S. Ghosh, S. Dutta, E. Gomes, D. Carroll, R. D'Agostino Jr., J. Olson, M. Guthold, W.H. Gmeiner, Increased heating efficiency and selective thermal ablation of malignant tissue with DNA-encased multiwalled carbon nanotubes, ACS Nano 3 (2009) 2667–2673.
[70] S.A. Love, M.A. Maurer-Jones, J.W. Thompson, Y.S. Lin, C.L. Haynes, Assessing nanoparticle toxicity, Annu Rev. Anal Chem. (Palo Alto Calif) 5 (2012) 181–205.
[71] A.I. Minchinton, I.F. Tannock, Drug penetration in solid tumours, Nat. Rev. Cancer 6 (2006) 583–592.
[72] Y. Wang, S. Tu, A.N. Pinchuk, M.P. Xiong, Active drug encapsulation and release kinetics from hydrogel-in-liposome nanoparticles, J. Colloid Interface Sci. 406 (2013) 247–255.
[73] H.S. Yoo, T.G. Park, Biodegradable polymeric micelles composed of doxorubicin conjugated PLGA-PEG block copolymer, J. Control. Release 70 (2001) 63–70.
[74] S. ST, S.E. McNeil, Nanotechnology safety concerns revisited, Toxicol. Sci. 101 (2008) 4–21.
[75] C. Zhang, N. Awasthi, M.A. Schwarz, S. Hinz, R.E. Schwarz, Superior antitumor activity of nanoparticle albumin-bound paclitaxel in experimental gastric cancer, PLoS One 8 (2013) e58037.
[76] A. Ahmad, S. Sheikh, S.M. Ali, M.U. Ahmad, M. Paithankar, D. Saptarishi, K. Maheshwari, K. Kumar, J. Singh, G. Patel, Development of aqueous based formulation of docetaxel: safety and pharmacokinetics in patients with advanced solid tumors, J. Nanomed Nanotechnol. 6 (1) (2015).
[77] A. Ahmad, S. Sheikh, R. Taran, S.P. Srivastav, K. Prasad, S.J. Rajappa, V. Kumar, M. Gopichand, M. Paithankar, M. Sharma, Therapeutic efficacy of a novel nanosomal docetaxel lipid suspension compared with taxotere in locally advanced or metastatic breast cancer patients, Clin. Breast Cancer 14 (2014) 177–181.
[78] S.P. Haen, P.L. Pereira, H.R. Salih, R. HG, C. Gouttefangeas, More than just tumor destruction: immunomodulation by thermal ablation of cancer, Clin. Dev. Immunol. 2011 (160250) (2011).

[79] R. Paliwal, R.J. Babu, S. Palakurthi, Nanomedicine scale-up technologies: feasibilities and challenges, AAPS Pharm Sci.Tech. 15 (2014) 1527–1534.
[80] S. Falsini, Sustainable strategies for large-scale nanotechnology manufacturing in the biomedical field, Green Chem. 20 (17) (2018) 3897–3907.
[81] Z. Yu, Q. Li, J. Wang, Y. Yu, Y. Wang, Q. Zhou, P. Li, Reactive oxygen species-related nanoparticle toxicity in the biomedical field, Nanoscale Res. Lett. 15 (115) (2020).
[82] A. Manke, L. Wang, Y. Rojanasakul, Mechanisms of nanoparticle-induced oxidative stress and toxicity, BioMed Res. Int. 2013 (2013) 942916-.
[83] S.-S. Qi, J.-H. Sun, H.-H. Yu, S.-Q. Yu, Co-delivery nanoparticles of anti-cancer drugs for improving chemotherapy efficacy, Drug Deliv. 24 (2017) 1909–1926.
[84] A.C. Anselmo, S. Mitragotri, Nanoparticles in the clinic: an update, Bioeng. Transl. Med. 4 (3) (Sep 5 2019) e10143.doi: 10.1002/btm2.10143.
[85] M.A. Obeid, R.J. Tate, A.B. Mullen, VA. Ferro, Chapter 8 - Lipid-based nanoparticles for cancer treatment, in: A.M. Grumezescu (Ed.), Lipid Nanocarriers for Drug Targeting, William Andrew Publishing, Norwich, New York, 2018, pp. 313–359.
[86] P. Sánchez-Moreno, H. Boulaiz, J.L. Ortega-Vinuesa, J.M. Peula-García, A. Aránega, Novel drug delivery system based on docetaxel-loaded nanocapsules as a therapeutic strategy against breast cancer cells, Int. J. Mol. Sci. 13 (2012) 4906–4919.
[87] S.M. Rafiyath, M. Rasul, B. Lee, G. Wei, G. Lamba, D. Liu, Comparison of safety and toxicity of liposomal doxorubicin vs. conventional anthracyclines: a meta-analysis, Exp. Hematol. Oncol. 1 (10) (2012).
[88] M. O'brien, N. Wigler, M. Inbar, R. Rosso, E. Grischke, A. Santoro, R. Catane, D. Kieback, P. Tomczak, S. Ackland, Reduced cardiotoxicity and comparable efficacy in a phase III trial of pegylated liposomal doxorubicin HCl (CAELYX™/Doxil®) versus conventional doxorubicin for first-line treatment of metastatic breast cancer, Ann. Oncol. 15 (2004) 440–449.
[89] J. Lao, J. Madani, T. Puértolas, M. Álvarez, A. Hernández, R. Pazo-Cid, A. Á, A. Antón Torres, Liposomal doxorubicin in the treatment of breast cancer patients: a review, J. Drug Deliv. 2013 (2013) 1–12. doi: 10.1155/2013/456409. Article ID 456409.
[90] A. Gabizon, F. Martin, Polyethylene glycol-coated (pegylated) liposomal doxorubicin, Drugs 54 (1997) 15–21.
[91] T. Šimůnek, M. Štěrba, O. Popelová, M. Adamcová, R. Hrdina, V. Geršl, Anthracycline-induced cardiotoxicity: overview of studies examining the roles of oxidative stress and free cellular iron, Pharmacol. Rep. 61 (2009) 154–171.
[92] A.A. Gabizon, Pegylated liposomal doxorubicin: metamorphosis of an old drug into a new form of chemotherapy, Cancer Investig. 19 (2001) 424–436.
[93] L. Zhang, N. Zhang, How nanotechnology can enhance docetaxel therapy, Int. J. Nanomedicine 8 (2013) 2927–2941.
[94] K.L. Hennenfent, R. Govindan, Novel formulations of taxanes: a review. Old wine in a new bottle? Ann. Oncol. 17 (2005) 735–749.
[95] P. Narayanan, P.S. Dattatreya, R. Prasanna, S. Subramanian, K. Jain, N.S. Somanath, N. Joshi, D. Bunger, M.A. Khan, A. Chaturvedi, et al., Efficacy and safety of nanosomal docetaxel lipid suspension-based chemotherapy in sarcoma: a multicenter, retrospective study, Sarcoma 2019 (2019) 3158590.
[96] M.A. Khan, N. Joshi, A. Chaturvedi, I. Ahmad, Letter to the editor: current advances in development of new docetaxel formulations, Expert Opin Drug Deliv. 16 (2019) 773–774.
[97] K. Greish, Enhanced permeability and retention (EPR) effect for anticancer nanomedicine drug targeting, Methods Mol. Biol. 624 (2010) 25–37.
[98] K. McKeage, Nanosomal Docetaxel Lipid Suspension: a guide to its use in cancer, Clin. Drug Investig. 37 (2017) 405–410.
[99] M. Ashraf, R. Sajjad, M. Khan, M. Shah, Y. Bhat, Z. Wani, 156P Efficacy and safety of a novel nanosomal docetaxel lipid suspension (NDLS) as an anti cancer agent-a retrospective study, Ann. Oncol. 27 (2016) ix46–ix51.
[100] R. Naik, M.A. Khan, Doceaqualip in a patient with prostate cancer who had an allergic reaction to conventional docetaxel: a case report, Mol. Clin. Oncol. 6 (2017) 341–343.
[101] R. Prasanna, D. Bunger, M.A. Khan, Efficacy and safety of DoceAqualip in a patient with locally advanced cervical cancer: a case report, Mol. Clin. Oncol. 8 (2018) 296–299.

[102] V.Vyas, N. Joshi, M. Khan, Novel docetaxel formulation (NDLS) in low cardiac reserve ovarian cancer, Open Access J. Cancer. Oncol. 2 (2018) 000122.
[103] S. Gupta, S.S. Pawar, D. Bunger, Successful downstaging of locally recurrent penile squamous cell carcinoma with neoadjuvant nanosomal docetaxel lipid suspension (NDLS) based regimen followed by curative surgery, BMJ Case Rep. 2017 (2017) bcr2017220686.
[104] A. Ahmad, S. Sheikh, R. Taran, S.P. Srivastav, K. Prasad, S.J. Rajappa, V. Kumar, M. Gopichand, M. Paithankar, M. Sharma, et al., Therapeutic efficacy of a novel nanosomal docetaxel lipid suspension compared with taxotere in locally advanced or metastatic breast cancer patients, Clin. Breast Cancer 14 (2014) 177–181.
[105] A. Murali, S. Gupta, D. Pendharkar, Efficacy and tolerability of nanoparticle docetaxel lipid suspension, J. Clin. Oncol. 36 (2018) e14542.
[106] H. Gelderblom, J. Verweij, K. Nooter, A. Sparreboom, Cremophor EL: the drawbacks and advantages of vehicle selection for drug formulation, Eur. J. Cancer 37 (2001) 1590–1598.
[107] K.L. Hennenfent, R. Govindan, Novel formulations of taxanes: a review. Old wine in a new bottle? Ann. Oncol. 17 (2006) 735–749.
[108] S. Alken, C.M. Kelly, Benefit risk assessment and update on the use of docetaxel in the management of breast cancer, Cancer Manag Res. 5 (2013) 357–365.
[109] E.K. Rowinsky, E.A. Eisenhauer, V. Chaudhry, A. SG, R.C. Donehower, Clinical toxicities encountered with paclitaxel (Taxol), Semin. Oncol. 20 (1993) 1–15.
[110] A.J. ten Tije, J. Verweij, W.J. Loos, A. Sparreboom, Pharmacological effects of formulation vehicles: implications for cancer chemotherapy, Clin. Pharmacokinet. 42 (2003) 665–685.
[111] L.A. Trissel, Pharmaceutical properties of paclitaxel and their effects on preparation and administration, Pharmacotherapy 17 (1997) 133s–139s.
[112] M.M. Jain, S.U. Gupte, S.G. Patil, A.B. Pathak, C.D. Deshmukh, N. Bhatt, C. Haritha, K. Govind Babu, S.A. Bondarde, R. Digumarti, et al., Paclitaxel injection concentrate for nanodispersion versus nab-paclitaxel in women with metastatic breast cancer: a multicenter, randomized, comparative phase II/III study, Breast Cancer Res. Treat. 156 (2016) 125–134.
[113] A.A. Ranade, D.A. Joshi, G.K. Phadke, P.P. Patil, R.B. Kasbekar, T.G. Apte, R.R. Dasare, S.D. Mengde, P.M. Parikh, G.S. Bhattacharyya, et al., Clinical and economic implications of the use of nanoparticle paclitaxel (Nanoxel) in India, Ann. Oncol. 24 (2013) v6–v12.
[114] A. Ahmad, S. Sheikh, S. Ali, M. Paithankar, A. Mehta, Nanosomal paclitaxel lipid suspension demonstrates higher response rates compared to paclitaxel in patients with metastatic breast cancer, J. Cancer Sci. Ther. 7 (2015) 116–120.
[115] Y.N. Lamb, L.J. Scott, Liposomal irminotecan: a review in metastatic pancreatic adenocarcinoma, Drugs 77 (2017) 785–792.
[116] A.H. Ko, Nanomedicine developments in the treatment of metastatic pancreatic cancer: focus on nanoliposomal irinotecan, Int. J. Nanomed. 11 (2016) 1225–1235.
[117] J.E. Frampton, Liposomal irinotecan: a review in metastatic pancreatic adenocarcinoma, Drugs 80 (2020) 1007–1018.

CHAPTER 17

Drug loaded nanomaterials for hematological malignancies diagnosis and enhanced targeted therapy

Priyanka Samal[a], Shahani Begum[b]
[a]Department of Clinical Hematology- Hemato-oncology & Stem Cell Transplant, IMS & SUM Hospital, Bhubaneswar, Odisha, India
[b]Plant Biotechnology Laboratory, Post Graduate Department of Botany, Utkal University, Bhubaneswar, Odisha, India

17.1 Introduction

Cancer is a group of highly heterogeneous disease. Malignant cells are characterized by self-sufficiency in growth signals, insensitivity to anti-growth signals, evasion of apoptosis, limitless replicative potential, sustained angiogenesis, tissue invasion, and metastasis. According to the World Health Organization (WHO), 18.1 million people had cancer, and 9.6 million people lost their lives from it in 2018 worldwide. These figures are likely to double by 2040. The incidence of lung, prostate, colorectal, stomach, and liver cancer is frequent in males, whereas breast, cervical, lung, and thyroid malignancies are common in females. Cancer Mortality is more in the developing and underdeveloped countries due to inequity in access to effective treatment, lack of awareness, and increased frequency of infections associated with cancer. Cancer Mortality is more in the developing and underdeveloped countries due to inequity in access to effective treatment, lack of awareness, and increased frequency of infections associated with cancer. Hematological malignancies are classified as tumors of Hematopoietic and lymphoid tissue by WHO. These include both liquid malignancies as well as solid malignancies, predominantly the Lymphomas [1,2] (Fig. 17.1). The etiology of these malignancies is influenced by host factors like inherited disorders, chromosomal abnormalities, immunodeficiencies, environmental factors, such as radiation, viruses, drugs, and chemicals [3–5]. Recently, the World Cancer Research Fund International presented Global Cancer Statistics in 2018, which revealed that leukemia contributed to about 2.3% of cancers (excluding non-melanoma skin cancer) in women while 2.8% in men.

The goal of chemotherapy is to destroy cancer cells. Traditional chemotherapies work by killing cells that divide rapidly. Nevertheless, as they wipe out fast-growing cancer cells, they can also damage fast-growing healthy cells, which results in significant side effects. 7+3 chemotherapy is the standard regimen for treating Acute Myeloid Leukemia and is considered one of the most toxic regimens. However, it has very little if any activity on the quiescent Leukemia stem cells (LSCs), which are very different

Advanced Nanomaterials for Point of Care Diagnosis and Therapy
DOI: https://doi.org/10.1016/B978-0-32-385725-3.00016-7

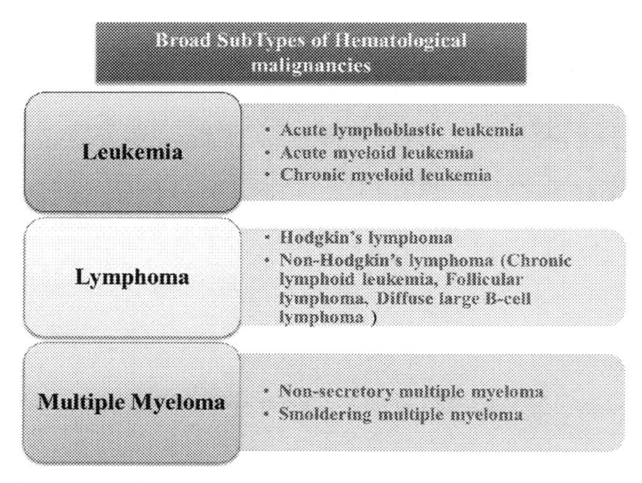

Fig. 17.1 ***Types and subtypes of hematological malignancies.***

from the normal and the blast cells. An LSC is a functionally defined entity arising from a normal stem cell following a mutation but retaining self-renewal properties, multipotency, and proliferation. It also has the unique property of drug efflux, which renders these cells resistant to chemotherapy and hence persistence of disease in spite of toxicity. Tumor-associated properties in the LSC could include mutations in the kinase domains, transcription factors, and tumor suppressors, or alterations in the growth and survival mechanisms mediated through NF-kappa B (NF-κB) or PI3 kinase, or changes in physiology, glucose metabolism, or responses to oxidative stress, to name a few.

Further, chemotherapies involving the administration of anthracyclins or radiation elevated the free radicals production in the body. Incapacitating reactions are reported during the high dose chemotherapy and stem cell transplantation, which emotionally weigh down patients and increase the risk of lugubriously [6,7]. Some of the potential drugs like nilotinib or sonidegib have been tested in mouse models of CML (chronic myeloid leukemia), and the results of the study demonstrated suppression of leukemia [8]. However, this pharmacodynamic approach was also associated with detrimental events like fatigue, nausea, spasm, and alopecia [9,10]. Additionally, studies have reported that stem cell transplantation patients also suffer cardiovascular complications [11]. The sub-normal prognosis and cancer diagnosis lead to an intensified mortality rate and minimal change in the treatment outcome. Imprecise drug-delivery is correlated with less accumulation, non-specificity, and brief half-life period of conjugated drugs. By 2030, the world might witness 13.1 million deaths from malignancies, as stated by WHO [12].

The replicative phase and M-phase (mitosis) are the targets of current conventional therapy. Different materials used in therapy for the transfer of medicine are polymer, lipids, inorganic carriers, polymeric hydrogels, and biomacromolecular scaffolds. Howbeit, the drug-resistant nature of malignant cells is charged for the certain drawbacks of the therapy, as mentioned earlier [13]. Hematological cancers are also being demonstrated to be treated by immunotherapy, which aims to elevate immune response by infecting cancer cells [14–16]. This approach's primary concern is the weak response of major fragment of patients to immunotherapy [17–20]. Administration of Il-2 in Immunotherapy causes biochemical abnormalities such as pyrexia, high blood pressure, and kidney ailments in patients [21–23]. 5-fluorouracil, a chemotherapy medicine used to treat cervical, breast, colon, and bladder cancer, is another cytotoxic prototype whose continuous administration causes toxic side effects. This anti-neoplastic drug damages the digestive tract, cardiac, nervous, and blood-forming system [24].

Nanoparticles, due to their remarkable size range (1–100 nm) and drug release attributes, are the preferred candidate for pharmaceutical application and therapy. The rapidly growing nanoscience has also ventured into the field of cancer prevention and treatment [25,26]. Nanocomposites morphology, surface charge, and other physicochemical properties can be altered to encapsulate hydrophobic or hydrophilic drugs to improve their efficacy and attain notable interactions with various functional groups present in biomedicine [27–29]. The success of nanoparticles as targeted drug carriers depends on their biochemical characteristics, thermal stability, optical properties, and non-toxicity [30–32]. Nanocomposites have the potentiality to adhere to cellular biomolecules like nucleic acids, proteins, and immunoglobulins. The Local delivery approach using nanocrystals excels in transferring a higher amount of chemotherapeutics at the malignant tissue site. This modernistic tool minimizes the harmful off-site effects in the body. The shortcomings of present immunotherapies also compel the researchers to evolve the innovative delivery program. Efforts are ongoing to target the nanocomposite conjugated drugs to those cells of the immune system, which invade tumor cells and generate an anticancer immune response in the blood [33]. Penetration of cytotoxic drugs in transformed tissue is an essential condition to promote the curative effect. Intravenous drug application leads to unnecessary extensive allocation of pharmacon in various organs resulting in insufficient concentration at the diseased site. Therefore, a potent drug shot is continuously administered to the patient, interfering with the body's physiological systems [34,35].

Biogenic metallic, inorganic, or polymeric nanoparticles are considered safe due to their biocompatibility and biodegradability. Polymeric nanoparticles are described as competent drug carrier vehicle with bettered non-polar drug delivery strategy to destined cells. The FDA has approved the production of some chemotherapeutics containing nanocomposites to treat breast cancer [36,37]. The unification of Cold atmospheric plasma biomedical technique and cancer nanotechnology aims to reduce the damage to non-cancerous cells [38,39]. Likewise, nanosize mesoporous structures with flexible pore size, shape, and considerable volume exhibit control release and targeted biomedicine delivery at the tumor site (Fig. 17.2) [40,41]. Fruitful use of drug-loaded graphene and graphene oxide nanocomposites in a controlled environment have also

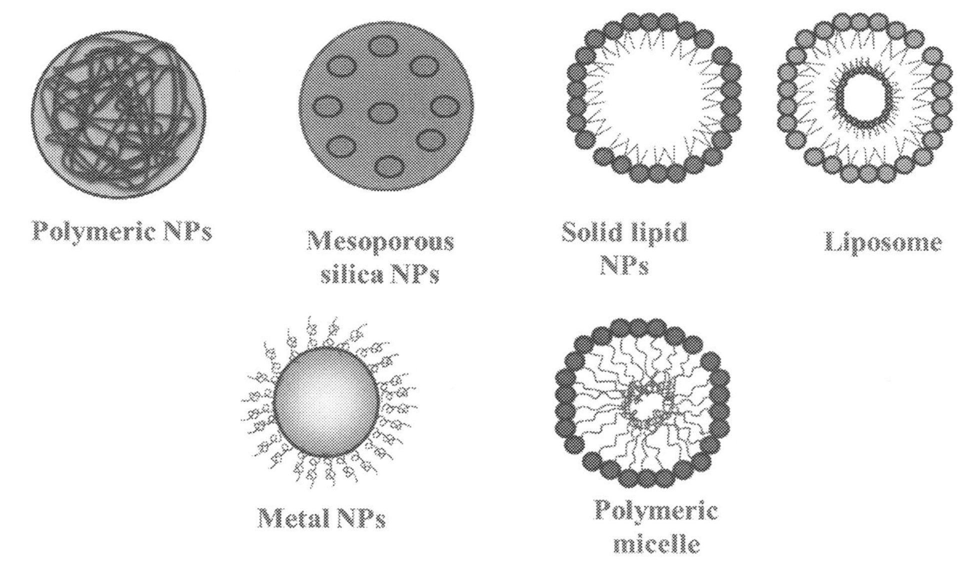

Fig. 17.2 ***Schematic illustration of types of nanocarriers used for targeted drug delivery.***

been reported for cancer diagnosis and therapy. In addition to exceptional drug-binding capacity, it is versatile, highly soluble, and adaptable. Non-covalent π–π interactions play a significant role in loading drugs onto graphene nanocomposites [42,43].

In this review, we discuss nanomaterials' advantages and their application in drug delivery system as carrier. We shall also emphasize the promising scientific impacts on cancer treatment and recent breakthroughs in nanocarrier use in hematological cancer diagnosis and therapeutics. Nano drug-based chemotherapy-associated side-effects and challenges will be addressed.

17.2 Different nanomaterials and their fabrication for targeted drug therapy

The concept of drug development faces a few major challenges in treating leukemia, such as a high tendency of disease recurrence and low survival rates, ranging between 38%–58% for the first few years. The use of nanotechnology can overcome the complications associated with standard conventional therapy. The engineered nanoparticles grant a different advantage to control drug release and enhance conjugated drugs' therapeutic effect. The nanocomposites could be organic, inorganic or hybrid. Delivery of engineered nanoparticles to bone marrow, neoplastic tissue, liver, or spleen occurs due to the EPR effect (enhanced permeation and retention) [44]. Tumor vessels with a dilated aperture between the endothelial cells, limited drainage structure, and anomalistic constitution permit scattering of nanomaterials outside the circulatory system and promotes its diffusion into tumor cell-matrix. pH responsiveness has also been utilized in various cancer treatments because normal tissue has pH 7.4, whereas tumor cells have a pH value of 5.7-7. Some biogenic or non-biogenic conjugated compounds are being investigated for their cytotoxicity against leukemia or lymphoma cells in preclinical models. Table 17.1 represents a list of nanomedicine being investigated for the treatment of hematological malignancies in recent times.

Curcumin is an herbal hydrophobic compound which is extracted from turmeric and nanodrugs containing it, is known to possess cytotoxic properties. Curcumin medical application is boosted when chitosan-based, alginate, magnetic and polymeric nanomaterials encapsulate it. This prepared it for slow drug release and targeted cancer therapy. Stabilization of drugs was much needed because curcumin was poorly absorbed, insoluble, and quickly degraded in our body [45]. Similarly, Citrate fabricated gold nanoparticles conjugated with mi-R221 inhibitor and an aptamer, was used to target nucleolin expressing leukemia [46]. In a cancerous cell, the nucleolin protein is distributed in both the nucleus and cytoplasm, which alter the metabolism of RNA and DNA. Mi-R221 modulated the mutated tumor suppressor genes like TP53, phosphatase and tensin homolog and cyclin-dependent kinase inhibitors, but no cytotoxicity was detected in healthy cells and blood-producing cells.

Another illustration demonstrated the therapeutic effect of conjugated lipid nanoconstruct against chronic lymphocytic leukemia ($CD38^+$ cells). In this report, anti-$CD38^+$ cells loaded lipid nanocarrier improved delivery of different microRNA. The lipid nanocarrier is primarily composed of phospholipid and cholesterol with a positive charge to achieve greater synergy with endogenous lipids and efficiently deliver miRNA. Nearly two-fold of micro-RNA variants were effectively trucked to hematologic malignant cells compared to unconjugated drugs. Nanoparticles exhibited high apoptotic behavior in vivo studies, and decreased expression of CDKs resulted in reduced expansion of CDK6 chronic lymphocytic leukemia [65]. Wu et al. investigated the toxicity of iron nanoparticles on red blood cells integrity. Magnetic Fe_3O_4 nanomaterials were loaded with daunorubicin (DNR) and assessed to study their role in hematologic cancers. The hemolyzation rate of these bioadaptable nanodrugs was neither more than 5%, nor did they cause any mutations in the tested cells, indicating insignificant destruction of RBC architecture.

Additionally, the liver kuffer cells adequately removed iron nanoparticles and organized the iron into ferritin [66]. Another newer approach, layer by layer dual-targeted nanocarrier was being tested to fight against drug-resistant malignancies. In this method, an apoptosis regulator called bcl-2 is the target which is down-regulated by a combination of siRNA technology and nanotechnology. While targeting, the siRNA is protected from nucleases enzymes as it is embedded within polyelectrolyte covering, which forms a laminate polymeric coat. The nanoparticles successfully transfer the anionic siRNA through the lipid bilayer and promote its interaction with destined cells as the outer layer consists of a ligand that binds CD44 glycoprotein, which is a surface antigen

Table 17.1 List of nanomedicine being investigated for the treatment of hematological malignancies.

Sl. no.	Company/product	Types of nanocarrier or conjugated nanomedicine	Drug	Cancer type	Current status	Reference
1	Doxil/calyx [Johnson & Johnson]	Pegylated Liposome	Doxorubicin	Multiple myeloma	2007 [Europe, Canada], FDA approved	[47]
2	Marqibo [Talon]	Liposome	Vincristine	Acute Lymphoid leukemia	2012 [USA], FDA approved	[48]
3	Oncaspar [Enzon/ Sigma-tau]	PEG protein conjugate	L-asparaginase	Leukemia	2006	[49]
4	CPX-351 [celator]	Liposome	Cytarabine/ daunorubicine	Acute Myeloid leukemia	Phase II/III	[50]
5	DCR-MYC (DCR-M1711) [Dicerna pharmaceuticals]	Lipid nanoparticle	RNAi targeting MYC protooncogene	Multiple Myeloma lymphoma	Phase 1	[51]
6	L-Annamycin [callisto]	Liposome	Annamycin	ALL, AML	Phase 1	[52]
7	PNT2258	Liposome	DNA oligonucleotide against bcl-2	B-cell lymphoma	Phase III	[53]
8	JVRS-100	Lipid nanoparticles	Plasmid DNA	Refractory leukemia	Phase 1	[54]
9	SNS01-T	Polyethylenimine nanoparticles	siRNA against EIF5A	Relapsed B-cell tumor	Phase I/II	[55]
10	Torque therapeutics	Immunotherapy nanomedicine	Nanoparticle – Functionalized antigen Primed T-cell therapy	lymphomas	Phase 1	[56]
11	Moderna therapeutic		Lipid nanoparticle–delivered mRNA encoding interleukins	lymphomas	Phase 1	[57]

Sl. no.	Company/product	Types of nanocarrier or conjugated nanomedicine	Drug	Cancer type	Current status	Reference
12	(Ibritumomabtiuxetan zevalin) [IDCE/ spectrum]	Antibody-drug conjugate	Yttrium-90 or indium-111	Non-Hodgkin's Lymphoma	Market	[58]
13	Elacytarabine (Clavis pharma)	Lipid–drug conjugate	Cytarabine	AML	Phase 3	[59]
14	Nanobins	liposome	Carfilzomib	Multiple myeloma	Preclinical	[60]
15	Folate and retinoic Acid grafted/dextran (FARA/DEX)	Polymeric micelle	Doxorubicin	AML	Preclinical	[61]
16	ABI-011(nab5404) [Abraxis bioscience/ Celgene]	Albumin nanoparticle	Thiocolchicine dimer (IDN 5404)	lymphoma	Phase 1	[62]
17	ONCO-TCS [Inex/ enzon]	Liposome	vincristine	Non-Hodgkin	Phase ii/iii	[63]
18	Liposomal Grb-2 [MD Anderson/biopat]	Liposome	Grb2 antisense nucleotide	Leukemia	Preclinical	[64]

expressed on cancerous cells [67]. Further, oligodeoxynucleotides (ODN) nanoparticles conjugated rituximab, or RIT (anti-cd20 antibody), has been designed to beat down the Tlr-9 function in CDK6 chronic lymphocytic leukemia (CLL) B-cells with the aid of G3139 BCL-2 antisense oligonucleotide. The experiment aimed to minimize Tlr-9 directed survival cell signals. Unmodified antisense oligonucleotide did not show any remarkable change in malignant cells' expression because C_pG-motif present in RNA-targeted pharmaceutical activates pro-survival effects. It was noted that free G3139 was unable to cause apoptosis in cancerous cells. CLL B cells are identified by excess translation of Bcl-2 and MCL- proteins, which positively leads to the drug-resistant phenomenon. A subnormal delivery rate has been held responsible for the fruitless action of G3139 in CLL B cells, which was counterbalanced by CD-20 targeted delivery of the antisense oligonucleotide. Nanoparticle-assisted distribution of G3139 lengthened its retention in the early endosomal compartment and RIT cause feeble expression of G3139; therefore, it cannot stimulate antiapoptotic proteins production instead caused gene silencing effects in leukemia affected cells [68].

Further, the tumor growth was effectively controlled by the employment of lipoic acid cross-linked hyaluronic acid nanoparticles encasing the antibiotic doxorubicin. Hyaluronic acid is a biogenic polysaccharide with specific advantages such as high solubility and great capability to interact with drugs. This exciting combination of the delivery vehicle and anticancer medicine displayed marked cytotoxicity against multiple myeloma and acute myeloid leukemia cells in athymic mice. Moreover, overexpression of CD44 in the malignant cells is held accountable for the fast entry of drug in tumor cells and easy drug release at the target site [69]. Notably, unconjugated doxorubicin treated mice exhibited low survival rates, and liver and heart were significantly damaged due to drug-induced systemic toxicity. It is evident that for uninterrupted metastasis, the tumor cells require new capillaries and blood vessels. Angiogenesis is maintained through the VEGF pathway, which ensures the upregulation of the formation of blood vessels. Mukherjee et al. has investigated the role of anti-VEGF loaded gold nanoparticles against B-cell chronic lymphoid leukemia (BCLL) survival. In this work, the increased expression of antiapoptotic proteins in tumor cells was suppressed by the targeted delivery of conjugated gold nanomedicine. The rate of apoptosis was found to be directly proportional to the concentration of nanoparticles [70]. Reports also advocate that leukemia stem cells featuring drug resistance and dynamic characteristics can be modulated by near-infrared fluoroprobe indocyanine green calcium (IGC) phosphosilicate nanoconjugates. CD117 and CD96 are the exceptional markers present on the surface of different cells but found in a significant amount in malignant cells. The targeted photodynamic therapy selectively affected the leukemia stem cells resulting in a decrease in their population. The elevated number of normal leukemia cells was gradually detected that were incompetent to proliferate and progress through the cell cycle [71].

17.3 Cancer diagnosis

Imaging of cancer is necessary for disease management because it helps to follow the treatment effects and predict design therapy accordingly. Non-invasive and safe diagnostic imaging with minimal tissue destruction for early detection of cancers is a suitable strategy for monitoring malignancies' progress. Most imaging operating procedures demand the interplay of EMR (electromagnetic radiation) with target/off-target tissue of the body. Recent studies have established that clinical imaging based on Raman spectroscopy and SPR are more powerful tools for carcinogenesis studies. It is imperative to mention that identifying and targeting biomarkers by nanostructured imaging probes in cancer diagnosis is a newer and promising approach.

17.4 Some investigated nanomaterials based imaging techniques

The characteristic ability to interact with light and intense electromagnetic gold nanoparticles' features makes it excellent signal enhancer and a contrast agent [72]. For efficient imaging of CLL cells, Nguyen et al. successfully demonstrated the enhanced Raman scattering using MGITC (Malachite green isothiocyanate) bounded gold nanoparticles. Electrostatic attractions helped establish the bond between MGITC (Raman tag) and nanocrystals, while the stability of the mixture was achieved by applying thiol-polyethyleneglycol. The modified nanomaterials were further conjugated to CD19 antibody, a surface biomarker glycoprotein present on all B-cells subsets. Vigorous and uniform SERS (surface-enhanced Raman scattering) signals were detected in the gold nanoconjugates variant compared to control with the significant enhancement factor of 2.4×10^3 in scattered signal [73]. Equally important SPR (surface Plasmon resonance) imaging biosensor performance was enhanced using nanosize gold particles to detect low miRNA concentrations. A micro-nucleic acid named miRNA-155 is reported to control the gene expression of tumor suppressor genes and protooncogenes. To be specific, due to loss of notch cell signaling, this miRNA is overexpressed in hematological cancers and affects bone marrow cells. With this in mind, a DNA probe with the ability to bind miRNA is joined to gold nanoparticles, which improves the sensitivity of SPR imaging technique for visualization of target miRNA-155. Notably, detection of even a low concentration of miRNA of about 45pM has been reported in this clinical diagnostic work [74].

Confined toxicological effects and intensification of the radiation dose of tungsten oxide nanomaterials revised the performance of radiotherapy. The experimental analysis carried out by Qiu et al. also concluded that the abovementioned nanowires could be used as excellent NIR (near Infra-red) photosensitizers for direct medical imaging operation and cancerous tissue destruction. The administration of tungsten oxide nanowires and RT (radiation therapy) and NIR laser led to maximum cellular damage in breast cancer cell line 4T1 and in tumor-bearing Balb/c mouse models.

Additionally, improved CT (computed tomography) imaging has been reported post-injection of nanowires in tumor-bearing mice. This described tool for targeted imaging is quite attractive and feasible [75]. Furthermore, Bismuth selenide nanoparticles are well documented to have a high photoelectric absorption coefficient, making them suitable for contrasting and photosensitizer agents. The nanoparticles were coated with PVP (polyvinylpyrrolidone), which acts as a protective layer during the delivery process. Its gradual accumulation at tumor site after being injected in cancer (cervical cancer cell line, U14) bearing mice enhanced 3D CT imaging sensitivity in *vivo*. No substantial pathological changes were noticed during the treatment and diagnosis. However, the liver was slightly affected, which surprisingly recovered after a few months of exposure [76]. Another ultrasensitive cancer diagnostic strategy has been investigated using magnetic gold nanoparticles conjugated to the anti-leukemia aptamer. This aptamer binds to tyrosine kinase protein (TK7), which is highly expressed on leukemia cells. The intercalator attached to the aptamer's hairpin structure releases as soon as it is combined with tumor cells. Replacement of intercalator with tested cancer cells at the aptamer site led to a decrease in electrochemical signal, amplified by a nitrogen-linked graphene nanosize conductor. Further, the above study also made it possible to distinguish whether the originated output signal was from specific interactions between the cells and gold-conjugated aptamer or any other non-specific interactions [77]. To sum up, Nanotechnology based-multiple imaging systems offer several advantages over present invasive techniques and, in the future, will have a great impact on cancer management.

17.5 Clearance and toxicological investigation of nanomaterials

Although nanomaterials assist in obtaining better resolution and high contrast imaging of normal and malignant tissue, it is also very reasonable to focus on its distribution in the circulatory system and rapid elimination from the body. Small diameter nanoparticles of sizes <5 nm are mostly eliminated by the renal system very easily [78], whereas particles greater than 50 nm are moderately removed by the liver metabolic process [79–82]. The renal filtration potency depends on various factors such as morphology and charge carried by nanomaterials. This filtration route is mainly suitable for the clearance of ultrasmall particles [83]. The EPR effect can be accelerated if nanoparticle-based probes are stable in the blood system and have a shorter half-life period. The medical policy approved by the U.S Food and Drug Administration states that the contrast agents used for diagnostic purposes must not be retained in the body for longer duration; otherwise, they might interfere with other imaging procedures.

A study was undertaken to decode the hydrodynamic threshold size and zeta potential for rapid renal clearance using fluorescent quantum dots. The quantum dots had a core of zinc or cadmium compounds with cationic, anionic, or neutral coating. It was indicated that the excretion route for the particles of size around 10 nm is through the hepatic system and then finally into feces. As the surface coating increases the hydrodynamic size by 15 nm, the large-sized quantum dots are found to be cornered in the liver, lungs, and spleen, but small quantum dots of hydrodynamic size 5-6 nm are quickly cleared and excreted in the bladder [84]. In addition, Cheng et al. developed PEGylated porphyrin nanoparticles to achieve the desired parity between imaging output and toxicity level resulting from improficient removal of nanomaterials from the organ system. The hydrodynamic diameter size of 10 nm of nanoparticles is considered best for renal clearance, and such small materials also do not accumulate in the reticuloendothelial system (RES). For this reason, the non-invasive technique positron emission tomography (PET) revealed strong signals from the kidney and bladder after few hours of injection, indicating diminished aggregation of contrast agents in blood circulation [85].

17.6 Challenges to cancer nanotechnology

Although nanotechnology has improved the distribution and targeting of anticancer drugs, the success of ligand coated nanomaterials as vectors depends on its stability and longer circulation time in the bloodstream near the target site. The RES accelerates the nanoparticle uptake and clearance even before the conjugated drug starts acting on tumor cells. In order to diminish superfluous nanoparticle removal by RES, the use of some coating agents like PEG (polyethylene glycol), polysaccharide PVOH (polyvinyl alcohol), and poloxamer (non-ionic triblock copolymer) have been preferred to lengthen the circulation time of nanomedicine. The applied layer is also engaged in controlling augmentation, opsonization, endocytosis, colloidal stability, and interaction with other serum proteins. Nevertheless, some of the coating agents do have abominable physicochemical properties such as they promote uncontrolled drug release over a short time due to an increase in the hydrodynamic size of nanocarrier and also leads to a rapid uptake of particles by the organs of RES system as coated layer interferes with the immune system of the body [86]. In contrast, low antigenic biodegradable polymers (hydroxyl branched PEG, polyglycolic acid, amino acid coating, zwitterions coating) could be employed to contravene the disadvantages of long-established coating agents.

Therapeutic efficiency also depends on the internalization of nanomedicine and its effective transport to the target site. The nanoparticles interact with extracellular and intracellular components through the biomolecules coated on nanoparticles while entering the biological environment [87]. These biomolecules form corona like structures around nanocrystals, which might guide nanocarrier to unspecific targets by modulating the immune system's cells and mechanisms. Proteins of corona can also affect the course of treatment by advancing the production of reactive oxygen species and lipids' peroxidation [88]. A thorough investigation of corona proteins is required to improve the functioning of nanotherapeutics. Another barrier is the heterogeneous nature of hematological cancer cells; this property is attributed to several genetic alterations and a dynamic tumor growth environment. During the course of disease, continuous genetic evolution and cell remodeling occur, which positively leads to malignancy relapse and may produce more aggressive and drug-resistant clones [89]. Further, the same type of cancer exhibit different epigenetic mutations, which curbs nanomaterials-based targeted drug therapy but the intervention of vascular mediator (anti-VEGF drugs), and blood flow modulator might lead to curative effects in such cases [90]. The low sensitivity and specificity of tumor biomarkers have limited the success of early diagnosis. Unfortunately, there has been very little success in the early detection of cancer-specific molecular abnormalities before tumor formation. Recent advancements in the nano-based imaging approach have revolutionized malignancies pre-determination, but the development cost of nano-diagnostic technique and cancer biosensor is high.

17.7 Conclusion and future perspective

In recent years, the nanoparticles system grants multiple drugs loading onto them due to their small size and high surface area [70]. Conjugated NPs detects the presence of surface biomarkers of malignant cells during molecular imaging of neoplastic tissue for the possible early detection of disease. High reproducibility, superior sensitivity, and cost-effective features of nanomaterials make it an attractive tool in developing cancer point-of-care program and treatment. However, much research is still required to overcome the abovementioned challenges associated with the application of nanomaterials for targeted cancer therapy. Drug-loaded functional nanomaterials showed better and intensified inhibition of malignant cells' growth with fast renal clearance and minimal side effects. Metal nanoparticles can sharpen tomography picture contrast without affecting the imaging technique's performance and use the magnetic field to reduce the drug's off-target distribution. Smart unification of nanotechnology in cancer point-of-care therapy in coming years will definitely replace time-consuming conventional methods and will be an appealing dais for blood cancer treatment. This wing of biotechnology has also been explored simultaneously for the nanoparticles-mediated delivery of microRNA (miRNA) and messenger RNA (mRNA) for translating therapeutic proteins [65]. Notably, this strategy is also considered suitable for the significant generation of vaccines for hematological cancer immunotherapy [91].

Thus this review highlighted the progress of research regarding the use of nanomaterials for targeted drug therapy against hematological cancers. We discuss here the development, fabrication, and recent applications of exceptional nanomaterials to upgrade the delivery of anticancer agents implying this area of research has excellent competence for clinical practices.

References

[1] Zentrum Für Krebsregisterdaten, Gesellschaft Der Epidemiologischen Krebsregister in Deutschland E.V., Editors: Krebs in Deutschland Für 2013/2014. 11. Ausgabe Ed, Robert Koch-Institut, Berlin, 2017.

[2] B. Barnes, K. Kraywinkel, E. Nowossadeck, I. Schönfeld, A. Starker, A. Wienecke, U. Wolf, Bericht zum Krebsgeschehen in Deutschland 2016. 2016, Robert Koch-Institut.

[3] N.S. Majhail, M.E. Flowers, K.K. Ness, M. Jagasia, P.A. Carpenter, M. Arora, et al., High prevalence of metabolic syndrome after allogeneic hematopoietic cell transplantation, Bone Marrow Transplant. 43 (2009) 49–54.

[4] C. Annaloro, P. Usardi, L. Airaghi, V. Giunta, S. Forti, A. Orsatti, et al., Prevalence of metabolic syndrome in long-term survivors of hematopoietic stem cell transplantation, Bone Marrow Transplant. 41 (2008) 797–804.

[5] M. Taskinen, U.M. Saarinen-Pihkala, L. Hovi, M. Lipsanen-Nyman, Impaired glucose tolerance and dyslipidaemia as late effects after bone-marrow transplantation in childhood, Lancet 356 (2000) 993–997.

[6] L.A. Bellm, J.B. Epstein, A. Rose-Ped, P. Martin, H.J. Fuchs, Patient reports of complications of bone marrow transplantation, Support. Care Cancer. 8 (2000) 33–39.

[7] C. Meier, S. Taubenheim, F. Lordick, A. Mehnert-Theuerkauf, H. Götze. Depression and anxiety in older patients with hematological cancer (70+) – Geriatric, social, cancer- and treatment-related associations. 5, 2020, J. Geriatr. Oncol., Vol. 11, pp. 828-835.

[8] D.A. Irvine, B. Zhang, R. Kinstrie, A. Tarafdar, H. Morrison, V.L. Campbel, et al., Deregulated hedgehog pathway signaling is inhibited by the smoothened antagonist LDE225 (Sonidegib) in chronic phase chronic myeloid leukaemia, Sci. Rep. 6 (2016) 25476.

[9] M.W. Kieran, J. Chisholm, M. Casanova, A.A. Brandes, I. Aerts, E. Bouffet, et al., Phase I study of oral sonidegib (LDE225) in pediatric brain and solid tumors and a phase II study in children and adults with relapsed medulloblastoma, Neuro. Oncol. 19 (2017) 1542–1552.

[10] A. Sekulic, M.R. Migden, A.E. Oro, L. Dirix, K.D. Lewis, J.D. Hainsworth, et al., Efficacy and safety of vismodegib in advanced basal-cell carcinoma, N. Engl. J. Med. 366 (2012) 2171–2179.

[11] E.J. Chow, B.A. Mueller, K.S. Baker, K.L. Cushing-Haugen, M.E. Flowers, P.J. Martin, et al., Cardiovascular hospitalizations and mortality among recipients of hematopoietic stem cell transplantation, Ann. Intern. Med. 155 (2011) 21–32.

[12] A.P. Singh, A. Biswas, A. Shukla, P. Maiti, Targeted therapy in chronic diseases using nanomaterial-based drug delivery vehicles, Signal Transduct. Target. Ther. 4 (2019) 33.

[13] S. Senapati, A.K. Mahanta, S. Kumar, P. Maiti, Controlled drug delivery vehicles for cancer treatment and their performance, Signal Transduct. Target Ther. 3 (2018) 7.

[14] B. Thomas, D. Coates, V. Tzeng, L. Baehner, A. Boxer, Treatment of hairy cell leukemia with recombinant alpha-interferon, Blood 68 (1986) 493–497.

[15] S. Rai., K. Ahmed, Interferon in the treatment of hairy-cell leukemia, Best Pract. Res. Clin. Haematol. 16 (2003) 69–81.

[16] S.A. Rosenberg, IL-2: the first effective immunotherapy for human cancer, J. Immunol. 192 (2014) 5451–5458.

[17] K.A. Hay, et al., Kinetics and biomarkers of severe cytokine release syndrome after CD19 chimeric antigen receptor–modified T cell therapy, Blood 130 (2017) 2295–2306.

[18] C. Schmidt, The benefits of immunotherapy combinations, Nature 552 (2018) 567–569.

[19] R.S. Day, E.S. Riley, Gold nanoparticle-mediated photothermal therapyapplications and opportunities for multimodal cancer treatment, Wiley Interdiscip. Rev. Nanomed. Nanobiotechnol. 9 (2017) 1449.

[20] S. Menon, S. Shin, G. Dy, Advances in cancer immunotherapy in solid tumors, Cancers (Basel) 8 (2016) 1–21.

[21] S. Lee, K. Margolin, Cytokines in cancer immunotherapy, Cancers (Basel) 3 (2011) 3856–3893.

[22] L. Milling, Y. Zhang, D.J. Irvine, Delivering safer immunotherapies for cancer, Adv. Drug Deliv. Rev. 114 (2017) 79–101.

[23] C.H. June, J.T. Warshauer, J.A. Bluestone, Is autoimmunity the Achilles' heel of cancer immunotherapy? Nat. Med. 23 (2017) 540–547.

[24] R.B. Harris, B.E. Diasio, Clinical pharmacology of 5-fluorouracil, Clin. Pharmacokinet. 16 (1989) 215–237.
[25] S. Senapati, et al., Layered double hydroxides as effective carrier for anticancer drugs and tailoring of release rate through interlayer anions, J. Control. Release 224 (2016) 186–198.
[26] N.K. Singh, et al., Nanostructure controlled anti-cancer drug delivery using poly (ε-caprolactone) based nanohybrids, J. Mater. Chem. 22 (2012) 17853–17863.
[27] D.K. Patel, et al., Influence of graphene on self-assembly of polyurethane and evaluation of its biomedical properties, Polymer 65 (2015) 183–192.
[28] S. Senapati, A.K. Mahanta, S. Kumar, P. Maiti, Controlled drug delivery vehicles for cancer treatment and their performance, Signal Transduct. Target. Ther. 3 (2018) 7.
[29] M. Wooley., KL. Elsabahy, Design of polymeric nanoparticles for biomedical delivery applications, Chem. Soc. Rev. 41 (2012) 2545–2561.
[30] E. Nemutlu, Ý. Eroðlu, H. Eroðlu, et al., In vitro release test of nano-drug delivery systems based on analytical and technological perspectives, Curr. Anal. Chem. 15 (2019) 373–409.
[31] F. Yang, P. Song, M. Ruan, et al., Recent progress in two-dimensional nanomaterials: synthesis, engineering, and applications, 2019, Flat Chem. 18 (2019) 100133.
[32] Z. Lin, G. Wu, L. Zhao, et al., Carbon nanomaterial-based biosensors: a review of design and applications, Nanotechnol. Mag. 13 (2019) 4–14.
[33] C. Wang, Y. Ye, Q. Hu, A. Bellotti, Z. Gu, Tailoring biomaterials for cancer immunotherapy: emerging trends and future outlook, Adv. Mater. 29 (2017) 1–24.
[34] A.I. Tannock, IF. Minchinton, Drug penetration in solid tumours, Nat. Rev. Cancer 6 (2006) 583–592.
[35] S.K. Sriraman, B. Aryasomayajula, V.P. Torchilin, Barriers to drug delivery in solid tumors, Tissue Barriers 2 (2014) 29528.
[36] J. Bushman, et al., Functionalized nanospheres for targeted delivery of paclitaxel, J. Control. Release 171 (2013) 315–321.
[37] Y. Kataoka., K. Matsumura, Preclinical and clinical studies of anticancer agent-incorporating polymer micelles, Cancer Sci. 100 (2009) 572–579.
[38] M. Wang, et al., Cold atmospheric plasma for selectively ablating metastatic breast cancer cells, PLoS One 8 (2013) e73741.
[39] M. Keidar, et al., Cold atmospheric plasma in cancer therapy, Phys. Plasmas 20 (2013) 057101.
[40] E. Sayed, et al., Porous inorganic drug delivery systems—a review, AAPS Pharm. Sci. Tech. 18 (2017) 1507–1525.
[41] M. Vallet-Regí, et al., Mesoporous silica nanoparticles for drug delivery: current insights, Molecules 23 (2017) 47.
[42] J. Liu, L. Cui, D. Losic, Graphene and graphene oxide as new nanocarriers for drug delivery applications, Acta Biomater. 9 (2013) 9243–9257.
[43] G. Shim, M.G. Kim, J.Y. Park, Y.K. Oh, Graphene-based nanosheets for delivery of chemotherapeutics and biological drugs, Adv. Drug. Deliv. Rev. 105 (2016) 205–227.
[44] D. Kalyane, N. Raval, R. Maheshwari, V. Tambe, K. Kalia, RK. Tekade, Employment of enhanced permeability and retention effect (EPR): Nanoparticle-based precision tools for targeting of therapeutic and diagnostic agent in cancer, Mater. Sci Eng. C 98 (2019) 1252–1276.
[45] Z. Mirzaie, M. Barati, M. Asadi Tokmedash, Anticancer drug delivery systems based on curcumin nanostructures: a review, Pharm. Chem. J. 54 (2020) 353–360.
[46] R. Deng, B. Ji, H. Yu, et al., Multifunctional gold nanoparticles overcome microRNA regulatory network mediated-multidrug resistant leukemia, Sci. Rep. 9 (2019) 5348.
[47] F.M. Muggia, J.D. Hainsworth, S. Jeffers, P. Miller, S. Groshen, M. Tan, et al., Phase II study of liposomal doxorubicin in refractory ovarian cancer: antitumor activity and toxicity modification by liposomal encapsulation, J. Clin. Oncol. 15 (1997) 987–993.
[48] M.A. Rodriguez, R. Pytlik, T. Kozak, M. Chhanabhai, R. Gascoyne, B. Lu, et al., Vincristine sulfate liposomes injection (Marqibo) in heavily pretreated patients with refractory aggressive non-Hodgkin lymphoma: report of the pivotal phase 2 stud. 2009, Cancer, Vol. 115, pp. 3475-3482.
[49] P.A. Dinndorf, J. Gootenberg, M.H. Cohen, P. Keegan, R. Pazdur, FDA drug approval summary: pegaspargase (oncaspar) for the first-line treatment of children with acute lymphoblastic leukemia (ALL), Oncologist 12 (2007) 991–998.

[50] E.J. Feldman, J.E. Lancet, J.E. Kolitz, E.K. Ritchie, G.J. Roboz, A.F. List, et al., First-inman study of CPX-351: a liposomal carrier containing cytarabine and daunorubicin in a fixed 5:1 molar ratio for the treatment of relapsed and refractory acute myeloid, J. Clin. Oncol. 29 (2011) 979–985.
[51] H. Dudek, D.H. Wong, R. Arvan, A. Shah, K. Wortham, B. Ying, et al., Knockdown of β-catenin with dicer-substrate siRNAs reduces liver tumor burden in vivo, Mol. Ther. J. Am. Soc. Gene Ther. 22 (2014) 92–101.
[52] D.J. Booser, F.J. Esteva, E. Rivera, V. Valero, L. Esparza-Guerra, W. Priebe, et al., Phase II study of liposomal annamycin in the treatment of doxorubicin-resistant breast cancer, Cancer Chemother. Pharmacol. 50 (2002) 6–8.
[53] ClinicalTrials.gov, US National Library of Medicine. https://clinicaltrials.gov/ct2/show/NCT01733238?term.
[54] 82. US National Library of Medicine. ClinicalTrials.gov, 2014. https://clinicaltrials.gov/ct2/show/NCT00860522?term.
[55] US National Library of Medicine. ClinicalTrials.gov, 2014. https://clinicaltrials.gov/ct2/show/NCT01435720?term.
[56] M.T. Stephan, J.J. Moon, S.H. Um, A. Bershteyn, D.J. Irvine, Therapeutic cell engineering with surface-conjugated synthetic nanoparticles, Nat. Med. 16 (2010) 1035–1041.
[57] S.L. Hewitt, et al., Durable anticancer immunity from intratumoral administration of IL-23, IL-36gamma, and OX40L mRNAs, Sci. Transl. Med. 11 (2019) 9143.
[58] F. Morschhauser, J. Radford, A. Van Hoof, U. Vitolo, P. Soubeyran, et al., Phase III trial of consolidation therapy with yttrium-90-ibritumomab tiuxetan compared with no additional therapy after first remission in advanced follicular lymphoma, J. Clin. Oncol. 26 (2008) 5156–5164.
[59] C.D. DiNardo, S. O'Brien, V.V. Gandhi, F. Ravandi, Elacytarabine (CP-4055) in the treatment of acute myeloid leukemia, Future Onco. 9 (2013) 1073–1082 NCT02736721.
[60] J.D. Ashley, C.J. Quinlan, V.A. Schroeder, M.A. Suckow, et al., Dual carfilzomib and doxorubicin-loaded liposomal nanoparticles for synergistic efficacy in multiple myeloma, Mol. Cancer Ther. 15 (2016) 1452–1459.
[61] J. Varshosaz, F. Hassanzadeh, H. Sadeghi Aliabadi, et al., Synthesis and characterization of folate-targeted dextran/retinoic acid micelles for doxorubicin delivery in acute leukemia, Biomed. Res. Int. (2014) 525684.
[62] R.J. Bernacki, J. Veith, P. Pera, B.J. Kennedy, et al., A novel nanoparticle albumin bound thiocolchicine dimer (nab-5404) with dual mechanisms of action on tubulin and topoisomerase-1: evaluation of in vitro and in vivo activity, Cancer Res. 65 (2005) 560.
[63] P.A. Dinndorf, J. Gootenberg, M.H. Cohen, P. Keegan, R. Pazdur, FDA drug approval summary: pegaspargase (oncaspar) for the first-line treatment of children with acute lymphoblastic leukemia (ALL), Oncologist 12 (2007) 991–998.
[64] A.M. Tari, Y. Gutierrez-Puente, G. Monaco, C. Stephens, T. Sun, M. Rosenblum, J. Belmont, R. Arlinghaus, G. Lopez-Berestein, Liposome-incorporated Grb2 antisense oligodeoxynucleotide increases the survival of mice bearing bcr-abl-positive leukemia xenografts, Int. J. Onco. 31 (2007) 1243–1250.
[65] J.A. Kulkarni, D. Witzigmann, S. Chen, et al., Lipid nanoparticle technology for clinical translation of siRNA therapeutics, Acc. Chem. Res. 52 (2019) 2435–2444.
[66] K. Briley-Saebo, A. Bjomeru, D. Grant, H. Ahlstrom, T. Berg, G.M. Kindberg, Hepatic cellular distribution and degradation of irom oxide nanoparticles following single intravenous injection in rats: Implications for magnetic resonance imaging, Cell Tissue Res. 316 (2004) 315–323.
[67] K.Y. Choi, S. Correa, M.J, J. Li, et al., Binary targeting of siRNA to hematologic cancer cells in vivo using layer-by-layer nanoparticles, Adv. Funct. Mater. 29 (2019) 1900018.
[68] B. Yu, Y. Mao, L. -Yuan Bai, S.E.M. Herman, et al., Targeted nanoparticle delivery overcomes off-target immunostimulatory effects of oligonucleotides and improves therapeutic efficacy in chronic lymphocytic leukemia, Blood 121 (1) (2013) 136–147.
[69] Y. Zhong, F. Meng, C. Deng, X. Mao, Z. Zhong, Targeted inhibition of human hematological cancers in vivo by doxorubicin encapsulated in smart lipoic acid-crosslinked hyaluronic acid nanoparticles, Drug Deliv. 24 (1) (2017) 1482–1490.

[70] P. Mukherjee, R. Bhattacharya, N. Bone, Y.K. Lee, et al., Potential therapeutic application of gold nanoparticles in B-chronic lymphocytic leukemia (BCLL): enhancing apoptosis, J. Nanobiotechnol. 5 (2007) 4.
[71] B.M. Barth, E.I. Altınoglu, S.S. Shanmugavelandy, et al., Targeted indocyanine-green-loaded calcium phosphosilicate nanoparticles for in vivo photodynamic therapy of leukemia, ACS Nano 5 (2011) 5325–5337.
[72] L. Liu, S. Jiang, L. Wang, Z. Zhang, G. Xie, Direct detection of microRNA-126 at a femtomolar level using a glassy carbon electrode modified with chitosan, graphene sheets, and a poly (amidoamine) dendrimer composite with gold and silver nanocluster, Microchim. Acta. 182 (2015) 77–84.
[73] C.T. Nguyen, J.T. Nguyen, S. Rutledge, et al., Detection of chronic lymphocytic leukemia cell surface markers using surface enhanced Raman scattering gold nanoparticles, Cancer Lett. 292 (2010) 91–97.
[74] K. Zeng, H. Li, Y. Peng, Gold nanoparticle enhanced surface plasmon resonance imaging of microRNA-155 using a functional nucleic acid-based amplification machine, Microchim. Acta. 184 (2017) 2637–2644.
[75] J. Qiu, Q. Xiao, X. Zheng, et al., Single W18O49 nanowires: A multifunctional nanoplatform for computed tomography imaging and photothermal/photodynamic/radiation synergistic cancer therapy, Nano Res. 8 (2015) 3580–3590.
[76] X. Zhang, J. Chen, Y. Min, et al., Metabolizable Bi 2 Se 3 nanoplates: biodistribution, toxicity, and uses for cancer radiation therapy and imaging, Adv. Funct. Mater. 24 (2014) 1718–1729.
[77] S.M. Mehrgardi, M.A. Khoshfetrat, Amplified detection of leukemia cancer cells using an aptamerconjugated gold-coated magnetic nanoparticles on a nitrogendoped graphene modified electrode, J. Bioelechem. 114 (2017) 24–32.
[78] H.S. Choi, W. Liu, P. Misra, et al., Renal clearance of quantum dots, Nat. Biotechnol. 25 (2007) 1165–1170.
[79] W.S. Cho, M. Cho, J. Jeong, et al., Size-dependent tissue kinetics of PEG-coated gold nanoparticles, Toxicol. Appl. Pharm. 245 (2010) 116–123.
[80] C.D. Walkey, J.B. Olsen, H. Guo, A. Emili, W.C. Chan. Nanoparticle size and surface chemistry determine serum protein adsorption and macrophage uptake, J. Am. Chem. Soc. 134 (2012) 2139–2147.
[81] M.C. Mancini, B.A. Kairdolf, A.M. Smith, S. Nie, Oxidative quenching and degradation of polymer-encapsulated quantum dots: new insights into the long-term fate and toxicity of nanocrystals in vivo, J. Am. Chem. Soc. 130 (2008) 10836–10837.
[82] Y.R.S. Yang, L.W. Chang, J.P. Wu, M.H. Tsai, H.J. Wang, et al., Persistent tissue kinetics and redistribution of nanoparticles, quantum dot 705, in mice: ICP-MS quantitative assessment, Environ. Health Perspect. 115 (2007) 1339–1343.
[83] M. Longmire, P.L. Choyke, H. Kobayashi, Clearance properties of nano-sized particles and molecules as imaging agents: considerations and caveats, Nanomed. 3 (2008) 703–717.
[84] H.S. Choi, W. Liu, P. Misra, et al., Renal clearance of nanoparticles, Nat. Biotechnol. 25 (2007) 1165–1170.
[85] L. Cheng, D. Jiang, A. Kamkaew, et al., Renal-clearable PEGylated porphyrin nanoparticles for image-guided photodynamic cancer therapy, Adv. Funct. Mater. (2017) 1702928.
[86] S. Huang., L. Guo, Nanoparticles escaping RES and endosome: challenges for siRNA delivery for cancer therapy, J. Nanomater. 2011 (2011) 742895.
[87] H. Liu, J. Zhang, X. Chen, X.S. Du, et al., Application of iron oxide nanoparticles in glioma imaging and therapy: from bench to bedside, Nanoscale 8 (2016) 7808–7826.
[88] S.A. Majetich, T. Wen, R.A. Booth, Functional magnetic nanoparticle assemblies: formation, collective behavior, and future directions, ACS Nano 5 (2011) 6081–6084.
[89] D.A. Landau, S.L. Carter, G. Getz, C.J. Wu, Clonal evolution in hematological malignancies and therapeutic implications, Leukemia 28 (2014) 34–43.
[90] A. Wicki, D. Witzigmann, V.K. Balasubramanian, J. Huwyler, Nanomedicine in cancer therapy: challenges, opportunities, and clinical applications, J. Control Release 200 (2015) 138–157.
[91] N. Pardi, K. Parkhouse, E. Kirkpatrick, M. McMahon, et al., Nucleoside modified mRNA immunization elicits influenza virus hemagglutinin stalk-specific antibodies, Nat. Commun. 9 (2018) 3361.

CHAPTER 18

DNA nanotechnology based point-of-care theranostics devices

Anjali Rajwar, Vinod Morya, Dhiraj Bhatia
Biological Engineering Discipline, Indian Institute of Technology Gandhinagar, Palaj, Gujarat, India

18.1 Introduction

Deoxyribonucleic acid, DNA being the biological material and has been explored for its biological properties and functions; until Nadrian Seeman suggested unorthodox structures based on its physical and self-assembling properties [1]. Structural robustness, appropriate persistent length, inherent biocompatibility, chemical addressability, self-assembly into predefined structures and dissociation capabilities in specific environmental condition makes it a natural material of choice for nanoscale construction of various nanodevices in 1D/2D/3D [2]. Structural DNA nanotechnology stemmed from Seeman's innovative proposal that DNA could be used as a physical material for the self-assembly of nanoscale structures. The journey of DNA nanotechnology started in 1980s when Nadrian Seeman got inspired from a picture by Dutch artist M.C. Escher's woodcut Depth and realized that the center of each fish in the picture as an idealized branch point of a six-arm junction. He hypothesized that interweaving different helical limbs to make stable artificial junctions taking inspiration from previously described replication forks and Holliday junctions. The initial designs in structural DNA nanotechnology were the replica of the motifs found in nature like the immobile Holliday junction which served as a milestone at the beginning of DNA nanotechnology. Ned developed structures that turned out to be the cornerstone in nanotechnology like a double crossover (DX) motifs, triple crossover (TX) and paranemic crossover (PX) which further increases the stability of the motif [3–5]. The DX and PX (parallel crossover) tiles have been extensively characterized by biochemistry and molecular biology techniques and are now the gold standards of DNA nanotechnology. These motifs offers tremendous advantage in that the dimensions of the tiles can be easily modulated while maintaining the junction geometry which is the prime necessity while building higher-order, more complex structures from DNA. DNA nanotechnology has travelled a long path from simple four-way junction [6,7] to 1D DNA wires [8], 2D DNA lattices designed using DNA origami (P. [9]), 3D DNA crystal [10], 3D polyhedral structures [11] (Fig. 18.1A–D). DNA-based scaffolds exhibit outstanding properties resulted due to the supramolecular interactions amongst complementary Watson–Crick base pairing

Advanced Nanomaterials for Point of Care Diagnosis and Therapy
DOI: https://doi.org/10.1016/B978-0-32-385725-3.00012-X

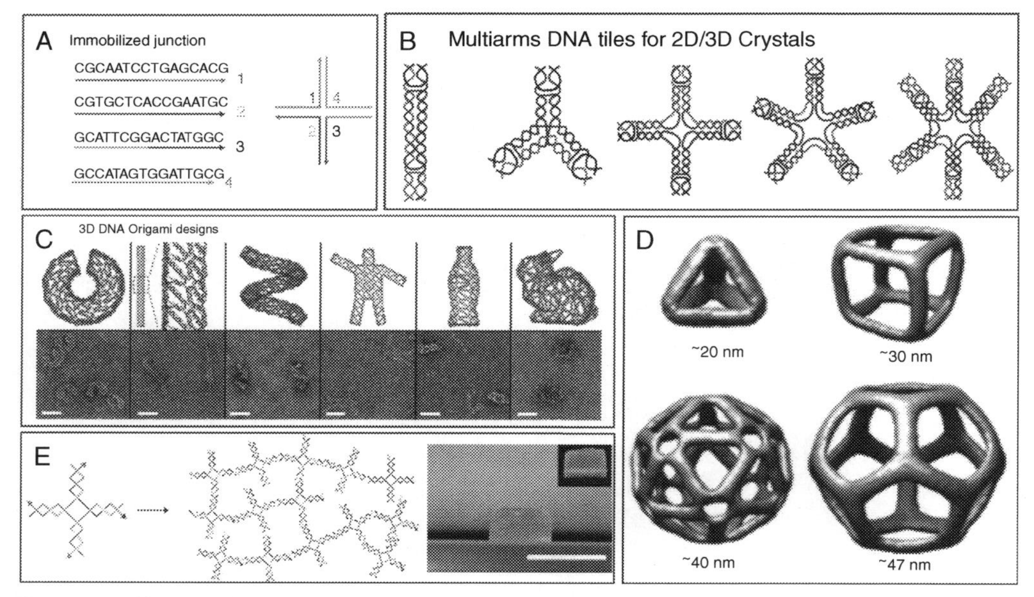

Fig. 18.1 ***Self-assembled DNA nanostructures.*** (A) four single stranded oligonucleotides were assembled into four-way junction, (B) multiple DNA strands of three, four, five, and six arms DNA junction with symmetric sticky arms can be extended into 2D DNA lattices and 3D DNA crystal. (C) Scaffolded assembly to form 3D DNA nanostructures [70]. *Reproduced with permission from Benson, E., Mohammed, A., Gardell, J. et al. DNA rendering of polyhedral meshes at the nanoscale. Nature. 523, 441–444 (2015).* (D) Cryo-EM reconstruction of DNA polyhedra with different geometries [11]. *Reproduced with permission from Zhang C., He, Y., et al. DNA self-assembly: from 2D to 3D. Faraday Discuss., 143, 221-233 (2009).* (E) Self-assembly of DNA hydrogel with X-shaped DNA monomer [41]. *Reproduced with permission from Um S., Lee J., Park N., et al. Enzyme-catalysed assembly of DNA hydrogel. Nat. Mater 5, 797–801 (2006).*

of individual DNA strands. Further, easy chemical modifications of DNA, making it compatible for coupling with multiple biomolecules, organic and inorganic nanoparticles, thus expanding its dimensions for biomedical and now clinical applications [12]. Multi-functional scaffolds have been realised by spatial arrangement of functional proteins (enzymes) on to the designed DNA structures [13]. Moreover, DNA nanodevices exhibit exceptionally high physicochemical stability that can be altered or programmed by external cues. Stability of DNA structures have been examined under different environmental conditions like high temperature, organic solvents, alkaline pH, UV exposure, etc. The results suggest DNA nanostructures can be used in a wider range of applications [14]. Stimuli-responsive behavior of designer DNA structures or functionalizable with other responsive materials triggered by external stimulus makes it suitable for sensing and other biomedical applications [15].

Other than DNA polyhedra and origami structures, DNA based hydrogels are proved as promising soft material building blocks for numerous applications including sensing, bioanalysis, therapeutic delivery, cell culture, *in vitro* translation leading to protein synthesis, mechanical devices, etc. [16]. Negatively charged DNA helix structure with polar sugar-phosphate backbone acts as highly hydrophilic polyelectrolyte. The hydrophilicity of DNA attracts the water molecules which allows the formation of gel like material [17,18]. DNA hydrogels are random 3D DNA polymer matrix which possesses the chemical properties of DNA but acts like hydrogel physically [19] (Fig. 18.1E). Matrix like structure of DNA hydrogel provides multiple sites and space for functionalization, thus makes it versatile supporting material.

Modern day lifestyle leaves truly little or no time for the health check-ups primarily because of the tedious and time-consuming diagnosis procedures. Additionally, in remote and underdeveloped regions, the cost and transportation are the limiting factors. Point-of-care (POC) devices are the miniaturized and low cost diagnostic systems used by any individual with minimal skills and resources provided with the device as a kit or manual. In past few decades, DNA nanotechnology based devices showed outstanding applications in sensing and diagnostics [20–23]. Given the above-mentioned properties of DNA and its adaptability with various chemical and biological entities; DNA nanotechnology based POC diagnostic devices can be realized with ease at very low coast. Herein, we cover the current trends in progress in DNA nanodevices toward applications as point-of-care diagnostics and preventive medicine. We conclude briefly about the challenges and shortcomings being faced by DNA nanodevices *in vivo* and also provide some of the prospects to translate these bench-side devices to bed-side applications.

18.2 Design and synthesis of DNA-based devices

The approach of constructing structures using DNA is based on Watson-Crick base pairing principle, where the arrangement of multiple DNA strands is shaped into unique designs in 1D/2D/3D [24]. The first DNA nanostructure to be realized was a immobile four-way junction created using four ssDNA was reported by Ned Seeman's group [7]. Later on, three, five, six, eight, and twelve-armed structures have been created using similar self-assembly strategy [25–27]. Larger functional DNA structures can be realized by programmed assembly of small branched DNA motifs or DNA tiles [28]. DNA motifs can be assembled using stick-ends into highly directive fashion to form two-dimensional (2D) and three-dimensional (3D) DNA crystals [10,29,30]. Initially used "tile" or "motifs" based approach suffered from some common drawbacks like the polyhedra are formed only at a low concentration of oligonucleotides as higher concentration resulted in giant supramolecular structures. DNA origami overcomes these drawbacks by using a single stranded DNA as a scaffold and folding it into any shape using multiple small, staple strands [9]. Multiple methods have been established for the construction of DNA nanodevices *in vitro* by exploiting the inherent properties of self-assembly and hybridization and DNA.

18.2.1 One pot assembly method

This is a single step synthesis in which all the component oligonucleotides are mixed together to hybridize into final structure [31]. This approach has been used for the simplistic 3D structures like tetrahedron and for complex structures like dodecahedron and Buckyball [32] (Fig. 18.2A).

18.2.2 Modular assembly method

This approach is used for synthesis of complex structure. In this method different modules are synthesised separately by simple hybridization methods and these modules are latter assembled into complete structure. DNA cube was the first structure to be synthesized by this approach [33]. DNA icosahedron and octahedron have also been synthesized using same approach [34,35] (Fig. 18.2B).

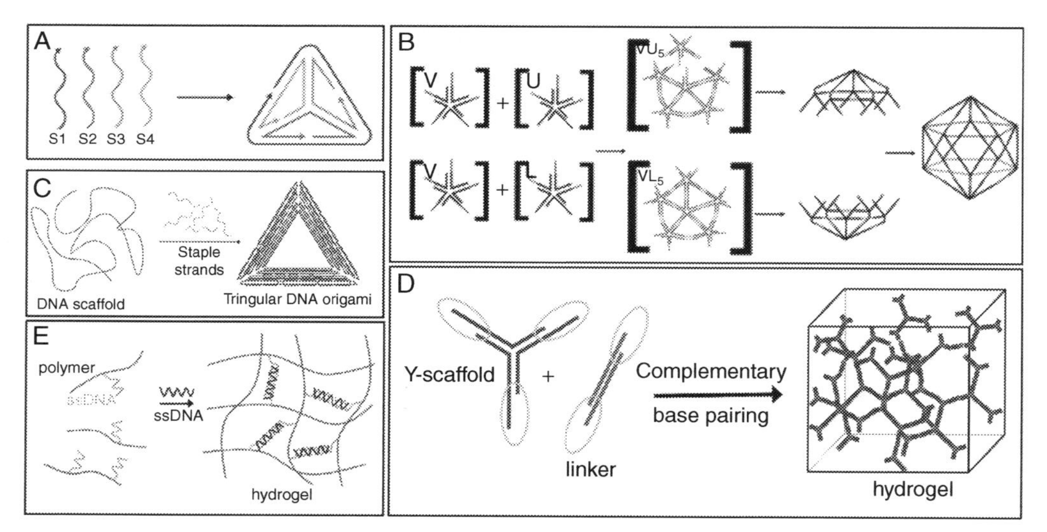

Fig. 18.2 ***Different modes of synthesis of DNA based nanostructures.*** (A) Single step synthesis of DNA tetrahedron (TD) which can be self-assembled using four single stranded oligonucleotides mixed together in equimolar ratios. The mixture is heated at higher temperature and subsequently cooled to form TD. (B) Modular assembly of DNA icosahedron (ID). DNA ID comprise to two half modules which are assembled from five-way junctions (V_5, U_5, and L_5). Half modules (VU_5 and VL_5) are mixed in equimolar ratio to form complete ID [34]. *Reproduced with permission from Bhatia, D., Mehtab, S., et al., Icosahedral DNA Nanocapsules by Modular Assembly. Angew. Chem. Int. Ed., 48: 4134-4137 (2009).* (C) Origami based synthesis of triangular DNA nanostructure. A large scaffold DNA mixed with small single stranded oligonucleotides, which hybridize with the scaffold DNA and fold into a predesigned structure. (D) Self-assembly of DNA hydrogel based on complimentary sticky ends and (E) synthesis of DNA cross-linked polyacrylamide hydrogel.

18.2.3 DNA origami-based assembly method

This approach was first introduced by Paul Rothemund in 2006 for designing large complex structures. In this method a large scaffold DNA is folded into desired shape with the help of multiple short staple strands (P. W. K. [36]). Since then, numerous 2D and 3D designs has been crafted using DNA origami strategy for various biomedical applications [37]. Apart from these 2D and 3D designs, polyhedral 3D structures have also been assembled from DNA motifs and tiles with similar bottom-up approach [38,39] (Fig. 18.2C).

18.3 DNA hydrogels

Hydrogels are mostly polymeric based scaffolds with enhanced capacity to absorb water and swell in dimensions. DNA hydrogels are highly branched DNA polymer based matrices. The properties of the hydrogel depend upon the components as well as formation method. The concentration and degree of cross-linking drives the density and porosity. The DNA hydrogels can be formed using various methods based on the desired biomedical applications. DNA hydrogels can be broadly classified into two types.

18.3.1 Hydrogels made only from DNA strands

A simple method involves mixing synthetic DNA motifs (monomers) like junctions in solution, which self-assembles into a defined 3D network mediated by complementary DNA strands [40]. These are reversible hydrogels and their stability depends upon the length of the complimentary strand (sticky end) and guanine-cytosine (GC) content. Enzymatic ligation makes them irreversible and more rigid [41]. 'Hybridization chain reaction' is a modular assembly approach to form DNA hydrogel, where strand displacement strategy is used to polymerise the DNA hairpin loops (monomers) [42]. In order to get large amount of DNA hydrogel, an in-situ synthesis method is used, that is, "rolling circle amplification" (RCA) [19]. DNA polymerase uses two circular ssDNA template to make multiple copies of monomer DNA strands. This strategy results into formation of bulk hydrogel with mechanical properties (Fig. 18.2D).

18.3.2 Hybrid DNA hydrogels

DNA is a soft biological material with some constraints like limited stability, and low yield, leading to limited applications. Therefore, hybrid DNA hydrogels are of much significance, where DNA is used as a function crosslinker or as a sensing motif (duplex DNA, aptamer, i-motif, G-quadruplex, etc.) [43–45]. These motifs are responsive to different stimulus like pH, biomolecules, metal ions, etc. DNA cross-linked synthetic polymers possess the stability and characteristics of the cross-linked polymer with function properties of DNA linker. ssDNA functionalized polyacrylamide hydrogel crosslinked with complimentary ssDNA is the simplest example of hybrid hydrogel (Fig. 18.2E). This responsive behavior proved to become backbone of various diagnostic devices leading to the translation of DNA nanodevices into devices for point-of-care diagnostics tools.

18.4 Characterization of DNA nanodevices

Basic characterization of DNA structures can be done by electro mobility shift assay (EMSA) [46,47]. EMSA is used to primarily identify the generation of high ordered structures of DNA in solution (Fig. 18.3A). The hydrodynamic size of DNA shapes or DNA nanogels can be determined by dynamic light scattering (DLS) or AFM [11,48] (Fig. 18.3B). The actual structure and morphology can be visualized by scanning electron microscope (SEM) or atomic force microscope (AFM) at nanometre scale [11,19,49] (Fig. 18.3C and D). The physical stability and thermodynamic properties of hydrogels can be demonstrated by rheological studies [50] (Fig. 18.3E). The length and G≡C content of sticky end determines the thermal stability of the hydrogels. Simulation studies helps in anticipation of structural, mechanical, and thermodynamics properties of DNA hydrogels [51]. The utility of DNA hydrogels in biological applications are closely related to their mechanical and physicochemical properties.

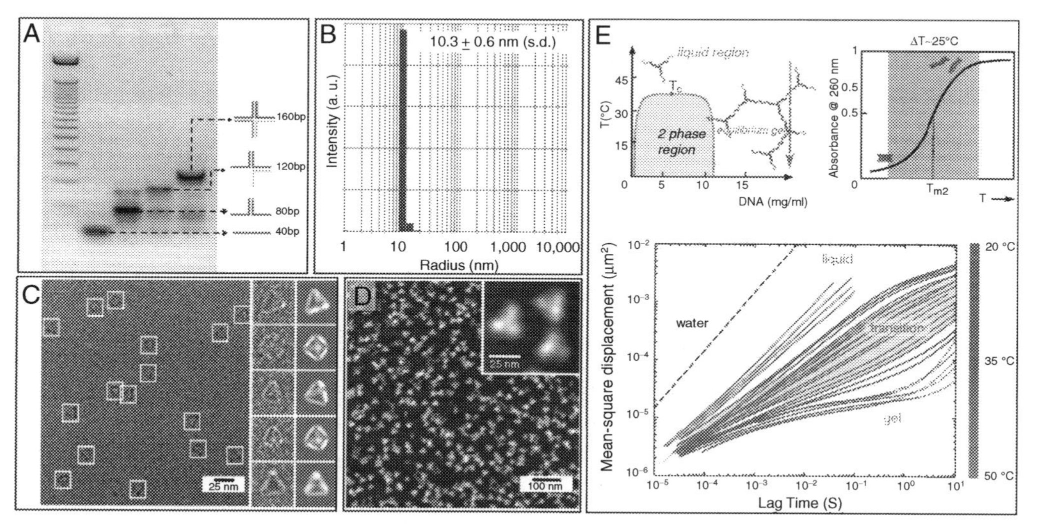

Fig. 18.3 (A) Gel retardation assay to confirm the formation of higher order structure [41]. *Reproduced with permission from Um S., Lee J., Park N, et al. Enzyme-catalysed assembly of DNA hydrogel. Nature Mater 5, 797–801 (2006).* (B–D) Size histogram of DNA tetrahedron measured by DLS; Cryo-EM; and AFM image to visualize the formation of defined structure [32]. *Reproduced with permission from He Y., Ye T., Su M., et al. Hierarchical self-assembly of DNA into symmetric supramolecular polyhedra. Nature 452, 198–201 (2008).* (E) Temperature stability and rheological studies of DNA hydrogels [50].

18.5 Current applications

DNA has unprecedented biocompatibility and chemical flexibility that make it a next generation biomaterial of choice for multiple purposes. DNA nanostructures designed either via modular self-assembly or through origami-based construction can serve as a potential biomaterial for disease diagnosis and drug delivery. The biocompatibility and stimuli-responsiveness lead to designing of DNA nanostructure for controlled drug delivery and biosensing purposes.

18.5.1 DNA nanodevices in therapeutics

DNA origami can serve as a scaffold which can be functionalized with multiple ligands like peptides, aptamer, protein, small molecules, etc. and can be used for diverse array of diagnostic applications *in vivo*. These 3D nanostructures can encapsulate multiple cargos within there cavity and release them at specific location in response to specific stimuli or trigger [52] (Fig. 18.4A). Douglas et al. have successfully designed origami-based DNA nanobot shaped like a hexagonal barrel. This nanorobot consists of two domains which are locked using aptamer-based mechanisms, these domains are attached covalently by single-stranded scaffold hinges. The aptamer complex is incorporated at both the ends and when both the aptamer recognizes their target it leads to dissociation of the nanorobot and the cargo is exposed [53]. Molecular payloads like metal nanoparticles or antibodies can be loaded inside the nanorobot. They have designed several lock combinations for different cell activation. These nanostructures provide an opportunity for selective regulation of cell sensing along with cargo loading capability. This study demonstrates the utility of nanorobots for tuning of specific cell behavior and have been used for cancer theragnostic. Li et al. have used this nanobot for tumor targeting by functionalizing it with the aptamer specific for nucleolin (protein highly expressed on cancer cells) (Fig. 18.4B). The nanobot was encapsulated with thrombin that can cause intravascular thrombosis resulting in tumor necrosis. This nanobot has shown efficient targeting and triggered release of thrombin specific to tumor site [54]. Maia Godonoga et al., have utilized DNA origami scaffold for diagnosing malarial infection. They have used rectangular DNA origami scaffold and functionalized it with 12 aptamers which are specific for Plasmodium Falciparum lactate dehydrogenase enzyme *(PfLDH)*, which is a well-known marker for malarial diagnosis [55,56]. The aptamer on the DNA origami capsule can interact with *PfLDH* resulting in conformational change which opens the capsule and signal can be detected. This approach can further be modified and can be used for designing "smart" vehicles for malarial diagnosis [55]. Vries et al., have shown that DNA coupled nanoparticles can serve as a better carrier system for the treatment of ocular infection. The limitation of the currently available ocular therapeutics is their low bioavailability that requires more frequent administration which might lead to biotoxicity [57–59]. They have used amphiphilic DNA bock copolymer

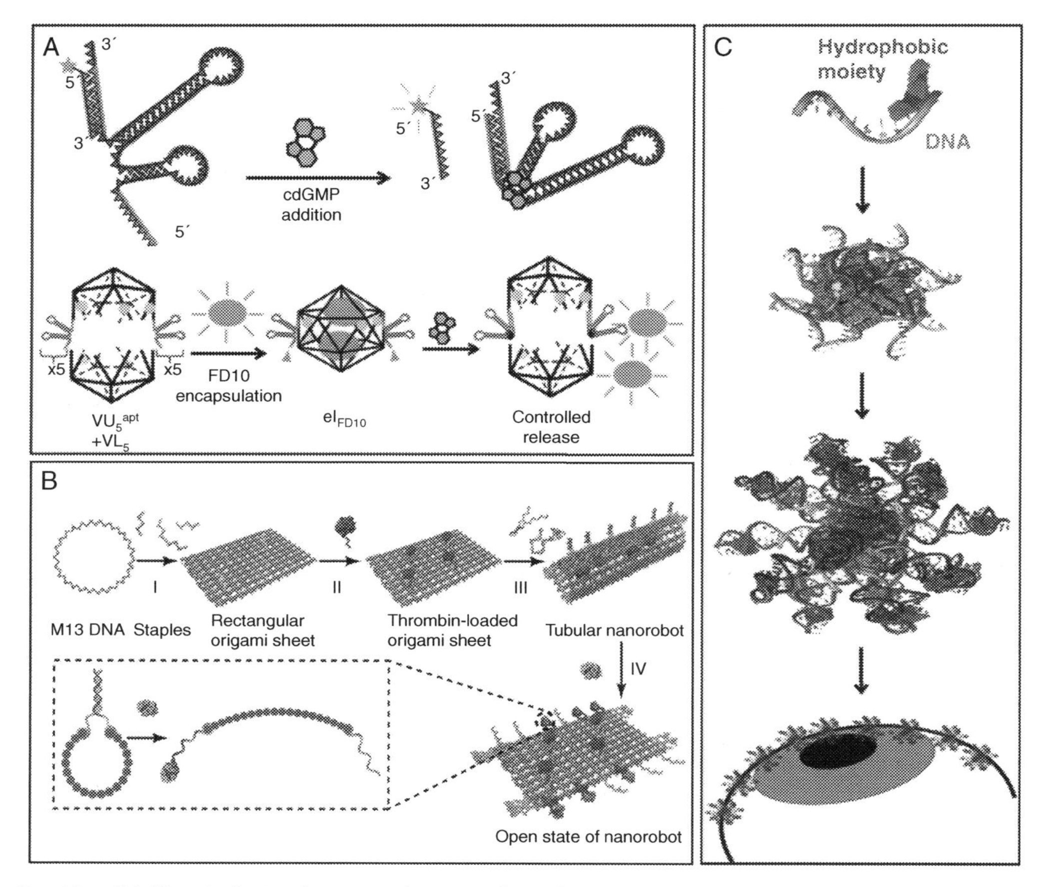

Fig. 18.4 (A) Chemical stimuli triggered cargo release from aptamer functionalized DNA icosahedron [52]. *Reproduced with permission from Banerjee, A., Bhatia, D., Controlled Release of Encapsulated Cargo from a DNA Icosahedron using a Chemical Trigger. Angew. Chem. Int. Ed., 52: 6854-6857 (2013).* (B) Schematic representation of the formation of "thrombinloaded nanorobot", and its reconfigurable open and closed state in response to nucleolin binding. (B, I) Formation of a rectangular DNA sheet by M13 phage genomic ssDNA cross-lined by staple strands. (B, II) Thrombin loading by hybridization of poly-T oligonucleotides (presented on thrombin molecules) with poly-A sequences (present on the surface of the DNA sheet). (B, III) Formation of thrombin-loaded, tubular DNA nanorobots by addition of the fasteners and aptamer strands. (B, IV) Opening and exposer of the encapsulated thrombin in response to presence of nucleolin [54]. *Reproduced with permission from Li, S., Jiang, Q., Liu, S. et al. A DNA nanorobot functions as a cancer therapeutic in response to a molecular trigger in vivo. Nat. Biotechnol. 36, 258–264 (2018).* (C) DNA nanoparticles for efficient ophthalmic drug delivery [61]. *Reproduced with permission from Willem de Vries, J., Schnichels, S., et al., DNA nanoparticles for ophthalmic drug delivery, Biomaterials. 157, 98-106 (2018).*

comprising of hydrophobic lipid core with extended DNA strands [60]. DNA-lipid based carrier system adhere efficiently on ocular surfaces enhancing the time for the active compounds to interact with the target tissue and thereby improving drug action (Fig. 18.4C). The drugs were attached via DNA and RNA aptamers for kanamycin and neomycin respectively at the 3' end of DNA strands. This system was administered in the form of eye drops at 20 μM concentration and volume of 30 μl. They observed that antibiotic loaded nanoparticles could remain adhered to the cornea for a period of 2 h whereas only antibiotics cannot be detected after 5mins showing DNA as an efficient tool for ophthalmic delivery [61].

18.5.2 DNA nanodevices in biosensing

DNA nanodevices have now been extensively explored for the detection of multiple analytes like pH, ions, nucleic acids, cells, viruses etc [62,63]. DNA and RNA can be conjugated with gold nanoparticles for colorimetric detection of different analyte [64] (Fig. 18.5A). Zhou et al. have recently developed simple and cost-effective DNA nanoswitches for the detection of viral RNA with high level of sensitivity. DNA nanoswitches comprise of a long scaffold DNA hybridized with single stranded oligonucleotides and detectors. The nanoswitches upon binding to viral RNA encounter confirmational changes from linear to looped which can be detected by gel electrophoresis [23] (Fig. 18.5B).

DNA-based hydrogels provide a platform to load and functionalize with various responsive motifs. There are plenty of application realized using DNA-based hydrogels in sensing and diagnosis. Mercury (Hg) is a highly toxic environmental pollutant, which has many adverse effects on our body. The traditional detection methods required massive and costly instruments like atomic absorption spectroscopy or inductively coupled plasma-mass spectrometry. Therefore, a comfortable and instrument-free technique can be of great use. Dave et al. reported an ultrasensitive detection of Hg^{2+} using polyacrylamide hydrogels incorporated with responsive thymine-rich DNA structure [65]. It's a fluorescence based visual detection technique, with the naked eye, detection limit is 10 nM Hg^{2+} in a 50 mL water sample (Fig. 18.5C).

DNA aptamer crosslinked hydrogels can show a sol-to-gel transition, which is a visual change. Aptamer can be designed for any target molecule using systematic evolution of ligands by exponential enrichment (SELEX) process. The potential application of these hydrogels is visual detection of any biomolecule of interest or can be used in selective release of load. For example, Yang et al. engineered a hydrogel based on aptamer-target interaction for qualitative detection of Adenosine [66]. The hydrogel matrix was crosslinked with Adenosine responsive aptamer and loaded with gold nanoparticles. When the hydrogel detects Adenosine in the solution it loosens up and releases the gold nanoparticles embedded in the hydrogel matrix. The release of gold nanoparticles can be seen in the solution. Similarly, a nonenzymatic system has been reported for visual detection of glucose [67].

Target-responsive DNA-based hydrogels are also developed to use with traditional digital glucometer for on-site detection of analyte. Chaoyang James Yang and his team designed glucoamylase-trapped aptamer-crosslinked hydrogel called "Sweet" hydrogel

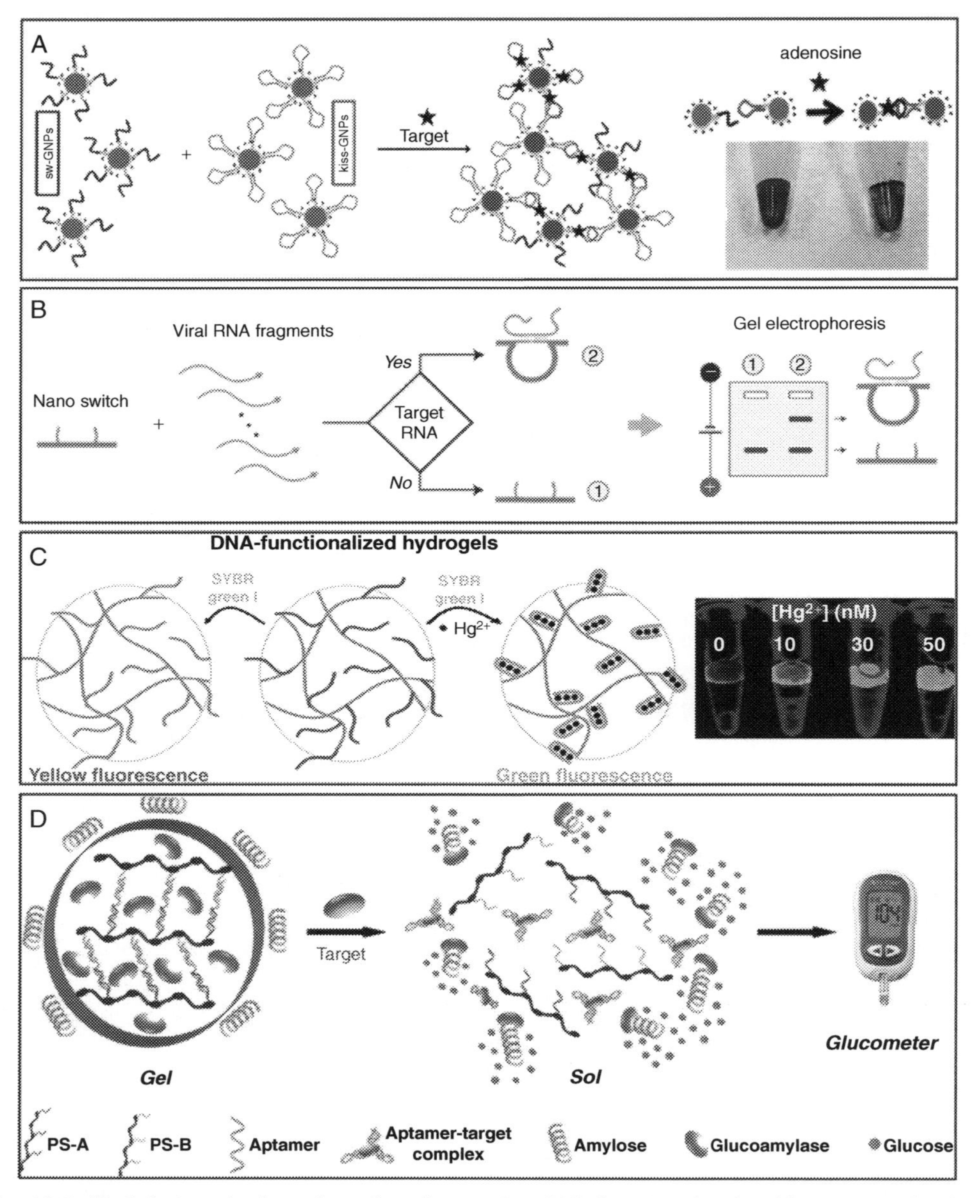

Fig. 18.5 (A) Colorimetric detection of analytes using DNA-functionalized gold nanoparticles [64]. *Reproduced with permission from Goux, E., Dausse, E., et al. A colorimetric nanosensor based on a selective target-responsive aptamer kissing complex. Nanoscale. 9, 4048–4052 (2017).* (B) Viral RNA detection by designer DNA switch, (C) mercury (Hg^{+2}) detection in water using DNA based-hydrogel [65]. *Reproduced with permission from Dave N., Chan M.Y., Regenerable DNA-functionalized hydrogels for ultrasensitive, instrument-free mercury(II) detection and removal in water. J. Am. Chem. Soc. 132, 12668–12673 (2010).* (E) Glucoamylase-trapped aptamer-crosslinked hydrogel ("sweet" hydrogel) for quantitative detection of non-glucose targets [68]. *Reproduced with permission from Yan L., Zhu Z., Target-responsive "sweet" hydrogel with glucometer readout for portable and quantitative detection of non-glucose targets. J. Am. Chem. Soc. 135, 3748–3751 (2013).*

for quantitative detection of non-glucose targets [68]. Here, the aptamer crosslinked the acrylamide polymer to form hydrogel and glucoamylase enzyme entrapped into the matrix. When hydrogel and amylose are mixed with a sample to analyze, the aptamer specifically binds to the analyte. It collapses the hydrogel to release the glucoamylase, which hydrolyze the amylose into glucose (Fig. 18.5D). The amount of glucose depends upon the concentration of the analyte. The glucose can be detected by any personal glucose meter (PGM), making it a proper POC diagnostic device. An Au-Pt core-shell nanoparticle encapsulated target-responsive hydrogel designed with volumetric bar-chart for quantitative readout was reported [69]. Target-responsive aptamer was used to release Au-Pt nanoparticles, these nanoparticles catalyses H_2O_2 and decompose into H_2O and O_2. The O_2 generated by catalysis creates pressure and pushes a dye solution placed in the volumetric bar to give a visual quantitative readout for the analyte. This kind of POC devices can be designed for any analyte using their respective aptamers.

18.6 Conclusion and future perspectives

Emerged as a genetic material at the beginning, interestingly, DNA has now metamorphosized into an attractive tool for designing tuneable nanostructures. Unique properties such as programmability and ready accessibility allow us to adopt DNA as template for designing desirable nanostructures. Earlier, "bottom-up" self-assembly technique was used for DNA nanostructure is achieved by the information concealed within the nucleotide sequence which guides them to form symmetrical nanostructures with desired dimensions. The biocompatibility and stimuli-responsiveness of DNA nanostructures endow broad spectrum application in bioimaging, cellular targeting, drug delivery, to start with. From the biomedical perspective, DNA nanostructure serves as a drug delivery carrier. A smart nanostructure helps in controlled drug delivery in vivo, with minimal doses administration to attenuate off-target effects. An ideal nanocarrier should have key features as programmability, high drug loading, target selectivity, biocompatibility. This makes DNA unique for this purpose. Using base pair complementarity, nanostructures with dictated size and geometries can be constructed. These nanostructures can be functionalized with multiple entities like targeting ligands (aptamer or antibodies) and bioimaging agent for theranostics purposes. Tuning these features allow, programmed drug delivery with specificity and controlled drug release. These very robust properties of DNA nanodevices poise them to be explored as next generation devices for point of care diagnosis in various healthcare related research and products.

Acknowledgments

We sincerely thank all the members of DB groups for critically reading the manuscript and their valuable feedback. VM and AR thank IITGN-MHRD, GoI for PhD fellowship. DB thanks SERB, GoI for Ramanujan Fellowship, BRNS-BARC, Gujcost and GSBTM for research grants and IITGN for start-up funds.

Conflict of interest

Authors declare no conflict of interest.

References

[1] N.C. Seeman, Nucleic acid junctions and lattices, J. Theor. Biol. 99 (1982) 237–247, https://doi.org/10.1016/0022-5193(82)90002-9.
[2] N.C. Seeman, DNA in a material world, Nature 421 (2003) 427–431, https://doi.org/10.1038/nature01406.
[3] T.J. Fu, N.C. Seeman, DNA double-crossover molecules, Biochemistry 32 (1993) 3211–3220, https://doi.org/10.1021/bi00064a003.
[4] T.H. LaBean, H. Yan, J. Kopatsch, F. Liu, E. Winfree, J.H. Reif, N.C. Seeman, Construction, analysis, ligation, and self-assembly of DNA triple crossover complexes, J. Am. Chem. Soc. 122 (2000) 1848–1860, https://doi.org/10.1021/ja993393e.
[5] Z. Shen, H. Yan, T. Wang, N.C. Seeman, Paranemic crossover DNA: a generalized holliday structure with applications in nanotechnology, J. Am. Chem. Soc. 126 (2004) 1666–1674, https://doi.org/10.1021/ja038381e.
[6] B.F. Eichman, J.M. Vargason, B.H.M. Mooers, P.S. Ho, The Holliday junction in an inverted repeat DNA sequence: sequence effects on the structure of four-way junctions, Proc. Natl. Acad. Sci. 97 (2000) 3971–3976, https://doi.org/10.1073/pnas.97.8.3971.
[7] N.R. Kallenbach, R.-I. Ma, N.C. Seeman, An immobile nucleic acid junction constructed from oligonucleotides, Nature 305 (1983) 829–831, https://doi.org/10.1038/305829a0.
[8] Y.A. Berlin, A.L. Burin, M.A. Ratner, DNA as a molecular wire, Superlattices Microstruct. 28 (2000) 241–252, https://doi.org/10.1006/spmi.2000.0915.
[9] P. Rothemund, Folding DNA to create nanoscale shapes and patterns, Nature 440 (2006) 297–302.
[10] J. Zheng, J.J. Birktoft, Y. Chen, T. Wang, R. Sha, P.E. Constantinou, S.L. Ginell, C. Mao, N.C. Seeman, From molecular to macroscopic via the rational design of a self-assembled 3D DNA crystal, Nature 461 (2009)a 74–77, https://doi.org/10.1038/nature08274.
[11] C. Zhang, Y. He, M. Su, S.H. Ko, T. Ye, Y. Leng, X. Sun, A.E. Ribbe, W. Jiang, C. Mao, DNA self-assembly: from 2D to 3D, Faraday Discuss 143 (2009) 221–233, https://doi.org/10.1039/B905313C.
[12] D.Y. Tam, P.K. Lo, Multifunctional DNA nanomaterials for biomedical applications, J. Nanomater. (2015), https://doi.org/10.1155/2015/765492.
[13] A.R. Chandrasekaran, Programmable DNA scaffolds for spatially-ordered protein assembly, Nanoscale 8 (2016) 4436–4446, https://doi.org/10.1039/C5NR08685J.
[14] H. Kim, S.P. Surwade, A. Powell, C. O'Donnell, H. Liu, Stability of DNA origami nanostructure under diverse chemical environments, Chem. Mater. 26 (2014) 5265–5273, https://doi.org/10.1021/cm5019663.
[15] W. Xu, W. He, Z. Du, L. Zhu, K. Huang, Y. Lu, Y. Luo, n.d. Functional nucleic acid nanomaterials: development, properties, and applications. Angew. Chem. Int. Ed. https://doi.org/10.1002/anie.201909927.
[16] F. Li, J. Tang, J. Geng, D. Luo, D. Yang, Polymeric DNA hydrogel: design, synthesis and applications, Prog. Polym. Sci. 98 (2019) 101163, https://doi.org/10.1016/j.progpolymsci.2019.101163.
[17] M.D. Frank-Kamenetskiĭ, V.V. Anshelevich, A.V. Lukashin, Polyelectrolyte model of DNA, Sov. Phys. Uspekhi 30 (1987) 317, https://doi.org/10.1070/PU1987v030n04ABEH002833.
[18] T. Ohnishi, Properties of double-stranded DNA as a polyelectrolyte, Biophys. J. 3 (1963) 459–468, https://doi.org/10.1016/S0006-3495(63)86831-9.
[19] J.B. Lee, S. Peng, D. Yang, Y.H. Roh, H. Funabashi, N. Park, E.J. Rice, L. Chen, R. Long, M. Wu, D. Luo, A mechanical metamaterial made from a DNA hydrogel, Nat. Nanotechnol. 7 (2012) 816–820, https://doi.org/10.1038/nnano.2012.211.
[20] Y. Du, S. Dong, Nucleic acid biosensors: recent advances and perspectives, Anal. Chem. 89 (2017) 189–215, https://doi.org/10.1021/acs.analchem.6b04190.
[21] J. Li, L. Mo, C.-H. Lu, T. Fu, H.-H. Yang, W. Tan, Functional nucleic acid-based hydrogels for bioanalytical and biomedical applications, Chem. Soc. Rev. 45 (2016) 1410–1431, https://doi.org/10.1039/c5cs00586h.

[22] C. Wiraja, D.C. Yeo, D.C.S. Lio, M. Zheng, C. Xu, Functional imaging with nucleic-acid-based sensors: technology, application and future healthcare prospects, Chem. Bio. Chem. 20 (2019) 437–450, https://doi.org/10.1002/cbic.201800430.
[23] L. Zhou, A.R. Chandrasekaran, J.A. Punnoose, G. Bonenfant, S. Charles, O. Levchenko, P. Badu, C. Cavaliere, C.T. Pager, K. Halvorsen, Programmable low-cost DNA-based platform for viral RNA detection, Sci. Adv. 6 (2020) eabc6246, https://doi.org/10.1126/sciadv.abc6246.
[24] J.D. Watson, F.H.C. Crick, Molecular structure of nucleic acids: a structure for deoxyribose nucleic acid, Nature 171 (1953) 737–738, https://doi.org/10.1038/171737a0.
[25] R. Assenberg, A. Weston, D.L.N. Cardy, K.R. Fox, Sequence-dependent folding of DNA three-way junctions, Nucleic Acids Res. 30 (2002) 5142–5150.
[26] X. Wang, N.C. Seeman, Assembly and characterization of 8-arm and 12-arm DNA branched junctions, J. Am. Chem. Soc. 129 (2007) 8169–8176, https://doi.org/10.1021/ja0693441.
[27] Y.L. Wang, J.E. Mueller, B. Kemper, N.C. Seeman, Assembly and characterization of five-arm and six-arm DNA branched junctions, Biochemistry 30 (1991) 5667–5674, https://doi.org/10.1021/bi00237a005.
[28] A. Heuer-Jungemann, T. Liedl, From DNA tiles to functional DNA materials, Trends Chem. 1 (2019) 799–814, https://doi.org/10.1016/j.trechm.2019.07.006.
[29] B. Ding, R. Sha, N.C. Seeman, Pseudohexagonal 2D DNA crystals from double crossover cohesion, J. Am. Chem. Soc. 126 (2004) 10230–10231, https://doi.org/10.1021/ja047486u.
[30] E. Winfree, F. Liu, L.A. Wenzler, N.C. Seeman, Design and self-assembly of two-dimensional DNA crystals, Nature 394 (1998) 539–544, https://doi.org/10.1038/28998.
[31] R.P. Goodman, R.M. Berry, A.J. Turberfield, The single-step synthesis of a DNA tetrahedron, Chem. Commun. (2004) 1372–1373, https://doi.org/10.1039/B402293A.
[32] Y. He, T. Ye, M. Su, C. Zhang, A.E. Ribbe, W. Jiang, C. Mao, Hierarchical self-assembly of DNA into symmetric supramolecular polyhedra, Nature 452 (2008) 198–201, https://doi.org/10.1038/nature06597.
[33] J. Chen, N.C. Seeman, Synthesis from DNA of a molecule with the connectivity of a cube, Nature 350 (1991) 631–633, https://doi.org/10.1038/350631a0.
[34] D. Bhatia, S. Mehtab, R. Krishnan, S.S. Indi, A. Basu, Y. Krishnan, Icosahedral DNA nanocapsules by modular assembly, Angew. Chem. Int. Ed. 48 (2009) 4134–4137, https://doi.org/10.1002/anie.200806000.
[35] Y. Zhang, N.C. Seeman, Construction of a DNA-truncated octahedron, J. Am. Chem. Soc. 116 (1994) 1661–1669, https://doi.org/10.1021/ja00084a006.
[36] P.W.K. Rothemund, Folding DNA to create nanoscale shapes and patterns, Nature 440 (2006) 297–302, https://doi.org/10.1038/nature04586.
[37] M. Endo, Y. Yang, H. Sugiyama, DNA origami technology for biomaterials applications, Biomater. Sci. 1 (2013) 347–360, https://doi.org/10.1039/C2BM00154C.
[38] X. Li, C. Zhang, C. Hao, C. Tian, G. Wang, C. Mao, DNA polyhedra with T-linkage, ACS Nano 6 (2012) 5138–5142, https://doi.org/10.1021/nn300813w.
[39] W. Qiu, X. Zhai, Y. Qiu, Architecture of platonic and archimedean polyhedral links, Sci. China Ser. B Chem. 51 (2008) 13–18, https://doi.org/10.1007/s11426-008-0018-3.
[40] Y. Xing, E. Cheng, Y. Yang, P. Chen, T. Zhang, Y. Sun, Z. Yang, D. Liu, Self-assembled DNA hydrogels with designable thermal and enzymatic responsiveness, Adv. Mater. 23 (2011) 1117–1121, https://doi.org/10.1002/adma.201003343.
[41] S.H. Um, J.B. Lee, N. Park, S.Y. Kwon, C.C. Umbach, D. Luo, Enzyme-catalysed assembly of DNA hydrogel, Nat. Mater. 5 (2006) 797–801, https://doi.org/10.1038/nmat1741.
[42] J. Wang, J. Chao, H. Liu, S. Su, L. Wang, W. Huang, I. Willner, C. Fan, Clamped Hybridization chain reactions for the self-assembly of patterned DNA hydrogels, Angew. Chem. Int. Ed. 56 (2017) 2171–2175, https://doi.org/10.1002/anie.201610125.
[43] W. Guo, C.-H. Lu, R. Orbach, F. Wang, X.-J. Qi, A. Cecconello, D. Seliktar, I. Willner, pH-stimulated DNA hydrogels exhibiting shape-memory properties, Adv. Mater. 27 (2015) 73–78, https://doi.org/10.1002/adma.201403702.
[44] C.-H. Lu, X.-J. Qi, R. Orbach, H.-H. Yang, I. Mironi-Harpaz, D. Seliktar, I. Willner, Switchable catalytic acrylamide hydrogels cross-linked by hemin/G-quadruplexes, Nano Lett. 13 (2013) 1298–1302, https://doi.org/10.1021/nl400078g.

[45] B.-C. Yin, B.-C. Ye, H. Wang, Z. Zhu, W. Tan, Colorimetric logic gates based on aptamer-crosslinked hydrogels, Chem. Commun. 48 (2012) 1248–1250, https://doi.org/10.1039/C1CC15639J.

[46] J.B. Lee, Y.H. Roh, S.H. Um, H. Funabashi, W. Cheng, J.J. Cha, P. Kiatwuthinon, D.A. Muller, D. Luo, Multifunctional nanoarchitectures from DNA-based ABC monomers, Nat. Nanotechnol. 4 (2009) 430–436, https://doi.org/10.1038/nnano.2009.93.

[47] Z. Liu, Y. Li, C. Tian, C. Mao, A smart DNA tetrahedron that isothermally assembles or dissociates in response to the solution pH value changes, Biomacromolecules 14 (2013) 1711–1714, https://doi.org/10.1021/bm400426f.

[48] H. Xue, F. Ding, J. Zhang, Y. Guo, X. Gao, J. Feng, X. Zhu, C. Zhang, DNA tetrahedron-based nanogels for siRNA delivery and gene silencing, Chem. Commun. 55 (2019) 4222–4225, https://doi.org/10.1039/C9CC00175A.

[49] B. Wei, M. Dai, P. Yin, Complex shapes self-assembled from single-stranded DNA tiles, Nature 485 (2012) 623–626, https://doi.org/10.1038/nature11075.

[50] Z. Xing, A. Caciagli, T. Cao, I. Stoev, M. Zupkauskas, T. O'Neill, T. Wenzel, R. Lamboll, D. Liu, E. Eiser, Microrheology of DNA hydrogels, Proc. Natl. Acad. Sci. 115 (2018) 8137–8142, https://doi.org/10.1073/pnas.1722206115.

[51] Z. Xing, C. Ness, D. Frenkel, E. Eiser, Structural and linear elastic properties of DNA hydrogels by coarse-grained simulation, Macromolecules 52 (2019) 504–512, https://doi.org/10.1021/acs.macromol.8b01948.

[52] A. Banerjee, D. Bhatia, A. Saminathan, S. Chakraborty, S. Kar, Y. Krishnan, Controlled release of encapsulated cargo from a DNA icosahedron using a chemical trigger, Angew. Chem. 125 (2013) 6992–6995, https://doi.org/10.1002/ange.201302759.

[53] S.M. Douglas, I. Bachelet, G.M. Church, A logic-gated nanorobot for targeted transport of molecular payloads, Science 335 (2012) 831–834, https://doi.org/10.1126/science.1214081.

[54] S. Li, Q. Jiang, S. Liu, Y. Zhang, Y. Tian, C. Song, J. Wang, Y. Zou, G.J. Anderson, J.-Y. Han, Y. Chang, Y. Liu, C. Zhang, L. Chen, G. Zhou, G. Nie, H. Yan, B. Ding, Y. Zhao, A DNA nanorobot functions as a cancer therapeutic in response to a molecular trigger in vivo, Nat. Biotechnol. 36 (2018) 258–264, https://doi.org/10.1038/nbt.4071.

[55] M. Godonoga, T.-Y. Lin, A. Oshima, K. Sumitomo, M.S.L. Tang, Y.-W. Cheung, A.B. Kinghorn, R.M. Dirkzwager, C. Zhou, A. Kuzuya, J.A. Tanner, J.G. Heddle, A DNA aptamer recognising a malaria protein biomarker can function as part of a DNA origami assembly, Sci. Rep. 6 (2016), https://doi.org/10.1038/srep21266.

[56] R. Piper, J. Lebras, L. Wentworth, A. Hunt-Cooke, S. Houzé, P. Chiodini, M. Makler, Immunocapture diagnostic assays for malaria using plasmodium lactate dehydrogenase (pLDH), Am. J. Trop. Med. Hyg. 60 (1999) 109–118, https://doi.org/10.4269/ajtmh.1999.60.109.

[57] A.C. Amrite, H.F. Edelhauser, U.B. Kompella, Modeling of corneal and retinal pharmacokinetics after periocular drug administration, Invest. Ophthalmol. Vis. Sci. 49 (2008) 320–332, https://doi.org/10.1167/iovs.07-0593.

[58] D.L. Budenz, A clinician's guide to the assessment and management of nonadherence in glaucoma, Ophthalmology 116 (2009) S43–S47, https://doi.org/10.1016/j.ophtha.2009.06.022.

[59] M.M. Hermann, A.M. Bron, C.P. Creuzot-Garcher, M. Diestelhorst, Measurement of adherence to brimonidine therapy for glaucoma using electronic monitoring, J. Glaucoma 20 (2011) 502–508, https://doi.org/10.1097/IJG.0b013e3181f3eb4a.

[60] F.E. Alemdaroglu, N.C. Alemdaroglu, P. Langguth, A. Herrmann, DNA block copolymer micelles – a combinatorial tool for cancer nanotechnology, Adv. Mater. 20 (2008) 899–902, https://doi.org/10.1002/adma.200700866.

[61] J. Willem de Vries, S. Schnichels, J. Hurst, L. Strudel, A. Gruszka, M. Kwak, K.-U. Bartz-Schmidt, M.S. Spitzer, A. Herrmann, DNA nanoparticles for ophthalmic drug delivery, Biomaterials 157 (2018) 98–106, https://doi.org/10.1016/j.biomaterials.2017.11.046.

[62] S. Modi, S.M. G, D. Goswami, G.D. Gupta, S. Mayor, Y. Krishnan, A DNA nanomachine that maps spatial and temporal pH changes inside living cells, Nat. Nanotechnol. 4 (2009) 325–330, https://doi.org/10.1038/nnano.2009.83.

[63] S. Saha, V. Prakash, S. Halder, K. Chakraborty, Y. Krishnan, A pH-independent DNA nanodevice for quantifying chloride transport in organelles of living cells, Nat. Nanotechnol. 10 (2015) 645–651, https://doi.org/10.1038/nnano.2015.130.

[64] E. Goux, E. Dausse, V. Guieu, L. Azéma, G. Durand, M. Henry, L. Choisnard, J.-J. Toulmé, C. Ravelet, E. Peyrin, A colorimetric nanosensor based on a selective target-responsive aptamer kissing complex, Nanoscale 9 (2017) 4048–4052, https://doi.org/10.1039/C7NR00612H.
[65] N. Dave, M.Y. Chan, P.-J.J. Huang, B.D. Smith, J. Liu, Regenerable DNA-functionalized hydrogels for ultrasensitive, instrument-free mercury(II) detection and removal in water, J. Am. Chem. Soc. 132 (2010) 12668–12673, https://doi.org/10.1021/ja106098j.
[66] H. Yang, H. Liu, H. Kang, W. Tan, Engineering target-responsive hydrogels based on aptamer−target interactions, J. Am. Chem. Soc. 130 (2008) 6320–6321, https://doi.org/10.1021/ja801339w.
[67] Y. Ma, Y. Mao, Y. An, T. Tian, H. Zhang, J. Yan, Z. Zhu, C.J. Yang, Target-responsive DNA hydrogel for non-enzymatic and visual detection of glucose, Analyst 143 (2018) 1679–1684, https://doi.org/10.1039/C8AN00010G.
[68] L. Yan, Z. Zhu, Y. Zou, Y. Huang, D. Liu, S. Jia, D. Xu, M. Wu, Y. Zhou, S. Zhou, C.J. Yang, Target-responsive "sweet" hydrogel with glucometer readout for portable and quantitative detection of non-glucose targets, J. Am. Chem. Soc. 135 (2013) 3748–3751, https://doi.org/10.1021/ja3114714.
[69] Z. Zhu, Z. Guan, S. Jia, Z. Lei, S. Lin, H. Zhang, Y. Ma, Z.-Q. Tian, C.J. Yang, Au@Pt nanoparticle encapsulated target-responsive hydrogel with volumetric bar-chart chip readout for quantitative point-of-care testing, Angew. Chem. Int. Ed. 53 (2014) 12503–12507, https://doi.org/10.1002/anie.201405995.
[70] E. Benson, A. Mohammed, J. Gardell, S. Masich, E. Czeizler, P. Orponen, B. Högberg, DNA rendering of polyhedral meshes at the nanoscale, Nature 523 (2015) 441–444, https://doi.org/10.1038/nature14586.
[71] D.D. Bhatia, C. Wunder, L. Johannes, Self-assembled, programmable DNA nanodevices for biological and biomedical applications. Chem. Bio. Chem. https://doi.org/10.1002/cbic.202000372.

CHAPTER 19

Current trends in theranostic approaches using nanotechnology for oral squamous cell carcinoma

Satya Ranjan Misra[a], Swagatika Panda[b], Neeta Mohanty[b]
[a]Department of Oral Medicine & Radiology, Institute of Dental Sciences, Siksha 'O' Ansuandhan deemed to be University, Bhubaneswar, Odisha, India
[b]Department of Oral Pathology & Microbiology, Institute of Dental Sciences, Siksha 'O' Ansuandhan deemed to be University, Bhubaneswar, Odisha, India

19.1 Introduction

Oral cancer is one of the common malignancies in the world and generally refers to oral squamous cell carcinoma (OSCC) which is the most frequently occurring epithelial malignancy in the oral cavity [1]. It may involve the lips, tongue, buccal mucosae, retromolar trigone, hard and soft palate, floor of the mouth or the poster part of the oral cavity extending into the tonsils and the oropharynx [2,3]. Although the oral cavity is one of the most accessible areas of the body which can be examined for any pathological changes, but sadly oral cancer is often diagnosed late which affects its prognosis and also increases the oral cancer burden [4]. The diagnostic delay can be attributed to various factors including the initial asymptomatic oral lesion and the lack of access to oral physicians [5]. One of the greatest therapeutic challenges is advanced OSCC which may be extremely difficult to treat once it spreads to the posteriorly to the base of the skull and inferiorly to the neck, compromising the quality of life in patients. The five-year survival rate also substantially decreases due to diagnostic delays with a dismal 50% [6]. Not only that owing to the anatomic location of the oral cavity, many vital functions can be severely impaired like swallowing, talking or even taste sensation [7]. Due to significant facial disfigurement following surgical intervention it also affects the patient's appearance, social well-being and despite a cure the morbidity associated with the disease has profound impact on the patient functionality [8].

Earlier the disease occurred in middle aged and elderly male patients with tobacco and alcohol abuse habits, but recent emerging trends of affliction of younger individuals, association of human papilloma virus (HPV), implication of unprecedented factors and even change in the gender predilection from males to females has only complicated the scenario [9]. It is being seen that even changes in circadian rhythms can aid in the initiation and progression of OSCC. Besides, HPV which has been associated with the cancer of the cervix has now emerged as a key factor in the causation of OSCC, especially in patients

Advanced Nanomaterials for Point of Care Diagnosis and Therapy
DOI: https://doi.org/10.1016/B978-0-32-385725-3.00003-9

who are not correlated to the traditional tissue abuse habits of tobacco and alcohol [10]. But the HPV associated oropharyngeal cancers have a comparatively better prognosis than the non-HPV associated ones, as they do not progressively metastasize leading to poor survival outcomes and respond better to standard cancer care protocols [11].

The customary methodologies for oral malignancies include radical surgery, which is the therapy of choice, ionizing radiation which is the common combination therapy, or a blend of radiotherapy, chemotherapy, and surgery; careful resection prompts perpetual distortion, modified self-appreciation and incapacitating physiological outcomes, considerable aestheticissues, and morbidity, while chemo and radio treatments increase toxic reactions, all influencing prosperity and personal satisfaction [12]. These chemotherapeutic agents are proficient for the treatment of the primary tumor yet are utilized with palliative purpose in advanced cases with metastatic disease, with critical and antagonistic impact on the quality of life [6,8]. Regardless of the advances in surgical procedure, chemotherapy, and radiotherapy for HNSCC therapy, the overall wellbeing for this disease has not been fundamentally improved throughout the most recent 50 years. Along with the aforementioned conventional treatment modalities, photodynamic therapy (PDT) and photothermal therapy (PTT) are also being used to eliminate OSCC [13]. Since, OSCC may result from immunosuppression (described by a lower total lymphocyte counts) that meddles with the patient's characteristic immune reaction, forestalling tumor cell recognition, Immunotherapy is also emerging as a promising treatment adjunct [14]. Accordingly, the advancement of novel methodologies or changes of current procedures is central to improve singular wellbeing results and endurance, while early tumor recognition is essential predictor of the prognosis.

Recently, nano-technology-based drug delivery systems have shown great potential in both the diagnosis and the treatment of OSCC, and each small development in this field is of great importance [15]. The emergence of nanotechnology in accelerating the efficacy of chemotherapeutic as well as radiotherapy is under research [16]. Even the use of nanotechnology for the diagnosis of OSCC and occult malignancy or oral potentially malignant disorders is being investigated with great interest [17].

19.2 Early diagnosis of oral cancer using nanotechnology

Since there is considerable morbidity associated with diagnostic delays in OSCC and even the overall survival rate is low, it is imperative to diagnose early to aim for a cure. Biopsy has been considered the gold standard for diagnosis and remains so [18]. However, due to invasive nature of the procedure, expertise involved and requirement of laboratory support, sometimes it is not feasible to take a biopsy even though early diagnosis is vital for good prognosis [19]. Therefore, researches these days are focused on the diagnostic applications using nanotechnology for OSCC due to their excellent optical properties [17].

El Sayed et al. constructed an antiepidermal growth factor receptor (anti-EGFR) antibody based conjugated gold nanoparticle system which used the principle of the resonant scatter visibility of gold nanoparticles, and then incubated it with normal and malignant cells [20]. When the surface plasmon oscillation occurs in these Gold nanoparticles, they resonantly scatter visible and near-infrared light [21]. The intensity of this light scatter depends on the size and the state of aggregation of the gold nanoparticles. Since, gold nanoparticles can cause intense and bright light scatter as compared to chemical fluorophores. In OPMDs and OSCC, a trans-membrane glycoprotein, EGFR is overexpressed and higher levels of EGFR expression correlates with the aggressiveness of the lesion as well as poorer prognosis, which if imaged by specific optical imaging using gold nanoparticles can be predictor of the general tumor behavior. It was found that gold nanoparticles could act as biosensors in live cells due to their unique optical properties as well as bind too the malignant cell surface with a specificity and homogeneity about six times more than normal epithelial cells, thereby serving as a molecular biosensor in the diagnosis of OSCC [20,22].

Another study by Kim et al. also used gold nanoparticles as a delivery system comprising of anti-EGFR antibody-conjugated PEGylated goldnanoparticles [23]. This system was injected into the inner layers of the oral epithelium using micro needling technique in hamsters induced with OSCC. An increase of 150% in the optical coherence tomography (OCT) contrast was observed in cases of oral carcinoma and the gold nanoparticles were better penetrated and distributed by this technique [17]. Hence, OCT images of oral dysplastic lesions enhanced by gold nanoparticles could aid in early oral cancer diagnosis.

Even Malhotra et al. in their study usingsingle wall carbon nanotube forests having attached capture antibodies for the detecting OSCC cells described it as an ultrasensitive electrochemicalimmunosensor [24]. An electrochemical sandwich immunoassay protocol was followed to successfully measure even extremely low and increased levels of interleukin-6 (IL-6) and it was proved accurate for a range of OSCC cells. Since IL-6 serves as a biomarker in OSCC, this system could be even used for early detection of IL-6 in OPMDs as well OSCC.

Since the tumor-associated biomarkers are even detectable before the appearance of clinical manifestations, they provide valuable information in the early stages of OSCC and with increase research aiming at molecular diagnosis of OSCC, incorporation of nanoparticle-based systems needs to be tried [25].

A study conducted by Kah et al. employing antibody conjugated gold nanoparticles provided excellent optical contrast in differentiating between oral cancerous cells from the normal epithelial cells [26]. As emphasized, these systems could detect malignant changes in the molecular level before their apparent clinical appearance. However, it was observed that even the salivary surface-enhanced Raman scattering (SERS) spectra from these densely packed gold nanoparticles films were significantly different in normal and malignant cells and could useful in early OSCC diagnosis [27].

Not only in the early diagnosis of the primary tumor, nanotechnology can even be used for detection of metastatic lymph nodes. Though the regional lymph node metastasis is detected by imaging modalities like contrast enhanced computed tomography (CECT), magnetic resonance imaging (MRI), ultrasonography (USG) and lymphoscintigraphy, no modality is superior than the other. Many a times more than one imaging modality has to be used and these modalities could still be plagued by lower detection sensitivity and poor resolution of the nodes [28]. Since there was no accurate single imaging modality, Tseng et al conducted a study using a formulation of 25 nm lipid calciumphosphate (LCP) nanoparticle havingbetter small interfering RNA (siRNA) delivery efficiency [29]. In this in vivo study the system preferentially concentrated in the lymph nodes following an intravenous injection, with and an enlarged, tumor-involved sentinel lymphnode visualized using single-photon emission computed tomography (SPECT). The study thereby proved the efficacy of LCP to achieve excellent imaging of nodal metastasis. Hence, nanotechnology opens up newer avenues to detect early stages of OSCC, and to increase the sensitivity and specificity of imaging is the main areas of research, although in nascent stages, it can go a long way in reducing the oral cancer burden.

19.3 Enhanced treatment of oral cancer using nanotechnology

Though surgery, radiotherapy or a combination of both surgery and radiotherapy is used for treating oral cancer, use of chemotherapy as an adjuvant or as a combination therapy has increased in recent years [30]. The anticancer drugs that are used may be used singly or in combination like cisplatin, 5-flurouracil, cetuximab, paclitaxel, bleomycin, docetaxel, methotrexate, and doxorubicin [31]. Although oral administration of cancer chemotherapeutic agents is convenient facilitating extended exposure to the drug but it also decreases the solubility of the drug in aqueous solution thereby lowering the bioavailability. Hence intravenous administration is the route of choice, which not only increases the bioavailability but also the absorption into the target tissues [32]. Since the absorption is better, the normal tissues are also affected and increase the morbidity of the patients [30]. So, the use of nanotechnology is gaining popularity to decrease the unwanted adverse effects of chemotherapeutic agents while keeping it efficacious. Nanotechnology is being used to ensure the availability of the drugs directly to the target tumor cells sparing the normal tissues, thereby eliminating or reducing the adverse effects [33]. Nanoparticles having a size of 100 nm or less with a hydrophilic surface escape being engulfed by the macrophages, so that a long circulation time is maintained. The tumor vascular architecture is defective and is highly vascular but with poor lymphatic drainage, allowing the nanodrugs to enter the malignant tissues easily, concentrate and depart slowly. This leads to better permeation as well as better retention of the chemotherapeutic agents [34]. Tumor angiogenesis has been a topic of extensive research as it plays a significant role in development of malignancy and vasculariZation facilitates the invasion and metastasis of malignancies. Nanotechnology helps in facilitation of the cancer chemotherapeutic agents in those areas which are more vascular, typically targeting the vascularized areas and controlling the carcinogenic process [35]. Not only chemotherapy, even brachytheapy, PDT, PTT, and gene therapy for OSCC are now employing nanotechnology-based systems and nanotechnology is trending in oral cancer therapeutics, while increasing the efficacy with lesser adverse effects [36].

19.4 Conclusion

In spite of numerous advances in oral cancer therapeutics, clinical, technological as well as therapeutic challenges remain which need to be overcome by a multidisciplinary and collaborative research. Advances in understanding of cancer biology and tremendous development in the biomaterial science and imaging technologies, it may be aptly said that oral cancer therapeutics is at a critical threshold awaiting a significant breakthrough. With increasing capabilities of the existent multi-functional nanoplatforms, integrating it with biomaterials in diagnostic imaging and cancer biology, in the near future we may essentially see not only theranostic applications of nanotechnology but overall advancements in oral cancer therapeutics. Since the main advantage of using nanotechnology is in its multifunctionality over the conventional approaches, therapeutic drugs, ligands, imaging label, functional fractions could be all combined with nano-conjugates ensuring both molecular imaging as well therapeutics. A special mention of the intriguing optical properties of gold nanoparticles which are useful in all kinds of theranostic applications in oral cancer, selectively detect and treat the malignant cells without affecting the normal tissues.

References

[1] R.L. Siegel, K.D. Miller, A. Jemal, Cancer statistics, 2016, CA Cancer J. Clin. 66 (1) (2016) 7–30.

[2] A. Jemal, F. Bray, M.M. Center, J. Ferlay, E. Ward, D. Forman, Global cancer statistics, CA Cancer J. Clin. 61 (2) (2011) 69–90.

[3] L.A. Torre, F. Bray, R.L. Siegel, J. Ferlay, J. Lortet-Tieulent, A. Jemal, Global cancer statistics, 2012, CA Cancer J. Clin. 65 (2) (2015) 87–108.

[4] S. Warnakulasuriya, Global epidemiology of oral and oropharyngeal cancer, Oral Oncol. 45 (4–5) (2009) 309–316.

[5] G.L. Marella, F. Raschellà, M. Solinas, P. Mutolo, S. Potenza, F. Milano, et al., The diagnostic delay of oral carcinoma, Ig E Sanita Pubblica 74 (3) (2018) 249–263.

[6] A. Le Campion, C.M.B. Ribeiro, R.R. Luiz, F.F. da Silva Júnior, H.C.S. Barros, C.B. dos Santos K de, et al., Low survival rates of oral and oropharyngeal squamous cell carcinoma [Internet], Int. J. Dent. 2017 (2017) e5815493, https://www.hindawi.com/journals/ijd/2017/5815493/ Accessed December 16, 2020.

[7] S. Núñez-González, J.A. Delgado-Ron, C. Gault, D. Simancas-Racines, Trends and spatial patterns of oral cancer mortality in Ecuador, 2001–2016 [Internet], Int. J. Dent. 2018 (2018) e6086595, https://www.hindawi.com/journals/ijd/2018/6086595/ Accessed December 16, 2020.

[8] S.B. Thavarool, G. Muttath, S. Nayanar, K. Duraisamy, P. Bhat, K. Shringarpure, et al., Improved survival among oral cancer patients: findings from a retrospective study at a tertiary care cancer centre in rural Kerala, India, World J. Surg. Oncol. 17 (1) (2019) 15.

[9] Prevalence of human papillomavirus in oral squamous cell carcinoma: A rural teaching hospital-based cross-sectional study D. Rajesh, S.M. Mohiyuddin, A.V. Kutty, S. Balakrishna - Indian J. Cancer [Internet]. https://www.indianjcancer.com/article.asp?issn=0019-509X;year=2017;volume=54;issue=3;spage=498;epage=501;aulast=Rajesh. (Accessed December 16, 2020).
[10] N. Sathish, X. Wang, Y. Yuan, Human papillomavirus (HPV)-associated oral cancers and treatment strategies, J. Dent. Res. 93 (7 Suppl) (2014) 29S–36S.
[11] Q. Zhang, Y. Chen, S.-.Q. Hu, Y.-.M. Pu, K. Zhang, Y-X. Wang, A HPV16-related prognostic indicator for head and neck squamous cell carcinoma, Ann. Transl. Med. 8 (22) (2020) 1492.
[12] J.P. Shah, Z. Gil, Current concepts in management of oral cancer – surgery, Oral Oncol. 45 (0) (2009) 394–401.
[13] MM. Al Qaraghuli, Biotherapeutic antibodies for the treatment of head and neck cancer: current approaches and future considerations of photothermal therapies, Front. Oncol. 10 (2020) 559596.
[14] Immunotherapy in oral cancer [Internet]. https://www.ncbi.nlm.nih.gov/pmc/articles/PMC6555318/. (Accessed December 17, 2020).
[15] M. Poonia, K. Ramalingam, S. Goyal, SK. Sidhu, Nanotechnology in oral cancer: a comprehensive review, J Oral Maxillofac Pathol. 21 (3) (2017) 407–414.
[16] Z. Ding, K. Sigdel, L. Yang, Y. Liu, M. Xuan, X. Wang, et al., Nanotechnology-based drug delivery systems for enhanced diagnosis and therapy of oral cancer, J. Mater. Chem. B 8 (38) (2020) 8781–8793.
[17] X.-.J. Chen, X.-.Q. Zhang, Q. Liu, J. Zhang, G. Zhou, Nanotechnology: a promising method for oral cancer detection and diagnosis, J. Nanobiotechnol. 16 (1) (2018) 52.
[18] Evaluation of excisional biopsy for stage I and IIsquamous cell carcinoma of the oral cavity | SpringerLink [Internet]. https://link.springer.com/article/10.1007/BF02628053. (Accessed December 16, 2020).
[19] S. Mazumder, S. Datta, J.G. Ray, K. Chaudhuri, R. Chatterjee, Liquid biopsy: miRNA as a potential biomarker in oral cancer, Cancer Epidemiol. 58 (2019) 137–145.
[20] I.H. El-Sayed, X. Huang, M.A. El-Sayed, Surface plasmon resonance scattering and absorption of anti-EGFR antibody conjugated gold nanoparticles in cancer diagnostics: applications in oral cancer, Nano Lett. 5 (5) (2005) 829–834.
[21] M.K. Bhalgat, R.P. Haugland, J.S. Pollack, S. Swan, RP. Haugland, Green- and red-fluorescent nanospheres for the detection of cell surface receptors by flow cytometry, J. Immunol. Methods 219 (1–2) (1998) 57–68.
[22] C. Medina, M.J. Santos-Martinez, A. Radomski, O.I. Corrigan, MW. Radomski, Nanoparticles: pharmacological and toxicological significance, Br. J. Pharmacol. 150 (5) (2007) 552–558.
[23] G.J. Kim, W. Kim, K.T. Kim, J.K. Lee, DNA damage and mitochondria dysfunction in cell apoptosis induced by nonthermal air plasma, Appl. Phys. Lett. 96 (2) (2010) 021502.
[24] R. Malhotra, V. Patel, J.P. Vaqué, J.S. Gutkind, JF. Rusling, Ultrasensitive electrochemical immunosensor for oral cancer biomarker IL-6 using carbon nanotube forest electrodes and multilabel amplification, Anal. Chem. 82 (8) (2010) 3118–3123.
[25] S.K. Arya, P. Estrela, Recent advances in enhancement strategies for electrochemical ELISA-based immunoassays for cancer biomarker detection, Sensors 18 (7) (2018) 1–45.
[26] Early diagnosis of oral cancer based on the surface plasmon resonance of gold nanoparticles [Internet]. https://www.ncbi.nlm.nih.gov/pmc/articles/PMC2676812/. (Accessed December 16, 2020).
[27] K. Niciński, J. Krajczewski, A. Kudelski, E. Witkowska, J. Trzcińska-Danielewicz, A. Girstun, et al., Detection of circulating tumor cells in blood by shell-isolated nanoparticle – enhanced Raman spectroscopy (SHINERS) in microfluidic device, Sci. Rep. 9 (2019) 1–14, https://www.ncbi.nlm.nih.gov/pmc/articles/PMC6592934/ Accessed December 18, 2020.
[28] P. Pałasz, Ł. Adamski, M. Górska-Chrząstek, A. Starzyńska, M. Studniarek, Contemporary diagnostic imaging of oral squamous cell carcinoma – a review of literature, Pol. J. Radiol. 82 (2017) 193–202.
[29] J. Li, Y.-.C. Chen, Y.-.C. Tseng, L. Huang, Biodegradable calcium phosphate nanoparticle with lipid coating for systemic siRNA delivery, J. Control Release Off. J. Control Release Soc. 142 (3) (2010) 416–421.
[30] L. Hartner, Chemotherapy for oral cancer, Dent. Clin. North Am. 62 (1) (2018) 87–97.
[31] F. Bootz, [Neoadjuvant radiochemotherapy for squamous cell carcinoma of the oral cavity], HNO 56 (2) (2008) 183–184.

[32] H. Kurita, T. Koike, H. Miyazawa, S. Uehara, H. Kobayashi, K. Kurashina, Retrospective analysis on prognostic impact of adjuvant chemotherapy in the patients with advanced and resectable oral squamous cell carcinoma, Gan. To. Kagaku. Ryoho. 33 (7) (2006) 915–921.
[33] T. Gupta, J. Singh, S. Kaur, S. Sandhu, G. Singh, I.P. Kaur, Enhancing bioavailability and stability of curcumin using solid lipid nanoparticles (CLEN): a covenant for its effectiveness, Front Bioeng. Biotechnol. 8 (2020) 879.
[34] R. Zein, W. Sharrouf, K. Selting, Physical properties of nanoparticles that result in improved cancer targeting, J. Oncol. 2020 (2020) 5194780.
[35] E. Pérez-Herrero, A. Fernández-Medarde, Advanced targeted therapies in cancer: drug nanocarriers, the future of chemotherapy, Eur. J. Pharm Biopharm. Off. J. Arbeitsgemeinschaft Pharm. Verfahrenstechnik EV 93 (2015) 52–79.
[36] M. Estanqueiro, M.H. Amaral, J. Conceição, JM. Sousa Lobo, Nanotechnological carriers for cancer chemotherapy: the state of the art, Colloids Surf. B Biointerfaces 126 (2015) 631–648.

CHAPTER 20

Advanced nanomaterials for point-of-care diagnosis and therapy

Sreejita Ghosh[a], Moupriya Nag[b], Dibyajit Lahiri[b], Dipro Mukherjee[b], Sayantani Garai[b], Rina Rani Ray[a]
[a]Microbiology Research Laboratory, Department of Biotechnology, Maulana Abul Kalam Azad University of Technology, Simhat, Haringhata, Nadia, West Bengal
[b]Department of Biotechnology, University of Engineering & Management, Kolkata

20.1 Introduction

Many advanced medicines and diagnostic systems have been reported to develop so far. Still, there are incidents of death due to lack of detection of diseases at proper time. Diseases like respiratory problems, ischemic heart diseases and diabetes are some of the major causes of death all over the world. However, these diseases can be prevented or proper therapies can be administered if detected on time.

Traditional methods like immunological assays or polymerase chain reactions (PCR) have some disadvantages like they give slow results, may not be specific and accurate, require skilled personnel and are expensive. Mainly, in the developing countries there are lack of skilled people and resources and so the use of such traditional diagnostic techniques is limited. Thus, the ideal diagnostic tools to use for the early detection of such diseases are the different nanobiomaterial based devices, which give rapid and accurate results, are more sensitive, specific, user friendly and cost effective. These rapid tools for disease identification also have the potential to emerge as point-of-care (PoC) diagnostic techniques. PoC testing systems refer to the testing and detection near or at the patient care site [1]. Nanomaterial based techniques are therefore considered as portable, affordable and robust in most of the developing countries. Nanomaterials are used because they have some striking optical, catalytic, electrical and mechanical properties unlike other biomaterials [2].

There also certain advanced nanomaterials developed for therapeutic purposes of various diseases such as cardiovascular diseases, neurodegenerative diseases and carcinomas. The vast application of nanomaterials in the field of biomedical engineering have assisted in various processes of gene therapy and targeted drug delivery or imaging tumor sites. Nanomaterial based therapies have many advantages in comparison to other therapies. They possesses increased bioavailability and permeability, enhanced penetration and retention capacities, stability, selective accumulation within the target site, efficient targeting, dose responses and lesser side effects. The pharmacokinetic profiles of nanomaterial therapeutics can also be modified by controlling the nanomaterial

Advanced Nanomaterials for Point of Care Diagnosis and Therapy
DOI: https://doi.org/10.1016/B978-0-323-85725-3.00010-6

shapes and sizes, surface modifications or development of multimodal nanomaterial therapies which are known as "combination therapy" or "theranostic," which is defined as the combination of therapeutic and diagnostic substances [3].

Over the recent years, nanomaterials have been widely used for potential PoC diagnosis and therapeutic purposes. Apart from the aforesaid advantages of nanomaterials, bioengineered nanomaterials have an inherent antioxidant property of themselves.

20.2 Point-of-care tests and diagnosis

Tests that can be carried out for rapid analysis and accurate detection of diseases near the patient are referred to as bedside or point-of-care tests (PoCT). It is essential that any PoCT must be user friendly and very simple so that even people who do not have laboratory or professional skills can use it and figure out its response. Also, the PoCT devices must be readily available and cost effective so that everyone everywhere can afford them. PoCT devices should be able to overcome changes in environmental conditions, must be able to respond against a particular analyte or parameter and not affected by others and differentiate between closely similar parameters. So far there have been many PoCT devices like glucose meters and pregnancy test kits that are being used on a large scale. However, with the development of nanobiotechnology, some nanomaterials are also being integrated within the different parts of PoCT test devices to more accurately monitor and detect various diseases [4].

20.3 Nanomaterial and its classifications

Nanomaterials are defined as nanostructured objects with a size of 100 nm or less than 100 nm. On the basis of nanostructure of these materials, they are considered as nanoparticles, nanogels, nanorods, nanowires, nanotubes, nanoribbons, and nanoscaffolds [5]. These nanomaterials are used for various diagnostic and therapeutic purposes because of their existence in different forms such as single or multiple aggregation or combined structures with various shapes. The different types of nanomaterials are classified as below (Tables 20.1 and 20.2).

Table 20.1 Classifications of nanomaterials [5].

Basis of classification	Classified types
Dimensions	Zero dimension (filaments, clusters, cluster assemblies) One dimension (nanotube, nanowire, nanofiber) Two dimension (nanofilms, nanoplates) Three dimension (quantum dots [QD], nanoparticles)
Pore size	Microporous (clay materials that occur naturally) Mesoporous (carbon mesoporous, SBA-15, MCM-48, MCM-41) Macroporous (porous glasses, porous gels, carbon microtubes)

Basis of classification	Classified types
Composition	Carbon-based nanotubes (CNT) (single-celled nanotube and multicelled nanotube) Fullerenes Metal based Hybrid (3D metal matrix and 2D lamellar) Nanowires (1D core shell and zero dimensional core shells)

Table 20.2 Descriptions of nanomaterials and their uses.

Type of nanomaterial	Description	Uses	References
Clusters	Small atomic or molecular agglomerates with molecules less than 2 nm diameter	In vitro diagnosis, bioimaginig, cancer treatment and surgeries and drug delivery system	[65]
Nanotubes	Also called carbon nanotubes in which molecules of carbon are present in hexagonal structure by formation of covalent bond with other carbon molecules	Diagnosis and treatment of malignant melanoma	[66]
Nanowires	Nanosized paths composed of CNT acting as channels for passage of electric current of very less amplitude	Measurement of RNA expression, levels of biomarkers and antigens measurement and detect gene mutations	[67]
Nanofibers	Network of porous mesh wires with striking interconnection between the pores	Regenerative medicine preparation and tissue engineering, dressing of wounds, therapeutic agent delivery, biosensor development	[68]
Nanogels (NG)	Hydrogels composed of covalently or non-covalently crosslinked polymer chains	Selective binding affinity various proteins and is still an emerging technology	[69]
Nanoshells	Gold coated minute structures that can absorb particular wavelengths of light	Used as molecular conjugates in drug delivery systems	Sharma et al., 2020
Quantum dots	Fluorescent and spherical nanocrystals made of semiconductor materials having a size of 2–8 nm	Cancer imaging and used as an alternative for fluorophores for biosensor development	[70]
Fullerenes	Carbon allotropes with 60 carbon atoms placed in icosahedron shape in a truncated fashion	Not much in use and still in the developmental stage	Bajwa et al., 2015
Metal-based nanomaterials	These are actually nanoparticles like silver and gold nanoparticles having higher compatibility and are nontoxic	Sunscreens and cosmetics manufacturing or production of antibacterial agents for therapeutic or diagnostic purposes	[71]

20.4 Applications of nanomaterials in point-of-care diagnosis

With respect to transducers, synthetic nanomaterials with a diameter of 1-100 nm are widely used due to their shape, size, biocompatibility, magnetism, fluorescence and thermal and electrical conductivity. There are various nanomaterials that are being used in PoC diagnosis and those nanomaterials are described in the following parts.

20.4.1 Spherical nanomaterials

Because of simple manufacture and preparation, spherical nanomaterials are predominantly used materials. On being attached to a biomarker, the spherical nanomaterials react in presence of particular pathological conditions to form an active nanosphere serving as a biological label to detect the presence of a pathogen or a given analyte. The different mechanisms [6] by which these nanomaterials can detect the presence of a pathogen or analyte are by measurement of the shift of wavelength peak due to accumulation of nanomaterials, surface plasmon response (radiation enhancement and optical modifications in nanomaterials because of disturbed dielectric constant caused by a molecule adsorption), increased secondary reactions of enzymes through quenching effect of nanomaterials (decreased intensity of fluorescent signal), electric impedance spectroscopy (modification of electrical resistivity in a medium) and electrochemical or electrical changes (if the nanomaterial has conductivity or catalyze an electrochemically identified reaction).

Out of all these spherical nanomaterials, gold nanoparticles (AuNP) are mostly used due to its increased bio-affinity, dark color (with small changes in surface diameter, wavelength varies widely) and catalytic features [7]. So, they are used in both electrochemical as well as optical PoC devices.

20.4.1.1 Optical point-of-care devices

Optical PoC devices are mainly lateral flow biosensors (LFB) [8] which are fabricated on paper substance such as pregnancy test kit. LFBs are based on the principle of thin layer immuno-affinity chromatography in which there is a "sandwich" formation by attachment of a primary biological marker (usually an antibody) with a paper surface (cellulose or nitrocellulose), an analyte (cells, proteins, bacteria, heavy metals or other molecules) and a secondary biological marker attached to a tag nanomaterial. Formation of this sandwiched portion initiates color formation in the "test zone" (site where the first biomarker attached) for a positive sample and if there is no sandwiched portion formed (no analyte in sample) then the "test zone" remains colorless for a negative sample. Latex beads are mostly used in LFB for pregnancy tests. Presence of AuNP in LFBs increases the sensitivity of the test because of the dark color of the nanomaterial due to its surface plasmon response [9]. Also, due to the small size of the nanomaterial, the ratio of label/analyte gets boosted. In this way, the nanomaterials

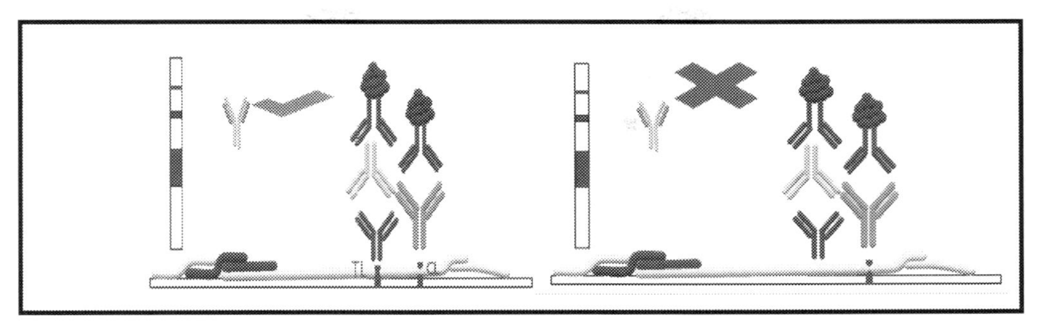

Fig. 20.1 ***Working mechanism of a LF strip and its components.*** (A) Positive and (B) Negative performances.

help in semiquantitative assays and this quantification can even be improved with more modifications (Fig. 20.1).

Apart from AuNP, other nanomaterials like silver nanoparticles (AgNP) are also used in LFBs. The wavelength variations by diameter modifications are enhanced in case of AgNP causing color tonalities. So, with AgNPs, multiplex tests can be carried out at the same time (different analyte detection in the same sample) obtaining various colors in different test zones [10]. Fluorescent nanomaterials such as quantum dots (QD) and upconverting phosphor reporters (made with expensive elements like Yttrium or Europium) can be used in LFB. Fluorescent signals cannot be read with naked eye and so an equipment is essential to read the resultant fluorescent signal.

Signal reading mechanisms for nanomaterials are present for paper substances, but these mechanisms need extra steps making the PoC device not much user friendly and increasing the chances of human error. Fu et al. [11] proposed a 2D paper network with LFB and other paper substrates for storage of enhancement reagents. In this PoC device, the user has to place the sample on the conjugated substrate, where the analyte is picked by the nanomaterials and then close the system followed by addition of water. At first, this device behaves as a traditional LFB; but after these enhancement reagents reach the zone of detection, the color of the nanomaterial becomes dark purple with increase in size. Against a light background, this dark color provides more contrast. Rodriguez et al., 2016 [12] also made use of an LFB for isolation, amplification and detection of nucleic acids in just one equipment free and more rapid than conventional devices. The system is composed of an LFB strip with extra detachable parts, a washing pad, a tab in order to stop the amplification reagent from being evaporated and few hydrophobic shields to stop DNA and other solvents from flowing to the LFB strip prematurely. For example, in a prototype designed by [13], a lab-on-a-syringe which is useful for urine collection and pumping the urine to paper pads arranged in cartridges connected serially. In the first cartridge, there is AuNP to capture the analyte which is a biomarker for cancer and a detection pad is present in the second cartridge. There is a nitrocellulose paper on detection pad to detect antibodies in a ring of wax and focussing and making the flow through antibodies. The interior portion of this ring turns more red with increased amount of analyte. Unlike the paper strip, the lab-on-a-syringe requires the user to control the flow making it less user friendly; but it also allows the incubation time modification of AuNP in presence of analyte improving sensitivity. A filter can even be combined with the syringe for matrix effect reduction.

Nanopaper called bacterial cellulose is a colorless substrate on which fluorescent nanomaterials can evoke a surface plasmonic response as an alternate way of ELISA [14]. These nanopapers can be fabricated in various forms like cuvette or spots on a piece of paper other than ELISA plates. Measurements such as quenching and fluorescence effects and surface plasmnonic resonance can be performed using these nanomaterials. Unlike nanopapers, hydrogels are also employed in various PoC diagnostic applications of which phenylboronic acid functionalized hydrogel. This system is reusable and finds application in filtering samples of urine and glucose retention.

Quantum dots (QD) are used in intense signalling for construction of multiplexed systems to detect blood infections by utilizing a barcode system [15]. Colorful QDs are encapsulated inside microbeads coated with conjugated antibodies specific against a different target and all these particles are mixed in a microfluidic system within the sample. Incubation within microbeads is electrokinetically controlled. Data was obtained by means of software based detection platform. This platform can individually detect wavelength of each QD thereby converting it to a barcode reading and measuring the quantity of each target. It was found that in presence of polyphenolic compounds there occurs a fluorescence quenching effect on graphene QDs [16]. The system comprised a 3D printed dark chamber provided with a UV LED drawing power from a mobile phone in order to excite the QDs. By means of wax-printed tracing spots on a strip of paper, these QDs were being physiosorbed within the spots to develop an ELISA plate-like system. On dropping polyphenolic compounds on these spots, the fluorescence gets quenched and therefore gets captured by the camera of a mobile phone. These captured images are then rapidly analyzed in the mobile phone through an app or computer software (Fig. 20.2).

20.4.1.2 Electrochemical point-of-care devices

Electrochemical diagnostic methods identify electrical signals evoked by a chemical change. Nanomaterials are used in these devices as commercialized inks to serve as electrodes. Nanomaterials such as CNT and nanoparticles like AuNP or AgNP are widely used in this application. For example, da [17] fabricated AgNPs to form a miniature and portable Ag/AgCl reference electrode with a stable and known potential. The electrode was printed on two different elastic substrates, polyethylene terepthallate (PET) or paper by means of a higher resolution piezoelectric inkjet material printer. Post printing, this

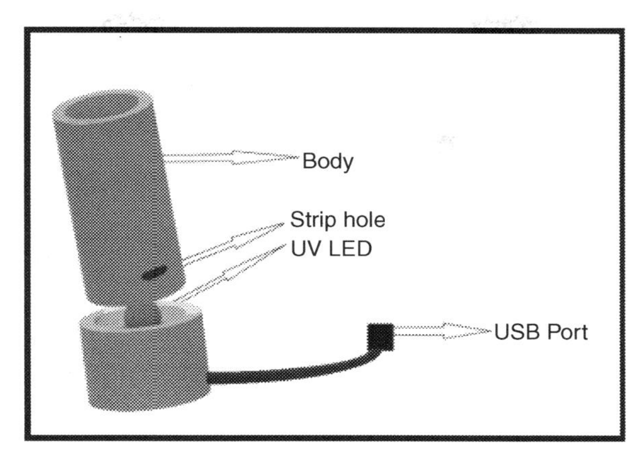

Fig. 20.2 ***Schematic representation of a 3D-printed device having QDs excited*** *via* ***UV-LED light and USB port directly attached to mobile phones.***

AgNP ink was restored at 120°C and then subjected to bleaching with sodium hypochlorite producing an Ag/AgCl mixture serving as pseudo reference electrode.

AuNP catalyzes reduction and such AuNPs are reported as transducers in these electrochemical assays because of their increased affinity for biomolecules [18]. There exists another reaction in the electrochemical assays catalyzed by AuNPs is the evolution of hydrogen reaction, that is, hydrogen gas formation by H^+ ion reduction. This reaction can be measured by magnetic beads (MB) or AuNPs and by DNA amplification systems on a screen consisting of printed carbon electrodes discriminating DNA of different hosts within the cells. AuNPs or MBs get attached to the magnified DNA through a magnet and this conjugate is placed over the working electrode for efficient signal measurement. Other nanomaterials like iridium oxide nanoparticles (IrO_2NP) are also used because of their catalytic properties [19] in water oxidation reaction, where oxygen is being produced from water and this finds application in impedimetric sensors. Impedance is defined as a method for the measurement of frequency modulations in dielectric medium near the nanomaterials. So, by the binding of IrO_2NP with biomarkers, the variations in the conductivity of the medium can be detected depending on whether the biomarker could capture the analyte.

Another paper-based PoC device was proposed by [20], in which there were 8 electrodes which are subjected to pretreatment with antibodies specific for a target analyte (Fig. 20.3). After sample addition to the electrodes, SiO_2 nanoparticles were also added to the electrodes. Here, the SiO_2 nanoparticles have been coated with antibodies that are specific for the given analyte and carries out a sandwich assay with horseradish peroxide, which initiates the electrochemical reaction. After this, electrodes are washed and again loaded with a reactive solution, which reinitiates the electrochemical reaction. One

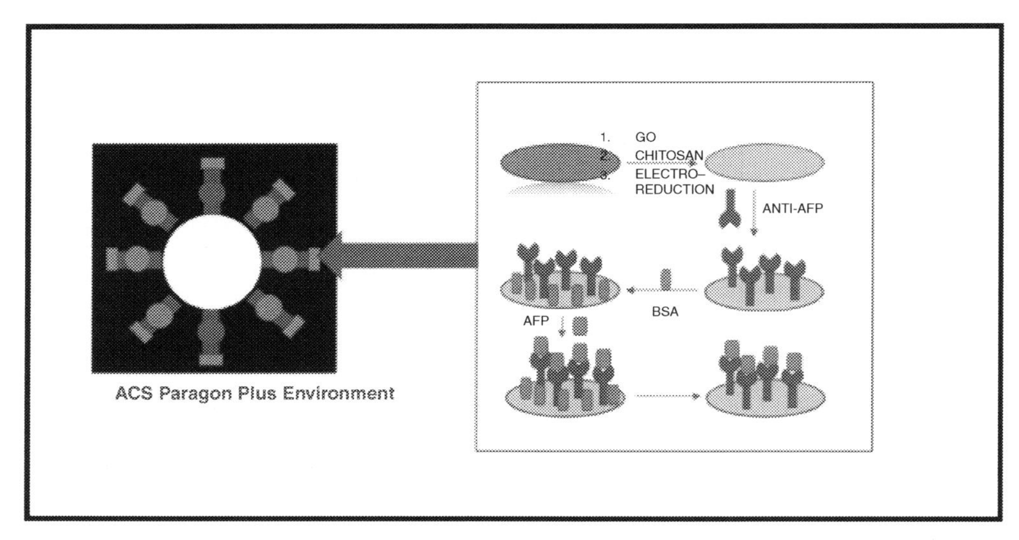

Fig. 20.3 ***Paper-based microfluidic electrochemical immunodevice integrated with nanobioprobes for detection of cancer biomarkers.***

more PoC device was fabricated by [21], in which an origami style proof-of-concept was utilized to detect AgNPs by oxidation with a gold working electrode. This device comprises 4 folded layers of paper, in which the first layer contains the inlet, outlet and the electrodes, the second and third layers contain a paper circuit demarcated by wax or a blue dye working as an indicator to stop the flow and the last layer contains a "sink" in order to redirect all the flow there.

20.4.1.3 Magnetic point-of-care devices

Magnetic nanoparticles (MNP) and MBs are used as supporters in PoC devices generally in the washing or amplification/preconcentration steps or for the analyte precipitation or some other nanomaterials on the electrode surface. MNPs are used as transducers in nuclear magnetic resonance (NMR) imaging. Since there is no interference due to nonmagnetic substances, this method is very sensitive because of low background signal. NMR also help to detect every tag in the detection zone in contrast to optical (only tags present on surfaces of substrates are detectable) or electrochemical (tags connected to the electrodes can only produce a signal) sensors. NMR detection is much rapid compared to other techniques. However, NMR is very expensive and may not be affordable by all users.

Liong et al. [22] developed a microfluidic PoC device for amplification of nucleic acids of a tuberculosis causing bacterium. This device can carry out different functions like incubation of DNA-MNPs, amplification of DNA, washing and detection by NMR. The device consists of 3 inlets for DNA, MNPs and washing buffer, mixing channels for incubation and an NMR coil to count the MNPs which is directly proportional to the initial DNA concentration. Although this device helps in rapid detection, the incubation and amplification processes increase the duration of the assay for more than 2 hours; but overall, this technique is a much rapid diagnosis of tuberculosis in comparison to the microscopic tests and cell culture assays. Chung et al. [23] devised another NMR microfluidic PoC device to detect the urine biomarker. Although this device has a complex matrix, the noise signal is less due to non-interference from non-magnetic substances with the readout.

20.4.2 1D nanomaterials

In single dimension nanomaterials, growth is directed in just one dimension in a linear fashion. Thickness of these nanomaterials is usually 1nm like single-walled carbon nanotubes (SWCNT) whereas their length can be much larger, about 100 μm or more. Absorption wavelengths by these nanomaterials are dependent on their shape (diameter and length). They are useful for the construction of large nanostructures because their increased length and reduced thickness impart them high mechanical strength. However, device designing using these nanomaterials is not easy as they require careful synthetic procedures for homogeneity determination.

Simplest 1D nanomaterial includes nanowires that are created from zero dimensional nanomaterials directly or by the use of various elements such as silicon, platinum, gold, silver and nickel. Introduction of nanowires (with copper, nickel or platinum) on paper substrates leads to the construction of electrodes for monitoring of electrocardiogram through tissue-electrode impedance working in dry conditions [24]. This device does not need any application of gels between the skin and electrodes as gels rapidly dry and is degraded with movement. Because of large surface area of the nanowires, good quality response (ranging between 100–1 kΩ) in the electrodes is obtained in the impedimetric measurements. These paper electrodes can even work well in an acidic battery.

Recently, SWCNT and CNT found wide applications in electrochemical PoC devices due to their high electrical and thermal conductivity. CNTs can also easily function with other nanomaterials (metallic oxides) enhancing their electrical properties apart from functioning with biological samples [25]. However, CNTs can be also used to fabricate screen printed electrodes (AgNP to fabricate reference electrodes and CNT to fabricate working and counter electrodes). In contrast to metal nanowires, CNT helps in fabricating large nanostructures with unique features like increased elasticity and flexibility and reduced weight because CNTs are hollow from inside. CNTs can even function properly in comparison to nanowires against variations in temperature as the coefficient of thermal expansion is very low in CNTs than that of the metals. On the other hand, it should be kept in mind that most of the reactions taking place on the CNT surfaces are irreversible thereby reducing the durability of the CNTs. Nemiroski et al. [26] integrated CNT made electrodes into the mobile phone system through audio jack of the mobile phone (since audio jack can send and receive information at the same time).

NFC sensing technology [27] helps in detecting antenna without an electrical power source. This technology can be found recently in door lock systems of hotels, smart phones, mail stamps, toys and also in metro stations. So, NFC tags are becoming cheaper everyday and can be easily fabricated and modified. SWCNT inside the circuit was replaced with NFC tag, which via a chemiresistive reaction can modify the circuit conductivity depending on the existence of various atmospheric compounds therefore making the device play as an on/off logic gate in the NFC tag. This technology is used in PoC application for detection of breath analytes or in body fluids like sweat and blood.

Interestingly, NFC can be used in chewing gums [28] by washing these chewing gums with a mixture of water and ethanol and then suspended in a solution containing CNTs. This mixture was then stretched further and pleated several times unidirectionally so that the CNTs can orient among themselves. This technology finds application in detection of humidity in a medium (mouth dryness caused by illness or specific medications or due to the damage to salivary glands or hormonal modifications) by sensing the electrical resistivity inside the gum. The CNTs can also be used as motion sensors in order to detect body movements and breathing in the users. These PoC diagnostic devices, through various chemical reactions, can detect specific targets like pulsations and dry mouth and biting problems. One disadvantage of such devices is that CNTs are considered to be a potential cytotoxic nanomaterial limiting its introduction within the mouth.

20.4.3 2D nanomaterials (single layer)

Two dimensional nanomaterials have the ability of expanding themselves in two directions and consist of a single layer of atoms or an ultrathin layer of few atoms. Although there are many inorganic 2D nanomaterials for use, graphene is the most commonly used 2D nanomaterial over the recent years. In spite of being made of a single layer of atoms, graphene is usually found in aggregations of multi-layered graphene. Optical, mechanical and electrical properties of graphene depend on the number of layers. There are also other properties like oxidation state, degree of purity and number of defects in the structure that have an effect on graphene and these parameters can be altered through some synthetic routes.

Graphenes, in combination with QDs behave as quenchers by silencing of the fluorescence signal emitted by QDs in presence of graphene nanomaterials. In a newly designed LFB system, 2 QD lines were dispensed on a paper substrate as test and control and here the first line could capture bacteria by presence of antibodies present on QDs [29]. After sample addition on LFB, graphene oxide (GO) solution (oxidized graphene consisting of hydroxyl and carboxyl groups and epoxy bridges) was added. GO would switch off the fluorescent signal emitted by QDs in test and control lines if there were no bacteria found in the given sample; but, when bacteria was captured in the test line, there would be a gap produced between QDs and GO leading to fluorescence emission.

In contrast to the traditional LFB, this device do not form false positives in the assay as a negative sample will always switch off both the lines and if the test line is affected by some exterior reasons, the control line will also get affected giving an invalid result but no false positives. The time taken for the assay is 1hour and is more than that taken by traditional LFB which commonly takes 5–10 min because of the extra GO addition and drying counting with the waiting time for the bacteria to flow through the strip.

In another PoC device, a nanopaper coated with GO was introduced with a complex antibody suspension attached with QD [30]. Here also, the quenching effect silences the fluorescence emitted by QDs. On addition of an analyte like protein or bacteria against which there are selective antibodies, a gap is formed between GO and QD suppressing the quenching effect and emitting fluorescence. This device has more advantages as it does not require any washing steps and is more rapid and portable. Hence, this can be considered as an efficient alternative of ELISA tests.

Because of its electrical properties, graphene can also be utilized to fabricate working electrodes [31]. Antibodies are conjugated with graphene surface through polymers by utilizing the chemistry of hydroxyl or carboxylic groups present in GO. Therefore, as graphene is planar, all the antibodies can orient perpendicularly to the layer of graphene in order to increase the analyte capturing probability [32]. Moreover, graphene electrodes are sensitive enough to allow small changes (impedimetric or electrochemical modifications) and label-free sensing on surfaces of the electrodes to be detected easily. The main disadvantage of graphene is that it requires very careful synthesis procedures to have reproducibility, mainly the structural defects and the number of layers.

An innovative PoC device was designed which was celled a solvent free method, in which GO was printed on various substrates utilizing patterns printed with wax through printing on nitrocellulose membrane by vacuum filtration [33]. GO stays within the areas that are not printed with wax and get transferred to the target substrate through pressure. In order to demonstrate the prospects of this procedure, there was printed a touch sensor, which works on the basis of the changes in resistance initiated by a finger on coming in contact with the circuit printed with GO. This touch sensor GO was connected to a power source and an LED, in which the sensor works as an ordinary switch turning the LED on/off. This PoC diagnostic device can replace touch screens that utilize contentious elements like rare-earth metals or indium and also in wearable flexible PoC devices like tattoo sensors on skin and contact sensing devices like motion or pressure sensors.

Apart from the features mentioned, graphene is also known for its electrochromic property that can be modulated to adjust the color reversibly depending on the amount of current applied. Various electrochromic flexible PoC diagnostic devices have been reported to develop. In one such PoC diagnostic device, there were two plastic substrates attached, whose one side was coated with graphene electrodes containing a liquid electrolyte within it [34]. On application of voltage through the electrodes, graphene becomes translucent as the voltage is sufficiently high to penetrate through the layers of graphene forming structural defects thereby fading its distinct black color. This device was designed using different graphene electrodes. The device is considered flexible since graphene can withstand a curvature of radius 1cm without getting damaged. Due to application of varying voltages on few electrodes, a chess pattern is created as transmittance increases on the graphene layers on increasing the current. The main advantage of using device is that it can efficiently convert electrochemical signals to optical responses *in situ* making it a more user friendly and wearable PoC diagnostic device.

20.4.4 3D nanomaterials

There exists a wide variety of 3D nanomaterials that vary in size and shape leading to various possibilities in the terminal sensing function. Nanomaterials that are not spherical possess similar feature like zero dimensional nanomaterials but shape modification may result in modification of absorbance, for example; color of gold nanosphere solution is different from that of nanotriangles or nanocubes even though their size and chemical composition are identical. Based on this principle, a plasmonic PoC diagnostic device was designed by Fu et al. [35]. The device can be moved from place to place and can be integrated in a mobile phone by utilizing the light sensor, which is found in most mobile phones nowadays in order to control the brightness of the screen or to enhance the photographic quality. This device contains a plate, where the sample is to be added and an LED for illuminating the sample on the plate to reach the light sensor. In order to examine the potential of this device, it was used to measure the plasmonic changes taking place on triangular silver nano prisms when cancer biomarker is absent. This method is called an indirect method of detection via shape transformation of nanomaterials by hydrogen peroxide. Three-dimensional silver nanomaterial like nano porous silver can also be used as a signal enhancer for labelling the metal ion carrier [36]. This PoC diagnostic device contains a multiplexed electrochemical origami paper system that utilizes nanoporous silver packed with various metal ions acting as labels for sensing any tumor growth. Screen printed carbon electrodes are used as the working and counter electrodes and a silver electrode is employed as a reference electrode. This PoC diagnostic paper device can be folded and can be coupled in an electric circuit (Fig. 20.4).

There are some nonspherical nanomaterials having electrochromic features like WO_3 nanoparticles that are used on paper substrates [37]. When electrochemically active bacteria are present, these nanomaterials change their color from yellow to blue. Hence, this PoC-based diagnostic device can be used as an easy-to-use sensor and it functions just like the ELISA but with no delayed steps or by not utilizing any intricate reagents as biological reagents have very short dates of expiration but these nanomaterials can survive for a longer time period even if they are kept at room temperature. Electro-chromic materials such as photonic crystals (PC) are made of many nanoparticles arranged in a crystalline manner. Such nanomaterials have great adaptive capacities in various sensors

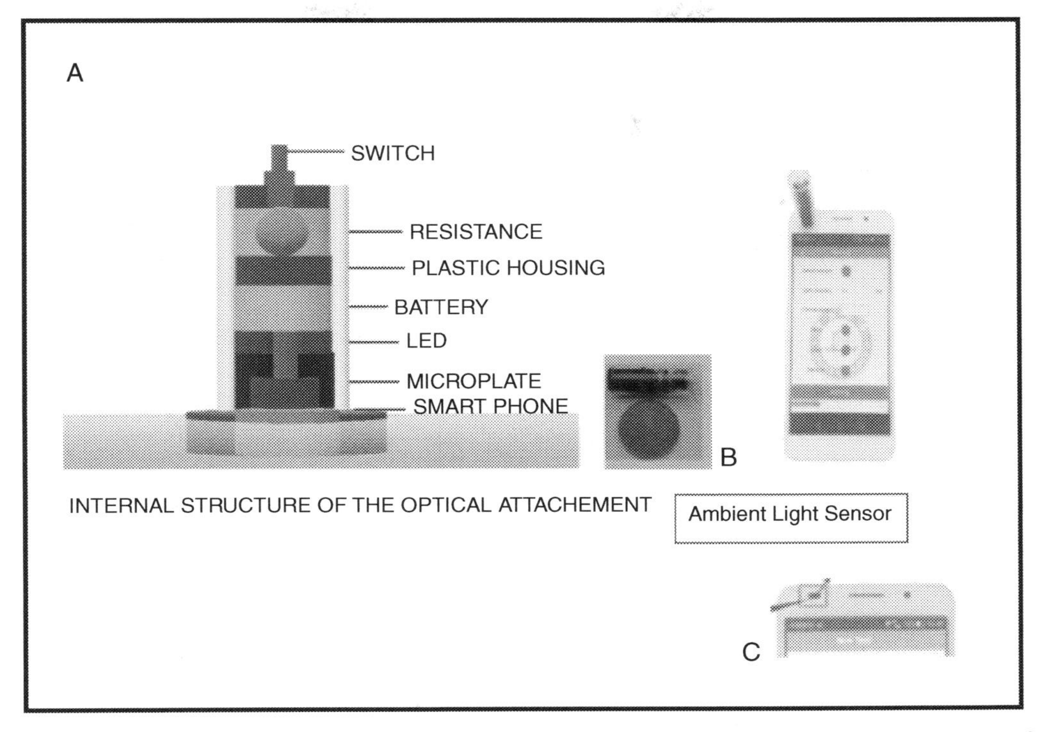

Fig. 20.4 ***Design of the smart phone-based plasmonic nanosensor.*** (A) Optics involved in the design of the internal structures. (B) Optical readout platform running on an Android-based smart phone.

because they are resistant to bending [38] and can even be used in microfluidic systems in polymeric channels as well as in paper substrates [39] that allow label-free sensing (change in color when an analyte passes near the PC or gets attached to the PC). These PCs can even be converted to wearable PoC stickers that can be stitched on clothes in order to warn the user in a zone of war about possible consequences of an expansive blast wave [40]. The PCs are broken down on exposure to extreme pressure as a result of a blast thereby changing its color and notifying the user that the explosion may have caused an invisible injury like injury to the brain or interior trauma.

20.5 Utilization of nanomaterials for point-of-care tests for infectious diseases

Even though many nanomaterial based diagnostic devices with increased sensitivity have been developed so far for detection of infectious diseases, still these devices require some external sources like optical, electrochemical and magnetic resonance sensing technology for detection of pathogenic agents in infectious diseases. In order to overcome these barriers, many nanodiagnostic strategies and technologies have been designed for the early and rapid detection of infectious diseases. Here, we will discuss about these POCT diagnostic devices.

20.5.1 Diagnostic magnetic resonance system

PoCT devices have been miniaturized with the help of nanotechnology and microtechnology. One such miniaturized diagnostic magnetic resonance (DMR) platform has been designed for the quantitative, rapid and multiplexed identification of pathogens in infectious diseases [41]. This DMR system has the ability to detect biological samples that are unprocessed by the utilization of magnetic nanomaterials as proximity sensor. This DMR sensor functions by a self-amplifying proximity test through magnetic nanoparticles. The nanoclusters formed are soluble and results in the reduction of bulk spin-spin relaxation time developed by the conjugations of target proteins or molecules and magnetic nanoparticles. The final result is recorded by NMR technique, which works rapidly than the surface structure based nanodevices, which require more time for sample preparation and pathogen targeting. This is why, DMR can be used to test urine, sputum or blood samples without any prior treatment or preparation procedures. This miniaturized DMR can also overcome the low NMR signal without magnetic fields and can detect bacteria and protein biomarkers altogether. Thus, this system with its microfabrication system and due to its low cost, high portability can be used for large scale detection of infectious diseases in the near future.

Miniaturized DMR platform is also used for rapid profiling of infectious agents directly in clinical samples [42]. Some omnipresent and particular novel magneto-DNA probes have been developed for sensing and specified targeting of bacterial 16s rRNA by nanoparticle hybridization assay with miniaturized DMR platforms. This specified detection of clinically valid infectious agents can lead to a single bacterial sensitivity. Overall, this system can rapidly profile infectious agents directly and can also detect a series of thirteen such agents within 2 h of time in a clinical sample. This system is advantageous over common culturing and real time PCR techniques in terms of cost and time.

20.5.2 Magnetic barcode assay platform

This nanodiagnostic technique is designed for the robust and sensitive detection of *M. tuberculosis* in tuberculosis infected patients [22]. This magnetic barcode assay system is designed for the rapid detection of nucleic acids from infectious pathogens within 2.5 h of time. All components of this nanodevice were integrated within a single micro fluidic chip. DNA-extraction procedure could be completed within 30 min in the off-chip method. The target sequence of DNA could be captured by polymer beads in combination with magnetic nanoprobes and capture DNA through complementary sequences. Amplification of target DNA sequence was done by asymmetric PCR followed by utilization of a portable NMR system for production of required signals. This diagnostic platform can supply analytical NMR signals for detection of nucleic acids of infectious pathogens. Apart from the DNA extraction step, all other steps were carried out in an integrated microfluidic system. After magnetic labeling and on-chip PCR amplification, the desired samples were concentrated and purified by a membrane filter followed by recording the final results on a portable NMR platform. The specificity and sensitivity of this diagnostic method have been proved by detecting bacteria in healthy people as well as in tuberculosis patients.

This diagnostic system can also be used for multiplexed detection of other infectious bacteria like *Klebsiella pneumonia*, methicillin-resistant *S. aureus* (MRSA) and *S. aureus*. In this PoCT platform, QD nanoparticles and magnetic particles were utilized and because of the changes in emission maxima of CdTe QDs altered by particular target specific bacterial genes like fnbA, mecA and wca, this PoCT system can detect very low pathogenic bacterial concentration of 10^2 CFU/mL [43]. These magnetic barcode systems of PoC diagnosis accounts for the efficient, sensitive, cost effective and fast detection of many infectious diseases.

20.5.3 Mobile phone-based microscopy of polarized light

Immunological diagnosis of infectious diseases develops a color for positive tests; but the signal is most of the times very weak for detectable readouts due to the limiting concentration of specific pathogens or very less quantity samples from infected people. Over the recent years, mobile phones embedded with a transmission polarized light microscopic system are used for PoC diagnosis of infectious diseases [44]. A cost effective and optical mobile phone based transmission polarized light microscopic technique has been fabricated for the detection of hemozoin in malaria. Hemozoin parasites are not easy to be observed and detected from other parasites through traditional microscopic systems even for the skilful technicians. So, the newly designed mobile phone based polarized light microscopic platform has high potentialities and high fidelity as well as very good optical resolution for detection of malarial parasites with considerable specificity and in a cost-effective and user-friendly way. In this technique, the 3D printed components were integrated into the camera of iPhone 5S mobile phone and the microscopic lens with a magnification of up to 40X having functions similar to that as the conventional large-scale laboratory microscope system. The characterization of the view field and its resolution, polarized and bright field malarial parasite imaging and polarized light produced by the nonmalarial parasite can be compared during sample testing to obtain the desired results [45]. A Leica polarized white light microscope was used as a positive reference for comparison.

One more successful application of mobile phone based polarized light microscopy platform was in specific and sensitive PoC diagnosis of blood-borne parasites such as *Loa loa* causing filaria [46]. This technique can diagnose the infectious filarial parasite on the basis of automatic quantification of *Loa loa* microfilariae in entire blood stream. The blood specimens could be obtained by a single finger prick and then collecting in a miniature glass capillary and so there is no requirement of traditional preparation of sample and staining methods. It was found that this diagnostic technique showed 100% sensitivity and 94% specificity from thirty-three actively infected *Loa loa* patients.

20.5.4 Mobile phone-based dongle system

Mobile phones have been increasingly used as PoC diagnostic device due to its user-friendliness, wireless processing of data, rapid computing action and reduced prices. Previously, mobile phones were only used for processing of data and its camera; but by using only these features of a mobile phone, it is not possible to meet the requirements of developing a PoC diagnostic device. Over the recent years, a new nanodiagnostic system duplicating a complete laboratory immunoassay in an accessory of a mobile phone has been designed [47]. This has proved to be the most successful PoC diagnostic technique to be used in clinical conditions in the near future. In this platform, a light and small dongle was fabricated to possess identical functions as a microplate ELISA making use of disposable plastic cassettes previously loaded with reagents. Silver ions and gold nanoparticles were used in amplification step in place of substrates and enzymes in ELISA. Every electronic, optical and mechanical function was carried out by using a mobile phone. All these features make this technique a novel one because it does not require an exterior source of energy or resources for sensing, unlike other nano diagnostic platforms. Blood from 96 patients suffering from active blood-borne infections by finger pricking was used for testing the detection ability of this mobile phone-based dongle system. A triplexed immunoassay by utilizing HIV antibody, *T. pallidum* specific antibodies against syphilis and non-*T. pallidum* antibodies for potential syphilis infections were carried out in a single test run in this technique. Test results were obtained within a span of 15min from this triplexed assay in a blinding experiment giving a detection specificity of 79–100% and sensitivity of 92%–100%. 97% patients prefer this mobile phone-based dongle platform for quick results by a single finger prick unlike the traditional ELISA methods.

Lately, a mobile phone-based dongle platform has been also designed to detect concentration of haemoglobin and to measure antibodies of HIV [48]. In this technique, an immunoassay device was enlarged for quantitative detection of concentration of haemoglobin and antibodies of HIV via colorimetric estimation of optical density of precipitation of silver over gold nanoparticles. It has been proved that immunoassay of HIV in presence of CHAPs detergent had 95% sensitivity and specificity. Such mobile phone-based dongle systems not only showed diagnostic abilities for syphilis and HIV in a real time PoCT diagnosis, but also provided a new pathway for developing unique nanodiagnostic systems with integrated microtechnology and nanotechnology and consumer electronics for PoCT of infectious diseases in the near future.

20.5.5 Paper-based point-of-care tests system

Unique features like portable, cost effective and disposable paper makes it the most useful material to be used for PoCT diagnosis. Various PoCT platforms based on paper have been extensively studied to diagnose infectious diseases. For multiplex detection of infectious pathogens, a unique paper-based PoC system in combination with multicolored silver nanoplates have been developed [10]. On the basis of multiplexed lateral flow platform, this paper-based PoCT system exhibited a high speed testing and increased capability for detection of multi-infectious pathogens by use of only one individual strip. These features make such systems suitable enough for the PoCT of infectious diseases because they are easy to fabricate, cost effective and are robust. This lateral flow paper-based PoCT device consists of an adsorbent wick, nitrocellulose membrane, conjugated pad (CP) and a sample pad. The CP was pre loaded with triangular plate-shaped silver nanoparticles that are conjugated with various antibodies, which help in producing simply differentiable colors because of size-tunable adsorption spectra of silver nanoparticles. Antibodies that could recognize Ebola virus, dengue virus NS1 protein, Zaire strain glycoprotein GP and yellow virus NS1 protein were utilized to conjugate with triangular plate-shaped silver nanoparticles for sandwich assay. This diagnostic platform could differentiate between various biomarkers of infectious diseases via different color readouts on the test line having a detection sensitivity of 150 ng/mL. Moreover, this nanomaterial-based PoCT system can carry out the multiplex analysis even in absence of any exterior source of excitation making this platform promising for PoC diagnosis of infectious diseases.

Although, this paper-based PoCT diagnostic platform is very promising and easy to use, there still exist certain limitations such as accuracy and detection sensitivity as to their application. For example, a commercially designed NanoSign HBs PoCT strip was tested for its capacity of HBsAg detection in patients infected with hepatitis B [49]. This PoC diagnostic platform was used to for screening of 297 serum samples from patients suffering with potential HBV infections. However, only 97.8% specificity and 73.7% detection sensitivity for HBsAg was analysed. Also, false negatives were customarily observed in a clinical environment, showing that the specificity and sensitivity of such paper-based PoCT diagnostic platform still needs improvement.

20.6 Therapeutic applications of nanomaterials

Although extensive efforts have been given for the therapeutic developments of various diseases, some loopholes still exist and cause a high mortality rates among the populations. Use of nanobiotechnology for rapid diagnosis and efficient treatment of certain diseases helps in lowering mortality rates of some lethal diseases. Various nanomaterial-based therapies to cure certain diseases like cancer, neurodegenerative diseases or brain tumors, oral biofilm infections, and COVID-19 have been developed so far but still there exists some challenges and limitations, which need to be overcome in order to use these therapies completely.

20.6.1 Nanomaterial-based cancer therapy

Actually the main challenge in cancer therapy is the destruction of tumor tissues without causing any harm to the other healthy organs and tissues. Conventional treatment procedures do not have this kind of specificity and can lead to acute and severe side effects. Therefore, development of targeted chemotherapy for selective destruction of cancer cells is always required. From this point of view, nanomaterial-based drug delivery systems (NM-DDSs) have already fuelled a standard shift in chemotherapy of cancer through toxicity reduction and at the same time keeping the therapeutic efficiency and also enhancing biocompatibility and safety. In comparison to otherwise healthy organs and tissues, cancer tissues or solid tumors have flawed vascular architecture, defective lymphatic drainage and reduced interstitial fluid uptake. By using this discrepancy, in combination with a long profile of blood circulation (carried out by introduction of stealth properties like PEGylation), NM-DDSs accumulate selectively within tumorous lesions and this phenomenon is commonly known as enhanced permeability and retention (EPR)-based passive targeting [50]. Different NM-based carriers of drugs are fabricated to use this passive targeting and many of such procedures have been proven to be successful enough in clinical trials. Researchers are trying to enhance tumor accumulation *via* active targeting through NM surface functionalization with ligands like antibodies, peptides and proteins, which can interact with overexpressed receptors on the tumor site leading to active targeting [51].

Recently there has been development in cancer stem cell (CSC) targeted drug delivery procedures [52]. As CSCs are classified as "promoters" of tumor growth, thus therapeutic targeting of CSCs can reduce tumor growth and development. This therapy is done for multiple types of tumors like leukemia and solid tumors restricting their metastasis and progression. Drugs like staurosporine (STS) are commonly used as an anticancer drug to treat malignant mesothalioma.

Another approach is gene therapy, which utilize genomic components as therapeutic molecules and possesess great potential in reduction of malignancies. Cancer is genetic disorder caused by gene mutations initiating uncontrolled cell division and thus cytotoxic therapy or gene therapy can be employed as a promising treatment for cancer development and tumor metastasis. *In-vivo* genetic therapeutic entity delivery into targeted cells and their subsequent cellular uptake can serve as a potential therapeutic process against cancer [53].

20.6.2 Nanomaterial-based treatment of brain diseases

Brain tumors are classified as significantly lethal as they are one of the leading causes of high mortality rates. Efficient treatments of brain tumor are still in the developing process and very limited due to the presence of an additional restriction called the blood-brain barrier (BBB). Till now, radiotherapy, surgical resection and chemotherapy are considered as some of the treatments of brain tumors like gliomas. With the development of nanobiotechnology, therapies like hyperthermia or gene therapy, drug or gene delivery and photodynamic therapy using nanomaterials have been so far used for certain brain tumor treatments.

20.6.2.1 Chemotherapy based on nanomaterials

Nanomaterial-based chemotherapy by rational surface engineering and meticulous size control of nanomaterials enable therapeutic agent delivery by passing the BBB effectively [54]. A dual targeting doxorubicin (DOX) liposome coupled with transferring and folate was developed to treat brain glioma. Moreover, it has been found that nanoparticle-coated chemotherapeutic drugs can more efficiently penetrate through the BBB and get internalized by the tumor tissues [55].

20.6.2.2 Gene therapy based on nanomaterials

Nanomaterial-based gene therapy approach has also shown a great efficiency in treating brain gliomas. Gene therapy is based on the principle of putting the corrected genetic material into target cells to reduce the symptoms of tumor progression. For example, a nano-delivery system consisting of multifunctional siRNA quantum dots (QDs) to selectively inhibit epidermal growth factor variant III (EGFV III) expression in glioma cells thereby downregulating the signalling pathway with great efficacy [56]. *In-vitro* studies indicated overexpression of EGFV III in cancer cells was decreased on increasing the coincubation time period of siRNA QDs. Over the recent years, gene therapy procedures have provided some initial promising results. However, most of these are exploratory researches and are not suitable for clinical use immediately.

20.6.2.3 Thermotherapy

Thermotherapy increases the temperature of body tissues for causing dysfunction of cells like induction of apoptosis, alteration of the cellular architecture and conformational changes in the DNA. Application of stereotactic injection containing magnetically active nanoparticles at the tumor site and then subsequent exposure to an alternating magnetic field (AMF) produced significant amount of heat into the tumor cells and tissues causing them to be dysfunctional leading to thermotherapy. Researches have been carried out to study the efficiency of iron oxide nanoparticles coupled with radiotherapy in treatment of recurring glioblastoma multiforme [57]. This approach caused mild side effects and no serious side effects thereby proving that coupling of thermotherapy with radiotherapy is more effective and less harmful than conventional chemotherapy.

20.6.2.4 Photodynamic therapy

Photodynamic therapy (PDT) is defined as non-invasive therapeutic procedure by utilization of photosensitizers, which are activated by light in order to produce reactive oxygen species (ROS) and eventually leading to tumor cell death. Nanomaterial-based photodynamic therapy exhibit targeted delivery of PDT drugs in the tumor tissues as well as efficient enhancement of efficacy of PDT. A new photodynamic nanoparticle was designed for treating brain tumors by coating of the polyacrylamide (PAA) core by PEG and molecular targeting groups [58]. MRI contrast agents and photosensitizers are encapsulated within the core that initiates cancer diagnosis, therapy as well as real-time monitoring simultaneously. This multidimensional approach provided a successful and multifunctional platform for treatment of brain tumor. Here, PEGylated AuNPs have been used to encapsulate photosensitizers and deliver them to the target tissue. These PEGylated AuNPs had exceptional water solubility, longer blood circulation profile and biocompatibility. Moreover, this platform led to the reduced uptake of AuNPs by the reticuloendothelial system (RES) and enhanced uptake by the cancer cells giving an improved therapeutic result for treating brain cancer.

20.6.3 Use of nanomaterials for treatment of oral biofilm infections

Nanomaterials have been developed as vehicles and embedded or loaded with photosensitizers (PSs) or the nanomaterials themselves serve as PSs. These nanomaterials are embedded with PSs since they contain several unique functions/properties such as: (i) restricted ability to release PSs, (ii) increased solubility in water, (iii) prevention of aggregation at higher concentrations, (iv) improved binding ability, easy uptake by bacteria and increased ROS yield, (v) restrictive antibacterial activity at the target site, (vi) increased penetration in bacterial cell wall, and (vii) widening the phototherapeutic platform for optimizing the depth of tissue penetration [59].

20.6.3.1 Fullerenes acting as PSs

Due to their extended π configuration, fullerenes can act as PS in treating periodontal diseases (PDT). Fullerene along with its derivatives possesses broad-spectrum antimicrobial photodynamic capability against bacteria as well as fungi. Fullerenes are known to cause disturbance in the cell wall structure of bacteria leading to the production of superoxide anions, O_2 and free radicals [10]. The cationic functional groups of fullerenes increase their antimicrobial activity in case of PDT. Over the recent years, a new amine group modification of fullerene called C_{70}-(ethylenediamine)$_8$ exhibited a good bactericidal activity against super bacteria and has the ability to counter bacterial infections as well as helps in wound healing procedures [60]. Adding potassium iodide further increases the photoactivity of fullerenes against Gram positive bacteria, Gram negative bacteria and fungus. Also aggregation of fullerenes possesses strong photocatalytic inactivation functionality.

20.6.3.2 Carbon nanotubes and graphene

Carbon nanotubes (CNTs) are considered as carbon-based scaffolds to encapsulate antimicrobial substances. CNTs can be modified to form ultrathin films with large surface area to immobile PSs followed by ROS generation after being illuminated. [61] designed a multi-walled carbon nanotube (MWCNT) in combination with protoporphyrin IX and found antimicrobial activity against PDT caused by *S. aureus* upon irradiation by visible light. When subjected to such treatments, it was found that the colonies formed by *S. aureus* significantly reduced by 15%–20% in comparison to the area occupied by biofilms (control) after being exposed to light for one hour.

Over the recent years, graphene has been used considerably to kill pathogenic microbes as these types of microbes are getting resistant toward the conventional treatments with each passing day. Graphene oxide (GO) has a wide surface area consisting of various functional groups having increased absorbance for near infra red (NIR) regions and well-dispersive capability in aqueous solution and thus GO can serve as an excellent platform for delivering PSs. Recently, delivery of PS is carried out on the basis of multifunctional nano systems for multimodelled therapies. In this system, polyethylene glycol (PEG) was loaded with chlorin e6 (Ce6) and was made functionalized through π-π stacking interactions and the delivery by GO will get enhanced to a significant level by intracellular shuttling of Ce6 thereby increasing their ability to kill cancer cells on being irradiated by visible light. Above all, combination of photothermal therapy (PTT) with slight heating locally by utilization of 808 nm laser increased GO-PEG-Ce6 cellular uptake by two times. Thus, GO can be used as a potential nanomaterial for targeting PDT and PTT.

20.6.3.3 Metal oxide nanoparticle

Titanium dioxide and zinc oxide are semiconductors having significant biocompatibility, optical properties as well as biodegradability. Electrons, on being excited by UV (ultraviolet) radiations jump from valence band to conductance band and then undergo electron transfer to produce ROS from oxygen in order to kill the microbes. Certain modifications on superficial level like doping of these metal oxides with other molecules can strongly increase their antibacterial efficacy. For example, fluorinated (F-doped) zinc oxide nanopowders possess photocatalytic antimicrobial activity against *S. aureus* and *E. coli* [62]. On being irradiated, F-doped zinc oxide nanopowders are very effective in stopping the microbial growth as compared to the undoped zinc oxide nanopowders. These doped nano powders showed an efficacy rate of 99.99% and 99.87% against *S. aureus* and *E. coli*, respectively.

Magnetic nanoparticles (MNPs) like ferric tetraoxide possess a peroxidase-like function by attachment to PSs and other surface molecules for enhancement of ROS production at the target sites to treat PDT. MNPs, which are multifunctional can be combined with vancomycin and PSs to form a composite network of reduced GO (rGO) iron oxide, which can induce a synergistic activity against methicillin-resistant *S. aureus* (MRSA).

20.6.3.4 Use of upconversion nanoparticles

Most of the PSs are excited by subjecting them to blue light with short wavelength or to UV radiation. However, these radiations actually lack significant tissue-penetrating capacity and may in turn cause harm to the human body. Near infrared (NIR) radiations having a wavelength ranging from 700–1100 nm possess a higher capability to penetrate deep into the tissues and also has low autofluorescence and therefore has no detrimental effects such as photodamage and phototoxicity. Upconversion is defined as a nonlinear optical procedure which includes sequentially absorbing two or more photons causing a short wavelength light emission as compared to the wavelength required for excitation (antistokes emission). UCNPs strongly absorb NIR converting it into photons with high energy and then it can be applied for the treatment of PDT.

Lanthanide ion doped solid-state materials are considered as UCNPs to obtain optical images and organic fluorophores to treat PDT. In general "core-shell" structures are applied along with various types of PSs against microbes causing PDT. A multifunctional nanosystem of UCNPs ($NaYF_4$: Mn/Yb/Er)/methylene blue (MB)/CuS-chitosan exhibited a strong bactericidal action through the synergistic functions of PDT/PTT [63]. Bactericidal efficacy of this nano system was found to be about 99% with wavelength of 980 nm at 1 W cm^{-2} for 20 min. Red light emitted (650–670 nm) by UCNPs was used to enhance the efficiency of energy transfer from UCNPs to MB. After this, copper sulfide (CuS) was added to produce a greater therapeutic index by means of the synergistic functions of PDT/PTT. In the last step, chitosan was used for grafting in order to stop aggregations of the particles thereby giving an appreciating biocompatibility as well as water solubility.

20.6.4 Antiviral activity of nanomaterials

Nanomaterials have been widely used as antiviral agents or as delivery systems of antiviral agents/compounds. Developed an aggregation consisting of titanium dioxide (TiO_2), colloid of silver, a binder and a dispersion stabilizer for their antibacterial, antifungal and antiviral activities. On being subjected to various antiviral tests, this aggregation showed a 100 times dilution of the concentration of its composition with an antiviral activity against porcine epidemic diarrhea virus (PEDV) and also transmissible gastroenteretitis virus (TGEV) with an efficacy of 99.99% or even higher. However, on 1000 fold dilution of composition concentration viral growth for PEDV was inhibited at a rate of 99.9% and that of TGEV was inhibited at a rate of 93%. Thus, it was proved from this study that the antiviral activity of the nanomaterial aggregate is solely dependent on composition concentration, that is, adjustment of dosage for desired level of inhibition.

20.6.5 Nanomaterials to treat autoimmune diseases

In case of autoimmune diseases, the immunity invades specific self-tissues by damaging their functional and structural compatibility. Nanomaterials have been designed for antigen presenting cell (APC) modulation and also to down regulate innate immune response thereby reinforcing adaptive immune response. For example, a study was carried out to pharmacologically treat experimental autoimmune encephalomyelitis (EAE), where liposomes were loaded with glucocorticoids and administered at comparatively lower doses than traditional glucocorticoid-based therapy [64]. The traditional particular antigen-based therapies for treating autoimmune diseases has a major limitation of facing the antigenic complexity in autoimmune diseases and therefore, these types of diseases require targeting the autoreactive T cells possessing a multifunctional characteristic. It has been found that peptide-loaded major histocompatibility complex (p-MHC) coated nanoparticles can enhance the level of CD4+ regulatory T-cells in low acidic conditions. Such nanoparticles on reaching the target tissue inhibit polyclonal autoimmune signals generated by a selective APC loaded with autoantigen. Newer nanocompounds will further be used for nano-based drug development for treatment of autoimmune diseases.

20.7 Conclusion

Day by day, PoC diagnostic tools are becoming an integral part of our everyday lives as they do not require any medical specialists to operate them and provide us with rapid and hassle-free detection of various diseases. Although most of the platforms for PoC diagnosis and treatment discussed in this study are only a proof-of-concept, it is clearly indicated that nanomaterials after being engineered with cutting-edge technologies can provide numerous advantages over the conventional diagnostic tools and disease therapies. Nanomaterial-based techniques, electrochemical and optical detection of diseases are mostly used because of their high sensitivity, simplicity and user-friendly nature. Various nanomaterial-based systems serve as effective treatment of certain diseases for which conventional therapeutic treatments are not much sufficient. Thus, this nanotechnology implemented with biotechnology is serving as an emerging technology for easy and rapid disease detection and treatment.

References

[1] V. Gubala, L.F. Harris, A.J. Ricco, M.X. Tan, D.E. Williams, Point of care diagnostics: status and future, Anal. Chem. 84 (2) (2012) 487–515, doi:10.1021/ac2030199.
[2] L. Sun, C. Zheng, T.J. Webster, Self-assembled peptide nanomaterials for biomedical applications: promises and pitfalls, Int. J. nanomedicine 12 (2016) 73–86, doi:10.2147/IJN.S117501.
[3] A.G. Arranja, V. Pathak, T. Lammers, Y. Shi, Tumor-targeted nanomedicines for cancer theranostics, Pharmacol. Res. 115 (2017) 87–95, doi:10.1016/j.phrs.2016.11.014.
[4] A. St John, C.P. Price, Existing and emerging technologies for point-of-care testing, Clin. Biochem. Rev. 35 (3) (2014) 155–167.

[5] J. Jeevanandam, A. Barhoum, Y.S. Chan, A. Dufresne, M.K. Danquah, Review on nanoparticles and nanostructured materials: history, sources, toxicity and regulations, Beilstein J. Nanotechnol. 9 (2018) 1050–1074, https://doi.org/10.3762/bjnano.9.98.
[6] M. Perfezou, A. Turner, A. Merkoci, Cancer detection using nanoparticle-based sensors, Chem. Soc. Rev. 41 (7) (2012) 2606–2622, doi:10.1039/c1cs15134g.
[7] M. Wuithschick, S. Witte, F. Kettemann, K. Rademann, J. Polte, Illustrating the formation of metal nanoparticles with a growth concept based on colloidal stability, Phys. Chem. Chem. Phys. 17 (30) (2015) 19895–19900, doi:10.1039/c5cp02219.
[8] D. Quesada-Gonzalez, A. Sena-Torralba, W.P. Wickasono, A.d. Escosora-Muniz, T.A. Ivandini, A. Merkoci, Iridium oxide (IV) nanoparticle-based lateral flow immunoassay, Biosens. Bioelectron. 132 (2019) 132–135, doi:10.1016/j.bios.2019.02.049.
[9] R.H. Shyu, H.F. Shyu, H.W. Liu, S.S. Tang, Colloidal gold-based immunochromatographic assay for detection of ricin, Toxicon 40 (3) (2002) 255–258, doi:10.1016/s0041-0101(01)00193-3.
[10] C.W. Yen, H.d. Puig, J.O. Tam, J. Gomez-Marquez, I. Bosch, K. Hamad-Schifferli, et al., Multicolored silver nanoparticles for multiplex disease diagnostics: distinguishing dengue, yellow fever and Ebola viruses, Lab Chip 15 (7) (2015) 1638–1641, doi:10.1039/c5lc00055f.
[11] E. Fu, T. Liang, P. Spicar-Mihalic, J. Houghtaling, S. Ramachandran, P. Yager, Two-dimensional paper network format that enables simple multistep assays for use in low-resource settings in the context of malaria antigen detection, Anal. Chem. 84 (10) (2012) 4574–4579, doi:10.1021/ac300689s.
[12] N.M. Rodriguez, W.S. Wong, L. Liu, R. Dewar, C.M. Klapperich, A fully integrated paperfluidic molecular diagnostic chip for the extraction, amplification, and detection of nucleic acids from clinical samples, Lab Chip 16 (4) (2016) 753–763, doi:10.1039/c5lc01392e.
[13] G.E. Pauli, A.d. Escosura-Muniz, C. Parolo, I.H. Bechtold, A. Merkoci, Lab-in-a-syringe using gold nanoparticles for rapid immunosensing of protein biomarkers, Lab Chip 15 (2) (2015) 399–405, doi:10.1039/c4lc01123f.
[14] E. Morales-Narvaez, H. Golmohammadi, T. Naghdi, H. Yousefi, U. Kostiv, D. Horak, et al., Nanopaper as an optical sensing platform, ACS Nano 9 (7) (2015) 7296–7305, doi:10.1021/acsnano.5b03097.
[15] J.M. Klostranec, Q. Xiang, G.A. Farcas, J.A. Lee, A. Rhee, E.I. Lafferty, Convergence of quantum dot barcodes with microfluidics and signal processing for multiplexed high-throughput infectious disease diagnostics, Nano Lett. 7 (9) (2007) 2812–2818, doi:10.1021/nl071415m.
[16] R. Alvarez-Diduk, J. Orozco, A. Merkoci, Paper strip-embedded graphene quantum dots: a screening device with a smartphone readout, Sci. Rep. 7 (1) (2017) 976, doi:10.1038/s41598-017-01134-3.
[17] E.T. Silva, S. Miserere, L.T. Kubota, A. Merkoci, Simple on-plastic/paper inkjet-printed solid-state Ag/AgCl pseudoreference electrode, Anal. Chem. 86 (21) (2014) 10531–10534, doi:10.1021/ac503029q.
[18] S.J. Park, T.A. Taton, C.A. Mirkin, Array-based electrical detection of DNA with nanoparticle probes, Science 295 (5559) (2002) 1503–1506, doi:10.1126/science.1067003.
[19] L. Rivas, A.d. Escosura-Muniz, L. Serrano, L. Altet, O. Francino, A. Sanchez, et al., Triple lines gold nanoparticle-based lateral flow assay for enhanced and simultaneous detection of Leishmania DNA and endogenous control, Nano Res. 8 (2015) 3704–3714, doi:10.1007/s12274-015-0870-3.
[20] Y. Wu, P. Xue, Y. Kang, K.M. Hui, Paper-based microfluidic electrochemical immunodevice integrated with nanobioprobes onto graphene film for ultrasensitive multiplexed detection of cancer biomarkers, Anal. Chem. 85 (18) (2013) 8661–8668, doi:10.1021/ac401445a.
[21] J.C. Cunningham, M.R. Kogan, Y.-J. Tsai, L. Luo, I. Richards, R.M. Crooks, Paper-based sensor for electrochemical detection of silver nanoparticle labels by galvanic exchange, ACS Sensors 1 (2016) 40–47, doi:10.1021/acssensors.5b00051.
[22] M. Liong, A.N. Hoang, J. Chung, N. Gural, C.B. Ford, C. Min, Magnetic barcode assay for genetic detection of pathogens, Nat. Commun. 4 (2013) 1752, doi:10.1038/ncomms2745.
[23] H.J. Chung, K.L. Pellegrini, J. Chung, K. Wanigasuriya, I. Jayawardene, K. Lee, Nanoparticle detection of urinary markers for point-of-care diagnosis of kidney injury, PLoS One 10 (7) (2015) 133417, doi:10.1371/journal.pone.0133417.
[24] P. Mostafalu, S. Sonkusale, A high-density nanowire electrode on paper for biomedical applications†, RSC Adv. 12 (2015) 1–3, doi:10.1039/C4RA12373E.
[25] M.A. Ali, P.R. Solanki, M.K. Patel, H. Dhayani, V.V. Agarwal, J. Renu, A highly efficient microfluidic nano biochip based on nanostructured nickel oxide, Nanoscale 5 (7) (2013) 2883–2891, doi:10.1039/c3nr33459g.

[26] A. Nemiroski, D.C. Christodouleas, J.W. Hennek, A.A. Kumar, E.J. Maxwell, M.T. Fernandez-Abedul, et al., Universal mobile electrochemical detector designed for use in resource-limited applications, Proc. Natl. Acad. Sci. U. S. A. 111 (33) (2014) 11984–11989, doi:10.1073/pnas.1405679111.
[27] J.M. Azzarelli, K.A. Mirica, J.B. Ravnsbaek, T.M. Swager, Wireless gas detection with a smartphone via rf communication, Proc. Natl. Acad. Sci. U. S. A. 111 (51) (2014) 18162–18166, doi:10.1073/pnas.1415403111.
[28] M. Darabi, A. Khosrozadeh, Q. Wang, M. Xing, Gum sensor: a stretchable, wearable, and foldable sensor based on carbon nanotube/chewing gum membrane, ACS Appl. Mater. Interfaces 7 (47) (2015) 26195–26205, doi:10.1021/acsami.5b08276.
[29] E. Morales-Narvaez, T. Naghdi, E. Zor, A. Merkoci, Photoluminescent lateral-flow immunoassay revealed by graphene oxide: highly sensitive paper-based pathogen detection, Anal. Chem. 87 (16) (2015) 8573–8577, doi:10.1021/acs.analchem.5b02383.
[30] N. Cheeveewattanagul, E. Morales-Narvez, A.R. Hassan, J.F. Bergua, W. Surareungchai, M. Somasundrum, Straightforward immunosensing platform based on graphene oxide-decorated nanopaper: a highly sensitive and fast biosensing approach, Adv. Funct. Mater. 27 (38) (2017) 1–8, doi:10.1002/adfm.201702741.
[31] S. Teixeira, R.S. Conlan, O.J. Guy, M.G. Sales, Label-free human chorionic gonadotropin detection at picogram levels using oriented antibodies bound to graphene screen-printed electrodes, J. Mater Chem. B 2 (13) (2014) 1852–1865, doi:10.1039/c3tb21235a.
[32] S.K. Tuteja, T. Duffield, S. Neethirajan, Graphene-based multiplexed disposable electrochemical biosensor for rapid on-farm monitoring of NEFA and βHBA dairy biomarkers, J. Mater Chem. B 5 (33) (2017) 6930–6940, doi:10.1039/c7tb01382e.
[33] A. Chamorro-Garcia, A. Merkoci, Nanobiosensors in diagnostics, Nanobiomedicine 3 (2016) 1–26, doi:10.1177/1849543516663574.
[34] E.O. Polat, O. Balci, N. Kakenov, H.B. Uzlu, C. Kocabas, R. Dahiya, Synthesis of large area graphene for high performance in flexible optoelectronic devices, Sci. Rep. 5 (2015) 1–10, doi:10.1038/srep16744.
[35] Q. Fu, Z. Wu, F. Xu, X. Li, C. Yao, M. Xu, A portable smart phone-based plasmonic nanosensor readout platform that measures transmitted light intensities of nanosubstrates using an ambient light sensor†, Lab Chip 6 (10) (2016) 1927–1933, doi:10.1039/c6lc00083e.
[36] W. Li, L. Li, S. Ge, X. Song, L. Ge, M. Yan, Multiplex electrochemical origami immunodevice based on cuboid silver-paper electrode and metal ions tagged nanoporous silver-chitosan, Biosens. Bioelectron. 56 (2014) 167–173, doi:10.1016/j.bios.2014.01.011.
[37] A.C. Marques, L. Santos, M.N. Costa, J.N. Dantas, P. Duarte, A. Goncalves, et al., Office paper platform for bioelectrochromic detection of electrochemically active bacteria using tungsten trioxide nanoprobes, Sci. Rep. 5 (2015) 1–7, doi:10.1038/srep09910.
[38] C.J. Choi, B.T. Cunningham, A 96-well microplate incorporating a replica molded microfluidic network integrated with photonic crystal biosensors for high throughput kinetic biomolecular interaction analysis, Lab Chip 7 (5) (2007) 550–556, doi:10.1039/b618584c.
[39] B.R. Schudel, C.J. Choi, B.T. Cunningham, P.J. Kenis, Microfluidic chip for combinatorial mixing and screening of assays, Lab Chip 9 (12) (2009) 1676–1680, doi:10.1039/b901999e.
[40] D.K. Cullen, Y. Xu, D.V. Reneer, K.D. Browne, J.W. Geddes, S. Yang, Color changing photonic crystals detect blast exposure, Neuroimage 54 (1) (2011) 37–44, doi:10.1016/j.neuroimage.2010.10.076.
[41] H. Lee, E. Sun, D. Ham, R. Weissleder, Chip-NMR biosensor for detection and molecular analysis of cells, Nat. Med. 14 (8) (2008) 869–874, doi:10.1038/nm.1711.
[42] H.J. Chung, C.M. Castro, H. Im, H. Lee, R. Weissleder, A magneto-DNAnanoparticle system for rapid detection and phenotyping of bacteria, Nature Nanotechnol. 8 (5) (2013) 369–375, doi:10.1038/nnano.2013.70.
[43] K. Cihalova, D. Hegerova, A.M. Jimenez, Antibody-free detection of infectious bacteria using quantum dots-based barcode assay, J. Pharm. Biomed. Anal. 134 (2017) 325–332, doi:10.1016/j.jpba.2016.10.025.
[44] Z.J. Smith, K. Chu, A.R. Espenson, et al., Cell-phone-based platform for biomedical device development and education applications, PLoS One 6 (3) (2011) e17150, doi:10.1371/journal.pone.0017150.
[45] Y. Wang, L. Yu, X. Kong, L. Sun, Application of nanodiagnostics in point-of-care tests for infectious diseases, Int J Nanomedicine 12 (2017) 4789–4803, doi:10.2147/IJN.S137338.

[46] M.V. D' Ambrosio, M. Bakalar, S. Bennuru, C. Reber, A. Skandarajah, L. Nilsson, Point-of-care quantification of blood-borne filarial parasites with a mobile phone microscope, Sci. Transl Med. 7 (286) (2015), doi:10.1126/scitranslmed.aaa3480.
[47] D. Zurovac, R.K. Sudoi, W.S. Akhwale, et al., The effect of mobile phonetext-message reminders on Kenyan health workers' adherence tomalaria treatment guidelines: a cluster randomised trial, Lancet 378 (9793) (2011) 795–803, doi:10.1016/S0140-6736(11)60783-6.
[48] T. Guo, R. Patnaik, K. Kuhlmann, A.J. Rai, S.K. Sia, Smartphone dongle for simultaneous measurement of hemoglobin concentration and detection of HIV antibodies, Lab Chip 15 (17) (2015) 3514–3520, doi:10.1039/c5lc00609k.
[49] R.G. Gish, J.A. Gutierrez, N. Navarro-Cazarez, A simple and inexpensive point-of-care test for hepatitis B surface antigen detection: serological and molecular evaluation, J. Viral Hepat. 21 (12) (2014) 905–908, doi:10.1111/jvh.12257.
[50] R.K. Jain, T. Stylianopoulos, Delivering nanomedicine to solid tumors, Nat. Rev. Clin. Oncol. 7 (11) (2010) 653–664, doi:10.1038/nrclinonc.2010.139.
[51] J. Shi, P.W. Kantoff, R. Wooster, O.C. Farokhzad, Cancer nanomedicine: progress, challenges and opportunities, Nat. Rev. Cancer 17 (1) (2017) 20–37, doi:10.1038/nrc.2016.108.
[52] A. Orza, D. Casciano, A. Biris, Nanomaterials for drug delivery to cancer stem cells, Drug Metab. Rev. 46 (2) (2014) 191–206, doi:10.3109/03602532.2014.900566.
[53] C.E. Thomas, A. Ehrhardt, M.A. Kay, Progress and problems with the use of viral vectors for gene therapy, Nat. Rev. Genet. 4 (5) (2003) 346–358, doi:10.1038/nrg1066.
[54] B. Auffinger, B. Thaci, P. Nigam, E. Rincon, C. Yu, M.S. Lesniak, New therapeutic approaches for malignant glioma: in search of the Rosetta stone, F1000 Med Rep 4 (18) (2012) 1–6, doi:10.3410/M4-18.
[55] X. Ying, H. Wen, W.L. Lu, J. Du, J. Guo, W. Tian, Y. Men, Y. Zhang, R.J. Li, T.Y. Yang, Dual targeting daunorubicin liposomes improve the therapeutic efficacy of brain glioma in animals, J. Control. Release 141 (2) (2010) 183–192, doi:10.1016/j.jconrel.2009.09.020.
[56] J. Li, B. Gu, Q. Meng, Z. Yan, H. Gao, X. Chen, X. Yang, W. Lu, The use of myristic acid as a ligand of polyethylenimine/DNA nanoparticles for targeted gene therapy of glioblastoma, Nanotechnology 22 (43) (2011) 435101, doi:10.1088/0957-4484/22/43/435101.
[57] K. Maier-Hauff, F. Ulrich, D. Nestler, H. Niehoff, P. Wust, B. Thiesen, H. Orawa, V. Budach, A. Jordan, Efficacy and safety of intratumoral thermotherapy using magnetic iron-oxide nanoparticles combined with external beam radiotherapy on patients with recurrent glioblastoma multiforme, J. Neuro-Oncol. 103 (2) (2011) 317–324, doi:10.1007/s11060-010-0389-0.
[58] R. Kopelman, Y-E.L. Koo, M. Philbert, B.A. Moffat, G.R. Reddy, P. McConville, D.E. Hall, T.L. Chenevert, M.S. Bhojani, S.M. Buck, Multifunctional nanoparticle platforms for in vivo MRI enhancement and photodynamic therapy of a rat brain cancer, J. Magn. Magn. Mater. 293 (1) (2005) 404–410, https://doi.org/10.1016/j.jmmm.2005.02.061.
[59] M. Qi, M. Chi, X. Sun, X. Xie, M.D. Weir, T.W. Oates, et al., Novel nanomaterial-based antibacterial lphotodynamic therapies to combat oral bacterial biofilms and infectious diseases, Int. J. Nanomedicine 14 (2019) 6937–6956, doi:10.2147/IJN.S212807.
[60] L. Huang, B. Bhayana, W. Xuan, et al., Comparison of two functionalized fullerenes for antimicrobial photodynamic inactivation: potentiation by potassium iodide and photochemical mechanisms, Photoch Photobio B 186 (2018) 197–206, doi:10.1016/j.jphotobiol.2018.07.027.
[61] I. Banerjee, D. Mondal, J. Martin, R.S. Kane, Photoactivated antimicrobial activity of carbon nanotube–porphyrin conjugates, Langmuir 26 (22) (2010) 17369–17374, doi:10.1021/la103298e.
[62] J. Podporska-Carroll, A. Myles, B. Quilty, et al., Antibacterial properties of F-doped ZnO visible light photocatalyst, J. Hazard. Mater. 324 (2017) 39–47, doi:10.1016/j.jhazmat.2015.12.038.
[63] M. Yin, Z. Li, E. Ju, et al., Multifunctional upconverting nanoparticles for near-infrared triggered and synergistic antibacterial resistance therapy, Chem. Commun. 50 (72) (2014) 10488–10490, doi:10.1039/C4CC04584J.
[64] N. Schweingruber, A. Haine, K. Tiede, A. Karabinskaya, J. van den Brandt, S. Wüst, J.M. Metselaar, R. Gold, J.P. Tuckermann, H.M. Reichardt, F. Lühder, Liposomal encapsulation of glucocorticoids alters their mode of action in the treatment of experimental autoimmune encephalomyelitis, J. Immunol 187 (2011) 4310–4318, https://doi.org/10.4049/jimmunol.1101604.

[65] P. Nasimi, M. Haidari, Medical use of nanoparticles: drug delivery and diagnosis diseases, Int. J. Green Nanotechnol. 1 (2013) 1–5, doi:10.1177/1943089213506978.
[66] N. Naderi, S.Y. Madani, E. Ferguson, A. Mosahebi, A.M. Seifalian, Carbon nanotubes in the diagnosis and treatment of malignant melanoma, Anticancer Agents Med. Chem. 13 (1) (2013) 171–185, doi: 10.2174/1871520611307010171.
[67] N. Shehada, J.C. Cancilla, J.S. Torrecilla, E.S. Pariente, G. Bronstrup, S. Christiansen, et al., Silicon nanowire sensors enable diagnosis of patients via exhaled breath, ACS Nano 10 (7) (2016) 7047–7057, doi:10.1021/acsnano.6b03127.
[68] R. Rosic, P. Kocbek, J. Pelipenko, J. Kristl, S. Baumgartner, Nanofibers and their biomedical use, Acta Pharm. 63 (3) (2013) 2304–2995, doi:10.2478/acph-2013-0024.
[69] H. Cho, U. Jammalamadaka, K. Tappa, Nanogels for pharmaceutical and biomedical applications and their fabrication using 3D printing techniques, Materials (Basel) 11 (2) (2018) 302, doi:10.3390/ma11020302.
[70] A. Maiti, S. Bhattacharyya, Review: quantum dots and application in medical science, Int. J. Chem. Chem. Eng. 3 (2) (2013) 37–42, https://www.ripublication.com/ijcce_spl/ijccev3n2spl_01.pdf.
[71] Y.H. Luo, L.W. Chang, P. Lin, Metal-based nanoparticles and the immune system: activation, inflammation, and potential applications, Biomed. Res. Int. 2015 (2015) 143720, doi:10.1155/2015/143720.

CHAPTER 21

Synthesis and applications of carbon nanomaterials-based sensors

Ravi Patel (Kumar)[a], Prakash Bobde[a], Vishal Singh (K.)[b], Deepak Panchal[c,d], Sukdeb Pal[c,d]

[a]Department of Research and Development, University of Petroleum & Energy Studies, Dehradun, Uttarakhand, India
[b]Department of HSE and Civil Engineering, University of Petroleum & Energy Studies, Dehradun, Uttarakhand, India
[c]Wastewater Technology Division, CSIR-National Environmental Engineering Research Institute (CSIR-NEERI), Nagpur, India
[d]Academy of Scientific and Innovative Research (AcSIR), Ghaziabad, India

21.1 Introduction

Nanomaterials are material science based approach to nanotechnology. On nanoscale the characterisation, fabrication and analysis of material is studied along with the morphological features [1–5]. The term nanosystem is referred when at least one dimension is reduced from 1 micrometre to less than 100 nm. The Nanomaterials are further classified based on their dimensions (D) such as zero dimensional, one dimensional, two dimensional and three dimensional [6]. 0D structure defines metallic, ceramic, quantum dots (QDs) and semiconducting nanoparticles [7–9], 1D structure defines as nanotubes, nanorods, and nanowires [8,10–12], 2D structure defines nano-textured surfaces, plates or thin films) [13–16] whereas 3D structure defines nanoterapods and nanocombs [17,18]. These dimensional structures exist as single, agglomerated or fused forms. The most commonly used nanomaterials are dendrimers, nanotubes, fullerenes and quantum dots which could exist as tubular, spherical, and irregular shapes. The advancement in nanoscience involves a basic understanding of nanoscale physical properties and emerging developments, both in manufacturing processes and in testing instruments, and interaction/exchange in different areas of research like physics, chemistry, materials science and technology, as Feynman suggests. Fundamental scientific work is currently focussed on exploring the potential of nanomaterials, which clearly vary from large objects of the same composition (e.g., bulk counterpart). There are variations throughout the bulk properties implied by size effects.

a) Chemical characteristics: catalysis and reactivity
b) Thermal properties: melting point and thermal conductivity
c) Mechanical characteristics: adhesion and strength
d) Optical characteristics: absorption, band structure and dispersion
e) Electrical properties: quantized conductance, coulomb charging and quantum tunneling current
f) Magnetic effect: paramagnetic outcome

Advanced Nanomaterials for Point of Care Diagnosis and Therapy
DOI: https://doi.org/10.1016/B978-0-323-85725-3.00019-2

Nanomaterials have unique physicochemical characteristics and have wide applications in several fields as catalysts, sensors, components of batteries, optoelectronic devices, and in agricultural and biomedical operations. The aim of the technology is to develop and control the form and size, size and form of nanomaterials and their composition. To study the nanoworld and nanotechnology observation tools such as transmission electron microscope (TEM), scanning electron microscope (SEM) and atomic force microscope AFM [19] and scanning near-field optical spectroscopy (SNOM) are generally required [20,21].

The nanomaterials which have high surface to volume ratio can be used in various sensors. A sensor diversity, flexibility, precision of extracted information, selectivity, and stability are the most critical characteristics of a sensor.

These carbon nanomaterials such as carbon nanotubes (CNT), nano/mesporous carbon and graphene can be used for electro-analytical purpose and they are unique in their way because of high stability, strong conductivity, low cost and noncomplex functionalization of the surface. Its nanostructures offer effective exposure for surface groups, leading to enhanced detection performance of environmental contaminants, to bind between analytic molecules and transduction material [22,23].

Synthesis of above shown carbon nanomaterials is achieved through various techniques depending upon the desired dimensionality and functionality of the synthesized nanomaterial and in this chapter we will discuss the synthesis, structure, properties, and applications of carbon-based nanomaterials.

21.2 Classification of nanosensors

Nanomaterials are being progressively studied due to their electrical, magnetic, and mechanical properties and have drawn interest of consumers in numerous applications such as biomedical, agriculture, gas sensors, batteries and optoelectronic instruments [24–28]. Due to a broad surface to volume ratio and a narrow diameter comparable to Debye thickness, nanomaterial with various sizes such as containment in two coordination systems provides even improved sensitivity for surface chemical reaction [29–32]. Two different methods, such as a bottom-up (chemical approach, electrochemical process, Sol-gel and solvothermal), can be used to synthesize nanomaterials. This means that the substance is created from the bottom, i.e. atom-by-atom, and top-down (kugel friction, laser ablation, lithography) [33–40]. The nanostructured materials used in nanosensors manufacturing include: nanoscale wires (high sensitivity detectability), polymer nanomaterials, carbon nanotubes, nanoparticles and thin films which is measured up to 100 nm [41]. Fig. 21.1 shows their classification based on energy source, structure and application.

21.2.1 Based on energy source

The nanosensors can be categorised into two types, that is, (1) active nanosensors such as thermistors and (2) passive nanosensors where no energy source such as thermocouple and piezoelectric nanomaterials. Classification of nanosensors based upon their responses to different stimuli are shown in Table 21.1.

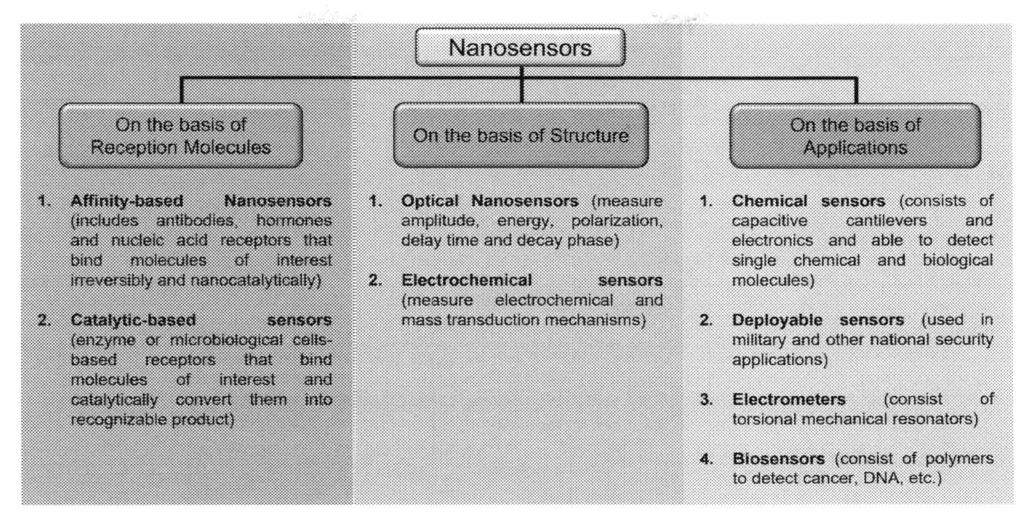

Fig. 21.1 *Classification of nanosensors.*

Table 21.1 Classification of nanosensors based upon their stimuli.

S. No.	Stimuli	Properties
1	Mechanical	Location, acceleration, tension, stress, strength, weight, mass, viscosity, momentum, torque, phase, polarization, velocity
2	Optical	Absorption, absorption, fluorescence, light, refractive index, light dispersion
3	Thermal	Flow, heat-specific temperature
4	Electrical	Load, current, potential, constant dielectric, conductivity
5	Magnetic	Load, current, potential, constant dielectric, conductivity
6	Chemical	Components (identities, concentrations, states)
7	Biological	Biomass (identities, ages, states)

21.2.2 Based on structure

21.2.2.1 Optical nanosensors

Chemical detection can be tracked by optical sensors. It relies on the nanomaterials optical properties. It can be used in different applications such as pharmaceutical, biotechnological, medical, environmental and human security industries. The first optical nanosensors reported in a fluorescent based nanoparticle polyacrylamide and was used in measuring pH [42]. Fluorescent sensors are particles which have minimum one binding element(s) and photosensitive unit(s) [43,44]. A fluorophorus receives light at a certain wavelength, followed by a quantum of light absorption which corresponds to the energy difference between ground and excited state [45,46]. The most fundamental type of optical nanosensor is the one used in the molecular fluorectuating colouring sensor within the cell as stated by Sasaki et al. [47]. The intrinsic chemical interference from the colouring of protein, cell sequestration and toxicity is a downside of free pigment. Another process is known as the mark of nanoparticles composed of an external reporter molecule [48,49]. In solid state the main distinction between the labelling of the nanoparticle and drying process is their fluidity and marked nanoparticles circulate easily and the reporter's molecules are in contact with the intracelled elements. The external component sensor style sensors are used for intracellular sensing, while retaining similar disadvantages of free fluorescent teeth as a signal derived from non-cellular receptor molecules [50,51].

(i) Fiber optic sensors: Fiber optic nanosensors can study major cell processes in vivo. Tan et al. [52] have the first fiber optic submicron nano sensor. As shown in Eq. 21.1 the association between the target molecule of (A) and receptor (R) is intended for physical and chemical interference, which can be transformed into a power signal [53–55]. This observable signal is then processed and transferred into the database by the optical sensor. Since the optical fiber arm is isolated from the environs and the affected region, the drawbacks of chemical contamination by the color-free approach are eliminated. The optical nanosensor also has the benefit of achieving the minimal degree of invasion.

$$R + A \rightarrow RA + \text{measurable signal} \tag{21.1}$$

21.2.2.2 Electromagnetic nanosensors

Referring to the detection process within the electromagnetic nanosensors there are two sensors which are categorised:

1. Monitoring by calculation of electric current.
2. Monitoring by calculation of magnetism.

It is a nonstandardized approach regarding the use of dyes in the calculation of electrical current as it involves the association of sulfide gas molecule and gold nanoparticle for obtaining result as investigated. Chromium electrode, gold electrode, source and drain in each sensing cell. A standard breakthrough of approx. between the two electrodes 40–60 nm was achieved. The nanoparticles are randomly placed around the vacuum. The formation of the sulfide shell prevents the "e" transition of the charges, namely so-called hopping phenomena, from one nanoparticle to another. This phenomenon was called jumping of electrons and it is calculated in applied electric field with the use of current and voltage in chrome and gold electrodes.

Magnetic measurement: These nanosensors have magnetism their relevant biomolecules with low femtomolar spectrum susceptibility are engineered to be observed. (e.g., virus). Magnetic nanosensors consist of magnetic oxide nanoparticles and they bind to the molecular objective they are intended for and further they form stability. Through magnetic resonance technique (MRI) it can be observed that their spin–spin relaxation time (T2) is decreasing.

21.2.2.3 Mechanical and/or vibrational nanosensors

Mechanical nanosensors have similar benefits for the identification of mechanical properties on a nanoscale relative to optical nanosensors and electromagnetic nanosensors [56]. Many kinds of mechanical nanosensors are available, including CNT-based fluidic shear-stress sensors and nanosensors. In the case of a vibrational and elastic property connected to a tapered cantilever, the earliest mechanical nano sensor was proposed by Binh et al. [57]. It was observed that the mechanical sensors plays an important role due to the nanoscale components and nano scale subassemblies in microelectronics.

21.2.3 Based on application

On the basis of application, the sensors are classified into four types:

(i) Chemical sensors

The study of a single chemical or molecule can be carried out with this form. Different types of optical chemical sensors are used to calculate certain properties such as pH and the concentration of ions.

(ii) Deployable nanosensors

These types of sensors are used in the military defense system. A compact light chemical detection device combined with a micro electromechanical detector for the sample collection and a micro concentration.

(iii) Electrometers

To pair the load on the mechanical part, it consists of a detector, mechanical resonator, electrode and gate electrode.

(iv) Biosensors

It is one of the most popular sensors for early cancer diagnosis and other illnesses. It can also facilitates the detection of unique DNA sequences [58]. The biosensors are generally regarded as a subcategory of chemical sensors since sensor platforms or transduction procedures are the same as in case of chemical sensors [59]. In different types of bio-sensing technology developed, field-effect transistors (FETs) have many advantages. The wearable nano-biosensors were analysed by Rai et al. [60]. This form can identify neural signs and detect abnormalities to diagnose targeted cardiovascular and neurological disorders.

Biosensors on-chip: Safe and cost-effective process points include the creation of micro-fluid biosensors (chip-on-chap biosensors) and biosensor technologies to make it possible to integrate chemical components into a single platform and to provide a new system for bio-sensing applications, including portabilities, disposability, real-time detection, unprecedented accuracy [61]. Das et al. [62] used this approach to detect nucleic acids (cfNAs) present at large blood levels in patients with cancer.

21.2.4 Classification of carbon-based nanosensors

The exceptional electrochemical properties of carbon nanomaterials such as graphene and nanotubes of carbon (CNTs) as well as crysteptic diamonds and carbon-like diamonds have led to their widely applied influence. Historically, electrode material for electrochemical detection has been used by a variety of materials from different sources including platinum, gold and carbon [63–67].

The study focuses, therefore, on the recent (<5 years) integration and usage for electrochemical sensing applications of carbon dependent nanomaterials, and their possible consequences. Various developments have been made in electrochemical sensors from nanomaterials such as high conductivity, large surface-to-volume ratios, and high electron mobility at room temperature. This analysis focuses on the introduction of carbon sensors use for detection of DNA, proteins, toxins, gases, ions metal, and immunosensors in different areas, as electrical sensors.

The result of C60 bucky ball discovered along with the emergency, alongside graphites and diamonds, of an additional carbohydrate structure, led to CNTs evolving by Iijima [68]. CNTs in many fields, including physics [69], chemistry [70–72] and materials science [73,74], have received tremendous popularity since they were discovered in the early 1990s and this curiosity has not yet waned.

The CNTs have a hexagonal "honeycomb" cylindrical form glasses made from units of sp2 carbon. This arrangement leads to a closed topology with nanometer diameters and micron lengths. CNT consists of two structural specified classes of nano-carbon mono (SWCNT) and multiwall (MWCNT) [75,76]. SWCNT's consist of a 1–2 nm graphite roller tube with a single graphite sheet, while MWCNT's are a result of the matrioshka-the nesting of numerous individual graphite cylinders exhibiting standard diameters ranging from 2 to 25 nm with the interlayer-like distance of around 0.34 nm [77].

CNTs have well-defined electronic characteristics, especially SWCNTs, and they exhibit similar properties to QDs and wires at low temperatures, including Coulomb blocking and single-electron loading [78–81].

Diamond has special features, since by eliminating oxygen, hydrogen or hydroxide the electric properties can be enhanced [82]. Diamond such as carbon is an amorphous or amorphous hydrogenated carbon species of which a high proportion, is not crystalline in nature. Thin film DLC free of Hydrogen is made from purified cathode vacuum, pulse laser deposition or from mass chosen deposition of ion beams [83–85].

Recently for some electrochemical uses, DLC films emerged as a field of great concern. This is because properties such as chemical inertness, low surface roughness, mechanical hardness, enhanced elastic module and their semiconductor existence with a tunable band scaping size from 1 to 4 eV (approximately) are achieved [86].

21.3 Study of carbon-based sensor

There has been a rising imperative to develop modern sensors with more complex characteristics. Enhanced sensitivity, quick response, more reliable, rapid recovery, low-cost, in situ analysis, reduced size and easy operation are some of the features needed to improve technological sensor devices [87]. A large variety of such as gas, heavy metal, humidity, biomolecules, and pressure sensing sensors are available. However, most of them are costly, require pre-treatment, exhibit difficult functioning and sluggish reaction, and have the inadequate detection, sensitivity and selectivity.

Nanotechnology has facilitated the most exciting updating of material properties in order to enhance the above described criteria, and provides important steps to correct flaws after traditional materials have been researched. Nanomaterials in the nanoscale range 1-100 nm, and their related structural components, such as bulk materials, change greatly in their properties rather than nanoscale [88].

Due to the exceptional characteristics, carbon-based materials are amongst most explored and used materials in the field of nanotechnology. Carbonaceous structures showed many advantages over commonly used materials due to their extraordinary physicochemical properties. Availability of simple manufacturing processes also results in a good yield of material with reduced densification defects. Moreover, these materials can be seen as an alternative to already existing costly electronic compounds, providing superior performance as an environmentally sustainable materials [89]. Therefore, carbon-based nanostructures have proven to be used as effective sensors and offer high-quality sensing efficiency.

In technical instruments, various carbon-based systems discovered several years ago are now being detected and introduced. The use of carbon nanomaterials (CNMs) such as carbon nanotubes (CNTs), fullerenes, graphene, carbon nanoparticles (CNPs) and their derivatives is growing exponentially in many fields of study, and the range and scope of their technologies are continuously increasing [90–99]. The potential applications of CNM in many research fields are therefore a core topic for assessing and evaluating their application during the manufacture of sensors [100–102]. Only atoms of sp^2 carbon that make up a complete hexagonal series of atoms compose this CNM.

Fullerenes are nanomaterials that have, owing to their composition and special properties, gathered pace in recent years. Fullerenes are derivatives of spherical carbon nanomaterials. This structure is composed of a closed shape of pentagonal and hexagonal carbon structures fused together, unlike other carbon structures. The fullerene nanomaterial's spherical structure gives this nanomaterial a wide surface area. For biosensors and sensor systems, this characteristic is a sought after attribute. The most beneficial aspect of the carbon structure's double bonds is that they can be changed as they can adapt to chemical reactions easily. Any of the chemical reactions that occur with nucleophilic attacks on the fullerene sphere will form active sides. Fullerens may be chemically modified with certain modifications. For biomolecule immobilisation or surface modification, chemical modification is necessary.

In a typical sp^2-bonded atomic-scale honeycomb (hexagonal) shape, carbon atoms are closely arranged in graphene and this shape is a solid base for other carbon-bonded sp^2 materials (allotropes) such as carbon nanotubes and fullerenes. The carbon nanotubes are cylindrical structures made of rolled-up graphene sheets and thus have a technically distinct structure. These structures can be broken into fragments such as single or multiple wells. Single well nanotubes were first reported in 1993 [76] as single-wall carbon nanotubes (SWCNTs), while multi-wall carbon nanotubes (MWCNTs) are those with more than one well and were first investigated by Iijima in 1991 [68].

Carbon can be applied to structures with entirely different characteristics in different ways. The sp2 carbon hybridisation creates a layered structure with a weak out-of-plan binding of van der Waals shape and solid boundary in-plane. During normal interlayer spacing some to some decades of concentration cylinders can be placed around the typical central void and MWCNTs are created. A difference (0.34 to 0.39 nm) in interlayer spacing was seen in the real-space analysis in multiwall nanotube materials. The inner diameter of the MWCNTs ranges from 2 to 20 nm depending on the number of layers and from 0.4 nm to a few nanometers. Normally, all MWCNT ends are trapped and half-fullerenic (pentagonal defects) dome-shaped clusters at ends are capped, ranging in axial scale from 1μm to a few centimetres. Half-fullerenous molécules have a feature to help close the tube at each end (pentagonal ring defect). SWCNT measurements, on the other hand, range from 0.4 to 2 to 3 nm and are typically volumetric micrometres. SWCNTs may usually come together and form bundles (ropes). To build a crystal-like structure [103] SWCNTs are hexagonally arranged in a packet system. Graphene is a single layer of carbon atoms and is closely wrapped in a hexagonal wave-pan. It is a carbon allotrop with a sp2 bonding atom plane with a molecular bond length of 0.142 nanometres. Graphite layers, each layered on top of another, have an interplanar spacing of 0.335 nanometers, form graphite. The powers ofVan der Waals, which can be solved by exfoliating graphite graphite, hold the different graphite levels of graphite together.

21.4 Synthetic methods of carbon nanomaterials in sensing

It is important to synthesise CNMs of specific structures with good synthesis efficiency in order to achieve electrochemical sensor applications. However, it is also appropriate to change the surface by chemical functionalisation to resolve the reduced solubility and compatibility/inertness/bonding of CNMs with other materials. The sensor of operation of CNMs requires a satisfactory composition and surface morphology.

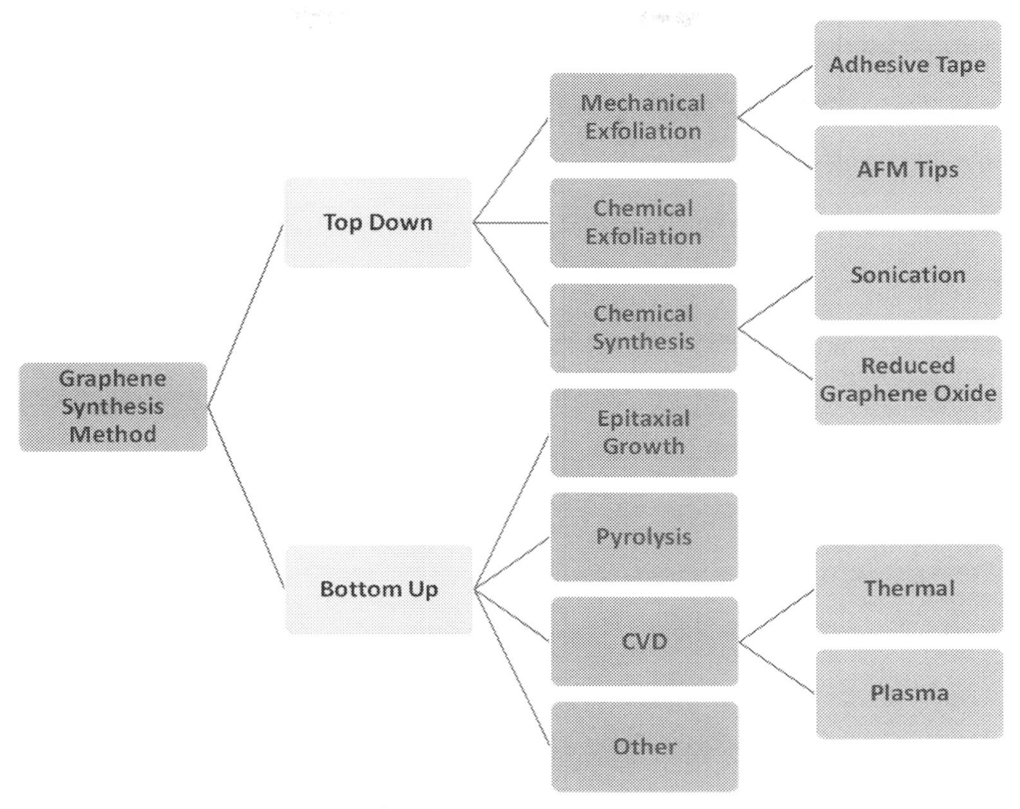

Fig. 21.2 *Different synthetic routes of graphene.*

21.4.1 Synthesis of graphene

Different methods for graphene synthesis have been developed in recent years (Fig. 21.2). The most commonly used methods today, however, are chemical and mechanical cleavage (exfoliation), chemical synthesis such chemical vapor deposition (CVD).

21.4.1.1 Mechanical exfoliation

The most unique and remarkable method for the extraction of single-layer graphene flakes on acceptable surfaces could be mechanical exfoliation. This is the first known graphene synthesis method. In nanotechnology, this is a top-down approach that induces vertical or horizontal stress on the substrate of layered structural materials. When monoatomic graphene layers are packed together by weak van der Waals forces, graphite is formed. Interparticle range and interlayer bond energy is 3.34Å and 2 eV/nm^2, respectively. To extract one mono-atomic layer from graphite, an external force of 300 nN/μm2 is required for mechanical glueing [104]. In practice, mechanical exfoliation or removal of sheets of graphite materials, including strongly arranged pyrolytic graphite (HOPG), natural graphite or single-crystal graphite, may achieve graphene sheets of varying thicknesses [105–109]. This exfoliation or removing can be carried out by using various types of materials such as electric field [110], scotch tape [111], ultrasonication [112] as well as transfer printing techniques [113,114]. In some experiments, HOPG has also been glued to the substrate either through epoxy resin as a standard adhesives [112,115] or SAMs [116] enhancing the yield of single and small graphene sheets.

Graphene is isolated from the graphite particle by adhesive tape in this micromechanical exfoliation process. The multiple-layer graphene exists on the film as the graphite has been extracted. By repetitive peeling, the multilayer graphene sheets are transformed into multiple flakes of a few-layers. Afterward, acetone substrate was used attaching the film facilitating separation. The resulting flakes differ significantly in thickness, and size ranging from nanometres to several tens of micrometres for substrate-based single-layer graphene. With only 2% absorption rate, single-layer graphene can be detected on SiO_2/Si under a light microscope, leading to interference effects [117].

21.4.1.2 Chemical exfoliation

Chemical approach for graphene synthesis is among the preferred technique creating a colloidal suspension that modifies graphene from graphite and graphite intercalation compounds. Various forms of polymer composites [118], transparent conductive electrodes [119], paper like material [120], energy storage materials [121] were utilized in the synthesis of graphene. Chemical exfoliation exhibit two-stage in the process; first begin with to maximise the interlayer spacing that decreases the interlayer van der Waals forces synthesizing intercalated graphene compounds (GICs) [122]. Then second, where graphene sheets are exfoliated into single layers through subsequent sonication or quick heating. Graphene oxide (GO) was synthesized by using Hummers method, utilizing very strong oxidants like $KMnO_4$, $NaNO_3$ in H_3PO_4 or H_2SO_4 for oxidation of graphite [123,124]. $NaBH_4$ was used as reducing agent to reduce graphene oxide. In recent years, many researchers used glucose [125], alkaline solutions [126], phenyl hydrazine [127], pyrrole [128], ascorbic acid [129], hydroxylamine [130], hydroquinone [131].

21.4.1.3 Pyrolysis of graphene

In the bottom-up process, the Solvothermal approach is also used as a graphene chemical synthesis. In the closed tank, the molar ratio of ethanol and sodium in this thermal reaction was 1:1. Sonication can be used to quickly remove the graphene sheets by pyrolating sodium ethoxide. Graphene sheets were made with dimensions of up to 10 μm. The spectroscopy of SAED, TEM and Raman was unabashed by the crystalline composition, the different layers, the graphic nature, the band structure [132]. A broad D-band, G-band, and IG/ID intensity ratio of about 1.16 is seen by Raman spectroscopy of the resulting layer, indicative of deficient graphene. Low-cost and easy-to-manufacture high-purity, low-temperature practical graphene were the benefits of the process. However, since it contained a large amount of flaws, the graphene quality was also not appropriate.

21.4.1.4 Chemical vapor deposition

Chemical vapor deposition involves the extraction and conversion of elements of precursor to a gaseous state. In CVD, the precursors are thermally disintegrated and diffused on substrates at high temperatures. A thin film of precursors (crystalline, solid, liquid, or gaseous) is deposited on the surface of the substrates. Different heavy metal substances such as Ruthenium [133], Copper [134], Nickel [135], Iridium [136] and Palladium [132] were used in the CVD process for the deposition of graphene. Graphite layer formation on Platinum by thermal chemical vapor deposition method was disclosed by Lang et al. [137]. In 2006, the first experiment of synthesis of graphene on nickel foil was performed through CVD using camphor as a starting material [138].

In a vacuum chamber that applies thin flam on the substrate surface, plasma-enhanced chemical vapor deposition (PECVD) creates plasma. The chemical reaction between gas molecules is involved. RF (AC frequency), microwave, and inductive coupling (electrical currents produced by electromagnetic induction) are used in the PECVD process. It is more practical for large-scale manufacturing applications as well as catalytic graphene-free production at comparatively low temperatures [139]. Within 5–40 minutes, plasma was processed. Complementary semiconductor metal-oxide (CMOS) systems require temperature minimization. The decreased temperature of PECVD has been widely used in the production of nanotubes and amorphous carbon during deposition [140–145].

21.4.1.5 Epitaxial growth of graphene

Epitaxial thermal growth is among the most known methods for the synthesis of graphene on a single crystalline silicon carbide (SiC) chip. The word "epitaxy" is a Greek term, and the prefix epi means "upon" or "over" and the taxis relates to "arrangement" or "order." When an epitaxial film is formed by single crystalline film deposition on a single crystalline substrate, the method known as epitaxial growth. There are two different surface-dependent epitaxial growth mechanisms, homo and hetero-epitaxial growth. If the deposited film on a surface is of the similar kind, it is homo-epitaxial, whereas when the film and substrate are different materials, it is known as hetero-epitaxial layer. For the electrical measurement of patterned epitaxial graphene, SiC was first used for the electrical measurement of patterned epitaxial graphene.

21.4.1.6 Other methods

Subrahmanyam et al. synthesized graphene by electron beam irradiation of PMMA nanofibers [146]. Arc discharge of graphite was also used for the synthesis of graphene [147]. Other methods such as thermal fusion of PAHs [148] and conversion of nano diamond to graphene have also been carried out [149].

21.4.2 Synthesis of carbon nanotubes

The CNTs output theory is clear. A combination of carbon feedstock, metal catalyst, and heat are used in all known processing techniques. Both SWNT manufacturing

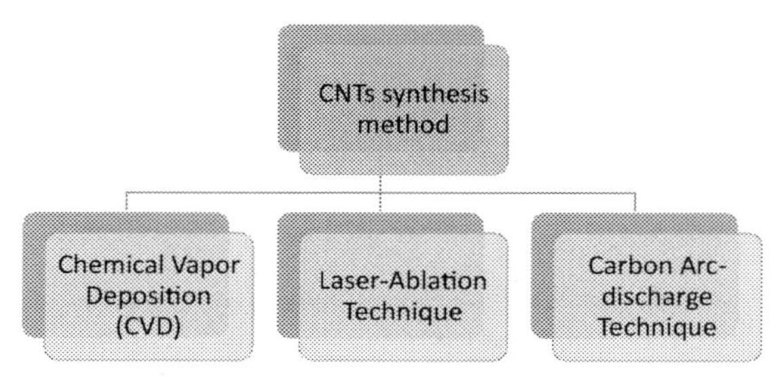

Fig. 21.3 ***Various synthetic methods of CNTs.***

techniques enable a metal catalyst, although arc discharge does not need any metal catalyst for manufacturing MWNTs. The processes of CNT production as summarized in Fig. 21.3 have been described in many books. Therefore, we concentrated on the characteristics of MWNTs and SWNTs obtained by different methods.

21.4.2.1 Chemical vapor deposition

Since the movement of cyanogen over red-hot porcelain in 1890 produced filamentous carbon, the decomposition of hydrocarbons using catalysis has been known to synthesize carbon fibres. Endo et al. [150] designed the floating catalyst reactor using catalyst particles of 10 nm diameter in the 1980s. This approach is a precursor to today's widely used aerosol-based CNT processing [151–157], whereas, transition metal catalyst such as Fe, Ni and Co facilitates the hydrocarbon pyrolysis process that produces SWNTs and MWNTs. MWNTs are created primarily in inert environment at low temperatures (300-800°C), whereas SWNTs need a mixture of an inert gas (Ar) and H2 and require higher temperatures (600-1150°C) [158–160]. A more complex catalyst preparation process is involved in producing DWNT using CVD techniques. In addition, DWNT samples also contain SWNTs and triple-walled CNTs [161].

21.4.2.2 Laser-ablation technique

There is a not much difference between the laser-ablation and arc discharge system. During both the methods, vaporisation of graphite targets produced the carbon atoms upon subsequent condensation. SWNTs are often produced when a laser vaporises catalyst-containing graphite targets, such as Ni, Co, and Pt [162–164]. A quartz tube containing graphite target is surrounded by a furnace at very high temperature (~1200°C). A continuous flow of gas such as Ar or He is maintained through the tube for passing the soot generated into a water-cooled Cu collector. The SWNTs are condensed as ropes or bundles containing individual SWNTs. There are also byproducts of encapsulated metal catalysts, such as amorphous carbon or pieces. The technique specifically favors the growth of SWNTs as there is no formation of MWNTs is observed. The material quality, weight, diameter, and distribution of chirality are believed to be equivalent to those of SWNTs developed by arc discharge.

21.4.2.3 Carbon arc-discharge technique

Ebbesen and Ajayan et al. demonstrated in 1992 that MWNTs can be formed by discharging a carbon arc. Two graphite electrodes are used to discharge the carbon arc, from which a direct current is emitted into an inert environment. The anode is swallowed and the cathode forms a cigar-like nanostructures. With a black soft inner core of MWNTs, polyhedral crystals, and amorphous carbon, the exterior shell of this deposit is gray and tough [165]. SWNTs can also be obtained, although they involve mixed metal catalysts that are incorporated into the anode, such as Fe: Co, Ni: Y [166]. SWNTs are found dispersed in the region as a smooth web-like substance after curving. Ando et al. demonstrated that for increasing macroscopic SWNT nets up to 20-30 cm in length, the arc evaporation of a graphite rod with a pure Fe catalyst in combination with a hydrogen and inert gas mixture can be used. Replacing He with H_2 results in a very narrow innermost tube of < 0.4 nm in MWNTs. By inserting B into the anode, the range of the MWNTs can be enhanced. The shape of zigzag MWNTs also tends to support Boron [167,168]. Loiseau et al. [169–172] and others showed that MWNTs were loaded with the addition of metals or their carbide yields. Arc-discharge driven MWNTs are usually 20 μm long with diameter of about 10 nm, as a rule of thumb. The wall count is reduced to 20-30. This process synthesizes the highly crystalline MWNTs with less defects in comparison to the other synthetic protocols. SWNTs created by the same technique appear in bundles [166]. Their length is about 1-2 nm. Because of the agglomeration of the SWNT packets, it remains hard to reliably calculate the SWNT length. A number of different chiralities are displayed by the SWNTs in the bundles. MWNT and SWNT arc-discharge samples usually contain large quantities of by-products such as polyhedral and amorphous carbon. Encapsulated metal catalyst particles are also found in samples of SWNTs.

21.4.3 Synthesis of fullerene

Unwin et al. used laser vaporization of carbon for the synthesis of fullerene, but very less quantity of fullerene was produce by using this method [173]. Fig. 21.4 summarizes the fullerene synthesis techniques.

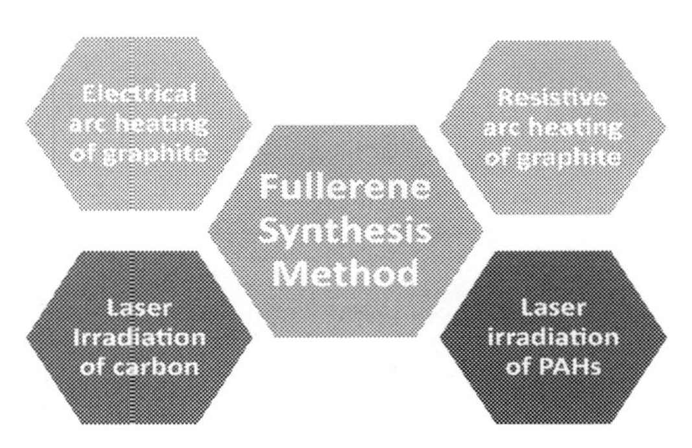

Fig. 21.4 ***Various synthetic routes of fullerene.***

21.4.3.1 Laser irradiation of carbon

In this process, fullerenes are created by a pulsed laser in a supersonic expansion nozzle that focuses within an inert atmosphere (helium) on a graphite target. This approach involves vaporising carbon from a revolving solid graphite disc into a high-density helium flow using a guided pulsed laser [174].

21.4.3.2 Electrical arc heating of graphite

Kratchmer et al. was used first time electrical arc heating of graphite for the synthesis of fullerene in 1900. The system includes producing an electric arc in an inert environment between graphite rods that creates a soft condensate (soot). Toluene extractible fullerenes form a portion of this fluffy condensate. The fullerenes in the soot are separated into a small volume of toluene by solvation [174,175]. A rotary evaporator was used to remove the toluene after extraction and a solid mixture consisting mainly of C_{60} with a limited number of higher fullerenes is exposed to a liquid chromatography system to achieve pure C_{60} [176–178].

21.4.3.3 Resistive arc heating of graphite

Heating of the heat resistive carbon rods causes the emission of a pale gray white cloud of dust, soot-like material composed of fullerenes. The product formed is collected from glass covers shielding the carbon rods. This process results into the evaporation heat resistive carbon rods upon heating in a small atmosphere filled with helium [179,180].

21.4.3.4 Laser irradiation of polycyclic aromatic hydrocarbons

As a means of introducing new homologues of the fullerenes class, controlled development of fullerenes was developed, which cannot be accomplished in good quantities by an unregulated graphite evaporation process. This synthetic protocol for fullerene focuses on polycyclic aromatic hydrocarbons (PAHs) that already have the carbon frameworks required.

21.5 Current applications

The major allotropic sources of carbon used in sensing device are graphene, graphene oxide, CNTs, and fullerene. The promising structure of carbon nanostructures facilitates the detection and quantification of various gases and their use in sensing applications have increased in recent time (Table 21.2). The features like high surface to volume ratio, hollow nanostructures are perfect for the adsorption and desorption of gas molecules. In addition, gas selectivity is improved by the prospect of functionalization. Carbon nanostructures also meets the requirement of effective gas sensor, such as high sensitivity and selectivity, rapid reaction and resonance recovery times and stability. Generally, variance in applied voltages and current in respect of the number of gas molecules adsorbed on

Table 21.2 Carbon-based nanosensor applications.

S. No.	Carbon type	Sensor	Application	References
1	CNTs	Electrochemical	Ions, metabolites, and protein biomarkers	[206]
2	MWCNTs	Cholesterol	Cholesterol in blood	[207]
3	f-CNTs	Electrochemical	Nitric oxide	[208]
4	f-MWCNTs	Electrochemical	Epinephrine	[209]
5	f-MWCNTs	Electronic	Rat striatum	[210]
6	SWCNTs	Electronic	Cellular nitric oxide	[211]
7	SWCNTs	Electronic	Epidermal Carcinoma cells	[212]
8	Pt/CNTs	Amperometric biosensor	Cysteine	[213]
9	CNT arrays	Amperometric biosensor	Biomarker	[214]
10	MWCNTs-bipolar electrode	Amperometric	Prostate-specific antigen	[215]
11	MWCNTs-P450	Amperometric	Breast cancer	[216]
12	D-(þ) galactose-SWCNTs	Electrochemical	Cancer marker galactin-3	[217]
13	SWCNTs/MWCNTs	Amperometric	Prostate cancer, PSA detection in blood samples	[218]
14	Au-Ag/MWCNTs	Amperometric	MGC-803 gastric cancer cells	[219]

the surface structure is the operating theory of the devices. During the interaction of molecules and sensors an electrical signal is produced. The detection and quantification of polluting gases, especially hydrogen sulfide (H_2S), carbon dioxide (CO_2), carbon monoxide (CO), methane (CH_4), nitrogen oxide (NO_x), and ammonia (NH_3) are among the major applications of carbonaceous materials as gas sensors.

A variety of electrochemical sensor platforms for medically and pharmaceutically essential compounds such as glucose, methynlongyoxal, nicostinamide adenine dinucleotide, acetaminophen and valacyclovir have been developed for the identification and quantification of. Recent advances have also opened the way for the synthesis of many new materials with desired moral characteristics and special physicochemical features in nanotechnology and materiel sciences [181].

The use of SWCNTs as sensing medium significantly reduces the ability of the carbon nanomaterial that is promising and various biological compounds are observed. In 0.1 M PBS (pH 7.4), Chen and colleagues demonstrated a SWCNT-based methylglyoxal sensors network, demonstrating that a sensor device was extremely precise and sensitive to methylglyoxal plasma levels for healthy individuals and patients with diabetes in a very highly sensitive way. Graphene is significantly younger than fullerenes in the family of carbon materials. Graphene is an atomically thin layer composed of two-dimension hybridised, hexagonally arranged carbon atoms [182]. Graphene is the perfect medium for anchoring metal nanoparticles for electrochemical sensing applications [183] due to the combination of attrible elements such as high surface areas, improved mobility of charging carriers and high stability. Chen and co-workers recently manufactured graphene and Au nanoparticle-rGO electrochemical sensors for sensitive detection of NADH and acetaminophen [184,185] using an electrochemical method. Buckypaper is a thin film (5–25 μm) constructed from lateral organised nanotubes networks, which offer a promising medium for manufacturing composites that reinforce highly condensed and aligned nanotubes. Macroscopic assemblies of carbohydrates, including carbohydrates, have been given significant attention recently to harness the characteristic properties of single carbohydrates on a macroscopical scale [186].

CNTs and graphs have an excellent conductivity that is used for the development of FET sensors for the identification of heavy metal ions. Kim et al. have developed the Hg^{2+} ion detecting nanosensors for a selective, sensitive, and fast single wall carbon nanotube (SWCNT), based on anomaly in the SWCNT conductance. The detection limit of the nanosensor for Hg^{2+} was 10 nM. In water, and with different metal ions (e.g. Cu^{2+}, Mg^{2+}, Pb^{2+}, Co^{2+}) selectivity to Hg^{2+} was provided.

The electron transfer between electrodes and electroactive species can be facilitated by CNTs and graphene. Many sensors for environmental detection of pollution based on CNT and graphene characteristics were created. The MWNTs have not only encouraged the transition, but also increased enzyme activity of AChE, in the manufacture of acetycholinesterase (AChE) dependent electrochemical sensors [187]. This led a rapid reaction and high sensitivity to the identification of a variety of pesticides by the MWNTs-based electrochemical sensor. 0.04 ppb dichlovrvos, and 0.1 ppb aqueous solution can be accurately detected by the electrode. In addition, the sensors were shown to be robust, reproducible and selective enough for actual samples to be detected. A gold-base organophosphate nanosheet nanoparticles (AuNP/cr-Gs) are co-designed by Li and Lin groups.

The Group of Mirkin has pioneered an onsite, real time qualitative, or semi-quantitative sensors dependent on color of AuNPs or AgNPs from red to blue, without complicating analytical instruments [188–190]. For eg, Liu et al. implemented Pb^{2+}-specific DNA-zymes as target recognition and DNA-based nanoparticles as signalling components, they engineered the Pb^{2+} colorimetric biometer, Lu et al. developed a lot of colorimetric sensors for ambient monitoring, such as heavy metal ion (Pb^{2+}, Hg^{2+}), radioactively detected metals (UO22 +). The substrate strand is closed in the presence of Pb^{2+} by the enzyme in the cooling phase that preserves reforming aggregates of nanoparticles and results in a red color. Thus Pb^{2+} can be detected and quantified by monitoring the sensor hue [191].

Colorimetric interface sensors can also be used for AgNPs. Chenet al. β-cyclodextrin synths (CD)-functional silver nanoparticles probes colorimetrically detect multiple isomers of aromatic compounds for the extremely stable, quantitative, sensitive, and naked eyes [192].

The specific mechanical qualities of CNTs allow CNTs to produce tissue engineering nanocomposites in reinforcement fillers [193–195]. Despite regulated osteogenic differentiation of human mesenchymal stems, CNTs induce differentiation of stem cells into osteogenic cell lines [196–200]. Mouse MSC-regulated proliferation is facilitated with a cell (hMSCs) that uses SWNT-structures without chemical-biological treatment [201]. Functional CNTs serve as ideal bone differentiation scaffolds [202]. Several studies have investigated the biocompatibility of CNTs, with findings that can also insert voluminous CNTs as possible products [203,204].

MWNTs have accelerated cell distinction relative to standard tissue culture plastic, particularly though external biochemical inductors are absent. When the tasks were equivalent to osteoblast and osteoclast [205].

21.6 Conclusions

Emerging technologies for carbon-based sensors have been addressed, with specific focus on bionanosystems, biochemical assay detectors and optical devices that focus on CNTs. The entire study reveals how CNTs and graphene sheets are especially suitable materials to many sensor applications, coupled with optical, radiofrequency and surface biofunctioning.

Acknowledgments

DP acknowledges the University Grant Commission, New Delhi, India for providing Senior Research Fellowship. The article is checked for plagiarism using the iThenticate software and recorded in the Knowledge Resource Center, CSIR-NEERI, Nagpur for anti-plagiarism (KRC No.: CSIR-NEERI/KRC/2020/NOV/WWTD/3).

References

[1] V.L. Colvin, The potential environmental impact of engineered nanomaterials, Nat. Biotechnol. 21 (10) (2003) 1166–1170.
[2] C.R. Martin, Nanomaterials: a membrane-based synthetic approach, Science 266 (5193) (1994) 1961–1966.
[3] C.R. Martin, Membrane-based synthesis of nanomaterials, Chem. Mater. 8 (8) (1996) 1739–1746.
[4] A. Nel, T. Xia, L. Mädler, N. Li, Toxic potential of materials at the nanolevel, Science 311 (5761) (2006) 622–627.
[5] G. Oberdörster, E. Oberdörster, J. Oberdörster, Nanotoxicology: an emerging discipline evolving from studies of ultrafine particles, Environ. Health Perspect 113 (2005) 823–839.
[6] V.V. Pokropivny, V.V. Skorokhod, New dimensionality classifications of nanostructures, Physica E: Low-dimensional Systems and nanostructures, 40, 2008 2521–2525.
[7] W.C.W. Chan, D.J. Maxwell, X. Gao, R.E. Bailey, M. Han, S. Nie, Luminescent quantum dots for multiplexed biological detection and imaging, Curr. Opin. Biotechnol. 13 (1) (2002) 40–46.
[8] M. Pashchanka, R.C. Hoffmann, A. Gurlo, J.J. Schneider, Molecular based, chimie douce approach to 0d and 1d indium oxide nanostructures. Evaluation of their sensing properties towards co and h 2, J. Mater. Chem. 20 (38) (2010) 8311–8319.

[9] E.Y.i. Sun, L. Josephson, R. Weissleder, "Clickable" nanoparticles for targeted imaging, Mol. Imaging 5 (2) (2006) 7290.2006. 00013.
[10] HW. Hillhouse, Development of double-gyroid nanowire arrays for photovoltaics, Proceedings of the 2011 AIChE annual meeting, 2011.
[11] Y.S. Zhao, H. Fu, A. Peng, Y. Ma, Q. Liao, J. Yao, Construction and optoelectronic properties of organic one-dimensional nanostructures, Acc. Chem. Res. 43 (3) (2010) 409–418.
[12] N. Chopra, V.G. Gavalas, L.G. Bachas, B.J. Hinds, Functional one-dimensional nanomaterials: applications in nanoscale biosensors, Anal. Lett. 40 (11) (2007) 2067–2096.
[13] Y. Wang, Z. Li, J. Wang, J. Li, Y. Lin, Graphene and graphene oxide: biofunctionalization and applications in biotechnology, Trends Biotechnol. 29 (5) (2011) 205–212.
[14] D.K. Yi, J.-H. Lee, J.A. Rogers, U. Paik, Two-dimensional nanohybridization of gold nanorods and polystyrene colloids, Appl. Phys. Lett. 94 (8) (2009) 084104.
[15] J.S. Chen, L.A. Archer, X.W.D. Lou, SnO2 hollow structures and TiO2 nanosheets for lithium-ion batteries, J. Mater. Chem. 21 (27) (2011) 9912–9924.
[16] A. Ciesielski, C.-A. Palma, M. Bonini, P. Samorì, Towards supramolecular engineering of functional nanomaterials: pre-programming multi-component 2D self-assembly at solid-liquid interfaces, Adv. Mater. 22 (32) (2010) 3506–3520.
[17] J. Lu, D. Yuan, J. Liu, W. Leng, T.E. Kopley, Three dimensional single-walled carbon nanotubes, Nano Lett. 8 (10) (2008) 3325–3329.
[18] H.S. Song, W.J. Zhang, C. Cheng, Y.B. Tang, L.B. Luo, X. Chen, C.Y. Luan, X.M. Meng, J.A. Zapien, N. Wang, Controllable fabrication of three-dimensional radial ZnO nanowire/silicon microrod hybrid architectures, Cryst. Growth Des. 11 (1) (2011) 147–153.
[19] G. Binnig, C.F. Quate, C.h. Gerber, Atomic force microscope, Phys. Rev. Lett. 56 (9) (1986) 930.
[20] E.H. Synge, XXXVIII. A suggested method for extending microscopic resolution into the ultra-microscopic region, London, Edinburgh Dublin Philos. Mag. J. Sci. 6 (35) (1928) 356–362.
[21] E. Betzig, J.K. Trautman, T.D. Harris, J.S. Weiner, R.L. Kostelak, Breaking the diffraction barrier: optical microscopy on a nanometric scale, Science 251 (5000) (1991) 1468–1470.
[22] W. Zhang, S. Zhu, R. Luque, S. Han, L. Hu, G. Xu, Recent development of carbon electrode materials and their bioanalytical and environmental applications, Chem. Soc. Rev. 45 (3) (2016) 715–752.
[23] P. Ramnani, N.M. Saucedo, A. Mulchandani, Carbon nanomaterial-based electrochemical biosensors for label-free sensing of environmental pollutants, Chemosphere 143 (2016) 85–98.
[24] B. Li, Y. Wang, Facile synthesis and enhanced photocatalytic performance of flower-like ZnO hierarchical microstructures, J. Phys. Chem. C 114 (2) (2010) 890–896.
[25] J.-H. Lee, Gas sensors using hierarchical and hollow oxide nanostructures: overview, Sens. Actuators B 140 (1) (2009) 319–336.
[26] Ü. Özgür, Y.a.I. Alivov, C. Liu, A. Teke, M.A.n. Reshchikov, S. Doğan, V. Avrutin, S.-J. ChoMorkoç, A comprehensive review of ZnO materials and devices, J. Appl. Phys. 98 (4) (2005) 11.
[27] S. Dutta, B.N. Ganguly, Characterization of ZnO nanoparticles grown in presence of Folic acid template, J. Nanobiotechnol. 10 (1) (2012) 29.
[28] A.-.M. Gurban, D. Burtan, L. Rotariu, C. Bala, Manganese oxide based screen-printed sensor for xenoestrogens detection, Sens. Actuators B 210 (2015) 273–280.
[29] J.G. Lu, P. Chang, Z. Fan, Quasi-one-dimensional metal oxide materials—synthesis, properties and applications, Mater. Sci. Eng. R: Rep. 52 (1-3) (2006) 49–91.
[30] A. Kolmakov, M. Moskovits, Chemical sensing and catalysis by one-dimensional metal-oxide nanostructures, Annu. Rev. Mater. Res. 34 (2004) 151–180.
[31] I.n.-S. Hwang, S.-J. Kim, J.-K.i. Choi, J. Choi, H. Ji, G.-.T. Kim, G. Cao, J.-.H. Lee, Synthesis and gas sensing characteristics of highly crystalline ZnO–SnO2 core–shell nanowires, Sens. Actuators B 148 (2) (2010) 595–600.
[32] X. Song, Z. Wang, Y. Liu, C.e. Wang, L. Li, A highly sensitive ethanol sensor based on mesoporous ZnO–SnO2 nanofibers, Nanotechnology 20 (7) (2009) 075501.
[33] S. Sharma, P. Chauhan, LPG gas sensing applications of SnO2/ZnO nanoparticles, Adv. Sci. Lett. 20 (5-6) (2014) 1198–1203.
[34] T. Arai, The study of the optical properties of conducting tin oxide films and their interpretation in terms of a tentative band scheme, J. Phys. Soc. Japan 15 (5) (1960) 916–927.

[35] X. Liu, J. Zhang, L. Wang, T. Yang, X. Guo, S. Wu, S. Wang, 3D hierarchically porous ZnO structures and their functionalization by Au nanoparticles for gas sensors, J. Mater. Chem. 21 (2) (2011) 349–356.
[36] K. Suri, S. Annapoorni, A.K. Sarkar, R.P. Tandon, Gas and humidity sensors based on iron oxide–polypyrrole nanocomposites, Sens. Actuators B 81 (2-3) (2002) 277–282.
[37] Y.X. Zhang, G.H. Li, Y.X. Jin, Y. Zhang, J. Zhang, L.D. Zhang, Hydrothermal synthesis and photoluminescence of TiO2 nanowires, Chem. Phys. Lett. 365 (3-4) (2002) 300–304.
[38] S.-.H. Keshmiri, M. Rezaee Rokn-Abadi, Enhancement of drift mobility of zinc oxide transparent-conducting films by a hydrogenation process, Thin Solid Films 382 (1-2) (2001) 230–234.
[39] V.R. Shinde, T.P. Gujar, C.D. Lokhande, R.S. Mane, S.-.H. Han, Use of chemically synthesized ZnO thin film as a liquefied petroleum gas sensor, Mater. Sci. Eng.: B 137 (1-3) (2007) 119–125.
[40] K.e. Yu, Y. Zhang, R. Xu, D. Jiang, L. Luo, Q. Li, Z. Zhu, W. Lu, Field emission behavior of cuboid zinc oxide nanorods on zinc-filled porous silicon, Solid State Commun. 133 (1) (2005) 43–47.
[41] R. Abdel-Karim, Y. Reda, A. Abdel-Fattah, Nanostructured materials-based nanosensors, J. Electrochem. Soc. 167 (3) (2020) 037554.
[42] V.L.P. Kumawat. Kalyani, Recent advancement in nanosensors with special reference to biomedical applications, J. Manag. Eng. Inf. Technol. 7 (3) (2020), https://doi.org/10.5281/zenodo.3926187.
[43] A.W. Czarnik, Fluorescent Chemosensors for Ion and Molecule Recognition, ACS Symposium Series, American Chemical Society, Washington, DC, 1992.
[44] S. Kulmala, J. Suomi, Current status of modern analytical luminescence methods, Anal. Chim. Acta 500 (1-2) (2003) 21–69.
[45] J.R. Lakowicz, Fluorophores. In: Principles of Fluorescence Spectroscopy, Springer, 1999, pp. 63–93.
[46] R.P. Haugland, Handbook of fluorescent probes and research chemicals, Molecular Probes, 8, Eugene, 1996.
[47] K. Sasaki, Z.-.Y. Shi, R. Kopelman, H. Masuhara, Three-dimensional pH microprobing with an optically-manipulated fluorescent particle, Chem. Lett. 25 (2) (1996) 141–142.
[48] J. Ji, N. Rosenzweig, I. Jones, Z. Rosenzweig, Molecular oxygen-sensitive fluorescent lipobeads for intracellular oxygen measurements in murine macrophages, Anal. Chem. 73 (15) (2001) 3521–3527.
[49] J. Ji, N. Rosenzweig, C. Griffin, Z. Rosenzweig, Synthesis and application of submicrometer fluorescence sensing particles for lysosomal pH measurements in murine macrophages, Anal. Chem. 72 (15) (2000) 3497–3503.
[50] Y.-.P. Kim, W.L. Daniel, Z. Xia, H. Xie, C.A. Mirkin, J. Rao, Bioluminescent nanosensors for protease detection based upon gold nanoparticle–luciferase conjugates, Chem. Commun. 46 (1) (2010) 76–78.
[51] M. Fehr, W.B. Frommer, S. Lalonde, Visualization of maltose uptake in living yeast cells by fluorescent nanosensors, Proc. Natl. Acad. Sci. 99 (15) (2002) 9846–9851.
[52] W. Tan, Z.-.Y. Shi, S. Smith, D. Birnbaum, R. Kopelman, Submicrometer intracellular chemical optical fiber sensors, Science 258 (5083) (1992) 778–781.
[53] T. Vo-Dinh, B. Cullum, Biosensors and biochips: advances in biological and medical diagnostics, Fresenius. J. Anal. Chem. 366 (6-7) (2000) 540–551.
[54] B.M. Cullum, G.D. Griffin, G.H. Miller, T. Vo-Dinh, Intracellular measurements in mammary carcinoma cells using fiber-optic nanosensors, Anal. Biochem. 277 (2000) 25–32, doi: https://doi.org/10.1006/abio.1999.4341.
[55] S.M. Mousavi, S.A. Hashemi, M. Zarei, A.M. Amani, A. Babapoor, Nanosensors for chemical and biological and medical applications, Med Chem (Los Angeles) 8 (8) (2018) 2161–0444.1000515.
[56] S. Agrawal, R. Prajapati, Nanosensors and their pharmaceutical applications: a review, Int. J. Pharm. Sci. Technol. 4 (2012) 1528–1535.
[57] Y.A. Romaniuk, S. Golovynskyi, A.P. Litvinchuk, D. Dong, Y. Lin, O.I. Datsenko, M. Bosi, L. Seravalli, I.S. Babichuk, V.O. Yukhymchuk, B. Li, J. Qu, Influence of anharmonicity and interlayer interaction on Raman spectra in mono- and few-layer MoS_2: A computational study, Phys. E: Low-Dimens. Syst. Nanostructures. (2021) 114999, https://doi.org/10.1016/j.physe.2021.114999.
[58] G. Doria, J. Conde, B. Veigas, L. Giestas, C. Almeida, M. Assunção, J. Rosa, P.V. Baptista, Noble metal nanoparticles for biosensing applications, Sensors 12 (2) (2012) 1657–1687.
[59] J.R. Stetter, W.R. Penrose, S. Yao, Sensors, chemical sensors, electrochemical sensors, and ECS, J. Electrochem. Soc. 150 (2) (2003) S11.
[60] P. Rai, S. Oh, P. Shyamkumar, M. Ramasamy, R.E. Harbaugh, V.K. Varadan, Nano-bio-textile sensors with mobile wireless platform for wearable health monitoring of neurological and cardiovascular disorders, J. Electrochem. Soc. 161 (2) (2013) B3116.

[61] J. Das, I. Ivanov, L. Montermini, J. Rak, E.H. Sargent, S.O. Kelley, An electrochemical clamp assay for direct, rapid analysis of circulating nucleic acids in serum, Nat. Chem. 7 (7) (2015) 569.
[62] J. Das, K.B. Cederquist, A.A. Zaragoza, P.E. Lee, E.H. Sargent, S.O. Kelley, An ultrasensitive universal detector based on neutralizer displacement, Nat. Chem. 4 (8) (2012) 642–648.
[63] O. Amor-Gutiérrez, E. Costa Rama, A. Costa-García, M.T. Fernández-Abedul, Paper-based maskless enzymatic sensor for glucose determination combining ink and wire electrodes, Biosens. Bioelectron. 93 (2017) 40–45.
[64] L.i. Fu, Y. Zheng, A. Wang, Poly (diallyldimethylammonium chloride) functionalized reduced graphene oxide based electrochemical sensing platform for luteolin determination, Int. J. Electrochem. Sci 10 (2015) 3518–3529.
[65] L. Nyholm, Electrochemical techniques for lab-on-a-chip applications, Analyst 130 (5) (2005) 599–605.
[66] A. Rahi, K. Karimian, H. Heli, Nanostructured materials in electroanalysis of pharmaceuticals, Anal. Biochem. 497 (2016) 39–47.
[67] N. Yang, G.M. Swain, X. Jiang, Nanocarbon electrochemistry and electroanalysis: current status and future perspectives, Electroanalysis 28 (1) (2016) 27–34.
[68] S. Iijima, Helical microtubules of graphitic carbon, Nature 354 (6348) (1991) 56–58.
[69] M.F.L. De Volder, S.H. Tawfick, R.H. Baughman, A. John Hart, Carbon nanotubes: present and future commercial applications, Science 339 (6119) (2013) 535–539.
[70] X. Pan, Z. Fan, W. Chen, Y. Ding, H. Luo, X. Bao, Enhanced ethanol production inside carbon-nanotube reactors containing catalytic particles, Nat. Mater. 6 (7) (2007) 507–511.
[71] M. Prato, K. Kostarelos, A. Bianco, Functionalized carbon nanotubes in drug design and discovery, Acc. Chem. Res. 41 (1) (2008) 60–68, doi:https://doi.org/10.1021/ar700089b.
[72] M. Trojanowicz, Analytical applications of carbon nanotubes: a review, TrAC Trends Anal. Chem. 25 (5) (2006) 480–489.
[73] R. Dastjerdi, M. Montazer, A review on the application of inorganic nano-structured materials in the modification of textiles: focus on anti-microbial properties, Colloids Surf. B 79 (1) (2010) 5–18.
[74] D.S. Hecht, L. Hu, G. Irvin, Emerging transparent electrodes based on thin films of carbon nanotubes, graphene, and metallic nanostructures, Adv. Mater. 23 (13) (2011) 1482–1513.
[75] S. Iijima, Carbon nanotubes: past, present, and future, Physica B 323 (1-4) (2002) 1–5.
[76] S. Iijima, T. Ichihashi, Single-shell carbon nanotubes of 1-nm diameter, Nature 363 (6430) (1993) 603–605.
[77] A. Qureshi, W.P. Kang, J.L. Davidson, Y. Gurbuz, Review on carbon-derived, solid-state, micro and nano sensors for electrochemical sensing applications, Diam. Relat. Mater. 18 (12) (2009) 1401–1420.
[78] Q. Cao, S.- Han, G.S. Tulevski, Y.u. Zhu, D.D. Lu, W. Haensch, Arrays of single-walled carbon nanotubes with full surface coverage for high-performance electronics, Nat. Nanotechnol. 8 (3) (2013) 180–186.
[79] J.E. Fischer, A.T. Johnson, Electronic properties of carbon nanotubes, Curr. Opin. Solid State Mater. Sci. 1 (4) (1999) 28–33.
[80] D. Jariwala, V.K. Sangwan, L.J. Lauhon, T.J. Marks, M.C. Hersam, Carbon nanomaterials for electronics, optoelectronics, photovoltaics, and sensing, Chem. Soc. Rev. 42 (7) (2013) 2824–2860.
[81] U.N. Maiti, W.J. Lee, J.u.M. Lee, Y. Oh, J.u.Y. Kim, J.i.E. Kim, J. Shim, T.H. Han, S.O. Kim, 25th anniversary article: chemically modified/doped carbon nanotubes & graphene for optimized nanostructures & nanodevices, Adv. Mater. 26 (1) (2014) 40–67.
[82] R.J. Nemanich, J.A. Carlisle, A. Hirata, K. Haenen, CVD diamond—research, applications, and challenges, MRS Bull. 39 (6) (2014) 490–494.
[83] K. Bewilogua, D. Hofmann, History of diamond-like carbon films—from first experiments to worldwide applications, Surf. Coat. Technol. 242 (2014) 214–225.
[84] H. Niakan, Q. Yang, J.A. Szpunar, Structure and properties of diamond-like carbon thin films synthesized by biased target ion beam deposition, Surf. Coat. Technol. 223 (2013) 11–16.
[85] M. Panda, R. Krishnan, T.R. Ravindran, A. Das, G. Mangamma, S. Dash, A.K. Tyagi, Spectroscopic studies on diamond like carbon films synthesized by pulsed laser ablation, AIP Conf. Proc. (2016).
[86] J. Vetter, 60 years of DLC coatings: historical highlights and technical review of cathodic arc processes to synthesize various DLC types, and their evolution for industrial applications, Surf. Coat. Technol. 257 (2014) 213–240.
[87] S.M. Aghaei, M.M. Monshi, I. Torres, S.M.J. Zeidi, I. Calizo, DFT study of adsorption behavior of NO, CO, NO2, and NH3 molecules on graphene-like BC3: a search for highly sensitive molecular sensor, Appl. Surf. Sci. 427 (2018) 326–333.

[88] W.G. Kreyling, M. Semmler-Behnke, Q. Chaudhry, A complementary definition of nanomaterial, Nano Today 5 (3) (2010) 165–168.
[89] L.-.M. Peng, Z. Zhang, S. Wang, Carbon nanotube electronics: recent advances, Mater. Today 17 (9) (2014) 433–442.
[90] A.I. El-Seesy, H. Hassan, Investigation of the effect of adding graphene oxide, graphene nanoplatelet, and multiwalled carbon nanotube additives with n-butanol-Jatropha methyl ester on a diesel engine performance, Renew. Energy 132 (2019) 558–574.
[91] W. Zhang, J. Zhang, Y. Lu, Stimulation of carbon nanomaterials on syntrophic oxidation of butyrate in sediment enrichments and a defined coculture, Sci. Rep. 8 (1) (2018) 1–13.
[92] R.S. Hebbar, A.M. Isloor, A.M. Asiri, Carbon nanotube-and graphene-based advanced membrane materials for desalination, Environ. Chem. Lett. 15 (4) (2017) 643–671.
[93] M.A. Voronkova, S. Luanpitpong, L.W. Rojanasakul, V. Castranova, C. Zoica Dinu, H. Riedel, Y. Rojanasakul, SOX9 regulates cancer stem-like properties and metastatic potential of single-walled carbon nanotube-exposed cells, Sci. Rep. 7 (1) (2017) 1–13.
[94] P.A. Rasheed, N. Sandhyarani, Carbon nanostructures as immobilization platform for DNA: a review on current progress in electrochemical DNA sensors, Biosens. Bioelectron. 97 (2017) 226–237.
[95] E. Heydari-Bafrooei, S. Askari, Ultrasensitive aptasensing of lysozyme by exploiting the synergistic effect of gold nanoparticle-modified reduced graphene oxide and MWCNTs in a chitosan matrix, Microchim. Acta 184 (9) (2017) 3405–3413, doi:https://doi.org/10.1007/s00604-017-2356-3.
[96] E. Heydari-Bafrooei, S. Askari, Electrocatalytic activity of MWCNT supported Pd nanoparticles and MoS2 nanoflowers for hydrogen evolution from acidic media, Int. J. Hydrogen Energy 42 (5) (2017) 2961–2969.
[97] E. Heydari-Bafrooei, N.S. Shamszadeh, Synergetic effect of CoNPs and graphene as cocatalysts for enhanced electrocatalytic hydrogen evolution activity of MoS 2, RSC Adv. 6 (98) (2016) 95979–95986.
[98] E. Heydari-Bafrooei, N.S. Shamszadeh, Electrochemical bioassay development for ultrasensitive aptasensing of prostate specific antigen, Biosens. Bioelectron. 91 (2017) 284–292.
[99] E. Heydari-Bafrooei, M. Amini, M.H. Ardakani, An electrochemical aptasensor based on TiO2/ MWCNT and a novel synthesized Schiff base nanocomposite for the ultrasensitive detection of thrombin, Biosens. Bioelectron. 85 (2016) 828–836.
[100] E. Heydari-Bafrooei, M. Amini, S. Saeednia, Electrochemical detection of DNA damage induced by Bleomycin in the presence of metal ions, J. Electroanal. Chem. 803 (2017) 104–110.
[101] S. Dong, L. Peng, W. Wei, T. Huang, Three MOF-templated carbon nanocomposites for potential platforms of enzyme immobilization with improved electrochemical performance, ACS Appl. Mater. Interfaces 10 (17) (2018) 14665–14672.
[102] F. Zhao, J. Wu, Y. Ying, Y. She, J. Wang, J. Ping, Carbon nanomaterial-enabled pesticide biosensors: design strategy, biosensing mechanism, and practical application, TrAC Trends Anal. Chem. 106 (2018) 62–83.
[103] L. Chico, V.H. Crespi, L.X. Benedict, S.G. Louie, M.L. Cohen, Pure carbon nanoscale devices: nanotube heterojunctions, Phys. Rev. Lett. 76 (6) (1996) 971.
[104] Y. Zhang, J.P. Small, W.V. Pontius, P. Kim, Fabrication and electric-field-dependent transport measurements of mesoscopic graphite devices, Appl. Phys. Lett. 86 (7) (2005) 073104.
[105] H. Hiura, T.W. Ebbesen, J. Fujita, K. Tanigaki, T. Takada, Role of sp 3 defect structures in graphite and carbon nanotubes, Nature 367 (6459) (1994) 148–151.
[106] T.W. Ebbesen, H. Hiura, Graphene in 3-dimensions: Towards graphite origami, Adv. Mater. 7 (6) (1995) 582–586.
[107] X. Lu, M. Yu, H. Huang, R.S. Ruoff, Tailoring graphite with the goal of achieving single sheets, Nanotechnology 10 (3) (1999) 269.
[108] T.M. Bernhardt, B. Kaiser, K. Rademann, Formation of superperiodic patterns on highly oriented pyrolytic graphite by manipulation of nanosized graphite sheets with the STM tip, Surf. Sci. 408 (1-3) (1998) 86–94.
[109] H.-.V. Roy, C. Kallinger, B. Marsen, K. Sattler, Manipulation of graphitic sheets using a tunneling microscope, J. Appl. Phys. 83 (9) (1998) 4695–4699.
[110] X. Liang, A.S.P. Chang, Y. Zhang, B.D. Harteneck, H. Choo, D.L. Olynick, S. Cabrini, Electrostatic force assisted exfoliation of prepatterned few-layer graphenes into device sites, Nano Lett. 9 (1) (2009) 467–472, doi:https://doi.org/10.1021/nl803512z.

[111] K.S. Novoselov, A.K. Geim, S.V. Morozov, D. Jiang, Y._. Zhang, S.V. Dubonos, I.V. Grigorieva, A.A. Firsov, Electric field effect in atomically thin carbon films, Science 306 (5696) (2004) 666–669.

[112] L. Ci, L.i. Song, D. Jariwala, A.L. Elias, W. Gao, M. Terrones, P.M. Ajayan, Graphene shape control by multistage cutting and transfer, Adv. Mater. 21 (44) (2009) 4487–4491.

[113] X. Liang, Z. Fu, S.Y. Chou, Graphene transistors fabricated via transfer-printing in device active-areas on large wafer, Nano Lett. 7 (12) (2007) 3840–3844.

[114] J.-H. Chen, M. Ishigami, C. Jang, D.R. Hines, M.S. Fuhrer, E.D. Williams, Printed graphene circuits, Adv. Mater. 19 (21) (2007) 3623–3627.

[115] V. Huc, N. Bendiab, N. Rosman, T. Ebbesen, C. Delacour, V. Bouchiat, Large and flat graphene flakes produced by epoxy bonding and reverse exfoliation of highly oriented pyrolytic graphite, Nanotechnology 19 (45) (2008) 455601.

[116] Li-H. Liu, M. Yan, Simple method for the covalent immobilization of graphene, Nano Lett. 9 (9) (2009) 3375–3378.

[117] C. Casiraghi, A. Hartschuh, E. Lidorikis, H. Qian, H. Harutyunyan, T. Gokus, K. Sergeevich Novoselov, A.C. Ferrari, Rayleigh imaging of graphene and graphene layers, Nano Lett. 7 (9) (2007) 2711–2717.

[118] S. Stankovich, D.A. Dikin, G.H.B. Dommett, K.M. Kohlhaas, E.J. Zimney, E.A. Stach, R.D. Piner, S.B.T. Nguyen, R.S. Ruoff, Graphene-based composite materials, Nature 442 (7100) (2006) 282–286.

[119] X. Wang, L. Zhi, K. Müllen, Transparent, conductive graphene electrodes for dye-sensitized solar cells, Nano Lett. 8 (1) (2008) 323–327.

[120] D.A. Dikin, S. Stankovich, E.J. Zimney, R.D. Piner, G.H.B. Dommett, G. Evmenenko, S.B.T. Nguyen, R.S. Ruoff, Preparation and characterization of graphene oxide paper, Nature 448 (7152) (2007) 457–460.

[121] M.D. Stoller, S. Park, Y. Zhu, J. An, R.S. Ruoff, Graphene-based ultracapacitors, Nano Lett. 8 (10) (2008) 3498–3502.

[122] Y.H. Wu, T. Yu, Z.X. Shen, Two-dimensional carbon nanostructures: Fundamental properties, synthesis, characterization, and potential applications, J. Appl. Phys. 108 (7) (2010) 10.

[123] W.S. Hummer Jr., R.E. Offeman, Preparation of graphitic oxide, J. Am. Chem. Soc. 80 (6) (1958) 1339.

[124] J. Wu, W. Pisula, K. Müllen, Graphenes as potential material for electronics, Chem. Rev. 107 (3) (2007) 718–747.

[125] C. Zhu, S. Guo, Y. Fang, S. Dong, Reducing sugar: new functional molecules for the green synthesis of graphene nanosheets, ACS Nano 4 (4) (2010) 2429–2437.

[126] X. Fan, W. Peng, Y. Li, X. Li, S. Wang, G. Zhang, F. Zhang, Deoxygenation of exfoliated graphite oxide under alkaline conditions: a green route to graphene preparation, Adv. Mater. 20 (23) (2008) 4490–4493.

[127] V.H. Pham, T.V. Cuong, T.-D. Nguyen-Phan, H.D. Pham, E.J. Kim, S.H. Hur, E.W. Shin, S. Kim, J.S. Chung, One-step synthesis of superior dispersion of chemically converted graphene in organic solvents, Chem. Commun. 46 (24) (2010) 4375–4377.

[128] C.A. Amarnath, C.E. Hong, N.H. Kim, B.-C. Ku, T. Kuila, J.H. Lee, Efficient synthesis of graphene sheets using pyrrole as a reducing agent, Carbon 49 (11) (2011) 3497–3502.

[129] J. Zhang, H. Yang, G. Shen, P. Cheng, J. Zhang, S. Guo, Reduction of graphene oxide via L-ascorbic acid, Chem. Commun. 46 (7) (2010) 1112–1114.

[130] X. Zhou, J. Zhang, H. Wu, H. Yang, J. Zhang, S. Guo, Reducing graphene oxide via hydroxylamine: a simple and efficient route to graphene, J. Phys. Chem. C 115 (24) (2011) 11957–11961.

[131] G. Wang, J. Yang, J. Park, X. Gou, B. Wang, H. Liu, J. Yao, Facile synthesis and characterization of graphene nanosheets, J. Phys. Chem. C 112 (22) (2008) 8192–8195.

[132] M. Choucair, P. Thordarson, J.A. Stride, Gram-scale production of graphene based on solvothermal synthesis and sonication, Nat. Nanotechnol. 4 (1) (2009) 30.

[133] P.W. Sutter, J.-I. Flege, E.A. Sutter, Epitaxial graphene on ruthenium, Nat. Mater. 7 (5) (2008) 406–411.

[134] A. Reina, X. Jia, J. Ho, D. Nezich, H. Son, V. Bulovic, M.S. Dresselhaus, J. Kong, Large area, few-layer graphene films on arbitrary substrates by chemical vapor deposition, Nano Lett. 9 (1) (2009) 30–35.

[135] K.S. Kim, Y. Zhao, H. Jang, S.Y. Lee, J.M. Kim, K.S. Kim, J.-H. Ahn, P. Kim, J.-Y. Choi, B.H. Hong, Large-scale pattern growth of graphene films for stretchable transparent electrodes, Nature 457 (7230) (2009) 706–710.

[136] J. Coraux, A.T.N. 'Diaye, C. Busse, T. Michely, Structural coherency of graphene on Ir (111), Nano Lett. 8 (2) (2008) 565–570.
[137] B. Lang, A LEED study of the deposition of carbon on platinum crystal surfaces, Surf. Sci. 53 (1) (1975) 317–329.
[138] P.R. Somani, S.P. Somani, M. Umeno, Planer nano-graphenes from camphor by CVD, Chem. Phys. Lett. 430 (1-3) (2006) 56–59.
[139] N.G. Shang, P. Papakonstantinou, M. McMullan, M. Chu, A. Stamboulis, A. Potenza, S.S. Dhesi, H. Marchetto, Catalyst-free efficient growth, orientation and biosensing properties of multilayer graphene nanoflake films with sharp edge planes, Adv. Funct. Mater. 18 (21) (2008) 3506–3514.
[140] M. Chhowalla, K.B.K. Teo, C. Ducati, N.L. Rupesinghe, G.A.J. Amaratunga, A.C. Ferrari, D. Roy, J. Robertson, W.I. Milne, Growth process conditions of vertically aligned carbon nanotubes using plasma enhanced chemical vapor deposition, J. Appl. Phys. 90 (10) (2001) 5308–5317.
[141] K.B.K. Teo, S.-B. Lee, M. Chhowalla, V. Semet, V.u. T. Binh, O. Groening, M. Castignolles, A. Loiseau, G. Pirio, P. Legagneux, Plasma enhanced chemical vapour deposition carbon nanotubes/nanofibres—how uniform do they grow? Nanotechnology 14 (2) (2003) 204.
[142] S. Hofmann, G. Csanyi, A.C. Ferrari, M.C. Payne, J. Robertson, Surface diffusion: the low activation energy path for nanotube growth, Phys. Rev. Lett. 95 (3) (2005) 036101.
[143] B.O. Boskovic, V. Stolojan, R.U.A. Khan, S. Haq, S.R.P. Silva, Large-area synthesis of carbon nanofibres at room temperature, Nat. Mater. 1 (3) (2002) 165–168.
[144] C. Casiraghi, A.C. Ferrari, R. Ohr, A.J. Flewitt, D.P. Chu, J. Robertson, Dynamic roughening of tetrahedral amorphous carbon, Phys. Rev. Lett. 91 (22) (2003) 226104.
[145] M. Moseler, P. Gumbsch, C. Casiraghi, A.C. Ferrari, J. Robertson, The ultrasmoothness of diamond-like carbon surfaces, Science 309 (5740) (2005) 1545–1548.
[146] K.S. Subrahmanyam, L.S. Panchakarla, A. Govindaraj, C.N.R. Rao, Simple method of preparing graphene flakes by an arc-discharge method, J. Phys. Chem. C 113 (11) (2009) 4257–4259.
[147] L.S. Panchakarla, A. Govindaraj, C.N.R. Rao, Boron-and nitrogen-doped carbon nanotubes and graphene, Inorg. Chim. Acta 363 (15) (2010) 4163–4174.
[148] Z.-S. Wu, W. Ren, L. Gao, J. Zhao, Z. Chen, B. Liu, D. Tang, B. Yu, C. Jiang, H.-M. Cheng, Synthesis of graphene sheets with high electrical conductivity and good thermal stability by hydrogen arc discharge exfoliation, ACS nNano 3 (2) (2009) 411–417.
[149] K.S. Subrahmanyam, S.R.C. Vivekchand, A. Govindaraj, C.N.R. Rao, A study of graphenes prepared by different methods: characterization, properties and solubilization, J. Mater. Chem. 18 (13) (2008) 1517–1523.
[150] M. Endo, Grow carbon fibers in the vapor phase, ChemTech September (1988) 568–576.
[151] N. Grobert, M. Mayne, M. Terrones, J. Sloan, R.E. Dunin-Borkowski, R. Kamalakaran, T. Seeger, H. Terrones, M. Rühle, D.R.M. Walton, Alloy nanowires: Invar inside carbon nanotubes, Chem. Commun. 5 (2001) 471–472, doi:https://doi.org/10.1039/B100190F.
[152] A.G. Nasibulin, A. Moisala, H. Jiang, E.I. Kauppinen, Carbon nanotube synthesis from alcohols by a novel aerosol method, J. Nanopart. Res. 8 (3-4) (2006) 465–475.
[153] B.K.i. Ku, M.S. Emery, A.D. Maynard, M.R. Stolzenburg, P.H. McMurry, In situ structure characterization of airborne carbon nanofibres by a tandem mobility–mass analysis, Nanotechnology 17 (14) (2006) 3613.
[154] M. Pinault, V. Pichot, H. Khodja, P. Launois, C. Reynaud, M. Mayne-L'Hermite, Evidence of sequential lift in growth of aligned multiwalled carbon nanotube multilayers, Nano Lett. 5 (12) (2005) 2394–2398.
[155] M. Mayne, N. Grobert, M. Terrones, R. Kamalakaran, M. Rühle, H.W. Kroto, D.R.M. Walton, Pyrolytic production of aligned carbon nanotubes from homogeneously dispersed benzene-based aerosols, Chem. Phys. Lett. 338 (2-3) (2001) 101–107.
[156] M. Reyes-Reyes, N. Grobert, R. Kamalakaran, T. Seeger, D. Golberg, M. Rühle, Y. Bando, H. Terrones, M. Terrones, Efficient encapsulation of gaseous nitrogen inside carbon nanotubes with bamboo-like structure using aerosol thermolysis, Chem. Phys. Lett. 396 (1-3) (2004) 167–173.
[157] R. Kamalakaran, F. Lupo, N. Grobert, D. Lozano-Castello, N.Y. Jin-Phillipp, M. Rühle, In-situ formation of carbon nanotubes in an alumina–nanotube composite by spray pyrolysis, Carbon 41 (14) (2003) 2737–2741.

[158] D. Kondo, S. Sato, Y. Awano, Low-temperature synthesis of single-walled carbon nanotubes with a narrow diameter distribution using size-classified catalyst nanoparticles, Chem. Phys. Lett. 422 (4-6) (2006) 481–487.
[159] G.F. Zhong, T. Iwasaki, H. Kawarada, Semi-quantitative study on the fabrication of densely packed and vertically aligned single-walled carbon nanotubes, Carbon 44 (10) (2006) 2009–2014.
[160] N.A. Kiselev, A.V. Krestinin, A.V. Raevskii, O.M. Zhigalina, G.I. Zvereva, M.B. Kislov, V.V. Artemov, Y.u.V. Grigoriev, J.L. Hutchison, Extreme-length carbon nanofilaments with single-walled nanotube cores grown by pyrolysis of methane or acetylene, Carbon 44 (11) (2006) 2289–2300.
[161] E. Flahaut, R. Bacsa, A. Peigney, C. Laurent, Gram-scale CCVD synthesis of double-walled carbon nanotubes, Chem. Commun. 12 (2003) 1442–1443.
[162] A.C. Dillon, P.A. Parilla, J.L. Alleman, J.D. Perkins, M.J. Heben, Controlling single-wall nanotube diameters with variation in laser pulse power, Chem. Phys. Lett. 316 (1-2) (2000) 13–18.
[163] N. Braidy, M.A. El Khakani, G.A. Botton, Single-wall carbon nanotubes synthesis by means of UV laser vaporization, Chem. Phys. Lett. 354 (1-2) (2002) 88–92.
[164] C.D. Scott, S. Arepalli, P. Nikolaev, R.E. Smalley, Growth mechanisms for single-wall carbon nanotubes in a laser-ablation process, Appl. Phys. A 72 (5) (2001) 573–580.
[165] T.W. Ebbesen, P.M. Ajayan, Large-scale synthesis of carbon nanotubes, Nature 358 (6383) (1992) 220–222.
[166] C. Journet, W.K. Maser, P. Bernier, A. Loiseau, M. Lamy de La Chapelle, S. Lefrant, P. Deniard, R. Lee, J.E. Fischer, Large-scale production of single-walled carbon nanotubes by the electric-arc technique, Nature 388 (6644) (1997) 756–758.
[167] X. Blase, J.-C. Charlier, A.D.e. Vita, R. Car, P.h. Redlich, M. Terrones, W.K. Hsu, H. Terrones, D.L. Carroll, P.M. Ajayan, Boron-mediated growth of long helicity-selected carbon nanotubes, Phys. Rev. Lett. 83 (24) (1999) 5078.
[168] L.-J. Li, M. Glerup, A.N. Khlobystov, J.G. Wiltshire, J.-L. Sauvajol, R.A. Taylor, R.J. Nicholas, The effects of nitrogen and boron doping on the optical emission and diameters of single-walled carbon nanotubes, Carbon 44 (13) (2006) 2752–2757.
[169] Z.Y. Wang, Z.B. Zhao, J.S. Qiu, Development of filling carbon nanotubes, Prog. Chem 18 (563) (2006) 533.
[170] A. Loiseau, F. Willaime, Filled and mixed nanotubes: from TEM studies to the growth mechanism within a phase-diagram approach, Appl. Surf. Sci. 164 (1-4) (2000) 227–240.
[171] N. Demoncy, O. Stephan, N. Brun, C. Colliex, A. Loiseau, H. Pascard, Filling carbon nanotubes with metals by the arc-discharge method: the key role of sulfur, Eur. Phys. J. B. 4 (2) (1998) 147–157.
[172] A. Loiseau, H. Pascard, Synthesis of long carbon nanotubes filled with Se, S, Sb and Ge by the arc method, Chem. Phys. Lett. 256 (3) (1996) 246–252.
[173] P. Unwin1990. Fullerenes (an overview), http://www.ch.ic.ac.uk/local/projects/unwin/Fullerenes.html. Google Scholar There is no corresponding record for this reference
[174] K.E. Geckeler, S. Samal, Syntheses and properties of macromolecular fullerenes, a review, Polym. Int. 48 (9) (1999) 743–757.
[175] A.M. Lopez, A. Mateo-Alonso, M. Prato, Materials chemistry of fullerene C 60 derivatives, J. Mater. Chem. 21 (5) (2011) 1305–1318.
[176] R.E. Smalley, Process for making fullerenes by the laser evaporation of carbon. Google Patents, 1994.
[177] P.P. Shanbogh, N.G. Sundaram, Fullerenes revisited, Resonance 20 (2) (2015) 123–135.
[178] R.F. Curl, R.E. Smalley, Probing C60, Science 242 (4881) (1988) 1017–1022.
[179] H. Ajie, M.M. Alvarez, S.J. Anz, R.D. Beck, F. Diederich, K. Fostiropoulos, D.R. Huffman, W. Kraetschmer, Y. Rubin, Characterization of the soluble all-carbon molecules C60 and C70, J. Phys. Chem. 94 (24) (1990) 8630–8633.
[180] J. Baggott, Great balls of carbon[from discovery in 1985 to latest applications], New Sci. 131 (1991) 34–38.
[181] T.A. Silva, H. Zanin, P.W. May, E.J. Corat, O. Fatibello-Filho, Electrochemical performance of porous diamond-like carbon electrodes for sensing hormones, neurotransmitters, and endocrine disruptors, ACS Appl. Mater. Interfaces 6 (23) (2014) 21086–21092.
[182] S. Chatterjee, J. Wen, A. Chen, Electrochemical determination of methylglyoxal as a biomarker in humanplasma, Biosens. Bioelectron. 42 (2013) 349–354.

[183] A.K. Wanekaya, Applications of nanoscale carbon-based materials in heavy metal sensing and detection, Analyst 136 (21) (2011) 4383–4391.
[184] F. Liu, G. Xiang, R. Yuan, X. Chen, F. Luo, D. Jiang, S. Huang, Y.i. Li, X. Pu, Procalcitonin sensitive detection based on graphene–gold nanocomposite film sensor platform and single-walled carbon nanohorns/hollow Pt chains complex as signal tags, Biosens. Bioelectron. 60 (2014) 210–217.
[185] M. Govindhan, M. Amiri, A. Chen, Au nanoparticle/graphene nanocomposite as a platform for the sensitive detection of NADH in human urine, Biosens. Bioelectron. 66 (2015) 474–480.
[186] B.-R. Adhikari, M. Govindhan, A. Chen, Sensitive detection of acetaminophen with graphene-based electrochemical sensor, Electrochim. Acta 162 (2015) 198–204.
[187] T.H. Kim, J. Lee, S. Hong, Highly selective environmental nanosensors based on anomalous response of carbon nanotube conductance to mercury ions, J. Phys. Chem. C 113 (45) (2009) 19393–19396.
[188] H. Chen, X. Zuo, S. Su, Z. Tang, A. Wu, S. Song, D. Zhang, C. Fan, An electrochemical sensor for pesticide assays based on carbon nanotube-enhanced acetycholinesterase activity, Analyst 133 (9) (2008) 1182–1186.
[189] N.L. Rosi, D.A. Giljohann, C. Shad Thaxton, A.K.R. Lytton-Jean, M.S.u. Han, C.A. Mirkin, Oligonucleotide-modified gold nanoparticles for intracellular gene regulation, Science 312 (5776) (2006) 1027–1030.
[190] J.-S. Lee, M.S.u. Han, C.A. Mirkin, Colorimetric detection of mercuric ion (Hg2+) in aqueous media using DNA-functionalized gold nanoparticles, Angew. Chem. Int. Ed. 46 (22) (2007) 4093–4096.
[191] C.A. Mirkin, R.L. Letsinger, R.C. Mucic, J.J. Storhoff, A DNA-based method for rationally assembling nanoparticles into macroscopic materials, Nature 382 (6592) (1996) 607–609.
[192] J. Liu, Y.i. Lu, Accelerated color change of gold nanoparticles assembled by DNAzymes for simple and fast colorimetric Pb2+ detection, J. Am. Chem. Soc. 126 (39) (2004) 12298–12305.
[193] X. Chen, S.G. Parker, G. Zou, W. Su, Q. Zhang, β-Cyclodextrin-functionalized silver nanoparticles for the naked eye detection of aromatic isomers, ACS Nano 4 (11) (2010) 6387–6394.
[194] B.S. Harrison, A. Atala, Carbon nanotube applications for tissue engineering, Biomaterials 28 (2) (2007) 344–353.
[195] H. Haniu, N. Saito, Y. Matsuda, T. Tsukahara, Y. Usui, N. Narita, K. Hara, K. Aoki, M. Shimizu, N. Ogihara, Basic potential of carbon nanotubes in tissue engineering applications, J. Nanomater. 2012 (2012), doi:https://doi.org/10.1155/2012/343747 343747.
[196] T.-I. Chao, S. Xiang, C.-S. Chen, W.-C. Chin, A.J. Nelson, C. Wang, J. Lu, Carbon nanotubes promote neuron differentiation from human embryonic stem cells, Biochem. Biophys. Res. Commun. 384 (4) (2009) 426–430.
[197] T.-I. Chao, S. Xiang, J.F. Lipstate, C. Wang, J. Lu, Poly (methacrylic acid)-grafted carbon nanotube scaffolds enhance differentiation of hESCs into neuronal cells, Adv. Mater. 22 (32) (2010) 3542–3547.
[198] C.-S. Chen, S. Soni, C. Le, M. Biasca, E. Farr, E.Y.T. Chen, W.-C. Chin, Human stem cell neuronal differentiation on silk-carbon nanotube composite, Nanoscale Res. Lett. 7 (1) (2012) 1–7.
[199] H. Kitahara, K. Yoshinori, T. Hiroko, A. Tsukasa, W. Fumio, I. Nobuo, Culture of ES cells and mesenchymal stem cells on carbon nanotube scaffolds, Nano Biomed. 2 (2) (2010) 81–92.
[200] M. Kalbacova, M. Kalbac, L. Dunsch, H. Kataura, U. Hempel, The study of the interaction of human mesenchymal stem cells and monocytes/macrophages with single-walled carbon nanotube films, Physica Status Solidi (b) 243 (13) (2006) 3514–3518.
[201] D. Cui, Z. Wang, H. Zhang, J. Ruan, C. Bao, H. Yang, A. Toru, H. Song, K. Wang, Single walled carbon nanotubes based regulation of proliferation and diffeneration of mouse embryonic stem cells, ECS Trans. 19 (13) (2009) 63.
[202] T.R. Nayak, L.i. Jian, L.C. Phua, H.K. Ho, Y. Ren, G. Pastorin, Thin films of functionalized multiwalled carbon nanotubes as suitable scaffold materials for stem cells proliferation and bone formation, ACS Nano 4 (12) (2010) 7717–7725.
[203] L. Chen, H. Cai, X.U. Guofu, Preparation of novel bulk carbon nano tubes dental materials for implant, Chinese J. Tissue Eng. Res. 12 (1) (2008) 180–184.
[204] W. Wang, O. Mamoru, W. Fumio, Y. Atsuro, Novel bulk carbon nanotube materials for implant by spark plasma sintering, Dent. Mater. J. 24 (4) (2005) 478–486.
[205] A.A. Kroustalli, S.N. Kourkouli, D.D. Deligianni, Cellular function and adhesion mechanisms of human bone marrow mesenchymal stem cells on multi-walled carbon nanotubes, Ann. Biomed. Eng. 41 (12) (2013) 2655–2665.

[206] Z. Wang, Z. Dai, Carbon nanomaterial-based electrochemical biosensors: an overview, Nanoscale 7 (15) (2015) 6420–6431.
[207] G. Li, J.M. Liao, G.Q. Hu, N.Z. Ma, P.J. Wu, Study of carbon nanotube modified biosensor for monitoring total cholesterol in blood, Biosens. Bioelectron. 20 (10) (2005) 2140–2144.
[208] R.M. Santos, M.S. Rodrigues, J. Laranjinha, R.M. Barbosa, Biomimetic sensor based on hemin/carbon nanotubes/chitosan modified microelectrode for nitric oxide measurement in the brain, Biosens. Bioelectron. 44 (2013) 152–159.
[209] B.B. Prasad, A. Prasad, M.P. Tiwari, R. Madhuri, Multiwalled carbon nanotubes bearing 'terminal monomeric unit' for the fabrication of epinephrine imprinted polymer-based electrochemical sensor, Biosens. Bioelectron. 45 (2013) 114–122.
[210] G.J. Kress, H.-J. Shu, A. Yu, A. Taylor, A. Benz, S. Harmon, S. Mennerick, Fast phasic release properties of dopamine studied with a channel biosensor, J. Neurosci. 34 (35) (2014) 11792–11802.
[211] J. Li, Q. Liu, Y. Liu, S. Liu, S. Yao, DNA biosensor based on chitosan film doped with carbon nanotubes, Anal. Biochem. 346 (1) (2005) 107–114.
[212] H. Jin, D.A. Heller, M. Kalbacova, J-Ho Kim, J. Zhang, A.A. Boghossian, N. Maheshri, M.S. Strano, Detection of single-molecule H_2O_2 signalling from epidermal growth factor receptor using fluorescent single-walled carbon nanotubes, 2010.
[213] S. Fei, J. Chen, S. Yao, G. Deng, D. He, Y. Kuang, Electrochemical behavior of L-cysteine and its detection at carbon nanotube electrode modified with platinum, Anal. Biochem. 339 (1) (2005) 29–35.
[214] Y. Tsujita, K. Maehashi, K. Matsumoto, M. Chikae, Y. Takamura, E. Tamiya, Microfluidic and label-free multi-immunosensors based on carbon nanotube microelectrodes, Japan. J. Appl. Phys. 48 (6S) (2009) 06FJ02.
[215] Q.-M. Feng, J.-B. Pan, H.-R. Zhang, J.-J. Xu, H.-Y. Chen, Disposable paper-based bipolar electrode for sensitive electrochemiluminescence detection of a cancer biomarker, Chem. Commun. 50 (75) (2014) 10949–10951.
[216] C. Baj-Rossi, G. De Micheli, S. Carrara, Electrochemical detection of anti-breast-cancer agents in human serum by cytochrome P450-coated carbon nanotubes, Sensors 12 (5) (2012) 6520–6537.
[217] Y.K. Park, B. Bold, W.K. Lee, M.H. Jeon, K.H. An, S.Y. Jeong, Y.K. Shim, D-(+)-galactose-conjugated single-walled carbon nanotubes as new chemical probes for electrochemical biosensors for the cancer marker galectin-3, Int. J. Mol. Sci. 12 (5) (2011) 2946–2957.
[218] B.N. Shobha, N.J.R. Muniraj, Design, modeling and performance analysis of carbon nanotube with DNA strands as biosensor for prostate cancer, Microsyst. Technol. 21 (4) (2015) 791–800.
[219] Y. Zhang, G. Gao, H. Liu, H. Fu, J. Fan, K. Wang, Y. Chen, B. Li, C. Zhang, X. Zhi, Identification of volatile biomarkers of gastric cancer cells and ultrasensitive electrochemical detection based on sensing interface of Au-Ag alloy coated MWCNTs, Theranostics 4 (2) (2014) 154.

CHAPTER 22

Nanomaterials for sensors: Synthesis and applications

Vinod Nandre[a], Yogesh Jadhav[b], Dwiti K. Das[c], Rashmi Ahire[a], Sougata Ghosh[d], Sandesh Jadkar[b], Kisan Kodam[a], Suresh Waghmode[a]
[a]Department of Chemistry, Savitribai Phule Pune University (Formerly Pune University), Pune, Maharashtra, India
[b]School of energy studies, Savitribai Phule Pune University (Formerly Pune University), Pune, Maharashtra, India
[c]National Centre for Nanoscience and Nanotechnology, University of Mumbai, Mumbai, Maharashtra, India
[d]Department of Microbiology, School of Science, RK University, Rajkot, Gujarat, India

22.1 Introduction

The advent of nanoscience and nanotechnology has been a boon to the world of material science, and the applications are immense. Nanotechnology deals with the nanomaterials which are the materials having at least one of the dimensions in the range of nanometres (nm, 10^{-9} m) [1,2]. As a general convention the dimensions of the material which are smaller than that of 100 nm is considered to be as nanomaterial. The interesting thing about these is that the materials have unique properties when they are in the nanoscale compared to their bulk counterparts. The high surface area to volume ratio and dominating quantum effects are the reason behind these properties [3]. This advancement in material science has enabled mankind to miniaturize many otherwise bulky devices and appliances, be it a transistor or an imaging system. The same effect can also be seen in Sensor Technology, while increasing the accuracy of the measurement or the sensitivity of the sensor [4].

The physicochemical properties of the material (like optical, electrical, chemical, strength, potential) are affected when the materials are transformed to nanoscale as compared to that of the bulk macroscale materials. Sensors have benefitted hugely from nanomaterials because of the low amount of sample required for the sensing. This reduces the problem which is faced because of lack of enough amount of sample to detect. Depending on the sensing method and the analyte, the limit of detection has also been getting lower, thanks to nanomaterials. So, in this chapter we will be discussing synthesis of nanomaterials and their application is sensor technology.

22.2 Sensors

Sensors are devices or modules which intend to detect changes in the environment (physical, chemical, physicochemical, biophysical, or biochemical) and then convert these into readable data/signals that can be recorded & analyzed. Sensors have

Advanced Nanomaterials for Point of Care Diagnosis and Therapy
DOI: https://doi.org/10.1016/B978-0-323-85725-3.00011-8

tremendous applications which spans from R & D level work to instruments of daily use. Sensors were built to capture/record the state or change in the various parameters to a comprehensible format/figure so that we humans can understand nature better, in due times the purpose and usage has expanded beyond few physical parameters. And in all these, nanotechnology has given us nanosensors, alternatively called as sensors based on nanomaterials. There are sensors of various kinds and for diverse applications, but here in this text, we are focusing mainly on nanosensors.

22.2.1 Sensors and their types

Sensors can be classified based on different factors, the main being the type of nanomaterials used to manufacture them. Mainly they are categorized as follows:

(i) Semiconductor based
(ii) Metal based
(iii) Carbon based
(iv) Organic materials based
(v) Biomaterials based/biosensors

Sensors are also classified on the basis of the type of analyte/parameter, which are:

a. Chemical sensors
b. Biosensors/biochemical sensors
c. Physical sensors/parametric sensors

22.2.1.1 Semiconductor-based sensors

Semiconductor-based sensors are the most commonly used sensors in electronics and imaging applications. As the name suggest, in this type of sensors, semiconductors are the type of material used. Due to the electrical properties of the semiconductors and the ability to control the doping level has allowed researchers to use these for sensing applications in various ways.

Nanotechnology has allowed the layers of materials to get thinner and also has achieved better sensitivity, range and accuracy. New techniques coupled with use of various novel materials is considered to be an attempt to displace the bulky conventional sensors for portable and inexpensive ones. Mostly 2D materials are typically used for semiconductor-based sensors.

Semiconductor materials are known for their doping capabilities which enable us to control their properties and use them for various detection/sensing techniques. Most of the semiconductor-based sensors rely on the p-n junction properties and capacitance of the materials. Voltage *vs* current characteristics at different conditions generally show uniqueness and thus that is exploited for sensor applications.

Silicon and Gallium are one of the most widely used sensor materials, other materials generally hail from the group III-V of the periodic table. But with advent of novel nanomaterials, metal oxides/nitrides/sulfides/selenides, etc., are also used for their semiconductor like properties. Along with the above-mentioned ones, certain ternary and quaternary materials are also utilized for sensing. The signal from the sensor is generally transduced by charge transfer in the form of a calibrated voltage/current signal.

Examples of semiconductor-based sensors are: complementary metal oxide semiconductor (CMOS) sensors, charge coupled devices (CCD), diode thermal sensors, etc.

22.2.1.2 Metal-based sensors

Here instead of semiconductors as the base materials, metals or metal alloys are used. With various approaches metal nanoparticles are synthesized and their unique properties are utilized to sense or detect the change in the environment. Electrical and magnetic parameters can be sensed using metal-based sensors. Depending on the reactivity or catalytic activity of the metal nanoparticles used in the metal-based sensors, it can be used to detect the analyte.

Noble metal nanoparticle (NP) like silver (Ag), gold (Au) and platinum (Pt) are known to show catalytic properties. It is important to note that control over the capping agents allows us to control their morphological features and thereby their physico-chemical and opto-electronic properties. AgNPs, AuNPs, etc., also exhibit colorimetric responses, thus have found application in colorimetric sensing/detection [5].

22.2.1.3 Carbon-based sensors

Carbon being a versatile material makes it one of the most widely used material in a wide variety of applications. Nanotechnology has given us materials like graphene, fullerenes, CNT (carbon nanotubes), carbon dots (carbon quantum dots), etc. [6–8]. Each of these have unique properties and find utilization in sensors of various kind, be it electronic sensors or chemical sensors or even sensors with biological sensing capabilities thanks to its non-toxic properties.

22.2.1.4 Organic materials-based sensors

For certain applications inorganic semiconductors or metal nanomaterials cannot be used, thus sensors with organic materials were explored. Organic polymer materials are used as the structural and support material while organometallic compounds are used as the sensing material.

22.2.1.5 Biomaterials-based sensors/biosensors

Bio Sensors refer to those sensors which can detect certain analyte utilizing biological component(s) like cell, proteins, nucleic acid, antibodies, other biomolecules, etc. Generally, a biosensor works in conjunction with a physiochemical sensor [4,9]. The "biomaterials" parts of the sensor (like the different biomolecules) are biologically derived components or artificial materials which mimic biological functions or conditions. The analyte generally interacts directly with the biological component of the sensor, and this step is usually referred as "biorecognition" [10].

Medical studies and diagnosis heavily depend on technologies in modern times; thus, biosensors have utmost importance. But unlike the other sensors discussed before, biosensors need a particular condition which is the condition of being non-toxic if the sensor needs to be in direct contact with a living being. Carbon-based sensors are at times tuned for the purpose. But here we are to discuss about various sensors which utilize biomolecules for the purpose of sensing. Biomolecules like proteins and nucleic acids are utilized because of their ability to specifically recognize the target analyte.

22.2.1.6 Chemical sensors

These are the analytical devices which measures/detects the change in chemical composition of the environment. Chemical sensors work the best in a fluid phase, that is, in liquid or gaseous phase, though there are ion sensors as well. Chemical sensors can be very specific depending on the analyte and the purpose of the sensor, thanks to the specificity of chemical reactions.

22.2.1.7 Biosensors/biochemical sensors

Biosensors are the devices that utilize biological or biomimetic materials to analyze the species. Biosensors have huge application in medical diagnosis, environmental studies, and biomedical research.

22.2.1.8 Physical sensors/parametric sensors

This category of sensors is used for detection or recording of physical or environmental parameters, like temperature, pressure, light etc. Semiconductor and metal-based sensors are widely used for these purposes.

22.2.2 Working mechanism

Sensors mainly have two components, the first being recognition of the species/parameter and the next step is that of signal transduction, which is essentially converting the detection/recognition into a comprehensible quantitative or qualitative form of reading.

The following are the classification based on the method of sensing:

(i) Optical/fluorescence sensors
(ii) Electrochemical sensors
(iii) Thermoelectric sensors
(iv) Electromagnetic sensors
(v) Sensors based on biochemical recognition/biosensors
(vi) Sensors based on direct physical interactions/change in physical parameters

22.2.2.1 Optical/fluorescence sensors

Optical sensors are based on the signal transduction that involves light (photons). It can be of various types like colorimetric, intensity dependent, fluorescent, or phosphorescent. These kinds of sensors can be implemented in two fundamental ways, one by showing fluorescence when the analyte is present and the other way is by fluorescence quenching, i.e., decrease in fluorescence when the analyte is present. Certain metal and metal oxide nanoparticles show fluorescence or phosphorescence, and thus find application in this technique of detection/sensing.

In a quenching based optical detection, the analyte (which is the quencher) decreases the intensity of the fluorescence of the sensing material (which is the nanomaterial for the case). If the amount of quenching is in a linear relation with that of the concentration of the analyte, then a good sensor can be built from that. If not, it can be simply used as a present or not detector.

22.2.2.2 Electrochemical sensors

These types of sensors involve measurement of current when a certain voltage is applied to the sample, and recording the graph and correlating it to certain standard of measurement. Electrochemical methods include various kinds of measurements, which are:

(i) Voltammetry
(ii) Potentiometry
(iii) Conductometry
(iv) Amperometry
(v) Impedimetric measurements

Voltammetry is the technique in which the potential is applied to the electrode with reference to a reference electrode and the current thus produced at the working electrode is measured.

In voltammetry or polarography, the potential ramp can be applied in various ways, some of which are shown in Fig. 22.2.

The common methods that fall under the voltammetry techniques are cyclic voltammetry (CV), linear sweep voltammetry (LSV), differential pulse voltammetry (DPV), square-wave voltammetry (SWV), etc. [11,12].

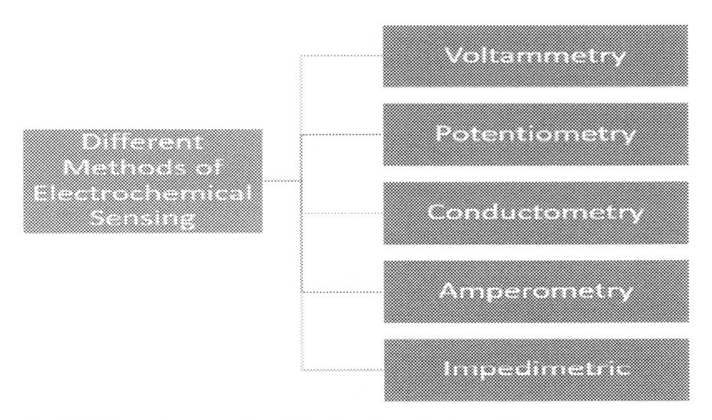

Fig. 22.1 ***Schematic of different methods of electrochemical sensing.***

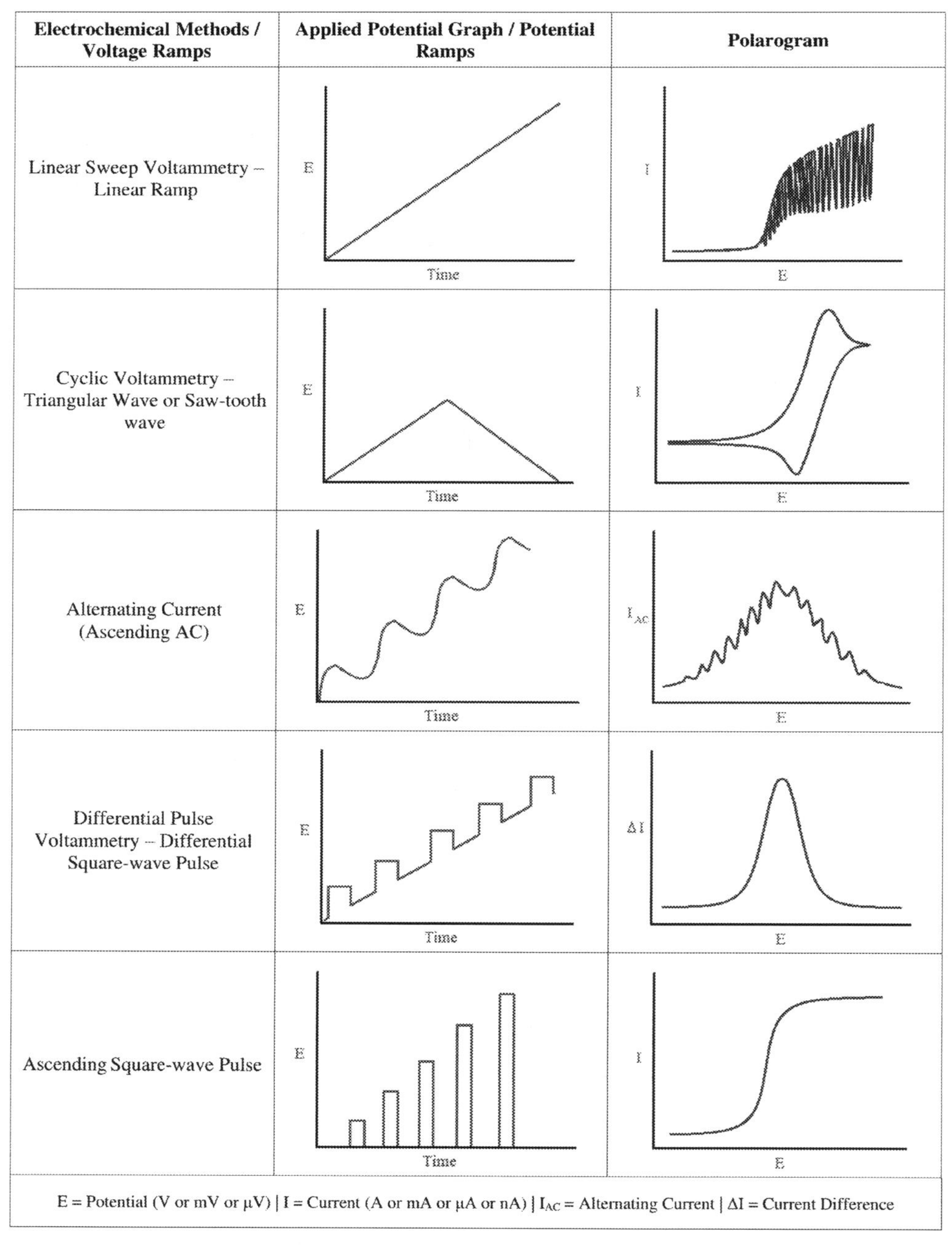

Fig. 22.2 ***Potential ramp and polarograms of different types used in electrochemistry.***

CV is one of the extensively used techniques among all the electrochemical techniques/studies, mostly used for initial fundamental studies, in which the potential is applied to the electrode in an increasing and decreasing fashion (triangular waveform) with reference to a reference electrode and the current thus produced is measured at the working electrode. The voltammograms thus generated reveals numerous qualitative and quantitative information on analytes. CV is also done on analytes to understand the thermodynamics and kinetics of the species undergoing electrochemical reactions.

22.2.2.3 Thermoelectric sensors

Thermoelectric sensors refer to the devices which sense the analyte on the basis of the thermal properties that can be transduced into a readable form. Most thermoelectric sensors consist of metallic or semi-conductive materials. A thermoelectric sensor essential measures the difference in voltage caused by difference in temperature across two of the terminals.

22.2.2.4 Electric and magnetic sensors

Almost all sensors can be considered as "electromagnetic sensors." But in this section we will be discussing specifically about sensors that detect or sense the electrical or magnetic properties of the environment or the analyte.

22.2.2.5 Sensors based on biochemical recognition/biosensors

As discussed previously in the earlier sections, Biosensors involve some sort of biochemical species for the recognition of the analyte. At times biological component can also be used for signal transduction. The schematic of an electrochemical sensor based on biorecognition is shown in Fig. 22.3.

Yan et al. (2018) had created sensor based on reduced graphene having polymer film of 5,10,15, 20-tetrakis (pentafluorophenyl)−21H,23H-porphyrin iron (III) chloride

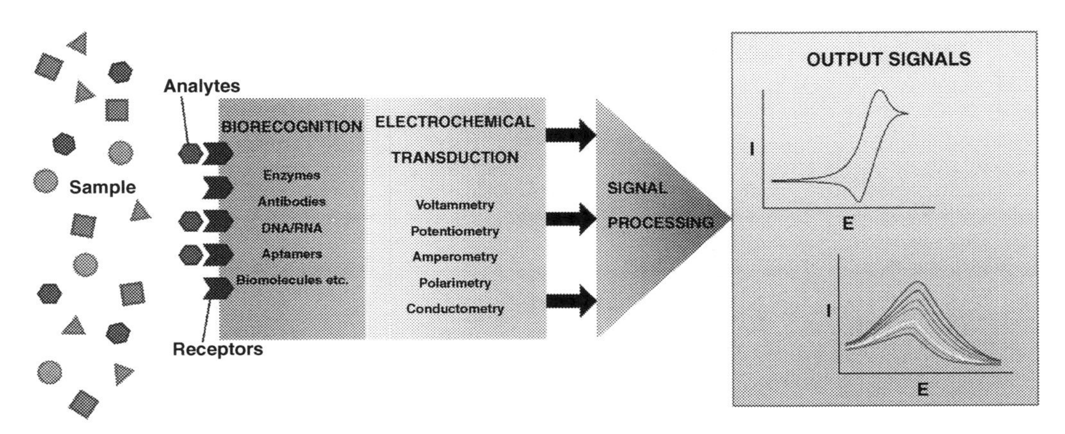

Fig. 22.3 *Schematic of an electrochemical biosensor [10].*

(FeTFPP) on reduced graphene oxide sheet over glassy carbon electrode (GCE) and achieved a detection limit of 0.023 μM [13].

22.2.2.6 Sensors based on direct physical interactions/change in physical parameters

Here, in this section we will be discussing about sensing all other physical interactions or parameters. We already have discussed thermal, electric, & magnetic parameters. Parameters like pressure, impact, force, viscosity, and pH will be discussed in this section of the chapter. Also, sensors for environmental applications are also there, some of which do not fall into the categories discussed earlier and based on physical or direct interaction (e.g., moisture sensor using moisture sensitive biomaterials).

22.3 Nanomaterials

The prefix nano comes from the Greek word for dwarf or atypically small. In the system of measurements, nano is used for representing something that is billionth (10^{-9}) of the base measurement. Nanomaterials refer to materials having at least one dimension of the structure lying in the range of 1 to 100 nm (10^{-9} meters). The nanomaterials are known to show distinctively different multifunctional properties from their bulky counterparts.

22.3.1 Nanomaterials and their classifications

Nanomaterials are classified based on the dimensions of the individual units of the nanoscale system,

(i) Zero dimensional (0D) or nanoparticles (NPs)
(ii) One dimensional (1D)
(iii) Two dimensional (2D)
(iv) Three dimensional (3D) or bulk nanostructured materials

22.3.1.1 Zero-dimensional (0D) nanomaterials or nanoparticles

Zero-dimensional nanomaterials refer to the nanomaterials where all the three dimensions of the unit are restricted to ≤ 100 nm. They are named so as they are point like units. 0D nanomaterials mainly consist of nanoparticles, quantum dots, fullerenes, etc. Nanoparticles can be metallic, like Au (gold) nanoparticles, or can be semiconductors like CdS or CdTe quantum dots. Nanoparticles can be of variety of shapes like spherical, cubic, octahedral, or polygonal with size range of 1–50 nm.

22.3.1.2 One-dimensional (1D) nanomaterials

One-dimensional nanomaterials refer to the nanomaterials where two of the three dimensions of the unit is restricted to ≤ 100 nm. These materials resemble lines or wires, thus the name one-dimensional nanomaterials. Examples of this category are nanowires, nanofibers, nanorods, nanotubes, CNT, etc.

22.3.1.3 Two-dimensional (2D) nanomaterials

Two-dimensional nanomaterials refer to the nanomaterials where one of the three dimensions of the unit is restricted to ≤ 100 nm. These are layered materials where the layer thickness is in the nanoscale, thus they look like sheets. Nanosheets, nanoribbons, nanoplates, graphene, hBN nanosheets are some examples of 2D nanomaterials.

22.3.1.4 Three dimensional (3D) or bulk nanostructured materials

The unit of this type of nanomaterials is not essentially restricted to 100 nm, but rather have features or appendages, which are at nanoscale. These mostly consist of nanocomposites or nanomaterials of different kind fused together for specific purposes. Nanotextured films, fullerene-nanotube composites, nanoflowers (nanocrystals) are some of the notable examples of these.

22.3.1.5 Methods for the synthesis

Synthesis of nanomaterials can be done using mainly the two different techniques of producing them.

(i) Top-down approaches
(ii) Bottom-up approaches

22.3.1.6 Top-down approaches

Top-down approaches of synthesis adopt various techniques, which essentially breaks down bulk materials into nanostructures using physical, chemical, electrochemical, or biologically assisted processes.

Various methods using this approach are ball milling, laser ablation, ultrasonication, etc. There are certain biological processes as well which can reduce certain materials into nanostructures.

22.3.1.7 Bottom-up approaches

Bottom-up approaches of synthesis involve assembly of molecules/atoms of the desired materials, forming nanostructures. The source materials can be in gas, liquid or solid state before it is incorporated into the desired nanostructure. The method of synthesis used can be either random or controlled. Methods involving controlled reactions give us a parametric control over what kind of geometry or shape is to be made. Examples of controlled processes for nanomaterial synthesis are ALD (atomic layer deposition), MBE (molecular beam epitaxy), self-limiting CVD (chemical vapor deposition), etc.

Synthesis methods that are termed random or chaotic are the ones where the parameters are difficult to control and the system is in a chaotic state. Though manipulating certain conditions (like reaction time, temperature, pressure, limiting agents, etc.) during the process can give a better distribution of the number of particles of a certain size or shape. Hydrothermal processes, pyrolysis processes, various chemical synthesis procedures would be the example for this.

Synthesis methods can also be classified using the basis of the routes involved in achieving the desired materials.

(i) Chemical routes
(ii) Physical routes
(iii) Biological routes

22.3.1.8 Chemical synthesis routes

Most commonly and unscalable route of synthesis is the chemical one, where materials chemistry is used to convert the desired materials into any of the nano form. The chemical routes include sol-gel method, hydrothermal method, reduction method, colloidal synthesis, CVD (chemical vapor deposition), ALD (atomic layer deposition), etc.

22.3.1.9 Physical synthesis routes

Physical routes utilized complex machineries to achieve nano level precision for the materials. Procedures like crystal-ALD, MBE, PVD.

22.3.1.10 Biological synthesis routes

Routes involving biological sources for materials or converting bulk materials into nanomaterials using certain biochemical process are put in this category. Biological routes broadly include, plant-based synthesis, fungi-based synthesis, bacteria-based synthesis, algae-based synthesis, etc.

22.4 Modification of nanomaterials as a function of size, shape, composition, doping

To utilize a certain material which is fine tuned for a said purpose may require a very specific set of properties, and thus it needs to be synthesized in a certain size, shape, and doping.

22.5 Applications of nanomaterials in sensor field

Here, in this section of the chapter, we will look at some of the examples of nanomaterials being used as sensors.

Semiconductor-based sensor – CMOS sensors
Metal-based sensor – gold NP based sensors
Carbon-based sensor – graphene-based sensors, CNT bases sensors
Metal + carbon sensor – graphene + metal NP sensor
Organic sensor – organic thin film transistor sensors
Biosensor – electrochemical neurosensors

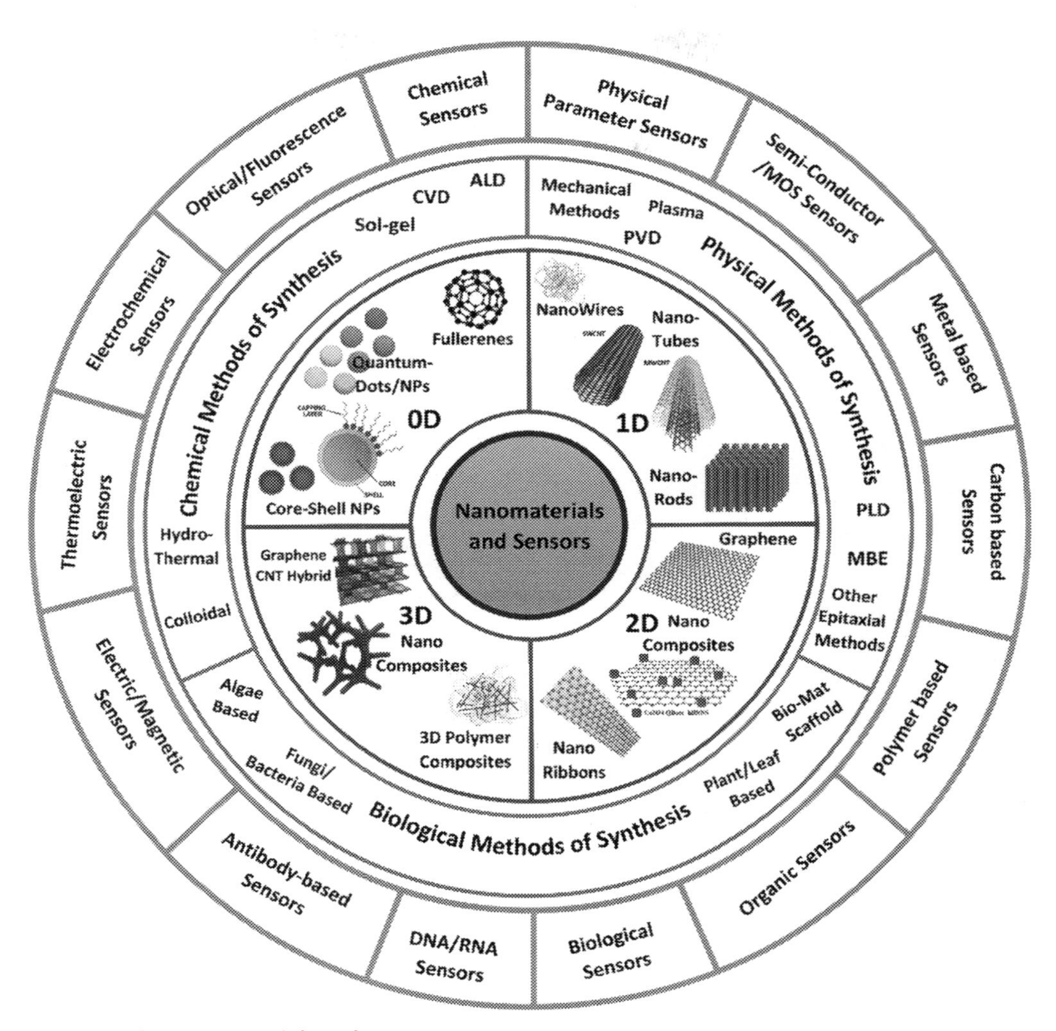

Fig. 22.4 *The nanomaterials and sensors.*

22.6 Summary and future perspectives

Nanoscience and nanotechnology is a vast and interdisciplinary field, and it is still giving us new kind of materials, and thus, we have a lot to expect in the near future. The pathway for research, analysis and diagnosis is through sensors only, as they are the way we humans can perceive nature both qualitatively and quantitatively. Fig. 22.4 summarizes the different nanomaterials and the various types of sensors which utilize the properties of the nanomaterials.

Table 22.1 Different type of sensors reported in literature.

Sr no.	Type of sensor	Materials used	Medium of analyte (solid, gas, liquid)	Analyte (to be sensed)	Method of sensing	Limit of detection reported (μM)	References
1	Electrochemical biosensor	Graphene, carbon fiber	Solution (liquid)	Dopamine (DA)	DPV (differential pulse voltammetry)	1.36	[14]
2	Electrochemical biosensor	Carbon dots	Solution (liquid)	Dopamine (DA)	DPV	0.01	[15]
3	Electrochemical biosensor	Reduced graphene oxide, MWCNT	Solution (liquid)	Dopamine (DA)	ECL (electrochemiluminescence)	0.067	[16]
4	Electrochemical biosensor	Graphene oxide	Solution (liquid)	DA	Amperometry	0.227	[17]
5	Electrochemical biosensor	Reduced graphene oxide	Solution (liquid)	Epinephrine (EP)	DPV	0.0012	[18]
6	Electrochemical biosensor	Graphene	Solution (liquid)	Norepinephrine (NE)	Square wave voltammetry	0.03	[19]
7	Electrochemical biosensor	Graphene	Solution (liquid)	Serotonin (5-HT)	Amperometry	0.0016	[20]
8	Electrochemical biosensor	Reduced graphene oxide	Solution (liquid)	5-HT	DPV	0.0001	[21]
9	Electrochemical biosensor	Gold nanoparticles	Solution (liquid)	Dopamine (DA)	DPV	0.0419	[22]
10	Electrochemical biosensor	Ag–Pt bimetallic nanoparticles	Solution (liquid)	Dopamine (DA)	DPV	0.11	[23]
11	Electrochemical biosensor	Pt3Ni nanoalloys	Solution (liquid)	Dopamine (DA)	Amperometry	0.01	[24]
12	Electrochemical biosensor	Fe_3O_4 nanorods	Solution (liquid)	DA	Amperometry	0.007	[25]
13	Electrochemical biosensor	Au-Pd nanocrystals	Solution (liquid)	EP	DPV	0.0012	[18]
14	Electrochemical biosensor	SnO_2	Solution (liquid)	DA	CV	0.0063	[26]
15	Electrochemical biosensor	Palladium nanoparticles	Solution (liquid)	NE	CV	0.1	[27]

Sr no.	Type of sensor	Materials used	Medium of analyte (solid, gas, liquid)	Analyte (to be sensed)	Method of sensing	Limit of detection reported (μM)	References
16	Electrochemical biosensor	SnO_2	Solution (liquid)	EP, NE	DPV	0.01, 0.006	[28]
17	Electrochemical biosensor	ZnO	Solution (liquid)	NE, 5-HT	SWV	0.2, 0.01	[29]
18	Electrochemical biosensor	WO_3 nanoparticles	Solution (liquid)	5-HT	DPV	0.0014	[30]
19	Electrochemical biosensor	Fe_3O_4 nanoparticles	Solution (liquid)	5-HT	DPV	0.08	[31]
20	Electrochemical biosensor	CuO	Solution (liquid)	Histamine (His)	Amperometry	0.33	[32]
21	Electrochemical biosensor	NiO	Solution (liquid)	Acetylcholine (ACh)	Amperometry	26.7	[33]
22	Electrochemical biosensor	Overoxidized poly-pyrrole	Solution (liquid)	Glutamate (Glu), DA	Amperometry	2.1, 0.062	[34]
23	Electrochemical biosensor	Poly (hydroquinone)	Solution (liquid)	DA	DPV	0.0419	[22]
24	Electrochemical biosensor	Poly(2,4,6-triaminopyrmidine)	Solution (liquid)	DA	DPV	0.017	[35]
25	Electrochemical biosensor	Poly(phenyl trimethoxysilane)	Solution (liquid)	NE	CV	0.1	[27]
26	Electrochemical biosensor	Polypyrrole/polyaniline	Solution (liquid)	Glu	Amperometry	0.0001	[36]
27	Electrochemical biosensor	Polypyrrole	Solution (liquid)	5-HT	SWV	0.0332	[37]
28	Electrochemical biosensor	Methacrylic acid	Solution (liquid)	His	CV	0.000074	[38]
29	Electrochemical biosensor	Poly (3-thiophenemalonic acid)	Solution (liquid)	Tryptamine	Amperometry	0.0417	[39]
30	Electrochemical biosensor	Polypyrrole nanotubes/gold nanoparticles	Solution (liquid)	EP	DPV	0.298	[40]

Acknowledgment

The author Vinod Nandre (SPPU-PDF/ST/BL/2019/0013) and Yogesh Jadhav (SPPU-PDF/ST/CH/2019/0004) sincerely acknowledged the financial support to Savitribai Phule Pune University (SPPU; formerly Pune University), Pune for providing the SPPU postdoctoral fellowship.

References

[1] A. Tuantranont, Applications of Nanomaterials in Sensors and Diagnostics, 14, Springer, Berlin Heidelberg, 2013.

[2] K. Levine (Ed.), Synthesis, Characterization and Modelling of Nano-Sized Structures, Nova Science Publishers, New York, 1990.

[3] Y. Shao, J. Wang, H. Wu, J. Liu, I.A. Aksay, Y. Lin, Graphene based electrochemical sensors and biosensors: a review, Electroanalysis 22 (10) (2010) 1027–1036, http://doi.org/10.1002/elan.200900571.

[4] J. Chattopadhyay, R. Srivastava (Eds.), Nanomaterials in Biomedical, Sensor and Energy Applications, Springer, Singapore, 2017.

[5] M.R. Willner, P.J. Vikesland, Nanomaterial enabled sensors for environmental contaminants, J. Nanobiotechnol. 16 (1) (2018) 1–16, http://doi.org/10.1186/s12951-018-0419-1.

[6] Y. Wu, Z. Shen, T. Yu (Eds.), Two-Dimensional Carbon Fundamental Properties, Synthesis, Characterization, and Applications, CRC Press, New York, 2014.

[7] Z. Farka, T. Juřík, D. Kovář, L. Trnková, P. Skládal, Nanoparticle-based immunochemical biosensors and assays: recent advances and challenges, Chem. Rev. 117 (15) (2017) 9973–10042, http://doi.org/10.1021/acs.chemrev.7b00037.

[8] C. Yang, M.E. Denno, P. Pyakurel, B.J. Venton, Recent trends in carbon nanomaterial-based electrochemical sensors for biomolecules: A review, Anal. Chim. Acta 887 (2015) 17–37, http://doi.org/10.1016/j.aca.2015.05.049.

[9] A. Chen, S. Chatterjee, Nanomaterials based electrochemical sensors for biomedical applications, Chem. Soc. Rev. 42 (12) (2013) 5425–5438, http://doi.org/10.1039/c3cs35518g.

[10] Y. Ou, A.M. Buchanan, C.E. Witt, P. Hashemi, Frontiers in electrochemical sensors for neurotransmitter detection: towards measuring neurotransmitters as chemical diagnostics for brain disorders, Anal. Methods 11 (21) (2019) 2738–2755, http://doi.org/10.1039/c9ay00055k.

[11] Z. Tavakolian-Ardakani, O. Hosu, C. Cristea, M. Mazloum-Ardakani, G. Marrazza, Latest trends in electrochemical sensors for neurotransmitters: A review, Sensors (Switzerland) 19 (9) (2019), http://doi.org/10.3390/s19092037.

[12] D.W. Kimmel, G. Leblanc, M.E. Meschievitz, D.E. Cliffel, Electrochemical sensors and biosensors, Anal. Chem. 84 (2) (2012) 685–707, http://doi.org/10.1021/ac202878q.

[13] X. Yan, et al., Catalytic activity of biomimetic model of cytochrome P450 in oxidation of dopamine, Talanta 179 (2017) 401–408, http://doi.org/10.1016/j.talanta.2017.11.038.

[14] J. Du, et al., Novel graphene flowers modified carbon fibers for simultaneous determination of ascorbic acid, dopamine and uric acid, Biosens. Bioelectron. 53 (2014) 220–224, http://doi.org/10.1016/j.bios.2013.09.064.

[15] Q. Huang, et al., A sensitive and reliable dopamine biosensor was developed based on the Au@carbon dots-chitosan composite film, Biosens. Bioelectron. 52 (2014) 277–280, http://doi.org/10.1016/j.bios.2013.09.003.

[16] D. Yuan, S. Chen, R. Yuan, J. Zhang, X. Liu, An ECL sensor for dopamine using reduced graphene oxide/multiwall carbon nanotubes/gold nanoparticles, Sensors Actuators, B Chem. 191 (2014) 415–420, http://doi.org/10.1016/j.snb.2013.10.013.

[17] A. Numan, M.M. Shahid, F.S. Omar, K. Ramesh, S. Ramesh, Facile fabrication of cobalt oxide nanograin-decorated reduced graphene oxide composite as ultrasensitive platform for dopamine detection, Sensors Actuators, B Chem. 238 (2017) 1043–1051, http://doi.org/10.1016/j.snb.2016.07.111.

[18] W. Dong, et al., Synthesis of tetrahexahedral Au-Pd core–shell nanocrystals and reduction of graphene oxide for the electrochemical detection of epinephrine, J. Colloid Interface Sci. 512 (2018) 812–818, http://doi.org/10.1016/j.jcis.2017.10.071.
[19] H.M. Moghaddam, H. Beitollahi, S. Tajik, H. Karimi Maleh, G.D. Noudeh, Simultaneous determination of norepinephrine, acetaminophen and tryptophan using a modified graphene nanosheets paste electrode, Res. Chem. Intermed. 41 (9) (2015) 6885–6896, http://doi.org/10.1007/s11164-014-1785-4.
[20] T.D. Thanh, J. Balamurugan, H. Van Hien, N.H. Kim, J.H. Lee, A novel sensitive sensor for serotonin based on high-quality of AuAg nanoalloy encapsulated graphene electrocatalyst, Biosens. Bioelectron. 96 (2017) 186–193, http://doi.org/10.1016/j.bios.2017.05.014.
[21] N.K. Sadanandhan, M. Cheriyathuchenaaramvalli, S.J. Devaki, A.R. Ravindranatha Menon, PEDOT-reduced graphene oxide-silver hybrid nanocomposite modified transducer for the detection of serotonin, J. Electroanal. Chem. 794 (2017) 244–253, http://doi.org/10.1016/j.jelechem.2017.04.027.
[22] X. Li, X. Lu, X. Kan, X. Li, X. Lu, X. Kan, 3D electrochemical sensor based on poly(hydroquinone)/gold nanoparticles/nickel foam for dopamine sensitive detection, J. Electroanal. Chem. 799 (2017) 451–458, http://doi.org/10.1016/j.jelechem.2017.06.047.
[23] Y. Huang, Y.E. Miao, S. Ji, W.W. Tjiu, T. Liu, Electrospun carbon nanofibers decorated with Ag-Pt bimetallic nanoparticles for selective detection of dopamine, ACS Appl. Mater. Interfaces 6 (15) (2014) 12449–12456, http://doi.org/10.1021/am502344p.
[24] G. Gao, Z. Zhang, K. Wang, Q. Yuan, X. Wang, One-pot synthesis of dendritic Pt3Ni nanoalloys as nonenzymatic electrochemical biosensors with high sensitivity and selectivity for dopamine detection, Nanoscale 9 (31) (2017) 10998–11003, http://doi.org/10.1039/c7nr03760k.
[25] J. Salamon, et al., One-pot synthesis of magnetite nanorods/graphene composites and its catalytic activity toward electrochemical detection of dopamine, Biosens. Bioelectron. 64 (2015) 269–276, http://doi.org/10.1016/j.bios.2014.08.085.
[26] P. Baraneedharan, S. Alexander, S. Ramaprabhu, One-step in situ hydrothermal preparation of graphene–SnO2 nanohybrid for superior dopamine detection, J. Appl. Electrochem. 46 (12) (2016) 1187–1197, http://doi.org/10.1007/s10800-016-1001-x.
[27] J. Chen, H. Huang, Y. Zeng, H. Tang, L. Li, A novel composite of molecularly imprinted polymer-coated PdNPs for electrochemical sensing norepinephrine, Biosens. Bioelectron. 65 (2015) 366–374, http://doi.org/10.1016/j.bios.2014.10.011.
[28] N. Lavanya, C. Sekar, Electrochemical sensor for simultaneous determination of epinephrine and norepinephrine based on cetyltrimethylammonium bromide assisted SnO2 nanoparticles, J. Electroanal. Chem. 801 (2017) 503–510, http://doi.org/10.1016/j.jelechem.2017.08.018.
[29] Y. Wang, et al., A disposable electrochemical sensor for simultaneous determination of norepinephrine and serotonin in rat cerebrospinal fluid based on MWNTs-ZnO/chitosan composites modified screen-printed electrode, Biosens. Bioelectron. 65 (2015) 31–38, http://doi.org/10.1016/j.bios.2014.09.099.
[30] A.C. Anithaa, K. Asokan, C. Sekar, Highly sensitive and selective serotonin sensor based on gamma ray irradiated tungsten trioxide nanoparticles, Sensors Actuators, B Chem. 238 (2017) 667–675, http://doi.org/10.1016/j.snb.2016.07.098.
[31] G. Ran, X. Chen, Y. Xia, Electrochemical detection of serotonin based on a poly(bromocresol green) film and Fe3O4 nanoparticles in a chitosan matrix, RSC Adv. 7 (4) (2017) 1847–1851, http://doi.org/10.1039/c6ra25639b.
[32] Y.T. Lin, C.H. Chen, M.S. Lin, Enzyme-free amperometric method for rapid determination of histamine by using surface oxide regeneration behavior of copper electrode, Sensors Actuators, B Chem 255 (2018) 2838–2843, http://doi.org/10.1016/j.snb.2017.09.101.
[33] N. Sattarahmady, H. Heli, R.D. Vais, An electrochemical acetylcholine sensor based on lichen-like nickel oxide nanostructure, Biosens. Bioelectron. 48 (2013) 197–202, http://doi.org/10.1016/j.bios.2013.04.001.
[34] T.T.C. Tseng, H.G. Monbouquette, Implantable microprobe with arrayed microsensors for combined amperometric monitoring of the neurotransmitters, glutamate and dopamine, J. Electroanal. Chem. 682 (2012) 141–146, http://doi.org/10.1016/j.jelechem.2012.07.014.
[35] E.A. Khudaish, et al., Sensitive and selective dopamine sensor based on novel conjugated polymer decorated with gold nanoparticles, J. Electroanal. Chem. 761 (2016) 80–88, http://doi.org/10.1016/j.jelechem.2015.12.011.

[36] B. Batra, S. Kumari, C.S. Pundir, Construction of glutamate biosensor based on covalent immobilization of glutmate oxidase on polypyrrole nanoparticles/polyaniline modified gold electrode, Enzyme Microb. Technol. 57 (2014) 69–77, http://doi.org/10.1016/j.enzmictec.2014.02.001.
[37] M. Tertiş, et al., Highly selective electrochemical detection of serotonin on polypyrrole and gold nanoparticles-based 3D architecture, Electrochem. commun. 75 (2017) 43–47, http://doi.org/10.1016/j.elecom.2016.12.015.
[38] M. Akhoundian, A. Rüter, S. Shinde, Ultratrace detection of histamine using a molecularly-imprinted polymer-based voltammetric sensor, Sensors (Switzerland) 17 (3) (2017), http://doi.org/10.3390/s17030645.
[39] X. Meng, et al., A molecularly imprinted electrochemical sensor based on gold nanoparticles and multiwalled carbon nanotube-chitosan for the detection of tryptamine, RSC Adv. 4 (73) (2014) 38649–38654, http://doi.org/10.1039/c4ra04503c.
[40] H. Mao, et al., Poly(ionic liquid) functionalized polypyrrole nanotubes supported gold nanoparticles: An efficient electrochemical sensor to detect epinephrine, Mater. Sci. Eng. C 75 (2017) 495–502, http://doi.org/10.1016/j.msec.2017.02.083.

CHAPTER 23

Nanomedicines as an alternative strategy for Fungal disease treatment

Swati Goswami, Vijay Kumar
Department of Microbiology, School of Science, RK University, Kasturbadham, Rajkot, Gujarat, India

23.1 Introduction

Fungal infections have an immeasurable impact on human health and the global economy. Superficial mycoses affect 20%–25% of the world's population [1]. It includes the infection of nails and skin infections such as ringworm and athlete's foot, predominantly caused by dermatophytes of family *Micosporum*, *Trichophyton,* and *Epidermophyton* species. Vulvovaginal candidiasis (thrush) affects 75% of females of childbearing age with a burden of recurrent infections in 5%–8% cases [2]. Invasive fungal infections are more severe with mortality of 1.5 million people per year [3]. Genera, including *Candida, Cryptococcus, Aspergillus,* and *Pneumocystis* are blameworthy for these lethal infections especially in immunocompromised patients and individuals with impaired immunity.

23.2 Fungi as human pathogens

Fungal species that human diseases differ from infections caused by other pathogens like bacteria or viral diseases in various ways. Fungal pathogens have many features in common with their host cells being eukaryotic, which impede the synthesis of antifungal drugs. Fungal tropism shows a huge diversity, as there are a wide range of cell types that are being attacked by fungi. This enables a single fungal pathogen capable of infecting multiple tissues in the same patient. Pathogenicity of fungi also depends on the host's immunological status. In addition to these, fungal pathogens have another unique feature that it can undergo morphogenic shifts during infection, this further enhances the infectivity. Despite the severity caused by fungal pathogens, it remained undervalued as major pathogens by both the public health officials and public. It was not earlier than the 1980s that invasive mycoses were widely acknowledged as medically important pathogens [4].

23.2.1 Dissemination of fungal diseases

Wide spread of fungal diseases usually indicates their ability to defy host defenses. This may occur due to various immune disorders or endocrinopathies. Fungal infections can

Advanced Nanomaterials for Point of Care Diagnosis and Therapy
DOI: https://doi.org/10.1016/B978-0-323-85725-3.00001-5

be treated and cured by productive management which requires a concerted effort to uncover and correct the underlying defects.

23.2.2 Host defense mechanisms

Our innate immune system resists fungal invasion at various levels such as skin that acts as the first line of defense. It offers barriers like pH variability that slows down fungal growth, competition with the normal bacterial flora, desiccated nature of the stratum corneum, naturally occurring long-chain unsaturated fatty acids and epithelial turn-over rate. Mucous membranes i.e. epithelium (respiratory tract, gastrointestinal tract, and vaginal vault) remain flooded with fluids, which contain antimicrobial substances along with ciliated cells that continuously help in removing foreign materials. Only a few fungi that survive these barriers, gain access to, colonize, and multiply in host tissues. Fungi enter the host tissues by inhalation or traumatic implantation. The severity of fungal disease is determined by a few curtail factors such as inoculum size, extent of tissue destruction, multiplication rate, and the host immunity [5–8].

23.3 Types of fungal infections

Most of the fungi that infect humans and cause disease are classified by tissue or organ levels that are primary sites of colonization. These are discussed below.

23.3.1 Superficial fungal infections

Superficial fungal infections only influence outermost layers of the skin stratum (*M. Furfur* and *Phaeoannellomyces werneckii*) or cuticle of the hair shaft (*Exophiala werneckii*); (*Piedraia hortae* and *Trichosporon beigelii*). These infections are typically aesthetic concerns and seldom cause the host to react immune (except occasionally *M. furfur* infections). Recently *T. beigelii* and *M. furfur*, especially immunosuppressed or otherwise weakened, were affected as opportunistic agents of the illness. These frequent germs infect patients through indoor catheters or intravenous lines accidently. Virtually nothing is known concerning the pathogenic mechanisms of these fungi [9].

23.3.2 Dermatophyte infections

The dermatophytes are a group of fungi that colonize skin, hair, and nails on the living host. These fungi contribute to greater invasive properties limited to the keratinized tissues thus causing mild scaling disorder to others which are generalized and highly inflammatory. Disease potential of these fungi depends on various host factors including the species of organism, immunologic status of the host, type of clothing worn, and type of footwear used in which trauma is important. These pathogens gain entry by releasing keratinase enzymes to metabolize the insoluble, tough fibrous protein. Factors like cell-mediated immunity and the presence of transferrin in serum inhibit fungal propagation to the deeper tissue layers and thus inhibit systemic diseases [10].

23.3.3 Subcutaneous mycoses

Subcutaneous mycoses is caused by the fungi that are abundant with low degree of infectivity. Although very little is known about their pathogenesis mechanism, they enter the host cell by traumatic implantation and survive by releasing proteolytic enzymes thus maintaining a facultative microaerophilic presence because of the reduced redox potential of the injured tissue. There is extensive tissue injury and purulent fluid production in eumycotic mycetoma, which exudes through various intercommunicating sinus tracts. Microabscesses are normal in chromoblastomycosis, however, as shown by the severe tissue reaction that characterizes the disease (pseudoepitheliomatous hyperplasia), the clinical presentation of the disease suggests a robust host response to the organism. Most of the fungi implicated in this category of disease exist in a hyphal morphology. The immunocompetent individuals have a high degree of innate resistance to disease. There is no information about mechanisms of pathogenesis [11].

23.3.4 Systemic mycoses

Blastomyces dermatitidis, *Coccidioides immitis*, *Histoplasma capsulatum*, *Paracoccidioides brasiliensis*, and *Penicillium marneffei* are dimorphic fungi which causes systemic mycoses by evading host immune system and successfully colonizing the host by surviving the elevated temperature of the body and either elude phagocytosis, neutralize the hostility they encounter, or adapt in a manner that will allow them to multiply. The primary site of infection is the respiratory tract. These fungi are dimorphic in nature thus changing from a mycelial to a unicellular morphology when they invade tissues, except *C. immitis* that forms spherules [12].

There is very little information about mechanisms of fungal pathogenicity, in contrast to what is known about molecular mechanisms of bacterial pathogenesis. Fungal pathogenesis is complex and involves the interplay of many factors [13].

23.4 Antifungal drugs and their mode of action

Currently, only four types of antifungal medications are available for the treatment of invasive fungal infections, including polyenes, pyrimidine analogs, echinocandins, and triazoles [14], whereas the fifth type, including allylamines, is used for the treatment of superficial dermatophyte infections [15]. Amphotericin B (AmB), a polyene has the ability to destabilize the membrane functions by binding to the ergosterol and acting as sterol [16]. Intrinsic toxic effects in humans by the use of AmB has been reported although this can be overcome by its liposome formulations [14]. Another drug molecule 5-fluorocytosine (5-FC), a pyrimidine analog, destabilizes nucleic acids (RNA, DNA) when metabolized by the fungal cells into fluorinated pyrimidines thus arresting its growth. For the treatment of *Cryptococcus* spp. meningitis 5-FC is used mainly in combination with AmB [17]. Catalytic subunit of β-1,3 glucan synthase is inhibited by Echinocandin, inhibiting the biosynthesis of the fungal cell wall [18]. The basic step that catalyzes lanosterol 145-007-demethylation in the biosynthesis of ergosterol is targeted at triazoles [14]. Fluconazole is commonly used due to its advantage of high oral availability and tolerability in patients.

23.4.1 Drug resistance mechanisms in fungi

Antifungal resistance mechanisms for most antifungal and fungal pathogens have been addressed at the molecular level. The processes are divided into multiple categories, including (1) reducing effective concentration by decreasing the concentrations of the intracellular drug by active efflux and by overexpression of the target drug. (2) The medication objectives change azoles and echinocandins. (3) Metabolic pathways are delayed because of loss or significant reduction in some functionalities in a certain route.

23.4.2 Multidrug resistance: A pattern of concern

It has been observed that fungal pathogens have developed resistance to single classes of drugs where different agents can exist. The classes of azoles (isavuconazole, fluconazole, voriconazole, itraconazole, and posaconazole) and echinocandins (anidulafungin, micafungin, and caspofungin) are few of the examples. Increased susceptibility to fungi infections and lack of appropriate treatment choices for infected patients are known to be one of the biggest problems in modern medicine. This suggests the need for new treatment approaches and the production of novel antifungal techniques. Different treatment approaches, including combination drug delivery, improvements in chemical composition and the use of drug transporters, are employed to overcome pathogenic strain resistances [19–21].

23.5 Nanomedicine/nanotherapy: An answer to antifungal resistance

The distinctive biological and physicochemical properties of nanomaterials make them riveting tools for developing modern antifungal methods [22,23]. Despite recent developments in antifungal chemotherapy with echinocandins and triazoles being less toxic and more efficient than traditional polyenes, the death risk of invasive fungal infections remains high, raising the need for new and more effective therapeutic methods [24].

The rate of discovery of antifungal drugs is unlikely to be sufficient for the future demands, since few drugs are currently being discovered. Here is a timeline presentation of approval of major antifungal drugs [25–28] (Fig. 23.1). It was not until the early 1970s, because of the risk of embolism, the intravenous administration of pharmaceutical suspensions was considered impractical.

Today, there is a consensus that nanotechnology represents a miniaturization of objects, as well as the preparation of nanomaterials with physical and chemical properties that drastically differ from those of bulk materials because they are on a nanoscale. As per current vista of microbial resistance and absence of new drugs, nanoparticles (NPs) emerged as the set of armories for the treatment of several diseases, including

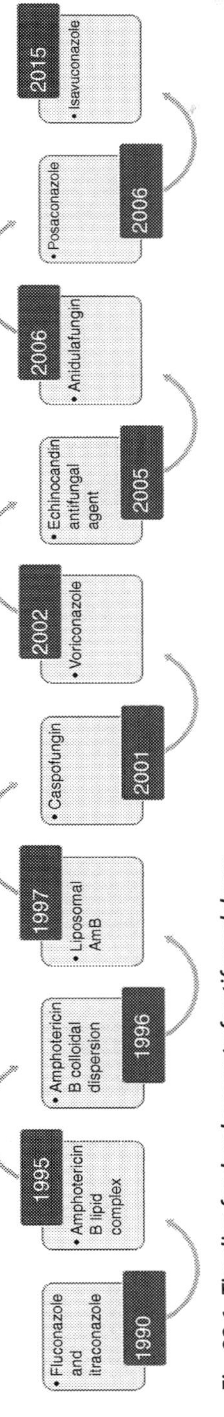

Fig. 23.1 ***Timeline for development of antifungal drugs.***

mycoses [81]. NPs are supramolecular ultradispersed structures with size ranging from 10 nm to 100 nm. The drugs required for delivery can be either encapsulated, entrapped, dissolved, or bound to a matrix of NPs, acting as a reservoir for particulate systems [29]. There are several studies supporting the increased efficacy of antifungal drugs incorporated into NPs for tackling fungal infections thus revolutionizing the drug delivery system using nanotechnology [3,30,31]. The nanomedicine or nanopharmaceutical suspension of NPs is currently used for the diagnosis, treatment and prevention of diseases. This improved the therapeutic index of various drugs by enhancing their efficacy by attacking desired tissues or cells and reducing toxicity. Delivery of drugs to the target site with the least toxicity is a matter of active effort for traditional antifungal drugs due to their low biodistribution, reduced potency, lack of selectivity, and toxicity to cells. These snags are well handled by the current drug delivery system, with NPs directly attacking compromised cells and tissues. The therapeutic ability of NPs as drug carriers depends on their form, hydrodynamic scale, surface chemistry, quantity, route of administration, period of stay in circulation and immune response, thereby exhibiting idiosyncratic biological and physicochemical properties for biomedical applications [32,158]. Drugs such as liposomes, inorganic and metal NPs, synthetic and natural polymers, silica and carbon materials, dendrimers and magnetic NPs (MNPs) are well carried by nanoscale or nanosized structures [33–35] (Fig. 23.2).

Nanoparticles are also used for topical and systemic antifungal drugs thus offering a lesser peril of systemic side effects while treating the skin infections. Silver, copper, gold,

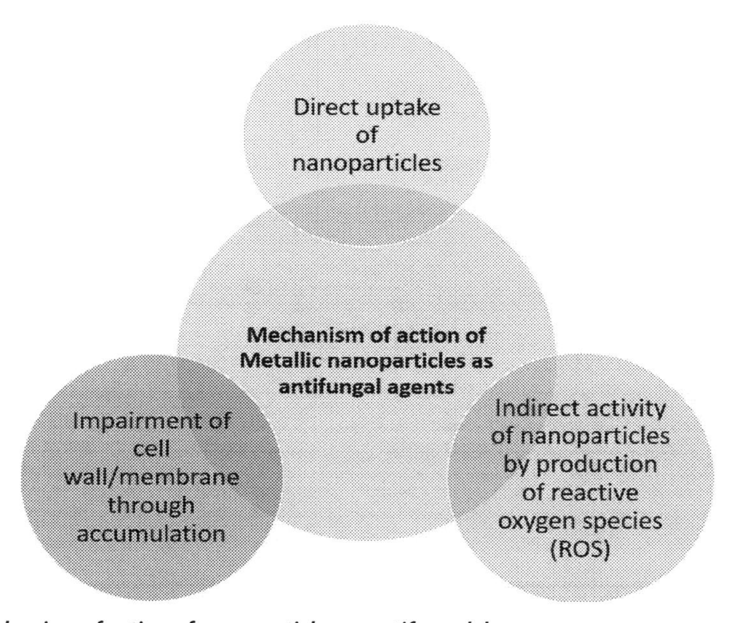

Fig. 23.2 ***Mechanism of action of nanoparticles as antifungal drugs.***

sulfur, titanium dioxide, zinc oxide, and some other NPs are designated for treating and managing the various skin disorders such as pityriasis versicolor, skin candidiasis and seborrhoeic dermatitis and tinea folliculitis. Treatment of skin infections by nanoparticle-based formulations have clinched weightage due to their increased skin permeability, targeted delivery and controlled release thereby increasing the bioavailability of active components with prolonged effect at the site of infection.

23.5.1 Metallic nanoparticles as antifungal agents

Metals present in nanoparticles can operate as catalysts, respond to biomolecules due to their very specialized surface, induce direct free radical formation, if they are exposed to the acidic environment or interact with organelles oxidative [36,37]. The inherent antibacterial activities of metallic nanoparticles have been employed to remove fungi which are diseases of humans and plants [38].

The specific processes of this activity are only speculated by three major paths, as seen in Fig. 23.1 [39]. Antimicrobial action is caused by the combination of these many paths [37]. ROS are oxygen-related by-products created under oxygenated circumstances of a material, such as superoxide anions, hydroxyl radicals and hydrogen peroxide that facilitate their interaction with biomolecules.

This can lead to a disruption between reactive species formation and the biological system's ability to detox or mend reactive mediums [39]. While cell defense somewhat minimizes the effects of ROS, overproduction of ROS can lead to oxidative stress and lipid peroxidation leading to membrane disintegration, mitochondrial malfunction, and DNA damage.

Cytotoxicity evaluations on nanoparticles, which owe their antimicrobial effects to ROS materials, are a toxic mechanism for humans, while preventing interactions and toxic reactions in men [39].

23.5.2 Mechanism of silver nanoparticles as an antifungal agent

Electrostatic attraction between the positively charged membrane of nanoparticles and the negatively charged microbial cellular membrane makes NPs of silver ions pivotal for the antimicrobial activity [39]. Due to the electrochemical potential of nanoparticles, they undergo dissolution processes thus leading to their separation into ions in the culture medium or within the microbial fluid [159]. Inhibitory action is caused by the accumulation of these ions in the interior or exterior of the microtubules resulting in the discontinuity within the cellular respiratory chains and damaging the microtubules [39]. The synthesis of adenosine triphosphate is hindered in the presence of Ag^+ as it has high affinity to thiol group in cysteine. Collapse of the proton motive force is also caused due to its binding with the transport proteins from the respiratory chain, causing the leak out of protons. Additionally, Ag^+ blocks the uptake of phosphate thus promoting the efflux of intracellular phosphate [59].

23.5.3 Mechanism of chitosan nanoparticles as an antifungal agent

Chitosan and its chemical derivatives possess ability to prolong the release of low-molecular-weight compounds to macromolecular drugs, mucoadhesive properties and in situ gelling performance which is used as key ingredient for drug delivery nanoformulations due to its mucoadhesive property, biodegradability, and biocompatibility [40]. *Candida* infections are well treated by chitosan NPs which is attributed to the presence of positively charged amino groups reacting with negatively charged groups of lipopolysaccharides and proteins on the surface of the microbial cells thus resulting in disintegration of the cell membrane [41]. This allows the binding of NPs with DNA followed by inhibiting mRNA and protein synthesis. Chitosan inhibits the activity of the growth-promoting enzymes thus affecting the sporulation and germination of fungal spores [41,42].

23.5.4 Mechanism of zinc oxide nanoparticles as an antifungal agent

Various fungal infections caused by *Aspergillus* and *Candida* along with dermatophyte infections can be treated by Zinc oxide nanoparticles (ZnONPs) [41]. Several studies revealed that antifungal drugs in presence of ZnONPs elevate the inhibitory activity thus reducing their toxicity due to decreased drug dosage and increased antifungal activity [43]. These NPs are the ray of hope as an engrossing and optimistic replacement to conventionally used preservatives in cosmetics in the upcoming days [44].

Dendrimers are one of the drug delivery tools which have an upper hand for complex therapy serving as both the adjunctive component of the dosage form and the drug carrier. Studies have reported that dendrimers also possess antifungal activity as demonstrated in Fig. 23.3 [45].

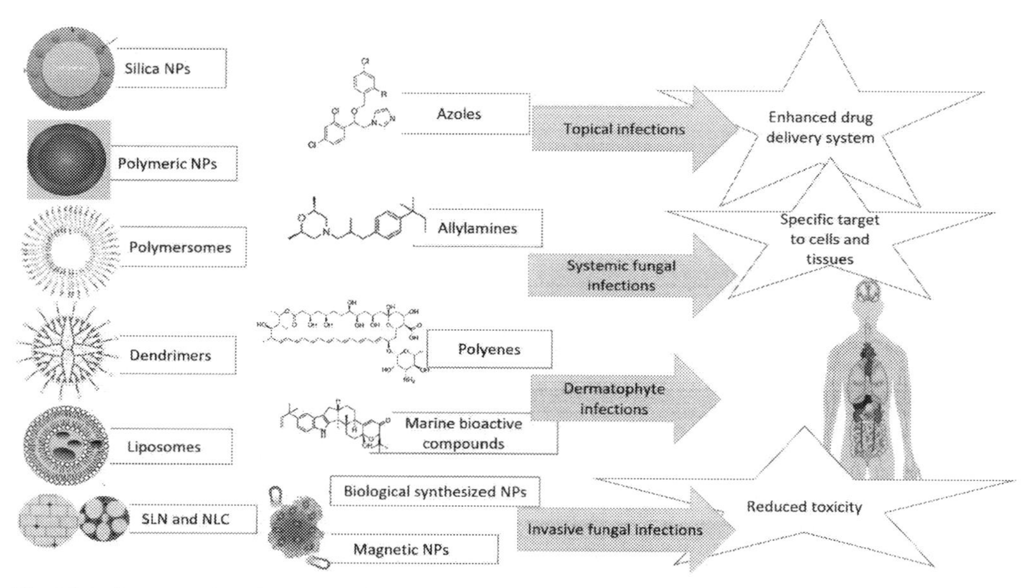

Fig. 23.3 ***Representing the nanotechnology-based new drug delivery systems.***

Because of their capacity to change and improve the pharmacokinetic and pharmacodynamic effects of medications, nanoparticles were used in pharmaceutical formulations. This is because they are able to boost the solubility and stability of the medications, to provide a regulated release of pharmaceuticals and to show biomass compatibility with tissues and cells [46,47]. Its size is also suitable with an intravascular injection and its large surface area is modified in ways that reduce systemic adverse effects, and increase therapeutic compliance with its decrease, so that it is delivered into a particular target by decreasing the usual dose and the frequency of administration [48,49]. This timely action is achievable as target ligands which allow for a preferential binding of particular types of cells may be included at a nanomolecular level with the use of antibodies and peptides on the transport surface [50–52]. Therefore, the creation of novel bioparticular drug systems, in particular nanoparticulate carriers, is a suitable technique for increasing traditional antifungal medications in therapeutic effectiveness, safety and compliance.

The below mentioned data in Table 23.1 represents an overview of the new delivery system for antifungal drugs and the drug chemical group, their route of administration, with their dosage form provided.

Table 23.1 Some of the novel drug delivery systems already developed for each antifungal drug.

Antifungal drugs	Novel drug delivery systems	Routes of administration	Dosage forms	References
Miconazole	Microemulsion	Topical	N.A.	[82]
	Solid lipid nanoparticles	Oral	N.A.	[83]
		Topical	Gel	[82]
	Liposomes	Topical	Gel	[84]
	Transethosomes	Topical	Gel	[85]
	Nanosponges	Vaginal	Gel	[86]
	Niosomes	Transdermal	Gel	[87]
	Nanoemulsion	Topical	N.A.	[88]
Econazole	Microemulsion	Percutaneous	N.A.	[89]
		Topical	Gel	[90]
	Solid lipid nanoparticles	Topical	Gel	[91]
	Nanostructured lipid carriers	Topical	Gel	[92]
	Liposomes	Topical	Gel	[93]
	Ethosomes	Topical	Gel	[94]
	Transethosomes	Transdermal	Gel	[95]
	Nanosponges	Topical	Hydrogel	[96]
	Niosomes	Transdermal	Gel	[97]
	Polymeric micelles	Topical	N.A.	[98]
	Nanoemulsion	Topical	N.A.	[99]

(Continued)

Table 23.1 (Cont'd)

Antifungal drugs	Novel drug delivery systems	Routes of administration	Dosage forms	References
Ketoconazole	Microemulsion	Oral	N.A.	[100]
	Solid lipid nanoparticles	Topical	Gel	[101]
	Nanostructured lipid carriers	Topical	Gel	[101]
	Liposomes	Topical	N.A.	[102]
	Spanlastics	Ocular	N.A.	[103]
	Dendrimers	Topical	Hydrogel	[45]
	Niosomes	Topical	Gel	[104]
Clotrimazole	Microemulsion	Buccal	Gel	[105]
		Vaginal	Gel	[77]
	Solid lipid nanoparticles	Topical	N.A.	[106]
	Nanostructured lipid carriers	Topical	N.A.	[106]
	Liposomes	Topical	Gel	[107]
	Ethosomes	Topical	Gel	[76]
	Transfersomes	Transdermal/ Topical	N.A.	[108]
	Nanosponges	Topical	Hydrogel	[109]
	Niosomes	Topical	Gel	[110]
	Polymeric emulgel	Topical	Gel	[111]
	Polymeric micelles	Topical	N.A.	[98]
Itraconazole	Microemulsion	Transdermal	N.A.	[112]
	Solid lipid nanoparticles	Ocular	N.A	[113]
	Nanostructured lipid carriers	Inhalation	N.A	[115]
	Liposomes	Topical	N.A.	[114]
	Polymersome	Intravenous	N.A	[115]
	Transfersomes	Transdermal	N.A	[116]
	Spanlastics	Ocular	N.A.	[117]
	Niosomes	Topical	N.A.	[118]
	Polymeric nanoparticles	Oral	N.A.	Pardeike et al., 2016, [78]
	Silica nanoparticles	Oral	N.A.	[119]
Fluconazole	Microemulsion	Vaginal	Gel	[77,120]
	Solid lipid nanoparticles	Topical	Gel	[121]
	Nanostructured lipid carriers	Oral	N.A.	[122]
	Liposomes	Intravitral	N.A.	[123]
	Ethosomes	Topical	Gel	[124]
	Spanlastics	Ocular	N.A.	[125]
	Microsponges	Topical	Gel	[79]
	Niosomes	Ocular	Gel	[126]
	Polymeric micelles	Topical	N.A.	[98]
	Polymeric amphiphilogel	Topical	Gel	[127]

Antifungal drugs	Novel drug delivery systems	Routes of administration	Dosage forms	References
Voriconazole	Microemulsion	Ocular	N.A.	[128]
	Solid lipid nanoparticles	Topical	Gel	[129]
	Ethosomes	Topical	N.A.	[130]
	Transethosomes	Topical	N.A.	[131]
	Polymeric nanoparticles	Ocular	N.A.	[132]
		Pulmonar	N.A.	[133]
Terbinafine	Solid lipid nanoparticles	Topical	N.A.	[134]
	Spanlastics	Transungual	N.A.	[135]
	Liposomes	Topical	Gel	[136]
	Polymeric chitosan nanoparticles	Topical	Hydrogel	[137]
Naftifine	Microemulsion	Topical	N.A.	[138]
	Niosomes	Topical	Gel	[139]
Butenafine	Microemulsion	Topical	Hydrogel	[140]
Amphotericin B	Microemulsion			
	Solid lipid nanoparticles	Oral	N.A.	[141]
	Nanostructured lipid carriers	Topical	N.A.	[142]
	Liposomes	Intravenous	N.A.	[143]
	Polymeric nanoparticles	Intravenous	N.A.	[144]
		Oral	N.A.	[145]
	Polymersomes	Oral	N.A	[146,147]
	Transfersomes	Topical	N.A.	[148]
	Micelles	Intravenous	N.A	[149]
	Silica nanoparticles	Intravenous	N.A	[150]
Nystatin	Solid lipid nanoparticles	Topical	N.A.	[151]
	Liposomes	Intravenous	N.A.	[152]
	Niosomes	Parenteral	N.A.	[153]
	Nanoemulsion	Topical	N.A.	[154]
Griseofulvin	Niosomes	Oral	N.A.	[155]
Ciclopirox	Niosomes	Topical	Gel	[156]
Caspofungin, Micafungin, Anidulafungin, Amorolfine		No nanotech studies yet released		[82]

23.6 Antifungal peptides as antifungal drugs/delivery system

There are a variety of disadvantages in existing antifungal agents that lower efficiency as therapeutic instruments against fungal infections. Only the presence of a few approved antifungal drug classes and the increased antifungal resistance make selective antifungal medication more complicated [53,54]. Azoles, polyes, and echinocandis, respectively, are the objectives of ergosterol synthesis, membrane ergosterol, and cell wall synthesis. When fungal pathogens are multi-resistant to antifungals, the low diversity of current processing mechanisms is a difficult situation [55]. In addition, current antifungal medications are concerned with adverse drug reactions, such as hypokalemia, infusion reactions, nephrotoxicity, hepatotoxicity, and gastrointestinal disorders, among others [53,56–58]. Any antifungal medicines affect typical eukaryotic targets both in human and pathogenic cells [55]. This makes it harder to create new and reliable antimicrobial therapies than antibacterial therapies. The AFPs have multiple benefits over the antifungal medicines that are specifically linked to their molecular targets and action mechanisms. AFPs are especially promising because they can identify several microbial targets, thus reducing the potential for resistance production [55], which is discussed in more detail in the following section. These microbial targets include fungal membranes, various cell wall components and molecules linked to physiological processes such as RNA, DNA and protein synthesis, and cell cycle [59,80]. AFPs are reported to target unique fungal compounds that include glucosylceramide, mannosyldiinositol phosphorylceramide, ergosterol-related enzymes or β-glucan synthesis, among others. This uniqueness of fungal killing mechanism along with no side effects translates into high pathogen selectivity and reduces the chance of cytotoxicity against mammalian cells but do not negates the possible cytotoxicity effects that may appear in rare cases [55,60]. As of now, three echinocandines, that is, anidulafungin, caspofungin, and micafungin have been permitted for use in Europe and the USA in past 13 years [61,160]. Synthetic killer peptides (KPs) are yet another promising antifungal peptides that have proven very promising antifungal effect barring cytotoxicity toward peripheral mononuclear blood cells in vitro or side consequences in murine studies. This peptide has demonstrated specific interaction with 1,3 β glucans in its dimeric form [62]. The reduced cytotoxicity properties of AFPs may be attributed to two reasons; firstly, compared with mammalian cell membranes, that are prominently neutral, there is a stronger interaction between the negatively charged fungal membrane (due to higher concentration of phosphatidylinositol and phosphatidic acid) and the cationic charges of the peptides (due to the high content of phosphatidylcholine). Second, some AFPs target membrane lipids that are unique to fungus but not found in mammalian cells, reducing toxicity [55,63]. The therapeutic activity of AMPs is multifaceted, according to growing evidence, and is not solely mediated by their direct antibacterial impact. Several publications and reviews have indicated that host defense peptides (HPDs), such as the defensins and cathelicidins families, have angiogeninc, immunomodulatory, and anti-inflammatory actions, as well as the ability to mobilize the adaptive immune response [62,64–68].

23.6.1 Resistance to antifungal peptides

C. albicans membrane remodeling has been linked to resistance to nonpeptide antifungal medicines, particularly ergosterol-sphingolipid-rich lipid rafts harboring multi-drug resistance (MDR) proteins connected to the membrane [69–72].

When it comes to AMP and AFP resistance, it's vital to know that, as previously said, these peptides primarily work through membrane contact, but that other ways of microbial suppression have also been proven [73]. When microorganisms, such as fungi, are exposed to antibiotics and antifungal medicines, they proliferate quickly and respond quickly. It's vital to remember, though, that cell membranes expand at a slower rate. Because of the quick and powerful action on membranes, as well as other inhibitory mechanisms displayed by AFPs, target microorganisms are less likely to develop de novo resistance [74]. Overuse of AFPs, as with other antimicrobial medicines now in use, may hasten the emergence of fungal resistance. Antifungal therapies are more difficult to use because of this difficulty. Indeed, widespread use of echinocandins in hospitals has resulted in an increase in the number of strains that have developed (secondary) resistance to these first-line antifungals, particularly among *C. glabratus* strains. Although there is a potential for high level resistance to AMPs, the risk is likely to be reduced in comparison with other antifungal products; however, it may depend on the administration of antimicrobial peptides [80]. Furthermore, while a reduced target concentration is required, the lack or change of a specific fungal target due to spontaneous mutation will result in resistance. However, such alterations to the conserved molecules may result in pathogen pathogenicity being lowered. Furthermore, combining antibiotics with other peptides, such as AMPs, would very certainly minimize the production of resistance [55,60,75,157].

23.7 Conclusion

The present chapter is focused on different nanoparticles and their formulations used for the treatment of various skin infections caused by fungi in humans. The different fungal pathogens and their diseases have shown continuous increase in severity and complexity among immunocompromised patients. Drugs presently available in the market possess additional side effects to human cells as both human and fungal cells are eukaryotic in nature. This has also led to increased incidences of fungal pathogens becoming drug resistant. Nanoparticle drug therapy has been looked upon as the need of the hour for existing fungal infections. Different nanoparticle therapeutic methods and their types along with its mechanism of actions were also discussed herein. This will give new insight to the researchers and pharma industries to explore in this field thus helping in prophylaxis and treatment to reach out to the largest number of patients.

References

[1] B. Havlickova, V.A. Czaika, M. Friedrich, Epidemiological trends in skin mycoses worldwide, Mycoses 51 (2008) 2–15.
[2] J.D. Sobel, Vulvovaginal candidosis, Lancet North Am. Ed. 369 (9577) (2007) 1961–1971.

[3] G.D. Brown, D.W. Denning, N.A. Gow, S.M. Levitz, M.G. Netea, T.C. White, Hidden killers: human fungal infections, Sci. Transl. Med. 4 (2012) 165rv13.
[4] M. Nucci, K.A. Marr, Emerging fungal diseases, Clin. Infect. Dis. 41 (4) (2005) 521–526.
[5] R.A. Cox, Immunosuppression by cell wall antigens of Coccidioides immitis, Rev. Infect. Dis. 10 (1988) S415–S418.
[6] L.G. Eissenberg, W.E. Goldman, P.H. Schlesinger, Histoplasma capsulatum modulates the acidification of phagolysosomes, J. Exp. Med. 177 (6) (1993) 1605–1611.
[7] G. Medoff, A. Painter, G.S. Kobayashi, Mycelial-to yeast-phase transitions of the dimorphic fungi Blastomyces dermatitidis and Paracoccidioides brasiliensis, J. Bacteriol. 169 (9) (1987) 4055–4060.
[8] D. Rotrosen, R.A. Calderone, J.E. Edwards Jr, Adherence of Candida species to host tissues and plastic surfaces, Rev. Infect. Dis. 8 (1) (1986) 73–85.
[9] B. Maresca, G.S. Kobayashi, Heat shock in dimorphic fungi, in: R.C. Matthews (Ed.), The Role of Heat Shock Proteins in Fungal Infection, RG Landes Co., Austin, Texas, 1995.
[10] G.S. Kobayashi, B. Maresca, Dimorphism in Histoplasma capsulatum: Study of cell differentiation and adaptation. In: Dimorphic Fungi in Biology and Medicine, Springer, Boston, MA, 1993, pp. 213–218.
[11] K.J. Kwon-Chung, J.C. Rhodes, Encapsulation and melanin formation as indicators of virulence in Cryptococcus neoformans, Infect. Immun. 51 (1) (1986) 218–223.
[12] B. Maresca, L. Carratù, G.S. Kobayashi, Morphological transition in the human fungal pathogen Histoplasma capulatum, Trends Microbiol. 2 (4) (1994) 110–114.
[13] G.S. Kobayashi, Disease mechanisms of fungi, Medical Microbiology, 4th edition (1996).
[14] D. Sanglard, F.C. Odds, Resistance of Candida species to antifungal agents: molecular mechanisms and clinical consequences, Lancet Infect. Dis. 2 (2) (2002) 73–85.
[15] C.J. Jessup, M.A. Ghannoum, N.S. Ryder, An evaluation of the in vitro activity of terbinafine, Med. Mycol. 38 (2) (2000) 155–159.
[16] T.M. Anderson, M.C. Clay, A.G. Cioffi, K. Diaz, Amphotericin forms an extramembranous and fungicidal sterol sponge, Nat. Chem. Biol. 10 (2014) 400–406.
[17] J.C. Nussbaum, A. Jackson, D. Namarika, J. Phulusa, J. Kenala, C. Kanyemba, J.N. Jarvis, S. Jaffar, M.C. Hosseinipour, D. Kamwendo, C.M. van der Horst, Combination flucytosine and high-dose fluconazole compared with fluconazole monotherapy for the treatment of cryptococcal meningitis: a randomized trial in Malawi, Clin. Infect. Dis. 50 (3) (2010) 338–344.
[18] M.C. Arendrup, J.L. Rodriguez-Tudela, C. Lass-Flörl, M. Cuenca-Estrella, J.P. Donnelly, W. Hope, European committee on antimicrobial susceptibility testing-subcommittee on antifungal susceptibility testing (EUCAST-AFST, 2011. EUCAST technical note on anidulafungin, Clin. Microbiol. Infect. 17 (11) (2011) E18–E20.
[19] S. Nami, A. Aghebati-Maleki, H. Morovati, L. Aghebati-Maleki, Current antifungal drugs and immunotherapeutic approaches as promising strategies to treatment of fungal diseases, Biomed. Pharmacother. 110 (2019) 857–868.
[20] L. Scorzoni, A.C. de Paula e Silva, C.M. Marcos, P.A. Assato, W.C. de Melo, H.C. de Oliveira, C.B. Costa-Orlandi, M.J. Mendes-Giannini, A.M. Fusco-Almeida, Antifungal therapy: new advances in the understanding and treatment of mycosis, Front. Microbiol. 8 (2017) 36.
[21] A.C. Souza, A.C. Amaral, Antifungal therapy for systemic mycosis and the nanobiotechnology era: improving efficacy, biodistribution and toxicity, Front. Microbiol. 8 (2017) 336.
[22] S. Khatry, N.S. Sirish, M. Sadanandam, Novel drug delivery systems for antifungal therapy, Int. J. Pharm. Pharm. Sci. 2 (4) (2010) 6–9.
[23] J.R. Perfect, Is there an emerging need for new antifungals? Expert Opin. Emerg. Drugs 21 (2) (2016) 129–131.
[24] G. Petrikkos, A. Skiada, Recent advances in antifungal chemotherapy, Int. J. Antimicrob. Agents 30 (2) (2007) 108–117.
[25] D. Allen, D. Wilson, R. Drew, J. Perfect, Azole antifungals: 35 years of invasive fungal infection management, Expert Rev. Anti Infect. Ther. 13 (6) (2015) 787–798.
[26] S. Ascioglu, K.A. Chan, Utilization and comparative effectiveness of caspofungin and voriconazole early after market approval in the US, PLoS One 9 (1) (2014) e83658.
[27] S. Chandwani, C. Wentworth, T.A. Burke, T.F. Patterson, Utilization and dosage pattern of echinocandins for treatment of fungal infections in US hospital practice, Curr. Med. Res. Opin. 25 (2) (2009) 385–393.

[28] P.L. McCormack, Isavuconazonium: first global approval, Drugs 75 (7) (2015) 817–822.
[29] G. Calixto, J. Bernegossi, B. Fonseca-Santos, M. Chorilli, Nanotechnology-based drug delivery systems for treatment of oral cancer: a review, Int. J. Nanomed. 9 (2014) 3719.
[30] E. Dorgan, D.W. Denning, R. McMullan, Burden of fungal disease in Ireland, J. Med. Microbiol. 64 (4) (2015) 423–426.
[31] J. Chander, A.M. Stchigel, A. Alastruey-Izquierdo, M. Jayant, K. Bala, H. Rani, U. Handa, R.S. Punia, U. Dalal, A.K. Attri, A. Monzon, Fungal necrotizing fasciitis, an emerging infectious disease caused by Apophysomyces (Mucorales), Rev. Iberoam. Micol. 32 (2) (2015) 93–98.
[32] A.Z. Wilczewska, K. Niemirowicz, K.H. Markiewicz, H. Car, Nanoparticles as drug delivery systems, Pharmacol. Rep. 64 (5) (2012) 1020–1037.
[33] I. Liakos, A.M. Grumezescu, A.M. Holban, Magnetite nanostructures as novel strategies for anti-infectious therapy, Molecules 19 (8) (2014) 12710–12726.
[34] J. Baumgartner, L. Bertinetti, M. Widdrat, A.M. Hirt, D. Faivre, Formation of magnetite nanoparticles at low temperature: from superparamagnetic to stable single domain particles, PLoS One 8 (3) (2013) e57070.
[35] A.R. Voltan, G. Quindós, K.P.M. Alarcón, A.M. Fusco-Almeida, M.J.S. Mendes-Giannini, M. Chorilli, Fungal diseases: could nanostructured drug delivery systems be a novel paradigm for therapy? Int. J. Nanomed. 11 (2016) 3715.
[36] A. Qidwai, R. Kumar, S.K. Shukla, A. Dikshit, Advances in biogenic nanoparticles and the mechanisms of antimicrobial effects, Indian J. Pharm. Sci. 80 (4) (2018) 592–603.
[37] Y.N. Slavin, J. Asnis, U.O. Häfeli, H. Bach, Metal nanoparticles: understanding the mechanisms behind antibacterial activity, J. Nanobiotechnol. 15 (1) (2017) 1–20.
[38] F. Asghari, Z. Jahanshiri, M. Imani, M. Shams-Ghahfarokhi, M. Razzaghi-Abyaneh, Antifungal nanomaterials: synthesis, properties, and applications. In: Nanobiomaterials in Antimicrobial Therapy, William Andrew Publishing, The Boulevard, Kidlington, Oxford, UK, 2016, pp. 343–383.
[39] M.D. Mashitah, Y. San Chan, J. Jason, Antimicrobial properties of nanobiomaterials and the mechanism. In: Nanobiomaterials in Antimicrobial Therapy, William Andrew Publishing, The Boulevard, Kidlington, Oxford, UK, 2016, pp. 261–312.
[40] N.L. Calvo, S. Sreekumar, L.A. Svetaz, M.C. Lamas, B.M. Moerschbacher, D. Leonardi, Design and characterization of chitosan nanoformulations for the delivery of antifungal agents, Int. J. Mol. Sci. 20 (15) (2019) 3686.
[41] K. Kon, M. Rai (Eds.), The Microbiology of Skin, Soft Tissue, Bone and Joint Infections, Academic Press, London, United States, 2017.
[42] F. Sousa, D. Ferreira, S. Reis, P. Costa, Current Insights on antifungal therapy: novel nanotechnology approaches for drug delivery systems and new drugs from natural sources, Pharmaceuticals 13 (9) (2020) 248.
[43] Q. Sun, J. Li, T. Le, Zinc oxide nanoparticle as a novel class of antifungal agents: current advances and future perspectives, J. Agric. Food Chem. 66 (43) (2018) 11209–11220.
[44] P. Singh, A. Nanda, Antimicrobial and antifungal potential of zinc oxide nanoparticles in comparison to conventional zinc oxide particles, J. Chem. Pharm. Res. 5 (11) (2013) 457–463.
[45] K. Winnicka, M. Wroblewska, P. Wieczorek, P.T. Sacha, E. Tryniszewska, Hydrogel of ketoconazole and PAMAM dendrimers: Formulation and antifungal activity, Molecules 17 (4) (2012) 4612–4624.
[46] P. Bhatt, R. Lalani, I. Vhora, S. Patil, J. Amrutiya, A. Misra, R. Mashru, Liposomes encapsulating native and cyclodextrin enclosed paclitaxel: Enhanced loading efficiency and its pharmacokinetic evaluation, Int. J. Pharm. 536 (1) (2018) 95–107.
[47] B.V.N. Nagavarma, H.K. Yadav, A.V.L.S. Ayaz, L.S. Vasudha, H.G. Shivakumar, Different techniques for preparation of polymeric nanoparticles-a review, Asian J. Pharm. Clin. Res. 5 (3) (2012) 16–23.
[48] J.H. Lee, Y. Yeo, Controlled drug release from pharmaceutical nanocarriers, Chem. Eng. Sci. 125 (2015) 75–84.
[49] S. D'Souza, A review of in vitro drug release test methods for nano-sized dosage forms, Advances in Pharmaceutics, 2014, Hindawi Publishing Corporation, Hindawi Limited Adam House, Third Floor1 Fitzroy Square London, W1T 5HF, United Kingdom, 2014.
[50] R. Goyal, L.K. Macri, H.M. Kaplan, J. Kohn, Nanoparticles and nanofibers for topical drug delivery, J. Control. Release 240 (2016) 77–92.

[51] A.T. Rangari, P. Ravikumar, Polymeric nanoparticles based topical drug delivery: an overview, Asian J. Biomed. Pharm. Sci. 5 (47) (2015) 5.
[52] R.A. Siegel, M.J. Rathbone, Overview of controlled release mechanisms. In: Fundamentals and applications of controlled release drug delivery, Springer, Boston, MA, 2012, pp. 19–43.
[53] P.G. Pappas, M.S. Lionakis, M.C. Arendrup, L. Ostrosky-Zeichner, B.J. Kullberg, Invasive candidiasis, Nat. Rev. Dis. Primers 4 (1) (2018) 1–20.
[54] M. Sanguinetti, B. Posteraro, C. Lass-Flörl, Antifungal drug resistance among Candida species: mechanisms and clinical impact, Mycoses 58 (2015) 2–13.
[55] M. Rautenbach, A.M. Troskie, J.A. Vosloo, Antifungal peptides: to be or not to be membrane active, Biochimie 130 (2016) 132–145.
[56] S.C.A. Chen, M.A. Slavin, T.C. Sorrell, Echinocandin antifungal drugs in fungal infections, Drugs 71 (1) (2011) 11–41.
[57] M.C.P. de Souza, A.G.D. Santos, A.M.M. Reis, Adverse drug reactions in patients receiving systemic antifungal therapy at a high-complexity hospital, J. Clin. Pharmacol. 56 (12) (2016) 1507–1515.
[58] I. Kyriakidis, A. Tragiannidis, S. Munchen, A.H. Groll, Clinical hepatotoxicity associated with antifungal agents, Expert Opin. Drug Saf. 16 (2) (2017) 149–165.
[59] M. Bondaryk, M. Staniszewska, P. Zielińska, Z. Urbańczyk-Lipkowska, Natural antimicrobial peptides as inspiration for design of a new generation antifungal compounds, J. Fungi 3 (3) (2017) 46.
[60] A. Matejuk, Q. Leng, M.D. Begum, M.C. Woodle, P. Scaria, S.T. Chou, A.J. Mixson, Peptide-based antifungal therapies against emerging infections, Drugs Future 35 (3) (2010) 197.
[61] M.W. Pound, M.L. Townsend, V. Dimondi, D. Wilson, R.H. Drew, Overview of treatment options for invasive fungal infections, Med. Mycol. 49 (6) (2011) 561–580.
[62] W. Magliani, S. Conti, T. Ciociola, L. Giovati, P.P. Zanello, T. Pertinhez, A. Spisni, L. Polonelli, Killer peptide: a novel paradigm of antimicrobial, antiviral and immunomodulatory auto-delivering drugs, Future Med. Chem. 3 (9) (2011) 1209–1231.
[63] L.T. Nguyen, E.F. Haney, H.J. Vogel, The expanding scope of antimicrobial peptide structures and their modes of action, Trends Biotechnol. 29 (9) (2011) 464–472.
[64] T. Hirsch, M. Metzig, A. Niederbichler, H.U. Steinau, E. Eriksson, L. Steinstraesser, Role of host defense peptides of the innate immune response in sepsis, Shock 30 (2) (2008) 117–126.
[65] I.N. Hsieh, K.L. Hartshorn, The role of antimicrobial peptides in influenza virus infection and their potential as antiviral and immunomodulatory therapy, Pharmaceuticals 9 (3) (2016) 53.
[66] Z. Li, R. Mao, D. Teng, Y. Hao, H. Chen, X. Wang, X. Wang, N. Yang, J. Wang, Antibacterial and immunomodulatory activities of insect defensins-DLP2 and DLP4 against multidrug-resistant Staphylococcus aureus, Sci. Rep. 7 (1) (2017) 1–16.
[67] T. Tran, R.A. Elliott, S.E. Taylor, M.C. Woodward, A self-administration of medications program to identify and address potential barriers to adherence in elderly patients, Ann. Pharmacother. 45 (2) (2011) 201–206.
[68] M. Zasloff, Antimicrobial peptides in health and disease, N. Engl. J. Med. 347 (15) (2002) 1199.
[69] K. Mukhopadhyay, T. Prasad, P. Saini, T.J. Pucadyil, A. Chattopadhyay, R. Prasad, Membrane sphingolipid-ergosterol interactions are important determinants of multidrug resistance in Candida albicans, Antimicrob. Agents Chemother. 48 (5) (2004) 1778–1787.
[70] R. Pasrija, T. Prasad, R. Prasad, Membrane raft lipid constituents affect drug susceptibilities of Candida albicans, Biochem. Soc. Trans. 33 (5) (2005) 1219–1223.
[71] R. Pasrija, S.L. Panwar, R. Prasad, Multidrug transporters CaCdr1p and CaMdr1p of Candida albicans display different lipid specificities: both ergosterol and sphingolipids are essential for targeting of CaCdr1p to membrane rafts, Antimicrob. Agents Chemother. 52 (2) (2008) 694–704.
[72] P. Shahi, W.S. Moye-Rowley, Coordinate control of lipid composition and drug transport activities is required for normal multidrug resistance in fungi, Biochim. Biophys. Acta 1794 (5) (2009) 852–859.
[73] M. Wu, E. Maier, R. Benz, R.E. Hancock, Mechanism of interaction of different classes of cationic antimicrobial peptides with planar bilayers and with the cytoplasmic membrane of Escherichia coli, Biochemistry 38 (22) (1999) 7235–7242.
[74] A.T. Yeung, S.L. Gellatly, R.E. Hancock, Multifunctional cationic host defence peptides and their clinical applications, Cell. Mol. Life Sci. 68 (13) (2011) 2161–2176.
[75] M. Fernández de Ullivarri, S. Arbulu, E. Garcia-Gutierrez, P.D. Cotter, Antifungal peptides as therapeutic agents, Front. Cell. Infect. Microbiol. 10 (2020) 105.

[76] N. Akhtar, K. Pathak, Cavamax W7 composite ethosomal gel of clotrimazole for improved topical delivery: development and comparison with ethosomal gel, AAPS Pharmscitech 13 (1) (2012) 344–355.
[77] Y.G. Bachhav, V.B. Patravale, Microemulsion based vaginal gel of fluconazole: formulation, in vitro and in vivo evaluation, Int. J. Pharm. 365 (1-2) (2009) 175–179.
[78] X. Ling, Z. Huang, J. Wang, J. Xie, M. Feng, Y. Chen, F. Abbas, J. Tu, J. Wu, C. Sun, Development of an itraconazole encapsulated polymeric nanoparticle platform for effective antifungal therapy, J. Mater. Chem. B 4 (10) (2016) 1787–1796.
[79] A. Moin, T.K. Deb, R.A.M. Osmani, R.R. Bhosale, U. Hani, Fabrication, characterization, and evaluation of microsponge delivery system for facilitated fungal therapy, J. Basic Clin. Pharm. 7 (2) (2016) 39.
[80] N.L. van der Weerden, M.R. Bleackley, M.A. Anderson, Properties and mechanisms of action of naturally occurring antifungal peptides, Cell. Mol. Life Sci. 70 (19) (2013) 3545–3570.
[81] V. Weissig, T.K. Pettinger, N. Murdock, Nanopharmaceuticals (part 1): products on the market, Int. J. Nanomed. 9 (2014) p.4357.
[82] I. Shahzadi, M.I. Masood, F. Chowdhary, A.A. Anjum, M.A. Nawaz, I. Maqsood, M.Q. Zaman, Microemulsion formulation for topical delivery of miconazole nitrate, Int. J. Pharm. Sci. Rev. Res. 24 (6) (2014) 30–36.
[83] B.M. Aljaeid, K.M. Hosny, Miconazole-loaded solid lipid nanoparticles: formulation and evaluation of a novel formula with high bioavailability and antifungal activity, Int. J. Nanomed. 11 (2016) 441.
[84] R.M. Elmoslemany, O.Y. Abdallah, L.K. El-Khordagui, N.M. Khalafallah, Propylene glycol liposomes as a topical delivery system for miconazole nitrate: comparison with conventional liposomes, AAPS Pharm. Sci. Tech. 13 (2) (2012) 723–731.
[85] M. Qushawy, A. Nasr, M. Abd-Alhaseeb, S. Swidan, Design, optimization and characterization of a transfersomal gel using miconazole nitrate for the treatment of candida skin infections, Pharmaceutics 10 (1) (2018) p.26.
[86] P.S. Kumar, N. Hematheerthani, J.V. Ratna, V. Saikishore, Design and characterization of miconazole nitrate loaded nanosponges containing vaginal gels, Int. J. Pharm. Ana. Res. 5 (3) (2016) 410–417.
[87] P.M. Firthouse, S.M. Halith, S.U. Wahab, M. Sirajudeen, S.K. Mohideen, Formulation and evaluation of miconazole niosomes, Int. J. Pharm. Tech. Res. 3 (2) (2011) 1019–1022.
[88] H.L. Maha, K.R. Sinaga, M. Masfria, Formulation and evaluation of miconazole nitrate nanoemulsion and cream, Asian J. Pharm. Clin. Res 11 (2018) 319–321.
[89] S. Ge, Y. Lin, H. Lu, Q. Li, J. He, B. Chen, C. Wu, Y. Xu, Percutaneous delivery of econazole using microemulsion as vehicle: formulation, evaluation and vesicle-skin interaction, Int. J. Pharm. 465 (1-2) (2014) 120–131.
[90] D. Evelyn, C.C. Wooi, J.R. Kumar, S. Muralidharan, S.A. Dhanaraj, Development and evaluation of microemulsion based gel (MBGs) containing econazole nitrate for nail fungal infection, J. Pharm. Res. 5 (4) (2012) 2385–2390.
[91] V. Sanna, E. Gavini, M. Cossu, G. Rassu, P. Giunchedi, Solid lipid nanoparticles (SLN) as carriers for the topical delivery of econazole nitrate: in-vitro characterization, ex-vivo and in-vivo studies, J. Pharm. Pharmacol. 59 (8) (2007) 1057–1064.
[92] L. Keshri, K. Pathak, Development of thermodynamically stable nanostructured lipid carrier system using central composite design for zero order permeation of econazole nitrate through epidermis, Pharm. Dev. Technol. 18 (3) (2013) 634–644.
[93] X.R. Qi, M.H. Liu, H.Y. Liu, Y. Maitani, T. Nagai, Topical econazole delivery using liposomal gel, STP Pharma Sci. 13 (4) (2003) 241–245.
[94] P. Verma, K. Pathak, Nanosized ethanolic vesicles loaded with econazole nitrate for the treatment of deep fungal infections through topical gel formulation, Nanomed. Nanotechnol. Biol. Med. 8 (4) (2012) 489–496.
[95] S. Verma, P. Utreja, Transethosomes of econazole nitrate for transdermal delivery: development, in-vitro characterization, and ex-vivo assessment, Pharm. Nanotechnol. 6 (3) (2018) 171–179.
[96] R. Sharma, R. Walker, K. Pathak, Evaluation of the kinetics and mechanism of drug release from econazole nitrate nanosponge loaded carbapol hydrogel, Indian J. Pharm. Educ. Res. 45 (1) (2011) 25–31.

[97] Y.P. Kumar, K.V. Kumar, R.R. Shekar, M. Ravi, V.S. Kishore, Formulation and evaluation of econazole niosomes. Sch. Acad, J. Pharm. 2 (4) (2013) 315–318.
[98] Y.G. Bachhav, K. Mondon, Y.N. Kalia, R. Gurny, M. Möller, Novel micelle formulations to increase cutaneous bioavailability of azole antifungals, J. Control. Release 153 (2) (2011) 126–132.
[99] M.P.Y. Piemi, D. Korner, S. Benita, J.P. Marty, Positively and negatively charged submicron emulsions for enhanced topical delivery of antifungal drugs, J. Control. Release 58 (2) (1999) 177–187.
[100] N. Tiwari, A. Sivakumar, A. Mukherjee, N. Chandrasekaran, Enhanced antifungal activity of Ketoconazole using rose oil based novel microemulsion formulation, J. Drug Deliv. Sci. Technol. 47 (2018) 434–444.
[101] E.B. Souto, R.H. Müller, SLN and NLC for topical delivery of ketoconazole, J. Microencapsul. 22 (5) (2005) 501–510.
[102] S. Ashe, D. Nayak, G. Tiwari, P.R. Rauta, B. Nayak, Development of liposome-encapsulated ketoconazole: formulation, characterisation and evaluation of pharmacological therapeutic efficacy, Micro Nano Lett. 10 (2) (2015) 126–129.
[103] S. Kakkar, I.P. Kaur, Spanlastics—a novel nanovesicular carrier system for ocular delivery, Int. J. Pharm. 413 (1-2) (2011) 202–210.
[104] S.B. Shirsand, M.S. Para, D. Nagendrakumar, K.M. Kanani, D. Keerthy, Formulation and evaluation of Ketoconazole niosomal gel drug delivery system, Int. J. Pharm. Investig. 2 (4) (2012) 201.
[105] J. Kaewbanjong, P. Wan Sia Heng, P. Boonme, Clotrimazole microemulsion and microemulsion-based gel: evaluation of buccal drug delivery and irritancy using chick chorioallantoic membrane as the model, J. Pharm. Pharmacol. 69 (12) (2017) 1716–1723.
[106] E.B. Souto, S.A. Wissing, C.M. Barbosa, R.H. Müller, Development of a controlled release formulation based on SLN and NLC for topical clotrimazole delivery, Int. J. Pharm. 278 (1) (2004) 71–77.
[107] M. Ning, Y. Guo, H. Pan, X. Chen, Z. Gu, Preparation, in vitro and in vivo evaluation of liposomal/niosomal gel delivery systems for clotrimazole, Drug Dev. Ind. Pharm. 31 (4-5) (2005) 375–383.
[108] R.G. Maheshwari, R.K. Tekade, P.A. Sharma, G. Darwhekar, A. Tyagi, R.P. Patel, D.K. Jain, Ethosomes and ultradeformable liposomes for transdermal delivery of clotrimazole: a comparative assessment, Saudi Pharm. J. 20 (2) (2012) 161–170.
[109] A.S. Kumar, P.S. Sheri, M.A. Kuriachan, Formulation and Evaluation of Antifungal Nanosponge Loaded Hydrogel for Topical Delivery, Int. J. Pharm. Pharm. Res. 13 (2018) 362–379.
[110] S.B. Shirsand, G.R. Kumar, G.G. Keshavshetti, S.S. Bushetti, P.V. Swamy, Formulation and evaluation of clotrimazole niosomal gel for topical application, RGUHS J Pharm Sci 5 (1) (2015) 32–38.
[111] G.E. Yassin, Formulation and evaluation of optimized clotrimazole emulgel formulations, J. Pharm. Res. Int. 4 (2014) 1014–1030.
[112] A. Chudasama, V. Patel, M. Nivsarkar, K. Vasu, C. Shishoo, Investigation of microemulsion system for transdermal delivery of itraconazole, J. Adv. Pharm. Technol. Res. 2 (1) (2011) 30.
[113] B. Mohanty, D.K. Majumdar, S.K. Mishra, A.K. Panda, S. Patnaik, Development and characterization of itraconazole-loaded solid lipid nanoparticles for ocular delivery, Pharm. Dev. Technol. 20 (4) (2015) 458–464.
[114] A.F.G. Leal, M.C. Leite, C.S.Q. Medeiros, I.M.F. Cavalcanti, A.G. Wanderley, N.S.S. Magalhães, R.P. Neves, Antifungal activity of a liposomal itraconazole formulation in experimental Aspergillus flavus keratitis with endophthalmitis, Mycopathologia 179 (3-4) (2015) 225–229.
[115] J. Pardeike, S. Weber, T. Haber, J. Wagner, H.P. Zarfl, H. Plank, A. Zimmer, Development of an itraconazole-loaded nanostructured lipid carrier (NLC) formulation for pulmonary application, Int. J. Pharm. 419 (1–2) (2011) 329–338.
[116] W.S. Zheng, X.Q. Fang, L.L. Wang, Y.J. Zhang, Preparation and quality assessment of itraconazole transfersomes, Int. J. Pharm. 436 (1-2) (2012) 291–298.
[117] A.N. ElMeshad, A.M. Mohsen, Enhanced corneal permeation and antimycotic activity of itraconazole against Candida albicans via a novel nanosystem vesicle, Drug Deliv. 23 (7) (2016) 2115–2123.
[118] V.D. Wagh, O.J. Deshmukh, Itraconazole niosomes drug delivery system and its antimycotic activity against Candida albicans, Int. Sch. Res. Notices 2012 (2012) 1–7, doi:10.5402/2012/653465. https://downloads.hindawi.com/archive/2012/653465.pdf. 653465.
[119] R. Mellaerts, R. Mols, J.A. Jammaer, C.A. Aerts, P. Annaert, J. Van Humbeeck, G. Van den Mooter, P. Augustijns, J.A. Martens, Increasing the oral bioavailability of the poorly water soluble drug itraconazole with ordered mesoporous silica, Eur. J. Pharm. Biopharm. 69 (1) (2008) 223–230.

[120] Y.G. Bachhav, V.B. Patravale, Microemulsion-based vaginal gel of clotrimazole: formulation, in vitro evaluation, and stability studies, Aaps Pharmscitech 10 (2) (2009) 476–481.
[121] S. El-Housiny, M.A. Shams Eldeen, Y.A. El-Attar, H.A. Salem, D. Attia, E.R. Bendas, M.A. El-Nabarawi, Fluconazole-loaded solid lipid nanoparticles topical gel for treatment of pityriasis versicolor: formulation and clinical study, Drug Deliv. 25 (1) (2018) 78–90.
[122] H.R. Kelidari, M. Moazeni, R. Babaei, M. Saeedi, J. Akbari, P.I. Parkoohi, M. Nabili, A.A. Gohar, K. Morteza-Semnani, A. Nokhodchi, Improved yeast delivery of fluconazole with a nanostructured lipid carrier system, Biomed. Pharmacother. 89 (2017) 83–88.
[123] S.K. Gupta, N. Dhingra, T. Velpandian, J. Jaiswal, Efficacy of fluconazole and liposome entrapped fluconazole for C. albicans induced experimental mycotic endophthalmitis in rabbit eyes, Acta Ophthalmol. Scand. 78 (4) (2000) 448–450.
[124] N. Indora, D. Kaushik, Design, development and evaluation of ethosomal gel of fluconazole for topical fungal infection, Int.J. Eng. Sci. Invention Res. Develop. 1 (8) (2015) 280–306.
[125] I.P. Kaur, C. Rana, M. Singh, S. Bhushan, H. Singh, S. Kakkar, Development and evaluation of novel surfactant-based elastic vesicular system for ocular delivery of fluconazole, J. Ocul. Pharmacol. Ther. 28 (5) (2012) 484–496.
[126] O.A.E.A. Soliman, E.A. Mohamed, N.A.A. Khatera, Enhanced ocular bioavailability of fluconazole from niosomal gels and microemulsions: Formulation, optimization, and in vitro–in vivo evaluation, Pharm. Dev. Technol. 24 (1) (2019) 48–62.
[127] S.K. Lalit, A.S. Panwar, G. Darwhekar, D.K. Jain, Formulation and evaluation of fluconazole amphiphilogel, Der Pharmacia Lettre 3 (5) (2011) 125–131.
[128] R. Kumar, V.R. Sinha, Preparation and optimization of voriconazole microemulsion for ocular delivery, Colloids Surf. B 117 (2014) 82–88.
[129] D.K. Pandurangan, P. Bodagala, V.K. Palanirajan, S. Govindaraj, Formulation and evaluation of voriconazole ophthalmic solid lipid nanoparticles in situ gel, Int. J. Pharm. Investig. 6 (1) (2016) 56.
[130] W. Faisal, G.M. Soliman, A.M. Hamdan, Enhanced skin deposition and delivery of voriconazole using ethosomal preparations, J. Liposome Res. 28 (1) (2018) 14–21.
[131] C.K. Song, P. Balakrishnan, C.K. Shim, S.J. Chung, S. Chong, D.D. Kim, A novel vesicular carrier, transethosome, for enhanced skin delivery of voriconazole: characterization and in vitro/in vivo evaluation, Colloids Surf. B 92 (2012) 299–304.
[132] E. Başaran, et al., Voriconazole incorporated polymeric nanoparticles for ocular application, Lat. Am. J. Pharm. 36 (10) (2017) 1983–1994.
[133] P.J. Das, P. Paul, B. Mukherjee, B. Mazumder, L. Mondal, R. Baishya, M.C. Debnath, K.S. Dey, Pulmonary delivery of voriconazole loaded nanoparticles providing a prolonged drug level in lungs: a promise for treating fungal infection, Mol. Pharm. 12 (8) (2015) 2651–2664.
[134] Y.C. Chen, D.Z. Liu, J.J. Liu, T.W. Chang, H.O. Ho, M.T. Sheu, Development of terbinafine solid lipid nanoparticles as a topical delivery system, Int. J. Nanomed. 7 (2012) 4409.
[135] N.I. Elsherif, R.N. Shamma, G. Abdelbary, Terbinafine hydrochloride trans-ungual delivery via nanovesicular systems: in vitro characterization and ex vivo evaluation, Aaps Pharmscitech 18 (2) (2017) 551–562.
[136] S. Tuncay Tanrıverdi, S. Hilmioğlu Polat, D. Yeşim Metin, G. Kandiloğlu, Ö. Özer, Terbinafine hydrochloride loaded liposome film formulation for treatment of onychomycosis: in vitro and in vivo evaluation, J. Liposome Res. 26 (2) (2016) 163–173.
[137] İ. Özcan, Ö. Abacı, A.H. Uztan, B. Aksu, H. Boyacıoğlu, T. Güneri, Ö. Özer, Enhanced topical delivery of terbinafine hydrochloride with chitosan hydrogels, Aaps Pharmscitech 10 (3) (2009) 1024–1031.
[138] M.S. Erdal, G. Özhan, M.C. Mat, Y. Özsoy, S. Güngör, Colloidal nanocarriers for the enhanced cutaneous delivery of naftifine: characterization studies and in vitro and in vivo evaluations, Int. J. Nanomed. 11 (2016) 1027.
[139] H.S. Barakat, I.A. Darwish, L.K. El-Khordagui, N.M. Khalafallah, Development of naftifine hydrochloride alcohol-free niosome gel, Drug Dev. Ind. Pharm. 35 (5) (2009) 631–637.
[140] A.B. Pillai, J.V. Nair, N.K. Gupta, S. Gupta, Microemulsion-loaded hydrogel formulation of butenafine hydrochloride for improved topical delivery, Arch. Dermatol. Res. 307 (7) (2015) 625–633.
[141] P. Jansook, Z. Fülöp, G.C. Ritthidej, Amphotericin B loaded solid lipid nanoparticles (SLNs) and nanostructured lipid carrier (NLCs): physicochemical and solid-solution state characterizations, Drug Dev. Ind. Pharm. 45 (4) (2019) 560–567.

[142] D. Butani, C. Yewale, A. Misra, Topical Amphotericin B solid lipid nanoparticles: Design and development, Colloids Surf. B 139 (2016) 17–24.
[143] R.J. Hay, Liposomal amphotericin B, AmBisome, J. Infect. 28 (1994) 35–43.
[144] A.C.O. Souza, A.D. Nascimento, N.M. De Vasconcelos, M.S. Jerônimo, I.M. Siqueira, L. R-Santos, D.O.S. Cintra, L.L. Fuscaldi, O.P. Junior, R. Titze-de-Almeida, M.F. Borin, Activity and in vivo tracking of Amphotericin B loaded PLGA nanoparticles, Eur. J. Med. Chem. 95 (2015) 267–276.
[145] J.L. Italia, M.N.V. Ravi Kumar, K.C. Carter, Evaluating the potential of polyester nanoparticles for per oral delivery of amphotericin B in treating visceral leishmaniasis, J. Biomed. Nanotechnol. 8 (4) (2012) 695–702.
[146] J.P. Jain, N. Kumar, Development of amphotericin B loaded polymersomes based on (PEG) 3-PLA co-polymers: factors affecting size and in vitro evaluation, Eur. J. Pharm. Sci. 40 (5) (2010) 456–465.
[147] X. Tang, H. Zhu, L. Sun, W. Hou, S. Cai, R. Zhang, F. Liu, Enhanced antifungal effects of amphotericin B-TPGS-b-(PCL-ran-PGA) nanoparticles in vitro and in vivo, Int. J. Nanomed. 9 (2014) 5403.
[148] A.P. Perez, M.J. Altube, P. Schilrreff, G. Apezteguia, F.S. Celes, S. Zacchino, C.I. de Oliveira, E.L. Romero, M.J. Morilla, Topical amphotericin B in ultradeformable liposomes: formulation, skin penetration study, antifungal and antileishmanial activity in vitro, Colloids Surf. B 139 (2016) 190–198.
[149] A.C. Moreno-Rodríguez, S. Torrado-Durán, G. Molero, J.J. García-Rodríguez, S. Torrado-Santiago, Efficacy and toxicity evaluation of new amphotericin B micelle systems for brain fungal infections, Int. J. Pharm. 494 (1) (2015) 17–22.
[150] A. Lykov, K. Gaidul, I. Goldina, V. Konenkov, V. Kozlov, N. Lyakhov, A. Dushkin, Silica nanoparticles as a basis for efficacy of antimicrobial drugs. In: Nanostructures for Antimicrobial Therapy, Elsevier, Amsterdam, Netherlands, 2017, pp. 551–575.
[151] R.M. Khalil, A.A. Abd El Rahman, M.A. Kassem, M.S. El Ridi, M.M. Abou Samra, G.E. Awad, S.S. Mansy, Preparation and in vivo assessment of nystatin-loaded solid lipid nanoparticles for topical delivery against cutaneous candidiasis, Int. J. Pharm. Pharm. Sci. 8 (7) (2014) 421–429.
[152] F. Offner, V. Krcmery, M. Boogaerts, C. Doyen, D. Engelhard, P. Ribaud, C. Cordonnier, B. de Pauw, S. Durrant, J.P. Marie, P. Moreau, Liposomal nystatin in patients with invasive aspergillosis refractory to or intolerant of amphotericin B, Antimicrob. Agents Chemother. 48 (12) (2004) 4808–4812.
[153] M.S. El-Ridy, A. Abdelbary, T. Essam, R.M. Abd EL-Salam, A.A. Aly Kassem, Niosomes as a potential drug delivery system for increasing the efficacy and safety of nystatin, Drug Dev. Ind. Pharm. 37 (12) (2011) 1491–1508.
[154] F. Fernández-Campos, B. Clares Naveros, O. Lopez Serrano, C. Alonso Merino, A.C. Calpena Campmany, Evaluation of novel nystatin nanoemulsion for skin candidosis infections, Mycoses 56 (1) (2013) 70–81.
[155] P.S. Jadon, V. Gajbhiye, R.S. Jadon, K.R. Gajbhiye, N. Ganesh, Enhanced oral bioavailability of griseofulvin via niosomes, Aaps Pharmscitech 10 (4) (2009) 1186–1192.
[156] S.B. Shirsand, G.G. Keshavshetti, Formulation and characterization of drug loaded niosomes for antifungal activity, Sper. J. Adv. Nov. Drug Deliv. 1 (2016) 12–17.
[157] L. Steinstraesser, U.M. Kraneburg, T. Hirsch, M. Kesting, H.U. Steinau, F. Jacobsen, S. Al-Benna, Host defense peptides as effector molecules of the innate immune response: a sledgehammer for drug resistance? Int. J. Mol. Sci. 10 (9) (2009) 3951–3970.
[158] V. Weissig, D. Guzman-Villanueva, Nanopharmaceuticals (part 2): products in the pipeline, Int. J. Nanomed. 10 (2014) 1245.
[159] A. Qidwai, S. Khan, S. Md, M. Fazil, S. Baboota, J.K. Narang, J. Ali, Nanostructured lipid carrier in photodynamic therapy for the treatment of basal-cell carcinoma, Drug Deliv. 23 (4) (2016) 1476–1485.
[160] N.D. Beyda, R.E. Lewis, K.W. Garey, Echinocandin resistance in Candida species: mechanisms of reduced susceptibility and therapeutic approaches, Ann. Pharmacother. 46 (7–8) (2012) 1086–1096.

CHAPTER 24

Technological advancement in nano diagnostics point of care test development for biomedical application

Anulipsa Priyadarshini[a], Tejaswini Sahoo[a], Deepak Senapati[a], Sabyasachi Parida[b], Rojalin Sahu[a]
[a]School of Applied Sciences, Kalinga Institute of Industrial Technology, Deemed to be University, Bhubaneswar, Odisha, India
[b]Pradyumna Bal Memorial Hospital, Kalinga Institute of Medical Sciences, Bhubaneswar, Odisha, India

24.1 Introduction

Research on nanotechnology advancement leads to the invention of different nanomaterials which are used in a wide variety of applications. Especially, sensors designed using nanomaterials show incredible performance in various fields like manufacturing industries, protection of the environment, medical diagnosis, and bioengineering because of its outstanding properties such as small size, high surface–volume ratio, etc. [1-3]. In this era of nanotechnology, the significant favorable impact has in the biomedical field. For example, the conventional methods used to diagnose different diseases always require a specialized laboratory, skilled persons, and measuring or detecting devices. These methods are costly and are not able to respond rapidly. So it is always desirable to have low cost, easy to use, real-time detection, fast response, and portable devices. These features can only be implemented by nanotechnology using different well-fabricated nanomaterials. Sensors are primarily used for detecting any changes in quantity and convert them into specific signals. For a suitable sensor, it should be highly sensitive, quickly respond, and stable. Nanosensors are applied for monitoring physical and chemical phenomena in regions difficult to reach, detecting biochemicals in cellular organelles, measuring nanoscopic particles in the industry and environment. Nanosensors are also used for national security purposes, in aerospace, integrated circuits, and many more. Different types of nanosensors are manufactured in several ways. Nanosensors can be made from metals (e.g., aluminum, cobalt, copper, gold, zinc) or metal oxides (e.g., iron oxide, zinc oxide, titanium oxide) based nanoparticles. It can also be made from carbon- or polymer-based nanoparticles. These nanoparticles are developed or improved by using various synthesis methods and also enhance the properties like optical, physical, chemical, electrical, and mechanical. It improves nanoparticle characterization and subsequent application.

Advanced Nanomaterials for Point of Care Diagnosis and Therapy
DOI: https://doi.org/10.1016/B978-0-323-85725-3.00022-2

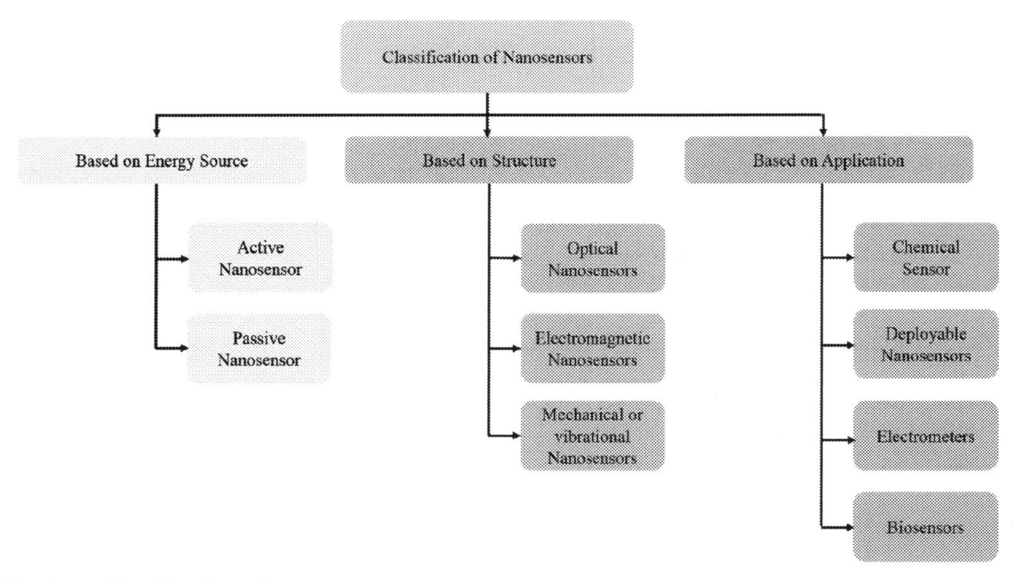

Fig. 24.1 ***Classification of nanosensors.***

24.2 Classification of nanosensors

According to the energy source, structure [4], and applications [5], the classification of nanosensors is illustrated in Fig. 24.1.

Active nanosensors: It requires no energy source to convert signals from one form to another form. They work on the energy conversion principle. Some of the best examples of active nanosensors are piezoelectric sensor and thermocouple.

Passive nanosensors: It requires additional or external energy source for their operation. Thermistor is an example of a passive nanosensor.

Optical nanosensors: It has the ability to monitor chemical analysis. These sensors are used in various fields like pharmaceuticals, chemical industries, biotechnology, and environmental sciences. According to the report, fluorescein-based nanosensor using polyacrylamide nanoparticle was the first optical sensor [6]. Fluorescent sensors are particles that include at least one binding component and photoactive units [7–8]. Fluorescent sensors are used for pH measurement. Sasaki et al. [9] reported fluorescent dye investigation within the cell, which is the most basic type of optical sensor. Molecular fluorescent dye is used to reduce cell disruption.

Electromagnetic nanosensors: This sensor is operated via two mechanisms, that is, by electrical current measurement and magnetism measurement.

Electrical current measurement: In the case of electrical current mechanism, a salient advantage is that it is the label-free methodology over dyes' use. The sensors developed using these mechanisms are used to detect and monitor different oxidase, such as glucose [10–12] and lactate [13].

Magnetism measurement: Magnetic nanosensors are basically fabricated by using magnetic nanoparticles. This sensor type is used in magnetic resonance imaging to distinguish between tissue types by magnetic resonance techniques. Apart from bioscience application, magnetic nanosensors are used in the electronics industries, which work on the basis of the magnetoresistance phenomenon [14].

Mechanical or vibrational nanosensors: It is used for detecting mechanical properties in nano range of any mechanical device. Binh et al. invented the first-ever mechanical nanosensor to monitor the elastic property of a tapered cantilever [15]. Some examples of this type of sensor are nanomechanical cantilever sensors and carbon nanotube (CNT)-based fluidic shear-stress sensors.

Chemical sensors: This type of sensor can detect a single chemical or biological molecule. Various chemical nanosensors were used to measure different chemical properties, such as pH and other entities.

Biosensors: It is the most commonly used sensor for detection of cancer earlier and other different diseases. It can also be utilized for the identification of particular sorts of DNA [16]. The biosensors are considered as a subset of chemical sensors in terms of transduction methods that are similar to those for chemical sensors.

Electrometers: Another type of nanosensor is the electrometer, which is a nanometer-scale mechanical electrometer comprised of a torsional mechanical resonator, a detection electrode, and a gate electrode, used to couple charge to the mechanical element.

Deployable nanosensors: It is used explicitly in the military or for other national security purposes (e.g., SnifferSTAR). It is a portable and straightforward chemical detection system comprised of a nanomaterial to collect samples and a microelectromechanical-based detector. This sensor would likely be used during the war to easily detect chemicals in the air without affecting human lives [17].

24.2.1 Working principle of biosensor, chemical sensor, and gas sensor

Biosensor: A biosensor can be defined as an analytical device containing a biological recognition system (e.g., enzymes) to target analytes and a transducer that converts the recognition into a detectable output signal. An analyte is a chemical compound (e.g., urea, glucose, pesticide) whose concentration has to be measured. A biosensor usually consist of three components (i) bioreceptor, (ii) transducer, and (iii) detector [18]. The major performance of biosensors depends on the selectivity and specificity of biological reactions. A bioreceptor (biological recognition system) contains an immobilized biocomponent that can detect specific target analytes [19]. These biosubstances are made up of enzymes, cells, antibodies, nucleic acids, etc. Then the transducer which basically converts one form of energy into another. The interaction between the analyte and bioreceptor produces a biochemical signal, which is converted into an electrical signal through the transducer. The received electrical signal is properly amplified, and the corresponding response can be read through the detector.

Chemical sensor: A chemical sensor can be defined as an analytical device that detects and converts chemical properties like pressure, concentration, etc. into an electrical signal to obtain quantitative or qualitative information about specific chemical components. Each chemical sensor contains two primary components: a chemical recognition system (receptor) and a transducer. The receptor function is to interact with the analyte molecules. Molecular sensing usually depends on recognizing molecular structure or its reactivity; this recognition aspect is called selectivity. The sensitivity and detection limit is related to the amount or concentration of the element or molecule to be analyzed. The transducer function is to convert the chemical response (nonelectrical quantity) to a detectable electrical signal. A transducer can be classified as electrical, optical, mass, or thermal sensors depending on the property to be determined. They are designed for detecting and responding to an analyte in the solid, gaseous, or liquid state [20]. Electrochemical sensors are particularly attractive compared to optical, mass, and thermal sensors because of their remarkable detection, ease of testing, and low cost [21].

Gas sensor: A gas sensor is a device that detects the concentration or presence of different toxic or explosive gases in the atmosphere. A suitable gas sensor can be characterized by high sensitivity, high-temperature tolerance, and short response/recovery time. There are different gas sensors, but metal oxide semiconductor-based sensors are most widely used in gas sensing applications. Semiconductor-based gas sensors are characterized by n-type nanomaterial in which majority charge carriers are electrons (−ve) and p-type nanomaterial in which holes (+ve) are majority charge carriers. The principle of sensing mostly relies on the change in electrical resistance or conductivity because of adsorption/desorption of experimental gases on surface of nanomaterials [22,23]. Sensor devices are composed of n-type nanomaterials; when comes in contact with air, the oxygen in air is adsorbed on the particle surface by capturing free electrons [24-26]. This will lead to form an electron depletion layer that increases the electron transport barrier between nanoparticles. Thus, it influences the conductance or resistance of nanomaterials-based devices. Three types of stable oxygen anions, namely O_2^-, O^-, O^{2-} generates during adsorption of oxygen on the surface of nanomaterials [27-29]. The complete reaction is given below in steps:

$$O_2(\text{gas}) \leftrightarrow O_2\ (\text{adsorbed})$$

$$O_2(\text{adsorbed}) + e^- \leftrightarrow O_2^-\ (\text{ads})$$

$$O_2^-\ (\text{ads}) + e^- \leftrightarrow 2O^-\ (\text{ads})$$

$$O^-\ (\text{ads}) + e \leftrightarrow O^{2-}\ (\text{ads})$$

When exposed to experimental gas, the gas molecules react with the adsorbed oxygen ions at the surface and release the trapped electrons. There is a reduction in the electron depletion barrier, and the electron transport barrier is also shrunk [30-33]. The reverse is true for p-type nanomaterials.

24.3 Synthesis

Nanotechnology has grown because of the distinct properties of nanomaterials. This can only be done by different synthesis methods. These methods make nanomaterials different from those of bulk materials. There are two general methods/approaches for the fabrication of nanostructured materials: (i) top-down approach and (ii) bottom-up approach. In the top-down approach, bulk materials break down into nanosized materials using different physical processes like crushing, milling, and grinding. It is a simpler method for the fabrication of nanomaterials. However, this is not suitable for preparing materials of uniform shape. As regular or smooth surfaces cannot be obtained, so it has a significant impact on physical properties and surface chemistry of nanomaterials. Mechanical milling, nanolithography, sputtering, and laser ablation are some of the most widely used for nanoparticle synthesis.

Mechanical milling: Mechanical milling is used to reduce the particle size and blending of particles in new phases. It consists of a high energy ball mill, along with a suitable milling medium and an inert atmosphere. The influencing factors in mechanical milling are plastic deformation that leads to particle shape; fracture leads to a decrease in particle size, and cold-welding leads to an increase in particle size. It can be used for large-scale production of nanoparticles. The demerit of this process is contamination due to steel balls and milling medium used. It also requires high energy for milling process [34].

Sputtering: Sputtering is the deposition of nanoparticles on a surface (anode) by ejecting particles from target (cathode) when ions are bombarded on the cathode. The main principle of sputtering is to use the energy of plasma (partially ionized gas) on the surface of cathode, to release the atoms of the material and deposit them on the surface (substrate). Sputter deposition is done in an evacuated vacuum chamber where sputtering gas is admitted and pressure is maintained relatively low [34]. The sputtering gas used in this process is usually an inert gas like argon, neon, krypton, etc. There are different types of the sputtering process such as direct current sputtering, RF sputtering, magnetron sputtering, and reactive sputtering.

Nanolithography: There are different types of nanolithographic processes, for example, electron-beam, optical, dip-pen, nanoimprint, plasmonic, and scanning probe lithography. Generally, there are three steps involved in this method (i) preparing of mask/resist to cover the surface, which is intended to be smooth, (ii) etching of uncovered portion, and (iii) placing of the desired pattern directly on the surface and producing a final product. This technique is suitable for nanofabrication of various semiconducting integrated circuits. The drawbacks of this method are that it is costly and requires complex equipment.

Laser ablation: In this method, nanoparticles or thick films are obtained using laser irradiation. The solid target material is immersed in a liquid solution and exposed to pulsed laser irradiation. In this process, Ti:sapphire (titanium-doped sapphire) laser, copper vapor lasers and Nd:YAG (neodymium-doped yttrium aluminum garnet) laser are used [35]. Some parameters like ablation time, laser frequency, and a liquid medium with or without surfactant influences characteristic of metal particle formed [36]. As a stable synthesis of nanoparticles can be achieved using water and organic solvents, so it does not have any need of a chemical or stabilizing agent. Hence, this process is referred as a "green process." The bottom-up approach refers to the nanomaterials build-up by joining smaller atoms or molecules. This method is the most frequently used for creating nanomaterials of uniform shape and size. It controls further particle growth. It gives less waste during processing; hence this approach is more economical. Some most commonly used top-down methods are discussed below.

Sol–gel method: This synthesis method for nanomaterials is trendy and is widely employed to fabricate different oxide materials. This method involves the evolution of the system by forming a colloidal suspension (sol) and gelation to create a continuous liquid phase (gel) system. The precursor for preparing these colloids consists of ions of metal alkoxides and alkoxy silanes. A catalyst is used to start the reaction and control pH. The initial material is processed to form a dispersible oxide and form a sol in contact with dilute acid or water. The liquid removal from the sol leads to gel formation and the oxides are produced by calcination of gel [37]. The reaction involved in this process is based on hydrolysis and condensation of metal alkoxides M(OR)z can be given as follows:

$$\mathrm{MOR} + \mathrm{H_2O} \rightarrow \mathrm{MOH} + \mathrm{ROH}\ (\mathrm{Hydrolysis})$$

$$MOH + ROM \rightarrow M - O - M + ROH\ (Condensation)$$

Hydrothermal method: This synthesis process is carried out in a reactor known as autoclave by the reaction in aqueous solution at high temperature and pressure. This synthesis process is widely used to prepare metal oxide nanoparticles which can be easily achieved through hydrothermal treatment of peptized precipitates of a precursor with water [38]. The advantage of this method is desired size and shape of nanocrystals can be prepared with high crystallinity. The disadvantage is that this process is expensive and difficult to control.

Chemical reduction method: This method involves the reduction of ionic salts by reducing agents in surfactant presence. Trisodium citrate, ethanol, ethylene glycol, glucose, hydrazine hydrate, and sodium borohydrate ($NaBH_4$), etc. are used as reducing agents for synthesis of metal nanoparticles [39]. Sometimes stabilizing agent and reducing agent are used together. Some major drawbacks of this method are toxicity, impurities, and high cost.

Chemical vapor deposition (CVD): CVD process is the deposition of a thin film of the target material (volatile precursors) on a substrate surface through a chemical reaction of gaseous molecules. The deposition process is carried out in a reaction chamber at ambient temperature. The quality of deposited materials mainly depends on the rate of reaction, temperature, and concentration of the precursor [40]. The advantage of CVD

Table 24.1 Different categories of nanomaterials synthesized from various methods [43].

Category	Method	Nanomaterials
Bottom-up	Hydrothermal	Metal oxide-based
	Sol–gel	Carbon-, metal-, and metal oxide-based
	Chemical reduction	Metal-based
	Chemical vapor deposition	Carbon- and metal-based
	Biosynthesis	Organic polymers- and metal-based
Top-down	Mechanical milling	Metal-, oxide-, and polymer-based
	Sputtering	Metal-based
	Nanolithography	Metal-based
	Laser ablation	Carbon-based and metal oxide-based

has controlled surface morphology and crystal structure. The major disadvantage of CVD is the chance of chemical hazards because of toxic and explosive precursor gases.

Biosynthesis: Biosynthesis is one of the most trending and adaptable technology for fabrication of nanoparticles. It is an eco-friendly and green approach for synthesizing nanoparticles that are biodegradable and nontoxic [41]. This method of synthesis uses biological agents (e.g., bacteria and fungi), plant extracts, etc. instead of chemical agents. Nanoparticles synthesized by biosynthesis show unique and improved properties and have a wide range of biomedical applications [42]. It is also very cost effective.

The different categories of nanomaterials synthesized from various methods are given in Table 24.1 [43].

24.4 Metal and noble metal nanomaterials as nanosensors

Metal nanomaterials present very special properties at nanoscale range from those of bulk materials which can be implemented in real-life applications. Various metals such as Au, Pt, Pd, Ag, Cu, Co, including rare earth metals have been employed for sensing. Among various metal nanoparticles, copper nanoparticles are attractive because of their catalytic, optical, and electrical conductivity properties, including easy availability and low cost. Copper nanoparticles also show the antibacterial and antifungal activities along with lesser cytotoxicity and anticancer properties, so it has many applications in biological and medicinal fields [44]. Here laser ablation method has been used for the synthesis of copper nanoparticles [45]. This method is green, easy, and chemical free. A copper plate (with 99.99% purity; Sigma-Aldrich) was taken and ablated in the presence of nitrogen gas flow to prevent the formation of copper oxide. The copper plate was immersed in walnut oil with a desired height to be operated by laser. The ablation was carried out via a pulsed Q-Switched Nd:YAG laser at duration of 5 ns, a repetition rate of 10 Hz at wavelength of 1064 nm. The laser beam was focused on the target with a 300 mm focal length lens for 5, 10, 20, 30, and 50 min at room temperature with spot size about 0.7 mm. The nanoparticles were observed after a few minutes during the

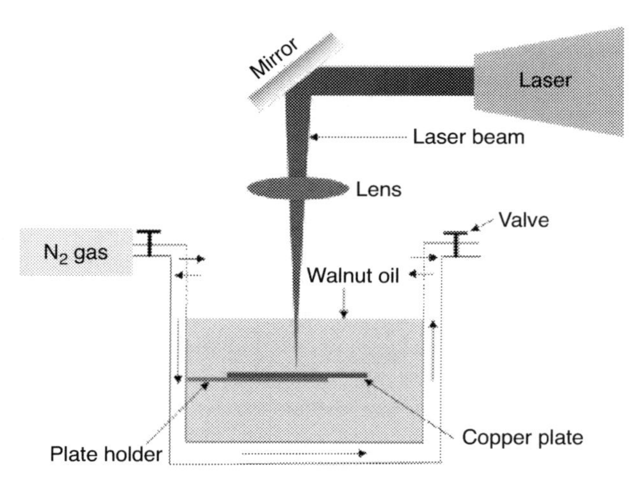

Fig. 24.2 ***Laser ablation synthesis process [45].***

ablation. To disperse the nanoparticles in the oil, a magnetic stirrer was used during the ablation process. The laser ablation process is given in Fig. 24.2 [45].

Another method for fabrication of copper nanoparticles is radio frequency (RF) sputtering method [46]. In this method, a copper target of cylindrical shape (99.99% purity) was used. Plasma is generated by RF capacitive discharge between two electrodes in a chamber. The copper target acts as a cathode which is connected to a high-frequency RF generator and anode is a substrate that is grounded. The cathode was connected to an RF source through a capacitor and an impedance matching network in series. The capacitor is a part of that impedance network. Argon gas (sputtering gas) is injected through a gas inlet valve. The gas passing through the grid forms a laminar flow between the electrode space, which is evacuated continuously by the pump. When RF potential is increased, an electrical breakdown has occurred in the gas and the plasma is ignited. Then the ions of the gas bombard the copper target during the ionization process and release the copper atoms. These sputtered copper atoms deposited on the substrate as a thin film. Noble metals are outstandingly resistant to oxidation and corrosion in moist air. These metals include ruthenium (Ru), palladium (Pd), silver (Ag), iridium (Ir), rhodium (Rh), osmium (Os), platinum (Pt), and gold (Au). Noble metal nanoparticles, especially of gold and silver, have been intensely explored by the scientific community owing to their spectacular properties. Gold and silver nanoparticles are used in nanobiotechnology, biosensor studies, visualization of cell structures [47], and targeted drug delivery [48]. Chemical reduction method has been used here for the synthesis of Au and Ag nanoparticles. For synthesis of Ag nanoparticles, 50 mL of 1 mM silver nitrate ($AgNO_3$) was dissolved in 500 mL of distilled water and heated to boiling temperature. To this solution, 5 mL of 1% trisodium citrate ($Na_3C_6H_5O_7$) was included drop by drop as a reducing agent. Solution was heated until the color change was noticeable (greenish). Then it was cooled to the room temperature. Here Ag nanoparticles are formed by the reduction of $AgNO_3$ because of trisodium citrate. The chemical reaction occurred is given below [49]:

$$4Ag^+ + C_6H_5O_7Na_3 + 2H_2O \rightarrow 4Ag^0 + C_6H_5O_7H_3 + 3Na^+ + H^+ + O_2 \uparrow$$

For synthesis of Au nanoparticles, 20 mL of 1 mM tetrachloroauric acid ($HAuCl_4$) was added to distilled water and heated to boiling temperature. To this solution, trisodium citrate of amount 2 mL (1 g dissolved in 100 mL of water) was added. Then a deep red solution was obtained by heating the solution further. In this reaction, Au^{+3} is reduced to Au^0 by reducing agent. The chemical reaction occurred is given below [50]:

$$2HAuCl_4 + 4Na_3C_6H_5O_7 \rightarrow H_2 + 2Au^0 + 4CO_2 + 4NaC_5H_5O_5 + 8NaCl$$

24.4.1 Applications

Glucose biosensors are commonly used in the food industries, waste water treatment, and diagnose blood sugar levels [51,52]. Though conventional enzymatic glucose sensors have evolved over three generations and are highly sensitive and selective, they are always affected by performance variability with the environment [53,54] and the enzymes used are denatured and inactive, resulting in poor performance and repeatability of enzymatic electrodes [55,56]. Hence, this problem can be avoided using nonenzymatic glucose biosensors by directly catalyzed oxidation to glucose on the electrode [57-59]. Glucose biosensors fabricated from copper nanoparticles (CuNPs) have received more interest because of their properties and higher catalytic activity toward glucose oxidation than other metal nanoparticles [60-62]. To reduce the aggregation and increase the dispersion of CuNPs, some carbon materials like CNTs and graphene are used as matrices [63-65]. Interestingly, nitrogen-doped carbon materials offer numerous active sites and large surface areas for the integration of CuNPs without any aggregation. So scientists were developed a nonenzymatic glucose biosensor using CuNPs/nitrogen-doped graphene oxide nanomaterials. The CuNPs/nitrogen-doped graphene oxide materials were coated on a disposable screen-printed carbon electrode for faster and more accurate sensing. The screen-printed carbon electrodes are inexpensive and nonrecyclable. From different studies, it was found that this biosensor shows high sensitivity (2500 $\mu A/mM/cm^2$) and also detects in low concentration (0.44 μM) [66].

Chemical oxygen demand (COD) is one of the chemical factors which is used for water quality assessment and pollution control in water and waste water [67]. COD is defined as the number of oxygen equivalents consumed in the oxidation of organic compounds using strong oxidizing agents like permanganate or dichromate [68-71]. The conventional method used for the evaluation of COD has several disadvantages like low precision, low sensitivity, high toxicity, and time-consuming [72]. So for high sensitivity, short analysis time, and easy handling, electrochemical sensors were developed for the determination of COD using Cu modified electrode. Despite of copper modified electrode, there are also number of different electrodes used for evaluation of COD. For the fabrication of the sensor, Cu nanoparticles were deposited on the surface of glassy carbon electrode (GCE) by electrodeposition method. Thus nano-Cu modified GCE (nano-Cu/GCE) was obtained. Experiments were performed using a conventional three-electrode system where nano-Cu/GCE, Ag/AgCl, and platinum wire were taken as working electrode, reference electrode, and counter electrode, respectively. Glycine was taken as sample organic compound in water for the evaluation of COD. From the different studies, electrochemical sensors using modified nano-Cu/GCE electrode were found to be highly sensitive with a detection limit as low as 1.7 mg/L because of the high electrocatalytic activity and stability [73].

24.5 Metal oxide nanomaterials as nanosensors

Metal oxide-based sensors fetch a great deal of attention for its numerous advantages like sensitive and fast detection and low cost. The physicochemical properties of metal oxides make the sensors very sensitive. So they can detect even if there is a small change in environment. Sensors fabricated from different metal oxides such as TiO_2, WO_3, and ZnO show high sensitivity and good stability in detecting different gases.

Titanium dioxide (TiO_2) nanoparticles are of great interest due to several applications such as optical filters, chemical sensors, photocatalytic degradation of various contaminants in waste water treatment, and solar cells. This wide range of applications is mainly due to its electronic and structural properties. TiO_2 exists mostly in three different crystal forms, namely anatase, rutile, and brookite. To make TiO_2 very effective, it should have certain properties such as suitable size and shape, crystallinity, and a good ratio of anatase to rutile. This is the reason why most researchers have been trying to use different methods to get particles with proper characteristics for different remediation and other applications. Among the methods used for synthesis of TiO_2 nanoparticles, sol–gel method is the most widely applied approach to prepare TiO_2 nanoparticles [74-76]. This method basically gives accessibility for synthesizing TiO_2 nanoparticles with different morphologies such as sheets, tubes, rods, wires, particles, mesoporous, and aerogels. Sol–gel method is generally based on the hydrolysis of titanium alkoxide. Here titanium tetra isopropoxide was taken as the precursor for the sol–gel method. Titanium tetra isopropoxide [$Ti(OCH(CH_3)_2)]_4$, 97% purity, Aldrich] was mixed with isopropanol [$(CH_3)_2CHOH$, 99.7% purity, Aldrich] in a dry atmosphere. The mixture was then added drop by drop into another mixture consisting of 12 mL deionized water and 10 mL isopropanol under constant stirring at 80°C. After 1 h, concentrated HNO_3 (0.8 mL) mixed with deionized water was added into titanium tetra isopropoxide solution and keep it under constant stirring at 60°C for 6 h highly viscous sol–gel was obtained. The prepared sol–gel was heated at 300°C for 2 h in the open atmosphere.

After annealing, the TiO_2 nanocrystalline powder was obtained. For further preparation of TiO_2 film, the prepared powder was added in the ratio of 1:10 of the solution of isopropanol. The TiO_2 nanoparticles deposited on titanium substrate using the dip coating method. However, the sol–gel method encounters some problems like weak anatase crystallinity and poor monodispersity. So in order to make pure anatase TiO_2 nanoparticles, electrochemical method was introduced [77]. In the overall process, the bulk metal is oxidized at the anode and the metal cations migrate to the cathode. The reduction takes place with formation of metal or metal oxide in the zero oxidation state. For synthesis of TiO_2, titanium metal sheet was taken as anode and a platinum sheet as the cathode. 0.01 M of tetra propyl ammonium bromide (TPAB) and acetonitrile/tetrahydrofuran in 4:1 ratio served as the electrolyte. Upon applying current, titanium dioxide clusters (>95%) were obtained which was stabilized by TPAB. Electrolysis was carried out in nitrogen atmosphere. TiO_2 nanoparticles were found to be white in color and separated out by a simple decantation process as these materials were insoluble in the solvent mixture used. Decanted solid product was washed with dry tetrahydrofuran for three to four times to remove excess TPAB and dried under vacuum. It was found that variation in current density affects the particle size of nanocluster in the electrolysis process.

Tungsten trioxide (WO_3) nanoparticles have attracted special interests among researchers due to its s strong adsorption within the solar spectrum (≤500 nm), stable physicochemical properties, as well as its resilience to photocorrosion. WO_3 is a metal oxide semiconductor with a wide bandgap and n-type conductivity. It is suitable for several applications such as photocatalysis, temperature sensor, electrochromic devices, and gas sensor. The amazing feature of crystalline tungsten oxide is its numerous polymorphs, like monoclinic, tetragonal, triclinic, hexagonal, cubic, and orthorhombic. Numerous methods have been reported to fabricate tungsten-trioxide nanoparticles in terms of physical and chemical processes. Hydrothermal synthesis [78] has various advantages such as simplicity and requires low handling temperature. It provides better uniformity and controls the shape and size of the structures. It is also cost effective. The optical, morphological, and crystalline properties of WO_3 nanostructures can be owned by changing reaction time, temperature, and precursor material. For the hydrothermal synthesis of WO_3 nanoparticles, 0.005 mol sodium tungstate ($Na_2WO_4 \cdot 2H_2O$, 99.9% purity, Aldrich), 0.025 mol sodium sulfate (Na_2SO_4, 99.2% purity, Aldrich), and 0.006 mol citric acid ($C_2H_8O_7 \cdot H_2O$, 99.9% Aldrich) were dissolved in distilled water of 50 mL. All chemicals were used without any further purification. When an apparent solution was formed, then 2 M of HCl was added dropwise to that solution. The solution was then transferred to a 100 mL PTFE chamber, placed inside a Teflon-lined autoclave and installed in the oven. The synthesis was done with different intervals of times at 180°C. Finally, the resulting precipitate was filtered and thoroughly washed with ethanol and distilled water. The end product was dried in the oven at 60°C for at least 10 h. As we know, tungsten trioxide has different polymorphs such as monoclinic, orthorhombic, tetragonal, triclinic, and hexagonal. Hexagonal structured WO_3 (h-WO_3) has gained more attention as an electrochromic material [79-80]. Therefore, it is possible to develop color-sensitive sensors using h-WO_3 in the presence of reducing gases. For the synthesizing of nanostructured h-WO_3, the acid precipitation method was used [81,82]. In this method, 1.17 g of sodium tungstate dihydrate (Na_2WO_4•$2H_2O$) is taken. Then it is dissolved in 17 mL of deionized water and cooled to 10°C. HCl solution of 8.4 mL is added in one dose to that solution. Then the solution mixture is cooled for about 20 h. The reaction occurs as follows:

$$Na_2WO_4 \cdot 2H_2O + H^+ = H_2WO_4 \cdot 2H_2O + Na^+$$

After that, the complete mixture turned into a colorless gel. The gel with some amount of water was added to a vessel and slowly stirred the mixture manually. The supernatant fluid was removed after centrifugation. The process of stirring, centrifuging, and removing supernatant liquid was repeated until H_2WO_4•H_2O was obtained and the precursor of final h-WO_3 powders. H_2WO_4•H_2O suspensions were transferred to hydrothermal dehydration under air at temperature 300°C and annealing time is 90 min.

$$H_2WO_4 \cdot 2H_2O \rightarrow WO_3 \cdot \frac{1}{3}H_2O \rightarrow h - WO_3$$

24.5.1 Applications

Metal oxide-based gas sensors show high sensitivity for detection and monitoring of different gases, including combustible, flammable, and toxic gases. WO_3 thin films and nanostructures are appeared to be very effective in detecting NO_2 gas because of its electrical properties and reactivity to oxidizing gases. For NO_2 gas sensor, an n-type WO_3 thin film was formed on an alumina substrate and it was fabricated with interdigital electrodes (IDEs). The IDEs were formed by evaporating an Au/In film on the surface of WO_3 material [83]. As NO_2 is an oxidizing gas, so it has the ability to accept electrons. When the WO_3 sensor is exposed to NO_2 gas, the resistance of sensor material increases as it donates electrons to NO_2. The sensor shows recovery times of 3 s and 89 s after exposure to NO_2 gas. This demonstrates the rapid reaction on the upper surface of the WO_3 sensor [84].

Biosensors using WO_3 nanoparticles can be possible due to its biocompatibility feature [85]. It has been seen that nanostructured WO_3 has direct accelerated interfacial electron transfer like TiO_2 and ZnO which can be used in implementing of nonmediated biosensors. Scientists developed a nitrite biosensor using indium tin oxide (ITO) glass electrodes, WO_3 nanoparticles, and cytochrome *c* nitrite reductase enzyme. Here, cytochrome *c* nitrite reductase was chosen for its high nitrite reduction activity, which may significantly impact biosensors' development for clinical diagnosis and pollution control. The dispersion of WO_3 was drop casted on the surface of ITO electrode and processed through annealing to obtain modified WO_3/ITO electrode. An electrochemical experiment was carried out in an electrochemical cell composed of modified WO_3/ITO, an Ag/AgCl and a platinum wire as working electrode, reference electrode, and counter electrode, respectively. Successful feedback was obtained by drop casting the enzyme solution together with ITO electrode and WO_3 nanoparticles dispersion as a result of high capacitive current and impedance of the material [86].

24.6 Carbon-based nanomaterials as nanosensors

Various carbon nanomaterials such as CNTs, graphene, fullerenes, and carbon nanofiber have their own unique physical, chemical, magnetic, optical, and electrical properties. These materials were used for various electroanalytical applications. Carbon nanomaterials can be classified according to their number of dimensions, that is, zero-dimensional nanoparticles (e.g., carbon dots which is also referred as graphene quantum dots), one-dimensional (e.g., CNTs), and two-dimensional materials (e.g., graphene) [87]. The different types of carbon nanomaterial are given in Fig. 24.3 [87].

Sumio Iijima discovered CNTs in 1991. CNTs are the tubular structures of rolled-up sheets of graphene comprising single wall and multiwall species. CNTs are one-dimensional nanomaterial possessed sp2 carbon units with many microns in length and some nanometers in diameter. Cylindrical fullerenes are called as CNTs. For synthesizing single-walled (SWCNTs) and multi-walled carbon nanotubes (MWCNTs), mainly three methods are used, that is, arc discharge, laser ablation, and CVD [88-90]. These processes take place in a vacuum or with process gases. Nanotubes were observed in 1991 during arc discharge process. Sumio Iijima, in 1991, first utilized the arc discharge method for synthesizing of fullerenes by using current of 100 amps. In the arc discharge method, a chamber is consisting of two electrodes, that is, cathode and anode, and plasma is created by passing current through the electrode. During this process, the

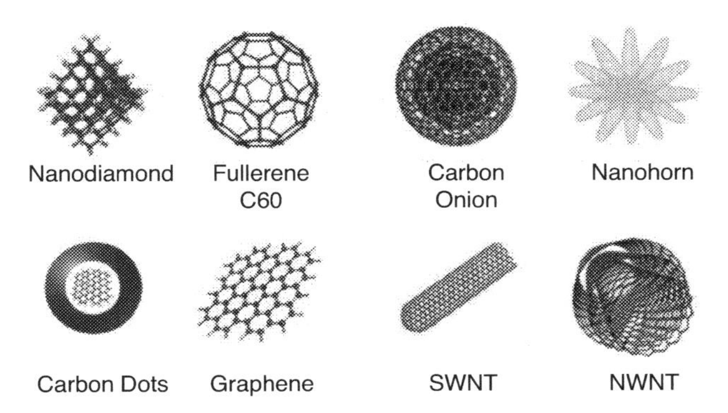

Fig. 24.3 ***Members of carbon nanomaterial family [87].***

cathode containing carbon sublimates because of the high-discharge temperatures. The first macroscopic production of CNTs was made in 1992 and the method used was the same as in 1991. In arc discharge method, the use of high temperatures (above 1700°C) for synthesizing of nanotubes generally results in the expansion of CNTs with fewer structural defects compared to other methods.

Guo et al. first introduced transition-metal/graphite composite rods' laser vaporization, which produced SWCNTs [90]. In laser ablation, a continuous or pulsed laser beam is introduced into a high-temperature reactor, which vaporizes a graphite target. Meanwhile, an inert gas is bled into the chamber. Graphite is used to create MWCNTs while graphite with metal catalyst particles (mixture of nickel and cobalt) is used to develop SWCNTs. Using this method, we can obtain a relatively high purity of SWCNTs with a controllable diameter determined by the reaction temperature. However, this method of synthesizing nanotubes is more expensive than either CVD or arc discharge.

CVD is a better synthesis process for the production of high-purity and high-yield CNT arrays at an average temperature. In the CVD process, using metal catalyst particles, a layer on the substrate is prepared and heated up to 700°C. For initial growth of nanotubes, two gases are bled into the reactor: a carbon-containing gas like methane, ethylene, acetylene, and a process gas like hydrogen, nitrogen, and ammonia. The growth in diameters of the nanotubes depends on the metal particles' dimensions, which can be controlled by the incorporation of the metal pattern. The carbon-containing gas is split up at the metal catalyst particle's surface and carried the carbon to the particle's edges, where it forms the nanotubes.

In all the above methods, some contaminants are always found when CNTs are created. Different acid treatments can remove these contaminants. These acid treatments may also introduce new types of contaminants that reduce the integrity and length of the nanotube.

Graphene is an allotrope of carbon which has sp2-hybridized carbon atoms combined by covalent bonds and it is closely packed in a honeycomb crystal lattice. Graphene has one atomic layer thick often called monolayer and has two-dimensional structure. It fetched appreciable attention for its remarkable properties at nanometer scale, including high thermal conductivity, super hydrophobicity, optical transmittance, high current density, and chemical inertness [91-92]. The first fabrication of graphene was done using graphite by micromechanical cleavage technique [93]. There are several types of graphene-like graphene nanoplatelets, graphene nanoribbons, graphene quantum dots, and graphene oxide. Graphite is simply a stack layer of several graphene sheets attached by van der Waals force. If this force is overcome then bonds will break and we can produce graphene from graphite sheets. The synthesization of single-layer graphite was tried by Lang et al. in 1975. He used thermal decomposition of carbon on single-crystal Pt substrates and showed single- and multi-layered graphite formation.

However, because of instability between properties of such sheets, he failed to classify the practical usage of the product. After a long period of time, graphene was discovered by Novoselov et al. in 2004 using mechanical exfoliation technique [93]. Mechanical exfoliation is generally a process of repeated peeling. Novoselov et al. used a 1 mm thickness highly oriented pyrolytic graphite sheet. There are also different graphitic materials such as single-crystal graphite, or natural graphite which can be used for production of graphene sheets. Several 5 μm deep mesas of area 0.3–3 mm^2 were formed by dry etching of the sheet in oxygen plasma. Then, the mesas were set on a photoresist and baked. For peeling off the layers from graphite sheet, the scotch tape was used. This peeling/exfoliation process can be done using different means such as scotch tape, ultrasonication, electric field, and even by transfer printing technique, etc. When thin scales attached to the photoresist were released in acetone and transferred to a Si substrate then single- to multi-layer sheets of graphene were found. This process was found to be easy and cheapest for producing high quality graphene sheets. Graphene synthesis using CVD method was first reported in 1975. In 2006, planar few-layer graphene was fabricated using CVD on Ni foil, taking camphor as the precursor material. First, camphor evaporated at 180°C and was subsequently decomposed at 700–850°C in another CVD furnace chamber. Here argon is used as a carrier gas to carry this process. After cooling at room temperature, graphene sheets of a few-layer were found in the Ni films. The first mono- to few-layer graphene were developed by using plasma-enhanced CVD (PECVD) method in 2004. In the PECVD process, [94] plasma is used for deposition of thin films. The plasma is generally formed by direct current or RF discharge between two electrodes and the interelectrode space is filled with the reacting gases. A typically PECVD process includes an external energy source for atoms or molecules' ionization, generating the plasma, a pressure reduction system for maintaining the plasma state, and a reaction chamber. The single- and multilayer of graphene was synthesized by PECVD on different substrates such as Mo, Si, Ti, Hf, Zr, Nb, Ta, W, Cu, SiO_2, Al_2O_3, and 304 stainless steel. A mixture of 5–100% CH_4 in H_2 with total pressure of 12 Pa inside the chamber was used for the production of graphene sheets by maintaining substrate temperature from 600°C to 900°C.

24.6.1 Applications

High sensitivity gas sensors are always essential to detect harmful gases. The principle of gas sensing involves the adsorption/desorption of gas molecules in sensing materials, so nanomaterials with a high surface-to-volume ratio and hollow structure are always preferred for storage and adsorption of gas molecules. Carbon nanomaterials like CNTs are found to be excellent for gas sensing applications. Gas sensors for the detection of ammonia (NH_3) gas were fabricated using MWCNTs [95]. For the fabrication NH_3 sensor, Au IDE electrode was sputtered on anodized aluminum substrate. The synthesized MWCNTs were sonicated in deionized water for several hours and coated on

IDE using spin coating method to create a sensor. By conducting experiments, it was observed that MWCNTs were highly sensitive to ammonia at room temperature 25°C. When the sensor is exposed to ammonia gas, donation of electrons occurred from NH_3 to CNTs, which increases the separation between the conduction band and the valence band. Therefore electrical resistance is increased. The increase in resistance proves that the CNTs are of p-type semiconductors. The sensitivity of ammonia sensor can be calculated by the equation: $(R_0 - R/R_0)*100\%$, where R_0 is sensors' resistance before exposed to NH_3 gas and R is sensors' resistance after exposed to NH_3 gas.

The derivatives of graphene, including graphene oxide, gained more attention due to its amazing thermal and electrical conductivities, good biocompatibility, large specific surface area, and low production cost. Like CNTs, extensively used in electrode materials, graphene also offers a new way of making electrochemical devices as it facilitates the transfer of electrons between electrodes and electroactive species. Electrochemical sensors were fabricated using graphene nanomaterials for detection of vanillin [96]. Vanillin ($C_8H_8O_3$) is one of the world-famous flavoring ingredients found mainly in *Vanillia planifolia*. The desirable flavor and aroma properties of vanilla have led to its widespread use in pharmaceuticals, foods, beverages, and perfumes. Vanillin is also used to inhibit the oxidation of human low-density lipoproteins, leading to lower rates of cardiac disease mortality, and used as an antisickling effect in sickle cell anemia sufferers. Therefore, controlling the quality of vanillin is very important. Several methods are used for determination of vanillin. These methods are costly and time-consuming. Vanillin detection using electrochemical sensor is used because of its simplicity, high sensitivity, and fast response. For cyclic voltammetric experiment, three electrode configuration system was used. The working electrode was graphene modified GCE, reference electrode was saturated calomel electrode (SCE) and auxiliary electrode was platinum wire. Graphene modified GCE was obtained by coating graphene and N,N-dimethylformamide mixed suspension with prepolished GCE and then drying under infrared lamp. Biscuits containing vanillin concentration (range of 6.0×10^{-7}–4.8×10^{-5} mol/L) were used as samples. The developed sensor was found to be with detection limit (5.6×10^{-8} mol/L) and satisfied recoveries from 97.9% to 103.5%.

24.7 Polymer nanomaterials as nanosensors

Since the discovery of electrically conducting polymers, it grabbed great attention of the scientific community. Conducting polymers such as polyaniline (PANI), polypyrrole (PPy), polythiophene (PTH), poly(p-phenylene) (PPP), poly(p-phenylenevinylene) (PPV), and their derivatives are a class of polymers containing a large resonation structure with many sp^2-carbon atoms that permits delocalized transport of charge carriers. Due to the inherent ability of these polymers to conduct electricity through charge delocalization they are called intrinsically conducting polymers. Intrinsic conducting polymers also referred as "synthetic metals" having a π-conjugated system in the polymeric chains. The polymer materials with combination of metals, metal oxides, CNTs or graphene for the fabrication of nanocomposites, improves its properties which can be utilized for chemical and biological sensing. Even desired properties are achieved with small quantities of nanomaterials, so the use of nanocomposites is always economical.

Among all conducting polymers, PANI is found to be an excellent polymer because of its extraordinary properties such as doping/dedoping effect, good environmental stability, electrical conductivity, etc. [97] including low cost, ease of synthesis, and promising applications in various fields. Synthesis of nanocomposites using PANI with other nanomaterials generally based on two routes: (i) one-step redox reactions where simultaneous polymerization of aniline and formation of nanoparticle takes place or (ii) in situ polymerization where presynthesized nanoparticles are mixed into the monomer solution followed by chemical or electrochemical polymerization.

Here we discussed about the synthesis of composites of acid-doped PANI with CNTs using in situ chemical oxidation polymerization [98]. For preparing the composites, first, different weight ratio of MWNTs were dissolved in 1.0 M HCl solutions and was ultrasonicated at 50°C for several hours. The above MWCNT/HCl suspension was then preserved in a refrigerator for an ice bath. Aniline monomer also dissolved in 1 M HCl solution and was then added to the MWNTs suspension. A 200 mL 1 M HCl solution containing 0.125 M ammonium persulfate was slowly added dropwise into the suspension and stirred with a magnetic stirrer at a reaction temperature 0–5°C by an ice bath. The reaction mixture was then kept in a refrigerator for 24 h. This results a black-green suspension which indicated the beginning of polymerization reaction of aniline monomer. The black-green suspension was then filtered and washed with deionized water and methanol until the impurities were removed. Finally, the precipitate was dried at 60°C for 24 h under vacuum. The same procedure was used to prepare PANI–SWCNT composites.

Another well-known method for synthesis of composites from conducting polymers is electrochemical polymerization. In this method cleaner products are produced than chemical methods because no additional chemicals such as oxidant and surfactant are used. This method provides better polymerization with a fine control of the initiation and termination steps. This polymerization includes three routes: (i) galvanostatic route (when applied a constant current), (ii) potentiostatic route (when applied a constant potential), (iii) potentiodynamic route (where current and potential vary with time).

For synthesis of PANI/MWCNT composite films [99], a commercial grade stainless-steel sheet was taken which was polished with successive grades of emery paper to a mild rough finish, washed free of abrasive particles and then air-dried. Electrochemical polymerization was carried in a three electrode cell where a stainless steel sheet as working electrode, a platinum plate as counter electrode, and a SCE as reference electrode.

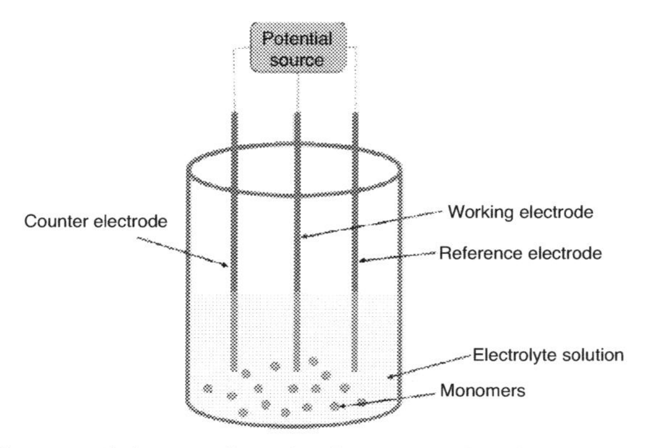

Fig. 24.4 ***Schematic diagram of electropolymerization process [100].***

These electrodes were dipped in the electrolyte containing monomer. Here the electrolyte is a mixture of 0.5 M H_2SO_4 + 0.325 M aniline (monomer) and MWCNT is added in different mixture ratios. Then the potential was applied between −0.2 and 1.0 V at a rate of 100 mV/s. MWCNT/PANI composite films were deposited on the working electrode's surface, which was then washed with deionized water and dried at 60°C for 1 day. Schematic diagram of electrochemical polymerization process is given in Fig. 24.4 [100].

24.7.1 Applications

In the case of gas sensors, PANI is a such conducting polymer that provides high sensitivity to different gases. PANI is used as active material in gas sensors to identify reducing gases such as CO, NH_3. It is seen that when PANI experiences volumetric change during reduction or oxidation, it becomes unstable mechanically. But the addition of suitable fillers like CNT or graphene can increase mechanical stability of PANI. The properties of PANI/SWCNT such as high conductivity and large active sites make it ideal for gas sensing application [101]. Gas sensor for the detection of carbon monoxide (CO) gas at room temperature was fabricated using PANI/SWCNT nanomaterials [102]. For fabrication of the sensor, A Ti/Au IDE electrode is evaporated on silicon substrate using thermal evaporator. The synthesized PANI/SWCNT film and IDE electrodes were fabricated by photolithography method. Upon exposure to CO gas, the charge transfer occurred from PANI/SWCNTs composite to the adsorbed CO gas. As PANI/SWCNTs have p-type semiconducting properties, the resistance is decreased when the number of holes increased in PANI/SWCNTs composite. PANI/SWCNTs gas sensor shows fast response time for the detection of CO gas. PANI is a suitable material for immobilization of biomolecules because its amine functional group can be utilized as a matrix for the crosslinking of various biomolecules. PANI polymer in combination with CNTs shows good stability, stability, good conductivity, and strong biomolecular interaction. So biosensors can be fabricated using PANI/CNTs for detection different biological compounds. Biosensors for the detection of cholesterol are discussed here [103]. Cholesterol level is an important parameter in clinical diagnosis. In general, cholesterol is measured by different other methods which are complicated and expensive. So the biosensor is most preferable because of its biochemical analysis, good selectivity, rapid response, and low cost. For the fabrication of cholesterol sensor, PANI/MWCNT film was electropolymerized on the IDE. IDEs were made by the lift-off process on the substrate of silicon using a pair of Pt metal. The enzyme cholesterol oxidase (ChOx) was immobilized in PANI/MWCNT film by cross-linking method. Finally, PANI/MWCNT/ChOx/IDE electrode was obtained. Potassium ferricyanide ($K_3[Fe(CN)_6]$) was taken as redox mediator. For cyclic voltammetric experiment purpose, three electrode configuration system was used where the working electrode was PANI/MWCNT/ChOx/IDE, reference electrode was SCE, and counter electrode was platinum wire. The sensitivity of the PANi-MWCNT biosensor was 464 $\mu A/mM/cm^2$. It is proposed that the presence of cholesterol was detected via change in signal transduction related to the redox group of FeCN as the redox mediator.

24.8 Bionanomaterials as nanosensors

Nanomaterials prepared by both physical and chemical methods using chemical reducing agents and protective agents have harmful effect on the nature, humans, and animals. These impacts can be minimized by biosynthesis or green methods. Biosynthesis method employs the use of biological agents including microbes, plants, and animals. It is also cost effective and environmental friendly. The shape and size of the nanoparticles can be controlled by this method. Biosynthesis of nanoparticles can be either of intra- or extracellular manner. Nanomaterials fabricated by biosynthesis method are defined as bionanomaterials. These bionanomaterials have potential application in different fields. Three main steps involved for biosynthesis of nanoparticles which must be estimated based on green chemistry perspectives, including (i) the choice of suitable solvent medium, (ii) the choice of environmentally benign reducing agent, and (iii) the choice of nontoxic substance for stabilization of nanoparticles [104]. Here synthesis of silver nanoparticles using mango leaf extract [105], gold nanoparticles using Locust bean gum (LBG) [106] and ZnO nanoparticles [107] have been described.

For the process of synthesis, first extract solution was prepared from mango leaf. The collected leaves were thoroughly washed using distilled water and dried at room temperature for 10 days under the shade. Then leaves were powdered using electrical grinder. The powdered leaves were added to 150 mL deionized water and heated for 30 min at 80°C under magnetic stirring. The final extract solution was cooled to room temperature and then filtered Whatman No. 1 filter paper. This prepared solution of mango leaves was used as the stabilizing and reducing agent in this synthesis process. Silver nitrate ($AgNO_3$) was added in deionized water. The extract solution of 2 mL was added in 25 mL of 1 mM silver nitrate solution. The mixture's color became reddish-brown under magnetic stirring at room temperature, indicating Ag nanoparticles' formation. For synthesis of gold (Au) nanoparticles, $HAuCl_4$ (chloroauric acid) and LBG extracted from the *Ceratonia silique* seeds were taken. All chemicals were taken without any further purification. First, 0.2 g of LBG was dissolved in 100 mL of distilled water. Then the mixture was heated at 80°C till 0.2% LBG solution was obtained. One millimolar of $HAuCl_4$ and 0.2% LBG were mixed in 1:4 ratio. The mixture was then autoclaved at 120°C and 15 psi pressure for 10 min to obtain Au nanoparticles. Zinc oxide (ZnO) nanoparticles were fabricated using aloevera plant extract. For this synthesis process, aloevera extract of an approximate amount was added dropwise to zinc nitrate ($(Zn(NO_3)_2)$ in 100 mL of deionized water and heated at 60°C for 4 h. The evaporation of water from the resulting solution took place by placing the solution on a hot plate. The sample material was then collected and annealed in a muffle furnace at 400°C for 4 h.

24.8.1 Applications

The biosynthesized Ag nanoparticles have gained more attention in biosensors because of their catalytic properties and high quantum electron transfer. Biosensor was fabricated using biosynthesized silver nanoparticles for detection of catechol by immobilization of the polyphenol oxidase (PPO) enzyme [108]. Catechol ($C_6H_6O_2$), 1,2-dihydroxybenzene/ 1,2-benzenediol) is generally found in many natural sources such as tea, vegetables, fruits, and other plants. Catechol quantification is important, due to its biological roles and environmental significance, in such topics as antioxidation, antivirus, toxicity, and carcinogenicity. The main challenge in developing enzyme-based electrochemical sensors is to find a suitable electrode for immobilization of the enzyme without any modification in its active sites. Silver nanoparticles have high conductivity and large surface area. So they can adsorb biomolecules such as proteins, cells and redox enzymes, without the loss of activity and thus maintaining their functional characteristics to a great extent. The PPO enzyme used in this sensor was extracted from Manilkara Z. (sapota) fruit. For fabrication of biosensor, electropolymerization of polypyrrole (PPy) on to a prefined graphite rod (Gr) was done using pyrrole monomer. Then biosynthesized Ag nanoparticles were added in deionized water. The solution mixture was then drops casted over Gr/PPy electrode and dried to obtain the Gr/PPy/AgNPs electrode. At last PPO enzyme was immobilized onto Gr/PPy/AgNPs electrode by drop casting method to get Gr/PPy/AgNPs/PPO electrode. For electrochemical experiments, the Gr/PPy/AgNPs/PPO electrode, SCE, and platinum wire were taken as working electrode, reference electrode, and counter electrode, respectively. The samples of green tea (Lipton brand) were collected for experiment purpose. From experimental studies, it was found that the sensor developed was able to detect catechol concentration in tea sample at low limit of 0.47 μM, with a sensitivity of 13.66 μA/mM/cm^2. The developed sensor can be useful in detection of catechol concentration for quality assessment of the tea samples in the tea industries to decide taste and aroma of the tea.

The zinc oxide nanoparticles have a wide band gap (3.4 eV) and high concentration of the active sites on the surface, which enables it to use in chemical and biological sensing applications. The biosynthesized ZnO nanoparticles were used to fabricate gas sensors for detection of various liquefied petroleum gas (LPG) concentrations [107]. For the gas sensor to be fabricated, a thin film is obtained by dissolving a quantity of ZnO nanoparticles in 1–2 drops of dimethyl formaldehyde to form a paste and depositing it on preprinted electrodes. When gas sensor was exposed to LPG, the gas molecules react with preadsorbed oxygen molecules adsorbed by n-type semiconductor material. Thus increasing the electron transport and consequently increasing the current. The measured current is converted to resistance to find the sensitivity of the sensor. The sensitivity can be calculated using the equation Ra-Rg/Ra, where Ra is sensor resistance in an atmospheric air and Rg is resistance in the presence of LPG gas. As a comparison, it was observed that biosynthesized ZnO sensors show similar sensing properties as of chemically synthesized ZnO sensors. But biosynthesized ZnO nanoparticles are preferred as it is cost effective and eco-friendly.

Electrochemical sensors fabricated using biosynthesized ZnO nanoparticles for the detection of ethanol (C_2H_5OH) have been reported [109]. For fabrication ethanol sensor, ZnONPs/PANI/GCE was taken as working electrode material. First, electropolymerization of PANI onto prepolished and purified GCE was done using aniline monomer. The resulting electrode PANI/GCE was then deposited on ZnO nanoparticles by electrodeposition method. The modified GCE was taken instead of simple GCE as it increases the conductance of ZnO nanoparticles. By conducting cyclic voltammetry experiments, it was observed that ZnONPs/PANI/GCE sensing material can detect ethanol in low limit (90 ppm). So this ZnO nanoparticles modified sensor could be used in real life application for the detection of ethanol in many alcoholic drinks where the levels of ethanol are usually greater than 2%.

24.9 Electrochemical sensors for biomedical application

Electrochemical sensors are generally a class of sensors which convert the information associated with electrochemical reactions (the reaction between an electrode and analyte) into an applicable qualitative or quantitative signal. Simply it can be said that, the change in electrical signal of desired electrochemical reactions occurred at an electrode surface are monitored in electrochemical sensors. The electrochemical signal is usually as a result of an applied potential, current, or frequency through an electrode. These sensors have wide range of commercial applications especially in biomedical field because of its several factors such as the use of the electron for signal acquisition, which is considered a clean model for analytical applications, with no generation of waste; miniaturization in portable devices (analyses with microvolumes of samples); fast analysis; noninvasive; and low production cost, allowing the popularization of these methods (e.g., as commercial glucose sensors).

Electrochemical gas sensors are used for the diagnosis of breast cancer, lung cancer, diabetes, halitosis, and renal failure via breath analyses. These sensors are able to detect biomarkers (an indicator of a particular disease state) usually present in the exhaled breath of diseased people. The detection of acetone using tungsten trioxide (WO_3) semiconductors is seen to give an excellent performance in terms of selectivity and sensitivity. Acetone is considered to be one of the biomarkers for monitoring of diabetes. Usually, diabetes patients are tend to have higher acetone levels in their breath than healthy people. The range of acetone level in the exhaled breath of diabetes people is in between 1.25 and 2.5 ppm. It has been seen that nanosensors fabricated using Si-doped WO_3 for detection of acetone shows highest response with a detection limit of 0.16 ppm [110]. Another serious disease is lung cancer which is more dangerous than breast cancer as the survival rate is low [111]. So for accurate and earlier detection of cancer SnO_2 gas sensors were used to detect the ethyl acetate level. Ethyl acetate is found in the breath of lung cancer patients which can be used as the biomarker. It is found that SnO_2 fabricated gas sensors show a low detection limit of 0.376 ppb with highest response [112]. Chronic kidney disease or renal disease is one of the most severe disease in which patients progressively lose the dialysis function of their kidneys (also referred as kidney failure). Hence, if the disease can be diagnosed in early stages, the progression can be curbed effectively by medicines. The risk and progression of chronic kidney disease is generally assessed by calculating the glomerular filtration rate. Serum creatinine is one of the established markers for glomerular filtration rate assessment. The concentration of creatinine in blood sera healthy person is 45–140 μM [113]. To detect the level of creatinine, electrochemical biosensors of enzymatic type are used. These biosensors show detection limit of 0.1 μM [114]. The foodborne diseases caused by various pathogens have become a global public health problem which affects everyone. There are many kinds of pathogens that are capable of producing toxins causing foodborne diseases [115,116], among them *Escherichia coli*, *Vibrio cholerae*, *Bacillus cereus*, *Staphylococcus aureus*, and *Clostridium perfringens* are common [117,118]. Human are likely to be infected with *E. coli* by drinking contaminated water or by eating food contaminated with feces [119-120]. So it is necessary for detection of *E. coli*. A simple, label-free electrochemical biosensor was developed using peroxidase-mimicking DNA enzyme to detect *E. coli* [121]. The first *Vibrio cholerae* electrochemical biosensor was constructed by Rao et al. [122]. Diseases like severe vomiting, diarrhea, water loss, and high mortality are caused by *Vibrio cholerae* [123-125].

24.10 Conclusion

Many types of nanosensors have been reviewed, categorized, and discussed according to energy source, structure, and materials. Many nanostructured materials applied for nanosensors were presented such as: metal, metal oxide, CNTs, graphene, polymers, and biomaterials. With the continuing progress in nanotechnology tools and increasing research on the nanoscale phenomena, one may expect further achievements in the field of nanosensors. This can be reached through the enhanced performance of existing nanosensors and newer nanosensors based on novel mechanisms.

References

[1] H. Akimoto, Global air quality and pollution, Science 302 (5651) (2003) 1716–1719.
[2] I.F. Akyildiz, W. Su, Y. Sankarasubramaniam, E. Cayirci, A survey on sensor networks, IEEE Commun. Mag. 40 (8) (2002) 102–114.
[3] F. Blais, Review of 20 years of range sensor development, J. Electron. Imaging 13 (1) (2004) 231–243.
[4] S. Agrawal, R. Prajapati, Nanosensors and their pharmaceutical applications: a review, Int. J. Pharm. Sci. Technol. 4 (2012) 1528–1535.
[5] R.K. Saini, L.P. Bagri, and A.K. Bajpai, in New Pesticides and Soil Sensors, A. Grumezescu (Ed.), Smart nanosensors for pesticide detection. Elsevier, Amsterdam, chapter 14, 2017, pp. 519–559.
[6] A.P. De Silva, H.Q. Nimal Gunaratne, T. Gunnlaugsson, A.J.M. Huxley, C.P. McCoy, J.T. Rademacher, T.E. Rice, Signaling recognition events with fluorescent sensors and switches, Chem. Rev. 97 (5) (1997) 1515–1566.
[7] D.H. Vance, A.W. Czarnik, Functional group convergency in a binuclear dephosphorylation reagent, J. Am. Chem. Soc. 115 (25) (1993) 12165–12166.
[8] S. Kulmala, J. Suomi, Current status of modern analytical luminescence methods, Anal. Chim. Acta 500 (1-2) (2003) 21–69.
[9] K. Sasaki, Z.-Y. Shi, R. Kopelman, H. Masuhara, Three-dimensional pH microprobing with an optically-manipulated fluorescent particle, Chem. Lett. 25 (2) (1996) 141–142.
[10] J. Wang, M. Musameh, Enzyme-dispersed carbon-nanotube electrodes: a needle microsensor for monitoring glucose, Analyst 128 (11) (2003) 1382–1385.
[11] J. Wang, M. Musameh, Y. Lin, Solubilization of carbon nanotubes by Nafion toward the preparation of amperometric biosensors, J. Am. Chem. Soc. 125 (9) (2003) 2408–2409.
[12] M. Gao, L. Dai, G.G. Wallace, Biosensors based on aligned carbon nanotubes coated with inherently conducting polymers, Electroanalysis 15 (13) (2003) 1089–1094.
[13] M.D. Rubianes, G.A. Rivas, Enzymatic biosensors based on carbon nanotubes paste electrodes, Electroanalysis 17 (1) (2005) 73–78.
[14] T.-C. Lim, S. Ramakrishna, A conceptual review of nanosensors, Zeits. Naturforsch. A 61 (7-8) (2006) 402–412.
[15] V.T. Binh, N. Garcia, A.L. Levanuyk, A mechanical nanosensor in the gigahertz range: where mechanics meets electronics, Surf. Sci. 301 (1-3) (1994) L224–L228.
[16] G. Doria, J. Conde, B. Veigas, L. Giestas, C. Almeida, M. Assunção, J. Rosa, P.V. Baptista, Noble metal nanoparticles for biosensing applications, Sensors 12 (2) (2012) 1657–1687.
[17] C.R. Yonzon, D.A. Stuart, X. Zhang, A.D. McFarland, C.L. Haynes, R.P.V. Duyne, Towards advanced chemical and biological nanosensors—an overview, Talanta 67 (3) (2005) 438–448.
[18] D.W.G. Morrison, M.R. Dokmeci, U.T.K.A.N. Demirci, A. Khademhosseini, Clinical applications of micro- and nanoscale biosensors, Biomed. Nanostruct. 1 (2008) 433–458.
[19] K. Kahn, K.W. Plaxco, Principles of biomolecular recognition. In: M. Zorob (Ed.), Recognition Receptors in Biosensors, Springer, New York, NY, 2010, pp. 3–45.

[20] J. Janata, Peer Reviewed: Centennial Retrospective on Chemical Sensors. Journal of the Brazilian Chemical Society; Institute of Chemistry, Universidade Estadual Paulista, CP 355, Araraquara - SP, Brazil, 2001, p. 150 A.
[21] N.R. Stradiotto, H. Yamanaka, M.V.B. Zanoni, Electrochemical sensors: a powerful tool in analytical chemistry, J. Braz. Chem. Soc. 14 (2) (2003) 159–173.
[22] N. Yamazoe, Toward innovations of gas sensor technology, Sens. Actuators B 108 (1-2) (2005) 2–14.
[23] S.K. Gupta, A. Joshi, M. Kaur, Development of gas sensors using ZnO nanostructures, J. Chem. Sci. 122 (1) (2010) 57–62.
[24] B. Shouli, C. Liangyuan, L. Dianqing, Y. Wensheng, Y. Pengcheng, L. Zhiyong, C. Aifan, C.C. Liu, Different morphologies of ZnO nanorods and their sensing property, Sens. Actuators B 146 (1) (2010) 129–137.
[25] M. Bagheri, A.A. Khodadadi, A.R. Mahjoub, Y. Mortazavi, Highly sensitive gallia-SnO_2 nanocomposite sensors to CO and ethanol in presence of methane, Sens. Actuators B 188 (2013) 45–52.
[26] Z.-M. Liao, H.-Z. Zhang, Y.-B. Zhou, J. Xu, J.-M. Zhang, D.-P. Yu, Surface effects on photoluminescence of single ZnO nanowires, Phys. Lett. A 372 (24) (2008) 4505–4509.
[27] B. Shouli, C. Liangyuan, L. Dianqing, Y. Wensheng, Y. Pengcheng, L. Zhiyong, C. Aifan, C.C. Liu, Different morphologies of ZnO nanorods and their sensing property, Sens. Actuators B 146 (1) (2010) 129–137.
[28] P. Rai, Y.-T. Yu, Citrate-assisted hydrothermal synthesis of single crystalline ZnO nanoparticles for gas sensor application, Sens. Actuators B 173 (2012) 58–65.
[29] O. Lupan, G. Chai, L. Chow, Novel hydrogen gas sensor based on single ZnO nanorod, Microelectron. Eng. 85 (11) (2008) 2220–2225.
[30] P. Romppainen, V. Lantto, The effect of microstructure on the height of potential energy barriers in porous tin dioxide gas sensors, J. Appl. Phys. 63 (10) (1988) 5159–5165.
[31] J.-I. Yang, H. Lim, S.-D. Han, Influence of binders on the sensing and electrical characteristics of WO_3-based gas sensors, Sens. Actuators B 60 (1) (1999) 71–77.
[32] Y.-D. Wang, X.-H. Wu, Q. Su, Y.-F. Li, Z.-L. Zhou, Ammonia-sensing characteristics of Pt and SiO_2 doped SnO_2 materials, Solid State Electron. 45 (2) (2001) 347–350.
[33] H.-J. Kim, J.-H. Lee, Highly sensitive and selective gas sensors using p-type oxide semiconductors: overview, Sens. Actuators B 192 (2014) 607–627.
[34] P.G. Jamkhande, N.W. Ghule, A.H. Bamer, M.G. Kalaskar, Metal nanoparticles synthesis: an overview on methods of preparation, advantages and disadvantages, and applications, J. Drug Deliv. Sci. Technol. 53 (2019) 101174.
[35] A.V. Simakin, V.V. Voronov, N.A. Kirichenko, G.A. Shafeev, Nanoparticles produced by laser ablation of solids in liquid environment, Appl. Phys. A 79 (4-6) (2004) 1127–1132.
[36] A. El-Nour, M.M. Kholoud, A'.A. Eftaiha, A. Al-Warthan, R.A.A. Ammar, Synthesis and applications of silver nanoparticles, Arab. J. Chem. 3 (3) (2010) 135–140.
[37] M. Parashar, V.K. Shukla, R. Singh, Metal oxides nanoparticles via sol–gel method: a review on synthesis, characterization and applications. J. Mater. Sci. Mater. Electron. 31 (5) (2020) 3729-3749.
[38] J. Yang, S. Mei, J.M.F. Ferreira, Hydrothermal synthesis of TiO_2 nanopowders from tetraalkylammonium hydroxide peptized sols, Mater. Sci. Eng.: C 15 (1-2) (2001) 183–185.
[39] S.M. Landage, A.I. Wasif, P. Dhuppe, Synthesis of nanosilver using chemical reduction methods, Int. J. Adv. Res. Eng. Appl. Sci. 3 (5) (2014) 14–22.
[40] C.S. Kim, K. Okuyama, K. Nakaso, M. Shimada, Direct measurement of nucleation and growth modes in titania nanoparticles generation by a CVD method, J. Chem. Eng. Jpn. 37 (11) (2004) 1379–1389.
[41] P. Kuppusamy, M.M. Yusoff, G.P. Maniam, N. Govindan, Biosynthesis of metallic nanoparticles using plant derivatives and their new avenues in pharmacological applications—an updated report, Saudi Pharm. J. 24 (4) (2016) 473–484.
[42] S. Hasan, A review on nanoparticles: their synthesis and types, Res. J. Recent Sci. 2277 (2015) 2502.

[43] A.M. Ealias, M.P. Saravanakumar, A review on the classification, characterisation, synthesis of nanoparticles and their application, IOP Conference Series on Material Science Engineering, vol. 263, 2017 032019.
[44] M.I. Din, F. Arshad, Z. Hussain, M. Mukhtar, Green adeptness in the synthesis and stabilization of copper nanoparticles: catalytic, antibacterial, cytotoxicity, and antioxidant activities, Nanoscale Res. Lett. 12 (1) (2017) 638.
[45] A.R. Sadrolhosseini, S.A. Rashid, A. Zakaria, K. Shameli, Green fabrication of copper nanoparticles dispersed in walnut oil using laser ablation technique, J. Nanomater. 2016 (2016) 1–7.
[46] Z.T. Nakysbekov, M.Z. Buranbayev, M.B. Aitzhanov, G.S. Suyundykova, M.T. Gabdullin, "Synthesis of copper nanoparticles by cathode sputtering in radio-frequency plasma." J. Nano Electron. Phys. 10 (2018) 1–4.
[47] L. Zhang, E. Wang, Metal nanoclusters: new fluorescent probes for sensors and bioimaging, Nano Today 9 (1) (2014) 132–157.
[48] P. Ghosh, G. Han, M. De, C.K. Kim, VM. Rotello, Gold nanoparticles in delivery applications, Adv. Drug. Deliv. Rev. 60 (11) (2008) 1307–1315.
[49] R. Nain, R.P. Chauhan, Colloidal synthesis of silver nanoparticles, Asian J. Chem. 21 (10) (2009) S113–S116.
[50] A.P.B. Dackiw, Induction and modulation of monocyte/macrophage tissue factor/fibrin deposition and TNF secretion in the microenvironment of inflammation: the role of tyrosine phosphorylation, National Library of Canada= Bibliothèque nationale du Canada, 1999.
[51] M.I. Din, F. Arshad, Z. Hussain, M. Mukhtar, Green adeptness in the synthesis and stabilization of copper nanoparticles: catalytic, antibacterial, cytotoxicity, and antioxidant activities, Nanoscale Res. Lett. 12 (1) (2017) 638.
[52] S. Ameen, M.S. Akhtar, H.S. Shin, Nanocages-augmented aligned polyaniline nanowires as unique platform for electrochemical non-enzymatic glucose biosensor, Appl. Catal. A 517 (2016) 21–29.
[53] I. Katakis, E. Domínguez, Characterization and stabilization of enzyme biosensors, TrAC Trends Anal. Chem. 14 (7) (1995) 310–319.
[54] K.E. Toghill, R.G. Compton, Electrochemical non-enzymatic glucose sensors: a perspective and an evaluation, Int. J. Electrochem. Sci. 5 (9) (2010) 1246–1301.
[55] Y.B. Vassilyev, O.A. Khazova, N.N. Nikolaeva, Kinetics and mechanism of glucose electrooxidation on different electrode-catalysts: part I. Adsorption and oxidation on platinum, J. Electroanal. Chem. Interfacial Electrochem. 196 (1) (1985) 105–125.
[56] B. Beden, F. Largeaud, K.B. Kokoh, C. Lamy, Fourier transform infrared reflectance spectroscopic investigation of the electrocatalytic oxidation of D-glucose: identification of reactive intermediates and reaction products, Electrochim. Acta 41 (5) (1996) 701–709.
[57] L.-M Lu, L. Zhang, F.-L. Qu, H.-X. Lu, X.-B. Zhang, Z.-S. Wu, S.-Y. Huan, Q.-A. Wang, G.-L. Shen, R.-Q. Yu, A nano-Ni based ultrasensitive nonenzymatic electrochemical sensor for glucose: enhancing sensitivity through a nanowire array strategy, Biosens. Bioelectron. 25 (1) (2009) 218–223.
[58] L. Meng, J. Jin, G. Yang, T. Lu, H. Zhang, C. Cai, Nonenzymatic electrochemical detection of glucose based on palladium—single-walled carbon nanotube hybrid nanostructures, Anal. Chem. 81 (17) (2009) 7271–7280.
[59] F. Xiao, F. Zhao, D. Mei, Z. Mo, B. Zeng, Nonenzymatic glucose sensor based on ultrasonic-electrodeposition of bimetallic PtM (M = Ru, Pd and Au) nanoparticles on carbon nanotubes–ionic liquid composite film, Biosens. Bioelectron. 24 (12) (2009) 3481–3486.
[60] R.M.A. Hameed, Amperometric glucose sensor based on nickel nanoparticles/carbon Vulcan XC-72R, Biosens. Bioelectron. 47 (2013) 248–257.
[61] T. Wang, Y. Yu, H. Tian, J. Hu, A novel non-enzymatic glucose sensor based on cobalt nanoparticles implantation-modified indium tin oxide electrode, Electroanalysis 26 (12) (2014) 2693–2700.
[62] J. Chen, W.-D. Zhang, J.-S. Ye, Nonenzymatic electrochemical glucose sensor based on MnO_2/MWNTs nanocomposite, Electrochem. Commun. 10 (9) (2008) 1268–1271.
[63] L. Mei, P. Zhang, J. Chen, D. Chen, Y. Quan, N. Gu, G. Zhang, R. Cui, Non-enzymatic sensing of glucose and hydrogen peroxide using a glassy carbon electrode modified with a nanocomposite

consisting of nanoporous copper, carbon black and nafion, Microchim. Acta 183 (4) (2016) 1359–1365.
[64] J. Luo, S. Jiang, H. Zhang, J. Jiang, X. Liu, A novel non-enzymatic glucose sensor based on Cu nanoparticle modified graphene sheets electrode, Anal. Chim. Acta 709 (2012) 47–53.
[65] H.-X. Wu, W.-M. Cao, Y. Li, G. Liu, Y. Wen, H.-F. Yang, S.-P. Yang, In situ growth of copper nanoparticles on multiwalled carbon nanotubes and their application as non-enzymatic glucose sensor materials, Electrochim. Acta 55 (11) (2010) 3734–3740.
[66] K. Sivasankar, K. Kohila Rani, S.-F. Wang, R. Devasenathipathy, C.-H. Lin, Copper nanoparticle and nitrogen doped graphite oxide based biosensor for the sensitive determination of glucose, Nanomaterials 8 (6) (2018) 429.
[67] Association of Official Agricultural Chemists, and W. Horwitz. Official Methods of Analysis, vol. 222. Washington, DC: Association of Official Analytical Chemists, 1975.
[68] W.A. Moore, W.W. Walker, Determination of low chemical oxygen demands of surface waters by dichromate oxidation, Anal. Chem. 28 (2) (1956) 164–167.
[69] K.-H. Lee, Y.-C. Kim, H. Suzuki, K. Ikebukuro, K. Hashimoto, I. Karube, Disposable chemical oxygen demand sensor using a microfabricated Clark-type oxygen electrode with a TiO_2 suspension solution, Electroanalysis 12 (16) (2000) 1334–1338.
[70] T. Korenaga, H.Y. Ikatsu, The determination of chemical oxygen demand in waste-waters with dichromate by flow injection analysis, Anal. Chim. Acta 141 (1982) 301–309.
[71] T. Korenaga, X. Zhou, K. Okada, T. Moriwake, S. Shinoda, Determination of chemical oxygen demand by a flow-injection method using cerium (IV) sulphate as oxidizing agent, Anal. Chim. Acta 272 (2) (1993) 237–244.
[72] K.-H. Lee, T. Ishikawa, S.J. McNiven, Y. Nomura, A. Hiratsuka, S. Sasaki, Y. Arikawa, I. Karube, Evaluation of chemical oxygen demand (COD) based on coulometric determination of electrochemical oxygen demand (EOD) using a surface oxidized copper electrode, Anal. Chim. Acta 398 (2-3) (1999) 161–171.
[73] I.H.A. Badr, H.H. Hassan, E. Hamed, A.M. Abdel-Aziz, Sensitive and green method for determination of chemical oxygen demand using a nano-copper based electrochemical sensor, Electroanalysis 29 (10) (2017) 2401–2409.
[74] A. Karami, Synthesis of TiO_2 nano powder by the sol-gel method and its use as a photocatalyst, J. Iran. Chem. Soc. 7 (2) (2010) S154–S160.
[75] C.B.D. Marien, C. Marchal, A. Koch, D. Robert, P. Drogui, Sol-gel synthesis of TiO_2 nanoparticles: effect of Pluronic P123 on particle's morphology and photocatalytic degradation of paraquat, Environ. Sci. Pollut. Res. 24 (14) (2017) 12582–12588.
[76] A. Sharma, R.K. Karn, S.K. Pandiyan, Synthesis of TiO_2 nanoparticles by sol-gel method and their characterization, J. Basic Appl. Eng. Res. 1 (9) (2014) 1–5.
[77] P. Anandgaonker, G. Kulkarni, S. Gaikwad, A. Rajbhoj, Synthesis of TiO_2 nanoparticles by electrochemical method and their antibacterial application, Arab. J. Chem. 12 (8) (2019) 1815–1822.
[78] H. Ahmadian, F. Shariatmadar Tehrani, M. Aliannezhadi, Hydrothermal synthesis and characterization of WO_3 nanostructures: effects of capping agent and pH, Mater. Res. Express 6 (10) (2019) 105024.
[79] L. Wang, J. Pfeifer, C. Balázsi, I.M. Szilágyi, P.I. Gouma, Nanostructured hexagonal tungsten oxides for ammonia sensing, Nanosensing: Materials, Devices, and Systems III, International Society for Optics and Photonics, Boston, MA, United States, 6769, (2007), p. 67690E. https://doi.org/10.1117/12.736679.
[80] L. Wang, J. Pfeifer, C. Balazsi, P.I. Gouma, Synthesis and sensing properties to NH_3 of hexagonal WO_3 metastable nanopowders, Mater. Manuf. Processes 22 (6) (2007) 773–776.
[81] C. Marcel, J.-M. Tarascon, An all-plastic WO_3 H_2O/polyaniline electrochromic device, Solid State Ion. 143 (1) (2001) 89–101.
[82] L. Wang, J. Pfeifer, C. Balazsi, P.I. Gouma, Synthesis and sensing properties to NH_3 of hexagonal WO_3 metastable nanopowders, Mater. Manuf. Processes 22 (6) (2007) 773–776.
[83] M. Yano, T. Iwata, S. Murakami, R. Kamei, T. Inoue, K. Koike, Gas sensing characteristics of a WO_3 thin film prepared by a sol-gel method, Multidiscip. Digital Publish. Inst. Proc. 2 (13) (2018) 723.

[84] R.N. Mulik, M.A. Chougule, G.D. Khuspe, V.B. Patil, Hydrothermal synthesis of tungsten oxide for the detection of NO_2 gas. In: Techno-Societal 2018, Springer, Cham, 2020, pp. 957–961.
[85] S.-J. Yuan, H. He, G.-P. Sheng, J.-J. Chen, Z.-H. Tong, Y.-Y. Cheng, W.-W. Li, Z.-Q. Lin, F. Zhang, H.-Q. Yu, A photometric high-throughput method for identification of electrochemically active bacteria using a WO_3 nanocluster probe, Sci. Rep. 3 (2013) 1315.
[86] L. Santos, C.M. Silveira, E. Elangovan, J.P. Neto, D. Nunes, L. Pereira, R. Martins, et al., Synthesis of WO_3 nanoparticles for biosensing applications, Sens. Actuators B 223 (2016) 186–194.
[87] F.R. Baptista, S.A. Belhout, S. Giordani, S.J. Quinn, Recent developments in carbon nanomaterial sensors, Chem. Soc. Rev. 44 (13) (2015) 4433–4453.
[88] T.W. Ebbesen, P.M. Ajayan, Large-scale synthesis of carbon nanotubes, Nature 358 (6383) (1992) 220–222.
[89] J. Kong, A.M. Cassell, H. Dai, Chemical vapor deposition of methane for single-walled carbon nanotubes, Chem. Phys. Lett. 292 (4-6) (1998) 567–574.
[90] T. Guo, P. Nikolaev, A. Thess, D.T. Colbert, R.E. Smalley, Catalytic growth of single-walled manotubes by laser vaporization, Chem. Phys. Lett. 243 (1-2) (1995) 49–54.
[91] J.-H. Chen, C. Jang, S. Xiao, M. Ishigami, M.S. Fuhrer, Intrinsic and extrinsic performance limits of graphene devices on SiO_2, Nat. Nanotechnol. 3 (4) (2008) 206–209.
[92] A.K. Geim, P. Kim, Carbon wonderland, Sci. Am. 298 (4) (2008) 90–97.
[93] K.S. Novoselov, A.K. Geim, S.V. Morozov, D. Jiang, Y. Zhang, S.V. Dubonos, I.V. Grigorieva, A.A. Firsov, Electric field effect in atomically thin carbon films, Science 306 (5696) (2004) 666–669.
[94] W. Choi, I. Lahiri, R. Seelaboyina, Y.S. Kang, Synthesis of graphene and its applications: a review, Crit. Rev. Solid State Mater. Sci. 35 (1) (2010) 52–71.
[95] M. Guo, K.-H. Wu, Y. Xu, R.-H. Wang, M. Pan, Multi-walled carbon nanotube-based gas sensor for NH_3 detection at room temperature, 2010 Fourth International Conference on Bioinformatics and Biomedical Engineering, IEEE, 2010, pp. 1–3.
[96] J. Peng, C. Hou, X. Hu, A graphene-based electrochemical sensor for sensitive detection of vanillin, Int. J. Electrochem. Sci. 7 (2) (2012) 1724–1733.
[97] E.T. Kang, K.G. Neoh, K.L. Tan, Polyaniline: a polymer with many interesting intrinsic redox states, Prog. Polym. Sci. 23 (2) (1998) 277–324.
[98] S. Ghatak, G. Chakraborty, A.K. Meikap, T. Woods, R. Babu, W.J. Blau, Synthesis and characterization of polyaniline/carbon nanotube composites, J. Appl. Polym. Sci. 119 (2) (2011) 1016–1025.
[99] J. Zhang, L.-B. Kong, B. Wang, Y.-C. Luo, L. Kang, In-situ electrochemical polymerization of multi-walled carbon nanotube/polyaniline composite films for electrochemical supercapacitors, Synth. Met. 159 (3-4) (2009) 260–266.
[100] P. Singh, S.K. Shukla, Advances in polyaniline-based nanocomposites, J. Mater. Sci. 4 (2020) 1–35.
[101] A. Roy, A. Ray, P. Sadhukhan, K. Naskar, G. Lal, R. Bhar, C. Sinha, S. Das, Polyaniline-multiwalled carbon nanotube (PANI-MWCNT): room temperature resistive carbon monoxide (CO) sensor, Synth. Met. 245 (2018) 182–189.
[102] I. Kim, K.-Y. Dong, B.-K. Ju, H.H. Choi, Gas sensor for CO and NH_3 using polyaniline/CNTs composite at room temperature, Tenth IEEE International Conference on Nanotechnology, IEEE, 2010, pp. 466–469.
[103] H.B. Nguyen, N.T. Nguyen, T.D. Nguyen, Portable cholesterol detection with polyaniline-carbon nanotube film based interdigitated electrodes, Adv. Nat. Sci.: Nanosci. Nanotechnol. 3 (1) (2012) 015004.
[104] K. Sahayaraj, Bionanomaterials: synthesis and applications, in: N. Joseph John (Ed.), Proceedings of the National Seminar on New Materials Research and Nano Technology, Govt. Arts College, Ooty, Tamil Nadu, 2012, pp. 24–29.
[105] F. Samari, H. Salehipoor, E. Eftekhar, S. Yousefinejad, Low-temperature biosynthesis of silver nanoparticles using mango leaf extract: catalytic effect, antioxidant properties, anticancer activity and application for colorimetric sensing, New J. Chem. 42 (19) (2018) 15905–15916.
[106] C.K. Tagad, K.S. Rajdeo, A. Kulkarni, P. More, R.C. Aiyer, S. Sabharwal, Green synthesis of polysaccharide stabilized gold nanoparticles: chemo catalytic and room temperature operable vapor sensing application, RSC Adv. 4 (46) (2014) 24014–24019.

[107] S. Goutham, S. Kaur, K. Kumar Sadasivuni, J. Kumar Bal, N. Jayarambabu, D. Santhosh Kumar, K.V. Rao, Nanostructured ZnO gas sensors obtained by green method and combustion technique, Mater. Sci. Semicond. Process. 57 (2017) 110–115.
[108] S. Sandeep, A.S. Santhosh, N. Kumara Swamy, G.S. Suresh, J.S. Melo, Detection of catechol using a biosensor based on biosynthesized silver nanoparticles and polyphenol oxidase enzymes, Portugaliae Electrochim. Acta 37 (4) (2019) 257–270.
[109] O.P. Emelda, M.I. Nyambura, M. Masikini, E. Iwuoha, Biosynthesised zinc oxide nanoparticles for ethanol chemical sensor, Int. J. Nano Res. 59 (2019) 94–104.
[110] A. Rydosz, Sensors for enhanced detection of acetone as a potential tool for noninvasive diabetes monitoring, Sensors 18 (7) (2018) 2298.
[111] World Health Organization. "World Health Organization Cancer Fact Sheet." (2009).
[112] Z. Khatoon, H. Fouad, H.-K. Seo, O.Y. Alothman, Z.A. Ansari, S.G. Ansari, Ethyl acetate chemical sensor as lung cancer biomarker detection based on doped nano-SnO_2 synthesized by sol-gel process, IEEE Sens. J. 21 (2020) 12504–12511.
[113] P. Kumar, R. Jaiwal, C.S. Pundir, An improved amperometric creatinine biosensor based on nanoparticles of creatininase, creatinase and sarcosine oxidase, Anal. Biochem. 537 (2017) 41–49.
[114] P. Kumar, M. Kamboj, R. Jaiwal, C.S. Pundir, Fabrication of an improved amperometric creatinine biosensor based on enzymes nanoparticles bound to Au electrode, Biomarkers 24 (8) (2019) 739–749.
[115] C.-Q. He, Y.-X. Liu, H.-M. Wang, P.-L. Hou, H.-B. He, N.-Z. Ding, New genetic mechanism, origin and population dynamic of bovine ephemeral fever virus, Vet. Microbiol. 182 (2016) 50–56.
[116] A.M. Ammar, A.M. Attia, N.K. Abd El-Aziz, M.I. Abd El Hamid, A.S. El-Demerdash, Class 1 integron and associated gene cassettes mediating multiple-drug resistance in some food borne pathogens, Int. Food Res. J. 23 (1) (2016) 332.
[117] S.P. Oliver, My, how time flies: foodborne pathogens and disease to begin its seventh year with publication frequency to increase in 2010, Foodborne Pathogens Dis. 7 (1) (2010) 1–2.
[118] J.Y. Huang, O.L. Henao, P.M. Griffin, D.J. Vugia, A.B. Cronquist, S. Hurd, M. Tobin-D'Angelo, et al., Infection with pathogens transmitted commonly through food and the effect of increasing use of culture-independent diagnostic tests on surveillance—Foodborne Diseases Active Surveillance Network, 10 US sites, 2012–2015, Morbid. Mortality Wkly. Rep. 65 (14) (2016) 368–371.
[119] S. Guo, M.Y.F. Tay, A.K. Thu, K.L.G. Seow, Y. Zhong, L.C. Ng, J. Schlundt, Conjugative IncX1 plasmid harboring colistin resistance gene mcr-5.1 in *Escherichia coli* isolated from chicken rice retailed in Singapore, Antimicrob. Agents Chemother. 63 (11) (2019).
[120] S. Zhang, G. Yang, Y. Huang, J. Zhang, L. Cui, Q. Wu, Prevalence and characterization of atypical enteropathogenic *Escherichia coli* isolated from retail foods in China, J. Food Prot. 81 (11) (2018) 1761–1767.
[121] H. Huang, M. Liu, X. Wang, W. Zhang, D.-P. Yang, L. Cui, X. Wang, Label-free 3D Ag nanoflower-based electrochemical immunosensor for the detection of *Escherichia coli* O157: H7 pathogens, Nanoscale Res. Lett. 11 (1) (2016) 507.
[122] M.K. Sharma, A.K. Goel, L. Singh, V.K. Rao, Immunological biosensor for detection ofVibrio cholerae O1 in environmental water samples, World J. Microbiol. Biotechnol. 22 (11) (2006) 1155–1159.
[123] F. Zhang, Y.H. Huang, S.Z. Liu, L. Zhang, B.T. Li, X.X. Zhao, Y. Fu, J.J. Liu, X.X. Zhang, *Pseudomonas reactans*, a bacterial strain isolated from the intestinal flora of *Blattella germanic* a with anti-*Beauveria bassiana* activity, Environ. Entomol. 42 (3) (2013) 453–459.
[124] Y.H. Huang, X.J. Wang, F. Zhang, X.B. Huo, R.S. Fu, J.J. Liu, W.B. Sun, D.M. Kang, X. Jing, The identification of a bacterial strain BGI-1 isolated from the intestinal flora of *Blattella germanica*, and its anti-entomopathogenic fungi activity, J. Econ. Entomol. 106 (1) (2013) 43–49.
[125] L. Li, H.-j. Yang, D. Liu, H.-b. He, C.-f. Wang, J.-f. Zhong, Y.-d. Gao, Z. Yanjun, Analysis of biofilms formation and associated genes detection in staphylococcus isolates from bovine mastitis, Int. J. Appl. Res. Vet. Med. 10 (1) (2012) 62.

CHAPTER 25

Smart and intelligent vehicles for drug delivery: Theranostic nanorobots

Vishakha Dave[a], Medha Pandya[b], Rakesh Rawal[c], S.P. Bhatnagar[a], Rasbindu Mehta[a]

[a]Department of Physics, Maharaja Krishnakumarsihji Bhavnagar University, Bhavnagar, Gujarat, India
[b]The KPES Science Collage, Maharaja Krishnakumarsihji Bhavnagar University, Bhavnagar, Gujarat, India
[c]Department of Life Sciences, University School of Sciences, Gujarat University, Ahmedabad, Gujarat, India

25.1 Introduction

The human race has experienced many pandemics in the past and has put forward an appropriate vaccine or cure. Diseases and viruses like H1N1 Influenza, plague, polio, smallpox, tuberculosis, rubella, rabies are almost curable. However, this is a never-ending process like human life evolves other micro and nanobacteria, viruses have found themselves to live in nature. Multidisciplinary areas like bionanotechnology, bioinformatics, biophysics, biochemistry, genetics, pharmacogenomics, proteomics have a lot of challenges and often hard to define. The development of Medical types of equipment like a microscope, X-ray, CT scanner, magnetic resonance imaging (MRI), Laser therapy, radiation therapy requires expertise in physics while to synthesis new drug composition, a chemist or a pharmacist plays a vital role. With advances in medical technologies, we have tackled death threats in the past few pandemics. Artificial intelligence and leading-edge technology have reduced the person's role in processes like diagnosis, surgery, and treatment. The hybrid smart and intelligent theranostic nanorobots are cutting-edge tools in the development of medicine, especially for the treatment of cancer and cardiac disease. Nanoparticle-based theranostic tools are arising as a promising model toward personalized Nanomedicine for disease- and patient-oriented diagnosis and treatment [1–4].

The most important breakthrough of nanomedicine is the development of nanorobots or nanobots. The idea of nanobots was mooted during the famous lecture by Feynman – a great theoretical physicist- 'there is a plenty room at the bottom: an invitation to enter a new field of physics'. In this lecture, Feynmann hinted possibility to use this nanotechnology in surgery. He says 'a friend of mine suggests a very interesting possibility for relatively small machines, He says that, although it is a wild idea, it would be interesting in surgery if you could swallow the surgeon. You put the mechanical surgeon inside the blood vessel and it goes into the heart and "looks" around. It finds out which valve is faulty one and takes a little knife and slices out [5]. This idea is very similar to the present-day development of nanorobots or nanobots. It helps to unaided diagnose

Advanced Nanomaterials for Point of Care Diagnosis and Therapy
DOI: https://doi.org/10.1016/B978-0-323-85725-3.00004-0

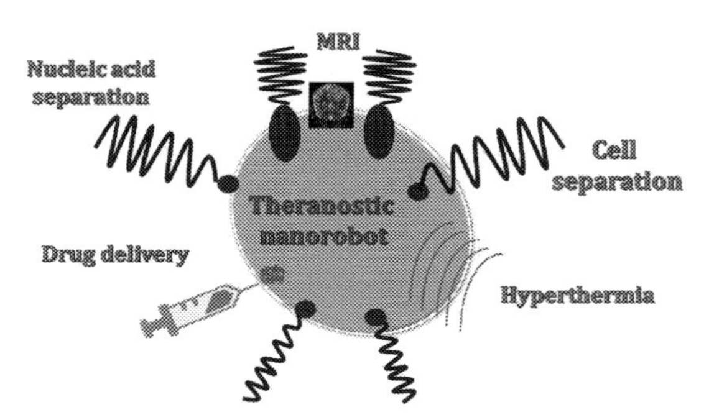

Fig. 25.1 ***Schematic of smart "all-in-one" theranostic nanorobot.***

and cure illnesses while the person lives a typical life. However, engineering smart "all-in-one" theranostic nanorobots (Fig. 25.1) for multifunction remains challenging. These self-propulsion synthetic nanorobots are made of three parts: nanoparticle as a core, stabilizing agent, and functioned group like a drug. Depending on their working principle, they are also known as nanoengines, nanorocket, nanobots include 3D-DNA nanomachines, nanoswimmers, bacteria nanobots, and sperm-like nanorobots [6]. We focus on the magnetic fluid based magnetic nanoparticle (MNP) theranostic nanorobots for biomedical applications.

25.2 Magnetic fluid as a smart material

Magnetic fluid is a stable suspension of single-domain magnetic nanoparticles (MNPs) of ferro- or ferrimagnetic material, primarily iron oxide. Magnetic nanoparticles are frequently coated with polymers, surfactants, or charged substances to form a stable colloidal suspension via steric and/or electrostatic repulsion. In the proximity of a magnetic field, they remain liquid and show novel physical phenomena like Rosensweig instability, phase transitions, levitation of non-magnetic material, magneto viscous effects and magneto-optical effects, etc. The usual methods for the preparation of magnetic fluids are chemical co-precipitation and size reduction [7]. Their properties can be manipulated by varying properties of magnetic nanoparticles, liquid carriers, and stabilizer.

Magnetic fluids are widely applied in technological and material science applications, e.g., compound separation, damping, pressure-tight rotary-shaft sealing, cooling agents for loudspeakers, and magnetic domain detection [8,9]. Some of the versatile properties of magnetic nanoparticles (MNPs) are mentioned below [10,11].

- Magnetic particles can be moved or retained at a precise location in the presence of a magnetic field applied externally. One can regulate their movement from afar.
- Magnetic fluid loses magnetism after the field pulls out.
- By applying the alternating magnetic field of the proper frequency, the particles can be heated through mechanisms, including hysteresis and relaxational losses.

- The single-domain particles act like tiny magnets, so it requires a moderate amount of magnetic field in most applications.
- Mechanical strength, chemical activity, and corrosion resistance are all physical features of nanoparticles.

The magnetic fluid which contains magnetic nanoparticles of the range of ~ 10 nm is emerging as theranostic nanorobot. Their exclusive chemical and physical properties are utilized for a variety of biomedical applications *in-vivo* and *in-vitro*, both, such as hyperthermia, drug release, tissue engineering, biomagnetic separations of biomolecules, and magnetic resonance (MR) imaging as contrast agents [12,13]. Here are some of the characteristics of MNPs that make them a promising tool for biomedical applications [11].

- Magnetic nanoparticles are smaller in size compared to a cell, a gene, and protein.
- They can be coated with biomolecules like protein, silane, starch, etc. As a result, it facilitates interaction with or bound to a biological entity. Thus a biomolecule of interest can be tagged or labeled.
- It exhibits less toxicity.

Water-based or biocompatible carrier liquids are used for the preparation of magnetic fluids in therapeutics. MNPs are functionalized with surface modification by binding agents like ligand, drug, or functional group. Magnetic fields can be used to extract and enrich labelled biological components such as bacteria, cells, nucleic acids, and proteins from complicated solutions such as blood. Both fractions can be completely retrieved after the process.

25.3 Physical properties of magnetic fluids

The magnetic fluid contains superparamagnetic soft ferrite particles. They exhibit different magnetic behavior as the particle size decreases. As the grain size of the magnetic particle goes from multi-domain to a single domain, the coercivity also changes. In a single domain region, it reaches a critical grain size value where the required coercive field becomes zero (Fig. 25.2C). At this point, the particles exhibit superparamagnetic behavior [14]. While preparing the stable magnetic fluid, these particles are suspended in a carrier liquid after being coated with a molecular layer of a dispersant. Forces like van der Waals, dipole-dipole interaction, and gravitational lead to the agglomerations and settling of the particles in a magnetic fluid. To achieve the stability of particles against magnetic field gradient, gravitational force, and dipole-dipole interaction, the criteria is to keep particle size in the range of 8–10 nm. It depends on thermal contribution and balance between attractive and repulsive forces. Particles can be coated with polar group adsorption or by adsorbing the opposite charges on the surface which create steric repulsive and electrostatic repulsive forces (Fig. 25.2(A,B)). There are two mechanisms to relax particles after removing the applied magnetic field. Relaxation caused by the particle rotation in the carrier liquid is known as Brownian relaxation. The mechanism

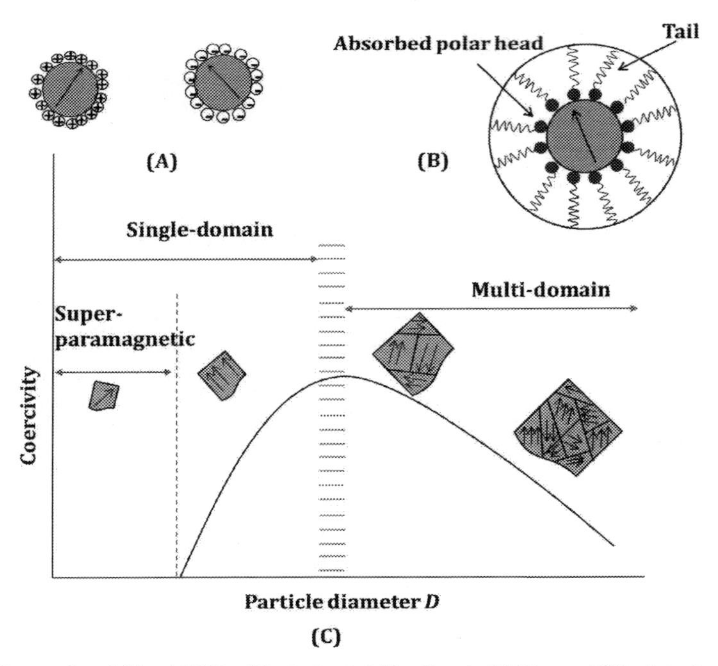

Fig. 25.2 ***(A) Charged stabilized MNPs, (B) steric stabilization in MNPs, and (C) plot of variation in particle diameter vs coercivity.***

in which the magnetic vector rotates within the particle is called the Néel relaxation. The following equations give the Néel (τ_N) and Brownian (τ_B) relaxation.

$$\tau_B = \frac{3V_H\eta_0}{kT}$$

$$\tau_N = \frac{1}{f_0}\exp\left(\frac{KV_M}{kT}\right)$$

$$\tau = \frac{\tau_B\tau_N}{\tau_B + \tau_N}$$

where V_H, η_0, k, T, f_0, K, τ and V_M and are the hydrodynamic volume of nanoparticles, the viscosity of the fluid, the Boltzmann constant, the temperature, the characteristic flipping frequency, the magnetic anisotropy constant, the effective relaxation time, and the volume of the nanoparticle, respectively. Both the relaxing time and the size of MNPs are strongly influenced by their size.

When MNPs are exposed in the alternating magnetic field, the required time for the reversal of the magnetic moment is less than that of the relaxation time. This realignment of magnetic moments results in heat generation [15]. The amount of heat dissipation is governed by the properties of MNPs and the frequency and amplitude of the applied magnetic field.

25.4 Engineering of magnetic fluid for biomedicine applications

Magnetic fluids have been fabricated based on their various applications. One can change the core, stabilizer, and functional groups to engineer them for their biomedical applications. Solution precipitation, decomposition, ultrasound chemistry, and microwave heating are all approaches for producing iron oxide nanoparticles [16–18]. To make them biocompatible, their surface engineering is required and need the additional coating. Coatings can also improve circulation time by stabilising water dispersion and protecting it from deterioration [19]. Coating layers can be applied to the surface of oxides by chemical or physical bonding [20] utilizing procedures including solvent evaporation and coupling agents [21] and can be done in situ or after synthesis. Iron oxide nanoparticles made up of magnetite (Fe_3O_4) and maghemite (γ-Fe_2O_3) are typically adopted in biomedical applications as an inorganic core. On the other side the organic shell/ coating functions to prevent agglomeration of core particles. It also provides the binding platform to drug molecules and limits opsonization.

25.4.1 Synthesis of magnetic fluid/magnetic nanoparticles

There are several methods for the preparation of MNPs and magnetic fluids mainly divided into three categories: biomineralization, physical method, and chemical method. Certain living organisms prepare magnetic nanoparticles, for example, the magnetotactic bacteria produce the magnetosomes. The magnetosomes are membranous structures made up of protein-coated magnetic particles. These particles serve as a magnetic compass for bacteria looking for a home in anaerobic zones at the ocean's bottom [22]. By creating the environment of their habitat in the lab, MNPs of a diameter of 20 to 45 nm are synthesized [23,24]. Apart from that, magnetic fluids are generally manufactured utilizing two approaches: size reduction and chemical procedures. A size reduction is a top-down approach in which the bulk size ferro/ferrimagnetic particles are reduced to the nanoscale by using high energy ball mills or attrition mills. In a wet grinding process, the carrier liquid and dispersant agent are also added with the MNPs to prepare stable magnetic fluid. Sol-gel, co-precipitation, sonochemical, microwave approach, microemulsion, thermal decomposition, electrochemical, aerosol/vapor phase route, and other bottom-up technologies are employed in biomedical applications. Because of their harmful effects on living organisms, toxic metals such as iron, nickel, and cobalt are avoided. The single or sub-domain superparamagnetic particles are attracted and form aggregates due to dipole-dipole interaction and Van der Waal forces. These forces lead to sedimentation. For the stability of MNPs, the mechanisms used are (1) steric stabilization and (2) the charge or Coulomb repulsion. In steric stabilization, short-chain polymers adsorb on the particles, and in charge stabilization, the particle surface is charged by treating it with an acidic or an alkaline medium. The MNPs are more stable in the steric stabilization method.

25.4.2 Binding drug and coating process of magnetic nanoparticles

Coating serves a variety of functions, including avoiding toxicity, stabilising particles against agglomeration, and providing a surface for conjugating ligands and medicinal molecules. Dextran, polyethylene glycol (PEG), PEG copolymers, and antifouling polymers are among the most researched surface coatings [25]. The binding mechanism of drugs with MNPs depends on their functionality and applications. The surface of MNP is functionalized by the hydrophilic polymers, and able to bind active molecules with chemical groups on the surface. In some cases, the NPs are capped with hydrophobic ligands. MNPs need to be moved from organic to aqueous solutions using a phase transfer agent. CTAB (hexadecyltrimethylammonium bromide) is a phase transfer agent for magnetic nanoparticles (MNPs, [26,27]). They also work as a labile ligand in the ligand exchange process with other biofunctional ligands. In another process, a ligand addition carries out on NP-ligand shell modification without removing the pre-existing ligands.

25.4.3 Magnetic core-shell designed

Core/shell MNPs are one of the most prevalent structures used. For bacterial cell identification, hybrid magnetic NPs (Fe/Fe_3O_4) with a large Fe core and a thin ferrite shell are created [28]. With little modification in the shell, mono-dispersed particles of $CoFe_2O_4$ and $MnFe_2O_4$ show high relaxation. They are useful for the detection of proteins and cancer cells [29]. The bimagnetic core/shell particles of Fe58Pt42/Fe_3O_4 exhibit remarkable changes in the coercive field as the particle size changes from 0.5 nm to 3 nm. By tuning the thickness of the core/shell of the particles, the magnetic properties can be tuned [30]. Ag NPs give an antimicrobial activity. Antifungal activities involving Ag-Fe_3O_4 core-shell particles were examined for Aspergillus glaucus [31]. The anticancer drug doxorubicin is bonded with a magnetic core and later encapsulated with triblock polymer (PEO-PLGA-PEO) for targeted drug delivery. These polymers are biodegradable and thermosensitive, which increases drug loading efficiency by 89% compared to conventional biocompatible polymers such as PEG and dextran [32]. The gold nanoparticles show Plasmon resonance by combining with MNPs. The Au-Fe combination core-shell and dumbbell NPs are synthesized [33]. Apart from the core-shell structure, the dumbbell metal oxide particles are also widely studied. Because of the Fe_3O_4 in Au-Fe_3O_4 NPs, the absorption peak is displaced from 520 to 538 nm. These dumbbell particles are also superparamagnetic like core-shell. These characteristics can be used as an optical/magnetic probe for diagnostic and therapeutic applications [34]. Particle morphology and chelating agent play an important role as their composite. Compared to spherical iron oxide particles, Octopod iron oxide NPs showed five times more ultrahigh transverse relaxivity, which affects the T2 image contrast capacity for MRI applications [35]. In MRI, T2 relaxation rates of nearby protons can be controlled by the effective radius of the nanomaterials. This study looked at manganese-doped iron oxide NPs with six distinct morphologies: cube, octopod, tetrahedra, spheres, rhombohedra, and plates with the same volume [36]. The induced magnetic field of NPs can be affected by a chelating agent. The presence of a chelating agent affects the relaxation time [37].

25.5 Nanorobots in diagnostics

It was stated earlier that nanorobots may be used to manipulate biological systems at the molecular level. Involving magnetic fluid in synthesizing nanorobots facilitates auto propulsion. Diagnostic gadgets, contrast agents, imaging methods, and enhanced therapeutic applications such as drug delivery vehicles have all benefited from the combination of magnetic nanoparticles with biology. The magnetic nanorobots are biocompatible, biodegradable, chemically stable, and nontoxic to the cells. The MNPs can be designed and modified for multifunctional applications and are promising diagnostic tools in medicine. Several theranostic applications are discussed here.

25.5.1 Molecular diagnostics by magnetic extraction

Molecular diagnostics incorporates a series of techniques that aim to detect deviations at the DNA, RNA, or protein levels. Thus, nucleic-acid-based testing has been developed as an inimitable diagnostic tool for prognosis, detection, diagnosis, and monitoring diseases [38,39]. Separation of nucleic acid is a step before multiple diagnostic and biochemical methods, including cloning, detection, amplification, hybridization, sequencing, and DNA synthesis. Nanorobot paved ways in diagnostics through magnetic separation methods for elusive and consistent captivities of specific proteins, genetic material, and other biomolecules. Recent research reports validate the yields of DNA isolated with the magnetic method which were superior or equivalent, a conventional techniques [40]. Further, magnetic handling is easy to control through the external magnetic field and does not require costly peripheral systems. Magnetic bead technology is one of the developing schemes for extracting nucleic acids. The technique includes the separation of genomic DNA and RNA from mixtures through complementary hybridization. Nowadays, functionalized MNPs or beads have been fixed with appropriate buffer systems for a speedy and effective extraction procedure [41,42]. In 2017, researchers developed the magnetic beads to extract nucleic acids from sputum to detect viral pathogens causing acute respiratory infections [43]. In 2020, the world experiencing a pandemic situation due to severe acute respiratory syndrome coronavirus -2 (SARS-CoV-2). Recently, for diagnostics for this virus, the magnetic beads based RNA extraction method is established [44]. Different types of functionalized MNPs commercially available for extractions of biomolecules are listed in Table 25.1.

Traditional methods of protein separations are tedious and can damage the target protein. In 1997, our research group reported that protein directly binds to MNPs [45]. Magnetic separation techniques have noticeable benefits over standard separation

Table 25.1 Commercially available magnetic nanoparticle based extraction kit.

Name of kit for biomolecular extraction	Application
NucleoMag RNA	Purification of total RNA
MagJET (Thermo Fisher scientific)	Plasmid DNA, genomic DNA, viral RNA isolation
Xpress DNA (Mag Genome)	For extraction of high quality genomic DNA from fresh, frozen and ethanol preserved mammalian tissues, fish tissues, and cell lines.

processes. These methods are simpler and do not require any specialized skill. A single test tube is used to execute all the stages of the cleaning procedure. Affluent filters, centrifuges, chromatography systems, or any additional equipment is not required. The magnetic affinity adsorption method contains the usage of particles with immobilized affinity ligands like antibodies, avidin-biotin, streptavidin, protein A and protein G. These immobilized ligands included in the MNPs can function as general solid phases leading to immobilize altered affinity ligands. The surface of MNPs can be modified by silanization [46]. The MNPs are encapsulated with different biopolymers. MNPs have exceptional bacterial and viral sensing abilities. The amine-modified magnetic nanoparticles were designed to remove harmful microorganisms from the environment [47]. Immobilized biotin labelled anti-E. coli antibodies with avidin coated MNPs and gold-coated magnetic spherical nanoparticles were employed for E.Coli separations [48]. Wu et al. developed antibody-conjugated Au/MNCs for the separation of Salmonella bacteria from milk [49]. Electrochemical bienzyme immune sensor based on magnetic beads for the detection of the bird flu virus (AIV) H9N2 [50]. The malarial plasmodium histidine-rich protein-2 detection was first time done with magneto immunoassay-based strategies using MNPs [51]. The conventional diagnostic methods for cancer detection are laborious and require higher volumes of samples. Galanzha et al. [52] devised a method for magnetically detecting circulating tumor cells (CTCs) in the bloodstream of mice, and were able to perform fast photoacoustic detection of CTCs as a result. This approach has a lot of applications in human metastasis prevention and basic tumor diagnostics. Fluorescent-magnetic-biotargeting multifunctional nanobioprobes can identify and isolate a variety of tumor cells [53]. Countless advancements have been achieved in the development of MNP-based biological detection, as well as a variety of MNP-based methodologies for investigating biological entities. Still, progress is required to alter the features of MNPs. New approaches, particularly in-vivo detection, will be required to make further advancements in this field.

25.5.2 Contrast agents for magnetic resonance imaging

L Paul has developed a non-invasive technique to detect tumors, and other defects in a human body named Magnetic Resonance Imaging (MRI) [54]. This technique is based on NMR (Nuclear Magnetic Resonance). NMR earlier called nuclear induction was first discovered in 1952 by F. Bloch and E.M. Purcell. When the nucleus (proton) experiences a strong magnetic field, the magnetic moment associate with it precesses around the magnetic field with a frequency called Larmor frequency. The applied magnetic field determines the value of this precessional frequency. Most magnetic moments align in the field direction when a magnetic field is applied, resulting in the longitudinal magnetization, or net magnetic vector in the field direction. These protons stay in a low energy state while some of the moments align in opposite directions remain in a high energy state. When a radiofrequency pulse interacts with this structure, some low energy parallel protons flip to a high energy state, lowering longitudinal magnetization. Protons synchronise and precess in phase during the process, causing the net magnetization vector to move toward the transverse phase, which is at a right angle to the primary magnetic field. This is referred to as transverse magnetization.

After the RF pulse is turned off, protons return to their low energy condition parallel to the primary magnetic field, transferring their energy to the surrounding lattice. This causes alterations in the longitudinal relaxation, which is known as T1 relaxation or spin-lattice relaxation. In the transverse axis, the protons that were in phase begin to dephase. T2 relaxation or spin-spin relaxation is the process through which transverse magnetization is reduced (Fig. 25.3). In NMR most commonly used nuclei are hydrogen atoms. The existence of the proton and the difference in relaxation durations (T1 and T2) of protons after being subjected to an external magnetic field are crucial to the operation of MRI. In human body, there are a large number of hydrogen nuclei in

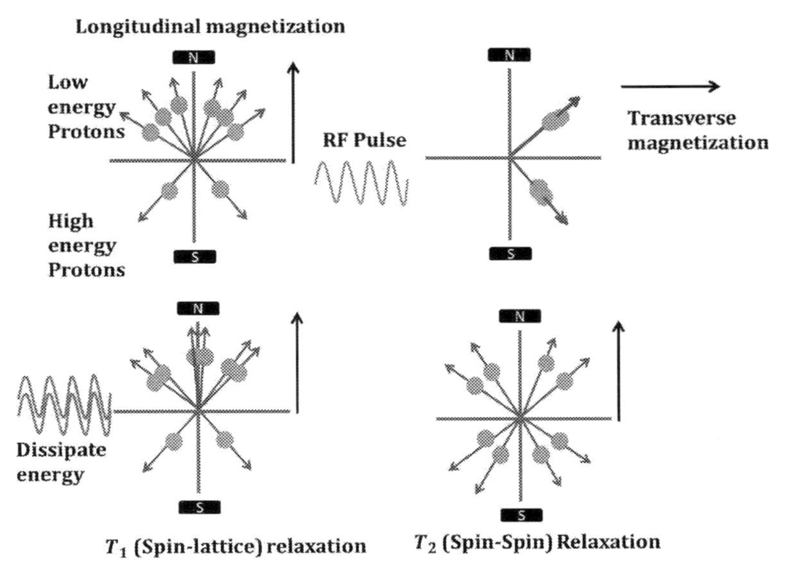

Fig. 25.3 *T_1 and T_2 relaxation mechanism.*

form of water, lipids, hydroxyapatite, carbohydrates, etc. This leads to a significant difference between T1 and T2 relaxation times. A good contrast agent is required to control relaxation times (T1 and T2) to get an enhanced MRI image

Paramagnetic gadolinium (Gd+3) is a common contrast agent. It has been discovered to be beneficial for brain tumor angiography. It has an effect on the local magnetic field and the relaxing of hydrogen protons. It helps the regional protons to flip back to their lower energy state quicker. The faster the longitudinal magnetization recovers, the stronger the T1 signal and the brighter that area looks on the T1 weighted picture [55,56]. Gadolinium is a heavy metal and quite toxic in the human body. If patients are having either diabetes, liver, or kidney diseases, gadolinium cannot be removed and gets accumulated in the body. In such cases, it is essential to use alternative ions. In this case, magnetic ions such as superparamagnetic iron oxide nanoparticles (SPIONS) or SPION MNP are useful. MNPs are more preferable since a drug can be easily bound to them. MRI with a drug-bound MNP can serve two purposes-MNP will enhance contrast and the drug can help to treat tumors or tissues [57]. Paramagnetic compounds give a bright contrast image resulting T1-weighted image, while superparamagnetic compounds give dark contrast images, resulting T2-weighted image. To improve relaxometry and low cytotoxicity, scientists have developed a T1/T2 hybrid contrast agent by encapsulation of superparamagnetic nanoparticles into an iron-based coordination polymer nanoparticles (CPP-Fe) as a T1-weighted signal [58]. The surface modification of the contrast agent by coating material also enriched their properties for MRI imaging. The superparamagnetic iron oxide nanoparticles (SPIONs) nanoparticles covered with amphiphilic alkyl-polyethyleneimine (PEI) show better performance for MRI imaging. However, it can cause cytotoxicity. Further modification with lactose, the new PEN-alkyl-PEI-lactobionic acid-coated MNPs works as a contrast agent for MRI [59]. To increase the functional advantages of stem cell-based stroke therapy, it is vital to investigate the long-term performance of mesenchymal stem cells (MSCs). Researchers used SPION-loaded cationic polymersomes to mark MSCs expressing green fluorescent protein (GFP) to investigate MRI's ability to reflect MSC survival, long-term destiny, and probable pathways [60]. It has been demonstrated that grafted cells are pre-labeled with SPION-loaded polymersome. The MRI technique can verify their biodistribution and migration. SPION in the range of 60 nm to 180 nm are used as approved clinical diagnosis MRI contrast agents. Two drugs, Feridex and Resovist are prepared using SPIO colloids with a coating of dextran or carboxy dextran respectively for MRI of the liver [61]. A recent study also showed the potentiality of nanosized magnetite and maghemite particles in MRI [62].

25.6 A drug delivery vehicle

Conventional drug system, while widely utilized, have many problems like low stability and less specificity that may be potentially overcome by modern methods. There are several factors increasing attention in the development of new devices, concepts, and techniques for drug delivery vehicles for deadly illnesses like cancer and infectious diseases. The MNPs have the potential to be used as an unconventional remedy since they offer excessive aids for targeted drug delivery straight to cancerous cells and enhance the therapeutic efficacy [63].

The drug delivery vehicles are primarily designed to carry the payload absorbed in their interior volume and on their surface. The physical and chemical features of the therapeutic substance influence the choice of vehicle type and the construction of the whole nanorobot.

25.6.1 Targeted drug delivery

The drug delivery system aims to modulate the pharmacokinetics and pharmacodynamics of drugs beneficially. Novel drug discovery system uses drug potency and controlled drug release and gives sustained therapeutic effects, it provides safety and also target a drug specifically to the target site. Till date, most of the nano vehicles are designed for cancer therapy. Using drugs in combination with specifically engineered drug delivery systems can help increase therapeutic efficacy and lower the risk of anti-cancer therapies causing systemic toxicity. The magnetically guided drug delivery systems for anticancer therapeutics are based on two different concepts of passive targeting and active targeting.

25.6.1.1 Passive targeting

Passive targeting techniques depend on architectural and functional differences between normal and tumor vasculature to enable selective medication accumulation at the tumor location [64]. The important aspect is that there is an increased permeability and retention effect (EPR) in many vascularized solid and metastatic tumor nodule tumours [65]. The EPR effect can be abused for so-called passive targeting of antitumor agents. The cause of EPR effect is leaky vasculature and impaired lymphatic drainage in solid tumors. Passive targeting by stimuli-sensitive vehicles may have an advantage over conventional systems that strained extreme treatment for oncogenic tumors. The hypothesis builds on the fact that acid pH and greater temperature generally occur in tumours than normal tissues [66]. Nano-sensitive stimuli can be manufactured to release the integrated medicine if exposed to certain "differentiated" tumor circumstances.

25.6.1.2 Active drug targeting

The advanced approach for tumor control is to actively target malignant cells with folate, lectins, antibodies, peptides and ligands supplied with MNPs through the therapy of MNPs or through interactions [67]. The activation of endothelial cells is evident during angiogenesis with an increased production of molecules of cellular adhesion and proteolytic enzymes. The vascular endothelium is therefore responsible for many cancer therapy objectives, including the endothelial cells, stromal components, which are extremely accessible to any circulating systems and may subsequently be employed to target drugs and drug carriers. Special biomarkers only occur in or on the tumor cells known as tumor-associated antigens (TAA). TAA can be exploited for medication targeting tumor cells using antibodies or ligands [68]. The design of the vectorized drug delivery system necessitates knowledge of surface receptors such as proteins, glycoproteins, and lipoproteins, which are more prevalent on malignant cells than on healthy cells. Recognizing ligands that bind strongly to these receptors, such as peptides, enzymes or enzyme inhibitors, saccharides, lectins, antibodies, and antibody fragments, is also significant.

25.6.1.3 Magnetic nanoparticles as drug delivery vehicle

In the 1970s, a new vehicle of transport for the delivery of drugs was manufactured to be used with an external magnetic field [69]. The systemic distribution of drug-carrying nanoparticles to particular areas in the body has been the subject of several investigations [70]. For effective drug delivery, nanoparticles must circulate in the blood for extended periods of time without obstructing arteries, veins, or capillaries, and the physical ingredients must be non-toxic, biocompatible, biodegradable, and otherwise suitable for human treatment [65]. MNPs were created for localized drug delivery, mostly in cancer patients. When the MNPs are injected into the bloodstream, a magnetic field is used to keep the particles in the tumor's specific area. It is one of the most promising cancer treatment options currently accessible. Over the last few years, the use of magnetic nanoparticles in drug targeting has skyrocketed. Nanoparticles can be regulated to convey the medication to the target location, fix the particles in place, release the medications, and perform magnetic drug targeting when utilized with an external magnetic field. MNP minimizes the likelihood of adverse effects in localized or targeted medication administration, which is frequent in traditional treatments like radiation treatment and chemotherapy.

Lubbe and colleagues conducted the first Phase 1 clinical study utilising magnetically focused drug delivery in 1996 [71]. Following that, other research groups have used various ways to synthesis magnetic drug carriers for drug delivery and MRI [72]. MNPs transport chemotherapeutic drugs such as doxorubicin (DOX), paclitaxel (PTX), Gemcitabine, Cisplatin, Docetaxel (Dtxl), Bortezomib, Artemisinin etc. [73] by attaching or encapsulation. The magnetic nanocarriers using chemotherapeutic agents and coting materials are listed in Table 25.2. The MNPs loaded with DOX were produced by FeRx, Inc., [74]. Alnis Biosciences produces a magnetic nanoparticle hydrogel containing chemotherapeutic drugs, Fe oxide colloids, and various targeting ligands [75]. Chemicell GmbH, Berlin, Germany, has launched TargetMAG-DOX nanoparticles with a multidomain magnetite core and a cross-linked starch matrix containing terminal cations [76]. Due to remarkable magnetic and biological properties, drug nanocarriers systems have been developed and tested *in-vitro* and *in-vivo* in the last two decades. Andhariya et al reported PEG functionalized OA/OL coated $Fe3O4$ nanoparticles drug delivery system for controlled release of doxorubicin [77]. Magnetic poly (ethyl-2-cyanoacrylate)

Table 25.2 Magnatic nanoparticles loaded with chemotherapeutic drugs and coating agents.

Drugs	Mechanism of action	Coating agents in different systems
Doxorubicin (DOX)	Intercalation into DNA and topoisomerase II inhibition	Dopamine/human serum albumin [102] Polyethylene glycol/polyethyleneimine [103] Heparin [77] Polyvinyl alcohol [104] Polymer [81] Polyethyleneimine [105] PEG [106] Chitosan [84,86,107] Phospholipid-PEG [108]
Paclitaxel (PTX)	Disruption of the normal tubule dynamics essential for cell division	Thermally cross-linked polymers/β-cyclodextrin [109] PEG/carboxymethylated β-cyclodextrin [80] Amphiphilic copolymer [110]
Gemcitabine (GEM)	DNA polymerase inhibition	Amphiphilic polymer [72] Dimercaptosuccinic acid [111,112]
Cisplatin	Crosslinking with purine nucleotides in DNA causes cell damage	Gold/Polyethylene glycol [82] Dextran/Human serum albumin [113]
Docetaxel (Dtxl)	Microtubule dynamics disruption	Pluronic F127 polymer/β-cyclodextrin [71]
Bortezomib	Proteasomal activity inhibition	Chitosan [83]
Artemisinin	The production of free radicals causes cell damage	Chitosan [114]

(PECA) nanoparticles carrying the anti-cancer medicines cisplatin and gemcitabine were created using an inter-facial polymerization process [78]. In-vivo testing of DOX-loaded mesoporous silica nanoparticles containing magnetite nanocrystals was performed. The in-vivo investigation established that the anticancer medicine was delivered to the tumor locations and that its activity was maintained throughout the process [79]. Yang and colleagues created the theranostic nanoplatform comprising histidine-tagged, cyan fluorescent protein (CFP)-capped magnetic mesoporous silica nanoparticles (MMSNs) [80]. These smart material have both drug delivery and cell imaging capabilities. Few peptides conjugated vacterized Theranostic nanorobot exhibited cytotoxicity for brest cancer and pancreas cancer cells [81–83] prepared gold-coated iron oxide nanoparticles bearing PEG linked anticancer drug cisplatin. With an external magnetic field, these nanorobots demonstrated cytotoxicity on human ovarian cancer cell lines A2780. The fluctuation in pH can be utilized to initiate the distribution of drugs onto nanocarriers via coating agents such as polymers or liposomes. The chitosan-coated magnetic nanoparticles are pH sensitive. The FDA approved proteasome inhibitor drug bortezomib coated with

Table 25.3 Commercially available theranostic nanoparticles.

Drug	Usage
Feraheme/Ferumoxytol	For the treatment of anemia caused by iron deficiency in adults with chronic renal disease (CKD)
Endorem	Contrast agent for MRI
Gastromark	Aqueous solution of silicone-coated superparamagnetic iron oxide for oral administration as a magnetic resonance imaging contrast material.
Combidex	MRI contrast agent to see lymph nodes in the pelvis
Lumiren	MRI contrast agent for bowel imaging
Radiogardase	Treatment with Prussian blue insoluble for thallium contamination
NanoTherm	Thermotherapy
Feridex IV	Liver and spleen imaging
Resovist	A clinically approved MRI contrast agent consisting of superpara-magnetic iron oxide

chitosan magnetic nanoparticles shows anti-cancer activity on HeLa and SiHa cells [84,85]. Adimoolam et al. recently reported a cytotoxic effect of chitosan-coated MNPs with pH-sensitive glutaraldehyde linkers for pH responsive delivery of doxorubicin in the intracellular components of human breast cancer (MCF-7) and ovarian cancer (SK-OV-3) cell lines in the intracellular components of human breast cancer (MCF-7) and ovarian cancer (SK-OV-3) cell lines in 2018 [86]. Some commercially available theranostic nanoparticles are listed in Table 25.3.

25.6.2 Hyperthermia

The temperature of cancerous cell regions of the body increases to 40–43°C during hyperthermia (HT) treatment. It can induce cancer cell death by amplifying cytotoxic effects via radiotherapy and chemotherapy [87]. Hyperthermia can be applied to a specific area of the body, a region of the body, or the entire body. Temperature increases have been achieved in a variety of ways over the last few decades, including the use of thermal chambers, hot water blankets, electromagnetic energy, perfusion of the limb or body cavity with heated fluids, ultrasound, and magnetic nanoparticles [87,88]. In 1957 Gilchrist et al, first time introduced magnetic hyperthermia, the magnetic nanoparticles inserted into lymphatic channels to heat malignant cells under an Alternating Magnetic Field (AMF) [89]. Jordan et al. presented that direct injection of magnetic nanoparticles into the tumor could be the upshot in the operative heating of tumors compared to other heating techniques like radiofrequency heating and ultrasound [90]. MNPs produce heat via the mechanism of hysteresis loss and relaxation losses, when it is exposed to an alternating magnetic field. The oral administration of MNPs is not feasible due to its large size and easy fecal excretion [91]. MNps can possibly be delivered to the tumor via intratumoral, intra-peritoneal, intra-arterial, intra-cavitary, and intravenous

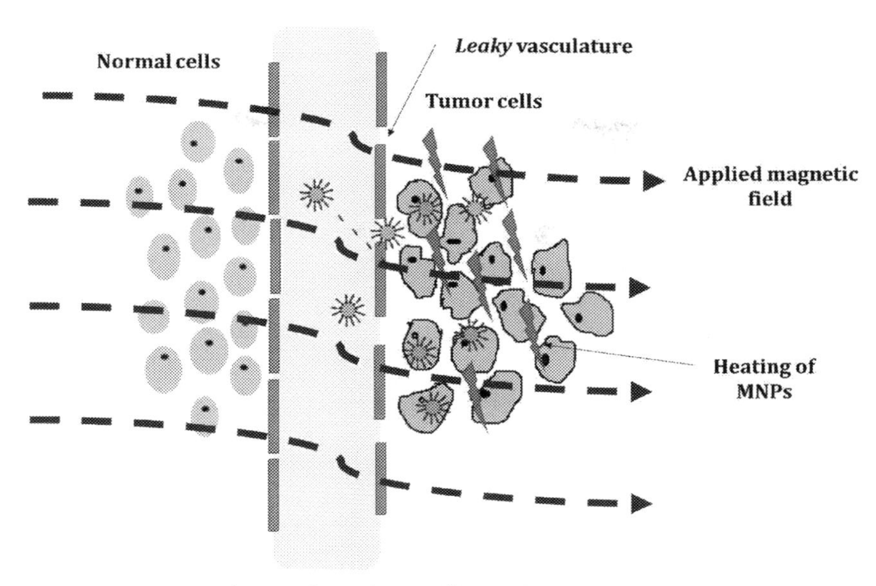

Fig. 25.4 ***Passive targeting of MNPs through EPR effect and magnetic hyperthermia.***

administration. Fig. 25.4 demonstrates that high-magnetism MNPs had a great heating effect and a lot of accumulation in the tumor due to enhanced permeability and retention (EPR) effects. As a result, employing magnetically enticed heat, tumour growth is effectively prevented within a short period of time.

Several studies reported that malignant cells efficiently uptake the MNPs than normal tissues [92]. The German company has developed aminosilane coated ferrofluid, an alternating magnetic field applicator NanoActivator and temperature simulation software NanoPlan. Numerous clinical trials performed on the influence of hyperthermia on glioblastoma and prostate cancers with subsequent application of AMF via the MagForce system.

Over the last decade, clinical trials for Glioblastoma (GBM) patients have used magnetic hyperthermia as a therapeutic approach. Initially, 14 patients with primary or recurrent GBM were treated in Phase 1 clinical trials with a combination of radiation therapy (RT) and hyperthermia. The treatment was performed twice a week for 60-min for an average of six sessions. After injection of the MNPs, a closed-end thermometry catheter was placed in the tumor area. The average 44.6 °C intratumoral temperature was extended. In the Phase II study, 59 patients with recurrent GBM were treated with MNPs heating in an alternating magnetic field combined with radiotherapy. In comparison to traditional therapies in the treatment of recurrent glioblastoma, the results showed that therapy with MNPs in combination with a lower radiation dose is non-toxic and effective, and leads to a more extensive diagnosis of first tumor recurrence. Table 25.4 summarizes some of the research into the efficiency of nanoparticle

Table 25.4 *In-vivo, in-vitro* and clinical trial studies for magnetic hyperthermia.

Experimental setup	Tumor site in patient/ cell line/animal	Magnetic nanoparticle composition	References
Clinical trials	Glioblastoma (14 patients)	SPION coated with aminosilane and distributed in water	[115]
Clinical trials	Glioblastoma (59 patients)	SPION coated with aminosilane and distributed in water	[116]
Clinical trials	Glioblastoma (3 patients)	SPION coated with aminosilane and distributed in water	[117]
Clinical trials	Prostate cancer (10 patients)	SPION coated with aminosilane and distributed in water	[118,119]
In-vitro & *In-vivo*	T-9 cell lines/F344 rats	Magnetite cationic liposomes	[120–123]
In-vitro & *In-vivo*	T-9 cell lines/F344 rats	Carboxymethyl-cellulose magnetite	[124]
In-vitro & *In-vivo*	U251-SP cell line/ Athymic nude mice	Magnetite cationic liposomes	[125]
In-vitro & *In-vivo*	C6 cell lines/Wistar rats	Maghemite	[126]
In-vitro & *In-vivo*	U251 cell lines/rats	Iron oxide	[127]
In-vitro	parasite Leishmania mexicana	Iron oxide	[128]
In-vitro	HeLa and MCF7 cell lines	Mn–Zn ferrite nanoparticles	[129]
In-vitro	Primary human glioblastoma cells	Dextran or aminosilane-coated magnetite	[92]
In-vivo & *Ex-vivo*	*Ex-vivo* human breast tissue and *in-vivo* tumors in mice	Magnetite nanoparticles	[130]
In-vivo	Tumours in mice	Coated magnetite	[131]
In-vivo	Mice	Magnetite nanoparticles coated with lipid membrane	[132]
In-vivo	Mice	Iron oxide coated with Dextran and PEG and conjugated to the Chimeric L6 antibody	[133]
In-vivo	In Mice	Ferromagnetic, dextran-coated nanoparticles	[134]
In-vivo	Mice	Ferromagnetic, dextran-coated nanoparticles	[135]
In-vivo	Mice	Magnetic nanoparticles based on iron are loaded into liposomes. Trastuzumab antibody-conjugated liposomes	[136]

hyperthermia induced single-domain ferromagnetic and superparamagnetic nanoparticles in various biological systems. Despite of this, certain improvements are needed to make MNP based hyperthermia clinically feasible.

25.7 Safety measurements and risks

We discuss various applications of magnetic nanoparticles as theranostic nanorobots. Tailoring MNPs and the Magnetic field are two major common factors in all applications. It is equally important to study their safety measures and risk in implementing them into clinical practices successfully. In ex-vivo and *in*-vitro studies, e.g. gene extraction doesn't involve a patient's body. Here, it is essential to study the effects like toxicity, hematocompatibility, biodegradation, immunogenicity, and pharmacokinetics, before the clinical trial of biocompatible magnetic nanoparticles. By modifications of the surface of MNPs, researchers have observed a reduced toxicity effect [93]. The cytotoxicity largely depends on nanoparticle size, shape, coating agents, composition, and surface area [94]. Large surface to volume ratio, large particle size, and concentration show higher cytotoxicity and unfavorable biological response [95,96]. In an MRI study, the patients' body experiences a high magnetic field. Even if human body parts are dia, para, ferri, ferro, or superparamagnetic, some restriction has been implicated for the application of high magnetic field gradient. Earlier 2Tesla and 4Tesla magnetic field gradients were considered as threshold limits [97]. Later the 8Tesla field was considered safe for the adults, but in *in-vivo* study, it reduces the human blood flow rate by 30% [98]. Companies like Feridex, Endorem, GastroMARK, Lumirem, Sinerem, or Resovist are producing Fe oxides superparamagnetic based contrast agents, while Magnevist, Dotarem, Gadovist, Teslacan are selling paramagnetic encapsulated in chelating agent-based contrast media. For the targeted drug delivery MagNaGel32 and FluidMAG and TargetMAG have commercialized the products [99]. In his article, Kostarelos stated that nanodevices must be toxicologically inert, degradable, or ejected from the patient's body, regardless of their self-propulsion and functionality [100]. By expelling these nanorobots from the body, we may reduce human toxicity, but then the question remains the same of environmental pollution with the wide-spread application of nanorobots in the environment. Researchers have developed smart synthetic micro/nanorobots for environmental sensing, monitoring, and remediation [101].

We have cited names of a few companies marketing materials/devices involving magnetic particles. This is only to cite a few examples. Authors neither suggest use nor otherwise endorse these products.

25.8 Conclusion and prospects

In this chapter, we have described the synthesis and applications of multifunctional nanoparticles of magnetic materials especially as theranostic nanorobots. These particles are biocompatible and using appropriate techniques so that drugs or biomolecules can easily be attached to these particles. Further, being superparamagnetic they can be controlled by an external magnetic field. These properties are promising for the development of novel applications in biotechnology and biomedicines. We have described four such applications namely, gene extraction, a contrast agent for MRI, targeted drug delivery, and hyperthermia for cancer treatment. Lastly, safety measurements and risk factors are also discussed.

A number of applications of theranostic nanomagnetic particles are found to be successful in-vivo, but the clinical applications are yet to be tried. A detailed study of cytotoxicity is required. Secondly, strong permanent magnets are required for targeted drug delivery systems. Present-day magnets can effectively generate fields only up to the depth of a few centimeters in body tissues. Nanoparticles with larger magnetic moments are also desirable. For these hybrid materials or alloys are being developed. Attaching magnetic particles with carbon nanotubes or other hollow nanotubes like halloysite can be useful as one can load more nanoparticles both onside and insides of the tubes. A good mathematical model can be very useful for successful trials. All these developments require a good multidisciplinary team of researchers. A day is not far when this very versatile material will be easily accessible even in a moderately good hospital.

References

[1] D.E. Lee, H. Koo, InC Sun, JuH Ryu, K. Kim, I.C. Kwon, Multifunctional nanoparticles for multimodal imaging and theragnosis, Chem. Soc. Rev. 41 (7) (2012) 2656–2672, https://doi.org/10.1039/c2cs15261d.

[2] J. Xie, S. Lee, X. Chen, Nanoparticle-based theranostic agents, Adv. Drug. Deliv. Rev. 62 (11) (2010) 1064–1079, https://doi.org/10.1016/j.addr.2010.07.009.

[3] KiY Choi, G. Liu, S. Lee, X. Chen, Theranostic nanoplatforms for simultaneous cancer imaging and therapy: current approaches and future perspectives, Nanoscale 4 (2) (2012) 330–342, https://doi.org/10.1039/c1nr11277e.

[4] B.T. Luk, R.H. Fang, L. Zhang, Lipid- and polymer-based nanostructures for cancer theranostics, Theranostics 2 (12) (2012) 1117–1126, https://doi.org/10.7150/thno.4381.

[5] P.R. Feynman, There's plenty of room at the bottom. In: Engineering and Science, CRC Press, Boca Raton, 1960, https://doi.org/10.1201/9780429500459.

[6] B. Gutierrez, C.V. Bermúdez, Y.R.C. Ureña, S.V. Chacón, J.V. Baudrit, Nanobots: development and future, Int. J. Biosens. Bioelectron. 2 (5) (2017), https://doi.org/10.15406/ijbsbe.2017.02.00037.

[7] R.E. Rosensweig, Ferrohydrodynamics, Cambridge University Press Cambridge, New York, Melbourne, 1985.

[8] R.V. Mehta, R.V. Upadhyay, Science and technology of ferrofluids, Curr. Sci. 76 (3) (1999) 305.

[9] K. Raj, B. Moskowitz, R. Casciari, Advances in ferrofluid technology, J. Magn. Magn. Mater. 149 (1–2) (1995) 174–180, https://doi.org/10.1016/0304-8853(95)00365-7.

[10] R.V. Mehta, Synthesis of magnetic nanoparticles and their dispersions with special reference to applications in biomedicine and biotechnology, Mater. Sci. Eng. C 79 (2016) (2017) 901–916, https://doi.org/10.1016/j.msec.2017.05.135.

[11] R.V. Mehta, Impact of magnetic nanomaterials on biotechnology and biomedicine, Mod Appl Bioequiv Availab 2 (2) (2017) 1–3, https://doi.org/10.19080/MABB.2017.02.555581.

[12] M. Lévy, C. Wilhelm, J.M. Siaugue, O. Horner, J.C. Bacri, F. Gazeau, Magnetically induced hyperthermia: size-dependent heating power of γ-Fe2O3 nanoparticles, J. Phys. Condensed Matter 20 (20) (2008), https://doi.org/10.1088/0953-8984/20/20/204133.

[13] F. Arteaga-Cardona, S. Hidalgo-Tobón, U. Pal, M.A. Méndez-Rojas, Ferrites as magnetic fluids for hyperthermia and mri contrast agents, AIP Conference Proceedings, 1747, 2016, https://doi.org/10.1063/1.4954118.

[14] B.D. Cullity, C.D. Graham, Introduction to Magnetic Materials, John Wiley & Sons, Piscataway, NJ, 2011.

[15] Q.A. Pankhurst, J. Connolly, S.K. Jones, J. Dobson, Applications of magnetic nanoparticles in biomedicine, J. Phys. D Appl. Phys. 36 (13) (2003) R167–R181.

[16] P. Tartaj, M. Del Puerto Morales, S. Veintemillas-Verdaguer, T. González-Carreño, C.J. Serna, The preparation of magnetic nanoparticles for applications in biomedicine, J. Phys. D 36 (13) (2003) R182–R197.

[17] J. Huang, H. Pen, Z. Xu, C. Yi, Magnetic Fe3O4/Poly(styrene-co-acrylamide) composite nanoparticles prepared by microwave-assisted emulsion polymerization, React. Funct. Polym. 68 (1) (2008) 332–339.

[18] J. Pinkas, V. Reichlova, R. Zboril, Z. Moravec, P. Bezdicka, J. Matejkova, Sonochemical synthesis of amorphous nanoscopic iron(III) oxide from Fe(Acac)3, Ultrason. Sonochem. 15 (3) (2008) 257–264.

[19] P. Dallas, A.B. Bourlinos, D. Niarchos, D. Petridis, Synthesis of tunable sized capped magnetic iron oxide nanoparticles highly soluble in organic solvents, J. Mater. Sci. 42 (13) (2007) 4996–5002.

[20] R.V. Ramanujan, Y.Y. Yeow, Synthesis and characterisation of polymer-coated metallic magnetic materials, Mater. Sci. Eng. C 25 (1) (2005) 39–41.

[21] J.L. Arias, V. Gallardo, S.A. Gómez-Lopera, R.C. Plaza, A.V. Delgado, Synthesis and characterization of poly(ethyl-2-cyanoacrylate) nanoparticles with a magnetic core, J. Control. Release 77 (3) (2001) 309–321.

[22] D.A. Bazylinski, A.J. Garratt-reed, R.B. Frankel, Electron microscopic studies of magnetosomes in magnetotactic bacteria, Microsc. Res. Tech. 401 (1994) 389–401.

[23] D. Faivre, D. Schu, Magnetotactic bacteria and magnetosomes, Chem. Rev. 108 (11) (2008) 4875–4898.

[24] M. Timko, M. Molcan, A. Hashim, A. Skumiel, M. Muller, G. Arkadiusz, J. Hubert, Hyperthermic effect in suspension of magnetosomes prepared by various methods, IEEE Trans. Magn. 49 (1) (2013) 250–254.

[25] P. Nezhad-Mokhtari, F. Salahpour-Anarjan, A. Rezanezhad, A. Akbarzadeh, Magnetic nanoparticles: an emergent platform for future cancer theranostics. In: Nanobiotechnology in Diagnosis, Drug Delivery, and Treatment, 2020 171–195.

[26] J. Kim, H.S. Kim, N. Lee, T. Kim, H. Kim, T. Yu, InC Song, W.K. Moon, T. Hyeon, Multifunctional uniform nanoparticles composed of a magnetite nanocrystal core and a mesoporous silica shell for magnetic resonance and fluorescence imaging and for drug delivery, Angew. Chem. 47 (44) (2008) 8438–8441, https://doi.org/10.1002/anie.200802469.

[27] N. Pazos-Pérez, Y. Gao, M. Hilgendorff, S. Irsen, J. Pérez-Juste, M. Spasova, M. Farle, LM. Liz-Marzán, M. Giersig, Magnetic - noble metal nanocomposites with morphology-dependent optical response, Chem. Mater. 19 (18) (2007) 4415–4422, https://doi.org/10.1021/cm070248o.

[28] H. Lee, T. JongYoon, R. Weissleder, Ultrasensitive detection of bacteria using core-shell nanoparticles and an NMR-filter system, Angew. Chem. Int. Ed. 48 (31) (2009) 5657–5660, https://doi.org/10.1002/anie.200901791.

[29] T.J. Yoon, H. Lee, H. Shao, R. Weissleder, Highly magnetic core-shell nanoparticles with a unique magnetization mechanism, Angew. Chem. Int. Ed. 50 (20) (2011) 4663–4666, https://doi.org/10.1002/anie.201100101.

[30] H. Zeng, J. Li, Z.L. Wang, J.P. Liu, S. Shouheng, Bimagnetic core/shell FePt/Fe3O4 nanoparticles, Nano Lett. 40 (1) (2004) 187–190.

[31] B. Chudasama, A.K. Vala, N. Andhariya, R.V. Upadhyay, R.V. Mehta, Antifungal activity of multifunctional Fe3O4–Ag nanocolloids, J. Magn. Magn. Mater. 323 (10) (2011) 1233–1237.

[32] N. Andhariya, B. Chudasama, R.V. Mehta, R.V. Upadhyay, Biodegradable thermoresponsive polymeric magnetic nanoparticles: a new drug delivery platform for doxorubicin, J. Nanopart. Res. 13 (4) (2011) 1677–1688.

[33] W. Shi, H. Zeng, Y. Ding, Y. Sahoo, TY. Ohulchanskyy, Z.L. Wang, M. Swihart, N. Paras, A general approach to binary and ternary hybrid nanocrystals, Nano Lett. 6 (4) (2006) 875–881.

[34] J. Jiang, H. Gu, H. Shao, E. Devlin, GC. Papaefthymiou, JY. Ying, Bifunctional Fe3O4–Ag heterodimer nanoparticles for two-photon fluorescence imaging and magnetic manipulation, Adv. Mater. 20 (23) (2008) 4403–4407.
[35] Z. Zhao, Z. Zhou, J. Bao, Z. Wang, J. Hu, X. Chi, K. Ni, et al., Octapod iron oxide nanoparticles as high-performance T 2 contrast agents for magnetic resonance imaging, Nat. Commun. 4 (2013) 1–7, https://doi.org/10.1038/ncomms3266.
[36] L. Yang, Z. Wang, L. Ma, Ao Li, J. Xin, R. Wei, H. Lin, R. Wang, Z. Chen, J. Gao, The roles of morphology on the relaxation rates of magnetic nanoparticles, ACS Nano 12 (5) (2018) 4605–4614, https://doi.org/10.1021/acsnano.8b01048.
[37] J. Zeng, L. Jing, Yi Hou, M. Jiao, R. Qiao, Q. Jia, C. Liu, F. Fang, H. Lei, M. Gao, Anchoring group effects of surface ligands on magnetic properties of Fe3O4 nanoparticles: towards high performance MRI contrast agents, Adv. Mater. 26 (17) (2014) 2694–2698, https://doi.org/10.1002/adma.201304744.
[38] P.B. Raghavendra, T. Pullaiah, Advances in Cell and Molecular Diagnostics, Academic Press, 2018.
[39] G.P. Patrinos, W.J. Ansorge, Molecular diagnostics: past, present, and future. Molecular Diagnostics, 3rd edition. Academic Press, 2010, pp. 1–11.
[40] Z.M. Saiyed, C.N. Ramchand, Extraction of Genomic DNA Using Magnetic Nanoparticles (Fe3O4) as a Solid-Phase Support, Am. J. Infect. Dis. 3 (4) (2007) 225–229, https://doi.org/10.3844/ajidsp.2007.225.229.
[41] MJ. Archer, B. Lin, Z. Wang, DA. Stenger, Magnetic bead-based solid phase for selective extraction of genomic DNA, Anal. Biochem. 355 (2) (2006) 285–297.
[42] S.M. Azimi, G. Nixon, J. Ahern, W. Balachandran, A magnetic bead-based DNA extraction and purification microfluidic device, Microfluid. Nanofluid. 11 (2) (2011) 157–165.
[43] H. He, R. Li, Yi Chen, P. Pan, W. Tong, X. Dong, Y. Chen, D. Yu, Integrated DNA and RNA extraction using magnetic beads from viral pathogens causing acute respiratory infections, Sci. Rep. 7 (2017) 45199.
[44] S. Klein, T.G. Müller, D. Khalid, V. Sonntag-Buck, A.-.M. Heuser, B. Glass, M. Meurer, et al., SARS-CoV-2 RNA extraction using magnetic beads for rapid large-scale testing by RT-QPCR and RT-LAMP, Viruses 12 (8) (2020) 863.
[45] R.V. Mehta, R.V. Upadhyay, S.W. Charles, C.N. Ramchand, Direct binding of protein to magnetic particles, Biotechnol. Tech. 11 (7) (1997) 493–496.
[46] T.I. Shabatina, O.I. Vernaya, V.P. Shabatin, M. Ya Melnikov, Magnetic nanoparticles for biomedical purposes: modern trends and prospects, Magnetochemistry 6 (3) (2020) 30.
[47] S. Zhan, Y. Yang, S. Zhiqiang, S. Junjun, Li Yi, Y. Shanshan, Z. Dandan, Efficient removal of pathogenic bacteria and viruses by multifunctional amine-modified magnetic nanoparticles, J. Hazard. Mater. 274 (2014) 115–123.
[48] L. Lu, X. Wang, C. Xiong, Li Yao, Recent advances in biological detection with magnetic nanoparticles as a useful tool, Sci. China Chem. 58 (5) (2015) 793–809.
[49] S. Wu, N. Duan, S. Zhao, F. CongCong, W. Zhouping, Simultaneous aptasensor for multiplex pathogenic bacteria detection based on multicolor upconversion nanoparticles labels, Anal. Chem. 86 (6) (2014) 3100–3107.
[50] C.-.H. Zhou, Y.-.M. Long, B.-.P. Qi, D.-.W. Pang, Z.-.L. Zhang, A magnetic bead-based bienzymatic electrochemical immunosensor for determination of H9N2 avian influenza virus, Electrochem. Commun. 31 (2013) 129–132.
[51] C. Souza, T.L M., Y. Hideko, S. Alegret, M.I. Pividori, Magneto immunoassays for plasmodium falciparum histidine-rich protein 2 related to malaria based on magnetic nanoparticles, Anal. Chem. 83 (14) (2011) 5570–5577.
[52] E.I. Galanzha, E.V. Shashkov, K. Thomas, K. Jin-Woo, Y. Lily, VP. Zharov, In vivo magnetic enrichment and multiplex photoacoustic detection of circulating tumour cells, Nat. Nanotechnol. 4 (12) (2009) 855–860.
[53] Er-Q Song, J. Hu, C.-.Y. Wen, Z.-.Q. Tian, Xu Yu, Z.-.L. Zhang, Y-Bo Shi, D.-.W. Pang, Fluorescent-magnetic-biotargeting multifunctional nanobioprobes for detecting and isolating multiple types of tumor cells, ACS Nano 5 (2) (2011) 761–770.
[54] P.C. Lauterbur, Image formation by induced local interactions: examples employing nuclear magnetic resonance, Nature 242 (5394) (1973) 190–191.

[55] P.L. Choyke, J.A. Frank, M.E. Girton, S.W. Inscoe, M.J. Carvlin, J.L. Black, H.A. Austin, A.J. Dwyer, Dynamic Gd-DTPA-enhanced MR imaging of the kidney: experimental results, Radiology 170 (3) (1989) 713–720.
[56] E.J. Rummeny, P. Reimer, W. Heindel, MR Imaging of the Body, 1st edn, Thieme Medical Publishers, New York, 2009.
[57] S. Laurent, D. Forge, M. Port, A. Roch, C. Robic, L. Vander Elst, RN. Muller, Magnetic iron oxide nanoparticles: synthesis, stabilization, vectorization, physicochemical characterizations, and biological applications, Chem. Rev. 108 (6) (2008) 2064–2110.
[58] M. Borges, S. Yu, A. Laromaine, A. Roig, S. Suárez-García, J. Lorenzo, D. Ruiz-Molina, F. Novio, Dual T 1/T 2 MRI contrast agent based on hybrid SPION@ coordination polymer nanoparticles, RSC Adv. 5 (105) (2015) 86779–86783.
[59] J. Du, W. Zhu, Li Yang, C. Wu, B. Lin, J. Wu, R. Jin, T. Shen, H. Ai, Reduction of polyethylenimine-coated iron oxide nanoparticles induced autophagy and cytotoxicity by lactosylation, Regen. Biomater. 3 (4) (2016) 223–229.
[60] X. Duan, L. Lu, Y. Wang, F. Zhang, J. Mao, M. Cao, B. Lin, X. Zhang, X. Shuai, J. Shen, The long-term fate of mesenchymal stem cells labeled with magnetic resonance imaging-visible polymersomes in cerebral ischemia, Int. J. Nanomed. 12 (2017) 6705.
[61] Yi-XJ Wang, Superparamagnetic iron oxide based MRI contrast agents: current status of clinical application, Quant. Imaging Med. Surg. 1 (1) (2011) 35.
[62] Y. Bao, J.A. Sherwood, Z. Sun, Magnetic iron oxide nanoparticles as T 1 contrast agents for magnetic resonance imaging, J. Mater. Chem. C 6 (6) (2018) 1280–1290.
[63] O. Veiseh, J. W. Gunn, M. Zhang, Design and fabrication of magnetic nanoparticles for targeted drug delivery and imaging, Adv. Drug. Deliv. Rev. 62 (3) (2010) 284–304.
[64] J.L.-S Au, S.H. Jang, M. GuillWientjes, Clinical aspects of drug delivery to tumors, J. Control. Release 78 (1–3) (2002) 81–95.
[65] V.P. Torchilin, Targeted pharmaceutical nanocarriers for cancer therapy and imaging, AAPS J. 9 (2) (2007) E128–E147.
[66] Z. Lin, VP. Torchilin, Stimulus-responsive nanopreparations for tumor targeting, Integr. Biol. 5 (1) (2013) 96–107.
[67] A.D. Friedman, S.E. Claypool, R. Liu, The smart targeting of nanoparticles, Curr. Pharm. Des. 19 (35) (2013) 6315–6329.
[68] S. Stevanovic, Identification of tumour-associated T-cell epitopes for vaccine development, Nat. Rev. Cancer 2 (7) (2002) 514.
[69] U. Zimmermann, G. Pilwat, Organ specific application of drugs by means of cellular capsule systems (author's transl), Z. Naturforsch. C J. Biosci. 31 (11–12) (1976) 732.
[70] K.J. Widder, AE. Senyei, DG. Scarpelli, Magnetic microspheres: a model system for site specific drug delivery in vivo, Proc. Soc. Exp. Biol. Med. 158 (2) (1978) 141–146.
[71] A.S. Lübbe, C. Bergemann, H. Riess, F. Schriever, P. Reichardt, K. Possinger, M. Matthias, B. Dörken, F.G Herrmann, Clinical experiences with magnetic drug targeting: a phase i study with 40-epidoxorubicin in 14 patients with advanced solid tumors, Cancer Res. 56 (1996) 4686–4693.
[72] P.K.B. Nagesh, N.R. Johnson, V. KN Boya, SF. Othman, P. Chowdhury, B. B. Hafeez Vahid Khalilzad-Sharghi, PSMA targeted docetaxel-loaded superparamagnetic iron oxide nanoparticles for prostate cancer, Colloids Surf. B 144 (2016) 8–20.
[73] G.Y. Lee, W.P. Qian, L. Wang, Y.A. Wang, CA. Staley, M. Satpathy, S. Nie, H. Mao, L. Yang, Theranostic nanoparticles with controlled release of gemcitabine for targeted therapy and MRI of pancreatic cancer, ACS Nano 7 (3) (2013) 2078–2089.
[74] S.C. Goodwin, C.A. Bittner, C.L. Peterson, G. Wong, Single-dose toxicity study of hepatic intra-arterial infusion of doxorubicin coupled to a novel magnetically targeted drug carrier, Toxicol. Sci. 60 (1) (2001) 177–183.
[75] C.J. Sunderland, M. Steiert, J.E. Talmadge, A.M. Derfus, S.E. Barry, Targeted nanoparticles for detecting and treating cancer, Drug Dev. Res. 67 (1) (2006) 70–93.
[76] U. Steinfeld, C. Pauli, N. Kaltz, C. Bergemann, H.-.H. Lee, T lymphocytes as potential therapeutic drug carrier for cancer treatment, Int. J. Pharm. 311 (1–2) (2006) 229–236.
[77] N. Andhariya, R. Mehta, R. Upadhyay, B. Chudasama, Folic acid conjugated magnetic drug delivery system for controlled release of doxorubicin, J. Nanopart. Res. 15 (1) (2013) 1416.

[78] Yi Yang, QFa Guo, J.R. Peng, J. Su, X.L. Lu, Y.X. Zhao, Z.Y. Qian, Doxorubicin-conjugated heparin-coated superparamagnetic iron oxide nanoparticles for combined anticancer drug delivery and magnetic resonance imaging, J. Biomed. Nanotechnol. 12 (11) (2016) 1963–1974.
[79] JiE Lee, N. Lee, H. Kim, J. Kim, S.H. Choi, J.H. Kim, T. Kim, Uniform mesoporous dye-doped silica nanoparticles decorated with multiple magnetite nanocrystals for simultaneous enhanced magnetic resonance imaging, fluorescence imaging, and drug delivery, J. Am. Chem. Soc. 132 (2) (2010) 552–557.
[80] X. Yang, Z. Li, M. Li, J. Ren, X. Qu, Fluorescent protein capped mesoporous nanoparticles for intracellular drug delivery and imaging, Chem. Eur. J. 19 (45) (2013) 15378–15383.
[81] Q. Mu, FM. Kievit, RJ. Kant, G. Lin, M. Jeon, M. Zhang, Anti-HER2/Neu peptide-conjugated iron oxide nanoparticles for targeted delivery of paclitaxel to breast cancer cells, Nanoscale 7 (43) (2015) 18010–18014.
[82] J.C. Leach, A. Wang, K. Ye, S. Jin, A RNA-DNA hybrid aptamer for nanoparticle-based prostate tumor targeted drug delivery, Int. J. Mol. Sci. 17 (3) (2016) 380.
[83] A.J. Wagstaff, S.D. Brown, M.R. Holden, G.E. Craig, J.A. Plumb, R.E. Brown, N.J. Wheate, N. Schreiter, W. Chrzanowski, Cisplatin drug delivery using gold-coated iron oxide nanoparticles for enhanced tumour targeting with external magnetic fields, Inorg. Chim. Acta 393 (2012) 328–333.
[84] G. Unsoy, R. Khodadust, S. Yalcin, P. Mutlu, U. Gunduz, Synthesis of doxorubicin loaded magnetic chitosan nanoparticles for pH responsive targeted drug delivery, Eur. J. Pharm. Sci. 62 (2014) 243–250.
[85] G. Unsoy, S. Yalcin, R. Khodadust, P. Mutlu, O. Onguru, U. Gunduz, Chitosan magnetic nanoparticles for pH responsive bortezomib release in cancer therapy, Biomed. Pharmacother. 68 (5) (2014) 641–648.
[86] M.G. Adimoolam, M.R. Nalam, N. Amreddy, M.V. Sunkara, A simple approach to design chitosan functionalized Fe3O4 nanoparticles for pH responsive delivery of doxorubicin for cancer therapy, J. Magn. Magn. Mater. 448 (2018) 199–207.
[87] P. Wust, B. Hildebrandt, G. Sreenivasa, B. Rau, J. Gellermann, H. Riess, R. Felix, P.M. Schlag, Hyperthermia in combined treatment of cancer, Lancet Oncol. 3 (8) (2002) 487–497.
[88] J.V. Zee, Heating the patient: a promising approach? Ann. Oncol. 13 (8) (2002) 1173–1184.
[89] R.K. Gilchrist, R. Medal, W.D. Shorey, J.C. Parrott, R.C. Hanselman, C.B. Taylor, Selective inductive heating of lymph nodes, Ann. Surg. 146 (4) (1957) 596.
[90] A. Jordan, P. Wust, H. Fähling, W. John, A. Hinz, R. Felix, Inductive heating of ferrimagnetic particles and magnetic fluids: physical evaluation of their potential for hyperthermia, Int. J. Hyperthermia 9 (1) (1993) 51–68.
[91] S. Chamorro, L. Gutiérrez, D. Verdoy María, P. Vaquero, G. Salas, Y. Luengo, A. Brenes, F.J. Teran, Safety assessment of chronic oral exposure to iron oxide nanoparticles, Nanotechnology 26 (20) (2015) 205101.
[92] A. Jordan, R. Scholz, P. Wust, H. Fähling, R. Felix, Magnetic fluid hyperthermia (MFH): cancer treatment with ac magnetic field induced excitation of biocompatible superparamagnetic nanoparticles, J. Magn. Magn. Mater. 201 (1–3) (1999) 413–419.
[93] S.I. Park, J.H. Lim, J.H. Kim, H.I. Yun, C.O. Kim, Toxicity estimation of magnetic fluids in a biological test, J. Magn. Magn. Mater. 304 (1) (2006) e406–e408.
[94] P.P. Macaroff, A.R. Simioni, Z.G.M. Lacava, E.C.D. Lima, P.C. Morais, A.C. Tedesco, Studies of cell toxicity and binding of magnetic nanoparticles with blood stream macromolecules, J. Appl. Phys. 99 (8) (2006) 08S102.
[95] R. Duncan, L. Izzo, Dendrimer biocompatibility and toxicity, Adv. Drug. Deliv. Rev. 57 (15) (2005) 2215–2237.
[96] H. Yin, H.P. Too, G.M. Chow, The effects of particle size and surface coating on the cytotoxicity of nickel ferrite, Biomaterials 26 (29) (2005) 5818–5826.
[97] J.F. Schenck, Physical interactions of static magnetic fields with living tissues, Prog. Biophys. Mol. Biol. 87 (2–3) (2005) 185–204.
[98] Y. Haik, V. Pai, C.-J. Chen, Apparent viscosity of human blood in a high static magnetic field, J. Magn. Magn. Mater. 225 (1–2) (2001) 180–186.
[99] M. Arruebo, M.R. Ibarra Rodrigo Fernández-Pacheco, J. Santamaría, Magnetic nanoparticles for drug delivery, Nano Today 2 (3) (2007) 22–32.
[100] K. Kostarelos, Nanorobots for medicine: how close are we? Nanomedicine 5 (3) (2010) 341–342.
[101] W. Gao, J. Wang, The environmental impact of micro/nanomachines: a review, Acs Nano 8 (4) (2014) 3170–3180.

[102] M. Johannsen, B. Thiesen, U. Gneveckow, K. Taymoorian, C.H. Cho, R. Scholz, N. Waldöfner, A. Jordan, S.A. Loening, P. Wust, Thermotherapy of prostate cancer using magnetic nanoparticles: feasibility, imaging, and three-dimensional temperature distribution, Eur. Urol. 52 (6) (2007) 1653–1662.

[103] Q. Quan, J. Xie, H. Gao, M. Yang, F. Zhang, G. Liu, X. Lin, HSA coated iron oxide nanoparticles as drug delivery vehicles for cancer therapy, Mol. Pharm. 8 (5) (2011) 1669–1676.

[104] Y. Huang, K. Mao, B. Zhang, Y. Zhao, Superparamagnetic iron oxide nanoparticles conjugated with folic acid for dual target-specific drug delivery and MRI in cancer theranostics, Mater. Sci. Eng. C 70 (2017) 763–771.

[105] M. Nadeem, M. Ahmad, M.S. Akhtar, A. Shaari, S. Riaz, S. Naseem, M. Masood, M.A. Saeed, Magnetic properties of polyvinyl alcohol and doxorubicine loaded iron oxide nanoparticles for anticancer drug delivery applications, PLoS One 11 (6) (2016) e0158084.

[106] F.M. Kievit, F.Y. Wang, C. Fang, H. Mok, K. Wang, JR. Silber, M. Zhang, R.G. Ellenbogen, Doxorubicin loaded iron oxide nanoparticles overcome multidrug resistance in cancer in vitro, J. Control. Release 152 (1) (2011) 76–83.

[107] J. Gautier, E. Munnier, A. Paillard, K. Hervé, L. Douziech-Eyrolles, M. Soucé, P. Dubois, I. Chourpa, A pharmaceutical study of doxorubicin-loaded PEGylated nanoparticles for magnetic drug targeting, Int. J. Pharm. 423 (1) (2012) 16–25.

[108] Y. Zou, P. Liu, C-He Liu, Xu-T Zhi, Doxorubicin-loaded mesoporous magnetic nanoparticles to induce apoptosis in breast cancer cells, Biomed. Pharmacother. 69 (2015) 355–360.

[109] CA. Quinto, P. Mohindra, S. Tong, G. Bao, Multifunctional superparamagnetic iron oxide nanoparticles for combined chemotherapy and hyperthermia cancer treatment, Nanoscale 7 (29) (2015) 12728–12736.

[110] H. Jeon, J. Kim, YMi Lee, J. Kim, J. Lee, W.C. Hyung, H. Park, Poly-paclitaxel/cyclodextrin-SPION nano-assembly for magnetically guided drug delivery system, J. Control. Release 231 (2016) 68–76.

[111] MdSU Ahmed, A.B. Salam, J. Jaynes, K. Willian, C. Yates, MO. Abdalla, T. Turner, Double-receptor-targeting multifunctional iron oxide nanoparticles drug delivery system for the treatment and imaging of prostate cancer, Int. J. Nanomed. 12 (2017) 6973.

[112] A. Aires, S.M. Ocampo, B.M. Simões, J.F.C.M.J. Rodríguez, P. Couleaud, K. Spence, Multifunctionalized iron oxide nanoparticles for selective drug delivery to CD44-positive cancer cells, Nanotechnology 27 (6) (2016) 065103.

[113] S. Trabulo, A. Aires, A. Aicher, AL. Cortajarena, C. Heeschen, Multifunctionalized iron oxide nanoparticles for selective targeting of pancreatic cancer cells, Biochim. Biophys. Acta Gen. Subj. 1861 (6) (2017) 1597–1605.

[114] H. Unterweger, R. Tietze, C. Janko, J. Zaloga, S. Lyer, S. Dürr, N. Taccardi, Development and characterization of magnetic iron oxide nanoparticles with a cisplatin-bearing polymer coating for targeted drug delivery, Int. J. Nanomed. 9 (2014) 3659.

[115] S. Natesan, C. Ponnusamy, S.S. Palaniappan, A. Sugumaran, S. Chelladurai, R. Palanichamy, Artemisinin loaded chitosan magnetic nanoparticles for the efficient targeting to the breast cancer, Int. J. Biol. Macromol. 104 (2017) 1853–1859.

[116] K. Maier-Hauff, R. Rothe, R. Scholz, U. Gneveckow, P. Wust, B. Thiesen, A. Feussner, Intracranial thermotherapy using magnetic nanoparticles combined with external beam radiotherapy: results of a feasibility study on patients with glioblastoma multiforme, J. Neurooncol. 81 (1) (2007) 53–60.

[117] K. Maier-Hauff, F. Ulrich, D. Nestler, H. Niehoff, P. Wust, B. Thiesen, H. Orawa, V. Budach, A. Jordan, Efficacy and safety of intratumoral thermotherapy using magnetic iron-oxide nanoparticles combined with external beam radiotherapy on patients with recurrent glioblastoma multiforme, J. Neurooncol. 103 (2) (2011) 317–324.

[118] V. Landeghem, KH. Frank, Post-mortem studies in glioblastoma patients treated with thermotherapy using magnetic nanoparticles, Biomaterials 30 (1) (2009) 52–57.

[119] M. Johannsen, U. Gneveckow, K. Taymoorian, B. Thiesen, N. Waldöfner, R. Scholz, K. Jung, A. Jordan, P. Wust, S.A. Loening, Morbidity and quality of life during thermotherapy using magnetic nanoparticles in locally recurrent prostate cancer: results of a prospective phase i trial, Int. J. Hyperthermia 23 (3) (2007) 315–323.

[120] M. Johannsen, U. Gneveckow, K. Taymoorian, B. Thiesen, N. Waldöfner, R. Scholz, K. Jung, A. Jordan, P. Wust, S.A. Loening, Intracellular hyperthermia for cancer using magnetite cationic liposomes: an in vivo study, Jpn. J. Cancer Res. 89 (4) (1998) 463–470.

[121] M. Yanase, M. Shinkai, H. Honda, J.Y.T. Wakabayashi, T. Kobayashi, Antitumor immunity induction by intracellular hyperthermia using magnetite cationic liposomes, Jpn. J. Cancer Res. 89 (7) (1998) 775–782.

[122] M. Shinkai, M. Yanase, H. Honda, J.Y.T. Wakabayashi, T. Kobayashi, Intracellular hyperthermia for cancer using magnetite cationic liposomes: in vitro study, Jpn. J. Cancer Res. 87 (11) (1996) 1179–1183.

[123] A. Ito, H. Honda, M. Shinkai, S. Saga, K. Yoshikawa, J. Yoshida, T. Wakabayashi, T. Kobayashi, Heat shock protein 70 expression induces antitumor immunity during intracellular hyperthermia using magnetite nanoparticles, Cancer Immunol. Immunother. 52 (2) (2003) 80–88.

[124] T. Ohno, A. Takemura, T. Wakabayashi, J. Yoshida, A. Ito, H. Honda, M. Shinkai, T. Kobayashi, Effective Solitary hyperthermia treatment of malignant glioma using stick type CMC-magnetite. In vivo study, J. Neurooncol. 56 (3) (2002) 233–239.

[125] B. Le, T.K.M. Shinkai, H. Honda, J.U.N. Yoshida, T. Kobayashi, T. Wakabayashi, Preparation of tumor-specific magnetoliposomes and their application for hyperthermia, J. Chem. Eng. Jpn. 34 (1) (2001) 66–72.

[126] I. Rabias, D. Tsitrouli, E. Karakosta, G. Diamantopoulos, T. Kehagias, D. Stamopoulos, M. Fardis, Rapid magnetic heating treatment by highly charged maghemite nanoparticles on wistar rats exocranial glioma tumors at microliter volume, Biomicrofluidics 4 (2) (2010) 024111.

[127] H. Xu, H. Zong, C. Ma, X. Ming, M. Shang, K. Li, X. He, L. Cao, Evaluation of nano-magnetic fluid on malignant glioma cells, Oncol. Lett. 13 (2) (2017) 677–680.

[128] S.L. Berry, K. Walker, C. Hoskins, N.D. Telling, H.P. Price, Nanoparticle-mediated magnetic hyperthermia is an effective method for killing the human-infective protozoan parasite leishmania mexicana in vitro, Sci. Rep. 9 (1) (2019) 1–9.

[129] A. Bhardwaj, K. Parekh, N. Jain, In vitro hyperthermic effect of magnetic fluid on cervical and breast cancer cells, Sci. Rep. 10 (1) (2020) 1–13.

[130] I. Hilger, W. Andra, R. Hergt, R. Hiergeist, H. Schubert, W.A. Kaiser, Electromagnetic heating of breast tumors in interventional radiology: in vitro and in vivo studies in human cadavers and mice, Radiology 218 (2) (2001) 570–575.

[131] I. Hilger, R. Hiergeist, R. Hergt, K. Winnefeld, H. Schubert, W.A. Kaiser, Thermal ablation of tumors using magnetic nanoparticles: an in vivo feasibility study, Invest. Radiol. 37 (10) (2002) 580–586.

[132] A. Ito, K. Tanaka, H. Honda, K. Kondo, M. Shinkai, T. Saida, K. Matsumoto, T. Kobayashi, Tumor regression by combined immunotherapy and hyperthermia using magnetic nanoparticles in an experimental subcutaneous murine melanoma, Cancer Sci. 94 (3) (2003) 308–313.

[133] S.J. DeNardo, L.A.M.G.L. DeNardo, A. Natarajan, A.R. Foreman, C. Gruettner, G.N. Adamson, R. Ivkov, Thermal dosimetry predictive of efficacy of 111In-ChL6 nanoparticle AMF–induced thermoablative therapy for human breast cancer in mice, J. Nucl. Med. 48 (3) (2007) 437–444.

[134] P.J. Hoopes, J.A. Tate, J.A. Ogden, R.R. Strawbridge, S.N. Fiering, A.A. Petryk, S.M. Cassim, et al., Assessment of intratumor non-antibody directed iron oxide nanoparticle hyperthermia cancer therapy and antibody directed IONP uptake in murine and human cells, Proc. SPIE Int. Soc. Opt. Eng. 7181 (2009) 71810.

[135] C.L. Dennis, A.J. Jackson, J.A. Borchers, P.J. Hoopes, R. Strawbridge, A.R. Foreman, J. Van Lierop, C. Grüttner, R. Ivkov, Nearly complete regression of tumors via collective behavior of magnetic nanoparticles in hyperthermia, Nanotechnology 20 (39) (2009) 395103.

[136] T. Kikumori, T. Kobayashi, M. Sawaki, T. Imai, Anti-cancer effect of hyperthermia on breast cancer by magnetite nanoparticle-loaded anti-HER2 immunoliposomes, Breast Cancer Res. Treat. 113 (3) (2009) 435.

Index

Page numbers followed by "*f*" and "*t*" indicate, figures and tables respectively.

O

P

Q

Printed in the United States
by Baker & Taylor Publisher Services